Global Change and Future Earth

Global Change and Future Earth is derived from the work of several programs of the International Union of Geodesy and Geophysics (IUGG). It demonstrates how multi- and interdisciplinary research outputs from the geoscience community can be applied to tackle the physical and societal impacts of climate change and contribute to the Future Earth program of the International Council for Science (now known as the International Science Council following a merger with the International Social Science Council). The volume brings together an international team of eminent researchers to provide authoritative reviews on the wide-ranging ramifications of climate change spanning eight key themes: planetary issues; geodetic issues; the Earth's fluid environment; regions of the Earth; urban environments; food security; risk, safety, and security; and climate change and global change. Covering the challenges faced by urban and rural areas, in both developed and developing countries, this volume provides an important resource for a global audience of graduate students and researchers from a broad range of disciplines, as well as policy advisers and practitioners.

Tom Beer is the Chair of the IUGG Commission on Climatic and Environmental Change (CCEC). He is an expert on environmental risk management, including greenhouse gas and air quality issues and their application to transport and health. Past positions include being President of the Australia–New Zealand chapter of the Society for Risk Analysis and President of the International Union of Geodesy and Geophysics (IUGG). He was part of the team that won the Commonwealth Scientific and Industrial Research Organisation (CSIRO) Chairman's Medal in 2000. Having been a panel member and lead author he was sent a certificate of appreciation by the Intergovernmental Panel on Climate Change (IPCC) when they were awarded half the Nobel Prize in peace in 2007.

Jianping Li is Professor at the State Key Laboratory of Earth Surface Processes and Resource Ecology (ESPRE) and the College of Global Change and Earth System Sciences (GCESS), both at Beijing Normal University. He is also Vice-Chair of the IUGG Union Commission on Climatic and Environmental Change (CEC) and Executive Secretary General of the International Commission of Climate (ICCL). His research interests include climatic dynamics and predictability, monsoons and annular modes. He is a co-editor of *Dynamics and Predictability of Large-Scale High-Impact Weather and Climate Events* (Cambridge University Press, 2016).

Keith Alverson is Director of the UN Environment International Environmental Technology Centre in Osaka, Japan. In the past he has served as Coordinator of the Freshwater, Land and Climate Branch of the Ecosystems Division of UN Environment in Nairobi, Kenya; Head of Ocean Observations and Services Section at the Intergovernmental Oceanographic Commission of UNESCO in Paris, France; and Executive Director of the Past Global Changes Project in Bern, Switzerland. He is an elected At-Large Member of the executive committee of the International Association of Meteorology and Atmospheric Physics, Executive Secretary of the IUGG Union Commission on Climatic and Environmental Change and past President of the International Commission of Climate.

Special Publications of the International Union of Geodesy and Geophysics Series

The International Union of Geodesy and Geophysics (IUGG) promotes and coordinates international scientific studies of Earth (physical, chemical, and mathematical) and its environment in space, aiming to apply this knowledge to societal needs such as mineral resources, mitigation of natural hazards, and environmental preservation. The series is copublished by the IUGG and Cambridge University Press, providing researchers and graduate students with authoritative insights into major scientific developments and state-of-the-art research.

TITLES IN PRINT IN THIS SERIES
A. Ismail-Zadeh, J. Urrutia-Fucugauchi, A. Kijko, K. Takeuchi, and I. Zaliapin, *Extreme Natural Hazards, Disaster Risks and Societal Implications*
J. Li, R. Swinbank, R. Grotjahn, and H. Volkert, *Dynamics and Predictability of Large-Scale, High-Impact Weather and Climate Events*
T. Beer, J. Li, and K. Alverson, *Global Change and Future Earth: The Geoscience Perspective*

Global Change and Future Earth: The Geoscience Perspective

EDITED BY

TOM BEER
IUGG Commission on Climatic and Environmental Change (CCEC)

JIANPING LI
College of Global Change and Earth System Sciences (GCESS), Beijing Normal University

KEITH ALVERSON
UN Environment International Environmental Technology Centre (UNEP IETC)

CAMBRIDGE
UNIVERSITY PRESS

CAMBRIDGE
UNIVERSITY PRESS

University Printing House, Cambridge CB2 8BS, United Kingdom

One Liberty Plaza, 20th Floor, New York, NY 10006, USA

477 Williamstown Road, Port Melbourne, VIC 3207, Australia

314–321, 3rd Floor, Plot 3, Splendor Forum, Jasola District Centre, New Delhi – 110025, India

79 Anson Road, #06–04/06, Singapore 079906

Cambridge University Press is part of the University of Cambridge.

It furthers the University's mission by disseminating knowledge in the pursuit of education, learning, and research at the highest international levels of excellence.

www.cambridge.org
Information on this title: www.cambridge.org/9781107171596
DOI: 10.1017/9781316761489

First published 2018

Printed in the United Kingdom by TJ International Ltd. Padstow Cornwall

A catalogue record for this publication is available from the British Library.

Library of Congress Cataloging-in-Publication Data
Names: Beer, Tom, editor.
Title: Global change and future earth : the geoscience perspective / edited by Tom Beer, IUGG Commission on Climatic and Environmental Change (CCEC), Jianping Li, College of Global Change and Earth System Sciences (GCESS) and Keith Alverson, UN Environment International Environmental Technology Centre, Osaka, Japan.
Description: Cambridge, United Kingdom ; New York, NY : Cambridge University Press, 2018. | Series: International union of geodesy and geophysics series | Includes bibliographical references and index.
Identifiers: LCCN 2018013796 | ISBN 9781107171596 (hardback)
Subjects: LCSH: Earth science–Research.
Classification: LCC QE40 .G56 2018 | DDC 550–dc23
LC record available at https://lccn.loc.gov/2018013796

ISBN 978-1-107-17159-6 Hardback

Contents

Contributors

Ian Allison
Antarctic Climate and Ecosystems Cooperative
Research Centre
Australia

Charles Bardeen
National Center for Atmospheric Research
United States of America

Sarah Beaven
University of Canterbury
New Zealand

Tom Beer
IUGG, Commission on Climatic and
Environmental Change
Australia

Julia Blanchard
Centre for Marine Socioecology
University of Tasmania
Australia

Harry L. Bryden
National Oceanography Centre, Southampton
University of Southampton
United Kingdom

Bruce M. Campbell
CGIAR
International Center for Tropical Agriculture (CIAT) and
University of Copenhagen
Denmark

Anny Cazenave
Laboratoire d'Etudes en Géophysique et
Océanographie Spatiales (LEGOS)
Observatoire Midi-Pyrénées
France

Jianli Chen
Center for Space Research
University of Texas, Austin
United States of America

William Cheung
Institute for the Fisheries and Oceans
University of British Columbia
Canada

İhsan Çiçek
Faculty of Language History and Geography
Ankara University
Turkey

Mustafa Coşkun
Turkish State Meteorological Service
Turkey

Athena Coustenis
Observatoire de Paris, CNRS
Université Paris Diderot
France

Gisela de Aragão Umbuzeiro
Faculdade de Tecnologia
UNICAMP
Brazil

Silvia Berlanga de Moraes Barros
School of Pharmaceutical Sciences
University of São Paulo
Brazil

Dhanush Dinesh
CGIAR
Climate Change, Agriculture and Food Security
(CCAFS)
and University of Copenhagen
Denmark

Ruiqiang Ding
Institute of Atmospheric Physics
Chinese Academy of Sciences
China

John C. Eichelberger
University of Alaska, Fairbanks
United States of America

Ibrahim Elmadfa
Past President of the International Union of Nutritional Sciences
Former Director of the Department of Nutritional Sciences
University of Vienna
Austria

Elaine M. Faustman
Institute for Risk Analysis and Risk Communication
University of Washington
United States of America

Heidi Foth
Martin Luther University
Germany

Eigil Friis-Christensen
IUGG, Commission on Climatic and
Environmental Change
Denmark

Elizabeth A. Fulton
CSIRO Oceans and Atmosphere and
Centre for Marine Socioecology, University of Tasmania
Australia

Tuyara N. Gavrilyeva
North-Eastern Federal University, Yakutsk
Siberian Branch of the Russian Academy of Sciences
Russia

Aakriti Grover
Department of Geography
School of Earth Sciences
Central University of Tamil Nadu
India

Regine Hock
Geophysical Institute
University of Alaska
United States of America

Sophia Huyer
CGIAR
Climate Change, Agriculture and Food Security Programme
(CCAFS) – WISAT
Canada

Salmaan H. Inayat-Hussain
Global Toxicology
Group Health, Safety, Security and Environment
Petroliam Nasional Berhad (PETRONAS)
Malaysia

Reiichiro Ishii
Research Institute for Humanity and Nature
Japan

Alik Ismail-Zadeh
Karlsruhe Institute of Technology, Applied Geosciences,
Karlsruhe, Germany; and
Russian Academy of Sciences, Institute of Earthquake
Prediction Theory and Mathematical Geophysics, Moscow,
Russia

Matt A. King
Surveying and Spatial Sciences
University of Tasmania
Australia

Thomas Kohl
Applied Geosciences, Geothermal Research
Karlsruhe Institute of Technology
Germany

Yekaterina Y. Kontar
Fletcher School of Law and Diplomacy
Tufts University
United States of America

Jianping Li
State Key Laboratory of Earth Surface Processes and
Resource Ecology, and College of Global Change
and Earth System Science, Beijing Normal
University, China
and Laboratory for Regional Oceanography and
Numerical Modeling
Qingdao National Laboratory for Marine Science and
Technology, China

Andrew N. Mackintosh
Antarctic Research Centre
Victoria University of Wellington
New Zealand

Hein Mallee
Research Institute for Humanity and Nature
Japan

Albert McGill
Future for Food
Melbourne, Victoria
Australia

Michael E. Meadows
Department of Environmental and Geographical Science
University of Cape Town
South Africa; and
School of Geographic Sciences
East China Normal University
Shanghai, China

Alexa Leonie Meyer
Department of Nutritional Sciences
University of Vienna
Austria

Thierry Meyrath
Mathematics Research Unit
University of Luxembourg
Luxembourg

Robert Mgendi
United Nations Environment Programme
Kenya

Michael Mills
National Center for Atmospheric Research
United States of America

Alberto Montanari
Department of Civil, Chemical, Environmental,
and Material Engineering
University of Bologna
Italy

Keto E. Mshigeni
Hubert Kairuki Memorial University
Tanzania

Richard Munang
United Nations Environment Programme
Kenya

Lawrence A. Mysak
Department of Atmospheric and Oceanic Sciences
McGill University
Canada

Julio Alejandro Navoni
Department of Clinical Chemistry
Federal University of Rio Grande do Norte
Brazil

Godwin D. Ndossi
Hubert Kairuki Memorial University
Tanzania

Hindumathi Palanisamy
Laboratoire d'Etudes en Géophysique et Océanographie
Spatiales (LEGOS)
Observatoire Midi-Pyrénées
France

Ligia Pérez-Cruz
Universidad Nacional Autónoma de México
Mexico

Éva Plagányi
CSIRO Oceans and Atmosphere and
Centre for Marine Socioecology, University of Tasmania
Australia

Christina Plainaki
ASI – Italian Space Agency
Italy

Alan Robock
Rutgers University
United States of America

Jane E. Rovins
Disaster Reduction & Resilience (DRR) Solutions, Ltd
Massey University, Joint Centre for Disaster Research
New Zealand

Ladislaus Rybach
Institute of Geophysics
ETH Zurich
Switzerland

Luis Schiesari
Environmental Management
University of São Paulo
Brazil

Anke Schneider
"The Future Ocean" Cluster of Excellence
Kiel University
Germany

Serhat Şensoy
Turkish State Meteorological Service
Turkey

Michael G. Sideris
Department of Geomatics Engineering
Schulich School of Engineering
Canada

R. B. Singh
Department of Geography, Delhi School of Economics
University of Delhi
India

Marissa N. Smith
Institute for Risk Analysis and Risk Communication
University of Washington
United States of America

Walter E. L. Spiess
Karlsruhe Institute of Technology
Germany

Cheng Sun
State Key Laboratory of Earth Surface Processes and
Resource Ecology, and College of Global Change and Earth
System Science
Beijing Normal University
China

Nikita I. Tananaev
Melnikov Permafrost Institute, Yakutsk
Siberian Branch of the Russian Academy of Sciences
Russia

Makoto Taniguchi
Research Institute for Humanity and Nature
Japan

Fred W. Taylor
Atmospheric, Oceanic, and Planetary Physics
University of Oxford
United Kingdom

Owen B. Toon
University of Colorado
United States of America

Sarah F. Trainor
Alaska Center for Climate Assessment and
Policy (ACCAP)
University of Alaska, Fairbanks
United States of America

Necla Türkoğlu
Faculty of Language History and Geography
Ankara University
Turkey

Jaime Urrutia-Fucugauchi
Universidad Nacional Autónoma de México
Mexico

Tonie van Dam
Faculty of Science, Technology, and Communication
University of Luxembourg
Luxembourg

Martin Visbeck
GEOMAR Helmholtz Centre for Ocean Research
Kiel
Germany

Mark L. Wahlqvist
Past President, International Union of Nutritional Sciences
Emeritus Professor and former Head of Medicine
Monash University, Melbourne, Australia and
Visiting Professor, National Health Research Institutes
Taiwan

Reg Watson
Institute for Marine and Antarctic Studies
University of Tasmania
Australia

Lili Xia
Rutgers University
United States of America

Tetsuzo Yasunari
Research Institute for Humanity and Nature
Japan

Preface

The Union Commission on Climatic and Environmental Change (CCEC) was established by the Executive Committee of the International Union of Geodesy and Geophysics (IUGG) in June 2012 in order to promote the advancement of scientific understanding of climatic and environmental change, to boost research in reducing uncertainties in climate and environmental models, to define criteria for collaborative transdisciplinary research on climate and environmental change, to fulfill the objectives of IUGG and its associations, to provide an all-Union perspective on climatic and environmental change, and to make available the knowledge and insights developed through scientific research to the benefit of society and planet Earth, including consideration of the science of global change, related vulnerability and impacts, and potential responses.

CCEC provides a focus for IUGG scientific expertise in climate- and environment-related areas across the breadth of all IUGG disciplines and associations. CCEC enables the breadth of IUGG expertise to be brought to bear at the global level through collaborating with, and underpinning, the work of ICSU and other international organizations. It also enables the geographic spread of IUGG expertise to be applied at the local level through involvement with national bodies in the organization of meetings and other activities.

Objectives

- To build scientific capacity for responsibly addressing the broad, multidisciplinary issues involved in climatic and environmental change;
- to provide useful information, understanding, and support to the public and governmental organizations;
- to interact and cooperate with outside activities that would benefit from the capabilities and resources of the IUGG Associations; and
- to strengthen links across the Scientific Associations within IUGG, to build new external links to organizations outside IUGG, to strengthen existing links to external organizations, and to promote IUGG's contribution to global change research.

Information about CCEC is also available on the main IUGG web page at

www.iugg.org/about/commissions/ccec.php.

IUGG has a long-established commission on climate (ICCL) that is one of the commissions of the International Association of Meteorology and Atmospheric Sciences (IAMAS) with an established membership that arranges activities and meetings that concentrate on the meteorological and atmospheric aspects of climate.

Figure 1 Speakers at the special session at the AGU Meeting of the Americas: Alik Ismail-Zadeh, Jaime Urrutia Fucugauchi, O. Brian Toon, Katya Kontar, Tom Beer, the late Mark Pagani.

Figure 2 Albert McGill, IUFoST, addresses the IUNS symposium. Mark Wahlqvist, past President of IUNS, is on the left.

Table 1 *A List of Scientific Meetings That Dealt with the Topic of Weather, Climate, and Food Security*

Location	Theme	Linked to
Symposium Cancun, Mexico	Weather, Climate, and Food Security: American and Global Perspectives	AGU Meeting of the Americas
Workshop Nairobi, Kenya	Weather, Climate, and Food Security: African and Global Perspectives	First Africa Food Security Conference
Workshop Brisbane, Australia	Weather, Climate, and Food Security: Regional and Global Perspectives	AOGS Annual Meeting
Open Forum Granada, Spain	Weather, Climate, and Food Security: Global Perspectives	General Assembly of the International Union of Nutrition Science (IUNS)

Figure 3 Opening Session of the meeting of authors. (Photo: Y. Kontar).

CCEC seeks to focus on relevant aspects of climate that correspond to the interests of the wider scientific community. CCEC focuses on both climate and environmental change within a transdisciplinary setting.

CCEC will "bring the Union's research expertise to bear on some of the most challenging and important environmental and societal challenges of the 21st century." Food security is obviously one of these challenges. Thus, CCEC was instrumental in persuading a consortium of international scientific unions to undertake an initiative on weather, climate, and food security. Specially targeted advanced research workshops and symposia (see Figure 1) listed in Table 1 drew together experts working on relevant issues.

When the International Council for Science (ICSU), as it was then known, initiated its new program called "Future Earth" by amalgamating a number of its interdisciplinary research efforts (www.futureearth.org/), CCEC sought to examine how activities could be included in Future Earth that had not hitherto been a part of the research efforts that were amalgamated. Selected distinguished scientists were approached, chosen on the basis that they had participated in the food security meetings listed in Table 1, or were involved in activities of international scientific unions (see Figure 2). There was an enthusiastic response to the idea of preparing a volume showing the scope of the possible geoscience perspective to (a) Future Earth as a research program and (b) the future of the Earth as it undergoes global change.

A workshop of the authors of this volume (Figure 3) took place at the Université du Luxembourg in Belval, Luxembourg, 21–22 October 2016. They agreed that this volume should be a monograph that seeks to expound how the geoscience community can assist Future Earth so as to improve the future of the Earth and that it should comprise international and interdisciplinary contributions around the subject of climate change and its impacts on natural disasters and food security around the globe.

The content and structure of the thirty chapters are built around eight key sections: Part I, Future Earth and Planetary Issues (Chapters 1 to 4); Part II, Future Earth and Geodetic Issues (Chapters 5 to 8); Part III, Future Earth and the Earth's Fluid Environment (Chapters 9 to 12); Part IV, Future Earth and Regions (Chapters 13 and 14); Part V, Future Earth and Urban Environments (Chapters 15 to 17); Part VI, Future Earth and Food Security (Chapters 18 to 22); Part VII, Future Earth, Risk, Safety and Security (Chapter 23 to 27); Part VIII, Future Earth, Climate Change and Global Change (Chapters 28 to 30).

We expect the readers of this book to be research scientists who subscribe to the vision of Future Earth with an interdisciplinary, multidisciplinary, or pluridisciplinary outlook. This includes geoscientists in the broadest sense – namely, those dealing with solid earth science and fluid earth science – as well as bioscientists and other members of the international scientific research community involved in examining the impacts of global change.

The Future Earth Web site notes that Future Earth is committed to building and connecting global knowledge to intensify the impact of research and find new ways to accelerate sustainable development. The success of Future Earth will depend on active collaboration among nations, disciplines, programs, researchers, and stakeholders to ensure that knowledge is generated in partnership with society and users of science. This book is intended for all those who share this vision and for those who engage in such collaboration.

We hope that you enjoy reading this book.

Acknowledgments

The editors wish to acknowledge the support of the International Union of Geodesy and Geophysics (IUGG), who provided funding to enable the authors to meet in Luxembourg. This volume has been a project of the IUGG Commission for Climatic and Environmental Change (CCEC), whose logo is shown in Figure 4.

Any project of this magnitude is reliant on the voluntary support of numerous people, many of whom rely on the implicit support of their home institution, the international scientific union to which they belong, and national and international funding agencies. We acknowledge them all.

In addition, the members of the IUGG Commission for Climatic and Environmental Change, under whose auspices this monograph has been prepared, have been helpful and supportive. Their names and photographs may be found at http://ccec-iugg.org/node/13.

In the preparatory phase of the monograph, the editors were assisted by an editorial advisory board consisting of

- Hans Volkert (Germany) – German Aerospace Centre (DLR) Institute of Atmospheric Physics
- Hubert Savenije (Netherlands) – Delft University of Technology
- Athena Coustenis (France) – Observatoire de Paris-Meudon, France
- Alberto Montanari (Italy) – University of Bologna, Italy
- Makoto Taniguchi (Japan) – Research Institute for Humanity and Nature, Japan
- John Burrows (Germany) – University of Bremen

We sincerely thank all authors of the book chapters for their hard work and their excellent contributions. The excellence of the final chapters was due, in part, to the peer review undergone by all the chapters and by the peer reviewers' willingness to undertake critical reviews and provide constructive comments and by the authors' willingness to heed these comments and revise their work accordingly. The chapter reviewers, whom we acknowledge and thank, were Ian Allison, Keith Alverson, Tom Beer, Harry Bryden, Bruce M. Campbell, Anny Cazenave, Athena Coustenis, Ibrahim Elmadfa, Elaine M. Faustman, Eigil Friis-Christensen, Elizabeth A. Fulton, Alik Ismail-Zadeh, David Johnston, Fumiko Kasuga, Yekaterina Y. Kontar, Jianping Li, Trevor McDougall, Albert McGill, Michael E. Meadows, Richard Munang, Lawrence A. Mysak, Godwin D. Ndossi, Jane Rovins, Ladislaus Rybach, Serhat Sensoy, Michael G. Sideris, Ramesh Singh, R. B. Singh, Walter Spiess, Makoto Taniguchi, Brian Toon, Tonie van Dam, and Mark L. Wahlqvist.

Special thanks go to the staff of Cambridge University Press – Susan Francis, Sarah Lambert, Zoë Pruce – for their guidance throughout the production of the book, and to Alik Ismail-Zadeh, the Editor-in-Chief of the IUGG Special Publication Series, for his guidance, contributions, and support.

Figure 4 Logo of the IUGG Commission for Climatic and Environmental Change. For an explanation of the symbolism see http://ccec-iugg.org/content/explanation-logo.

Abbreviations

AR5	Fifth Assessment Report of IPCC
AR6	Sixth Assessment Report of IPCC
ASTAAG	Asia Science, Technology and Academia Advisory Group
CCEC	Commission on Climatic and Environmental Change (of IUGG)
CCl	Commission for Climatology of WMO
CFC	chlorofluorocarbon
CH_4	methane
CLIVAR	CLImate VARiability and predictability (Core Project on Climate and Ocean)
CMIP5	Coupled Model Inter-Comparison Project Phase 5
CO_2	carbon dioxide
CO_3^{2-}	carbonate
COBSEA	Coordinating Body on the Seas of East Asia
COP	conference of parties
DIVERSITAS	an international program of biodiversity science
DRR	disaster risk reduction
EEZ	Exclusive Economic Zone
FAO	Food and Agricultural Organization
GCM	Global Climate Model, and also General Circulation Model
GDP	gross domestic product
GHG	greenhouse gas
GOOS	Global Ocean Observing System
H^+	hydrogen ion
H_2CO_3	carbonic acid
H_2S	hydrogen sulfide
HCO_3^-	bicarbonate
HELCOM	Helsinki Commission
IAMAS	International Association of Meteorology and Atmospheric Sciences
IAG	International Association of Geodesy
ICSU	International Council for Science, now known as the International Science Council
IDMC	Internal Displacement Monitoring Centre
IDP	Internally Displaced People
IGBP	International Geosphere-Biosphere Programme
IHDP	International Human Dimensions Programme
IMO	International Maritime Organization
IOC	Intergovernmental Oceanographic Commission
IOM	International Organization for Migration
IPBES	Intergovernmental Platform on Biodiversity and Ecosystem Services
IPCC	Intergovernmental Panel on Climate Change
IRDR	Integrated Research on Disaster Risk
ISA	International Seabed Authority
ISSC	International Social Science Council (now merged with ICSU to become ISC)
IUFoST	International Union of Food Science and Technology
IUGG	International Union of Geodesy and Geophysics
IUNS	International Union of Nutrition Sciences
KAN	Knowledge-Action Network
LDC	least developed country
MDG	Millennium Development Goal
MPA	marine protected area
MSP	marine spatial planning
N_2O	nitrous oxide
NAPA	National Adaptation Programme of Action
Ocean KAN	Ocean Knowledge-Action Network
OSPAR	Oslo-Paris Convention
OWG	open working group
ppm	part per million
RCP	representative concentration pathway scenario
ROPME	Regional Organisation for Protection of the Marine Environment
RSP	Regional Seas Programme
SCOR	Scientific Committee on Ocean Research
SDG	sustainable development goal
SDSN	Sustainable Development Solutions Network

SIDS	Small Island Developing State
SLR	sea level rise
SO_2	sulfur dioxide
SREX	Special Report on Managing the Risks of Extreme Events and Disaster Advance Climate Change Adaptation of IPCC
SSP	Shared Socio-Economic Pathway Scenario
UN	United Nations
UNCLOS	United Nations Convention of the Law of the Sea
UNEP	United Nations Environmental Programme
UNESCO	United Nations Educational, Scientific and Cultural Organization
UNFCCC	United Nations Framework Convention on Climate Change
UNHCR	United Nations High Commissioner for Refugees
UNISDR	United Nations Office for Disaster Risk Reduction
URL	Uniform Resource Locator
UV	ultraviolet radiation
WCRP	World Climate Research Programme
WHO	World Health Organization
WMO	World Meteorological Organization

PART I

Future Earth and Planetary Issues

1
International Drivers to Study Climatic and Environmental Change: A Challenge to Scientific Unions

Tom Beer

1.1 Climatic Change

An English language purist will point out that the word 'climate' is a noun whereas the adjective from this word is 'climatic' so that the correct term for what is popularly called 'climate change' is actually 'climatic change'. The Springer scientific journal *Climatic Change*[1] uses correct English. Thus, even though this chapter will deal with climatic change and environmental change, the term climate change is synonymous, and will be used interchangeably, with climatic change.

Though the term climate change has become standard usage, there is less standardisation in relation to the term that is used to cover the *effects* of climate change on the ambient and social environment, generally through studies of impacts, adaptation and vulnerability. This chapter will use the term 'environmental change' but the international research community also uses the term 'global change' or 'global environmental change' for the same phenomenon.

Climatic change as a scientific discipline includes studies of processes, detection and attribution (which in this context means separating out the influences of climatic variability from the influences of climatic change) and projections of climatic change into the future and its effects on sea level and on coasts. Climatic change research can also involve detailed measurement or modelling of the greenhouse gas (GHG) in the atmosphere. In this context, the phrase 'detection and attribution' is used to determine the nations or industries responsible for the emissions of particular gases.

This chapter will examine the international political drivers in relation to climate change, the international inter-governmental drivers, the international scientific response and the role of scientific unions focussing on sustainability and then determine where the organisational and research gaps lie.

1.2 International Political Drivers in Relation to Climate Change

The General Assembly of the United Nations called for a Framework Convention on Climate Change (UNFCCC) in 1990, a call that was supported at the Second World Climate Conference in 1990 so that the convention was finally adopted in New York in May 1992. It was opened for signatures at the Intergovernmental Conference on Sustainable Development, held in Rio de Janeiro in 1992. More than 175 countries have ratified the UNFCCC. It is called a 'framework' convention because it leaves the details of implementation to be worked out later, by a series of protocols or other legal devices or agreements to be adopted by the member countries at the Conference of Parties (COP).

The first of these was the 1997 Kyoto Protocol in which, for the period from 2008 to 2012, greenhouse gas reduction measures were agreed. The scope of the protocol was extended until 2020 with the Doha Amendment to that protocol in 2012.[2]

The objective of the UNFCCC is stated in Article 2 to be:

> the stabilisation of greenhouse gas concentrations in the atmosphere at a level that would prevent dangerous anthropogenic interference with the climate system. Such a level should be achieved within a timeframe sufficient to all ecosystems to adapt naturally to climate change, to ensure that food production is not threatened, and to enable economic development to proceed in a sustainable manner.

The UNFCCC contained no binding commitments on emissions levels, but it did lay down some general principles and objectives to shape future negotiations on these commitments. Pittock (2009) notes that these included that

- developed countries (most members of the Organisation for Economic Cooperation and

Development, OECD), plus many former communist states undergoing transition to a market economy), should take the lead with abatement measures;
- the climate and economic vulnerabilities of developing states should be recognised;
- abatement should be consistent with sustainable development and not infringe the goals of an open and supportive international economy.

These provisions, and negotiations towards their implementation, led to argument between member countries (termed 'parties' to the Convention), especially over the contents and implementation of the Kyoto Protocol adopted in 1997. These debates were originally clouded by uncertainties as to the actual risk from climate change, and uncertainties regarding the costs of impacts and of abatement measures. By the time of the 2015 COP in Paris there was general agreement that a global mean temperature rise of 2 degrees Celsius (°C) would lead to unacceptable warming so that the Paris Agreement[3] (French: *L'accord de Paris*) was an agreement to set a goal of limiting global warming to less than 2 degrees Celsius (°C) compared to pre-industrial levels. The agreement calls for zero net anthropogenic greenhouse gas emissions to be reached during the second half of the twenty-first century. In the final adopted version of the Paris Agreement, the parties will also 'pursue efforts to' limit the temperature increase to 1.5 °C. The 1.5 °C goal will require zero emission sometime between 2030 and 2050. However, no detailed timetable or country-specific goals for emissions were incorporated into the Paris Agreement – as opposed to the previous Kyoto Protocol.

1.3 International Inter-Governmental Drivers

How does one set up a formal process to incorporate scientific advice into the international political process? This question came to the fore in relation to the Montreal Protocol. From the previous information about the UNFCCC it is apparent that in the UN system a protocol relates to a convention. One of the most successful, if not *the* most successful, scientific protocols was the Montreal Protocol on Substances That Deplete the Ozone Layer,[4] which is a protocol to the Vienna Convention for the Protection of the Ozone Layer, an international treaty designed to protect the ozone layer by phasing out the production of numerous substances that are responsible for ozone depletion. It was agreed on 16 September 1987.

Due to its widespread adoption and implementation the Montreal Protocol been hailed as an example of exceptional international co-operation, with UN Secretary-General Kofi Annan, in his presentation to the Millennium Assembly of the United Nations in September 2000, saying that 'perhaps the single most successful international agreement to date has been the Montreal Protocol'. In comparison, effective burden sharing and solution proposals mitigating regional conflicts of interest have been among the success factors for the ozone depletion challenge, where global regulation based on the Kyoto Protocol has failed to be. The two ozone treaties (the Vienna Convention and the Montreal Protocol) have been ratified by 197 parties, which includes 196 states and the European Union, making them the first universally ratified treaties in United Nations history.

In the case of the ozone layer it took only thirteen years for science to be incorporated into international policy. The science that led to the Montreal Protocol (Molina and Rowland, 1974) was published in 1974, after the discovery that chlorofluorocarbon (CFC) molecules were stable enough to remain in the atmosphere until they were transported into the middle of the stratosphere, where they would finally (after an average of 50–100 years for two common CFCs) be broken down by ultraviolet radiation, releasing a chlorine atom, which would then react with the ozone in the stratosphere

Then, Farman et al. (1985) noted abnormally low ozone concentrations above Halley Bay near the South Pole. They speculated that this was connected to increased levels of CFCs in the atmosphere. Subsequent work confirmed the reality of the Antarctic losses and established them to be significant. The impact of these studies, the metaphor 'ozone hole', and the colourful visual representation in a time lapse animation proved shocking enough for negotiators in Montreal to take the issue seriously.

The science in relation to climate change was less clear-cut, and some climate change sceptics would claim that it is still less clear-cut. Because of this the bodies responsible for administration of the UNFCCC – the World Meteorological Organization (WMO) and the United Nations Environment Programme (UNEP) – established the Intergovernmental Panel on Climate Change (IPCC) to assess scientific, technical and socio-economic information concerning climate change, its potential effects and options for adaptation and mitigation.

1.3.1 The Work of the IPCC

As implied previously, the major task of the IPCC[5] is to produce assessments. These provide a scientific basis for governments at all levels to develop climate-related

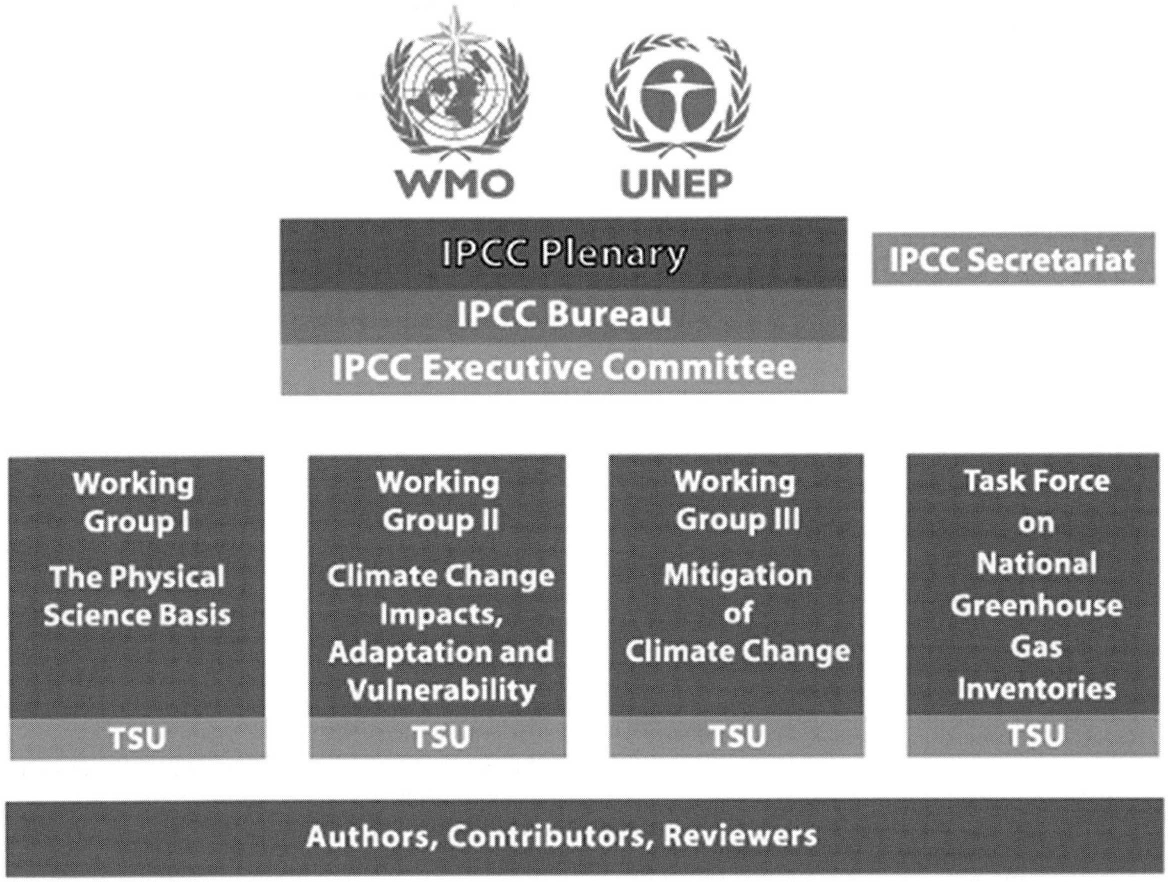

Figure 1.1 Diagram illustrating the structure of the Intergovernmental Panel on Climate Change (reproduced from www.ipcc.ch/organization/organization_structure.shtml with permission of the IPCC).

policies, and the IPCC assessments underlie negotiations at the various UNFCCC conferences. The assessments are policy-relevant but not policy-prescriptive: they may present projections of future climate change based on different scenarios and the risks that climate change poses and discuss thc implications of response options, but they do not tell policymakers what actions to take.

According to Article 14 of the Paris Agreement, Nationally Determined Contributions (NDCs) will be revised on a five-year basis through a global stocktaking mechanism established under the UNFCCC, which will start in 2023. The IPCC sixth assessment report, known as AR6, is to be finished by 2022; hence, all elements of the AR6 will be available for consideration by the global stocktake in 2023, and the IPCC will be in a position to provide the policy-relevant scientific input to the global stocktaking.

The IPCC also produces special reports, which are an assessment on a specific issue, and methodology reports that provide practical guidelines for the preparation of greenhouse gas inventories.

The IPCC scientific work is currently organized by three working groups and a task force (Figure 1.1). They are assisted by technical support units (TSUs), which are hosted and financially supported by the government of the developed country co-chair of that working group or task force. Working Group I deals with The Physical Science Basis of Climate Change, Working Group II with Climate Change Impacts, Adaptation and Vulnerability and Working Group III with Mitigation of Climate Change. The main objective of the Task Force on National Greenhouse Gas Inventories is to develop and refine a methodology for the calculation and reporting of national greenhouse gas emissions and removals.

1.3.1.1 Special Reports

The IPCC has produced various special reports and continues to do so. During the Sixth Assessment Report (AR6) cycle there will be three special reports: one on the impacts of global warming of 1.5 °C above pre-industrial levels, a second on climate change and oceans and the cryosphereand a third on climate change, desertification, land degradation, sustainable land management, food security and greenhouse gas fluxes in terrestrial ecosystems. Recently completed special reports include the *Special Report on Renewable Energy Sources and Climate Change Mitigation* (SRREN) and the *Special Report on Managing Risks of Extreme Events and Disasters to Advance Climate Change Adaptation* (SREX).

The work of the IPCC provides an outstanding example of risk assessment in action. Beer (1997) noted that

> the IPCC process is an ongoing risk assessment of the consequences of climate change. But the noteworthy aspect of it is that, for a problem of such global scale and long temporal duration, the appropriate assessment involves: international cooperation; the involvement of numerous scientists, technologists, environmentalists, economists, politicians, lawyers and other groupings, and as a result takes time and money. Yet there is no doubt that a quick assessment undertaken by a small focused group would fail to win the consensus support that concerted action requires and fail to spread ownership of the consensus across a sufficiently wide cross-section of the community.

From the political perspective, the IPCC process has been a great success that was acknowledged in 2007 when the Nobel Committee awarded one-half of the Nobel Prize in peace to the IPCC. However, from the scientific perspective, and especially from the viewpoint of an individual scientist, it is constrained by its inter-governmental nature. The various IPCC assessment reports are reviews of the scientific literature. The IPCC process does not exist to generate the scientific literature but to review and assess it.

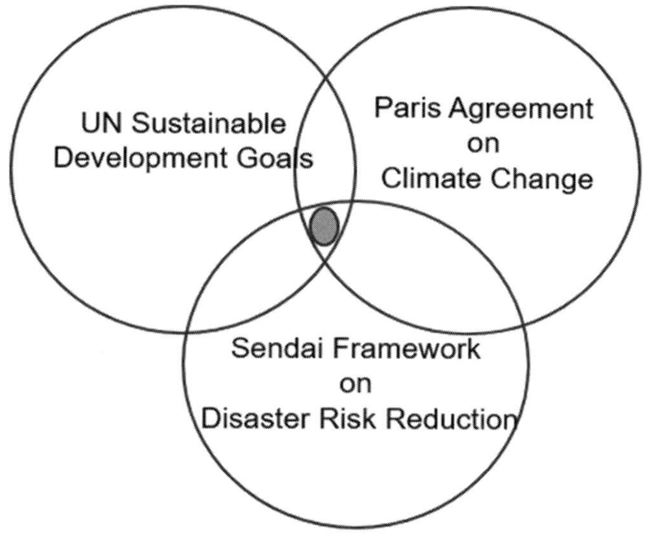

Figure 1.2 Relations between the major international drivers. The spot at the intersection of the three circles is meant to indicate that the greatest research potential for scientific unions is to be found at this point.

1.4 Other International Drivers

The year 2015 was an important one. The Paris agreement has already been mentioned. In addition, two other relevant international agreements were achieved. In terms of disaster risk reduction, assembled delegates approved the Sendai Framework for Disaster Risk Reduction,[6] whereas in terms of sustainability, at a UN summit in September 2015 world leaders adopted the Sustainable Development Goals,[7] discussed in more detail later on, which officially came into force on 1 January 2016. Over the next fifteen years, with these new goals that universally apply to all, countries will mobilize efforts to end all forms of poverty, fight inequalities and tackle climate change, while ensuring that no one is left behind.

This chapter, and indeed this whole volume, is predicated on the assumption that international scientific union members of the International Council for Science (ICSU) can produce worthwhile and significant achievements if they collaborate in areas that span two or more of these international agreements, as illustrated diagrammatically in Figure 1.2. ICSU itself has felt that the best way to meet the existing and future challenges facing science and society is to merge with the International Social Sciences Council (ISSC). In October 2017 the membership of ICSU agreed to such a merger.

1.5 The International Scientific Response

The international process that has been established to assist researchers to generate the scientific literature is the World Climate Research Programme (WCRP),[8] an international scientific research programme jointly sponsored by the World Meteorological Organization (WMO), the International Council for Science (ICSU) and the Intergovernmental Oceanographic Commission (IOC) of UNESCO.

The WCRP mission is to facilitate analysis and prediction of Earth system variability and change for use in an increasing range of practical applications of direct relevance, benefit and value to society. The two overarching objectives of the WCRP are

- to determine the predictability of climate, and
- to determine the effect of human activities on climate.

Recent progress in the understanding of climate system variability and change makes it possible to gauge its predictability, and to use this predictive knowledge in developing adaptation and mitigation strategies. Such strategies assist global communities in responding to the impacts of climate variability and change on major social and economic sectors including food security, energy and transport, environment, health and water resources.

The main foci of WCRP research are:

- observing changes in the components of the Earth system (atmosphere, oceans, land and cryosphere) and in the interfaces between these components;
- improving our knowledge and understanding of global and regional climate variability and change, and of the mechanisms responsible for this change;
- assessing and attributing significant trends in global and regional climates;
- developing and improving numerical models that are capable of simulating and assessing the climate system for a wide range of space and time scales and
- investigating the sensitivity of the climate system to natural and human-induced forcing and estimating the changes resulting from specific disturbing influences.

In actual fact, WCRP was established in 1980 – a testament to the prescience of the scientific community. Its mandate is to study the climate, rather than climate change in particular. Accordingly, once it became apparent to the scientific community that what was needed was scientific research targeted at the science, effects, impacts and human response to climate change, then-new scientific programmes were established. The International Geosphere Biosphere Programme (IGBP) was set up in 1986 to study the total Earth system, the changes that are occurring and the manner in which changes are influenced by human actions. The International Human Dimensions Programme (IHDP) was set up in 1996 to address the coupled human-natural system in the context of global environmental change (GEC) and to examine ways in which individuals and societies

- contribute to global environmental change,
- are influenced by global environmental change and mitigate and adapt to global environmental change.

A fourth programme, DIVERSITAS, established in 1991, belongs to a family of four global change programmes (IGBP, WCRP, IHDP and DIVERSITAS) that have ICSU as a common sponsor. DIVERSITAS existed to promote integrative biodiversity science, linking biological, ecological and social disciplines in an effort to produce socially relevant new knowledge and to provide the scientific basis for an understanding of biodiversity loss, and to draw out the implications for policies for conservation and sustainable use of biodiversity.

At the end of 2014, the projects under IGBP, IHDP and DIVERSITAS were all transferred into the new global change research programme Future Earth,[9] that aims to study – along with WCRP – the Anthropocene. 'Anthropocene' is a term (Crutzen and Stoermer, 2000) coined in relation to humans affecting the Earth and its geology. Conversely – or perhaps in cyclical tandem – changes in our Earth and its geology affect humans, physically and emotionally.

All of these initiatives entrained many working scientists into research programmes and projects to examine global change and provided an impetus for many national governments to become involved with such research. Nevertheless, one could argue that the initiatives really occurred at a level above that of the normal researcher. The programmes were initiatives of the International Council for Science (ICSU), which is a body to which most scientists do not feel any attachment. Scientists feel attachment to their own scientific union. Chemists identify with the International Union of Pure and Applied Chemistry. Physicists identify with the International Union of Pure and Applied Physics. Earth Scientists identify with the International Union of Geodesy and Geophysics. This is normal. The national and international analogy is with the United Nations. Citizens identify with their own country, even if they happen to be involved in a programme for one of the United Nations agencies.

1.6 The Role of Scientific Unions

The upsurge of interest in global change issues strengthened the resolve of the ICSU Unions involved in the sciences that study the Earth to work together. They formed an alliance (Figure 1.3) known as the GeoUnions,[10] involving the International Astronomical Union (IAU), the International Geographical Union (IGU), the International Union for Quaternary Research (INQUA), the International Society for Photogrammetry and Remote Sensing (ISPRS), the International Union of Geodesy and Geophysics

Figure 1.3 ICSU GeoUnions (reprinted from the ICSU GeoUnions website (with permission from A. Ismail-Zadeh)).

(IUGG), the International Union of Geological Sciences (IUGS), the International Union of Soil Sciences (IUSS), the International Union of Radio Science (URSI) and a later member, the International Cartographic Association (ICA).

One of the issues for the GeoUnions, separately and collectively, is how to respond to the international activity that has been generated by initiation of Future Earth. The reason is that by combining the existing programmes[11] of IGBP, IHDP and DIVERSITAS there was no obvious niche for the establishment of new research programmes.

In a document addressing this issue, De Mulder et al. (2015) note that 'as Future Earth is intended to be a "global research platform designed to provide knowledge needed to support transformations towards sustainability" all four spheres of System Earth should be included, the atmosphere, the hydrosphere, the biosphere and the geosphere. In Future Earth priorities as given in Future Earth's Strategic Research Agenda, the authors observe a distinct lack of attention to the geosphere and its processes.'

Since then, Future Earth developed a mechanism to incorporate new activities in the form of Knowledge Action Networks,[12] with the following eight topics: Water–Energy–Food Nexus, Oceans, Transformations, Natural Assets, Sustainable Development Goals, Cities, Health and Finance & Economics.

Even though it may be difficult for the GeoUnions, as a collective, to find a niche in Future Earth, various initiatives of the individual unions are either already part of Future Earth or were established to foster such interaction. For example, The International Global Atmospheric Chemistry (IGAC) project[13] is a non-profit organisation created in the late 1980s to address growing international concerns over rapid changes observed in Earth's atmosphere.

IGAC developed under joint sponsorship of the Commission on Atmospheric Chemistry and Global Pollution (CACGP) of the International Association of Meteorology and Atmospheric Sciences (IAMAS) and the International Geosphere-Biosphere Programme (IGBP) and became part of Future Earth when IGBP was incorporated into Future Earth. IAMAS is one of the eight constituent scientific associations of IUGG.

Within IUGG, there are two separate commissions that exist to deal with climate change. Within IAMAS itself there exists the International Commission on Climate (ICCL[14]), a group with a long history of bringing together physical scientists who study climate. In 2012, IUGG established a union commission – the Commission on Climatic and Environmental Change (CCEC [15]) to

a build scientific capacity for responsibly addressing the broad, multi-disciplinary issues involved in climatic and environmental change;
b provide useful information, understanding and support to the public and governmental organisations;
c interact and cooperate with outside activities that would benefit from the capabilities and resources of the IUGG Associations; and
d strengthen links across the scientific associations within IUGG, to build new external links to organisations outside IUGG, to strengthen existing links to external organisations and to promote IUGG's contribution to global change research.

Figure 1.4 Illustration of the seventeen UN Sustainable Development Goals.

1.7 Sustainability

Sustainable development ('sustainability' for short) has been defined in many ways; the most frequently quoted definition is from Our Common Future, also known as the Brundtland Report: 'Sustainable development is development that meets the needs of the present without compromising the ability of future generations to meet their own needs.' The report continues: 'It contains within it two key concepts: the concept of "needs", in particular the essential needs of the world's poor, to which overriding priority should be given; and the idea of limitations imposed by the state of technology and social organization on the environment's ability to meet present and future needs'.

The United Nations has entered a fifteen-year development goal package, from 2016 to 2030, known as the 17 Sustainable Development Goals (SDGs), that are illustrated in Figure 1.4. These show an understanding of the bi-directional impacts between humans and Earth. The original MDGs were included, goals relating to the environment were better defined and the importance of globally shared and delegated roles and responsibilities was further highlighted. The December 2015 Paris Agreement includes also acknowledgement 'that climate change is a common concern of humankind' and of an 'intrinsic relationship' between climatic change and poverty, hunger, equality, health, education, etc.

The scientific community has, to date

- Undertaken a review of the targets (ICSU, ISSC 2015). There are 17 Goals but 169 targets;
- Produced a working paper on a framework for SDG interactions, with a commentary that was published in *Nature*[16] (Nilsson et al., 2016);
- Produced a report (International Council for Science, 2017) on SDG interactions focussing on the detailed interactions for Goals 2 (Hunger), 3 (Health), 7 (Energy) and 14 (Water).

1.7.1 Quantifying Sustainability

The ecological footprint is probably the best known way of quantifying sustainability. The method measures thc biologically productive land and water area required to produce the resources consumed and to assimilate the wastes generated using prevailing technology. This area, called the Ecological Footprint, represents the fraction of the biosphere necessary to maintain the current material throughput of the human economy under current management and production practices (Wackernagel et al. 2002; Amekudzi et al. 2015).

These calculations indicate that at present humanity uses the equivalent of 1.6 planets to provide the resources we use and absorb our waste. This means it now takes the Earth one year and six months to

Table 1.1 *How the International Scientific Association That Comprises the International Union of Geodesy and Geophysics (IUGG) Can Assist in Analysis of the Sustainable Development Goals*

Mapping IUGG/Associatious to the UN Sustainable Development Goals

Goal	IUGG	IACS	IAG	IAGA	IAHS	IAMAS	IAPSO	IASPEI	IAVCEI
1.5	CCEC, GRC	x	x	x	x	x	x	x	x
2.4	CCEC, GRC	x	x		x	x	x	x	x
3.9					x	x			
4.7	x	x	x	x	x	x	x	x	x
4.b	x	x	x	x	x	x	x	x	x
5.5	x	x	x	x	x	x	x	x	x
6.3–6.7		x				x			
7.a					x	x	x	x	x
9.5–9.6	x	x	x	x	x	x	x	x	x
9.6	x	x	x	x	x	x	x	x	x
11.5	GRC	x	x	x	x	x	x	x	x
11.6	CCEC					x			
11.b	GRC	x	x	x	x	x	x	x	x
13.1	CCEC, GRC	x	x	x	x	x	x	x	x
13.3	CCEC	x			x	x	x		
14.1–14.3							x		
14.a, 14.c							x		
15.1		x			x	x		x	
15.3	GRC				x	x			
16.6–16.7	x	x	x	x	x	x	x	x	x
17.6	x	x	x	x	x	x	x	x	x
17.16	x								
17.18	UCDI		x	x			x		

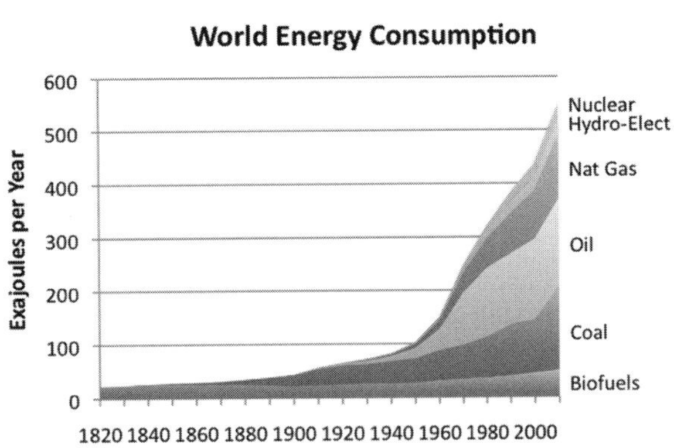

Figure 1.5 World Energy Consumption (from https://ourfiniteworld.com/2012/03/12/world-energy-consumption-since-1820-in-charts/).

regenerate what we use in a year.[17] If we re-examine this issue from an energetic perspective, in 2008 the world consumed, through fossil fuel burning, 48.5×10^{19} J. Over a year the sun radiates 40.24×10^{23} J on the Earth's surface. If we assume that 1 per cent of this is available for photosynthesis, then in energetic terms the fossil fuel consumption is 1 per cent of the energy available for photosynthesis. However if energy consumption continues to grow as shown in Figure 1.5, then eventually it will exceed available. Haberl et al. (2007) introduced the concept of 'human appropriation of the world's net primary productivity (HANPP)', as an indicator of the amount and intensity of land use by humans. It is measured as a percentage of total potential vegetation, which in turn is a measure of the incoming solar radiation and has been estimated to have been between 23 and 45 per cent around the year 2000.

Hall et al. (2014) show that the energy return on energy invested in fossil fuel production is declining, indicating that more and more effort is required to extract the fossil fuel needed by the world.

1.7.2 The Role of Geodesists and Geophysicists

Table 1.1 shows how the international geophysical community can assist in analysis of the Sustainable Development Goals. The particular goals and targets that bring together climatic and environmental change (shown as CCEC in Table 1.1) are:

Goal 1 End poverty in all its forms everywhere

1.5 By 2030, build the resilience of the poor and those in vulnerable situations and reduce their exposure and vulnerability to climate-related extreme events and other economic, social and environmental shocks and disasters

Goal 2 End hunger, achieve food security and improved nutrition and promote sustainable agriculture

2.4 By 2030, ensure sustainable food production systems and implement resilient agricultural practices that increase productivity and production, that help maintain ecosystems, that strengthen capacity for adaptation to climate change, extreme weather, drought, flooding and other disasters and that progressively improve land and soil quality

Goal 11 Make cities and human settlements inclusive, safe, resilient and sustainable

11.6 By 2030, reduce the adverse per capita environmental impact of cities, including by paying special attention to air quality and municipal and other waste management

11.B By 2020, substantially increase the number of cities and human settlements adopting and implementing integrated policies and plans towards inclusion, resource efficiency, mitigation and adaptation to climate change, resilience to disasters, and develop and implement, in line with the Sendai Framework for Disaster Risk Reduction 2015–2030, holistic disaster risk management at all levels

Goal 13 Take urgent action to combat climate change and its impacts

13.1 Strengthen resilience and adaptive capacity to climate-related hazards and natural disasters in all countries

13.3 Improve education, awareness-raising and human and institutional capacity on climate change mitigation, adaptation, impact reduction and early warning

1.7.3 Neo-Cornucopianism and Malthusianism

All nations rely on agriculture for a large proportion of food production and, increasingly, as a source of energy. in the late 1700s the economist Thomas Malthus suggested that agricultural production was a major constraint on population growth. However, the increase of land available to agriculture, coupled with improvements in productivity, have seen agricultural output outpace population growth, especially since the 'Green Revolution' that commenced in the mid-twentieth century. The Green Revolution has contributed to a significant reduction in the proportion of the global population that is undernourished.

1.7.3.1 Food

Concerning the use of solar energy, the Malthusian prognosis about population outrunning food output on a global scale has not been realized, principally because the application of science and technology has provided new ways to increase production. Plant breeding is one of many examples that show a striking increase in performance far in excess of anything that could have been foreseen three hundred years ago by Malthus. Gains in cereal production, not least in Europe, one of the world's major grain producers, have progressed linearly over the last fifty years. The success of modern intensive agriculture, however, has occurred simultaneously with the loss of land to urbanisation and has incurred soil degradation, erosion, desertification and a rising dependence on irrigation. The picture that emerges is one of unsustainable production. Modern bioengineering is being used to develop high-yielding crops that can be grown in more efficient systems that require fewer chemical applications and produce lower levels of environmental pollutants. In Europe the successful adoption of these procedures is not yet assured. Ethical debates about safety, utility, fairness, naturalness, equity and the 'slippery slope' mean that, whereas Europe has invented and developed much of the science and many of the skills, it has yet to come to terms with the public acceptability of biotechnology.

The Earth's population, currently 7.2 billion, is expected to rise at a rapid rate over the next forty years. Current projections state that the Earth will need to support 9.6 billion people by the year 2050, a figure that climbs to nearly 11 billion by 2100 (United Nations, 2012). At the same time, most people envision a future Earth with a greater average standard of living than currently exists – and, as a result, greater consumption of planetary resources. How can this population growth be achieved in a manner that is sustainable from an economic, social and environmental perspective?

Odum (1973) pointed out that fossil fuels, and in particular oil, have allowed the Green Revolution to take place. Firstly, fossil fuels provide the energy to run the tractors that sow the seeds, the harvesters that harvest it, the trucks that distribute it and the factories that process it. Secondly, fossil fuels are used to provide the high temperatures and pressure required for the Haber process that is used to create nitrogenous fertiliser.

These interactions have led a number of people to suggest that world population and energy sources are correlated, and if there is a decline in energy availability, then there will be a decline in population. [18] Whether this is, or is not, the case is still to be determined. However, a general ecological principle is that the size of a biological population is determined by the limiting nutrient. Whether, in the case of human populations, that limiter is food or energy or water (or something else) is not clear because of the interrelations between the three. Energy availability determines water availability and food availability. In addition, the three are geographically dependent.

Pimental and Pimental (2008) have examined the environmental toll of modern food production. Every megajoule (MJ) of industrialised food, as consumed in developed countries, requires approximately 10 MJ of fossil fuel energy to plant, fertilise, harvest, ship and process the food. Instead of using energy units, an alternative way of undertaking such life-cycle calculations is to calculate the greenhouse gases emitted. Grant and Beer (2008), for example, found that the production of 400 g of maize-derived corn chips emits 530 g of carbon dioxide equivalents.

Though world food production is sufficient to feed the world's present population, provided that wastage can be eliminated, the difficulties of distribution mean that there remain areas of the world, especially in Africa, where food is in short supply (National Research Council, 2014). Energy availability is also geographically heterogeneous. The middle east has abundant supplies of oil, whereas Japan, for example, does not. And water availability is so critical that humans find it difficult to live in deserts unless they can tap into groundwater resources, or divert water from rivers flowing through or near the desert.

1.8 Gaps

The Union Commission on Climatic and Environmental Change (CCEC, as mentioned in Table 1.1) was established by the Executive Committee of IUGG in 2012 in order to promote the advancement of scientific understanding of climatic and environmental change, to boost research in reducing uncertainties in climate and environmental models, to define criteria for collaborative trans-disciplinary research on climate and environmental change, to fulfil the objectives of IUGG and its associations, to provide an all-Union perspective on climatic and environmental change and to make available the knowledge and insights developed through scientific research to the benefit of society and planet Earth, including consideration of the science of global change, related vulnerability and impacts and potential responses.

One of its objectives is to strengthen links across the scientific associations within IUGG, to build new external links to organisations outside IUGG, to strengthen existing links to external organisations and to promote IUGG's contribution to global change research.

As shown in Table 1.2, there are three key UN initiatives that deal with climatic and environmental change – climate change, sustainability and disasters. Geodesy and geophysics can provide the underpinning science to enable other disciplines and other practitioners to examine the long-term issues that are involved.

One of the niche areas that the commission identified as a gap relates to the issue of weather, climate and food security. Climate change and food security was examined in detail in chapter 7 of the IPCC Fifth Assessment Report Working Group 2 report.[19] The link between climate change and disasters was examined in detail in the IPCC Special Report on Extremes.[20] However, the food security issues that occur in the aftermath of a disaster appear to have been inadequately studied and constitute a research gap that could be addressed by the Geounions, working co-operatively with other ICSU unions and the ICSU programme Integrated Research on Disaster Risk.

1.9 Summary

Climate Change has now become a significant entity in both scientific and political discourse. This has led

Table 1.2 *Key UN Programmes and Their Internal and External Scientific Responses*

Political imperative	Internal UN Scientific Response	International Scientific Response by ICSU
Climate change (UNFCCC)	Inter-Governmental Panel on Climate Change	WCRP
Sustainable Development (UNDESA)	Technology Facilitation Mechanism	Future Earth
Sendai Framework (UNISDR)	Science and Technology Advisory Group	IRDR

many organisations to reflect on the best way that their existing members can respond to these new and significant challenges.

Institutional responses can be categorised into three:

1 Develop a new body – either by merger or by creation of a new body
2 Collaborate with like-minded individuals
3 Develop a new programme

1.9.1 Develop a New Body – Either by Merger or by Creation of a New Body

This option has been adopted by ICSU both for itself (in its plans to merge with ISSC) and for its subsidiary bodies such as Future Earth, which merged three of the existing research programmes of ICSU. This also appears to have been the option adopted by UN-ISDR when seeking to determine a mechanism for the receipt of scientific advice. They set up a new body called the Science and Technology Advisory Group.

1.9.2 Collaborate with Like-Minded Individuals

This option has been favoured by many of the scientific unions composing ICSU and is most notable in the Geo-Unions grouping. Such an approach can be successful, but fails to deal with the more difficult issue – how to collaborate with those that are not necessarily like-minded. Is this possible? Is it desirable? Is it necessary? It is certainly difficult, but we maintain that it is necessary in order to study and ideally to solve many of the problems involved in sustainability. The hope is that sustained dialogue of individuals who are initially not like-minded will lead to a final convergence of like-mindedness.

1.9.3 Develop a New Program

The World Meteorological Organisation (WMO), which is a co-sponsor of the WCRP, decided that a means to operationalise climate outputs to their membership of national meteorological services would be to develop a Global Framework for Climate Services (GFCS)[21] to link the priority areas of agriculture and food security, disaster risk reduction, energy, health and water.

These approaches are not mutually exclusive. Developing a new programme generally requires collaboration with a large group of potential participants. A successful program may achieve such success that it becomes a new permanent body. The common thread is that they are all different ways to meet the recent challenges arising from the international drivers of climate change, disasters and sustainability at the international governmental level, and at the international scientific research level.

Internet Resources

[1] http://link.springer.com/journal/10584
[2] https://en.wikipedia.org/wiki/Paris_Agreement
[3] http://unfccc.int/paris_agreement/items/9485.php
[4] http://ozone.unep.org/en/treaties-and-decisions/montreal-protocol-substances-deplete-ozone-layer
[5] www.ipcc.ch/
[6] www.unisdr.org/we/coordinate/sendai-framework
[7] www.un.org/sustainabledevelopment/sustainable-development-goals/
[8] www.wcrp-climate.org/
[9] www.futureearth.org/
[10] http://icsu-geounions.org/
[11] www.futureearth.org/projects
[12] www.futureearth.org/knowledge-action-networks
[13] www.futureearth.org/projects/igac-international-global-atmospheric-chemistry
[14] www.iccl-iamas.net/
[15] http://ccec-iugg.org/
[16] www.nature.com/news/policy-map-the-interactions-between-sustainable-development-goals-1.20075
[17] www.footprintnetwork.org/en/index.php/GFN/page/world_footprint/
[18] https://ourfiniteworld.com/2012/03/12/world-energy-consumption-since-1820-in-charts/
[19] www.ipcc.ch/report/ar5/wg2/
[20] www.ipcc.ch/publications_and_data/publications_and_data_reports.shtml#2
[21] www.wmo.int/gfcs/

References

Amekudzi, A., M. Khayesi, and C. J. Khisty (2015) Sustainable development footprint: a framework for assessing sustainable development risks and opportunities in time and space, *Int. J. Sustainable Development* 18, 1/2: 9–40.

Beer, T. (1997) Strategic risk management: a case study of climate change, *World Resource Review* 9: 151–164.

Brundtland, G. H. (1987) Report of the World Commission on Environment and Development: Our Common Future (Transmitted to the UN General Assembly as an Annex to document A/42/427 – Development and International Co-Operation: Environment), available at: www.un-documents.net/wced-ocf.htm.

Crutzen, P. J. and E. F. Stoermer (2000) The 'Anthropocene' *Global Change Newsletter* 41: 17–18.

De Mulder, E. F. J., R. Oberhänsli, Q. Cheng, S. Cloetingh, R. Finkelman, J. Martinez-Frías H. Gupta, M. Meadows, A. Mogessie, and W. Meng (2015) GeoSciences for Future Earth Research, available at: http://icsu-geounions.org/files/Geosci_FE.pdf.

Farman, J. C., B. G. Gardiner, and J. D. Shanklin (1985) Large losses of total ozone in Antarctica reveal seasonal ClO^x/NO^x interaction, *Nature* 315: 207–210.

Grant, T. and T. Beer (2008) Life-cycle assessment of greenhouse gas emissions from irrigated maize and their significance in the value chain, *Aust. J. Experimental Agriculture* 48: 375–381.

Haberl, H., K.-H. Erb, F. Krausmann, V. Gaube, A. Bondeau, C. Plutzar, S. Gingrich, W. Lucht, and M. Fischer-Kowalski (2007) Quantifying and mapping the human appropriation of net primary production in Earth's terrestrial ecosystems. *Proceedings of the National Academy of Sciences* 104: 12942–12947.

Hall, C. A. S., Jessica G. Lambert, and Stephen B. Balogh (2014) EROI of different fuels and the implications for society, *Energy Policy* 64: 141–152. DOI: http://dx.doi.org/10.1016/j.enpol.2013.05.049, available at: www.sciencedirect.com/science/article/pii/S0301421513003856.

ICSU, ISSC (2015) *Review of the Sustainable Development Goals: The Science Perspective*, International Council for Science (ICSU), Paris.

International Council for Science (ICSU) (2017) *A Guide to SDG Interactions: From Science to Implementation*, D. J. Griggs, M. Nilsson, A. Stevance, D. McCollum [eds.],) International Council for Science (ICSU), Paris DOI: 10.24948/2017.01.

Molina, Mario and F. Sherwood Rowland (1974) Stratospheric sink for chlorofluoromethanes: chlorine atomic catalyzed destruction of ozone, *Nature* 249: 810–812.

National Research Council (2014) Can Earth's and Society's Systems Meet the Needs of 10 Billion People? Summary of a Workshop, M. Mellody, Rapporteur, Board on Environmental Change and Society and Committee on Population, Division of Behavioral and Social Sciences and Education; Board on Life Sciences, Division on Earth and Life Studies, National Academies Press, Washington, DC.

Nilsson, M., D. Griggs, and M. Visbeck (2016) Map the interactions between Sustainable Development Goals, *Nature* 534: 320–322, available at: www.nature.com/news/policy-map-the-interactions-between-sustainable-development-goals-1.20075.

Odum, H. T. (1973) Energy, ecology and economics, *Ambio* 2(6): 220–227.

Pimental, D. and M. H. Pimental (2008) *Food, Energy and Society*, 3rd edn. CRC Press, Boca Raton, FL.

Pittock, A. B. (2009) *Climate Change: The Science, Impacts and Solutions*, 2nd edn. CSIRO, Clayton, Australia.

United Nations (2012) *World Population Prospects: The 2012 Revision*, United Nations, New York.

Wackernagel, M., C. Monfreda, and D. Deumling, (2002) *Ecological Footprint of Nations*, Sustainability Issue Brief, Redefining Progress, California.

2

Future Earth and Expected Mega Changes

Serhat Şensoy, Mustafa Coşkun, Necla Türkoğlu, and İhsan Çiçek

2.1 Introduction

The Earth was formed about 4.5 billion years ago. We can't say exactly what the Earth's daily weather was like in any particular place thousands or millions of years ago. But we know more or less what the Earth's climate was like in the past because of some proxy evidence that remains in rocks, ice, trees, corals and fossils. These clues tell us that the Earth's climate has been changed many times before. Over the last half-million years, our climate has experienced long ice ages regularly punctuated by brief warm periods called interglacials. Atmospheric carbon dioxide closely matches the cycle. Interglacials come along approximately every 100,000 years. This is called the Milankovitch cycles, brought on by changes in the Earth's orbit. There are three main changes to the orbit: the shape of the orbit around the Sun (eccentricity) varies from an ellipse to a more circular shape on a cycle of about 100,000 years. The Earth's axis is tilted relative to the Sun at around 23°. This tilt oscillates between 22.5° and 24.5° in a 41,000-year cycle (obliquity). It is currently 23.4° and decreasing at a rate of about 0.013° per century. The third and final of the Milankovitch cycles is Earth's precession.[1] As the Earth spins around its axis, the axis wobbles from pointing towards the North Star to pointing at the star Vega, which determines climatic patterns on Earth through orbital forcing. This top-like wobble, or precession, has a periodicity of 23,000 years. As a result we have had ice ages when most of the planet was covered in ice and there have also been much warmer periods despite CO_2 levels being lower than today (Barnola et al., 2003). The severity of hot and cold periods varies depending on the combination of these three cycles (see Figure 2.1). Today's CO_2 concentrations now exceed 400 parts per million (ppm).[2]

Our climate system results from complex, non-linear interactions among the atmosphere, cryosphere, biosphere, hydrosphere and lithosphere in response to a combination of external and internal forcing. The dominant external input is solar energy. Occasionally volcanic eruptions and impacts by extra-terrestrial bodies can cause major disruptions. The atmosphere is largely transparent to shortwave solar radiation, some of which is reflected back to space, especially from clouds and snow cover. Generally, the solar heating of the surface is high in the tropics and low near the poles, but the distribution of different types of land and water surfaces causes considerable regional variations. Some of the surface energy is used for biological processes, but the major proportion is redistributed by longwave radiation, conduction and convection by direct heating of the overlying atmosphere and by evaporation of water. Water vapour and other atmospheric GHGs such as carbon dioxide (CO_2), methane (CH_4), chlorofluorocarbons (CFCs) and nitrous oxide (N_2O) and clouds trap some of the outgoing longwave radiation to provide a "natural greenhouse effect", so that the Earth becomes warm enough to make life possible on it.

Natural factors (solar and volcanic activities, orbital changes etc.) have changed the Earth's climate in the past. But today's climate change is different than before. According to Intergovernmental Panel on Climate Change (IPCC), it's extremely likely (probability of 95 to 100 per cent) that human influence was the dominant cause of global warming between 1951 and 2010 (IPCC, 2014). Natural forces cannot explain today's warming. Human activities such as increased atmospheric concentrations of GHG due to massive use of fossil fuels, land use changes and deforestation are the main causes of today's climate change.

2.2 Observed Changes in the Climate System

Findings obtained from the synthesis report of the Fifth Assessment Report (AR5) of IPCC are as follows:

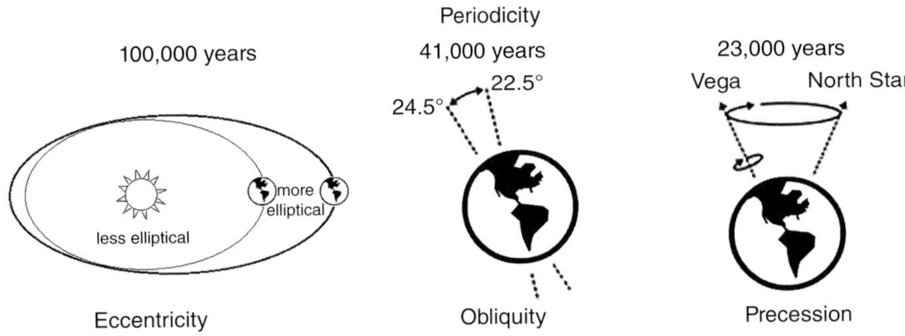

Figure 2.1 Milankovitch cycles.

- Human influence on the climate system is clear, and recent anthropogenic emissions of greenhouse gases are the highest in history. Recent climate changes have had widespread impacts on human and natural systems.
- Warming of the climate system is unequivocal. Combined land and ocean surface temperature data demonstrate a 0.85 °C linear trend over the period 1880–2012, as the amounts of snow and ice have diminished, and sea level has risen.
- Each of the last three decades has been successively warmer at the Earth's surface than any preceding decade since 1850. In the Northern Hemisphere, 1983–2012 was likely the warmest 30-year period of the last 1,400 years. Ocean warming dominates the increase in energy stored in the climate system, accounting for more than 90 per cent of the energy accumulated between 1971 and 2010.
- Over the last two decades, the Greenland and Antarctic ice sheets have been losing mass, glaciers have continued to shrink almost worldwide and Arctic sea ice and Northern Hemisphere spring snow cover have continued to decrease in extent.
- The rate of sea level rise since the mid-nineteenth century has been larger than the previous two millennia. Over the period 1901–2010, global mean sea level rose by 19 cm.
- Carbon dioxide concentrations have increased by 40 per cent since pre-industrial times, primarily from fossil fuel emissions and secondarily from land use change emissions. The ocean has absorbed about 30 per cent of the emitted anthropogenic CO_2, causing ocean acidification.
- Relative to 1850–1900, global surface temperature change through the end of the twenty-first century (2081–2100) is projected to exceed 2 °C under certain emission scenarios known as representative concentration pathway (RCP) scenarios RCP6.0 and RCP8.5.
- Great changes in many extreme weather and climate events have been observed since 1950. Some of these changes have been linked to human influences, including a decrease in cold temperature extremes, an increase in warm temperature extremes, an increase in extreme high sea levels and an increase in the number of heavy precipitation events in some of the regions.
- Climate change will affect food security and will force displacement of populations.
- Temperature increases are linearly related to carbon emissions. It is estimated that a temperature increase of 2 °C will occur when 2,900 Gt CO_2 has been emitted. Of our carbon budget (1900 Gt CO_2) 65 per cent) was already used between 1870 and 2011. After additional 1,000 Gt, Paris Agreement 2.0 °C goal will be exceeded.
- There is a need to act quickly to prevent destructive results. Measures exist to achieve the substantial emissions reductions required to limit likely warming to 2.0 °C (40–70 per cent reduction in GHG globally by 2050 and near-zero GHG in 2100).

More than thirty-six thousand papers have indicated that climate change has a medium to high probability of major and widespread impacts (Figure 2.2) on physical (glaciers, ice, permafrost, rivers, lakes, floods/drought and coastal), biological (terrestrial, marine and wildfire) and human systems (food, health and economics) on almost all continents, polar regions and small islands.

2.3 Future Climate and Expected Mega Changes

Scientists use computer models of the climate system to understand and to project future climate changes better. These models simulate changes based on a set of scenarios of anthropogenic forcing. Representative concentration pathway (RCP) scenarios were used in the Coupled Model Inter-Comparison Project Phase 5 (CMIP5) and IPCC AR5. The scenarios are gradually

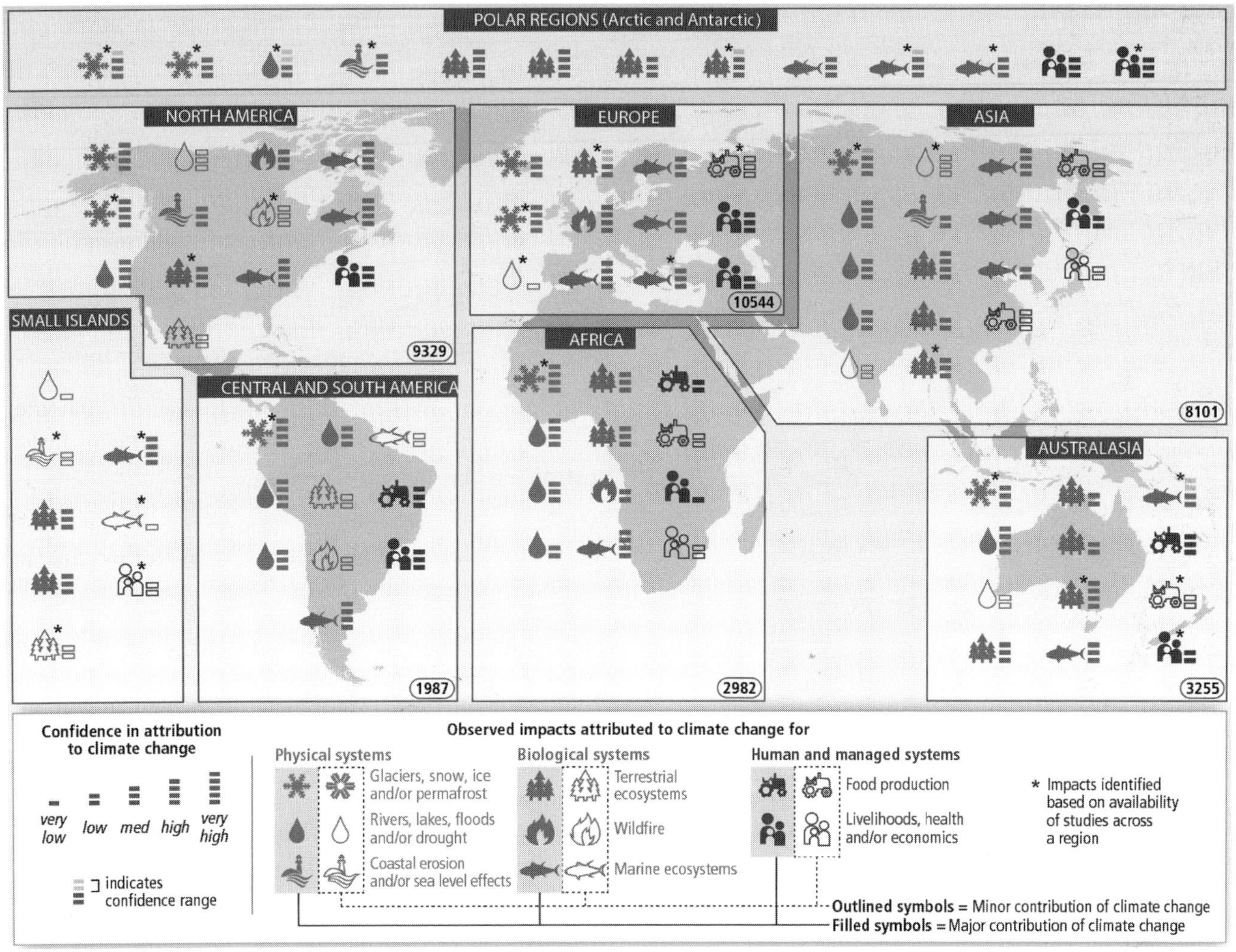

Figure 2.2 Widespread impacts attributed to climate change based on the available scientific literature (IPCC, 2014).

improving. New Shared Socio-Economic Pathway (SSP) scenarios will be used in future. New climate model projections with SSP scenarios under the CMIP6 project will be available within the 2018–2020 timeframe. AR6 is to be finished by 2022 in order to provide comprehensive, policy-relevant scientific inputs to the global stocktake mechanism established under the UNFCCC.

According to the IPCC AR5 Synthesis Report published in 2014, continued emission of greenhouse gases will cause further warming and long-lasting changes in all components of the climate system, increasing the likelihood of severe, pervasive and irreversible detrimental impacts for people and ecosystems. Cumulative emissions of CO_2 will largely determine global mean surface warming by the late twenty-first century and beyond. Future changes are expected to include a warmer atmosphere, a warmer and more acidic ocean, higher sea levels and changes in precipitation patterns. Even if greenhouse emissions stopped overnight, there will be additional warming by 0.5–1.0 °C.

2.3.1 Expected Future Changes in the Earth's Environment in Some Future Years

The following data are compiled from IPCC (2014) unless noted otherwise.

2020

- Heavy rainfall and flash floods will increase across Europe,
- low rainfall will reduce agricultural yields up to 50 per cent in some areas,
- world population will reach 7.6 billion people (U.S. Census Bureau).

2030

- Of the world's coral reef 18 per cent will probably be dead as a result of warmer and more acidic oceans,
- world population will reach 8.3 billion people (U.S. Census Bureau),
- equatorial mountains' glaciers will disappear because of warming temperatures.

2040

- The Arctic Sea could be ice-free in the summer and winter ice depth may shrink drastically.

2050

- Small alpine glaciers will disappear completely while large glaciers will shrink by 30–70 per cent,
- one-fourth of the world's plant and animal species could face extinction,
- world population will reach 9.6 billion people.[3]

2070

- Electricity production for the world's existing hydropower stations will decrease because of glaciers disappearing and increased drought; maximum impact will be in Europe (6 per cent) and the Mediterranean (50 per cent),
- warmer and drier conditions will lead to more frequent heatwaves, more frequent droughts and more forest fires, especially in Mediterranean regions.

2080

- Up to 100 million people will experience coastal flooding each year; 20 per cent of the world's population live in river basins that will be hit by increasing floods. The most vulnerable are the densely populated low-lying areas;
- water shortage will affect about 1–3 billion people, 600 million of whom will lack food.

2100

- Atmospheric CO_2 levels will be 850 ppm (highest of the last 650,000 years),
- ocean pH levels will decrease by 0.5 pH unit to reach the lowest value of the last 20 million years,
- permafrost will emit more CO_2 into the atmosphere than it absorbs as a result of large-scale thawing,
- of Earth's land 39 per cent will be in new climate zones,
- the species extinction rate of all plants and animals will reach one-fourth,
- global temperature will increase 4 °C according to RCP8.5 scenarios,
- world population will reach 10.9 billion people.[3]

2200

- Global surface temperature increase will reach 6 °C with RCP8.5,
- atmospheric CO_2 concentration will reach 1,800 ppm with RCP8.5,
- sea level rise will be more than 2 meters.

2300

- Global surface temperature increase will reach 7 °C with RCP8.5,
- atmospheric CO_2 concentration will reach 2,000 ppm with RCP8.5,
- sea level rise will be more than 3.6 meters.

2.4 What Are Mega Changes?

Mega changes are severe, pervasive and in some cases irreversible detrimental impacts due to climate change. Past climate changes led to extinction of many species, population migrations and great alterations in the land surface and the ocean circulation. The speed of current climate change is faster than that of most of the past changes. So it's expected that future mega changes will be more severe and pervasive and have irreversible impacts. Expected mega changes can be listed as follows:

1 Increased displacement of people due to:
   - Widespread drought and undermined food security,
   - increased poverty and hunger,
   - national security implications.

2 Reduced surface and groundwater resources and water crisis:
   - melting of Himalayan glaciers.

3 Increased extreme events

4 Irreversible changes:
   - Warmer temperature,
   - sea level rise,
   - changing geography,
   - ocean acidification,
   - extinction of species,
   - loss of Amazon rain forest.

There will also be some important (but not mega) changes:

- Reduced permafrost extent,
- exacerbated human health problems.

2.4.1 Increased Displacement of People

According to IPCC reports published since 1990, one of the great effects of climate change will be the migration of people. Nowadays this outcome is known as the climate refugee problem. But, according to the United Nations High Commissioner for Refugees (UNHCR), people displaced across borders as a result of climate change may not meet the formal definition of a refugee, which is "someone who has been forced to escape his/her country because of persecution, war or violence". In 2015, there were 65.3 million "forcibly displaced" people;[4] the term includes those formally defined as refugees as well as those who do not meet the formal definition. Also there are millions of internally displaced people (IDPs) who have not crossed a border. The International Organization for Migration (IOM) defines those affected by environmental displacement as a "person or group of persons who predominantly for reason of sudden or progressive change in the environment that adversely affects their lives or living conditions, are obliged to leave their habitual homes, or choose to do so, either temporarily or permanently and who move either within their country or abroad".[5]

Which climatic factors cause migrations?

- Intensification of extreme events such as hurricanes and cyclones that destroy housing and livelihoods and cause people to relocate;
- increased warming and drought that affect agricultural production (crops, livestock), reducing clean water and food security;
- rising sea levels, which will cause coastal areas to be uninhabitable; more than 600 million people live in low elevations;
- competition over natural resources, which creates the risk of violent conflict.

People try to adapt to these situations but many of them will move to another place to survive. According to UNHCR, 26.4 million people are displaced every year since 2008 by extreme events. The Internal Displacement Monitoring Centre (IDMC) declared that in 2015 alone, 27.8 million were displaced by conflict, violence and disaster.[6] This means that almost every second person has been displaced by a disaster. Drought in Somalia from 2011 to 2012, floods in Pakistan between 2010 and 2012 and earthquake in Nepal in 2015 left very many people without a house, clean water and basic supplies. In the Philippines 4 million people were displaced by the Haiyan typhoon, which occurred in 2013.

Mobility is a form of adaptation to climate change, as recognized by UNFCCC's COP 16 in Cancun, in December 2010. The resolution identified three types of mobility: climate change–induced displacement, migration and planned relocation. The term 'displacement' implies coercion, as people are forced to leave their home, while 'migration' is assumed to be voluntary. Adaptation and Disaster Risk Reduction (DRR) can involve steps to reduce the need for people to migrate. National Adaptation Programmes of Action (NAPA) are the principal mechanisms through which least developed countries (LDCs) identify adaptation needs, though the programmes which they provide are a very limited response to the adaptation challenge, focusing primarily on 'climate-proofing' through small-scale projects that fail to integrate adaptation.

Relocation would pose security risks that could provoke violence or even conflict. Small island developing states (SIDSs) that are affected by sea level rise are examples of nations that need planned relocation including international movement. As resources become scarcer, there will be increasing pressure to migrate to other countries. Because of the predictions of the impacts of climate change, there is a potential that the entire land surface could disappear inSIDSs such as Tuvalu, Kiribati, Samoa, Maldives, Mauritius, Solomon and Marshall Islands.

The Tuvalu–New Zealand agreement is a good example of international movement. Tuvalu is located in the southwest Pacific Ocean. It's one of the world's smallest and most isolated island states. The total land area is 26 square kilometres and population is about eleven thousand people. Average elevation above sea level of the islands is less than 2 metres. The land is very low-lying and highest elevation is 4.6 metres above sea level on Niulakita.There are no streams or rivers. Total rainfall is more than 3000 millimetres. Tuvaluans depend on fishing, but fishing is at risk because of coral bleaching. Tuvaluans are also at great risk of sea level rise. Because of these risks Tuvalu signed an agreement with New Zealand within the scope of the climate change adaptation project. New Zealand has agreed to a thirty-year immigration program taking seventy-five Tuvaluans per year if Tuvaluans would accept status as climate refugees and could resettle in New Zealand. This implies the irreplaceable loss of the island's traditional skill and knowledge.[7]

2.4.1.1 Widespread Drought and Undermined Food Security

The human population of the world is growing very rapidly. It's expected that world population[3] will reach 9.6 billion by 2050. Additional population implies acceleration in demand (more food, more clean water and more housing). But agricultural areas are shrinking. Human pressure is altering the global environment and the changes are most extreme at the continental surface: forests are transformed into agricultural land;

cities sprawl ever more widely even in the flood valleys. Human activities cause, or contribute to, desertification, deforestation, salinization and soil erosion but also reforestation, irrigation and landscape management. Human-induced land-use change, and consequent land-cover change, are inescapable features of current and future environments.

According to IPCC projections, in the coming decades climate change will unleash mega droughts lasting ten years or more especially in the Mediterranean, Australia, South Africa, South America and Caribbean. The impacts of climate change on key economic sectors, such as agriculture and water, can have profound effects on food security. Climate change – mostly drought – is already affecting the global agricultural supply and will likely drive up food prices. Climate change will affect food security in all five dimensions: availability, accessibility, utilization, production and distribution channels, changing purchasing power and market flows. Its impacts will be short term, resulting from more frequent and more intense extreme events and long term, caused by changing temperature and precipitation patterns.

About 384 million people fish in Indonesia, Philippines, Malaysia, Papua New Guinea, the Solomon Islands and Timor-Leste, a region known as the Coral Triangle. Scientists say that ocean temperatures have risen and continue to rise enough to affect fragile ecosystems of coral reefs, which have been bleaching and dying in recent decades. If people are not able to find fish, this will force them to be displaced. By comparison Syria (with a population of twenty-one million) caused the world's greatest refugee crisis so far with 4.9 million refugees fleeing in 2015; the movement of 384 million people due to the failure of their fisheries would be a great worldwide crisis.[8] Agriculture-based livelihood systems are already vulnerable to food insecurity; they are at immediate risk of increased crop failure, new patterns of pests and diseases, lack of appropriate seeds and planting material and loss of livestock. People living on the coastal areas and flood plains and in mountains, drylands and arctic regions are most at risk. As an indirect effect, low-income people in urban areas will be at risk of food insecurity. The food system will also be affected through possible internal and international migration, conflicts and civil unrest triggered by climate change. Climate change adaptation and mitigation measures need to be integrated into the overall production and delivery mechanism (Figure 2.3).

A recent study finds that sub-Saharan Africa will need to more than triple crop production by 2050 in order to feed its growing population. This hard-to-reach target will become more difficult in places like southern Africa and the western Sahel, where water supply is projected to fall[9].

A new World Bank report suggests that water scarcity may reduce GDP by 6 per cent in the tropics and mid-latitudes by 2050[10].

2.4.1.2 Increased Poverty and Hunger

'No poverty' and 'zero hunger' are the first two items of the Sustainable Development Goals (SDGs).[11] Climate change makes agricultural production more difficult by changing precipitation regimes (causing floods or droughts) and increasing temperature and CO_2 concentration in the air, soil and water. In addition, there will be an increase in the frequency and magnitude of extreme events such as storms, hail and floods. They have severe impacts on food production. In developing countries 80 per cent of people continue their livelihood by agriculture.

There is risk of crop failure due to storm, hurricanes, typhoons, floods or drought increase as a result of climate change. Climate change can push more than 100 million additional people into extreme poverty and hunger by 2030. Generally poor people are at high risk from climate-related shocks. Crop yield loss[12] could reach 5 per cent in 2030 and 30 percent in 2080. As a result of weather extremes, food often becomes more expensive. Drought and floods frequently lead to crop failures, which in turn affect prices. If there is no food, it must be imported. Even today one of eleven people suffers from hunger in Somalia, Ethiopia and other countries. About 800 million of them live in developing countries. Approximately 2 billion people suffer from micronutrient deficiency, which is hidden hunger.

Climate change–induced risks to agricultural production are as follows:

- Dried soil due to heatwaves, drought and rising temperatures: dried soil is very open to soil erosion and desertification, and generally difficult cultivation;
- changed precipitation patterns, which will cause more negative effects on rain-fed agriculture;
- plant growth condition change due to temperature rise and water scarcity, preventing many plants from growing at their previous location; Increase in pests and diseases as a result of temperature changes and potential livestock breeding problems;
- reduced cultivation area: the expected rise in sea levels threatening flooding and contributing to salinization of soil and groundwater; continued increase in dry regions.

As a result of climate change, agricultural land areas will be reduced, and their yields will decrease.

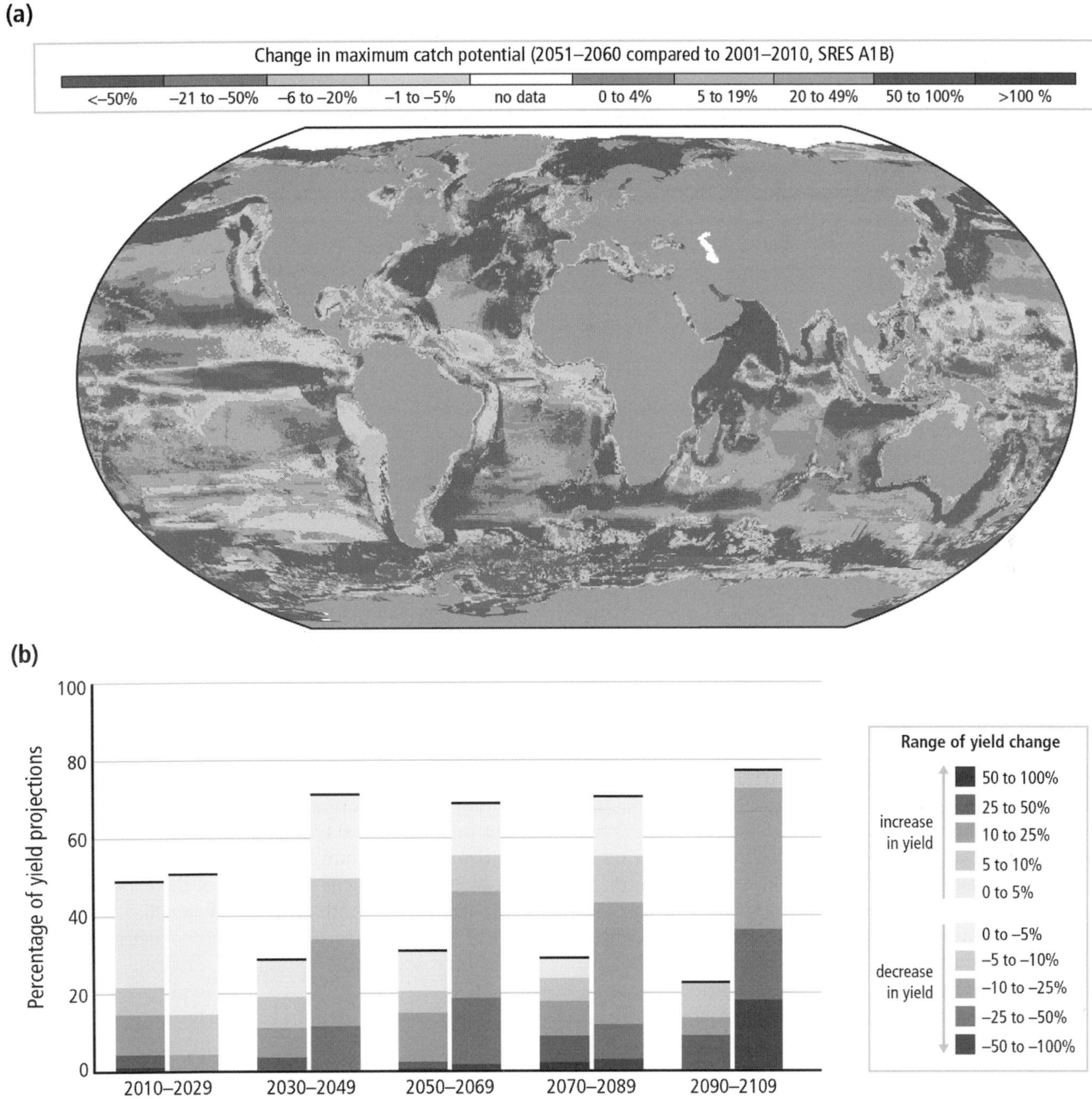

Figure 2.3 Climate change poses risks for food production: (a) change in maximum fish catch potential of ~1,000 species (2051–2060 compared to 2001–2010, SRES A1B, 2 °C warming); (b) summary of projected changes in crop yields (mostly wheat, maize, rice and soy) due to climate change over the twenty-first century. The figure combines 1,090 data points from crop model projections, covering different emission scenarios, tropical and temperate regions and adaptation and no-adaptation cases. Changes are relative to late twentieth-century levels (IPCC, 2014). For the color version, please refer to the plate section.

According to the World Bank, in 2050 between 2 and 5 billion people will not be adequately nourished. In order to combat hunger and poverty, sustainable agriculture must be expedited.

2.4.1.3 National Security Implications

Today it's accepted that climate change is an urgent and growing threat to national security because of its negative impacts on humans and the environment. Intensity and frequency of extreme events such as heat waves, droughts, wildfires, heavy rainfall and floods, sea level rise, ocean heating and acidifying and hurricanes and storms surges are increasing because of climate change. These effects are disrupting people's lives, property, infrastructure, economy and ecology, causing food insecurity, poverty, conflicts and refugee flows. On the other hand, impacts of climate change could directly affect the nation's critical infrastructure. Coastal regions are at risk through rising sea levels, higher storm surge and increased erosion. Higher temperatures could buckle railways, damage roads and strain power systems. In turn, the global economy suffers, compounding the growing costs of preparing and restoring infrastructure, which will cause decrease in the GDP. Pervasive and long-lasting drought can diminish crop yields. Also coral reefs, a source for fishing, will cease to exist by mid-century. These are not only environmental problems but national security problems. Decreasing water supply in the mid-latitudes may also contribute to destabilizing nations, adding to pre-existing political tensions, conflict and migration.

Climate change will have real impacts on the military and their missions. The military could be called to help people and civil authorities in cases of disasters. Weather has always affected military operations. Climate change impacts may exacerbate existing stressors, poverty, environmental degradation, terrorist activity and political instability. These are the very critical national security problems that threaten overall stability.

Unpredictable instability has become the "new normal," and this trend will continue for the foreseeable future. Extreme weather, climate change, environmental degradation, rising demand for food and water, poor policy decisions and inadequate infrastructure will magnify this instability. Consequences of psychosocial impacts caused by climate change include increase in violence, intergroup conflict, displacement and relocation and socioeconomic disparities. According to research, a positive relationship has been found between heat and violence: increasing temperature will cause more aggression.

In conclusion, climate change is predicted to strain economies and societies around the world, placing an additional burden on already-vulnerable nations, especially in the Mediterranean, Australia, the Caribbean, South Africa and Asia. Increasing population, increasing urbanization, increasing energy demand, decreasing water resources and decreasing food security will put pressure on national security. Coastal regions are vulnerable to rising sea levels and other impacts.

2.4.2 Reduced Surface and Groundwater Resources, Water Crisis

Water is an environmental, political, cultural, historical and economic natural resource. It's vital for human life. Clean water and sanitation are the sixth item of the sustainable development goals (SDGs).[11] A changing climate means less rain and lower water supplies in regions where many people live and much of the planet's food is produced: the mid-latitudes of the Northern and Southern Hemispheres, including the United States, southern Europe and parts of the Middle East, southern Africa, Australia and Chile. According to IPCC AR5, consecutive dry days will be increased by twenty to twenty-five days in those regions. Rain-fed agriculture, which many poor people depend upon, will also suffer as a result of reduced precipitation in the mid-latitude regions.

Generally all surface water such as streams, lakes, reservoirs and wetlands interacts with groundwater. These interactions can occur in many ways. In many situations surface water bodies gain water from the groundwater system; in others the surface water body is a source of groundwater. Surface water is hydrologically connected to groundwater. Climate change projections show significant rainfall reduction in the Mediterranean basin, which will cause a decrease in surface water and groundwater resource in the region (IPCC, 2014). The Water Risk Atlas of World Resource Institute shows overall water risk, which identifies areas with higher exposure to water-related risks.[13]

Another water crisis source are transboundary rivers. There are 276 transboundary rivers in the world; 148 countries include territory within one or more transboundary rivers.[14] Because of the very few agreements between user countries, water has become a security issue. There are many factors that may potentially increase conflict among riparian countries. Increasing demand for water may lead to a 40 per cent gap between water demand and supply by 2030. This gap will pose a threatening water crisis.[15]

In karstic regions, for example, in Konya-Karaman province in Turkey, collapse sinkhole growth is accelerated by excessive use of water and reduction of groundwater. Accordingly, agriculture and settlement areas are at risk (Dogan and Yılmaz, 2011). There is a risk that this effect will increase in the future.

2.4.2.1 Melting of Himalayan Glaciers

Glaciers have continued to shrink almost worldwide. The Himalaya-Hindu Kush mountain range and the Tibetan Plateau are sometimes known as the Third Pole because their ice fields contain the largest reserve of fresh water outside the polar regions. They cover 5 million square kilometres and the average elevation is 4000 metres. The Himalayan Plateau is important because it provides drinking and irrigation water that sustains 1.5 billion people in downstream countries including India, Nepal, China and Bangladesh. Although there was a projection that Himalayan glaciers could disappear by 2035 in the WGII report of IPCC 2007, this projection was not included in the AR4 summary for policymakers indicating that "widespread mass losses from glaciers and reductions in snow cover over recent decades are projected to accelerate throughout the 21st century, reducing water availability, hydropower potential, and changing seasonality of flows in regions supplied by meltwater from major mountain ranges (e.g. Hindu-Kush, Himalaya, Andes), where more than one-sixth of the world population currently lives.[16]

2.4.3 Increased Extreme Events

Extreme weather and climate events are typically rare. But climate change causes their frequency and intensity to increase; 95 per cent of extreme events are meteorological, hydrological and climatological.[17] Each year the impacts of extreme events around the Earth give rise to multiple casualties and significant damage to property and infrastructure, with adverse economic consequences for communities that can persist for many years. As trends in extreme temperature events react more sensitively to climate change than mean climate, they have a more intense impact on natural and human systems (Aguilar et al., 2009). Given their importance, it is essential to understand how and why weather and climate extremes have changed in the past. Changes in climate extremes are usually investigated by using climate extreme indices based on absolute or percentile-based thresholds (Zhang et al., 2005; Sensoy et al., 2007; Aguilar et al., 2009).

In its Special Report on Managing the Risks of Extreme Events and Disasters to Advance Climate Change Adaptation (SREX), the IPCC (IPCC, 2012) defines an extreme weather or climate event as "the occurrence of a value of a weather or climate variable above (or below) a threshold value near the upper (or lower) ends of the range of observed values of the variable".

Many papers which contributed to IPCC ARs, found that projected temperature extremes demonstrated significant warming during the second half of the twentieth century. The IPCC has already predicted that risks associated with extreme events will continue to increase as the global mean temperature rises. A changing climate leads to changes in the frequency, intensity, spatial extent, duration and timing of extreme weather and climate events.

Changes in extremes can be linked to changes in the mean. Small change in the mean value can cause more extreme events (IPCC, 2012). According to the study carried out by Fischer and Knutti (2015), heat extremes that previously occurred only once every 1,000 days are now happening four to five times more often. The number of heat extremes will increase to 27 per 1,000 days, if warming reaches 2 °C, this will increase to 62 per 1,000 days, if warming reaches 3 °C. This is a dramatic increase of heat extremes (Fischer and Knutti, 2015).

Climate extreme indices have also been used to compute and analyse projected future climate conditions based on simulations by Global Climate Models (GCMs) (Tebaldi et al., 2006). Summer temperature extremes are predicted to increase while extreme cold temperatures will become more moderate. The effects of heat waves can include an increase in morbidity and mortality rates, rising stress on agricultural resources and a strain on infrastructure. The Global Climate Risk Index of Germanwatch shows which countries suffer most from extreme weather events.[18]

2.4.4 Irreversible Changes

In its latest climate change report, the Intergovernmental Panel on Climate Change (IPCC) warned of certain "irreversible" impacts of climate change, which were likely to occur if carbon emissions weren't cut by the end of the century. Even if we stop burning all fossil fuels tomorrow, the carbon that's already in the atmosphere will not disappear. Some of the irreversible effects due to climate change are as follows:

2.4.4.1 Warmer Temperatures

We have already observed global warming over the last several decades. Future temperatures are expected to change further. Because of the long carbon life cycle in the atmosphere, greenhouse gas causes an effectively permanent increase in global temperatures. The IPCC's latest report states, "Warming caused by CO_2 emissions is effectively irreversible over multi-century timescales unless measures are taken to remove CO_2 from the atmosphere."

2.4.4.2 Sea-Level Rise

The IPCC writes, "It is virtually certain that global mean sea level rise will continue for many centuries beyond 2100, due to increased ocean warming and increased loss of mass from glaciers and ice sheets." The Greenland and Antarctic ice sheets, which both contain very large amounts of ice, are two locations that scientists are especially worried about because of their potential to cause large amounts of sea level rise. Total loss of the Greenland ice sheet would produce an associated 7-metre sea level rise (IPCC, 2014. Not only SIDS but also lowland coastal areas could remain underwater because of sea level rise (SLR).[19] SLRS causes the disappearance of islands in the Indian and Pacific Oceans, whose elevation is usually less than 3 metres. Some lowland inhabited islands are at risk of disappearance. (Some migratory birds die when they cannot find them on their flight route.) Today this process is affecting inhabited islands less than 2 metres in height, and it will affect islands, causing problems such as relocation of residents of these islands.

2.4.4.3 Changing Geography

One consequence of long-term SLR is a permanently altered geographic landscape. We're already starting to see its effects, as coastlines are shrinking and islands around the world are slowly being washed away.[20] Even within this century, SLR could erase some SIDS from the map.

2.4.4.4 Ocean Acidification

The ocean is a sink for carbon dioxide from the atmosphere, and chemical reactions make the water more acidic. This is harmful for certain types of marine organisms, including coral and some plankton, which survive by creating protective calcium carbonate shells around their bodies. The more acidic the ocean becomes, the harder it is for these animals to complete the chemical processes required to make their shells. IPCC warns that the acidification could continue for centuries.

2.4.4.5 Extinction of Species

Scientists predict that climate change will soon be a major player in the extinction of wildlife as it changes habitats and alters ecosystems around the globe. Many scientists already believe Earth is at the brink of a mass extinction, its sixth in the planet's history. Recent research has suggested that 26 per cent of all mammals, 13 per cent of birds, and 41 percent of amphibians are already in danger, and many of these jeopardized species will be even more vulnerable as changing landscapes, shifting food and water availability and rising temperatures threaten their ability to survive.[21]

Today's rapid loss of species is estimated to be 1,000–10,000 times higher than the natural. Experts calculate that between 0.01 and 0.1 per cent of species will become extinct every year. If the upper estimate of species number is true (100 million), then between 10,000 and 100,000 species are becoming extinct per year. Some scientists are estimating that half of all our plants and animals will be extinct by 2100. This is a biodiversity crisis.[22]

2.4.4.6 Loss of Amazon Rain Forest

About 20 per cent of oxygen is produced in Amazon rain forests and it's one of the tipping elements in the Earth climate system.[23] The term "tipping point" commonly refers to a critical threshold at which a tiny perturbation can qualitatively alter the state or development of a system. The term "tipping element" describes large-scale components of the Earth system that may pass a tipping point.

Projected reductions in rainfall show that large parts of the Amazon region are at risk. The forest has been under pressure from the timber industry, agriculture and effects of climate change. The new study examined what would happen if the dry seasons intensified in the Amazon. More than half of the Amazon's tree species are then threatened with extinction. Reduced rainfall increases the risk of forest dieback. In addition, forest loss can intensify drought. More droughts can lead to less forest. Forest losses can reach 25– 38 per cent of the Amazon. This would turn the forest into a different savannah environment (Zemp et al., 2017).[24] In the Amazon, the 2010 drought was worse than the rare 2005 dry spell.[25]

2.4.4.7 Reduced Permafrost Extent

According to IPCC, the near-surface permafrost extent at high northern latitudes will be reduced by global mean surface temperature increases. Warming of permafrost, thawing of ground ice and development of thermokarst terrains (mud and lakes) have been documented over the past several decades. Near-surface permafrost (upper 3.5 metres) is projected to decrease by an amount that ranges from 37 per cent (RCP2.6) to 81 per cent (RCP8.5). In developed areas of the Arctic, continuation of such changes may lead to costly damage to human infrastructure. This will make an extra contribution to global warming by causing the release of greenhouse gases (especially methane) stored in frozen soils. In addition, a decrease in permafrost areas will cause a decrease in albedo, which will cause more energy absorption and warming in terrestrial areas.

2.4.4.8 *Exacerbated Human Health Problem*

According to IPCC, health impacts include greater likelihood of injury and death due to more intense heat waves and fires and increased risks from foodborne and waterborne diseases. Risks of undernutrition in poor regions will increase. Risks from vector-borne diseases are projected to increase generally with warming.

The researchers suggest that future air quality could decline without increased regulations to reduce greenhouse gas emissions, and the development and deployment of new emission abatement technologies. Evidence from modelling studies suggests that climate change is likely to increase concentrations of tropospheric ozone, one of the most important urban air pollutants responsible for respiratory problems. Under this assumption, rapid reductions of emissions from fossil fuel burning are needed to protect the health of both current and future generations (Ebi and McGregor 2008).

According to the World Health Organization (WHO), 7 million premature deaths annually are linked to air pollution. Exceptionally severe smog in Beijing and other mega cities is a major public health problem at present.[26] Outdoor air pollution–caused deaths are broken down by disease as follows:

- 40 percent ischemic heart disease,
- 40 per cent stroke,
- 11 per cent chronic obstructive pulmonary disease (COPD),
- 6 per cent lung cancer, and
- 3 per cent acute lower respiratory infections in children.

At the global level, warming of 2–3 degrees Celsius could increase the number of people at risk for malaria[12] to 5 per cent and for diarrhoea up to 10 per cent by 2030.

2.5 Summary

Mega changes are severe, pervasive and in some cases irreversible detrimental impacts due to climate change. Our world has been in the course of various geological cycles and climatic changes for about 4.5 billion years. But the speed of current climate change is faster than that of most past changes. Past climate changes led to extinction of many species, population migrations and great alterations in the land surface and the ocean circulation. So it's expected that climate change–attributed mega changes will be more severe, pervasive and have irreversible impacts.

There are substantially more impacts in recent decades now attributed to climate change. More than thirty-six thousand papers have indicated with medium to high confidence (IPCC, 2014) that climate change has major and widespread impacts on physical, biological and human systems on almost all continents.

According to the Fifth Assessment Report (AR5) of the Intergovernmental Panel on Climate Change (IPCC), recent anthropogenic emissions of greenhouse gas are the highest of last 400,000 years. Today's CO_2 concentrations now exceed 400 ppm. Human influence on the climate system is clear. The latest global climate report of the WMO confirms that 2016 was the warmest year on record with a remarkable 1.1 °C above the pre-industrial period (WMO, 2017).

Despite some international efforts coordinated by the UNFCCC (Kyoto Protocol, Paris Agreement), greenhouse gas emissions continue to rise and this will cause further warming and long-lasting changes in all components of the climate system. Climate change will amplify existing risks and create new risks for natural and human systems (IPCC, 2014).

Climate change–attributed expected mega changes are as follows:

- Increased displacement of people due to widespread drought and undermined food security, increased poverty and hunger, refugee crises and national security implications;
- reduced surface and groundwater resources, melting of Himalayan glaciers and water crisis;
- increased frequency and intensity of extreme events;
- irreversible changes such as warmer temperature, sea level rise, changing geography, ocean acidification, extinction of species and loss of Amazon rain forest.

There will also be some important (but not mega) changes such as reduced permafrost extent and exacerbated human health problems.

Limiting climate change is essential to protect physical, biological and human systems from the severe, pervasive, irreversible and widespread impacts. This will require international consensus and efforts for substantial and sustained reductions in greenhouse gas emissions together with adaptation and mitigation measures. Otherwise we will not be able to leave a good world for future generations.

Acknowledgements

The authors would like to acknowledge Professor Tom Beer for the kind invitation to contribute to this Cambridge University Press book as well as his excellent organization and leadership. The authors thank the lead authors, contributing authors and review editors. The authors also thank the Cambridge University staff for their kind cooperation.

Internet Resources

[1] www.indiana.edu/~geol105/images/gaia_chapter_4/milankovitch.htm, 11.06.2017
[2] www.co2.earth/, 28.03.2017
[3] https://ourworldindata.org/future-population-growth, 15.04.2018, 21.10.2016
[4] www.unhcr.org/figures-at-a-glance.html, 28.03.2017
[5] www.iom.int/, 28.03.2017
[6] www.internal-displacement.org/, 31.03.2017
[7] https://germanwatch.org/download/klak/fb-tuv-e.pdf, 31.08.2016
[8] www.youtube.com/watch?v=YZXBawQGONs, 28.03.2017
[9] www.ifpri.org/news-release/study-africa-south-sahara-will-need-increased-crop-yields-trade-meet-rising-food-demand
[10] www.worldbank.org/en/topic/water/publication/high-and-dry-climate-change-water-and-the-economy?CID=WAT_TT_Water_EN_EXT, 31.08.2016
[11] https://sustainabledevelopment.un.org/?menu=1300, 31.03.2017
[12] www.worldbank.org/en/news/feature/2015/11/08/rapid-climate-informed-development-needed-to-keep-climate-change-from-pushing-more-than-100-million-people-into-poverty-by-2030, 29.03.2017
[13] www.wri.org/applications/maps/aqueduct-atlas, 31.08.2016
[14] www.unwater.org/statistics/statistics-detail/pt/c/211763, 29.03.2017
[15] The Water Resources Group, Background, Impact and the Way Forward URL: www3.weforum.org/docs/WEF/WRG_Background_Impact_and_Way_Forward.pdf
[16] www.ipcc.ch/publications_and_data/ar4/syr/en/mains3-3-1.html, 31.03.2017
[17] www.munichre.com/site/touch-naturalhazards/get/documents_E2080665585/mr/assetpool.shared/Documents/5_Touch/_NatCatService/Focus_analyses/1980-2014-Loss-events-worldwide.pdf, 21.10.2016
[18] https://germanwatch.org/fr/download/13503.pdf, 31.08.2016
[19] www.youtube.com/watch?v=Unyu4JpzXJ4, 31.03.2017
[20] www.theguardian.com/environment/2014/dec/04/in-30-years-maybe-kiribati-will-disappear-climate-change, 11.06.2017
[21] www.cmscoms.com/?p=2852, 31.03.2017
[22] http://wwf.panda.org/about_our_earth/biodiversity/biodiversity/, 31.03.2017
[23] www.blueplanetbiomes.org/amazon.htm, 31.03.2017
[24] www.independent.co.uk/environment/amazon-rainforest-drought-deforestation-jungle-death-spiral-potsdam-institute-a7627931.html, 28.03.2017
[25] www.livescience.com/30085-amazon-drought-2010-climate-change-110203.html, 30.03.2017
[26] www.who.int/mediacentre/news/releases/2014/air-pollution/en/, 31.03.2017

References

Aguilar, E., Aziz Barry, A., Brunet, M., Ekang, L., Fernandes, A., Massoukina, M., Mbah, J., Mhanda, A., do Nascimento, D. J., Peterson, T. C., Thamba Umba, O., Tomou, M., Zhang, X. (2009). Changes in temperature and precipitation extremes in western central Africa, Guinea Conakry, and Zimbabwe 1955–2006, *J. Geophys. Res.*, 114, D02115.

Barnola, J. M., Raynaud, D., Lorius, C., Barkov, N. I. (2003). Historical Carbon Dioxide Record from the Vostok Ice Core. In: *Trends: A Compendium of Data on Global Change*. Carbon Dioxide Information Analysis Center, Oak Ridge National Laboratory, U.S. Department of Energy, Oak Ridge, TN. http://cdiac.ess-dive.lbl.gov/trends/co2/vostok.html.

Ebi, K. L., McGregor, G. (2008). Climate change, tropospheric ozone and particulate matter and health impacts. *Environmental Health Perspectives.* 116(11), 1449–1456.

Dogan, U., Yılmaz, M. (2011). Natural and induced sinkholes of the Obruk Plateau and Karapınar-Hotamıs Plain, Turkey. *Journal of Asian Earth Sciences* 40, 496–508.

Fischer, E. M., Knutti, R. (2015), Anthropogenic contribution to global occurrence of heavy precipitation and high-temperature extremes, *Nature Climate Change* 5, 560–564, DOI:10.1038/nclimate2617.

IPCC (2012). Summary for Policymakers. In: *Managing the Risks of Extreme Events and Disasters to Advance Climate Change Adaptation* [Field, C. B., V. Barros, T. F. Stocker, D. Qin, D. J. Dokken, K. L. Ebi, M. D. Mastrandrea, K. J. Mach, G.-K. Plattner, S. K. Allen, M. Tignor, and P. M. Midgley (eds.)]. A Special Report of Working Groups I and II of the Intergovernmental Panel on Climate Change. Cambridge University Press, Cambridge and New York, pp. 1–19.

IPCC (2014). Climate Change 2014: Synthesis Report. Contribution of Working Groups I, II and III to the Fifth Assessment Report of the Intergovernmental Panel on Climate Change [Core Writing Team, R. K. Pachauri and L. A. Meyer (eds.)]. IPCC, Geneva, Switzerland, 151 pp.

Sensoy, S., Peterson, T. C., Alexander, L. V., Zhang, X. B. (2007). Enhancing Middle East climate change monitoring and indexes, *Bulletin of the American Meteorological Society*, 88, 1249–1254.

Tebaldi, C., Hayhoe, K., Arblaster, J. M., Meehl, G. A. (2006). Going to the extremes: an inter- comparison of model simulated historical and future changes in extreme events, Clim. *Change* 79, 185–211.

WMO (2017). Statement on the State of the Global Climate in 2016, World Meteorological Organization (WMO)-No. 1189.

Zhang, X., Aguilar, E., Sensoy, S., Melkonyan, H., Tagiyeva, U., Ahmed, N., Kutaladze, N., Rahimzadeh, F., Taghipour, A., Hantosh, T. H., Albert, P., Semawi, M. (2005). Trends in Middle East climate extreme indices from 1950 to 2003. *J. Geophys. Res*, 110, D22104, DOI: 10.1029/2005JD006181.

Zemp, D. C., Schleussner, C., Barbosa, H. M. J., Hirota, M., Montade, V., Sampaio, G., Staal, A., Wang-Erlandsson, L., Rammig, R. (2017). Self-amplified Amazon forest loss due to vegetation-atmosphere feedbacks. *Nature Communications* 8, Article number: 14681 (2017), DOI:10.1038/ncomms14681.

3
Global Change, Space Weather, and Climate

Eigil Friis-Christensen

3.1 Introduction

The concern about our environment and specifically about climate has made us increasingly aware of both the limiting conditions of our planet and the effect on the planet of human activity. Most discussed in science as well as in the public is of course the changing climate and the implications this may have for society. Therefore, now more than three decades ago in 1986, the International Council of Scientific Unions, a coordinating body of national science organisations, launched the International Geosphere-Biosphere Programme (IGBP) lasting until 2015 with the aim of conducting a research programme to look at the total Earth system, the changes that are occurring and the manner in which changes are influenced by human actions. Global change refers to planetary-scale changes in the Earth system. The system consists of the land, oceans, atmosphere, polar regions, life, the planet's natural cycles and deep Earth processes. All these constituent parts influence one another.

Our planet is, however, not a self-contained unit. It is part of our solar system and subject to the forces that rule here with the Sun as the main factor controlling the energy that makes our Earth a living planet. The varying orbit of the Earth around the Sun is the major cause of long-term changes in global climate, the systematic change between ice ages and warm periods (Imbrie et al. 1993), which has had a crucial influence on the evolution of humanity and our civilisation. But our Sun is a living star, most directly illustrated by the dark spots on the surface that change their location and distribution from day to day. They were described in Chinese books more than one thousand years ago and have been systematically studied since Galileo in the seventeenth century invented the telescope and used it to look at the Sun and the sunspots. Besides the life-giving radiation in the visible light the Sun is the source of radiation in the UV and X-ray range as well as the source of high-energy particles, magnetic fields and plasma (known as the solar wind) that continuously impinge on our planet and constitute a risk for life and society. All the planets are exposed to the radiation and particles from the Sun but the effect on the various planets is very different depending on their distance to the Sun, their magnetic fields and their atmospheres (Plainaki et al. 2016) (see also Chapter 4 for more details). Fortunately, during ordinary conditions, both the Earth's atmosphere and its magnetic field provide the necessary protection for human life as we know it.

One visible and for the observer overwhelming manifestation of these processes are the aurorae (see Figure 3.1), which during the history of mankind have been associated with a wealth of religious myths, in the absence of any physical explanation. But most of the processes related to the interaction between electric and magnetic fields and particles are not directly visible and they were therefore only discovered after people started to use electromagnetic applications systematically for communication purposes, for example, via radio. The connection to the occurrence of aurorae was soon realised, although not understood, whereas the relationship to processes on the Sun's surface had much more difficulty in being accepted among scientists. One of the early promoters of the idea that processes on the surface of the Sun were causing near-Earth effects like aurorae and magnetic storms was the Norwegian scientist Kristian Birkeland (1867–1917). One of his ideas, that the aurorae are associated with electric currents along the magnetic field lines, is today a fundamental concept in our understanding of the physics of the upper atmosphere but was not accepted, and was sometimes even ridiculed, by many mainstream scientists because the idea conflicted with the fundamental concept of empty space or vacuum between the Sun and the Earth. Consequently, his ideas were only definitively accepted when the existence of such currents could be documented through direct measurements of their

Figure 3.1 Photo of aurora from Greenland (Craig Heinselman). For the color version, please refer to the plate section.

magnetic field obtained from spacecraft in 1967. Even then, this interpretation of the measurements was not broadly accepted until it was confirmed by new observations (Armstrong and Zmuda 1973).

Today we realise that the conditions in our own environment are very dependent on the processes on the solar surface and in the interplanetary space between the Sun and the Earth. We describe these solar controlled conditions as "Space Weather". The term encompasses the various processes that connect the physics of the Sun with that of the Earth's environment. In this definition is buried the science of a very complex system involving the interaction between variations on the solar surface and the electrodynamic environment of the Earth, controlled by its magnetic field. Still we have not fully understood in detail the multitude of physical processes that are involved. However, in terms of the last fifty years of extensive spacecraft observations both in the near-Earth space environment and in the solar system we now have a basic understanding of the physics that is involved. This allows us to start making predictions relevant for society at large, but it has also made us realise the enormous forces that are released during solar eruptions and the risk that our society faces, in particular in the technologically developed part of it.

Although the perceived effects of space weather processes have been increasing, particularly during the space age and associated technological development, all the processes are natural consequences of very fundamental properties of the Sun and our planet, determining the interactions between the Earth's environment (magnetosphere, atmosphere) and the local interplanetary space. These interactions are the main agent of space weather resulting in associated features (e.g., auroras) and different atmospheric effects (e.g., ionization).

In the following sections the two main factors controlling the physics of space weather, the Sun and its heliosphere and the Earth's magnetic field, will be briefly described.

3.2 The Sun and the Heliosphere

For a long time it was believed that the Sun is just the reliable, constant and perpetual source of energy for life on Earth. This is emphasised by the term the "solar constant", indicating the integrated solar power over all wavelengths at the position of the Earth, which is approximately 1,368 Wm^{-2}. Although one could observe the varying distribution of sunspots, and although ground telescopes could identify associated variable features on the surface, the dynamic and explosive nature of solar activity was not really revealed until space craft outside the Earth's atmosphere could observe in detail the true dynamics of the solar surface, both regarding the solar plasma and the radiation in various wavelengths. These observations have contributed significantly to understanding the puzzling but systematic approximate eleven-year cycle of solar activity first observed by the systematic rise and fall in the occurrence of spots on the Sun's surface. These sunspots are dark areas on the solar surface with very strong magnetic fields. Although the approximate eleven-year period can be understood in terms of magnetic fields in the outermost convection zone of the Sun and the latitudinal gradient in rotation rate, which continually twists the magnetic field tubes until they break, it has been very difficult to predict quantitatively the amplitude and the timing of this periodic variation in sunspot numbers. Actually, however, the true solar magnetic cycle is about twenty-two years since the polarity of the solar magnetic fields changes every sunspot cycle.

The recent abundance of solar surface observations, starting with the dedicated solar and heliospheric observatory, the joint ESA-NASA spacecraft, SOHO (see http://soho.esac.esa.int/), has definitively demonstrated the true dynamic nature and the overwhelming power of the solar surface activity in terms of emission of energetic particles, X-ray radiation, magnetic fields and enormous emissions of matter in the form of the so-called coronal mass ejections (CMEs), which we now know are causing the geomagnetic storms on Earth that represent a dominant component of space weather.

The yearly sunspot number can only be reliably traced back to the invention of the telescope at the beginning of the seventeenth century. But the varying solar activity can actually be traced further back in time through the radioactive ^{14}C isotope in organic material.

This isotope is one of the naturally occurring isotopes of carbon in the atmosphere and is formed when high-energy particles from our galaxy, the so-called cosmic rays, interact with nitrogen in the atmosphere. The flux of the cosmic rays is modulated by the heliospheric magnetic field, which again varies with the solar activity. Ground-based neutron monitors were installed as part of an international effort and have since the 1950s contributed significantly to our understanding of the galactic cosmic ray modulation. Because the half-life of ^{14}C is around 5,730 years, there is a clear limitation regarding how far back in time one can determine the solar activity. However, one can go even further back in time using another radioactive isotope, ^{10}Be, created by cosmic rays in the atmosphere. This isotope has a half-life of 1.39×10^6 years and is recorded in the ice caps in Greenland and Antarctica. Examination of ice core data has revealed a detailed time series of these solar activity proxies during the last ninety-four hundred years (McCracken et al. 2013). The time series has demonstrated the existence of distinct periodic signals; the most conspicuous ones apart from the well-known eleven- year period are periods of 87,208 and 978 years. No physical explanation of these periodicities has gained universal acceptance, but while energy considerations clearly show that the planets cannot be the direct cause of the solar activity, it is an open question whether the planets can perturb the operation of the solar dynamo. Abreu et al. (2012) report an excellent agreement between the long-term cycles in solar activity, and the periodicities in the planetary torque, mainly from the Jovian planets, and also find that some periodicities remain phase-locked over ninety-four hundred years. On these observations they suggest that the long-term solar magnetic activity could be modulated by planetary effects. Stefani et al. (2016) find that planetary tidal forces may act as small external pace setters and invoke oscillation in the so-called alpha effect in the solar dynamo. Every 11.07 years, the Sun and the planets Venus, Earth and Jupiter are aligned and the associated tidal force could cause the polarity reversal of the solar magnetic field and possibly the twenty-two-year cycle of the solar dynamo. The idea that the other planets indirectly, via the Sun, have an influence on our environment adds just another dimension to the concept of global change.

3.3 The Magnetic Field of the Earth

The geomagnetic field is created deep inside our planet. This is the field that determines the shape of the magnetosphere, which is the environment around the Earth in which the Earth's magnetic field is controlling the motion of the electrically charged particles. Inside the magnetosphere we are protected from the continuous flow of charged particles constituting the solar wind. The geomagnetic field is dominated by a dipole field, which, however, is offset relative to the Earth's centre. This results in anomalies like the significantly reduced magnetic field strength in the so-called South Atlantic anomaly (SAA). The exact form of the magnetosphere is always changing as a result of the interaction between the varying solar wind and the Earth's magnetic field.

Whereas the Sun's magnetic field carried away from the Sun by the solar wind is changing at all time scales, the Earth's interior magnetic field shows only slow changes over years, the secular variation. Nevertheless, the changes in the magnetic field are large and occasionally, on average every 200,000 years, lead to complete reversals of the magnetic field. The last reversal, however, took place more than 750,000 years ago. About 95% of the observed magnetic field on the Earth's surface is being produced and sustained by a dynamo process involving turbulent motions of molten iron in the outer core, below 3000-km depth. Its dominant dipole component, however, is currently decreasing at a rate presumably ten times faster than the one at which it would naturally decay, were the dynamo to be switched off. It has decreased by nearly 9% over the last 175 years (Finlay et al. 2016), a figure comparable to that seen at times of magnetic reversals (Hulot et al. 2002). However, in some regions, as in the South Atlantic anomaly, the field has decreased even more. This region characterised by a weak magnetic field is also the region where most spacecraft failures occur during severe space weather events and magnetic storms. Understanding how this weakening shield is going to evolve in the future is a very important issue, which is one of the main science objectives of the advanced three-satellite mission, the ESA Swarm: the Earth's magnetic field and environment explorers launched in 2013 (Friis-Christensen et al. 2006; Friis-Christensen et al. 2009).

For many purposes in the fields of ionospheric, magnetospheric and cosmic ray physics and in studies of crustal fields, particularly in exploration geophysics, an internationally produced and agreed global model of the main field and its secular variation, the International Geomagnetic Reference Field (IGRF), is widely used. The IGRF, first developed in 1969, is sponsored by the International Association of Geomagnetism and Aeronomy (IAGA). It is a series of mathematical models of the Earth's magnetic field and its annual rate of change. It is a product of a collaborative effort between magnetic field modelling

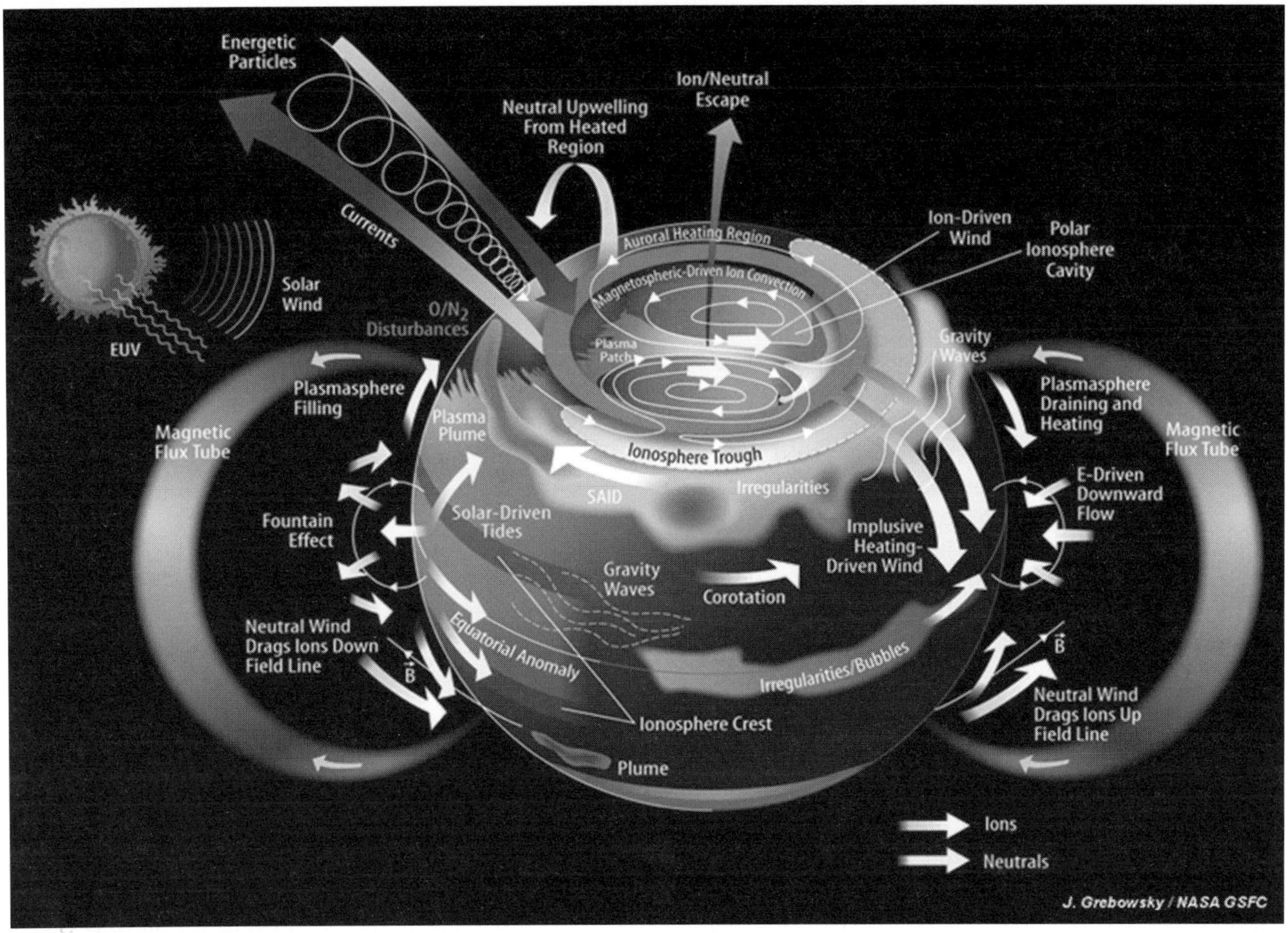

Figure 3.2 The magnetosphere comprising a number of different regions, each characterised by specific particle populations, particle drifts and magnetic fields (courtesy of J. Grebowsky). For the color version, please refer to the plate section.

groups and institutes involved in collecting and disseminating magnetic field data from observatories, surveys and satellites (Maus et al. 2005).

But in addition to the magnetic field from inside the Earth there exists a smaller but highly varying contribution from electric currents in the ionosphere and in the magnetosphere related to the dynamic solar wind.

3.4 Interaction between the Sun and the Earth's Magnetic Field

The Earth's magnetic field is an effective shield against charged particles impinging from outer space onto the Earth. The cocoon-like magnetosphere diverts most of the solar wind around the Earth with a bow shock defining the boundary between the magnetosphere and the ambient medium. The outermost layer of the magnetosphere is the magnetosheath consisting mainly of solar wind plasma. The particles entering into the magnetosphere are guided by the geomagnetic field and form fundamental and self-organising structures such as the radiation belts and the ring current as illustrated in Figure 3.2. Among the fundamental structures are the polar cusps, regions in the magnetosphere separating the closed magnetic field lines at the front of the magnetosphere from the open magnetic field lines that have one end connected to the Earth while the other end is part of the solar wind magnetic field. The polar cusps are the regions of most direct entry of magnetosheath plasma into the magnetosphere. Even low-energy particles may here follow the geomagnetic field lines, which converge to relatively small regions in the high-latitude day-side ionosphere where they heat the upper atmosphere. This results in an expanding atmosphere causing increased drag on low Earth-orbiting spacecraft, particularly strong during high solar activity (Schlegel et al. 2005).

The charge separation built up in the magnetosphere is neutralised through currents along the magnetic field lines. These currents close in the conducting ionosphere and are manifested by a dawn-dusk auroral zone potential difference of 15 kV during magnetically quiet conditions but may reach 150 kV during magnetically disturbed conditions. The field-aligned currents are a major link in the process of transferring energy from the solar wind into the upper atmosphere. The energy transferred is particularly large in the night-side auroral zone during the polar substorms where the auroral electrojets carry currents of the order of millions of amperes.

Geomagnetic substorms are local or regional phenomena that take place at regular time intervals and typically last one or two hours. Substorms are part of a natural and continuous process related to the unloading of energy collected in the night-side magnetosphere caused by the interaction between the solar wind and the Earth's magnetic field on the day side. But the dominant component of space weather is undoubtedly the geomagnetic storm (Gonzalez et al. 1994), which is a global phenomenon encompassing the whole magnetosphere. It is also the component that was identified long before it had any direct effect on technology, namely, as a period of large variations in the compass needle closely connected with the occurrence of bright aurora well equator-ward of the auroral zones. The large, systematic and worldwide component of the geomagnetic storm is the ring current, a westward current around the Earth's magnetic equator located in a band at a distance of 3 to 5 R_E (Earth radii). The current is carried by ions; mainly protons but also alpha particles and oxygen ions contribute to the current. The resulting magnetic field is opposite to the Earth's magnetic field, which implies a decrease in the horizontal north component of the magnetic field at the surface. Already early in the history of geomagnetic research this specific feature of the geomagnetic storm had been identified and an index, the disturbance field-storm (Dst) index had been defined in order to characterise the magnitude of geomagnetic storms.

The relationship between auroral substorms and geomagnetic storms is not fully understood, partly because it is difficult to study an auroral substorm that occurs during a geomagnetic storm, which is accompanied by a wide range of geophysical processes. Partamies et al. (2011) conclude that the energy dissipated during a substorm, even at the end of a prolonged storm recovery, is very large compared to the energy content of a typical isolated substorm.

The magnetic storm is the result of large transfer of energy and momentum from the heliosphere to the magnetosphere typically associated with a large explosive ejection of mass from the solar surface, the CME. CMEs are a natural phenomenon initiated at the solar surface associated with the strong magnetic fields also responsible for solar flares. But whereas solar flares create high-energy particles, the CME consists of large masses of solar plasma particles. To have an effect on Earth it is a requirement that the CME occur at a location on the Sun that faces the position of the Earth in order for the plasma masses to hit the Earth. Furthermore, to have a large effect, the magnetic field within the CME has to be oppositely directed to the Earth magnetic field where the two magnetic fields meet. Therefore, a southward-directed field in the CME facilitates the physical processes involved in transferring energy from the CME to the Earth's magnetosphere through magnetic reconnection and associated particle acceleration.

3.5 Space Weather Effects on Technology

Space weather processes have their origin in regions of strong magnetic fields at the solar surface associated with sunspots. When the magnetic field lines cross and reconnect, the stored energy is released in an explosion known as a solar flare. Solar flares occur frequently, at the maximum of the eleven-year sunspot cycle amounting to several per day. They are categorised in different classes corresponding to the released energy. The largest, X-class flares, cause long-lasting radiation storms in the upper atmosphere, and trigger radio blackouts. The high-energy charged particles, the so-called solar energetic particles (SEPs) with energy greater than 10 MeV/nucleon are dangerous for astronauts in space and may damage electronic equipment. They reach the Earth much faster than the solar wind plasma, in minutes to hours, depending on their energy. They penetrate easily to the ionosphere in the polar caps, the regions around the geomagnetic poles where the geomagnetic field is weakest and nearly vertical. Here they create high levels of ionisation that enhances absorption of radio signals. This can degrade and even completely black out HF communications along trans-polar aviation routes, requiring aircraft flying these routes to be re-directed to lower latitudes. Also the navigation signals from Global Positioning System (GPS) satellites can be affected, resulting in a reduction of the positioning accuracy. Various effects on society are sketched in Figure 3.3. The most energetic SEP events corresponding to energies greater than 500 MeV/nucleon may also affect the lower atmosphere as a result of cascading processes similar to the effect of

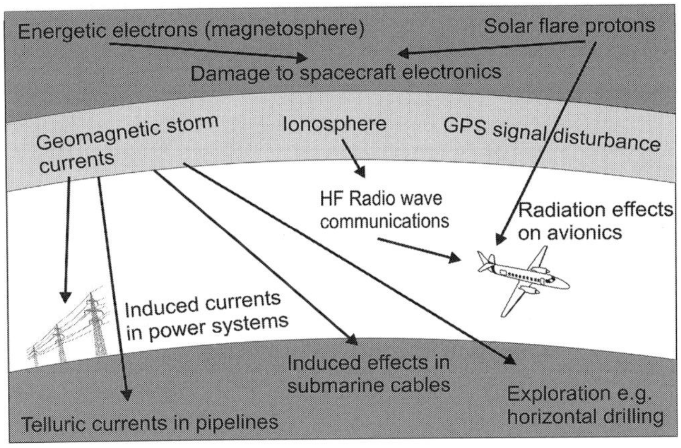

Figure 3.3 Space weather effects on technological systems.

the galactic cosmic rays. These events are known as ground level events (GLEs) and have been extensively studied through the data from the worldwide Neutron Monitor Network.

During magnetic storms the varying magnetic field due to electric currents in the auroral ionosphere may induce large voltages and associated currents in large-scale conducting structures on the ground. This may damage transformers connected to electric power grids and oil and gas pipelines may suffer from increased corrosion, which can be difficult to identify before catastrophic results may occur (see Figure 3.3).

Radiation damage to spacecraft and radiation exposure to humans in space are a matter of increasing concern as society becomes more dependent on infrastructure in space. Exposure of spacecraft to energetic particles during energetic particle events and radiation belt enhancements can cause temporary operational anomalies, damage critical electronics, degrade solar arrays and blind optical systems such as cameras and star trackers. The operational problems relate to charged particle of different types and energies. The most energetic ones are the ions that constitute the galactic cosmic ray (GCR) flux. These particles have energies larger than 100 MeV and are less modulated by the Earth's magnetic field and can therefore penetrate down to the surface. However, the GCR flux is subject to a large modulation by the heliospheric magnetic field, which, although much weaker than the Earth's field, works over huge distances. This effect therefore displays a distinct eleven-year cycle, with the largest flux at solar minimum where the heliospheric magnetic field is smallest. The low geomagnetic field strength in the South Atlantic anomaly allows high-energy particles from the radiation belts to penetrate deep into the upper atmosphere creating intense radiation accompanied by increased risk of satellite instrument failures (Heirtzler et al. 2002).

The effects of space weather on modern technological systems are well described in both the technical literature and newspapers. However, it may not always be realised that the effects on, for example, electric power lines may have enormous consequences for a number of different aspects of society including communication systems, provision of potable water, carrying out of banking and financing and execution of government services (Baker et al. 2014). A much-cited event is the collapse within ninety seconds of northeastern Canada's Hydro-Quebec power grid during the large geomagnetic storm of March 1989. Millions of people were left without electricity for up to nine hours. This event exemplifies the dramatic impact of extreme space weather events but also demonstrates how vulnerable our modern society has become because it depends more and more on a technology-based interconnected infrastructure, not only in space but also on the ground.

This event, although the most damaging, is, however, not the only geomagnetic superstorm that has occurred during the recent period. Seventeen major flares erupted on the Sun between 19 October and 5 November 2003, including perhaps the most intense solar flare ever measured directly, a huge X28 flare, causing an extreme radio blackout on 4 November (Gopalswamy et al. 2005).

However, the largest observed solar flare and subsequent geomagnetic storm occurred in 1859. Although this event, the Carrington event, took place when modern electricity-based technology was still rather primitive, it nevertheless had considerable and damaging effects on the recently invented and implemented electric telegraph system.

The gained understanding of the physics involved in recent severe space weather events has illustrated that extreme events such as the Carrington event are certainly likely to occur again some time in the future and will have a much larger damaging impact on many parts of our society including electric power blackout and associated cascading disruptive effects through dependent systems such as transportation, communication, banking, financing and government services. And this may lead to further disruption of necessary functions in our society such as supply of drinking water, food and medicine.

The increased concern prompted the Space Studies Board of the National Academies in the United States to establish a committee on the Societal and Economic Impacts of Severe Space Weather Events. The committee was charged to convene a public workshop that would feature invited presentations and discussion to assess the nation's current and future ability to manage the effects of space weather events and their societal and economic impacts. The workshop was held in Washington, D.C., on 22–23 May 2008. Participants were drawn from a broad cross section of those interested in or directly affected by severe space weather events, including government agencies and industry as well as private vendors of space weather services. The workshop provided an initial forum for gathering information on specific space weather effects and on the status and unmet challenges of forecasting. A comprehensive report summarised the workshop proceedings but was not intended to offer any recommendations. Instead, the objective of the workshop was to help gather information and identify issues for analysis in possible follow-on studies that could provide recommendations on future space weather programs, resource needs and interagency coordination to improve services and knowledge for those affected by space weather.

Recently, the European Union published the results of a two-day summit on the impact of extreme space weather on critical infrastructures (Krausmann et al. 2016). One of the main results is that an extreme space weather event may overwhelm a single nation's response capacity and that a strategic plan should be developed to define the roles of the key players in Europe.

Extreme space weather events are categorised as low-frequency/high-consequence events. It has been argued (Riley 2012) that a Carrington event will likely happen once every century. This makes the associated problems quite different from the problems raised by conventional and frequently experienced events.

But can we risk space weather events that are even more powerful? The NASA Solar Terrestrial Relations Observatory (STEREO) mission consists of two spacecraft, one located ahead in the Earth orbit, the other trailing behind. With this pair of viewpoints the mission is able to see the structure and evolution of solar storms as they leave the Sun and move out through space. This means that we are able to observe also solar storms that do not hit the Earth. A very powerful solar storm occurred on 23 July 2012. This solar storm was focused away from the Earth and was measured most directly by the STEREO-A spacecraft. This allowed to establish the timing and strength of the solar outputs of the event (Baker et al. 2013), and had the CME been striking the Earth, the world would have witnessed a storm larger (possibly much larger) than the 1859 Carrington event. This most likely would have had devastating consequences for many technological systems (Ngwira et al. 2013).

The observations illustrate that such events are real and that they occur now, even though the solar activity is not particularly high and that our society also needs to prepare for this kind of "global change".

3.6 The Climate System

Our climate is the result of the most complex physical system on Earth, apart from, perhaps, the motion of the electrical currents in the Earth's fluid core, which creates the varying magnetic field. Both systems are so complex that use of numerical models is the only way of describing, quantitatively, the temporal evolution, and observations provide the only way to verify the results. This complex system consists of different components (atmosphere, biosphere, cryosphere, hydrosphere, lithosphere and recently also anthroposphere), which are all highly interlinked through many concurrent operating processes on a wide range of temporal and spatial scales.

For the climate system, as for any system, the ability to predict in a satisfactory manner is dependent on understanding the physics that is involved in the multitude of processes that take place. By studying carefully past climate, by modelling the basic physical processes and by validating the models by observations, new climate knowledge can be achieved. The climate projections of variability and change and the physical understanding of the system rest on the application of advanced climate system models. On the other hand, it is not possible to model all the physical processes that are involved, in particular because some of the processes take place on scales well below the grid-cell size that is feasible within the available computer power. Furthermore, many of the small-scale processes, in particular those related to the formation of clouds, need to be better understood before they can be fully incorporated in the numerical models. The dominant factor in climate change relates to the complex physical processes determined by the difference in energy received from the solar radiation and the energy radiated away from the Earth. These processes involve the spectrum of the solar radiation, the composition of the atmosphere, including clouds, and the Earth's surface properties determining the reflection of incoming radiation.

Today, research in climate change is dominated by the effort to understand the effect of the rising level of greenhouse gasses in the atmosphere, in particular the anthropogenic CO_2. However, it is important to realise that many factors play a role and that the estimation of the effect of the anthropogenic contribution cannot be done properly without a better understanding of the natural climate processes that have always been and are still in play. Indeed, poor understanding of natural climate variations may be the largest obstacle to decreasing the current uncertainty in prediction of future climate.

Large unknown sources and sinks of energy are associated with the distribution of heat stored in the oceans. Observing the heat energy and the variations in the ocean currents transporting this heat, both horizontally and vertically, would provide a much more meaningful measure of global warming than just the global mean temperature, as discussed by Victor and Kennel (2014) and Balmaseda et al. (2013). Another potential driver of natural changes in our climate beyond the slowly varying orbital forcing is the variation in solar energy output and corresponding effects in the Earth's atmosphere.

3.7 Sun-Climate Relationship

Solar activity variations are probably the most dominating and systematic cause of climate variations after the last glaciation and a number of correlations between solar activity and various climatic variables or climate proxies have been reported. Already Herschel (1796) had noted an association between sunspot activity and the prices of wheat. More recently, Eddy (1977) undertook a comprehensive analysis and in particular noted the coincidence between the little ice age in the seventeenth century and the Maunder minimum in sunspot activity. Eddy's pioneering study was followed by a number of studies confirming a link between solar activity and climate: Labitzke and Van-Loon (1988), Friis-Christensen and Lassen (1991), Lassen and Friis-Christensen (1995), Soon et al. (1996), Bond et al. (2001), Neff et al. (2001) and Solanki and Fligge (2002).

Such correlations have been found on many time scales, including the 11-year Schwabe period, the 87-year Gleissberg period and the 978-year Eddy period. A persistent correlation over a three thousand-year period is illustrated in Figure 3.4, which shows a striking similarity between the $\delta^{18}O$ and $\Delta^{14}C$ data indicating that solar or GCR activity controlled the strength of the Indian Ocean monsoon. Increased GCR intensity is associated with a weakening of the monsoon (decreased rainfall) and a southerly shift of the intertropical convergence zone, ITCZ. In correlating longer time series based on solar activity proxies it must be realised that these proxies, based on recorded ^{14}C and ^{10}Be time series, may themselves be affected by climate variations. Fortunately, these two records, which are based on different isotopes, are not influenced by the same parts of the climate system; the ^{14}C concentration of the atmosphere is influenced by changes in the global carbon cycle (e.g., variations in the uptake of carbon in tree rings) whereas ^{10}Be concentrations are influenced by the varying amount and distribution of precipitation. Therefore, since these two processes in the climate system are so fundamentally different, considerable confidence can be placed in the records when they show similar time variations.

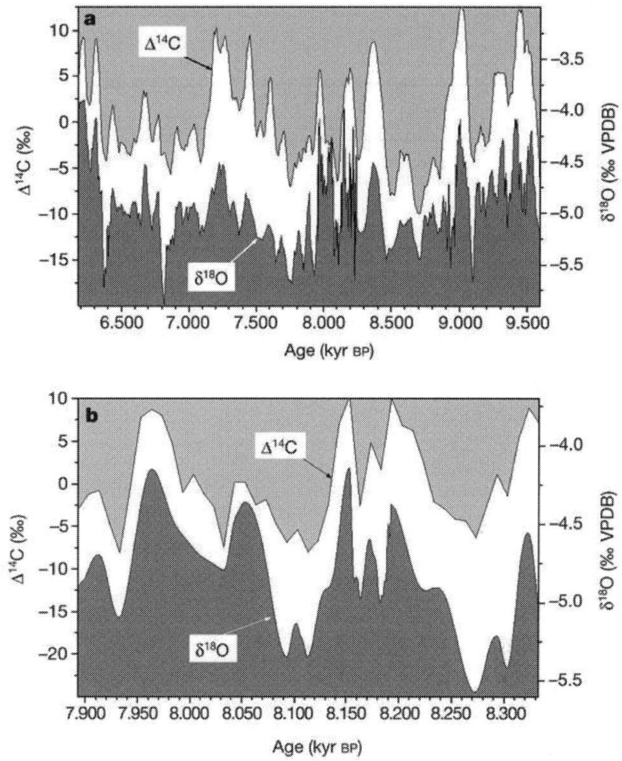

Figure 3.4 Profiles of $\delta^{18}O$ from a U-Th-dated stalagmite from a cave in Oman, together with $\Delta^{14}C$ from tree rings, for a) the 3.4-ky period from 9.6 to 6.2 ky BP (before present) and b) the 430-y period from 8.33 to 7.9 ky BP (Neff et al. 2001).

During modern times, when instrumental temperature measurements exist with a reasonable coverage, it was demonstrated that the Northern Hemisphere temperature over land from 1860 to 1985 very closely followed the solar activity (Friis-Christensen and Lassen 1991). In an updated analysis Lassen and Friis-Christensen (2000) realised that after 1985 the global surface temperature continued to rise whereas there was a decrease in solar activity. They therefore concluded that it was necessary to include additional physical processes to explain the change in global temperature, including the possibility of anthropogenic forcing. Lockwood and Frohlich (2007) concluded that the preceding thirty years of global surface temperature rise was primarily due to the increased amount of CO^2 in the atmosphere. However, observations show that the global mean temperature has not increased between the large El Niño in 1997–1998 and the recent one occurring in 2015–2016. This "hiatus" in global warming has been difficult to explain, in particular because it took place during an interval characterised by the fastest increase in global atmospheric CO_2 content.

Although it is difficult, in view of the abundance of reported correlations, to argue against a causal link between solar activity and climate, the observed correlations are not sufficient to determine whether such a link exists. In order to achieve this a physical theory is needed. Without a theory that can be verified following normal scientific methodology and practice nothing can be concluded regarding the causal link. But correlations, if significant, may indicate where to look for explanations. Of course, the first obvious hypothesis would be that the total solar irradiation changes with solar activity. This is easy to confirm on the basis of observations from spacecraft. During the eleven-year solar cycle the total solar irradiation changes about 0.1%. This effect is easy to incorporate in numerical climate models, but changes of this order are far too small to be comparable to the climate effect of the changing CO_2 content. A change of 0.1% is also too small to explain the observed correlation between solar activity and climate parameters on decadal to centennial time scales. Therefore, if the correlations are believed to represent a physical mechanism, this must imply a way, directly or indirectly, to amplify the solar effect.

The atmospheric temperature record displays large year-to-year variations due to internal variability, which are normally classified as "noise". To extract the eleven-year period signal from the noise has been difficult and therefore Shaviv (2008) suggested looking at the global sea level rise instead. The global sea level is supposed to rise in response to an increased surface temperature because of the expansion of the oceans. As a result of the large heat capacity of the oceans the global sea level is less influenced by short period changes than the atmosphere. In the study the global sea level record from 1900 to the present was examined. A clear eleven-year cycle in the rate of change of global sea level was found, consistent with a forcing from the varying solar irradiation. The observed effect, however, is about five times larger than expected from just total solar irradiance variations during the eleven-year solar cycle.

It is well known that during the solar cycle other solar parameters may vary more than the total irradiance. The UV part of the radiation varies much more, on the order of 20%. The available energy in the radiation in these wavelengths is, however, much too low to have direct effect on the atmosphere. But these wavelengths do have a considerable effect on the chemistry in the stratosphere, in particular on the ozone concentration, which could modify vertical transports between the stratosphere and the troposphere. Modelling studies can simulate the decadal variation of an effect in the troposphere, but the global effects are small. However, some of these results indicate that there may be a regional effect by changing the location of the atmospheric circulation. This could imply colder winters in Europe as at times observed during low solar activity (Lockwood et al. 2010).

Another parameter changing significantly during the eleven-year solar cycle is the ionisation in the troposphere. In the upper atmosphere, in particular in the ionosphere, the ionisation is caused by the extra UV EUV radiation from the Sun, but this radiation is not able to penetrate down to the troposphere. Nevertheless, a small ionisation does exist in the troposphere. This is caused by the cosmic rays mainly coming from our galaxy, the GCRs consisting of very high-energy particles. The GCR flux is negatively correlated with the solar activity because the larger heliospheric magnetic field at solar maximum allows fewer GCR particles to enter the Earth's atmosphere. Even before the space age Ney (1959) pointed out that the ionisation in the troposphere due to cosmic rays did show a very large variation during the solar cycle.

This feature was examined by Svensmark and Friis-Christensen (1997), who demonstrated that the global cloud cover during two solar cycles is well correlated with the cosmic ray flux, meaning an increased cloud cover at sunspot minimum where the cosmic ray flux is largest. Later, Marsh and Svensmark (2000) refined the analysis with additional data and found that it was actually the low-altitude clouds that were correlated with the cosmic rays. This was an important result because it gave a hint regarding a possible physical mechanism, which could explain the correlation. If only low clouds are influenced it has to be a mechanism involving liquid water as opposed to high clouds, which consist of ice particles.

This observation is crucial because one of the least understood processes in climate research is the formation of clouds. Clouds, on the other hand, are also one of the most efficient means of changing the radiation from the Sun. If ionisation can change the cloud coverage, this provides a very efficient way of changing the amount of solar irradiation reaching the Earth's surface. Higher solar activity implies lower cosmic ray flux and hence reduced cloud cover and increased solar radiation heating.

Cloud measurement, and in particular identification of low clouds, is very difficult to do from spacecraft

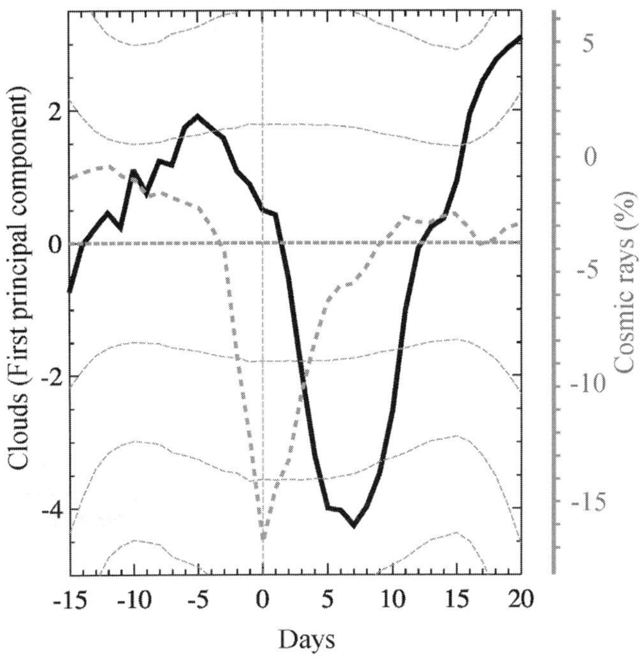

Figure 3.5 Statistical common disturbance in clouds (first principal component) based on three cloud satellite data sets (ISCCP, MODIS and SSM/I) superposed for the five strongest Forbush decreases (black) curve. The grey curve is the change (in percentage) of cosmic rays superposed for the same five events. The thin lines denote 1 to 3 standard deviations (adapted from Svensmark et al. 2016).

when middle and/or high clouds coexist. This could be a reason why the found correlation could not be reconstructed during later solar cycles. This has obviously caused some difficulty in the hypothesis's gaining general acceptance (Gray et al. 2010). However, studies of aerosol and various clouds parameters show a significant effect during intense solar storms accompanied the by so-called Forbush decreases (FDs), temporary large decreases in the observed cosmic ray flux lasting several days, as shown in Figure 3.5 (Svensmark et al. 2016). This figure shows the collected effect of FDs on various cloud parameters measured from satellites.

But a correlation is not sufficient. The physical mechanism has to be found and explained and possibly used to make predictions, which is the ultimate test of a physical theory. An effect of ionised particles on the creation of ultra-fine aerosols, which could evolve into cloud condensation nuclei, would indicate that the cosmic rays are the missing link between solar activity variations and climate. In laboratory experiments it was demonstrated that the presence of ions contributes strongly to the formation (Svensmark et al. 2007) and growth (Svensmark et al. 2013) of aerosols. Independent laboratory experiments (Kirkby et al. 2011, 2016) confirmed the role of ions in aerosol nucleation, but although the laboratory experiments suggest that there is an effect of cosmic rays on cloud formation in the troposphere, a complete physical model is still lacking. Therefore, it is widely debated whether the effect is large enough to have a significant effect on climate (Laken et al. 2012, Dunne et al. 2016).

The hypothesis of an effect on cloud formation of cosmic ray particles can explain many of the observations of the effects on climate of changes in solar activity. This hypothesis, however, has also been suggested as an explanation for large glacial periods in the Earth's geological history during the last 500 My (Shaviv 2002). Such periods could be associated with increased cosmic ray flux related to our solar system crossing the galactic spiral arms with increased supernovae production (Svensmark 2012).

The various laboratory experiments indicate that important physical processes related to the formation of clouds are still unexplained and therefore not included in the current numerical climate models, which are used to predict future changes in climate. It is conceivable that such a missing component in climate models could explain some of the existing discrepancies between model predictions and observations regarding climate change.

3.8 Conclusions

A well-developed infrastructure is crucial for the operation of our modern society. There is therefore a concern about the vulnerability of the critical parts of our infrastructure to the effects of natural hazards. In particular, because of the growing dependence on technological systems including power grids, telecommunication, avionics and navigation systems the potentially disastrous impacts of extreme space weather events on our infrastructure are becoming increasingly evident.

Our understanding of the physical processes involved in the interaction between the Earth's environment (magnetosphere, ionosphere and atmosphere) and the solar radiation, high-energy particles and solar wind plasma is restricted by the limited number of simultaneous observations in the huge region that we call geospace. Furthermore, the dynamics of these processes will never in a foreseeable future allow a comprehensive physical description of any single space weather event.

Nevertheless, research based on the increased number and quality of observations of the Sun's surface during the latest decades has elucidated some very fundamental physical space weather processes that affect life and technological systems – in space as well as on the ground. This means that although we are still far from predicting very well the conditions on the Sun we now have a sufficiently good understanding of the physics, which allows us to start making predictions relevant for society at large. The increased understanding of the physics has made us realise the enormous forces that are released during solar eruptions and the risk that our society faces, in particular in the technologically developed part of it.

The understanding we have gained has demonstrated that two main issues have to be dealt with: increased prediction capability and strategic plans, not only on a national level but in terms of internationally coordinated plans. It is not a question of whether, but of when a monster solar storm will hit the Earth with possibly catastrophic implications.

An important aspect of the Sun–Earth connection is the possibility of a link between solar variations and climate. If such a link exists and can be quantified, it has significant implications regarding the possibility of predicting climate changes, natural and man-made and mitigating the societal effects.

References

Abreu, Jose A, Jürg Beer, Antonio Ferriz-Mas, Kenneth G McCracken, and Friedhelm Steinhilber. Is there a planetary influence on solar activity? *Astronomy & Astrophysics*, 548:A88, 2012.

Armstrong, James C, and AJ Zmuda. Triaxial magnetic measurements of field-aligned currents at 800 kilometers in the auroral region: Initial results. *Journal of Geophysical Research*, 78 (28): 6802–6807, 1973.

Baker, Daniel N, X Li, A Pulkkinen, CM Ngwira, ML Mays, AB Galvin, and KDC Simunac. A major solar eruptive event in July 2012: Defining extreme space weather scenarios. *Space Weather*, 11 (10): 585–591, 2013.

Baker, Daniel N, Jamie M Jackson, and Lauren K Thompson. Predicting and mitigating socio-economic impacts of extreme space weather: Benefits of improved forecasts. *Extreme Natural Hazards, Disaster Risks and Societal Implications*, 1: 113, 2014.

Balmaseda, Magdalena A, Kevin E Trenberth, and Erland Källén. Distinctive climate signals in reanalysis of global ocean heat content. *Geophysical Research Letters*, 40 (9): 1754–1759, 2013.

Bond, Gerard, Bernd Kromer, Juerg Beer, Raimund Muscheler, Michael N Evans, William Showers, Sharon Hoffmann, Rusty Lotti-Bond, Irka Hajdas, and Georges Bonani. Persistent solar influence on north atlantic climate during the holocene. *Science*, 294 (5549): 2130–2136, 2001.

Dunne, Eimear M, Hamish Gordon, Andreas Kürten, João Almeida, Jonathan Duplissy, Christina Williamson, Ismael K Ortega, Kirsty J Pringle, Alexey Adamov, Urs Baltensperger et al. Global atmospheric particle formation from cern cloud measurements. *Science*, 354 (6316): 1119–1124, 2016.

Eddy, John A. Climate and the changing sun. *Climatic Change*, 1 (2): 173–190, 1977.

Finlay, Christopher C, Julien Aubert, and Nicolas Gillet. Gyre-driven decay of the earth's magnetic dipole. *Nature communications*, 7, 2016.

Friis-Christensen, E, and K Lassen. Length of the solar cycle: An indicator of solar activity closely associated with climate. *Science*, 254 (5032): 698–700, 1991.

Friis-Christensen, Eigil, H Lühr, and Gauthier Hulot. Swarm: A constellation to study the earth's magnetic field. *Earth, Planets and Space*, 58 (4): 351–358, 2006.

Friis-Christensen, Eigil, Hermann Lühr, Gauthier Hulot, Roger Haagmans, and Michael Purucker. *Geomagnetic Research from Space*. *Eos*, 90 (25): 213–214, 2009.

Gonzalez, WD, JA Joselyn, Y Kamide, HW Kroehl, G Rostoker, BT Tsurutani, and VM Vasyliunas. What is a geomagnetic storm? *Journal of Geophysical Research: Space Physics*, 99 (A4): 5771–5792, 1994.

Gopalswamy, N, L Barbieri, EW Cliver, G Lu, SP Plunkett, and RM Skoug. Introduction to violent sun-earth connection events of October–November 2003. *Journal of Geophysical Research: Space Physics*, 110 (A9), 2005. https://doi.org/10.1029/2005JA011268.

Gray, Lesley J, Jürg Beer, Marvin Geller, Joanna D Haigh, Michael Lockwood, Katja Matthes, Ulrich Cubasch, Dominik Fleitmann, G Harrison, L Hood et al. Solar influences on climate. *Reviews of Geophysics*, 48 (4), 2010.

Heirtzler, James R, Joe H Allen, and Daniel C Wilkinson. Ever-present south Atlantic anomaly damages spacecraft. *EOS, Transactions American Geophysical Union*, 83 (15): 165–169, 2002.

Herschel, W. Some remarks on the stability of the light of the sun. *Philos. Trans. Roy. Soc., London*, 166: 1796, 1796.

Hulot, Gauthier, Céline Eymin, Benoît Langlais, Mioara Mandea, and Nils Olsen. Small-scale structure of the geodynamo inferred from oersted and magsat satellite data. *Nature*, 4160 (6881): 620–623, 2002.

Imbrie, J, André Berger, EA Boyle, SC Clemens, A Duffy, WR Howard, G Kukla, J Kutzbach, DG Martinson, A McIntyre et al. On the structure and origin of major glaciation cycles 2. The 100,000-year cycle. *Paleoceanography*, 8 (6): 699–735, 1993.

Kirkby, Jasper, Joachim Curtius, João Almeida, Eimear Dunne, Jonathan Duplissy, Sebastian Ehrhart, Alessandro Franchin, Stéphanie Gagné, Luisa Ickes, Andreas Kürten et al. Role of sulphuric acid, ammonia and galactic cosmic rays in atmospheric aerosol nucleation. *Nature*, 476 (7361): 429–433, 2011.

Kirkby, Jasper, Jonathan Duplissy, Kamalika Sengupta, Carla Frege, Hamish Gordon, Christina Williamson, Martin Heinritzi, Mario Simon, Chao Yan, João Almeida et al. Ion-induced nucleation of pure biogenic particles. *Nature*, 533 (7604): 521–526, 2016.

Krausmann, E, E Andersson, M Gibbs, and W Murtagh. Space weather and critical infrastructures: Findings and outlooks. *EUR* 28237 *EN*, 2016. doi: 10.2788/152877.

Labitzke, Karin, and Harry Van Loon. Associations between the 11-year solar cycle, the QBO and the atmosphere. Part I: The troposphere and stratosphere in the northern hemisphere in winter. *Journal of Atmospheric and Terrestrial Physics*, 50 (3): 197–206, 1988.

Laken, Benjamin A, Enric Palle, Jasa Calogovic, and Eimear M Dunne. A cosmic ray-climate link and cloud observations. *J. Space Weath. Space Clim.*, 2: A18, 2012. doi: 10.1051/swsc/2012018.

Lassen, K, and E Friis-Christensen. Reply [to solar cycle lengths and climate: A reference revisited]. *Journal of Geophysical Research: Space Physics*, 105 (A12): 27493–27495, 2000. ISSN 2156–2202. doi: 10.1029/2000JA900067. URL http://dx.doi.org/10.1029/2000JA900067.

Lassen, Knud, and Eigil Friis-Christensen. Variability of the solar cycle length during the past five centuries and the apparent association with terrestrial climate. *Journal of Atmospheric and Terrestrial Physics*, 57 (8): 835–845, 1995.

Lockwood, Mike, and Claus Fröhlich. Recent oppositely directed trends in solar climate forcings and the global mean surface air temperature. In *Proceedings of the Royal Society of London A: Mathematical, Physical and Engineering Sciences*, Volume 463, pages 2447–2460. The Royal Society, 2007.

Lockwood, Mike, Richard Giles Harrison, T Woollings, and Sami K Solanki. Are cold winters in Europe associated with low solar activity? *Environmental Research Letters*, 5 (2): 024001, 2010.

Marsh, Nigel, and Henrik Svensmark. Cosmic rays, clouds, and climate. *Space Science Reviews*, 94 (1–2): 215–230, 2000.

Maus, Stefan, Susan MacMillan, T Chernova, S Choi, D Dater, V Golovkov, V Lesur, F Lowes, H Lühr, W Mai et al. The 10th-generation international geomagnetic reference field. *Geophysical Journal International*, 161 (3): 561–565, 2005.

McCracken, K G, J Beer, F Steinhilber, and J Abreu. A phenomenological study of the cosmic ray variations over the past 9400 years, and their implications regarding solar activity and the solar dynamo. *Solar Physics*, 286 (2): 609–627, 2013. ISSN 1573–093X. doi: 10.1007/s11207–013–0265–0. URL http://dx.doi.org/10.1007/s11207–013–0265–0.

Neff, U, SJ Burns, A Mangini, M Mudelsee, D Fleitmann, and A Matter. Strong coherence between solar variability and the monsoon in Oman between 9 and 6 kyr ago. *Nature*, 411 (6835): 290–293, 2001.

Ney, Edward P. Cosmic radiation and the weather. *Nature*, 183: 451–452, 1959.

Ngwira, Chigomezyo M, Antti Pulkkinen, M Leila Mays, Maria M Kuznetsova, AB Galvin, Kristin Simunac, Daniel N Baker, Xinlin Li, Yihua Zheng, and Alex Glocer. Simulation of the 23 July 2012 extreme space weather event: What if this extremely rare cme was earth directed? *Space Weather*, 11 (12): 671–679, 2013.

Partamies, N, L Juusola, E Tanskanen, K Kauristie, JM Weygand, and Y Ogawa. Substorms during different storm phases. In *Annales Geophysicae*, 29, 2031–2043.

Plainaki, Christina, Jean Lilensten, Aikaterini Radioti, Maria Andriopoulou, Anna Milillo, Tom A Nordheim, Iannis Dandouras, Athena Coustenis, Davide Grassi, Valeria Mangano et al. Planetary space weather: Scientific aspects and future perspectives. *Journal of Space Weather and Space Climate*, 6: A31, 2016.

Riley, Pete. On the probability of occurrence of extreme space weather events. *Space Weather*, 10 (2), 2012.

Schlegel, Kristian, Hermann Lühr, J-P St Maurice, Geoff Crowley, and Chris Hackert. Thermospheric density structures over the polar regions observed with champ. *Annales Geophysicae*, 23, 1659–1672, 2005.

Shaviv, Nir J. Publisher's note: Cosmic ray diffusion from the galactic spiral arms, iron meteorites, and a possible climatic connection [*Phys. Rev. Lett.* 89, 051102 (2002)]. *Phys. Rev. Lett.*, 89: 089901, Aug 2002. doi: 10.1103/PhysRevLett.89.089901. URL http://link.aps.org/doi/10.1103/PhysRevLett.89.089901.

Shaviv, Nir J. Using the oceans as a calorimeter to quantify the solar radiative forcing. *Journal of Geophysical Research: Space Physics*, 113 (A11), 2008.

Solanki, SK and M Fligge. Solar irradiance variations and climate. *Journal of Atmospheric and Solar-Terrestrial Physics*, 64 (5): 677–685, 2002.

Soon, WH, ES Posmentier, and SL Baliunas. Inference of solar irradiance variability from terrestrial temperature changes, 1880–1993: An astrophysical application of the sun-climate connection. *The Astrophysical Journal*, 472 (2): 891, 1996.

Stefani, F, A Giesecke, N Weber, and T Weier. Synchronized helicity oscillations: A link between planetary tides and the solar cycle? *Solar Physics*, 1–16, 2016. ISSN 1573–093X. doi: 10.1007/s11207–016–0968–0. URL http://dx.doi.org/10.1007/s11207-016-0968-0.

Svensmark, Henrik. Evidence of nearby supernovae affecting life on earth. *Monthly Notices of the Royal Astronomical Society*, 423 (2): 1234–1253, 2012. ISSN 1365–2966. doi: 10.1111/j.1365–2966.2012.20953.x. URL http://dx.doi.org/10.1111/j.1365–2966.2012.20953.x.

Svensmark, Henrik, and Eigil Friis-Christensen. Variation of cosmic ray flux and global cloud coverage – a missing link in solar-climate relationships. *Journal of Atmospheric and Solar-Terrestrial Physics*, 59 (11): 1225–1232, 1997.

Svensmark, Henrik, Jens Olaf P Pedersen, Nigel D Marsh, Martin B Enghoff, and Ulrik I Uggerhøj. Experimental evidence for the role of ions in particle nucleation under atmospheric conditions. In *Proceedings of the Royal Society of London A: Mathematical, Physical and Engineering Sciences*, Volume 463, pages 385–396. The Royal Society, 2007.

Svensmark, Henrik, Martin B Enghoff, and Jens Olaf Pepke Pedersen. Response of cloud condensation nuclei (>50nm) to changes in ion-nucleation. *Physics Letters A*, 377 (37): 2343–2347, 2013.

Svensmark, J, MB Enghoff, NJ Shaviv, and Henrik Svensmark. The response of clouds and aerosols to cosmic ray decreases. *Journal of Geophysical Research: Space Physics*, 121(9): 8152–8181, 2016.

Victor, David G, and Charles F Kennel. Climate policy: ditch the 2 °C warming goal. *Nature*, 1219(9), 8147–8151.

4

Climate Issues from the Planetary Perspective and Insights for the Earth

Athena Coustenis, Fred W. Taylor, and Christina Plainaki

4.1 Introduction

The laws of physics, in particular those that govern the formation, evolution and behaviour of any planetary body, are the same everywhere, and only the boundary conditions and evolutionary history are different. Hence, in order to understand climate, it is instructive to compare that of the Earth with climate on other planets, especially those with substantial atmospheres. From a fuller understanding of the climatology within the Sun's planetary family, and eventually that of the planets in other stellar systems, we may better understand our own, and develop predictive models for the future Earth. With this as our objective, we focus on Earth's neighbours in the inner solar system, Mars and Venus, with the addition of Saturn's large and Earth-like satellite Titan.

The inclusion of Titan may seem surprising to anyone new to the subject of comparative planetology; however, Titan is considered to be an 'honorary' terrestrial planet since its solid body is larger than the planet Mercury and it is surrounded by a dense atmosphere showing a full range of quite Earth-like climatic behaviour (Coustenis and Taylor, 1999; 2008). Mercury itself has a negligible atmosphere but, because of its proximity to the Sun, extreme environmental conditions leading to unique Sun–planet coupling effects take place. This kind of coupling has implications for the climate and its evolution on all planets, including the Earth.

Like the Earth, Venus, Mars and Titan are dynamic entities with atmospheres, surfaces and interiors evolving significantly over time. To understand this evolution requires information gathered from a variety of sources, especially deep space missions, to be applied to produce time-dependent models that incorporate all relevant physical and chemical processes. The climates of the planets are different not only because of their distances from the Sun, but also because of large variations in their atmospheric compositions and densities that affect the energy balance in the atmosphere at different levels. The resulting temperature differences drive the circulation of the atmosphere in ways that are also affected by parameters such as the planetary rotation rate and axial inclination (obliquity).

Key to any understanding of climate (conveniently represented in simple terms on an Earth-like planet by the mean surface temperature) is the component contributed by the atmosphere to the radiative energy balance at the surface, the so-called greenhouse effect. Gases such as carbon dioxide, methane and water vapour are essentially transparent to visible light but absorb infrared radiation very strongly. Thus they tend to trap solar energy, with the result that the atmosphere and surface are warmer than would otherwise be the case. At the same time, scattering by molecules and cloud or haze particles in the atmosphere tend to have a net cooling effect, preventing some sunlight from reaching the surface and reflecting a fraction back to space. This fraction, known as the albedo, has moderate values for Earth, Mars and Titan (about 30%, 20% and 27% respectively), while Venus's perpetual cloud cover of highly reflective sulphuric acid droplets contributes to a much higher albedo of about 76%.

The balance between heating and cooling effects and their variations give each of the planets its distinctive temperature distribution and atmospheric circulation, which together lead to interesting potential couplings between the biosphere and climate. On each planet, this can vary with time as the atmospheric composition, and other factors, such as solar activity, change. The net effect has important implications for the climate and potential habitability not just of the Earth and its neighbours in the solar system, but also of the more Earth-like exoplanets.

Past studies have demonstrated that space weather can be the cause of climate variations and that there is a link between solar activity and climate at the Earth (see Chapter 1.3 for an extended discussion and references). The search for similar correlations at other planets could be important for our overall understanding of the role of the Sun in planetary climate evolution. Moreover, the investigation of internally driven space weather at the outer planets and its relation with upper atmospheric phenomena (e.g., energy deposition) could provide an additional component to our understanding of the space weather–weather relationship and to the overall energy balance in the system.

4.2 The Terrestrial Planets

The Earth is one of a family of four rocky planets that orbit relatively close to the Sun. As with the H_2–He-dominated bodies of the giant planets and the tenuous atmospheres of most of their satellites, it is usual to disregard Mercury when looking for analogues to the Earth's surface environment, because its atmosphere is so thin as to be virtually non-existent and without any significant role in determining the surface energy balance. On the other hand Earth's nearest planetary neighbours, Venus, closer to the Sun, and Mars, more distant, both exhibit a great deal of Earth-like climatic behaviour and important comparisons can be made, for instance between the mean temperature profiles (Figure 4.1) of their atmospheres.

We can also compare the general circulation of the atmosphere on these three planetary neighbours (Figure 4.2). While they are superficially quite different, on closer investigation it is not difficult to understand the mean differences and seasonal changes in terms of simple circulation models that take account of the variations in basic parameters such as solar heating, albedo, rotation rate and atmospheric composition and density.

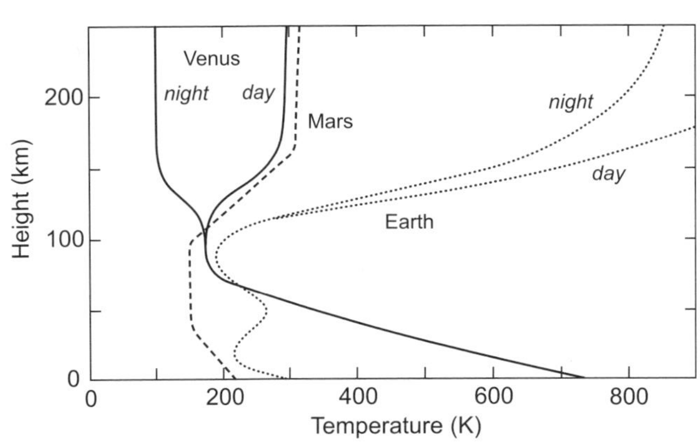

Figure 4.1 Models of the mean temperature profiles on Venus, Earth and Mars (Taylor, 2010).

For example, Venus has a very slow solid-body rotation rate with a period of 243 days, and almost no axial obliquity, while on Mars both of these parameters are much closer to the values that apply to the Earth. Thus equator-to-polar Hadley cell circulations are potentially stable on Venus, but not on its two siblings. Mars, with its thin atmosphere, has very large diurnal temperature variations while Venus has none and Earth is somewhere in between. Some of the details are much harder to explain, such as the dust storm seasons on Mars and the rapid zonal super-rotation on Venus, which have no real counterpart on the Earth. Topics like these, along with local meteorological phenomena, remain within the focus of current research.

4.2.1 The Current Climate of Venus and Contrast with Earth

The surface of Venus is very hot. At its landing site in 1975, the spacecraft Venera 9 measured a temperature of 758 K, which is hotter than a domestic self-cleaning oven, and indeed at this temperature soft metals such as lead and tin are molten. This state of affairs cannot be simply attributed to proximity to the Sun; at 240 K, the radiometric temperature of Venus (that of a blackbody in equilibrium with the incoming solar energy) is actually rather less than that of the Earth. The reason is that the very high albedo of the planet, which is due to its highly reflective cloud cover, is more than enough to offset the difference in orbital radius of the two planets.

This being so, it was long expected, in the centuries before the start of the space age, that Venus would have Earth-like conditions at its surface. The planets are

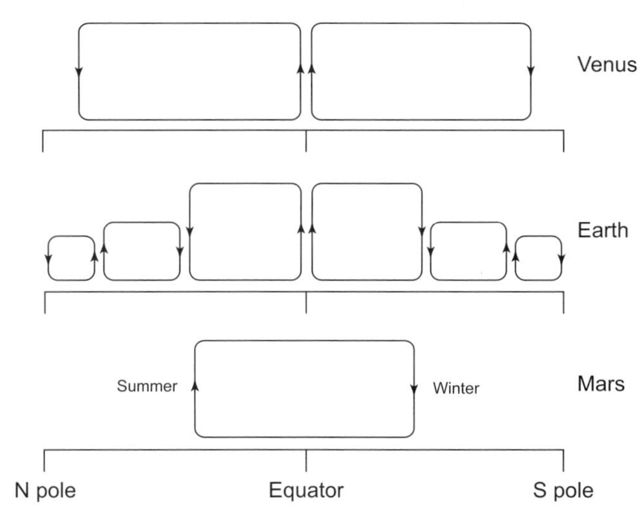

Figure 4.2 Schematics of the zonal mean general circulation in the atmospheres of Venus, Earth and Mars (Taylor, 2010).

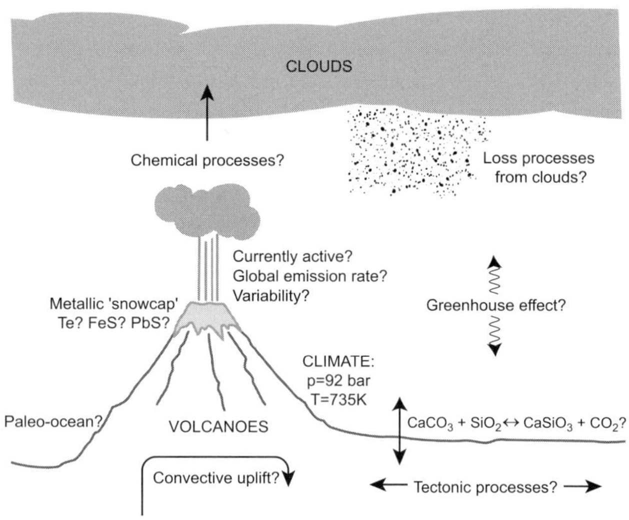

Figure 4.3 Some of the key processes affecting climate on Venus: volcanism, cloud formation and atmosphere–surface interactions are among the most important (Taylor, 2014).

about the same size, and the cloudiness of Venus, coupled with its closeness to Earth, suggested to early observers and theoreticians the presence of large amounts of water, so that oceans and a related hydrological cycle could be expected. Svante Arrhenius, during the early 1900s, weighed up all of the theoretical and observational evidence and concluded that Venus conditions could be warm, wet and suitable for life, if not indigenous then potentially capable of being provided by the arrival of colonists from Earth.

The very first successful interplanetary probe, Mariner 2 in 1964, confirmed suspicions – originating with radio telescope observations of higher-than-expected fluxes from Venus in the microwave region of the spectrum – that the surface and lower atmosphere are actually much hotter, and drier, than their terrestrial equivalents. They also hinted at the reason, which is that Venus has a much more massive atmosphere than Earth, and is therefore much more affected by 'greenhouse' heating of the surface (Figure 4.3). This effect is exacerbated by the fact that Venus's atmospheric composition, as found by later missions, including Soviet landers, is composed of around 96% carbon dioxide, which is of course a powerful greenhouse gas.

The vital question of how Earth and Venus evolved to be so different in climate is currently a hot research topic. Most experts accept that 'evolved' is the key term; it is broadly accepted that the two planets were quite similar after formation, particularly in internal structure and composition, including the amount of water present. Whether Venus was ever cool enough to have had liquid oceans like those on Earth is moot; current thinking (e.g., Taylor and Grinspoon, 2009; Taylor and Hunten, 2014) suggests that it is more likely that the water was present as steam, raising the surface pressure and the temperature far above its present value in the distant past. Either way, the water is no longer present except in trace amounts, and seems to have been lost by dissociation in the upper atmosphere because of the high levels of solar UV irradiation on Venus. To allow significant concentrations of water to reach the altitudes where dissociation could proceed rapidly, Venus must have had a warmer stratosphere at that time and so avoided the 'cold trap' that helps to retain water on the Earth. The highly enriched ratio of deuterium to hydrogen on Venus is often cited as powerful evidence for this hypothesis.

If Venus dried out relatively early in its evolution, this could account for two other significant differences from Earth. One is the retention of huge amounts of carbon dioxide in the atmosphere, in contrast to Earth, where processing through dissolution in the oceans and ingestion by plant life have removed more than 99% of the primordial inventory of atmospheric carbon dioxide. The other is the dryness of the crust on Venus, which apparently has suppressed large-scale plate tectonics to the extent that the primary route for the escape of interior heat is through volcanic activity rather than convection. These two processes could account together for the high surface pressure, nearly one hundred times Earth's, and also for the very high abundance of sulphur compounds on Venus. Sulphur dioxide is more than a hundred times over-abundant in the troposphere when compared to chemical equilibrium models, and the clouds are only about 20% water with the rest mostly sulphuric acid. On Earth, equivalent amounts of carbon and sulphur have been trapped in surface minerals such as carbonates and sulphates.

In the present day, although gases are removed from the atmosphere much more slowly on dry Venus than on wet Earth, there will still be a steady loss and some form of replenishment would be needed for equilibrium. This is especially the case for sulphur dioxide, which is more than one hundred times over-abundant on Venus compared to chemical equilibrium models. Here, the explanation seems to be that Venus has a high level of current volcanic activity, at least equal to that of the Earth and possibly greatly exceeding it. The latter might be expected, since the dryness of the solid crust on Venus has apparently suppressed plate tectonics, at least on the largest scale, and it may be that volcanic venting takes its place as the main mechanism releasing heat from the interior. The dryness and rigidity of the crust may also have a role in suppressing magnetic field generation on Venus, thus helping to explain

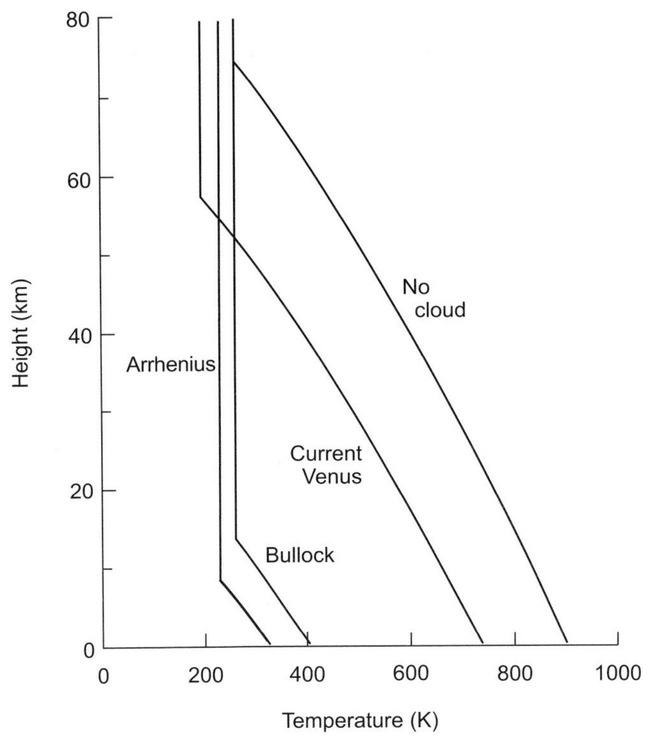

Figure 4.4 Some possible model temperature profiles for the atmosphere of Venus at various epochs in its history (Taylor, 2010).

another major difference between the two superficially identical planets.

In the distant future, when all volcanic activity has subsided, Venus may be expected to lose its sulphuric acid clouds, and eventually most of its atmospheric carbon dioxide as well, leaving a mainly nitrogen atmosphere more like the Earth's. Figure 4.4 shows model predictions for the temperatures that might result under these scenarios, compared with the present day and with the conditions envisioned by Arrhenius more than a century ago.

4.2.2 Current Climate of Mars Compared to Earth

The atmosphere of Mars is, like that of Venus, very dry compared to Earth's, but in this case the main reason is less a shortage of water than the low prevailing temperatures. In fact, most of the atmosphere of Mars is saturated with water most of the time, but since the surface exceeds the freezing point of H_2O only near the sub-solar point in summer, and then only for a relatively short time, the equivalent depth if the vapour could be condensed as liquid is only a few millimetres. However, a huge amount of water is trapped in the polar caps, and below the surface at lower latitudes, as

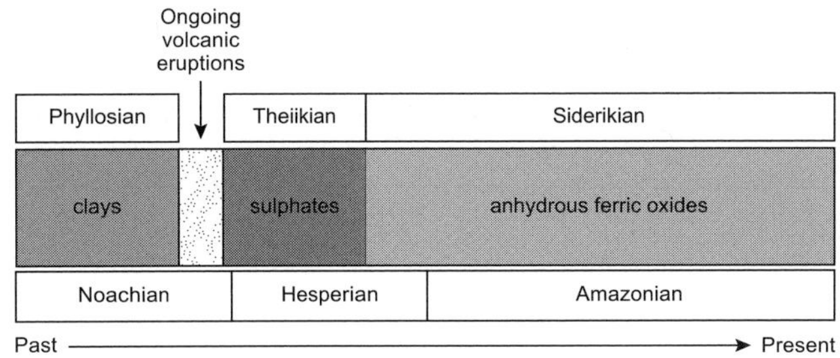

Figure 4.5 A summary of the climate history of Mars over the lifetime of the planets, showing alternative names proposed by researchers for the principal epochs and key formation processes (Taylor, 2009 and references therein).

permafrost and ice. Some of it may be liquid in places where residual volcanism gives rise to warm sub-surface aquifers that appear to break through in places, producing short-lived flows that leave visible traces in some crater walls.

Liquid water is not stable under the conditions on Mars, in particular the low atmospheric pressure that is less than 1% of that on Earth at an average of about 6 mb. The surface pressure varies with season by more than a third, because temperatures at the winter pole fall below the condensation point for carbon dioxide, which like that of Venus is the main constituent (about 96%) of the atmosphere. A temporary polar cap of dry ice forms over the permanent water ice cap during the polar night, to evaporate in the spring before it starts to form again at the other pole.

The low density of the Martian atmosphere results in only a very modest greenhouse effect, worth just a few degrees in surface temperature enhancement, compared to more than 30 degrees on Earth and around 500° on Venus. The main component of the Martian 'greenhouse' is, in fact, under most conditions, not carbon dioxide but airborne dust. The variability of the dust amount leads to differential warming of the surface that can provide feedback leading to strong winds. These in turn raise more dust and generate huge dust storms that, without rain and oceans to help to take out the airborne dust, regularly cover the entire planet.

There is ample evidence on present-day Mars for surface water in the past, implying a much thicker atmosphere during the so-called Noachian or Phillosian period (Figure 4.5). There is also plenty of evidence for widespread volcanic activity on early Mars. No doubt these two factors are linked, with volcanically exhaled gases adding carbon dioxide and other greenhouse gases to the young atmosphere, which was then able to maintain higher pressures and average temperatures above the freezing point of water for perhaps as long as the first two billion years of the planet's history. This would be long enough, as a consequence of what we know about the early Earth, for life to have developed on Mars, at least in simple aquatic forms. Whether it did or not is a key focus of current research with mobile laboratories on the surface of Mars, to include in the future the capacity to explore specific sites where liquid water may be anticipated to occur near the surface where it can be accessed with drills and tested for microbial life.

Meanwhile, considerable progress in understanding the history of the Martian climate is being made with orbiting spacecraft carrying high-resolution imagers, infrared and gamma ray sensors, along with robotic geologists on the surface. In particular the time scales for the change between the earlier, more Earth-like climate and the dry desert found at present is becoming clearer. The meteorology of the modern atmosphere, including seasonal circulation changes and the genesis and evolution of the global dust storms, is also being observed in sufficient detail that general circulation models used for the Earth's atmosphere have been successfully adapted to simulate the common aspects of weather behaviour on both planets.

4.2.3 Early Climates of Earth, Mars and Venus

It seems likely that all three terrestrial planets were characterised by denser and wetter atmospheres in the first billion years or so of their existence. On Earth, much of the carbon dioxide now locked up in surface minerals would have been in the atmosphere, leading to much higher surface pressures and temperatures than now. The water presently in the oceans would then have been in the atmosphere as vapour, further increasing the surface pressure to hundreds or even thousands of bars. Venus would have been in a similar state, but losing water more rapidly, and consequently losing CO_2 less rapidly later, than the Earth.

At some point the Earth's surface became cool enough for water to condense, which would have accelerated the removal of carbon dioxide and other soluble gases such as ammonia. Along with this, the rate at which further CO_2 and water were added would gradually have slowed as volcanic activity declined. The composition of the present-day terrestrial atmosphere is still affected by volcanism, but to a much lesser degree than on Venus, where the high sulphur content of the lower atmosphere and clouds, in particular, testifies to sustained and substantial exhalation from the interior. On Mars, in contrast, with its smaller radius, volcanic activity has all but ceased completely. As loss processes, including meteoritic bombardment and solar wind erosion, gradually gained the upper hand, Mars's atmosphere became too thin to support liquid water and the current frozen époque began.

Controversial evidence for isolated and time-dependent methane plumes on Mars has led to a search for regions with low levels of residual volcanic emissions, and if any are found they will provide further clues to the climate history as well as the likely locations of the potentially life-bearing aquifers. There is a possibility, probably remote but presently unquantifiable, that the methane itself is of biological origin.

4.2.4 Common Evolutionary Processes

All three terrestrial planets have secondary atmospheres, that is to say the high temperatures during and after formation, along with meteoritic bombardment and chemical combination, would have left the solid bodies without gaseous envelopes in some very early phase. The present atmospheres then accumulated as a result of exhalation of gases from the interior, primarily as volcanism, augmented to an unknown degree by icy material rich in volatiles that arrived as cometary objects, large and small. Attempts to reconstruct the history of these processes with the evidence available to us now depend heavily on the abundances and isotopic ratios found in the less reactive species such as nitrogen, but especially in the noble gases argon, neon, krypton and xenon (Figure 4.6). This history is still far from complete, partly because accurate measurements of some of the less common isotopes are still not

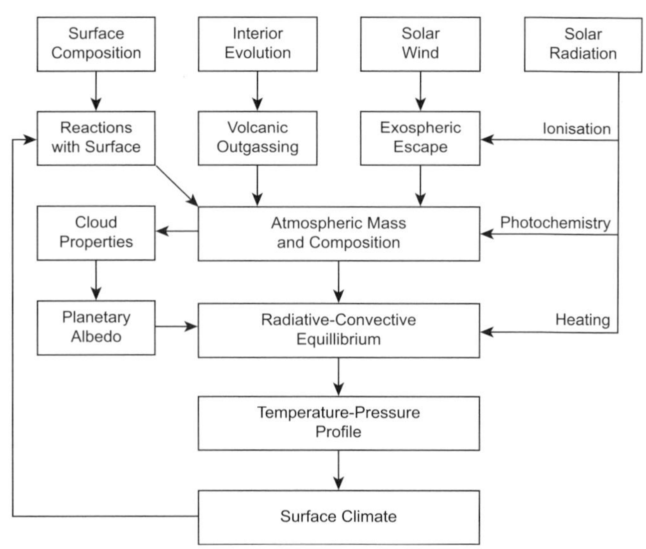

Figure 4.6 A summary of the key physical processes common to climate evolution on all three terrestrial planets.

available for all three planets, and partly because no satisfactory theory has been found to explain all of the measurements we do have. Sophisticated mass spectrometer measurements, carried out in situ on Mars and Venus to augment inferences from remote sensing and measurements on meteorites, will be a focus of future research.

Volcanism on Earth shows large variations in activity and composition of the exhaled gases. Mars, while obviously active in the past, shows few signs of current activity beyond the likely existence of some warm aquifers below the surface, so identifying contributions to the atmosphere from reservoirs in the interior is a goal for the very long term. Venusian volcanism is probably very active today, but hard to monitor because of the very cloudy atmosphere and the inhospitable surface conditions, although realistic ways in which this might be done in the relatively near future are being discussed.

The influx of icy material as comets and other debris containing volatiles is difficult to quantify, even at the present time. Some evidence has been presented for a steady arrival of a large number of small ('house-sized') icy objects into Earth's atmosphere from space on a daily basis, possibly amounting to a large flux of water and other volatiles, although the amounts involved are controversial. Whatever the flux is, it must be common to the other planets too, and it is likely to have been much greater in the past when the solar system contained a lot more randomly distributed material that has since been mostly swept up by the planets. However, it would take a million large comets, each ten times the mass of Halley and consisting mainly of water ice, to deliver in a series of catastrophic events the water found in the Earth's present-day oceans. This is not inconceivable, however; the apparently very extensive deposits of ice in the polar regions of Mercury and the Moon may be of cometary origin, even on these airless bodies, which can retain only a small fraction of any water that arrives.

Earth's present-day magnetic field has usually been described as protecting the atmosphere from being stripped away by the flow of very energetic particles from the Sun (the 'solar wind'). Venus has no such protection, and yet retains an atmosphere that is much thicker than Earth's. The loss rates measured recently by spacecraft such as *Venus Express* do not show loss rates that are vastly greater than Earth's, although getting completely comprehensive data, given the vast volumes of space and time involved in characterising the complex and variable phenomena involved, is an on-going task. It has, however, been suggested that, rather than protecting the planet from atmospheric erosion, the Earth's relatively strong field actually focuses the solar wind into high latitudes and increases the loss to space overall. On Mars, there seems to have been a strong field in the past that has now waned. It is tempting, but premature, to assume that the loss of the field contributed to the thinning of Mars's atmosphere over the history of the planet.

4.3 Titan

4.3.1 An Earth-Like World in the Outer Solar System

Titan, a large body about half the size of the Earth, is a very complex world as the diagram presented graphically in Figure 4.7 shows. It is the only one we know of today, beyond our own planet, to possess a thick nitrogen-dominated organic-laden atmosphere, a geologically complex active surface with liquid hydrocarbon lakes, and most probably a sub-surface ocean. The physical processes within this world, for instance, the methane cycle in the atmosphere, the formation of lakes, dunes, craters and mountains on the surface, and the interior structure, are very closely related to what we have on our planet and their study may help

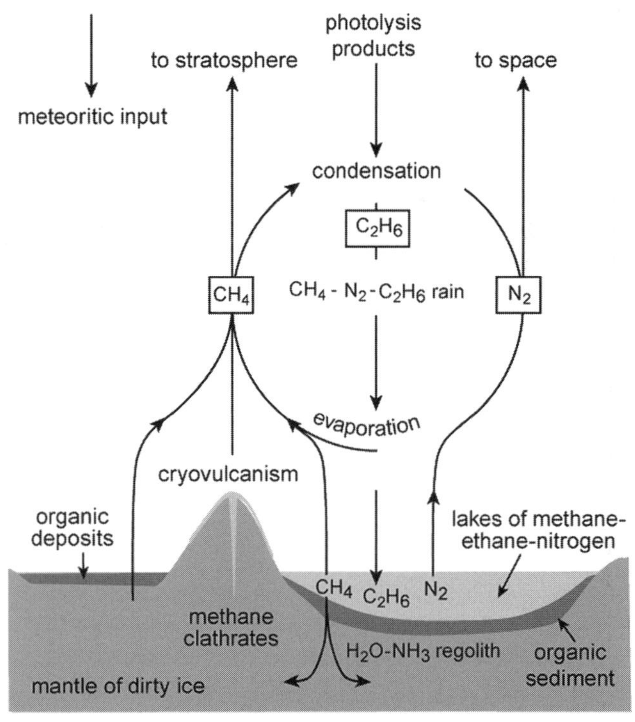

Figure 4.7 Titan's methane cycle. The physical processes remain poorly understood, including the nature and distribution of methane sources and the seasonal methane cycle in the atmosphere, the formation of lakes, dunes, craters and mountains on the surface and the interior structure (from Coustenis and Taylor, 2008).

to understand better the processes that also occur on Earth. In particular Titan's meteorology can give insights to our own climate, even though the time constants are very different.

Titan's thick and extended atmosphere is mainly composed of molecular nitrogen (97%) and methane (1.5%), with a surface temperature of $T = 94$ K and pressure $P= 1.5$ bar. In the upper atmosphere, complex photochemistry involving C, H, N and O species leads to the formation of hydrocarbons, nitriles and oxygen compounds, some of which condense to form a ubiquitous orange haze. Under the conditions found in Titan's lower atmosphere, CH_4 and C_2H_6 can condense and produce thick cumuliform clouds, precipitation and surface liquid deposits in a manner analogous to Earth's hydrological cycle. The surface itself has many familiar aspects to Earth, with lakes, seas, dunes and channels shaped by weather but formed from different material.

We know that 90% of the solar radiation arriving at Titan is absorbed by the haze in the higher atmospheric layers, leaving only 10% to reach the surface (that is only 0.1% of the solar flux arriving at Earth). At the surface of Titan, the dense atmosphere with high concentrations of nitrogen, methane and hydrogen (symmetrical molecules that normally do not have a significant greenhouse effect on Earth) raises the temperature by about 21 K. This effect is counter-balanced by an 'anti-greenhouse' effect, which reduces the surface temperature by about 9 K (McKay et al. 1991) because it allows the infrared radiation to escape from the optically thick upper atmospheric layers (Figure 4.8). Since the anti-greenhouse effect on Titan is half as strong as the greenhouse effect, the net balance is a higher temperature by 12 K with respect to the expected effective temperature at the surface in the absence of an atmosphere (Figure 4.8). This effect produces a temperature inversion at the tropopause of Titan around 40 km in altitude, with the temperature increasing with altitude above this level.

A version of the anti-greenhouse effect may have occurred on our planet in the past in the form of a huge impact that may have filled the atmosphere with material forming a haze as extensive as that on Titan. The resulting change in surface temperature is thought to have caused the disappearance of the dinosaurs about 65 million years ago. Something similar could happen in the future if a nuclear war filled the atmosphere with soot-like particles, again with destructive consequences (the so-called nuclear winter).

4.3.2 Seasonal Effects and Meteorology at 10 AU

As on Earth, on Titan we are observing an active hydrologic cycle subjected to seasonal and longer-term changes. By investigating such a close Earth analogue in terms of processes, but with sufficient differences in working fluids and time scales, we can rigorously test the depth of our understanding of planetary climate. Titan follows Saturn on its course around the Sun in roughly 29.5 years, so a season on Titan lasts about 7.5 years.

Since 2004, instruments on the *Cassini* Saturn orbiter have consistently observed clouds at Titan's mid-southern latitudes, while some images also show clouds close to the moon's equator. These observations provide evidence of a seasonal shift of Titan's weather systems to low latitudes following the Saturnian system's August 2009 equinox. During equinox, the sun lies directly over the equator, and both hemispheres receive equal heating from the Sun. The maximum illumination switches hemispheres during the transition to spring in the northern hemisphere and to fall in the

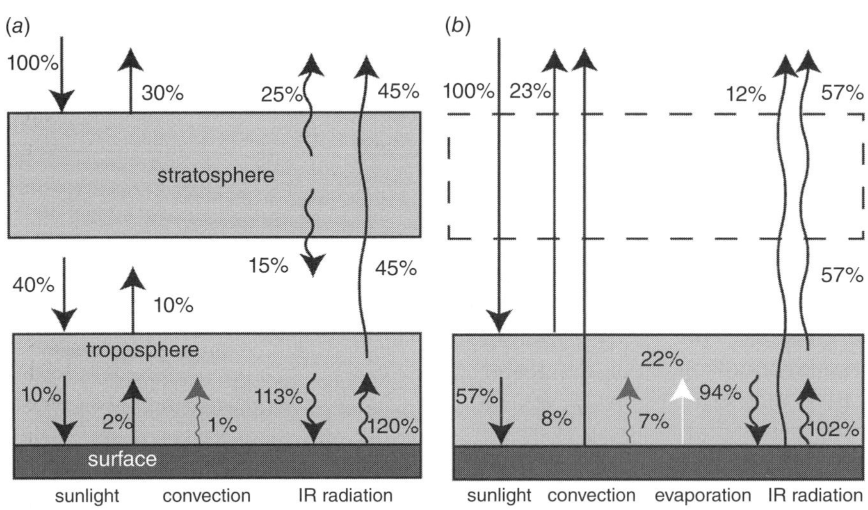

Figure 4.8 Diagram of the energy balance of the troposphere and lower stratosphere on Titan (left) and on Earth (right). The flux units are the percentage of the globally averaged solar radiation at the top of Titan's and Earth's atmospheres, which are 342 and 3.7 Wm^{-2}, respectively (from Griffith, 2009, after McKay et al. 1991).

Figure 4.9 An artist's impression of the change in observed atmospheric effects on Titan before, during and after equinox in 2009. The Titan globes also provide an impression of the detached haze layer that extends all around the moon (blue) (credit: NASA). For the colour version, please refer to the plate section.

southern hemisphere. In 2004, during Titan's late southern summer, extensive cloud systems were common in Titan's south polar region, but since 2005, southern polar cloud systems have been observed infrequently and after another year extensive near-equatorial clouds have been seen (Figure 4.9).

During the first years of *Cassini* 's exploration of the Saturnian system, Titan sported a 'hood' of dense white haze (white in Figure 4.9) in a vortex above its north pole, along with a high-altitude 'hot spot' (red in Figure 4.9). During this time the north pole was pointed away from the Sun. After the equinox, the north pole tilted towards the Sun, signalling the arrival of spring, while the southern hemisphere tilted away from the Sun and moved into autumn. After equinox and until 2011 there was still a significant build up of trace gases over the north pole, but the vortex and hot spot had almost disappeared. Instead, similar features began developing at the south pole, which are still present today. These observations are interpreted as a large-scale reversal in the single pole-to-pole atmospheric circulation cell of Titan immediately after equinox, with an upwelling of gases in the summer hemisphere and a corresponding downwelling in the winter hemisphere.

The cloud patterns observed from late southern summer to early southern fall on Titan suggest that Titan's global atmospheric circulation is influenced by both the atmosphere and the surface. The temperature of the surface responds more rapidly to changes in illumination than does the cold, thick atmosphere. The outbreaks of clouds that are observed may be the Titan equivalent of what creates Earth's tropical rain forest climates, even though the delayed reaction to the change of seasons and the apparently sudden shift are more reminiscent of the behaviour over Earth's tropical oceans than over tropical land areas.

4.4 Space Weather in the Solar System

Space weather describes the physical and phenomenological state of planetary (or satellite) natural space environments and the variability determined by solar activity and the dynamics of the interplanetary medium and the magnetosphere in which the solar system body may be embedded. Understanding the properties of the interactions of plasma and photons with exospheric and atmospheric environments in different places in the solar system is of key importance for understanding atmospheric loss processes and for nowcasting and forecasting planetary space weather during future solar system missions (Plainaki et al., 2014; 2016a).

4.4.1 Planetary Space Weather Agents, Analogy and Contrast with Earth

Space weather in the inner solar system is determined predominantly by the energy release from the Sun, in the form of solar photon flux, solar wind and high-speed stream events, coronal mass ejections (CMEs) and solar energetic particles (SEPs). The magnetic storm, the most powerful manifestation of circum-terrestrial space

Table 4.1 *Characteristic Parameters of Solar System Magnetospheres Determining Planetary Space Weather*

Solar system body	Surface magnetic field at the equator (T)	Dipole tilt[1]	Plasma sources[2]	Plasma composition (main species)	Plasma motion	Lifetime
Neptune	$0.14 \cdot 10^{-4}$	-47°	Triton (solar wind?)	N^+, H^+	Rotation (and solar wind driven?)	Days
Uranus	$0.23 \cdot 10^{-4}$	-59°	Ionosphere (solar wind)	H^+	Solar wind driven and rotation	1–30 Days
Saturn	$0.22 \cdot 10^{-4}$	-0°	Enceladus (rings, Tethys, Dione, Titan, solar wind)	H_2O^+, O^+, H^+	Rotation	30–50 Days
Jupiter	$4.28 \cdot 10^{-4}$	-9.6°	Io, Europa (ionosphere)	O^{n+}, S^{n+}, H^+	Rotation	20–80 Days
Mercury	$2 \cdot 10^{-7}$	+14°	Solar wind	H^+	Solar wind driven	Minutes
Ganymede	$7.19 \cdot 10^{-7}$	176°	Jupiter's magnetosphere (Ganymede's exosphere)	O^{n+}, S^{n+}, H^+	Corotation (Jupiter's magnetosphere driven)	Min-hours

[1]Angle between the magnetic and rotation axes.
[2]The dominant plasma sources are indicated. In parenthesis we indicate the believed secondary plasma sources.
Adapted from Plainaki et al. (2016a).

weather, is the result of the large transfer of energy and momentum from the heliosphere to the magnetosphere typically that is associated with CMEs (this is discussed more fully in Chapter 1.3). The presence of energetic particles in the inner solar system can be the result of acceleration either at the flare reconnection sites on the Sun or at the shocks driven by CMEs propagating through the solar corona and in interplanetary space. In the outer solar system, the space weather conditions around planets or satellites are determined by the balance between the incoming energy of solar or galactic origin and the internal sources in the form of plasma, neutral particles and fast planetary rotation. In both cases, the related space weather phenomena are evidenced in aurorae, ionization/dissociation of atmospheric constituents, atmosphere heating and escape. Plasma populations can be of solar, planetary or satellite origin and they can be described in general by thermal or power-law energy distributions. The energetic particle populations originate either from sources external to the planetary system (e.g. GCRs, SEPs) or from mechanisms taking place within the system such as plasma acceleration or charge exchange. The respective space weather phenomena vary, in principle, significantly in both time scale and spatial extent (Plainaki et al., 2016a).

The development of space weather phenomena at a planet or within a planetary system is strongly conditioned by the planetary bodies' magnetic fields (where present). In Table 4.1, we present the parameters determining the characteristics of plasma sources/losses and plasma circulation, which are responsible for space weather phenomena in different magnetospheres of our solar system. Ganymede is the only known satellite possessing an intrinsic magnetic field, located within a giant planet's magnetosphere. Mercury's magnetic dipole moment is offset northward from the planet's centre by 0.2 Mercury radii and tilted with respect to the planet's rotation axis by 14°. All planets in the outer solar system possess magnetic fields, which have approximately either a dipole form (e.g. Jupiter and Saturn, with tilt angles 9.6° and 0°, respectively) or a multipole form (e.g. Uranus and Neptune, with tilt angles 59° and 47°, respectively). The mechanisms through which the solar wind enters a planetary magnetosphere determine the characteristics of the space weather phenomena in that environment in terms of plasma circulation around the body, precipitation to the upper atmosphere and related auroral emissions.

Numerous solar-terrestrial studies based on theoretical models and observations have enriched our knowledge of the interaction between plasma of solar origin and the Earth's magnetosphere and atmosphere. The triggering process for space weather phenomena related to plasma–atmosphere interactions is reconnection, which is the breaking and topological rearrangement of magnetic field lines in a plasma, resulting in the conversion of magnetic field energy into plasma particle energy. During reconnection acceleration and heating of plasma occur at the expense of magnetic energy. The Earth's intrinsic magnetic field deflects the solar wind at the magnetopause location, a region where magnetic reconnection is likely to occur, allowing for the solar plasma to enter in the terrestrial magnetosphere. A necessary condition for reconnection is that the interplanetary magnetic field has an anti-parallel component to the planetary magnetic field at this location.

In the magnetospheres of rapidly rotating planets with strong magnetic fields (e.g. Jupiter, Saturn), plasma motions are dominated by planetary rotation (see Table 4.1). The icy moons located inside magnetospheres of this kind supply heavy ions such as water group species, oxygen, sulphur and nitrogen. Because of the fast planet rotation, such ions are concentrated near the centrifugal equator in a disk-like current sheet and associated plasma sheet. Internal plasma sources such as lunar volcanism and plume activity (e.g. at Io and Enceladus, respectively) determine the characteristics of the interactions within the planetary systems of Jupiter and Saturn, respectively.

Considering planetary systems at farther distances from the Sun our knowledge is rather limited and many open questions related to space weather exist. For example, whereas it is believed that solar wind could have an important role in driving the plasma motion at Uranus, at Neptune the distribution of plasma is generally interpreted as indicative of Triton acting as a major source.

In smaller magnetospheres around slowly rotating planets (e.g. Mercury), it is mainly the solar wind that controls the plasma motions. As a consequence, plasma and possibly energetic particles of solar origin become dominant space weather agents at Mercury.

4.4.2 The Significant Role of the Sun in Space Weather at the Terrestrial Planets

Space weather at the terrestrial planets lacking an intrinsic magnetic field (Venus, Mars) has direct effects in the bodies' atmospheres, even the deep atmospheric layers. Equivalently to the terrestrial case, space weather at Venus could be the driver for numerous processes related to atmospheric cascades and energy deposition. SEP fluxes may produce ionization at low altitudes in the Venusian atmosphere with characteristics that depend on the actual energy spectrum of the incoming particles. During ordinary conditions the maximum of the atmosphere ionization rate is expected at the altitude of ~63–64 km, but in the case of a SEP event the maximum appears at higher altitudes. Recent studies related to space weather at Venus showed that the maximum ionization rate during SEP events is

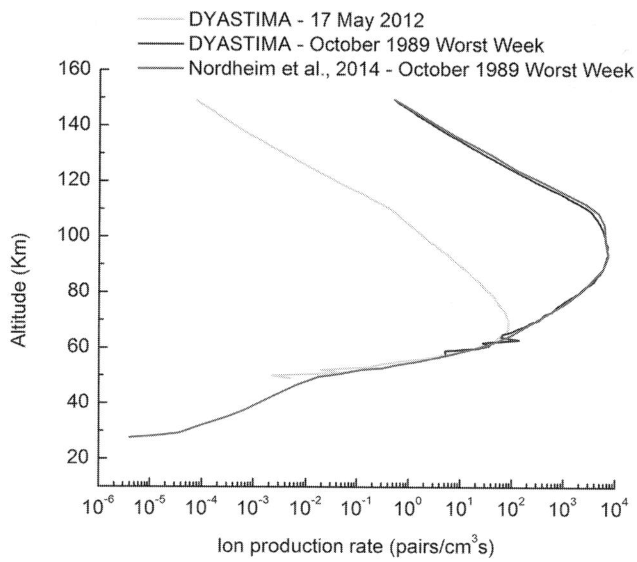

Figure 4.10 Solar energetic particles impacting the Venusian atmosphere can initiate space weather phenomena such as the ionization of the atmospheric molecules. The profile of the ionization rate varies significantly with the event. During conditions similar to those of the October 1989 and May 2012 events, the maximum of the ionization rate in the Venusian atmosphere is expected in relatively high altitudes and its intensity can be up to 105 times higher than the background (from Plainaki et al. 2016b). For the colour version, please refer to the plate section.

expected to be higher than the background ionization because of GCRs by a factor ranging between 10^2 and $\sim 10^5$, during the event's main phase (Plainaki et al., 2016b). Figure 4.10 shows the expected profiles of the ionization rates in the Venusian atmosphere during different periods of intense solar activity. Observational evidence for the effects of space weather at the terrestrial planets exists for Mars. The radar sounder MARSIS found some strong signal attenuation after SEP events. Moreover, one of the possible explanations for the disturbances found by the radio science experiment MaRS was the SEP precipitation. The MAVEN mission currently operating around Mars is expected to provide important new information about space weather at the red planet.

Among the inner solar system planets, Mercury is the one most subject to space weather since it is tightly linked to the Sun's activity. The weak intrinsic magnetic field of Mercury supports a small but dynamic magnetosphere hardly able to protect the planet from the Sun's action. Magnetosphere, exosphere and surface are tightly linked at Mercury and, therefore, interactions facilitate exchange of both material and energy. The induction effect due to the planet's conductive interior plays an important role in determining the global structure of Mercury's magnetosphere. The solar wind entering the magnetosphere of Mercury through the cusps (the regions separating the closed magnetic field lines at the front of the magnetosphere from the open magnetic field lines that have one end connected to the planet and the other end connected to the solar wind magnetic field) can either circulate or reach the surface. The region of impact corresponds, at a first approximation, to the footprint of the open magnetic field lines. The MESSENGER/FIPS instrument provided observational evidence for the precipitation of proton fluxes to the planet's surface in the region below the planet's northern cusps (Zurbuchen et al., 2011). The ESA/JAXA BepiColombo mission with its particle analysis package and magnetometer is expected to provide important additional knowledge on space weather at Mercury (Plainaki et al., 2016a).

4.4.3 Space Weather Phenomena in the Outer Solar System

Like the Earth, the main auroral emissions at Saturn appear to be associated with solar wind–magnetosphere interactions, since in both cases their morphology responds to solar wind changes. However, at Jupiter, the auroral dynamics are mainly driven by internal plasma sources such as Io volcanism, and fast rotation, and in particular to the magnetosphere–ionosphere coupling current system associated with the breakdown of rigid corotation in the middle magnetosphere. Persistent dim regions (discontinuities) of the main emission are possibly related to solar wind driven magnetospheric convection.

In future missions to the Jupiter system and the other outer planets, understanding space weather in the vicinity of planetary satellites will be an important part of the scientific requirements to achieve the mission's objectives (see Plainaki et al., 2016a) and strategies to assure spacecraft protection. The orbits of the icy satellites of Jupiter move up and down the planet's plasma sheet and as a consequence different surface ion-precipitation patterns exist (see Figure 4.11 referring to Ganymede). As a result of the plasma sheet oscillation, space weather phenomena at Ganymede are expected to manifest themselves in aurorae and variations of the spatial distribution of the moon's exosphere. The time variability in Io's volcanic activity could also result in the modification of the Jovian plasma properties in the region near the icy satellites.

The charged energetic particle populations play a significant role in the magnetosphere–moon interactions, having distributions that vary in both long and short terms. Although the intrinsic magnetic field

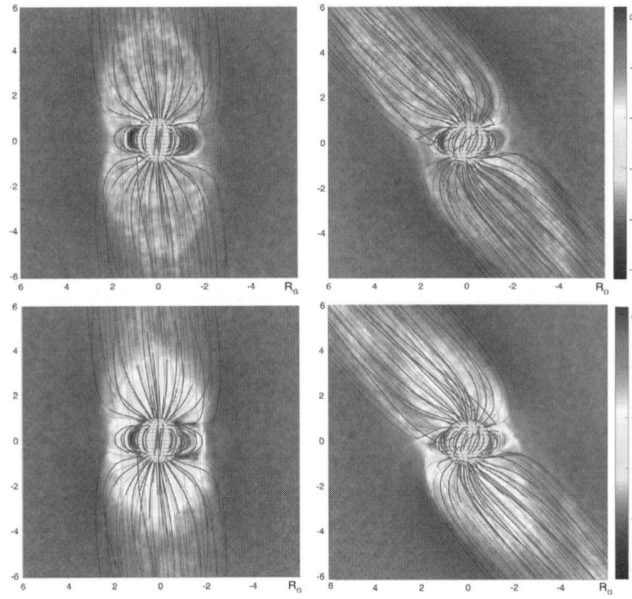

Figure 4.11 O^+ circulation in Ganymede's magnetosphere. The two upper panels refer to 10-keV O^+ and the two lower ones to 1-keV O^+. The panels on the left refer to orbital phases when Ganymede is close to the centre of Jupiter's plasma sheet (simulation consistent with the G8 Galileo flyby), whereas the panels on the right to those when Ganymede is above the plasma sheet (simulation consistent with the G2 Galileo flyby). The magnetic field configurations are given by the Jia et al. (2009) model (credit: Stefano Massetti). For the colour version, please refer to the plate section.

of Ganymede, reconnecting with the external Jovian magnetic field, partially shields the surface from ion impact, high ion precipitation rates, leading to intense sputtering and radiolysis effects, are expected to occur near the open-closed magnetic field boundary regions (the closed field lines are the ones that start and finish on the planet or moon's surface; see Figure 4.11). The spatial distribution of an icy moon's exosphere depends also on the illumination of the body, since the surface temperature is responsible for the efficiency of the source processes of radiolysis and sublimation. In an analogous way, the variability in the plasma electron density distribution around the moon during different orbital phases and/or different periods of plasma injections from Io affects the exosphere loss rates and determines its actual morphology. The moon exosphere source-loss balance determining the actual space weather conditions, consequently, depends on a complex pattern of plasma/photon interactions with the moon's exosphere and surface.

At Europa, the dominant exospheric species is O_2 because of its non-sticking property: lacking sufficient energy to overcome Europa's gravity, it bounces multiple times on the icy surface until it is lost to space through ionization, charge exchange or dissociation processes. As a result, a relatively dense exospheric envelope with a thickness of some hundreds of kilometres,consisting mainly of thermal O_2 molecules, accumulates around the moon. Space weather phenomena are guided by the magnetospheric electron spectrum and spatial distribution determining the morphology of the auroral emissions. Moreover, the exosphere generation and loss depend on the actual energetic particle and plasma distributions, respectively, around the moon. The density profile of the main exospheric species O_2 is subject to variations due to the varying impinging plasma and energetic particle distributions along Europa's orbit around Jupiter.

4.5 Conclusions: Implications from Planetary Studies for Future Earth Climate

Terrestrial planet climate studies provide a platform for the discussion of the climate changes currently affecting the Earth, which have generated wide concern about a decline in habitability as the population grows with increasingly harmful effects on the environment. The apparent stability of the Earth's climate over the last few thousand years has been shown to be deceptive, since it was preceded by huge changes taking place in the past, sometimes (during the most recent terrestrial ice ages, for example) on very short time scales of centuries or even decades. We now know that dramatic changes in climate, and potential habitability, have also taken place on Mars, and probably on Venus and Titan as well, although the time scales there remain uncertain.

Amelioration of the current problem on Earth requires a very complete understanding of the processes involved (radiative transfer in a water–carbon dioxide–cloudy 'greenhouse' atmosphere, for example) and the factors that affect atmospheric composition and surface pressure (volcanism, pollution, etc.). Theoretical methods and computer models of great sophistication do exist, of course, but are in constant need of refinement and testing before long-term (here meaning fifty or a hundred years rather than millennia or eons) forecasts can be relied upon as the basis for a political response. Applying such models to Venus and Mars not only allows us to develop detailed scenarios and possible histories for those extreme variants of the terrestrial situation, it shows up deficiencies in our understanding that could make important differences to climate forecast for the near-future of life on Earth.

The fundamental process, which is warming of the surface by altering the energy balance with the atmosphere through changes in atmospheric composition, is common to all four of the Earth-like planetary bodies. For instance, the 'greenhouse' theory can account for the much-higher-than-expected mean surface temperature on Venus, and the evident importance of dust aerosol on Mars for regulating the climate there. Both have direct analogues on Earth, and are well-enough understood that similar models are able to account for the behaviour of all three planets in general terms, although many important details remain unresolved. Similarly, the atmospheric circulations on the three worlds are obviously versions of the same paradigm, namely, a thin layer of gas on a rotating solid spheroid, heated by the Sun; and yet the observed behaviours are very different and we still do not have a single model that explains everything, as we should have if we understood all of the physics involved.

Model inter-comparisons that work quite well for present-day Earth, Venus, Mars and Titan raise the prospect of extrapolating the model descriptions of climate into the past, and into the future. Obviously, the level of detail must decrease in this sort of exercise if it is to have any credibility at the current state of the art, but equally obviously progress in this area is vital not just for understanding the evolution of the solar system but also for improving the relatively short-term climate forecasts on which much depends for life and civilisation on Earth.

On Venus, the possibility of a wetter, although probably not cooler, environment in the distant past has already been mentioned. More intriguingly, theoretical models suggest that the planet may gradually lose much of its thick carbon dioxide atmosphere in the future, when outgassing from the interior mainly through volcanoes declines to something more like the state on current Mars. In these predictions, enough nitrogen remains to maintain a tropical Earth-like regime that is similar to that originally expected before space age studies revealed a hot, sulphurous 'greenhouse' state maintained by what is probably a higher level of volcanic activity than the Earth, although this remains to be quantified by future space missions.

On Mars, the perspective over long periods of time suggests a scenario in which an Earth-like climate prevailed billions of years ago but then cooled to the frozen desert environment we now see there. Most likely, this change was driven by loss of most of the early atmosphere where the surface pressure was much closer to that of present-day Earth, rather than less than 1% as it is today. The loss processes undoubtedly included solar wind erosion, especially after Mars lost its global magnetic field as the interior cooled; decline in the large-scale exhalation of carbon dioxide, water vapour and other gases from volcanoes for the same basic reason and blast erosion during impacts by meteoritic debris, to which Mars may have been particularly vulnerable because of its orbit near the asteroid belt and its relatively low gravitation. Exactly what happened on Mars over time remains a keen research topic, and the subject of detailed mission planning to investigate the geological record more fully on and below the surface. The topic is particularly intriguing since there is an unknown, but probably better than zero, probability that life evolved during the early, more hospitable phase of the climate. If it did, evidence should be present and there may well be simple life forms still present in liquid aquifers below the surface, warmed by residual volcanic activity too weak now to produce much in the way of gaseous or lava eruptions as it once did.

The evolution of the solar luminosity makes it almost unavoidable that, in around a billion years from now, the tropopause temperature of the Earth will be elevated to the point that loss of the Earth's oceans by evaporation and then stratospheric photolysis of the water vapour will occur over a time scale of the order a hundred million years, the same process that removed most of Venus's water inventory in the distant past. From that point on, surface liquid water will be limited to the poles, will have a short photochemical lifetime, and if persistent will be resupplied by mantle/deep crustal water. At the equator, occasional intense rainstorms might scour out fluvial features, but the principal sedimentary deposits will be sand dunes.

The closest analogue in our solar system to such a future Earth is Titan (Figure 4.12). By viewing the current situation on Titan, where surface conditions preclude stable equatorial/mid-latitude oceans and the liquid on the surface is limited to the poles, while at the equator, occasional intense rainstorms scour out fluvial features, but the principal sedimentary deposits are sand dunes, one may be taking a glimpse into Earth's future evolution. Indeed, by studying Titan and how the various geologic and meteorological processes interact on that body, we have a picture of the Earth's future 'simulated' on a planetary scale, albeit by different materials. Future progress will depend heavily on new and more detailed investigations of the Titan climate system and its commonalities. These aspects of Titan, along with the fragility and changeability of the climate of our own home world, are essential to study more closely and monitor in the future by space- and ground-based observations, extending the rich input we currently have, in particular from the *Cassini-Huygens* mission.

On Venus, we can look forward to deep surveys using high-pressure aerobots that will be able, inter alia, to analyse the frequency, volume and composition of volcanic plumes. The return of core samples from the surface is also in prospect, and will provide time histories and data on chemical interactions between the atmosphere and the surface. Probably before that, we can expect to have cores from the polar caps on Mars that will provide much more detailed histories of past climate changes. There is also scope for the recent studies of strata in exposed

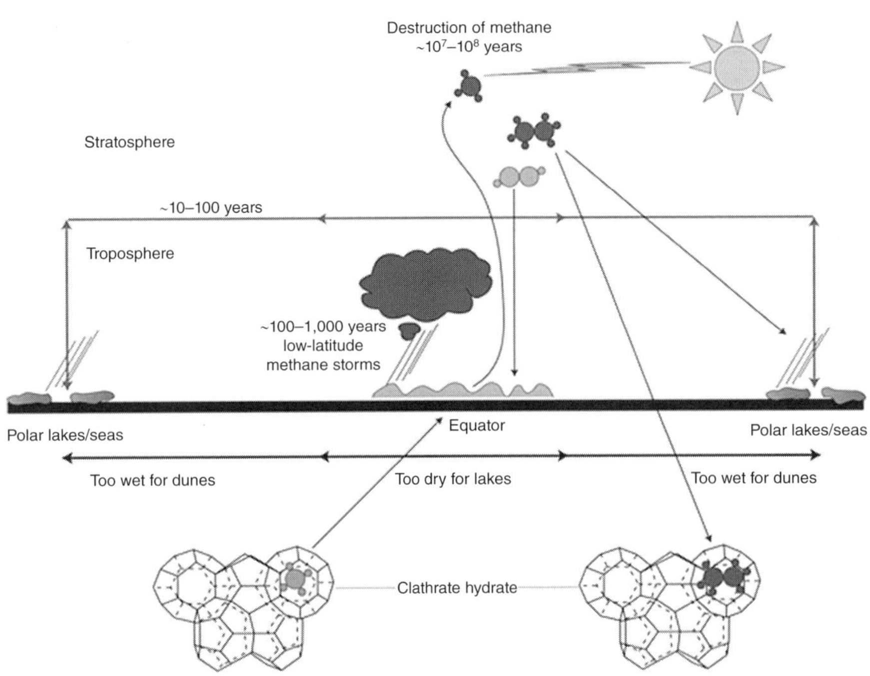

Figure 4.12 A schematic of Titan's methane cycle and its evolution with time, which may be representative of a future Earth (from Lunine and Atreya 2008).

rock faces using roving robotic geology laboratories to be greatly extended, including improved access to difficult sites such as vertical cliffs and sub-surface aquifers.

Expeditions and in situ laboratories staffed by humans are a perennial goal that may be inching closer. These have been extensively studied for Mars and progress is mostly a question of summoning the will (and, of course, the money) to proceed. For Venus, a floating laboratory near the cloud tops would enjoy an Earth-like temperature and pressure environment and could direct and enable sophisticated robotic studies of the deep atmosphere and surface. Again, the technology to do this is within reach.

A comparative investigation among the results obtained through observations and/or modelling of different environments in the solar system could provide important information useful for the interpretation of space weather phenomena at the Earth, leading to a more in-depth understanding of circum-terrestrial space weather phenomena. A very interesting example is the auroral substorm, a process initiated on the night side of the Earth's magnetosphere in which a significant amount of energy derived from the solar wind–magnetosphere interaction is deposited in the auroral ionosphere. Despite substantial progress in the last fifty years, there are still details of the related physical mechanisms that are not understood, such as how the unloaded energy produces a secondary dynamo, which powers the expansion phase of the magnetospheric substorm. Equivalent phenomena in other planetary systems include the evidence of night side reconnection occurring at Jupiter and Saturn, but with different characteristic times from those at Earth, and driven predominantly by internal processes. These are not yet understood either, but their investigation may be of importance for interpreting the substorm physical mechanisms at our own planet, providing a more complete frame for the study of the related physics and their dependence on different planetary properties.

References

Coustenis, A., and Taylor, F. W. (1999). *Titan, the Earth-like Moon*, Singapore: World Scientific.

Coustenis, A., and Taylor, F. W. (2008). *Titan: Exploring an Earthlike World*, Singapore: World Scientific.

Griffith, C. A. (2009). Storms, polar deposits and the methane cycle in Titan's atmosphere. *Phil. Trans. R. Soc. A*, 367, 713–728.

Jia, X., Walker, R. J., Kivelson, M. G. et al. (2009). Properties of Ganymede's magnetosphere inferred from improved three-dimensional MHD simulations. *J. Geophys. Res. [Space Phys.]*, 114, A09209, DOI: 10.1029/2009JA014375.

Lilensten, J., Coates, A. J., Dehant, V., et al. (2014). What characterizes planetary space weather? *Astron. Astrophys. Rev.*, 22, 79, DOI: 10.1007/s00159–014-0079–6.

Lunine, J. I. and Atreya, S. (2008). The methane cycle on Titan. *Nature Geoscience* 1, 159–164.

McKay, C. P., Pollack, J. B. and Courtin, R. (1991) The greenhouse and anti-greenhouse effects on Titan. *Science* 253, 1118–1121, DOI:10.1126/science.11538492.

Plainaki, C., Lilensten, J., Radioti, A. et al. (2016a). Planetary Space Weather: Scientific Aspects and Future Perspectives. *J. Space Weather Space Climate*, 6, A31, DOI:10.1051/swsc/2016024.

Plainaki, C., Paschalis, P., Grassi, D., Mavromichalaki, H. and Andriopoulou, M. (2016b). Solar energetic particle interactions with the Venusian atmosphere. *Ann. Geophys.*, 34, 595–608, DOI:10.5194/angeo-34–595-2016b.

Taylor, F. W. (2009). *The Scientific Exploration of Mars*, Cambridge University Press.

Taylor F. W. (2010). *Planetary Atmospheres*. Oxford University Press.

Taylor, F. W. (2014). *The Scientific Exploration of Venus*. Cambridge University Press.

Taylor, F. W. and Grinspoon, D. H. (2009). The Climate Evolution of Venus. *J. Geophys. Res.*, 114, 2156–2202.

Taylor, F. W. and Hunten, D. M. (2014). Venus: Atmosphere. In *Encyclopaedia of the Solar System*, ed. T. Spohn, D. Breuer, & T. V. Johnson, Elsevier, pp. 305–322.

Zurbuchen, T. H., Raines, J. M. and Slavin, J. A. et al. (2011). MESSENGER observations of the spatial distribution of planetary ions near Mercury. *Science*, 333 (6051), 1862–1865.

PART II

Future Earth and Geodetic Issues

5
Satellite Remote Sensing of Hydrological Change

Alberto Montanari and Michael G. Sideris

5.1 Introduction – Monitoring the Water Balance

The future of the planet Earth is tightly related to the future of its environment, which is in turn shaped by geological, climatic and human events. Geological events include tectonic evolution, volcanic eruptions and landslides. Climatic events include precipitation and runoff, formation of glaciers and snowmelt, floods and droughts. Climate is also the driver for erosion and sedimentation, which are considered geological events. Human events include agricultural and industrial evolution; water impoundment and diversion; human-induced floods, like those due to dam failure or occupation of perifluvial areas; contaminant release; groundwater extraction and many others. Human and climatic events, and some geological events, are strictly interlinked and therefore their future evolution can be fully understood only if such a connection is taken into account.

The links among climate, humans and the environment are mainly established through water, which is an essential element for life. Water is the main driver for climate and societal evolution, and is also a driver for several geological events. In fact, the ancient Greeks recognised water as one of the four essential elements (see Figure 5.1), and water is mentioned in the scriptures of several religions. Challenges associated with water have marked human history and will be more and more prominent at the global level in the coming years.

Water is a fluid and a moving element that continuously travels along the paths of the water cycle. The water cycle is studied by the science of hydrology, in its numerous and interconnected branches, which include surface hydrology, groundwater hydrology, mountain hydrology, eco-hydrology, chemical hydrology, hydroinformatics, hydrometeorology, surface hydrology, drainage basin management, water quality, sociohydrology and water resources management (see Sivapalan et al., 2012; Sivapalan et al., 2014; Konar et al., 2016).

Although the water cycle is a natural process, in many parts of the Earth it is now heavily influenced by human activities. Figure 5.2 shows a pictorial diagram of the human-impacted water cycle. It is clear that the human intervention, the so-called engineered hydrology that is superimposed on natural hydrology (Evenson et al., 2013), is necessary for managing water resources but, at the same time, is threatening the sustainable use of water and therefore its future.

Indeed, the main challenge for humans at present is to manage and exploit the environmental resources, and in particular water, in a sustainable way, meeting human development goals while ensuring the ability of natural systems to provide the natural resources and ecosystem services for the needs of future generations. The available renewable freshwater resources are limited at the global level and are unevenly distributed in space and time (Oki and Kanae, 2006). The current global rate of freshwater withdrawal is well below the upper limit of available freshwater resources, but about 2 billion people today live in river basins where groundwater withdrawal exceeds the recharge rate. In 2000, the Millennium Declaration was signed by the Member States of the United Nations, which led to the identification of the Millennium Development Goals (MDGs; see www.un.org/waterforlifedecade/water_at_rio.shtml). These included the target to halve, by 2015, the proportion of people without sustainable access to safe drinking water and basic sanitation. The target was met at the global level and locally by 147 countries, but much still remains to be done as about 700 million people still lack a safe access to water (UNICEF and World Health Organization, 2015).

To reduce further the number or people living in water scarcity conditions, in the face of rising development and global change, it is necessary to devise an

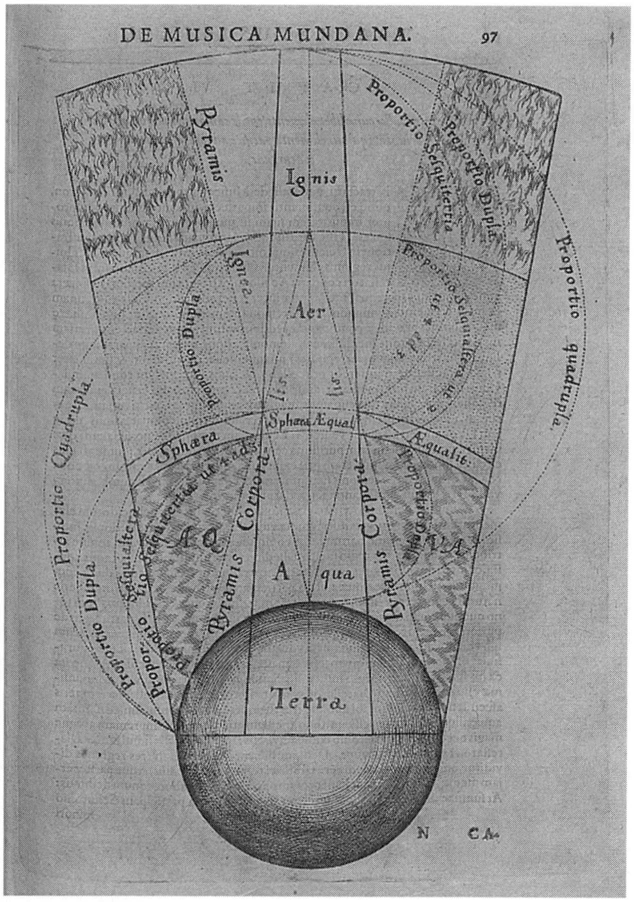

Quelle: Deutsche Fotothek

Figure 5.1 A diagram from Robert Fludd (1574–1637) showing the segment of the macrocosm with the elemental spheres of terra (earth), aqua (water), aer (air), and ignis (fire). (By Deutsche Fotothek, Public Domain; https://commons.wikimedia.org/w/index.php?curid=6479056).

improved strategy for the global assessment and management of water resources. The key condition to achieve this ambitious target is to improve our understanding of how the water cycle works and in particular to improve our capability to monitor and estimate water resources availability (Wheater and Gober, 2015; Lettenmaier et al., 2015). Several processes of the water cycle, depicted in Figure 5.2, are still a mystery: in many catchments we still do not know how water reaches the river and how the groundwater storage is replenished. Therefore, hydrologists must better understand and model water fluxes in a human-impacted environment. Such effort has to be made with a new vision, to profit from the tremendously enhanced international and interdisciplinary cooperation, increasing computational power and new monitoring techniques from space. New theories, new methods, new thinking, as well as the capability to look proactively to the future, are required. Above all, new forms of scientific research are needed to profit from global cooperation of the actors involved in water resources planning and management at the global level, by joining the efforts of scientists, policy makers, decision makers and governmental organizations. The present contribution aims to present an overview of such opportunities, and in particular the global monitoring methods employing data collected by Earth observation satellites.

5.2 The Panta Rhei Scientific Decade of the International Association of Hydrological Sciences: Change in Hydrology and Society

The awareness of the pivotal role that humans are playing in shaping societal development and its impacts has motivated the scientific community to concentrate research efforts on the interaction between humans and natural environment and resources. While the attention of researchers has been dedicated for centuries to natural processes, the geosciences of the twenty-first century are now specifically focusing on human impact. The interest in hydrology and society has been catalyzed by the International Association of Hydrological Sciences (IAHS), a non-profit, non-governmental scientific organization committed to serving the science of hydrology and the worldwide community of hydrologists. IAHS supports scientific decades that aim to encourage hydrological research around the globe under a common theme. After the success of the Prediction in Ungauged Basins decade (Sivapalan et al., 2003; Hracowitz et al., 2013), which mainly focused on pristine catchments, IAHS launched in 2013 the scientific decade Panta Rhei – Change in Hydrology and Society, which is concentrating on human-impacted waters.

Panta Rhei promotes scientific awareness and research by stimulating the aggregation of scientists in working groups that are focusing on research themes. Panta Rhei now counts more than thirty working groups comprising more than four hundred scientists. The purpose of Panta Rhei is to reach an improved interpretation of the processes governing the water cycle by focusing on their changing dynamics in connection with rapidly changing human systems (Montanari et al., 2013; see Figure 5.3). The practical aim is to improve our capability to make predictions of water resources dynamics to support sustainable societal

Figure 5.2 A diagram of the water cycle showing the linkages between "natural hydrology" and "engineered hydrology" (after Evenson et al., 2013).

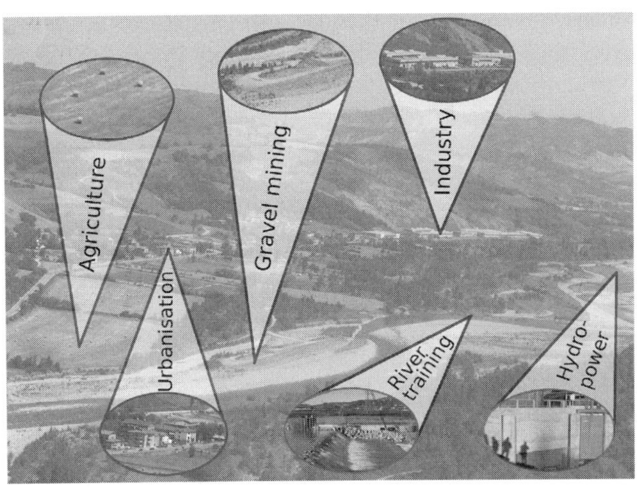

Figure 5.3 Interaction between hydrology and human systems (after Montanari, A., Young, G., Savenije, H. H. G. et al., 2013).

development in a changing environment. The concept implies a focus on hydrological systems as a changing interface between environment and society, whose dynamics are essential to determine water security, human safety and development, and to set priorities for environmental management. Panta Rhei will devise innovative theoretical blueprints for the representation of processes including change, and will focus on advanced monitoring and data analysis techniques. The assumption that is made is that change is resulting from the superimposition of natural variability and human-induced effects. Their interaction is critical for deciphering the feedbacks on the environment and hydrological systems. Interdisciplinarity will be sought by increased efforts to bridge with the socio–economic sciences and geosciences in general.

There is a final and very important premise, identified by the community, for the success of Panta Rhei. The research challenges for hydrology in the next ten years should be tackled through a collective effort, therefore emphasizing the key role that scientific associations must play. Cooperation among researchers through science initiatives, exchange programmes and virtual laboratories is fundamental for the long-term success of hydrological science as well as education and growth of the community. Hydrology must be dealt with by using collective and inclusive discussion and cooperation, while preserving the value of individual ideas and contributions. There are many countries in the world whose hydrological features are poorly known, and Panta Rhei can contribute to fill this knowledge gap and to strengthen scientific expertise in areas that have an immense potential in terms of research contribution and community building.

Panta Rhei recognises the value of the innovative information provided by global monitoring through satellite remote sensing, which can also enable us to decipher the human impact on the hydrological cycle (see, for instance, Ceola et al., 2014; Ceola et al., 2015). In the following sections, we will demonstrate this by discussing the contributions of satellite data to the solution of the well-known water balance equation.

5.3 The Water Balance under Human Impact

Water resources assessment is classically carried out by applying the mass balance equation to a control volume that corresponds to a given water system. For assessing water resources, the mass balance equation is mostly applied to catchments, but in some cases one may refer to a water supply system or an irrigation system to define the control volume. Water is frequently assumed to be incompressible and therefore the mass balance equation reduces to a volume balance. The resulting equation is called the "water balance equation" or "hydrological budget".

The hydrological budget during a time interval dt for a catchment not affected by human impact can be written in the form

$$P = Q_s + Q_g + dS/dt + E + \varepsilon \quad (5.1)$$

where P is the volume rate of precipitation during dt (positive, and expressed in L^3/T); Q_s and Q_g (positive, and expressed in L^3/T) are the surface and subsurface, respectively, average discharges during dt; S (positive, and expressed in L^3) is the water storage in the control volume; E (positive, and expressed in L^3/T) is the volume rate of evaporation during dt, which includes transpiration; and ε (positive or negative, and expressed in L^3/T) is an error term, which is necessary to close the balance in view of the random nature of variables in Eq. (5.1).

The application of Eq. (5.1) requires the definition of the volume rates of precipitation and evaporation, the surface and subsurface discharge and the catchment storage. All of the variables are affected by uncertainty and therefore should be treated as random variables. In fact, precipitation is usually monitored by using ground observations, which are discrete in space and time. Given the relevant variability affecting precipitation, uncertainty is introduced when quantifying the term P. Evaporation is usually estimated empirically and therefore approximately, depending on meteorological variables. Surface and subsurface discharges cannot be directly measured and therefore are often approximately inferred depending on other hydrological variables, such as river and/or groundwater levels. To provide an example, river discharge data for large rivers are likely to be affected by a 20% error even when measurements are taken in optimal conditions (Di Baldassarre and Montanari, 2009).

Furthermore, uncertainty is inflated in the presence of human impact. For a human-impacted catchment, the water balance equation can be written as

$$P = Q_s + Q_g + dS/dt + E - W + D + \varepsilon \quad (5.2)$$

where W is the volume rate of consumptive (i.e., not originating return flow) human water withdrawal from the catchment (positive, and expressed in L^3/T) and D is the volume rate of human-diverted water from locations outside the catchment (positive, and expressed in L^3/T) (see Konar et al., 2016). W and D are also affected by relevant uncertainty. In fact, water withdrawal may occur from surface water and groundwater, and may be not measured. Furthermore, the return flow can be evaluated only empirically in most cases. Water diversions may occur through infrastructure and in this case they are usually measured, but may also occur through virtual water transfer, namely, the water volume required to produce commodities traded to an importing or exporting nation or region.

There is a prolific and recent scientific literature highlighting the relevance of virtual water flow and its estimation (see, for instance, Tamea et al., 2014), which requires the use of advanced and interdisciplinary monitoring techniques.

The sum of the preceding uncertainties introduces several approximations in the estimation of water resources availability, which may therefore be unreliable for technical purposes, thus preventing effective management of water resources and planning of future developments. Two solutions may be adopted and integrated to resolve the preceding problem, which are described in the following sections.

5.3.1 Reducing Uncertainty of Water Budget through Hydrological Modelling

One possibility to reduce the uncertainty in Eqs. (5.1) and (5.2) is to assume that some of their variables are unknown, therefore introducing mathematical relationships for their computation that are associated to the water balance equation. The resulting system of mathematical relationships, where the number of equations is equal to the number of unknowns, constitutes a hydrological model. For instance, an energy balance equation may be introduced to relate the surface and subsurface discharge to the catchment storage. Of course these relationships are themselves affected by uncertainty and therefore their use should be carefully evaluated on the basis of the problem at hand. A large set of hydrological models can be found in the literature, which mainly refer to undisturbed catchments. However, in the past three decades the development of water resources management as an independent discipline has produced intensive research efforts to quantify the terms W and D, which were in most cases considered as exogenous terms. In fact, the impacts of society on hydrology are well documented by studies focusing on, e.g., the impacts of dams and water quality decline associated with human activities, and the impacts of urbanization, climate change and pollution (Poff and Zimmerman, 2010; Beck et al., 2012). However, the impacts of hydrology on society (i.e., in terms of human practices, politics, economics, migration and human geographies, among others) have historically been the subject of disciplines other than hydrology (see, e.g., Barnes and Alatout, 2012; Winiwarter et al., 2013; Linton and Budds, 2014).

Recently, several researchers have put emphasis on the need to incorporate the human impact into hydrological models adequately with an innovative approach, taking into account the two-way interaction between hydrological and human systems. These efforts led to defining a new branch of hydrology, termed socio-hydrology (Sivapalan et al., 2012; Sivapalan et al., 2014). Socio-hydrology aims to model coupled and interlinked changes in hydrology and society. Within the context of water resources management, the research question is to interpret how the future society will affect water availability, water quality and water distribution. How to detect and model such feedbacks reliably and how to identify effective indicators for the interactions between humans and hydrology are still open challenges. Refining reliable model representations of these interactions may provide insights and reduce the uncertainty associated with water resources assessment. Understanding and modelling the two-way coupling of human–water systems requires an interdisciplinary approach (Macleod et al., 2007; Braden et al., 2009; Hamilton et al., 2015). Socio-hydrology has recently been the subject of intensive debate (see Di Baldassarre et al., 2015; Montanari, 2015; Sivapalan, 2015; Troy et al., 2015; Loucks, 2015; Gober and Wheater, 2015). In particular, the capability of socio-hydrological models to capture human behaviour has been questioned, pointing out the inherently random and unpredictable nature of human decisions. However, it is definitely accepted by the scientific community that the human impact should be properly taken into account, by removing the assumption that humans are simply an exogenous forcing into undisturbed catchments.

5.3.2 Reducing Uncertainty of Water Budget through Improved Monitoring

One of the key opportunities for gaining a better understanding of water fluxes on the Earth's surface and subsurface is offered by new monitoring techniques and in particular by remote sensing through satellites. The increasing number of missions dedicated to Earth observation provides a wealth of data at the global scale and therefore unprecedented ways forward for research. Indeed, remotely sensed data are potentially able to provide a spatially distributed overview of hydrological fluxes and storage at the global level. In fact, Earth observation is now providing unprecedented opportunities to observe land use, precipitation, soil moisture, evaporation, river discharge and catchment storage (for an overview, see Lettenmaier et al., 2015). Therefore, intensive research activities have been recently undertaken to study solutions for incorporating the preceding information into hydrological models. There are still relevant problems to resolve, which are mainly due to the uncertainty of the remotely sensed information and the challenges to integrate it with ground observations.

Satellite-borne sensors can provide space measurements for the water cycle and in particular for the elements in Eq. (5.1). For example, near- and thermal-infrared and radar can be utilized for precipitation and evaporation, passive microwave and radar for soil moisture and surface water monitoring, and gravity for total groundwater content monitoring In order to show how, and which, satellite missions contribute to the water balance, we rewrite Eq. (5.1) in a slightly different form, which combines the surface and subsurface discharges into a combined discharge Q $(= Q_s + Q_g)$ and separates the water storage S into its soil moisture S_{SM}, surface water S_{SW} and groundwater S_{GW} constituents ($S = S_{SM} + S_{SW} + S_{GW}$):

$$P = Q + E + dS_{SM}/dt + dS_{SW}/dt + dS_{GW}/dt + e \quad (5.3)$$

We can now link each term of Eq. (5.3) to specific satellite missions by use of Figure 5.4 (Cherchali and

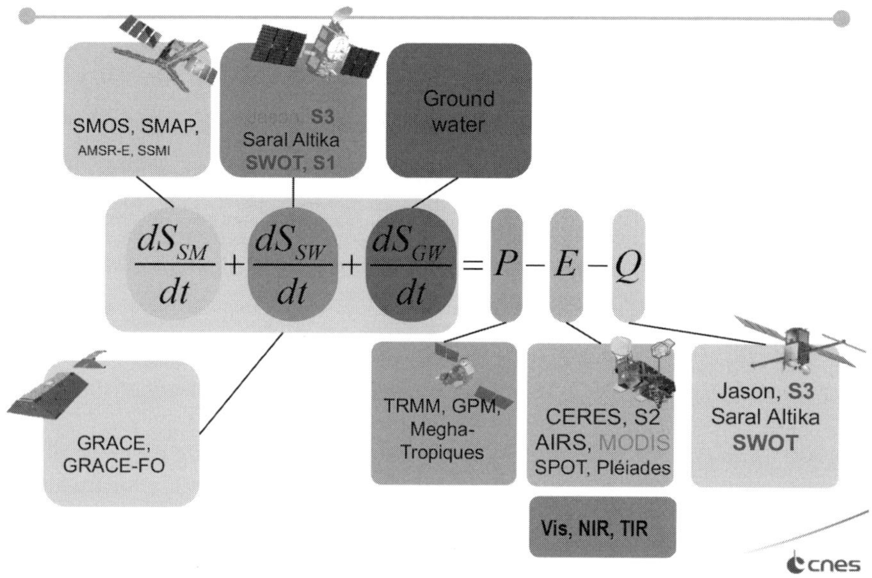

Figure 5.4 Satellite observations of the water cycle components (after Cherchali and Gosset, 2016).

Gosset, 2016). It is clear from this figure that more than one mission may contribute to each component in Eq. (5.3), that some missions can only observe an aggregate of components (e.g., GRACE can "see" only the total *dS*/*dt*) and that not all components can be observed from space. Because of orbit and other satellite mission limitations, the components that are observable from space cannot always be observed with the desired accuracy and spatial and temporal resolution.

In the sequel, and because of lack of space, we will briefly describe some but not all of these satellite missions. For a very good summary of practically all satellite missions contributing to the space observation of the water cycle, the interested reader is referred to the CEOS Water Constellation Feasibility Study Report (2016).

5.4 Monitoring Water Balance Components by Satellite Missions

5.4.1 Monitoring Water Storage: The GRACE Missions

The US and German space agencies NASA and DLR launched the Gravity Recovery and Climate Experiment (GRACE) twin satellite mission on 17 March 2002. Since then, and until October 2017, it had been observing the changes in the Earth's gravity field by making accurate measurements of the (approximately 220-km) distance between the two satellites, using a microwave ranging system and GPS. For example, when there is a decrease in gravity ahead of the pair, the front satellite slows down and the distance between the pair decreases, while when the decreased gravity is between the pair their distance increases. Relative velocities smaller than 1 μm/s can be detected.

The difference between two GRACE-derived global gravity fields represents a time-variable component, typically available on a monthly basis. Although on land GRACE monitors changes in all of the water stored there, i.e., the change in total water storage TWS (groundwater, surface water, soil moisture and snow) at monthly and longer time scales, these gravity changes primarily reflect the spatiotemporal evolution of groundwater (the *dS*/*dt* term in Eqs. 5.1 and 5.2) over large regions. Their vertical extent is measured in centimetres of equivalent water thickness, which is denoted as equivalent water height (EWH). Figure 5.5 shows the EWH trend, and it is not difficult to identify hydrological changes related to, e.g., groundwater depletion in northwestern Australia, California, China, India, the Middle East and South Africa; Orinoco and midwestern US floods; droughts in Argentina and the southwestern United States; the Greenland and Alaska ice sheet and glacier melting; precipitation increases in southern India and in high-latitude areas, and so on.

The main limitations of GRACE include its limited spatial resolution, the use of smoothing filters to reduce the striping effects caused by the orbit geometry (which further decreases the spatial resolution), leakage effects (water mass variations in neighbouring areas leak into the desired region, and part of the desired signal leaks outside the region, in particular along coastlines) and the monthly sampling rate, which limits its temporal resolution and hence its ability to identify signals that have shorter time scales (Wouters et al., 2014).

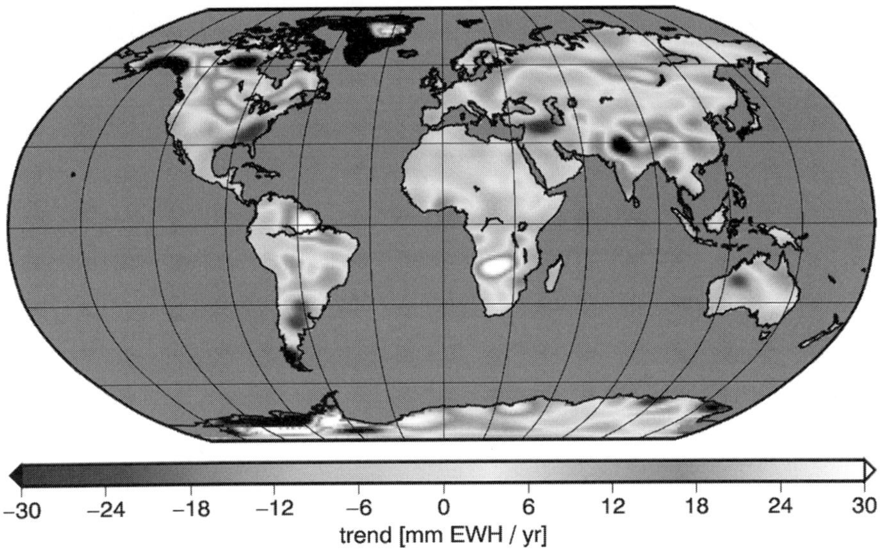

Figure 5.5 Average EWH trend for the period 2003–2012 from GRACE CSR-RL05 time series (after Eicker et al., 2016). For the color version, please refer to the plate section.

For hydrological purposes, since GRACE measures the entire water column as one measurement, the separation of dS/dt into its three constituents dS_{SM}/dt, dS_{SW}/dt and dS_{GW}/dt in Eq. (5.3) requires a combination of GRACE data with in situ measurements, land/hydrological models (such as the Global Land Data Assimilation System GLDAS; see, e.g., Rodell et al., 2004 and Syed et al., 2008), geophysical models (e.g., for glacial isostatic adjustment, see Peltier at al., 2015) and data from other satellite missions (such as those described later), which introduce their own limitations and errors in the assimilation procedure.

The one enormous advantage of GRACE is that it enabled, for the first time, the estimation of the terrestrial water storage at large scales and in remote regions. Therefore it has been used to monitor the terrestrial water cycle (Eicker at al., 2016), including floods (Reager and Famiglietti, 2009), droughts (Thomas et al., 2014), river basin water storage changes (Reager et al., 2014), continental aquifer and water storage changes in poorly monitored regions (Ramillien et al., 2014) and groundwater depletion (Rodell et al., 2009) and other anthropogenic regional water changes of high societal impact (such as identifying regions where irrigation using aquifers' water is non-sustainable). To understand such hydrological temporal changes and their impacts fully, it is imperative that satellite data be provided in uninterrupted, long-term time series, and therefore future GRACE-type missions are already planned by the space agencies.

On May 17, 2018, the GRACE follow-on (GRACE-FO) mission was launched to continue observations made by GRACE but using a more advanced and much more accurate laser technology, in addition to the microwave system, to measure the inter-satellite distance. The nominal mission life is five years but could possibly be extended to seven years (end of 2024). To ensure the continuity of gravity field monitoring from space, the GRACE-II mission is planned for launch in the late 2020s. Measurement accuracy will be further improved by use of a laser-based satellite-to-satellite interferometer and a drag-free propulsion system. In addition, the temporal and the spatial resolution should be improved, the latter from the current 400 km down to 100 km. Given that the resolution required for local aquifer monitoring is 1 km, GRACE-II could provide a tremendous boost not only to regional but also to local hydrology studies.

5.4.2 Monitoring Water Levels and Surface Discharges: The SWOT Mission

Research to infer river and lake levels and associated river discharges (the Q term in Eq. 5.3) from satellite data has been carried out since the 1960s by Buchanan and Somers (1969). The opportunity to monitor surface water (the dS_{SW}/dt term in Eq. 5.3) at the global level is providing exciting perspectives for improving water resources management and planning. The main problems and challenges are associated to the spatial discretization of the satellite images, which allows monitoring large water bodies only, and the transformation from river water levels to river discharges, which requires the availability of rating curves that are seldom available. A great improvement in measurement accuracy for water levels has been obtained in the past two decades thanks to progress in altimeter instrumentation and improvement in the precision of satellite orbit calculations.

The first satellite measurements of inland waters have been acquired from the TOPEX/Poseidon mission, which had been designed primarily to acquire ocean surface topography data (Birkett et al., 2002; Lettenmaier et al., 2016). The main challenges that have been found are related to the revisit times, which typically exceed one week, and the spacing of the tracks, which are typically located hundreds of kilometres apart. These limitations made the collection of regular time series at specified river locations possible in a few cases only. Improvements have been obtained with the successors of the TOPEX/Poseidon mission, namely, the Jason-1, Jason-2 and Jason-3 missions, which are a joint effort by the NASA (National Aeronautics and Space Administration, United States) and CNES (Centre National d'Etudes Spatiales, France) space agencies. The Jason missions were designed to measure climate change through precise measurements of global sea level changes. The availability of additional satellites provided more data at finer spatial and temporal resolution. However, river monitoring remained confined to large watercourses such as the Amazon River, the Nile River and a few others.

Exciting future perspectives are conveyed by the Surface Water Ocean Topography (SWOT) mission, a planned joint effort by NASA and CNES. SWOT is planned to make the first global survey of Earth's surface water and will produce new estimates of water storage in lakes, reservoirs and wetlands. SWOT will be fundamentally different from previous missions as it will deliver images of surface water height, rather than one-dimensional track returns. In fact, the SWOT instrument will be a wide swath altimeter that will provide maps of water elevation and extent at an unprecedented spatial resolution (50–100 m) and precision (centimetre vertical accuracy when averaged over areas of 1 km^2; Durand et al., 2010). Besides the preceding estimates of water level, SWOT will be able to deliver images of inundated areas, therefore providing fundamental additional information for the calibration and application of inundation models. Although SWOT is scheduled for launch not before 2020, several studies have been carried out to assess the usefulness of SWOT data for hydrological applications (see, for instance, Biancamaria et al., 2011; Durand et al., 2008; Yoon et al., 2012). At the global scale, Biancamaria et al. (2010) characterized the error of discharge estimates in rivers wider than 50 m using rating curves, with a focus on the SWOT temporal sampling. They also provided a comprehensive review of SWOT applications in hydrology.

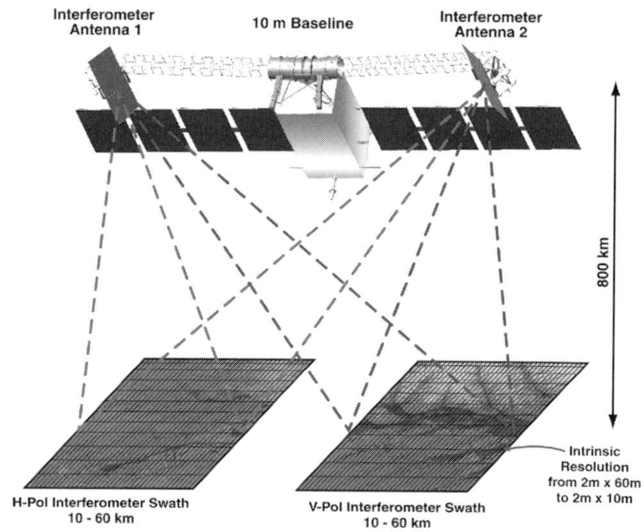

Figure 5.6 Two-dimensional mapping through swath altimetry (by JPL/NASA [Public domain], via Wikimedia Commons).

The innovative contribution of the SWOT mission is expected to be twofold. First of all, it is planned that an opportunistic use of SWOT images will be possible to decipher better the development and the impact of extreme events like floods and droughts. Second, SWOT will effect a paradigm change in technology, from nadir pointing altimeters (essentially all previous satellite altimeters) to swath altimetry. The swath altimeter will transform the sampling of surface topography from one-dimensional, along-track profiling to two-dimensional mapping (see Figure 5.6). It is likely that swath altimetry will become a standard in future missions.

5.4.3 Monitoring Soil Moisture: The SMOS and SMAP Missions

Soil moisture data are critical for improving the dS_{SM}/dt term in the water balance (Eq. 5.3). Since its launch on 2 November 2009, ESA's Soil Moisture and Ocean Salinity (SMOS) mission uses a microwave imaging radiometer and has been observing land soil moisture with specific accuracy (4% volumetric soil moisture), spatial resolution (35–50 km), spatial coverage (global) and temporal coverage (one to three days). The mission's main advantage for hydrological studies is that it has been providing novel accurate data globally, with a high acquisition rate. Its disadvantages stem from the low spatial resolution of these data and the challenges to integrate satellite information into hydrological models.

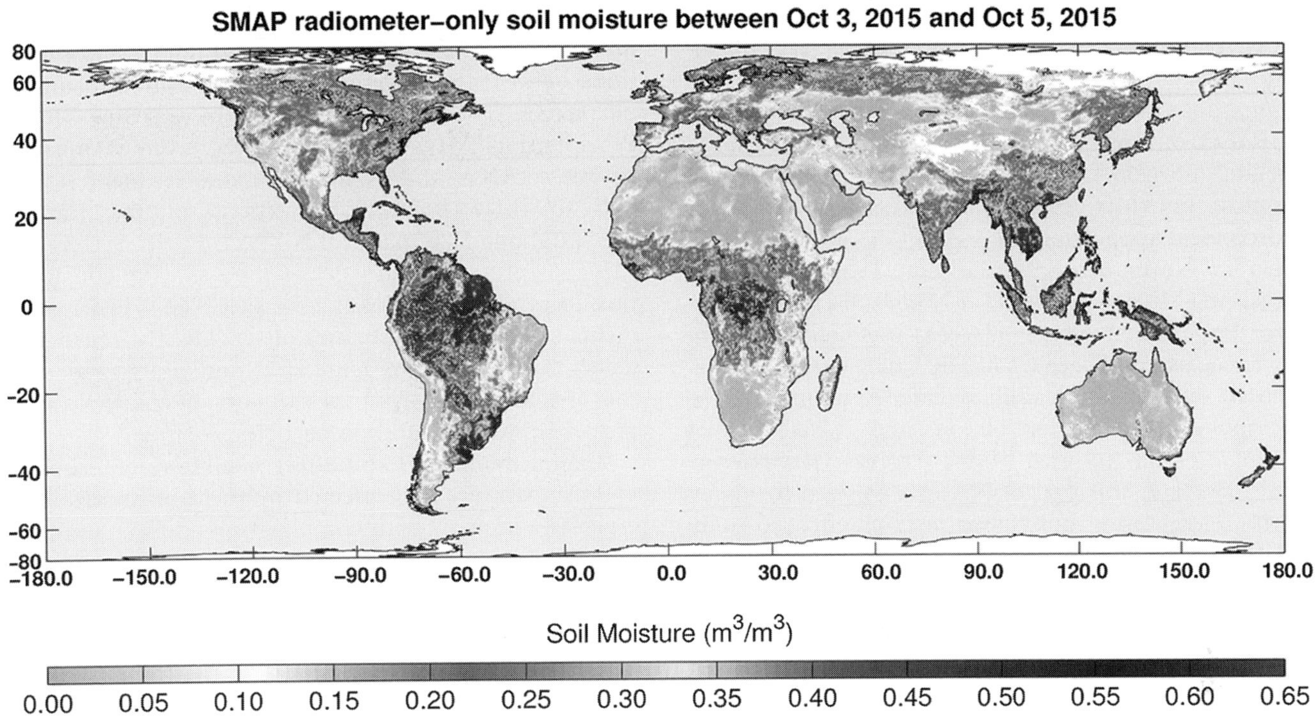

Figure 5.7 SMAP radiometer-only soil moisture between 3 and 5 October 2015 (courtesy NASA/JPL-Caltech, available at https://smap.jpl.nasa.gov/news/1250/smap-soil-moisture-data-available-at-nasa-national-snow-and-ice-data-center/). For the color version, please refer to the plate section.

SMOS was joined in 2014 by the NASA Soil Moisture Active Passive (SMAP) mission, which was designed to measure globally, among other things, the moisture in the top 5 cm of soil with three-day temporal resolution. The missions provide soil moisture and freeze/thaw measurements from space. Soil moisture content is mapped via a combination of radar and radiometer measurements with 40- to 10-km spatial resolution, while the freeze/thaw mapping is accomplished using unique properties of the radar system's measurements with 3-km spatial resolution (http://smap.jpl.nasa.gov). SMAP soil moisture measurements have also allowed the separation of TWS measured by GRACE into its soil moisture and groundwater components (Tapley et al., 2016).

These two missions have demonstrated the feasibility and value of obtaining global soil moisture from space (Figure 5.7) by use of L-band passive microwave measurements; see, e.g., Kerr et al. (2010) and Entekhabi et al. (2010). They have also contributed to the separation/identification of the three constituents of the TWS observed by GRACE. Their primary limitation is the relatively coarse (40 km) spatial resolution. Despite the need for long-term continuous observations, there is unfortunately no confirmed follow-on SMOS or SMAP mission, thus raising the risk of space observation gaps in the beginning of the next decade. Nevertheless, other current (GCOM-W, ASCAT, MetOp-A/B and Radarsat-2) and planned (Sentinel-1A/1B, SAOCOM-1A/B, NISAR) missions will be contributing space observations for soil moisture estimation.

5.4.4 Monitoring Precipitation: The TRMM and GPM Missions

Precipitation (*P* in Eqs. 5.1 to 5.3) is the main water flux from the atmospheric to the terrestrial part of the water cycle. It originates from clouds in the form of snow, rain, freezing rain, sleet or hail. Precipitation is characterized by a relevant variability in space and time. For this reason, quantifying precipitation is a challenging yet extremely important task, as precipitation is the main input for most hydrological analyses and predictions, and engineering design. It frequently happens, especially in developing countries, that precipitation cannot be adequately observed at the required spatial and temporal resolutions. Therefore, researchers are dedicating relevant efforts to devise innovative monitoring solutions. An interesting option is the use of satellites (Maggioni et al., 2016; Lettenmaier et al., 2015). Sensors and algorithms that allow satellite-based estimation of precipitation over land have been developed since the mid-1970s

(Kidd and Huffman, 2011; Gebregiorgis and Hossain, 2014). The first operational satellite precipitation estimates were developed using visible (VIS) or infrared (IR) observations of cloud-top temperature from geostationary (GEO) or Low Earth Orbiting (LEO) satellites. GEO satellites provide frequent observations (every 15–30 minutes), but their accuracy was often too limited for hydrological applications.

On 27 November 1997, the joint Tropical Rainfall Measuring Mission (TRMM) of NASA and Japan Aerospace Exploration Agency (JAXA) was launched from the Tanegashima Space Centre in Tanegashima, Japan. TRMM was a research satellite designed to improve the description of the distribution and variability of precipitation in the area of the Tropics (Kummerow et al., 1998). It was part of NASA's Mission to Planet Earth, a long-term, coordinated research effort to study the Earth as a global system. The science objective of TRMM mainly related to hydrology was to advance our understanding of the global energy and water cycles (Kummerov et al., 2001). This target was reached by providing distributions of rainfall and latent heating over the global Tropics, understanding how changes in tropical rainfall influence global circulation, understanding rainfall variability at monthly and longer time scales, evaluating diurnal variability of tropical rainfall globally and having a space-based system for rainfall measurements.

TRMM has shown the importance of taking measurements at different times of day to improve observations of weather systems and real-time monitoring of hurricanes. The successor of TRMM was the Global Precipitation Measurement (GPM), a joint mission between JAXA and NASA to make frequent (every two to three hours) observations of the Earth's precipitation. It is part of NASA's Earth Systematic Missions program and works with a satellite constellation to provide full global coverage. The GPM instruments extend the capabilities of the TRMM sensors to detect falling snow, measure light rain and provide, for the first time, quantitative estimates of microphysical properties of precipitation particles.

Several limitations still affect satellite rainfall observation. Data are still affected by relevant uncertainty, especially in areas with complex topography. Snowfall is still difficult to detect, because of the very diverse shape and radiative properties of snowflakes. However, despite these limitations, satellite rainfall estimation is a promising perspective for the future. In fact, several applications of the TRMM and GPM products have been presented in the literature. A relevant example has been provided by Wu et al. (2014), who used TRMM data for providing global real-time flood forecasting. Figure 5.8 presents a global map of landslide potential estimated by using GPM data, combined with an

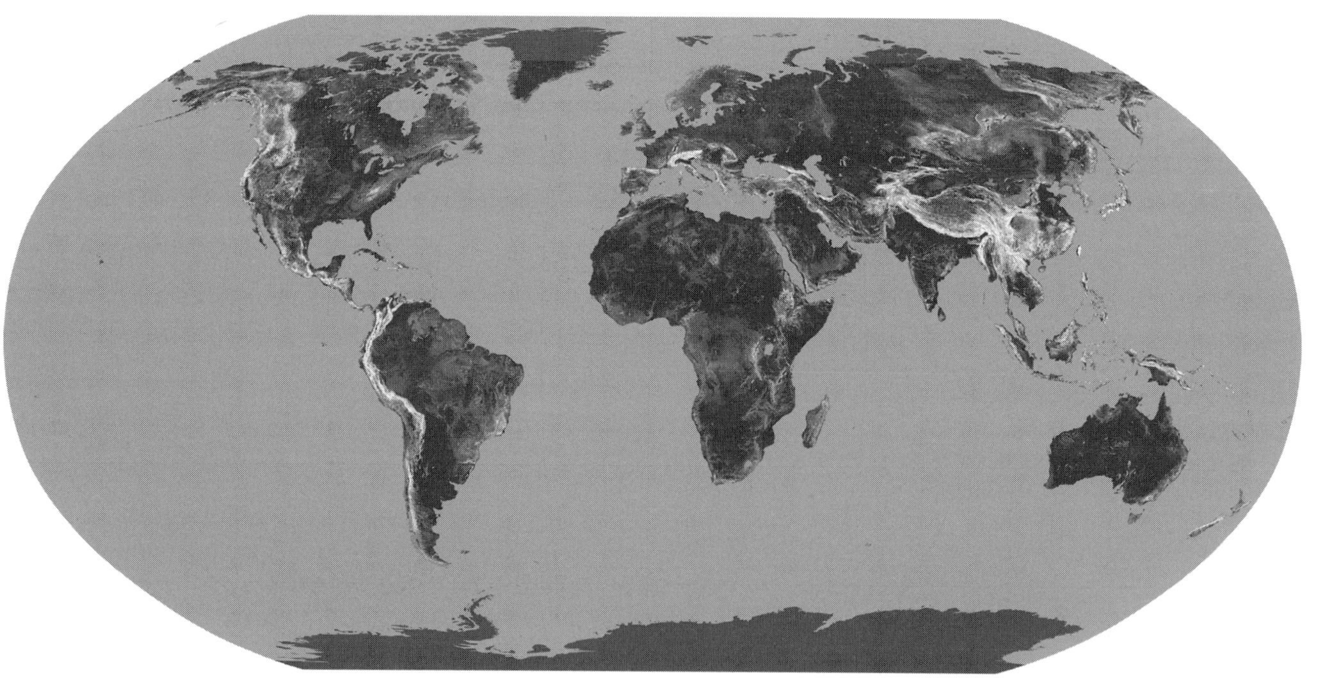

Figure 5.8 Global map of landslide potential estimated by using GPM data, combined with an assessment of the underlying susceptibility of terrain. Lighter areas are exposed to higher risk (NASA Earth Observatory images by Jesse Allen, using landslide susceptibility data (Stanley and Kirschbaum, 2017); topographic data from the Shuttle Radar Topography Mission (SRTM)).

assessment of the underlying susceptibility of terrain (Stanley and Kirschbaum, 2017).

5.4.5 Monitoring Evaporation: The Terra (CERES, MODIS) Mission

Evaporation-transpiration (E term in Eqs. 5.1 to 5.3), comprising evaporation from soils and transpiration from plants, is the second-largest component (after precipitation) of the global terrestrial water cycle. Satellite remote sensing has been employed in order to map globally, though not measure directly, evaporation-transpiration (ET) on the Earth's surface (see, e.g., Kustas and Norman, 1996; Anderson et al., 2011). NASA's Terra satellite caries five instruments (ASTER, CERES, MODIS, MISR and MOPITT). The CERES (Clouds and the Earth's Radiant Energy System) instruments aboard Terra measure the Earth's total energy radiation budget, while the MODIS (Moderate Resolution Imaging Spectroradiometer) instrument measures a large set of parameters and provides information on the water vapour content in the atmosphere with 1-km spatial and one-day temporal resolution. Land ET can then be evaluated as the residual of surface energy balance.

ET is important for a wide range of applications, ranging from assessing water use in irrigation in agricultural systems (Kisekka et al., 2010; Linquist et al., 2015) and water management (Vanino et al., 2015) to droughts (Mu et al., 2013) and climate change (Vinukollu et al., 2011). The extent of these applications ranges from local to global coverage. The use of these satellite missions is the only practical means of collecting spatially well-distributed ET data over large areas or globally (see Figure 5.9), even if their temporal resolution is limited by the infrequent acquisition of medium- to high-resolution satellite images. MODIS, for example, has moderate spatial and temporal resolution. Improvements in both the spatial and temporal resolution can then be obtained by assimilating in situ measurement, by using interpolation models and by combining with data from other ET contributing missions (e.g., LANDSAT for high spatial (40-m) and low temporal (16-day) resolution, and geostationary satellites for low spatial (3–10 km) and high temporal (<1 hour) resolution and in particular those providing land surface temperature, which is indispensable for remotely estimating ET.

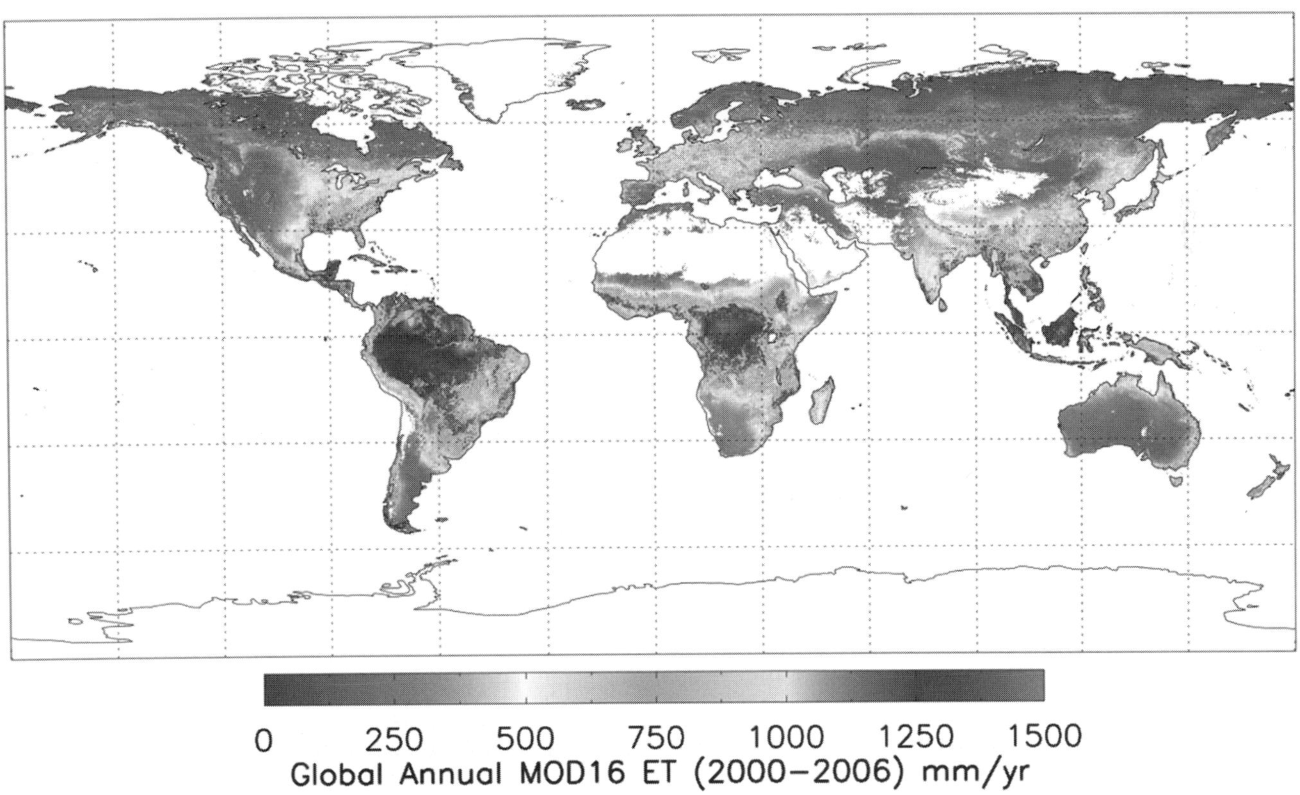

Figure 5.9 Global average ET over the 2000–2006 period (courtesy Numerical Terradynamic Simulation Group (NTSG), Univ. of Montana, MODIS Global Evapotranspiration Project (MOD16) www.ntsg.umt.edu/project/mod16.php). For the color version, please refer to the plate section.

5.5 Concluding Remarks

New monitoring techniques, and in particular remotely sensed data from satellites, are offering exciting opportunities for observing hydrological processes across a wide range of spatial and temporal scales. This original information offers unprecedented perspectives to improve global-scale modelling, and to gain a comprehensive understanding and mapping of water resource availability and water threats. Although uncertainties in global scale modelling are still a challenge, they are reduced by use of the growing global data sets retrieved through satellites. Working on larger spatial scales offers new ways forward to resolve global water problems.

For the preceding reason, new measurement techniques and new data processing will be primary subjects of research in hydrology in the near future. There is an increasing need to support hydrological studies with more refined knowledge, and enhanced theories, allowing us to profit from the new generation of spatially distributed and remotely sensed information. There will be a need to bridge the scale gap between local problems and the global challenges they are inducing. Water resources are frequently managed at the local scale of a single catchment or water resource system, but solutions need to be devised with a global vision, to take into account water transfers that are occurring at increasing spatial scales (e.g., through virtual water trade). Global information and a global vision will be key to anticipating the challenges and scientific trends for water resource management in the twenty-first century.

Water will become more and more the regulator of social dynamics, including economy, politics, immigration and social tensions. Virtual connections (including water sharing) around the world will increase and will make global analyses more and more important. Water will play a fundamental role in ensuring a peaceful future for humanity, provided scientists will be proactive enough to advise politicians effectively towards equity in water security and solving the water–food–energy nexus. To this end, long-term advanced monitoring of the global water cycle will be one of the keys to opening the doors to the future of humanity. It is therefore hoped that the space agencies will plan future satellite missions, which will extend the data sets obtained by the currently available satellites and improve the accuracy, resolution and latency of the observations contributing to the estimation of the various components in the water balance equation.

References

Anderson, M. C., Kustas, W., Norman, J. M., Hain, C. R., Mecikalski, J. R., Schultz, L., González-Dugo, M. P., Cammalleri, C., d'Urso, G., Pimstein, A. and Gao, F. (2011). Mapping daily evapotranspiration at field to continental scales using geostationary and polar orbiting satellite imagery. *Hydrol. Earth Syst. Sci.*, **15**, 223–239. doi: 10.5194/hess-15-223-2011.

Barnes, J. and Alatout, S. (2012). Water worlds: introduction to the special issue of *Social Studies of Science*. *Social Studies of Science*, **42**(4), 483–488. doi: 10.1177/0306312712448524.

Beck, M. W., Claassen, A. H. and Hundt, P. J. (2012). Environmental and livelihood impacts of dams: common lessons across development gradients that challenge sustainability. *International Journal of River Basin Management*, **10**(1), 73–92.

Biancamaria, S., Andreadis, K. M., Durand, M., Clark, E. A., Rodriguez, E. N., Mognard, M., Alsdorf, D. E., Lettenmaier, D. P. and Oudin, Y. (2010). Preliminary characterization of SWOT hydrology error budget and global capabilities. *IEEE J. Selec. Top. Appl. Earth Obs. Remote Sens.*, **3**, 6–19. doi:10.1109/JSTARS.2009.2034614.

Biancamaria, S., Lettenmaier, D. P., and Pavelsky, T. M. (2016). The SWOT Mission and its capabilities for land hydrology. *Surv. Geophys.*, **37**(2), 307–337. doi:10.1007/s10712-015-9346-y.

Biancamaria, S., Durand, M., Andreadis, K. M., Bates, P. D., Boone, A., Mognard, N. M., Rodriguez, E., Alsdorf, D. E., Lettenmaier, D. P. and Clark, E. A. (2011). Assimilation of virtual wide swath altimetry to improve Arctic river modeling. *Remote Sens. Environ.*, **115**, 373–381. doi:10.1016/j.rse.2010.09.008.

Birkett, C. M., Mertes, L. A. K., Dunne, T., Costa, M. H. and Jasinski, M. J. (2002). Surface water dynamics in the Amazon Basin: application of satellite radar altimetry. *J. Geophys. Res. Atmos.*, **107**(D20), 8059. doi:10.1029/2001JD000609.

Braden, J. B., Brown, D. G., Dozier, J., Gober, P., Hughes, S. M., Maidment, D. R., Schneider, S. L., Schultz, P. W., Shortle, J. S., Swallow, S. K. and Werner, C. M. (2009). Social science in a water observing system. *Water Resour. Res.*, **45**(11), W11301. doi:10.1029/2009WR008216.

Buchanan, T. J. and Somers, W. P. (1969). Discharge measurements at gaging stations. *U.S. Geological Survey Techniques of Water-Resources Investigations*, book 3, chap. A8, 65 pp. Available at http://pubs.usgs.gov/twri/twri3a8/.

Ceola, S., Monttanari, A., Krueger, T. et al. (2016). Adaptation of water resources systems to changing society and environment: a statement by the International Association of Hydrological Sciences. *Hydrological Sciences Journal*, **61**(16), 2803–2817. doi:10.1080/02626667.2016.1230674.

Ceola, S., Laio, F. and Montanari, A. (2014). Satellite night-time lights reveal increasing human exposure to floods worldwide. *Geophys. Res. Lett.*, **41**(20), 7184–7190.

Ceola, S., Laio, F., and Montanari, A. (2015). Human-impacted waters: New perspectives from global high-resolution monitoring. *Water Resour. Res.*, **51**(9), 7064–7079.

CEOS (2016). CEOS Water Constellation Feasibility Study Report. Available at http://ceos.org/document_management/Meetings/Plenary/30/Documents/2.3_Ishida_WSIST_Water Constellation FS report_V1.pdf.

Cherchali, S. and Gosset, M. (2016). Towards operational hydrology from space. Invited presentation at the *8th IPWG and 5th IWSSM Joint Workshop*, Bologna, October 3–7, 2016. Available at www.isac.cnr.it/~ipwg/meetings/bologna-2016/Bologna2016_Orals/12-1_Cherchali.pdf.

Di Baldassarre, G. and Montanari, A. (2009). Uncertainty in river discharge observations: a quantitative analysis. *Hydrol. Earth Syst. Sci.*, **13**(6), 913–921. doi:10.5194/hess-13-913-2009.

Di Baldassarre, G., Viglione, A., Carr, G., Kuil, L., Yan, K., Brandimarte, L. and Blöschl, G. (2015). Debates–perspectives on socio-hydrology: Capturing feedbacks between physical and social processes. *Water Resour. Res.*, **51**. doi:10.1002/2014WR016416.

Durand, M., Rodriguez, E., Alsdorf, D. E. and Trigg, M. (2010). Estimating river depth from remote sensing swath interferometry measurements of river height, slope and width. *IEEE J. Sel. Top. Appl. Earth Observ. Remote Sens.*, **3**(1), 20–31. doi:10.1109/JSTARS.2009.2033453.

Durand, M., Andreadis, K. M., Alsdorf, D. E., Lettenmaier, D. P., Moller, D. and Wilson, M. (2008). Estimation of bathymetric depth and slope from data assimilation of swath altimetry into a hydrodynamic model. *Geophys. Res. Lett.*, **35**, L20401. doi:10.1029/ 2008GL034150.

Eicker, A., Forootan, E., Springer, A., Longuevergne, L. and Kusche, J. (2016). Does GRACE see the terrestrial water cycle "intensifying"? *J. Geophys. Res. Atmos.*, **121**(2), 733–745. doi:10.1002/2015JD023808.

Entekhabi, D., Njoku, E. G., O'Neill, P. E., Kellogg, K. H., Crow, W. T., Edelstein, W. N., Entin, J. K., Goodman, S. D., Jackson, T. J., Johnson, J., Kimball, J., Piepmeier, J. R., Koster, R. D., Martin, N., McDonald, K. C., Moghaddam, M., Moran, S., Reichle, R., Shi, J. C., Spencer, M. W., Thurman, S. W., Tsang, L. and Van Zyl, J. (2010). The Soil Moisture Active Passive (SMAP) Mission. *Proceedings of the IEEE*, **98**(5), 704–716. doi:10.1109/JPROC.2010.2043918.

Evenson, E. J., Orndorff, R. C., Blome, C. D., Böhlke, J. K., Hershberger, P. K., Langenheim, V. E., McCabe, G. J., Morlock, S. E., Reeves, H. W., Verdin, J. P., Weyers, H. S., and Wood, T. M. (2013). U.S. Geological Survey water science strategy – observing, understanding, predicting, and delivering water science to the Nation. *U.S. Geological Survey Circular* 1383–G, 49 pp. Available at https://pubs.usgs.gov/circ/1383g/circ1383-G.pdf.

Fu, L.-L. and Ubelmann, C. (2014). On the transition from profile altimeter to swath altimeter for observing global ocean surface topography. *Journal of Atmospheric and Oceanic Technology*, **31**(2), 560–568. doi:10.1175/JTECH-D-13-00109.1.

Gebregiorgis, A. and Hossain, F. (2014). Making satellite precipitation data work for the developing world. *IEEE Geosci. Remote Sens. Mag.*, 2(2), 24–36. doi:10.1109/MGRS.2014.2317561.

Gober, P. and Wheater, H. S. (2015). Debates – perspectives on socio-hydrology: Modeling flood risk as a public policy problem. *Water Resour. Res.*, **51**, 4782–4788. doi:10.1002/2015WR016945.

Hamilton, S. H., El Sawah, S., Guillaume, J. H. A., Jakeman, A. J. and Pierce, S. A. (2015). Integrated assessment and modelling: Overview and synthesis of salient dimensions. *Environmental Modelling and Software*, **64**, 215–229. doi:10.1016/j.envsoft.2014.12.005.

Hrachowitz, M., Savenije, H. H. G., Blöschl, G. et al. (2013). A decade of Predictions in Ungauged Basins (PUB) – a review. *Hydrol. Sci. J.*, **58**(6), 1198–1255. doi:10.1080/02626667.2013.803183.

Kerr, Y. H., Waldteufel, P., Wigneron, J.-P., Delwart, S., Cabot, F., Boutin, J., Escorihuela, M.-J., Font, J., Reul, N., Gruhier, C., Juglea, S. E., Drinkwater, M. R., Hahne, A., Martín-Neira, M. and Mecklenburg, S. (2010). The SMOS Mission: new tool for monitoring key elements of the global water cycle. *Proceedings of the IEEE*, **98**(5), 666–687.

Kidd, C. and Huffman, G. (2011). Global precipitation measurement. *Meteorol. Appl.*, **18**, 334–353. doi:10.1002/met.284.

Kisekka, I., Migliaccio, K. W., Dukes, M. D., Schaffer, B. and Crane, J. H. (2010). Evapotranspiration-based irrigation scheduling and physiological response in a carambola (Averrhoa Carambola L.) orchard. *Applied Engineering in Agriculture*, **26**(3), 373–380.

Konar, M., Evans, T. P., Levy, M., Scott, C. A., Troy, T. J., Vörösmarty, C. J. and Sivapalan, M. (2016). Water resources sustainability in a globalizing world: who uses the water? *Hydrological Processes*, **30**(18), 3330–3336. doi:10.1002/hyp.10843.

Kummerow, C, Simpson, J., Thiele, O. et al. (2000). The Status of the Tropical Rainfall Measuring Mission (TRMM) after two years in orbit. *Journal of Applied Meteorology*, **39**, 1965–1982. doi:10.1175/1520-0450(2001)040<1965:TSOTTR>2.0.CO;2.

Kummerow, C., Barnes, W., Kozu, T., Shiue, J. and Simpson. J. (1998). The Tropical Rainfall Measuring Mission (TRMM) sensor package. *J. Atmos. Ocean Technol.*, **15**, 809–817. doi:10.1175/1520-0426(1998)015<0809:TTRMMT>2.0.CO;2.

Kustas, W. P. and Norman, J. M. (1996). Use of remote sensing for evapo-transpiration monitoring over land surfaces. *Hydrol. Sci. J.*, **41**, 495–516. doi:10.1080/02626669609491522.

Lettenmaier, D. P., Alsdorf, D., Dozier, J., Huffman, G. J., Pan, M. and Wood, E. F. (2015). Inroads of remote sensing into hydrologic science during the WRR era. *Water Resour. Res.*, **51**, 7309–7342. doi:10.1002/2015WR017616.

Linquist, B., Snyder, R., Anderson, F. et al. (2015). Water balances and evapotranspiration in water- and dry-seeded rice systems. *Irrig. Sci.*, **33**, 37–385. doi: 10.1007/s00271-015-0474-4.

Linton, J. and Budds, J. (2014). The hydrosocial cycle: defining and mobilizing a relational-dialectical approach to water. *Geoforum*, **57**, 170–180. doi: 10.1016/j.geoforum.2013.10.008.

Loucks, D. P. (2015). Debates–perspectives on socio-hydrology: simulating hydrologic-human interactions. *Water Resour. Res.*, **51**, 4789–4794. doi:10.1002/2015WR017002.

Macleod, C. J. A., Scholefield, D. and Haygarth, P. M. (2007). Integration for sustainable catchment management. *Science of the Total Environment*, **373**(2–3), 591–602.

Maggioni, V., Meyers, P. C. and Robinson, M. D. (2016). A review of Merged High-Resolution Satellite Precipitation Product Accuracy during the Tropical Rainfall Measuring Mission (TRMM) Era. *Journal of Hydrometeorology*, **17**(4), 1101–1117.

Montanari, A., Young, G., Savenije, H. H. G. et al. (2013). "Panta Rhei – everything flows": Change in hydrology and society – The IAHS Scientific Decade 2013–2022. *Hydrol. Sci. J.*, **58**(6), 1256–1275. doi:10.1080/02626667.2013.809088.

Montanari, A., 2015. Debates—Perspectives on socio-hydrology: introduction. *Water Resour. Res.*, **51**, 4768–4769. doi:10.1002/2015WR017430.

Mu, Q., Zhao, M., Kimball, J. S., McDowell, N. G. and Running, S. W. (2013). A remotely sensed global terrestrial drought severity index. *Bulletin of the American Meteorological Society*, (**94**)1, 83–98. doi:10.1175/BAMS-D-11-00213.1.

Munier, S., Polebistki, A., Brown, C., Belaud, G. and Lettenmaier, D. P. (2015). SWOT data assimilation for operational reservoir management on the upper Niger River Basin. *Water Resour. Res.*, **51**, 554–575. doi:10.1002/2014WR016157.

Oki, T. and Kanae, S. (2006). Global hydrological cycles and world water resources. *Science*, **313**(5790), 1068–1072. doi:10.1126/science.1128845.

Peltier, W. R., Argus, D. F. and Drummond, R. (2015). Space geodesy constrains ice-age terminal deglaciation: the global ICE-6G_C (VM5a) model. *J. Geophys. Res. Solid Earth*, **120**(1), 450–487. doi:10.1002/2014JB011176.

Poff, N. L. and Zimmerman, J. H. K. (2010). Ecological responses to altered flow regimes: a literature review to inform the science and management of environmental flows. *Freshwater Biology*, **55** (1), 194–205. doi:10.1111/j.1365-2427.2009.02272.x.

Ramillien, G., Frappart, F. and Seoane, L. (2014). Application of the regional water mass variations from GRACE satellite gravimetry to large-scale water management in Africa. *Remote Sens.*, **2014**(6), 7379–7405. doi:10.3390/rs6087379.

Reager, J. T. and Famiglietti, J. S. (2009). Global terrestrial water storage capacity and flood potential using GRACE. *Geophys. Res. Lett.*, **36**, L23402. doi:10.1029/2009GL040826.

Reager, J. T., Thomas, B. F. and Famiglietti, J. S. (2014). River basin flood potential inferred using GRACE gravity observations at several months lead time. *Nature Geoscience*, **7**, 588–592. doi:10.1038/ngeo2203.

Rodell, M., Houser, P. R., Jambor, U., Gottschalck, J., Mitchell, K., Meng, C.-J., Arsenault, K., Cosgrove, B., Radakovich, J., Bosilovich, M., Entin, J. K., Walker, J. P., Lohmann, D. and Toll, D. (2004). The Global Land Data Assimilation System. *Bull. Amer. Meteor. Soc.*, **85**(3), 381–394.

Rodell, M., Velicogna, I. and Famiglietti, J. S. (2009). Satellite-based estimates of groundwater depletion in India. *Nature*, **460**, 999–1002. doi:10.1038/nature08238.

Sivapalan, M. (2015). Debates—Perspectives on socio-hydrology: Changing water systems and the "tyranny of small problems" – socio-hydrology. *Water Resour. Res.*, **51**(6), 4795–4805. doi:10.1002/2015WR017080.

Sivapalan, M., Takeuchi, K. S., Franks, W., Gupta, V. K., Karambiri, H., Lakshmi, V., Liang, X., McDonnell, J. J., Mendiondo, E. M., O'Connell, P. E., Oki, T., Pomeroy, J. W., Schertzer, D., Uhlenbrook, S. and Zehe, E. (2003). IAHS Decade on Predictions in Ungauged Basins (PUB), 2003–2012: shaping an exciting future for the hydrological sciences. *Hydrological Sciences Journal*, **48** (6), 857–880. doi:10.1623/hysj.48.6.857.51421.

Sivapalan, M., Konar, M., Srinivasan, V., Chhatre, A., Wutich, A., Scott, C. A., Wescoat, J. L. and Rodríguez-Iturbe, I. (2014), Socio-hydrology: use-inspired water sustainability science for the Anthropocene. *Earth's Future*, **2**(4), 225–230. doi:10.1002/2013EF000164.

Sivapalan, M., Savenije, H. H. and Blöschl, G. (2012). Socio-hydrology: a new science of people and water. *Hydrological Processes*, **26**(8), 1270–1276.

Stanley, T. and Kirschbaum, D. B. (2017). A heuristic approach to global landslide susceptibility mapping. *Natural Hazards*, **87**(1), 145–164.

Syed, T. H., Famiglietti, J. S., Rodell, M., Chen, J. and Wilson, C. R. (2008). Analysis of terrestrial water storage changes from GRACE and GLDAS. *Water Resour. Res.*, **44**(2), W02433. doi:10.1029/2006WR005779.

Tamea, S., Carr, J. A., Laio, F., and Ridolfi, L. (2014). Drivers of the virtual water trade. *Water Resour. Res.*, **50**(1), 17–28.

Tapley, B., Rodell, M. and H. Save (2016). Droughts and floods as viewed by the GRACE mission and synergies with the SMAP data. *4th SMAP Applications Workshop and Tutorial,* Austin, TX, April 4–5, 2016. Presentation #16, available online at www.regonline.com/builder/site/tab3.aspx?EventID=18239.

Thomas, A. C., Reager, J. T., Famiglietti, J. S. and Rodell M. (2014). A GRACE-based water storage deficit approach for hydrological drought characterization. *Geophys. Res. Lett.*, **41**, 1537–1445. doi:10.1002/2014GL059323.

Troy, T. J., Pavao-Zuckerman, M. and Evans, T. P. (2015). Debates–perspectives on socio-hydrology: socio-hydrologic modeling: tradeoffs, hypothesis testing, and validation. *Water Resour. Res.*, **51**, 4806–4814. doi:10.1002/2015WR017046.

UNICEF and World Health Organization (2015). Progress on Sanitation and Drinking Water – 2015 update and MDG assessment. Available online at www.who.int/water_sanitation_health/monitoring/jmp-2015-update/en/.

Vanino, S., Nino, P., De Michele, C., Falanga Bolognesi S. and Pulighe, G. (2015). Earth observation for improving irrigation water management: a Case-study from Apulia region in Italy. *Agriculture and Agricultural Science Procedia*, **4**, 99–107. doi:10.1016/j.aaspro.2015.03.012.

Vinukollu, R. K., Wood, E. F., Ferguson, G. R. and Fisher, J. B. (2011). Global estimates of evapotranspiration for climate studies using multi-sensor remote sensing data: Evaluation of three process-based approaches. *Remote Sensing of Environment*, **115**(3), 801–823.

Wheater, H. S. and Gober, P. (2015). Water security and the science agenda. *Water Resour. Res.*, **51**, 5406–5424. doi:10.1002/2015WR016892.

Winiwarter, V., Schmid, M. and Dressel, G. (2013). Looking at half a millennium of co-existence: the Danube in Vienna as a socio-natural site. *Water History*, **5**(2), 101–119.

Wouters, B., Bonin, J. A., Chambers, D. P., Riva, R. E. M., Sasgen, I. and Wahr, J. (2014). GRACE, time-varying gravity, Earth system dynamics and climate change. *Rep. Prog. Phys.*, **77**, 116801, 41 pp. doi:10.1088/0034-4885/77/11/116801.

Wu, H., Adler, R. F., Tian, Y., Huffman, G. J., Li, H., & Wang, J. (2014). Real-time global flood estimation using satellite-based precipitation and a coupled land surface and routing model. *Water Resources Research*, 50(3), 2693–2717.

Yoon, Y., Durand, M., Merry, C. J., Clark, E. A., Andreadis, K. M. and Alsdorf, D. E. (2012). Estimating river bathymetry from data assimilation of synthetic SWOT measurements. *J. Hydrol.*, **464–465**, 363–375. doi:10.1016/j.jhydrol.2012.07.028.

6

Geodetic Observations as a Monitor of Climate Change

Tonie van Dam, Jianli Chen, and Thierry Meyrath

6.1 Introduction

The solid Earth is constantly changing. The subdaily to long-term changes in the distribution of environmental mass (atmospheric, oceanic, water, and ice) have an effect on the shape, the rotation, and the gravity field of the Earth. These changes can be observed by geodetic techniques. Geodesy is the mathematical branch of the Earth sciences that deals with the measurement of the Earth's shape, its gravitational field, and its rotation. These observations represent the three pillars of geodesy.

As the Earth's surface temperature responds to increasing concentrations of greenhouse gases in the atmosphere, so also do the shape, rotation, and mass distribution of environmental fluids on the Earth's surface (atmospheric, oceanic, water, and ice). Geodesy allows us to observe the results of climate warming indirectly.

Some of the most obvious effects of climate warming that can be observed by geodesy include sea level rise, changes in the thickness and extent of ice cover, changes in the water cycle, and changes in ocean dynamics.

In this chapter, we review how each of the three pillars of geodesy provides evidence for and insight into the effects of climate warming.

6.2 Shape of the Earth

The Earth's shape changes on time scales ranging from subdaily to decadal and longer. In addition, the changes in the Earth's shape can happen over wavelengths ranging from the radius of the Earth to tens of kilometers. The range of time scales and wavelengths of surface displacements reflects the large variety of processes that cause it to change including tidal forcing, the redistribution of surface mass (atmosphere, hydrosphere, cryosphere), tectonics, and glacial isostatic adjustment.

When it comes to detecting the effects of climate change, the Earth's shape changes as a result of the redistribution of surface mass. For example, increasing climate warming causes the ice sheets and glaciers to melt, increases evaporation at the equator, influences the pattern and the magnitude of precipitation, and increases sea level. All of these changes can be thought of as a redistribution of water mass (in all its forms).

At time scales less than about one hundred years, the Earth responds elastically to changes in surface mass. An elastic response means that the Earth responds linearly and immediately to the mass change. To demonstrate this property, Figure 6.1 shows the change of the height of the Earth's surface in response to redistributions of global water mass as predicted by GLDAS. GLDAS is a model of continental water storage that ingests satellite- and ground-based observational data products and uses advanced land surface modeling and data assimilation techniques, to generate optimal fields of land surface states and fluxes (Rodell et al., 2004). The observations are GPS observations of vertical surface motions. The predicted displacements are determined by convolving the global redistribution in water storage from GLDAS with mass loading Green's functions (Farrell, 1972). (For more detail the reader should see van Dam and Wahr, (1987) and van Dam et al., (2001).) Figure 6.1 shows the observed GPS vertical (in black) at four global sites: 1) Irkutsk, Siberia; 2) Potsdam, Germany; 3) Alice Springs, Australia; and 4) North Liberty, Iowa, United States. Also shown in the figure is the predicted vertical displacement from GLDAS (red). The two curves track each other very well indicating that a single GPS station can be used to observe changes in regional water storage. Thus, we can track changes in the water cycle by analyzing GPS time series from a global network of stations.

On the other hand, observations of surface displacements can be used to infer surface mass changes. In the first global inversion of its type, Blewitt et al. (2001) used five years of three-dimensional GPS observations

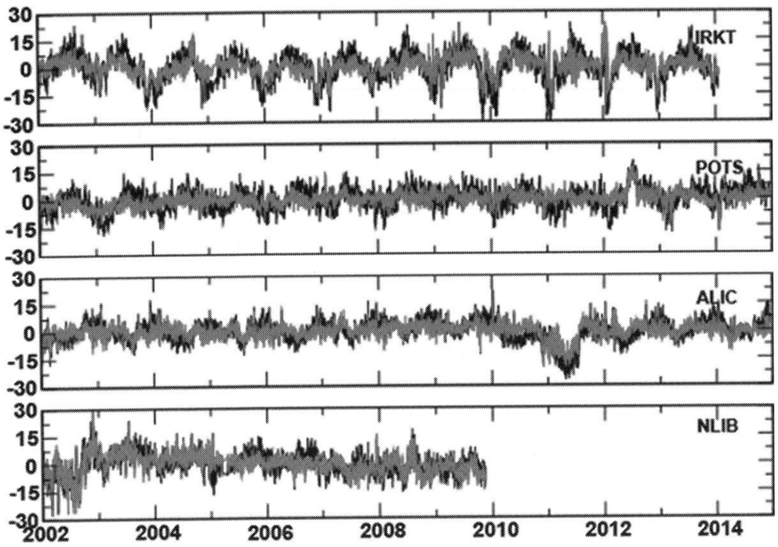

Figure 6.1 Vertical displacements of the Earth's surface in millimeters. The black lines are GPS observations of displacement; the grey lines are predicted displacements from the GLDAS water storage model. The stations from top to bottom are 1) Irkutsk, Siberia; 2) Potsdam, Germany; 3) Alice Springs, Australia; and 4) North Liberty, Iowa, United States.

at approximately sixty globally distributed sites to determine the degree-surface displacement of the Earth. These global displacements were inferred to be driven by annual and semiannual cycles of hemisphere-to-hemisphere mass exchange, primarily water mass. The estimated load coefficients correspond to a determination of the motion of the center of mass (CM) of the total Earth system including load with respect to the center of figure (CF) of the solid Earth surface (often referred to as geocenter motion).

The inversion of Blewitt et al. (2001) is, however, a theoretically underdetermined problem. In simple terms, if the data used to invert for a truncated spherical harmonic load series are not equally distributed and are not fully global (i.e., what about the breaks where we have ocean) the inversion will be subject to aliasing errors (Wu et al., 2002). This problem has been addressed by subsequent researchers who include estimates of the surface displacements over the ocean using models of global ocean bottom pressure and/or Gravity Recovery and Climate Experiment (GRACE) satellite gravity data (to be discussed later). Figure 6.2, taken from Wu et al. (2003), shows a global inversion of two hundred near-evenly spaced GPS sites for the annual water mass loading signal. This work demonstrates how multiannual time series of GPS observations can be used to infer long-wavelength, long–time scale changes in water mass.

In regions where the GPS network is particularly dense, e.g., the western United States because of the Plate Boundary Observatory, a network of more than eleven hundred permanent, continuously operating GPS stations, higher spatial resolution information can be obtained. Figure 6.3a shows the GPS vertical surface displacements due to water storage variations in the California Central Valley between April and October. Figure 6.3b shows the inversion results in units of equivalent water thickness. Finally, Figure 6.3c shows the results of a composite water storage model. Comparing Figure 6.3a and c demonstrates the potential of high-density GPS networks to determine changes in water storage.

GRACE satellite gravity observations have been used to infer that California's Central Valley lost 20 km^3 of groundwater from 2003 to 2010, for a loss rate of 3 km^3/yr (Famiglietti et al., 2011). However, the strength of GPS for this type of application is that GPS has the ability to resolve solid Earth's elastic response to surface loading at higher spatial resolution than GRACE and in this example constraining the distribution of snow and water.

In a similar manner, more detailed inversions of the GPS data over California's central basin have been published (Amos et al., 2014; Borsa et al., 2014), as well as inversions in Alaska (Fu et al., 2012), the Himalaya (Fu and Freymueller, 2012; Chanard et al., 2014), the Amazon (Fu et al., 2013), and many other regions of the world to determine long-term changes in water storage.

GPS observations have also been used to determine mass loss over the Greenland ice sheet. Figure 6.4 shows the long-term uplift velocity of stations in the southeast of Greenland (Bevis et al., 2012). What is interesting about this image is that the GPS uplift velocities vary between a maximum of 30.5 mm/yr and 3.6 mm/yr. The higher velocities are from stations located near large glaciers. Climate warming increases the

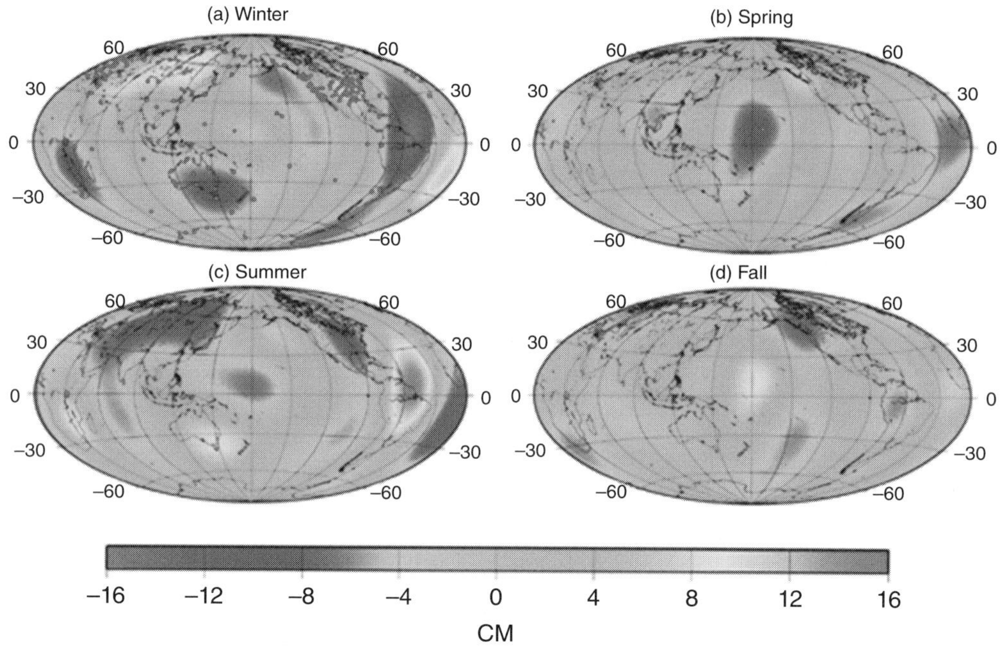

Figure 6.2 Inversion of three-dimensional displacements from two hundred global sites (red circles) over the continents plus estimates of surface displacements from ocean bottom pressure over the oceans. Inversion up to spherical harmonic degree and order 5, annual surface mass variation in centimeter equivalent water thickness. (a) Winter at mid-January, (b) spring at mid-April, (c) summer at mid-July, (d) fall at mid-October (from figure 1 of Wu et al., 2003). For the color version, please refer to the plate section.

temperature of the oceans melting the marine terminating glaciers from below and from the front edge allowing the glaciers to move faster. Thus, the very high velocities indicate mass loss due to glacier dynamics and the smaller velocities are more indicative of the longer-wavelength changes in ice mass on the ice sheet.

This section has shown (1) that the Earth responds to environmental mass loading; (2) that these observations can provide estimates of the changes in the environmental surface mass; and (3) that observations of changes in the shape of the Earth can be used to monitor or investigate long-wavelength changes in this surface mass. This dataset is very different from the methods used for investigating changes in surface mass, mainly ice and water mass. These data can provide a long-wavelength, long–time scale information that can be used tocomplement interpretation of other, more traditional types of ice and water observations. In the next section, we discuss how gravity, a second geodetic pillar, observations can be used to monitor surface mass changes.

6.3 The Earth's Gravity Field

In this section, we will report on how satellite gravity observations have been and are being used to monitor changes in the global distribution of water and ice mass, changes that may be linked to long-term climate change.

6.3.1 Satellite Gravity

The gravity field, the second pillar in geodesy, is defined by the mass distribution in the Earth system, including the solid Earth and its geophysical fluid envelope (atmosphere, ocean, hydrosphere, and cryosphere). The static gravity field is mainly controlled by mass distribution in the solid Earth (the crust, mantle, and core). The gravity anomalies (see Figure 6.5) more or less reflect the land topography, the ocean bathymetry, and the internal density structure of the Earth. However, temporal variations in the gravity field are mainly driven by air and water mass redistributions within the Earth climate system (i.e., atmosphere, ocean, hydrosphere, and cryosphere) (Wahr, et al., 1998; Cazenave and Chen, 2010), plus solid Earth contributions at long-term time scales (hundreds to thousands of years) in certain locations due to the glacial isostatic adjustment (GIA) effect, a slow viscoelastic response of the Earth crust after the last glacial maximum (LGM) (Peltier, 2004; Ivins and James, 2005).

The geodetically observed time-variable gravity field provides a unique tool for studying mass redistributions

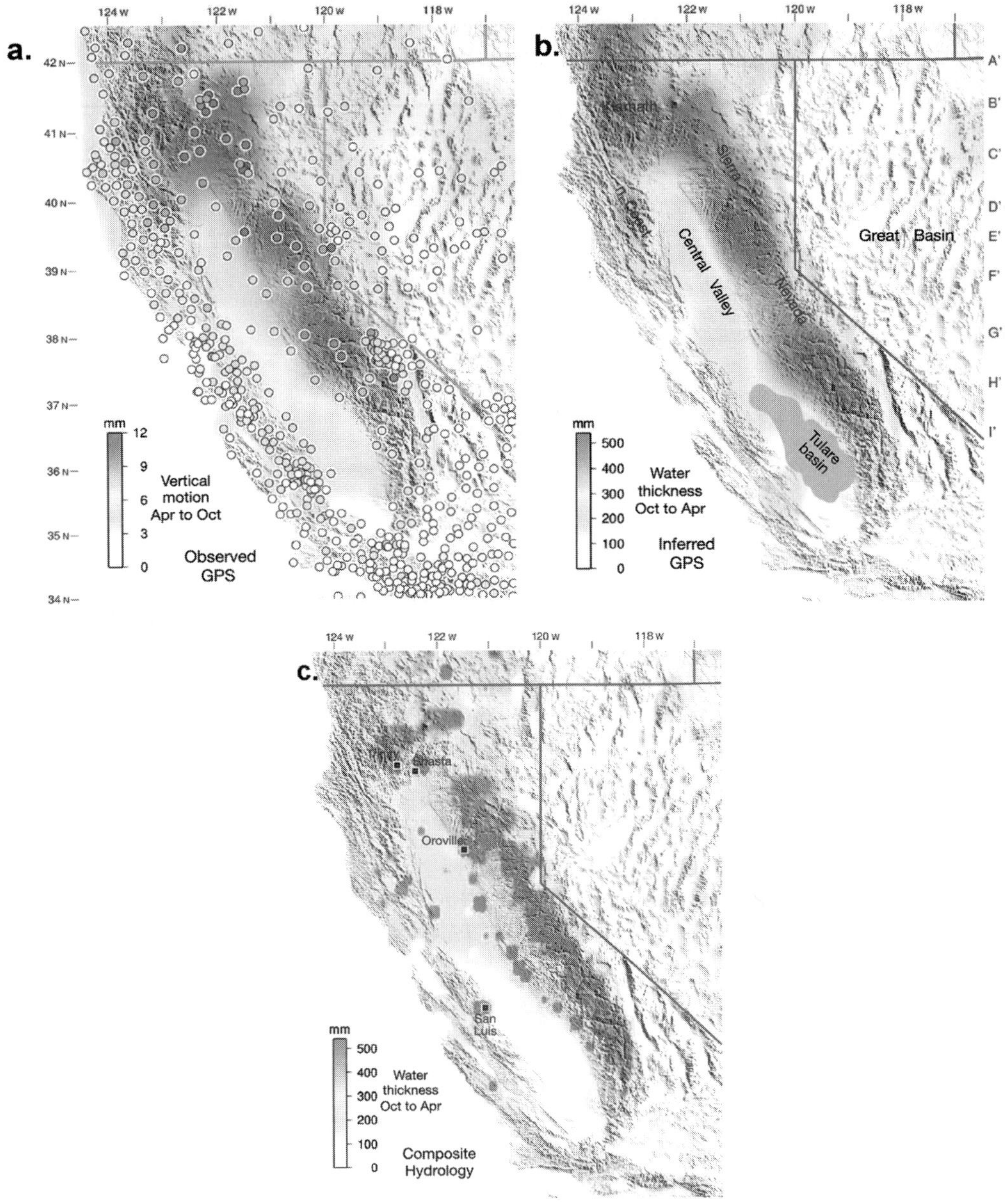

Figure 6.3 A comparison of surface uplift (a), inverted estimate of the annual change in water storage (b), and water storage change derived from a composite model of water storage (c) (composite image derived from figures 2 and 3 of Argus et al., 2014). For the color version, please refer to the plate section.

in the Earth system, especially those associated with the climate system, such as terrestrial water storage change, groundwater depletion, flood and drought, polar ice sheet melting, and sea level rise (Cazenave and Chen, 2010). Observed time-variable gravity changes represent an important tool for studying solid Earth geophysical variations, including quantifying GIA uplifting rates (e.g., Tamisiea et al., 2007) and seismic deformation from major earthquakes (e.g., Han et al., 2006). However, determining the temporal variations of the gravity field with adequate accuracy has been challenging. Historically, the Earth's gravity field has been measured by gravimeters (either absolute or relative), at fixed stations or through field surveys.

Combined with other geodetic observations, these historical in situ gravity measurements can help provide an approximate estimation of the Earth's static gravity field, with a limited spatial resolution. Because of the lack of adequate in situ measurements, it is simply impossible to determine temporal variations of the gravity field even on large regional scales (not to say continental or global scales). Since the mid-1970s, developments in satellite geodesy have begun to provide new tools for measuring temporal variations of the

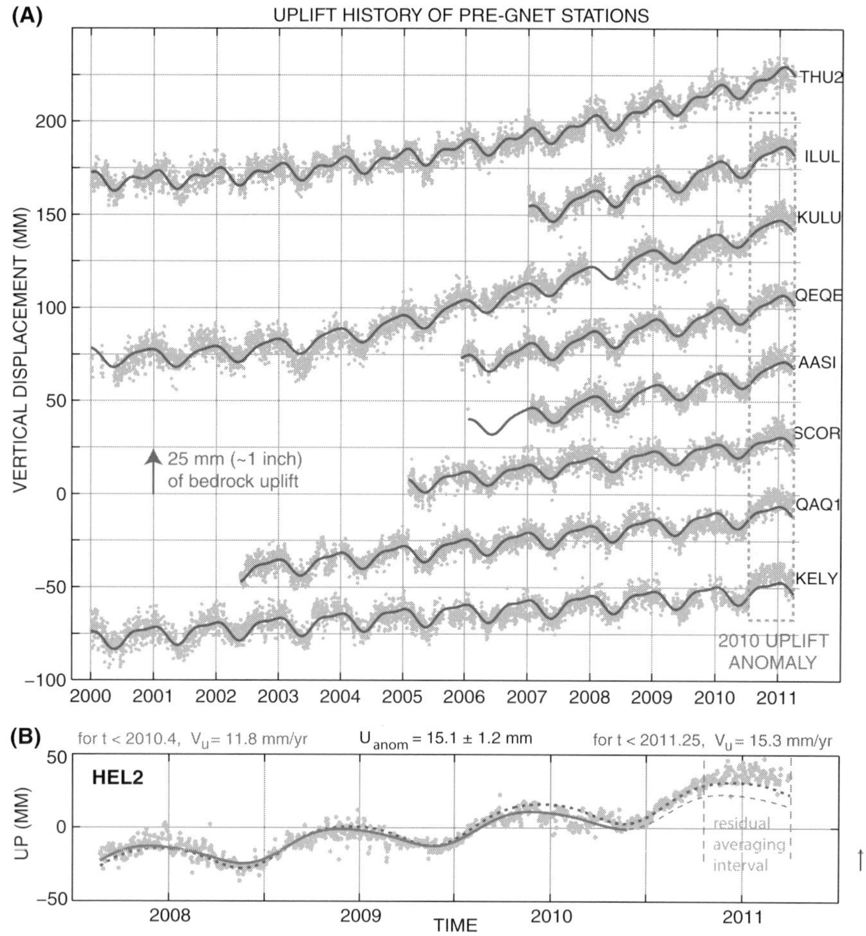

Figure 6.4 Velocity trends determined by GPS located on bedrock stations around the Greenland ice sheet (from figure 1 of Bevis et al., (2012)).

Earth's gravity field from space, starting with the satellite laser ranging (SLR) instrument on board the Lageo-1 satellite (which was launched in 1976).

Satellite geodesy reached a new milestone in 2002 when the GRACE satellite gravity mission was launched. GRACE allows us to measure temporal variations in the global gravity field with unprecedented accuracy, enabling us to study large-scale mass redistribution, and in the climate system (Tapley et al., 2004; Cazenave and Chen, 2010).

6.3.2 SatelliteLaser Ranging

Satellite laser ranging (SLR) is a space geodetic technique that measures the time it takes an ultrashort (<500 psec) laser pulse to travel between an SLR ground station and a retroreflector-equipped satellite and back. This technique provides instantaneous range measurements with the precision of millimeter level, which can be accumulated to provide accurate measurement of orbits and a host of other important scientific data (Pearlman et al., 2002).

Since the mid-1970s, SLR tracking of the orbits of tens of satellites at different altitudes and orbit inclinations (see Figure 6.6) has gradually improved knowledge of Earth's static (i.e., mean) gravity field. Traditionally, gravity field recovery from tracking data has been an iterative approach, which compares a numerically integrated orbit from an initial gravity field and other force models with observations of the satellite motion from ground-based stations. Accumulation of decades of tracking data from a large number of geodetic satellites in different orbital configurations led to a succession of gravity field solutions of increased precision and resolution (Cheng and Tapley, 1999). Moreover, the combination of these satellite-based solutions with surface measurements and satellite altimetry–based gravity anomalies (from ocean geoid height measurements) has provided global gravity field solutions of ever better quality (see Figure 6.5) (Pavlis et al., 2012).

SLR has significantly contributed to our understanding of the Earth's climate system, through accurately measuring low-degree gravitational changes (e.g., Yoder et al., 1983; Cheng and Tapley, 1999). These

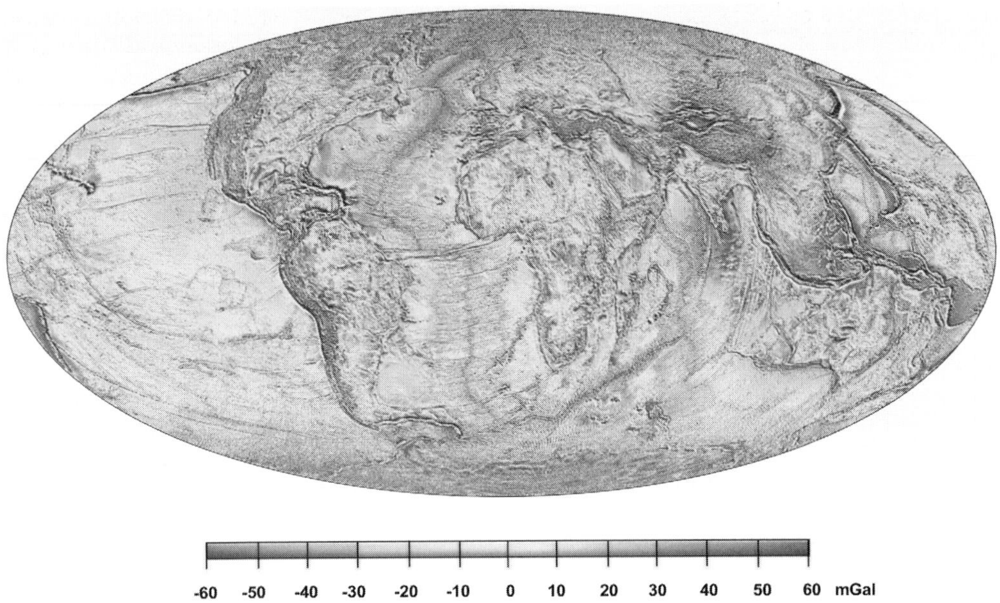

Figure 6.5 The static component of Earth's gravity field based on the EGM08 model (Pavlis et al., 2012). The figure shows gravity anomalies in mGal (modified from an original map in Cazenave and Chen, 2010). For the color version, please refer to the plate section.

low-degree gravitational changes provide an observational constraint on the global scale mass redistributions in the climate system. Among SLR observed low-degree gravitational changes, the degree-2 zonal terms ΔC20 (also termed as ΔJ2 by a scale factor), believed to be best determined, have been subject to extensive studies over the past few decades (e.g., Yoder et al., 1983; Cox and Chao, 1998; Dickey et al., 2002; Chen et al., 2003; Cheng and Tapley, 2004).

The C20 coefficient represents the Earth oblateness, which is sensitive to polar ice sheet ice melting and global sea level rise. The more-than-four-decade-long (1976 to the present) record of SLR ΔC20 time series has played an important role in helping us better understand the Earth's climate change, especially mass transport from high latitudes (the Antarctic and Greenland ice sheets and Arctic glaciers) to the world's ocean (Cox and Chao, 1998; Dickey et al., 2002; Chen et al., 2003).

SLR is also regarded as the most accurate technique currently available to determine the geocenter motion, which represents variation of the mass center of the Earth system with respect to the terrestrial reference frame (or vice versa, i.e., variation of the terrestrial reference frame with respect to the mass center of the Earth system) (Chen et al., 1999; Collilieux et al., 2009). SLR's ability to measure low-degree gravitational variations and to monitor geocenter motion, together with its capability of monitoring vertical motion of station sites, makes it a unique technique for modeling and evaluating long-term climate change. An improved reference system is important for better determining GIA, sea level change, and polar ice sheet mass change.

6.3.3 GRACE Satellite Gravimetry

The GRACE satellite mission is a dedicated twin satellite gravity mission, jointly sponsored by NASA and DLR (German Aerospace Center) under the NASA Earth System Science Pathfinder Program (Tapley et al., 2004). GRACE utilizes a state-of-the-art technique to observe variations of Earth's gravity by tracking the intersatellite range and range rate between two coplanar, low-altitude satellites (Figure 6.7, GRACE A and B) via a K-band ranging (KBR) system (Tapley et al., 2004). The GRACE Science Data System uses the KBR range and range rate data, along with ancillary data, to estimate a new gravity field every month in the form of corrections to a well-defined background gravity model used in the data processing procedure (Bettadpur, 2012).

Monthly GRACE global gravity solutions are provided by three GRACE data processing centers: the Center for Space Research (CSR) at the University of Texas at Austin, Geoforschungszentrum (GFZ) in Potsdam, and the NASA Jet Propulsion Laboratory (JPL). GRACE solutions are distributed through the NASA PODAAC archive (http://podaac.jpl.nasa.gov/grace/). GRACE monthly solutions are generally expressed in the form of fully normalized spherical harmonic coefficients, up to a maximum degree and order (typically

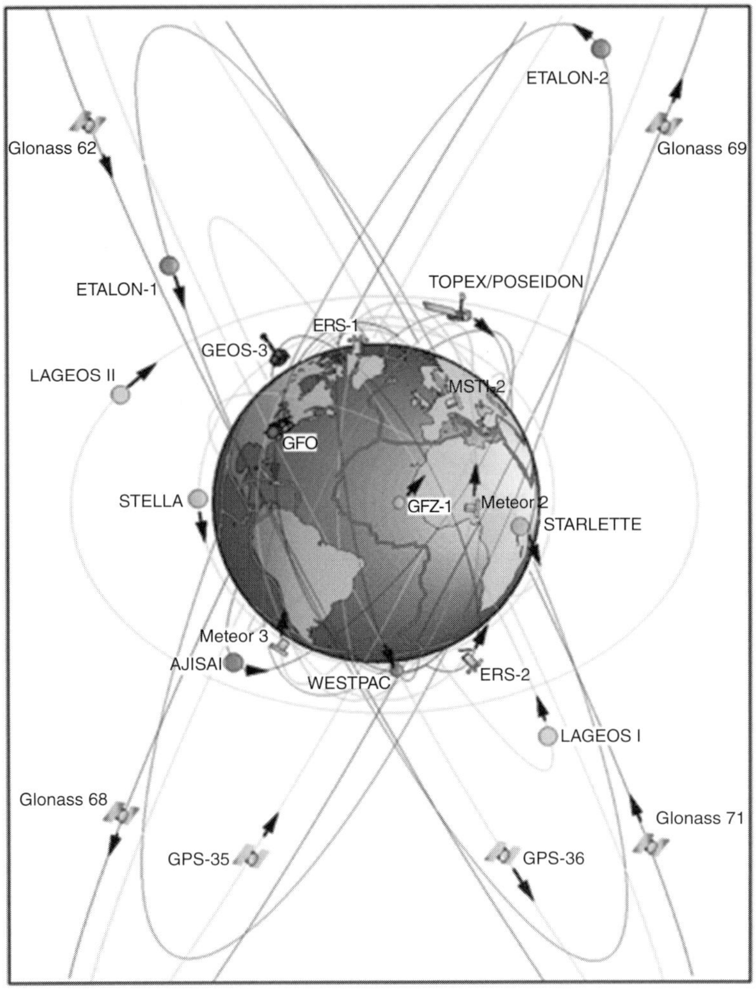

Figure 6.6 Constellation of satellites supported by SLR (source: NASA International Laser Ranging Service). For the color version, please refer to the plate section.

between 60 and 100). Atmospheric and oceanic mass redistributions and tidal effects are removed during GRACE data processing using atmospheric and ocean circulation models (e.g., Bettadpur, 2012; Dale et al., 2013).

GRACE observed time-variable gravity changes represent three-dimensional (3D) integrated contributions from mass redistribution in the Earth system. However, the inversion of mass variations from time-variable gravity solutions is nonunique (Chao, 2005); that means without constraints from other independent observations or assumptions, we are not able to compute 3D mass variations in the Earth system using GRACE time-variable gravity solutions alone. However, at decadal or shorter time scales, we can assume that mass variations mainly occur on the Earth surface in the Earth's geophysical fluid envelope (atmosphere, ocean, hydrosphere, and cryosphere). Under this assumption, GRACE monthly gravity spherical harmonic (SH) coefficients Clm and Slm can be readily converted into surface mass change (usually expressed in equivalent water height) as in Wahr et al. (1998).

GRACE high-degree and high-order SH coefficients are dominated by noise, especially spatially correlated noise (longitudinal stripes). Spatial filtering and/or smoothing is needed in order to reveal meaningful mass change signals such as those arising from ice melting and water redistribution (Wahr et al., 1998). The longitudinal stripes originate from the existence of correlation between the even and odd degree pairs of GRACE SH coefficients of a given order, and can be fairly effectively removed through the so-called decorrelation filtering (Swenson and Wahr, 2006). After decorrelation filtering, a certain degree of Gaussian smoothing is still needed to suppress residual noise further.

Even though GRACE monthly global gravity SH solutions have been the dominant form of GRACE data used worldwide, GRACE mass concentration (mascon) solutions are another data representation that

Figure 6.7 GRACE includes twin satellites orbiting the Earth at an initial altitude of ~500 km in a near-polar orbit (89° inclination). GRACE was launched on 17 March 2002, with a design lifetime of five years and is now in its fifteenth year (source: NASA/CSR).

can be useful in global and regional mass change studies and are being developed by a number of institutes, including NASA JPL (Watkins et al., 2015; Wiese et al., 2016), UT CSR (Save et al., 2016), and NASA GSFC (Luthcke et al., 2013). GRACE mascon solutions do not require decorrelation filtering or Gaussian smoothing, and offer improved spatial resolution. The JPL and CSR mascon solutions are currently available to the public ftp://podaac.jpl.nasa.gov/allData/tellus/L3/mascon/RL05/).

GRACE is in the fifteenth year of its mission, and it is still acquiring excellent data. The GRACE mission provides a revolutionary means for monitoring and quantifying large-scale mass redistributions and movements within the Earth system (e.g., Cazenave and Chen, 2010). Over the past decade, studies using GRACE gravity data have greatly enhanced our understanding of mass balance of polar ice sheets (e.g., Velicogna and Wahr, 2006; 2013; Luthcke, et al., 2006; Chen et al., 2006; 2009) and mountain glaciers (e.g., Tamisiea et al., 2005; Chen et al., 2007; Lutchke et al., 2008), terrestrial water and groundwater storage changes (e.g., Wahr et al., 2004, Rodell et al., 2009; Famiglietti et al., 2011), and sea level change (e.g., Chambers et al., 2004; Leuliette and Miller, 2009; Chen et al., 2013).

The more than fourteen-year record of GRACE time-variable gravity measurements has also provided a unique tool for studying solid Earth geophysical variations, including improving quantification of GIA uplifting rates, especially in regions with limited geological constraints and GPS measurements, such as the Antarctic (Yamamoto et al., 2011; Gunter, et al., 2014) and coseismic and postseismic deformation from major earthquakes, especially those with epicenters in the ocean (Han et al., 2014; 2016; Chen et al., 2007; Li et al., 2014; 2016).

Figure 6.8 shows GRACE observed long-term mass change rates (in centimeters per year of equivalent water height) over the ten-year period 2003–2012, based on the CSR RL05 GRACE monthly gravity solutions, with decorrelation filtering and 300-km Gaussian smoothing applied. The $\Delta C20$ terms have been replaced by SLR solutions based on the same GRACE RL05 standards (Cheng and Ries, 2012). GRACE has captured some interesting long-term mass change signals from either climate change or solid Earth geophysical processes, including dominant ice losses over the Greenland ice sheets and the Canadian Arctic Archipelago (CAA) and the Amundsen Sea Embayment (ASE) coastal region and Antarctic Peninsula in the West Antarctic (regions A and B marked on the plot). GRACE has also captured evident ice losses from mountain glaciers in southern Alaska (region C) and the Patagonia Ice Field (region D). The dominant mass increase in the Hudson Bay area (region E) in North America represents the Earth crust uplift from GIA effect. The mass loss in northwestern India (region F) is apparently due to excessive groundwater depletion for agricultural irrigations (Rodell et al., 2009). An interesting observation is the mass increase in the tropical Indian Ocean, which in turn apparently reflects sea floor deformation due to the gigantic 2004 Sumatra–Andaman earthquake (Mw 9.2). The increasing trend appears to result from the combination of coseismic uplift and postseismic relaxation (Han et al., 2016). Another interesting signal is the mass increase along the coastal regions in East Antarctica, which is attributed to increased snow accumulation during the period (Boening et al., 2012).

Despite the tremendous success of GRACE satellite gravimetry in studying mass redistribution and movement in various components of the Earth system, there are still some major challenges in correctly using and interpreting GRACE data, because of the coarse spatial resolution (a few hundred kilometers at best) and limited accuracy (Wahr et al., 2006). Major error sources include (but are not limited to) 1) residual error in GRACE measurements after filtering and smoothing, 2) leakage error due to truncation and spatial filtering and/or smoothing of GRACE data, 3)

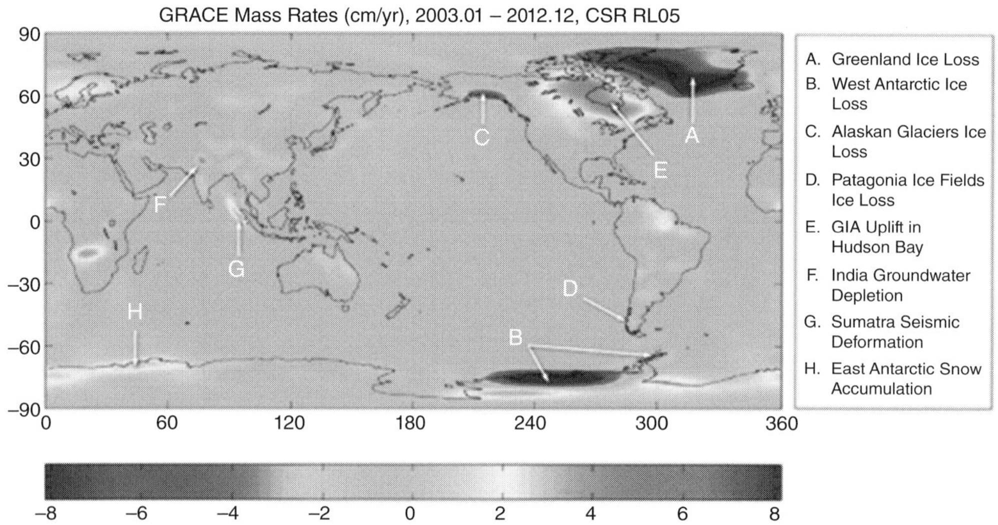

Figure 6.8 GRACE-observed long-term mass change rates (in centimeters per year of equivalent water height) over the ten-year period 2003–2012, based on the CSR RL05 monthly gravity solutions, with decorrelation filtering and 300-km Gaussian smoothing applied. Some long-term mass change signals from climate change and solid Earth geophysical processes are marked by white arrows and labeled on the right. For the color version, please refer to the plate section.

uncertainty of low-degree gravity coefficients and geocenter terms, 4) uncertainty of atmospheric and oceanic models used in the GRACE dealiasing process, and 5) uncertainty of other geophysical models (e.g., GIA) that need to be removed. While GIA is generally regarded as the largest error source in GRACE Antarctic mass balance estimates (Velicogna and Wahr, 2006), the leakage effect could be the largest error source to many other studies related to climate change.

Depending on the size or location of the signal, the applied truncation of GRACE SH and spatial filtering can significantly reduce the amplitude of the signal, and spread the signal to a much broader region, resulting in the so-called leakage effect (Landerer and Swenson, 2012; Chen et al., 2015). How to reduce the leakage bias correctly and recover the true signal has been challenging. Scale factors (rescaling) can be used to correct or reduced the leakage biases (Chen et al., 2005; Velicogna and Wahr, 2006; Landerer and Swenson, 2012). For hydrological applications, scale factors can be estimated from land surface model (LSM) TWS storage predictions that have been represented as truncated SH expansions and filtered (Landerer and Swenson, 2012). However, the scaling factor method may be suited to study of seasonal signals, but not to longer-term variability, because LSMs are thought not to replicate longer-term behavior well. The problem is complicated by anthropogenic pumping, which is not an element of many LSMs. It may be possible to improve the rescaling method by constructing a mass change model that reasonably represents both seasonal and long-term mass changes in a particular region.

Forward modeling has also proven useful for correcting or reducing leakage bias, especially when source locations of mass change signals are known (e.g., Chen et al., 2006, 2015; Wouters et al., 2008). Forward modeling is an iterative process that finds a mass change (or mass change rate) that agrees with GRACE data, but is confined to regions constrained by known geography. Forward modeling accounts for all processing steps such as truncation and Gaussian smoothing, and seeks the optimal solutions through minimizing differences between forward model solutions and GRACE observations. Even though GRACE mascon solutions offer better spatial resolution (than SH solutions), leakage error and bias still exist (to a lesser extent, though) (Wiese et al., 2016; Save et al., 2016). Quantification of mascon solutions leakage bias can be more challenging.

The GRACE Follow-On (FO) mission, launched in 2018, promises to continue satellite gravimetry observations well into the future. The GRACE and GRACE FO missions are expected to provide at least more than two (or even three) decades of time-variable gravity measurements, which will be a valuable data resource for studying various components of the Earth climate change at broad time scales. Continuous improvements of GRACE data processing methods and background geophysical models are also expected, which will help improve the accuracy and spatial resolution of GRACE measurements and enable a wider range of applications of satellite gravity observations.

In this section, we've described the some of the satellite gravity techniques that have been used to provide information on the long-wavelength changes in water and ice on the surface of the Earth. These changes are related to climate change over the last two decades or more and could not have been acquired in any other way.

6.4 Earth Rotation

In this section, we describe the third geodetic pillar, Earth rotation, and how changes in these parameters can be used to infer climate-driven changes in the distribution of mass on the surface of the Earth.

The rotation of the Earth is not constant, and both the position of its rotation axis and the rotation rate undergo change due to various processes and on different time scales. The discussion of the Earth's variable rotation is often divided into three parts: 1) variations of the rotation axis with respect to inertial space, the so-called precession and nutation; 2) motions of the rotation axis with respect to the Earth's crust, usually referred to as polar motion (PM); and 3) fluctuations in the rotation rate, which are directly related to variations in the length of day (LOD). While precession and nutation are mostly caused by the gravitational torques exerted by the Sun and the Moon, PM and LOD changes are excited by dynamical processes occurring in the Earth system, such as the redistribution of mass within the surficial layers of the Earth, including the atmosphere, the oceans, and the continental water storage. Hence, these latter quantities are directly affected by variations in the Earth's climate, so that the present discussion focuses on PM and LOD variability, while precession and nutation are not considered here. For more complete descriptions of the Earth's variable rotation, we refer the reader to the classical monographs of Munk and MacDonald (1960) and Lambeck (1980) or to the overview articles of Gross (2007), Seitz and Schuh (2010), and Schindelegger et al. (2013).

Fluctuations in the Earth's rotation are generally studied by applying the principle of conservation of angular momentum to the Earth system. This principle shows that in the absence of external torques, the Earth's angular velocity vector will be affected by changes in the mass distribution within and on the Earth, as these processes lead to variations in the Earth's inertia tensor. Furthermore, the motion of mass relative to the solid body of the Earth induces relative angular momentum and will therefore also influence the Earth's rotation, as the total angular momentum of the system is conserved. Hence, mass redistribution and movement in the Earth system result in fluctuations in the angular velocity vector, including directional changes that lead to PM, as well as changes in magnitude, inducing variations in LOD.

From the 1970s of the twentieth century onward, space geodesy has become an increasingly important tool to explore the Earth, and nowadays the Earth's rotation is permanently monitored to high accuracy by means of techniques such as Very Long Baseline Interferometry (VLBI), the Global Positioning System (GPS), Satellite Laser Ranging (SLR), Lunar Laser Ranging (LLR), and Doppler Orbitography and Radiopositioning Integrated by Satellite (DORIS). Figure 6.9 shows daily values of the Earth's PM from 1962 to 2016, taken from the widely used EOP 08 C04 series, which is derived from a combination of different space geodetic measurement techniques and published by the International Earth Rotation and Reference Systems Service (IERS). The units are milliarcseconds (mas), where one milliarcsecond roughly equals three centimeters on the Earth's surface, and note that by convention, the x-axis is oriented in the direction of the Greenwich meridian, while the y-axis points positively toward 90° western longitude.

The main features in the Earth's PM are a yearly periodic motion that is largely driven by seasonal mass redistribution within the surficial layers of the Earth, as well as an oscillation having a period of about 433 days, which is a free rotational mode of the Earth (e.g., Lambeck (1980); Gross et al. (2003)). Initially predicted by L. Euler in 1765, it was discovered later by the astronomer S. Chandler in 1891 and is therefore known as the Chandler Oscillation (or Chandler Wobble). The superposition of these two periodic motions results in a beat with a period of about 6.3 years and a maximum amplitude of about nine meters on the Earth's surface. We also mention the presence of a secular trend in the direction of ~76°–79° western longitude in the PM series, generally believed to be related to glacial isostatic adjustment.

Figure 6.10 displays daily variations in the LOD from the EOP 08 C04 series given in milliseconds and representing deviations of the effective LOD from the nominal value of 86400 s. Again, we see strong seasonal signals, related to geophysical processes in the Earth's surficial layers, particularly to atmospheric effects (e.g., Gross et al. (2004)). The high-frequency signals are mostly due to solid Earth and ocean tides, where the latter are relatively small. We further note the presence of strong decadal-scale fluctuations that are not related to mass redistribution, but rather the result of interactions between the Earth's core and mantle (e.g., Gross et al. (2005)).

Because of the relation between mass redistribution processes within and on the Earth and fluctuations in the Earth's rotation, measurements of the Earth's variable rotation can be used as an indicator of changes in the Earth's climate, and space geodetic observations of PM and LOD offer a unique possibility to detect and explore large-scale climate-related mass movements in the Earth system. This has been recognized fpr a long time and numerous studies have investigated atmospheric, oceanic, and hydrological effects on the Earth's rotation (e.g., Wahr (1982, 1983); Barnes et al. (1983); Eubanks (1993); Jochmann and Greiner-Mai (1996), Chen et al. (2000), Höpfner (2001), Gross et al. (2003; 2004; 2005); Dobslaw et al. (2010); Chen et al. (2012);

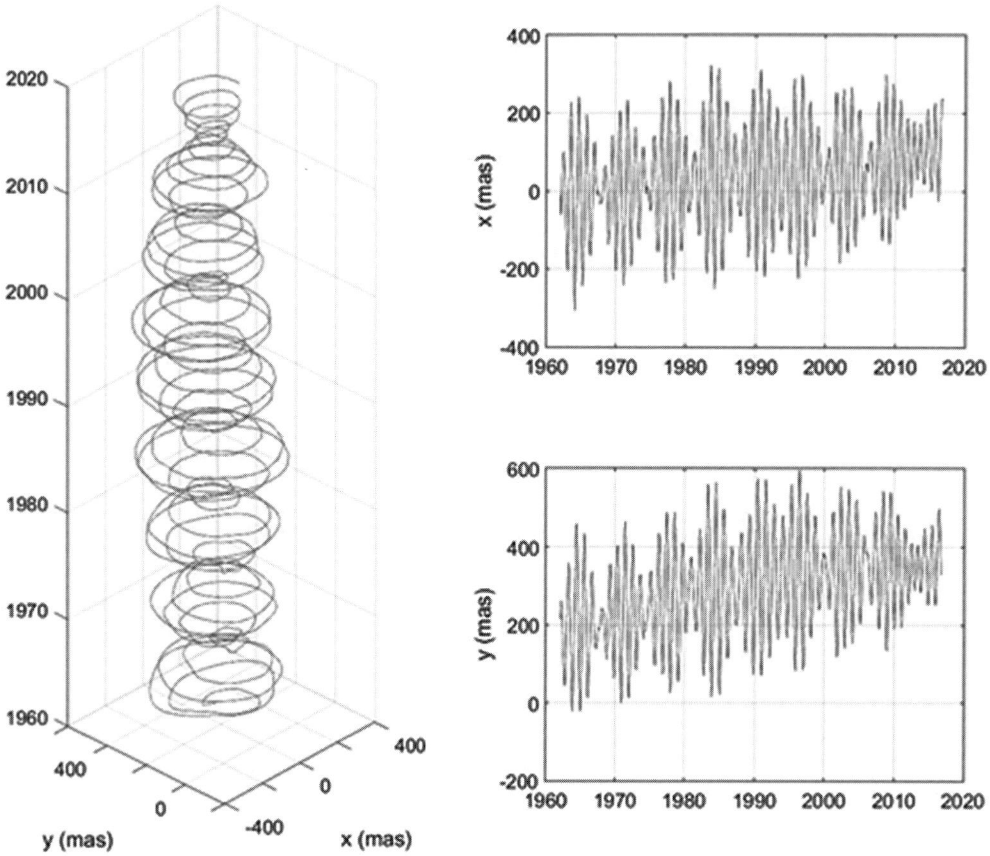

Figure 6.9 The Earth's PM for the period 1962–2016 (left) in three dimensions and (right) separately for the x- and y-components.

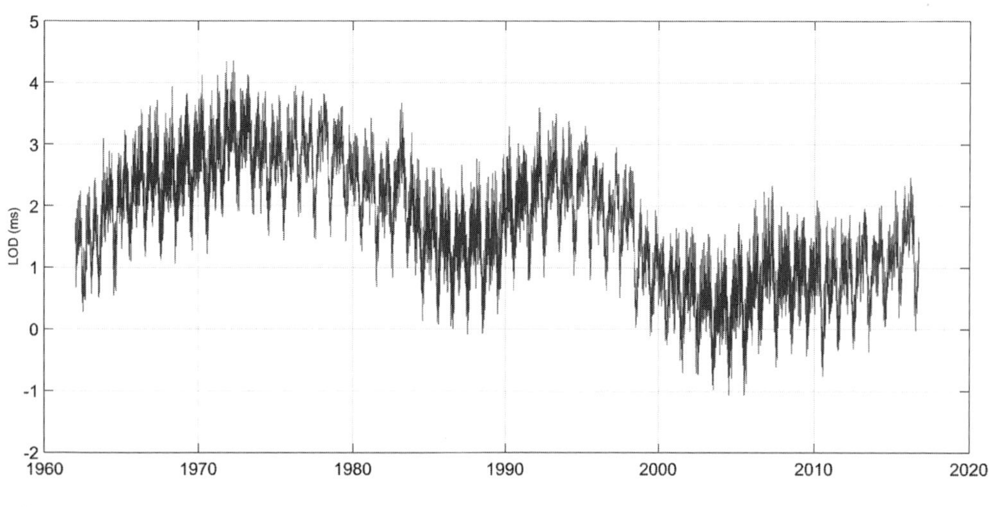

Figure 6.10 Variations in the Earth's LOD for the period 1962–2016.

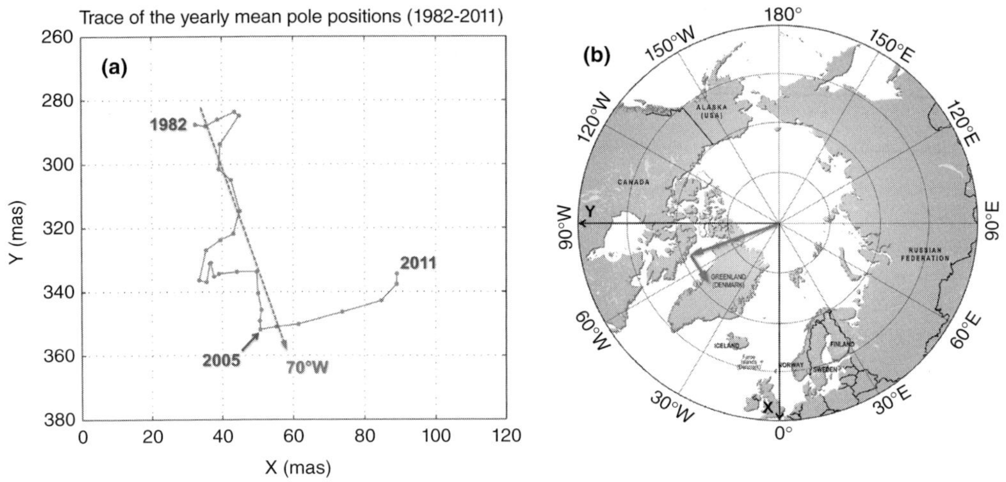

Figure 6.11 (a) Yearly mean pole positions for 1982–2011 in the left-handed (x, y) coordinate system. The arrow indicates the average drift direction of the mean pole over the last century (b) Illustration of the mean pole drift directions on the Earth surface. The long arrow represents the last century long-term drift, and the short arrow represents the direction for the period 2006–2011. Note that annual variations and Chandler Wobble were removed before calculating yearly means (taken from Chen et al. (2013)).

Zotov et al. (2016)). In recent times, the study of the Earth's variable rotation has also been shown to be a valuable tool to monitor the effects of global warming.

For instance, space geodetic observations of PM revealed that the annual average pole position began drifting toward the east around 2005, in contrast to the drift direction that was observed over the past century (see Figure 6.11). Combining these observations with data from the GRACE satellite mission, Chen et al. (2013) could show that the abrupt eastward turn of the mean pole position can largely be attributed to accelerated melting of polar ice sheets and mountain glaciers, as well as to related sea level rise, thus clearly demonstrating that global climate warming affects the Earth's rotation. A similar, but relatively smaller change in the long-term PM rate around 1992 was reported by Roy and Peltier (2011), who also suggested a connection to polar ice sheet melting. Further studies investigating the effects of present-day change in glacier and polar ice mass on long-term and secular PM were carried out earlier by Trupin (1992), Trupin et al. (1993), James and Ivins (1997), and Nakada and Okuno (2003), in which some of these authors additionally considered the impact of ice melting effects on LOD variability. The close link between PM and climate-related mass redistribution was recently confirmed in a study by Adhikari and Ivins (2016), where it was shown that all the main features in observed PM for the period 2003–2015 can be explained by global-scale continent–ocean mass transport. We also mention that correlations between PM and global temperature changes were reported by Huang et al. (2016).

However, the Earth's rotation rate is also believed to be affected by modifications of the global circulation in the atmosphere and oceans due to the anthropogenic increase in greenhouse gas concentrations in the Earth's atmosphere. For instance, using the output of different coupled ocean/atmosphere models, de Viron et al. (2002) concluded that an increase in LOD of the order of one microsecond per year can be expected as a result of changes in the mean zonal wind caused by a rise in atmospheric CO_2 concentrations. The effects of rising CO_2 concentrations on the Earth's rotation rate were further explored by Rosen and Gutowski (1992) and Räisänen (2003), while Abarca del Rio (1999) reported a linear correspondence between global surface temperature increase and LOD variability. All these studies suggest that measurements of variations in LOD may be used as an indirect indicator of global warming.

The relation between mass redistribution and mass movement within the Earth system and fluctuations in the Earth's rotation makes it possible to use space geodetic measurements of PM and LOD to investigate variations in the Earth's climate. This has become increasingly important in times of global warming, and many studies have demonstrated that by precisely monitoring the Earth's rotation, space geodesy provides a valuable tool to study the effects of global climate change.

In this section, we have provided a short overview of the types of changes in Earth rotation that are affected by long-term climate change. Thus, we can use these observations to infer changes in climate.

6.5 Conclusions

In this chapter, we have introduced the three observational pillars of geodetic science, namely, measuring the shape, the gravity field, and the rotation. We have discussed how these observations have been affected by climate change. Thus, we can use these contemporary observations to monitor long-wavelength mass redistributions that arise from climate warming.

There are many other types of geodetic observations and their interpretations that we were not able to present because of space restrictions. New types of Global Navigations Satellite Systems (GNSS) are being developed, which will allow us to discriminate between the technique errors in GNSS and environmental signals that might in fact be correlated. Because of the success of the GRACE satellite mission and the fact that the mission has continued far past the expected lifetime, the satellite gravity community has pushed for the development of a GRACE Follow-On mission to be launched as soon as possible. In addition, new insights into data processing techniques should allow us to extract even more information (higher spatial and temporal resolution) from the GRACE observations in the future. These and many other advances in geodesy will insure that geodetic observations will continue to contribute to monitoring mass redistributions that result from global change.

References

Abarca del Rio, R. (1999), The influence of global warming in Earth rotation speed, *Ann. Geophys.*, 17, p. 806–811, doi:10.1007/s00585-999-0806-x.

Adhikari, S., and E. R. Ivins (2016), Climate-driven polar motion: 2003–2015, *Science Advances* 2(4), doi: 10.1126/sciadv.1501693.

Amos, C.B., R. Burgmann, I. A. Johanson, and G. Blewitt (2014) Uplift and seismicity driven by groundwater depletion in central California, *Nature* 509, p. 483–486, doi:10.1038/nature13275.

Argus, D. F., Y. Fu, and F. W. Landerer (2014), Seasonal variation in total water storage in California inferred from GPS observations of vertical land motion, *Geophys. Res. Lett.*, 41, doi:10.1002/2014GL059570.

Barnes, R. T. H., R. Hide, A. A. White, and C. Wilson (1983), Atmospheric angular momentum fluctuations, length-of-day changes and polar motion, *Proc. R. Soc. Lond., A*, 387, p. 31–73.

Bevis, M., J. Wahr, S. A. Khan, F. B. Madsen, A. Brown, M. Willis, E. Kendrick, P. Knudsen, J. E. Box, T. van Dam, D. J. Caccamise II, B. Johns, T. Nylen, R. Abbot, S. White, J. Miner, R. Forsberg, H. Zhou, J. Wang, T. Wilson, D. Bromwich, and O Francis (2012), Bedrock displacements in Greenland manifest ice mass variations, climate cycles and climate change, *Proc. Natl Acad. Sci. USA*, 109, p. 11944–11948, doi:10.1073/pnas.1204664109.

Bettadpur, S. (2012), *CSR Level-2 Processing Standards Document for Product Release 05, GRACE 327–742, The GRACE Project, Center for Space Research*, University of Texas at Austin.

Blewitt, G., D. Lavallee, P. Clarke, and K. Nurutdinov (2001), A new global mode of Earth deformation: Seasonal cycle detected, *Science*, 294, p. 2342–2345.

Boening, C., M. Lebsock, F. Landerer, and G. Stephens (2012), Snowfall-driven mass change on the East Antarctic ice sheet, *Geophys. Res. Lett.*, 39, doi:10.1029/2012GL053316.

Borsa, A. A., D. C. Agnew, and D. R. Cayan (2014), Ongoing drought-induced uplift in the western United States, *Science*, 345, p. 1587–1590.

Cazenave A., and J. Chen (2010), Time-variable gravity from space and present-day mass redistribution in theEarth system, *Earth and Planetary Science Letters*, 298 (3), p. 263–274, ISSN 0012–821X, doi: 10.1016/j.epsl.2010.07.035.

Chambers, D. P., J. Wahr, and R. S. Nerem (2004), Preliminary observations of global ocean mass variations with GRACE, *Geophys. Res. Lett.*, 31, doi:10.1029/2004GL020461.

Chanard, K., J. P. Avouac, G. Ramillien, and J. Genrich (2014), Modeling deformation induced by seasonal variations of continental water in the Himalaya region: Sensitivity to Earth elastic structure, *J. Geophys. Res. Solid Earth*, 119, doi:10.1002/2013JB010451.

Chao, B. F., (2005), On inversion for mass distribution from global (time-variable) gravity field. *Journal of Geodynamics*, 39, p. 223–230.

Chen, J. L., C. R. Wilson, B. F. Chao, C. K. Shum, and B. D. Tapley (2000), Hydrological and oceanic excitations to polar motion and length-of-day variation, *Geophys. J. Int.*, 141(1), p. 149–156.

Chen, J. L., C. R. Wilson, X. G. Hu, B. D. Taple (2003), Large-scale mass redistribution in the oceans, 1993–2001, *Geophy. Res. Lett.*, 30(20), 2024, doi:10.1029/2003GL018048.

Chen, J. L., C. R. Wilson, and B. D. Tapley (2006), Satellite gravity measurements confirm accelerated melting of Greenland ice sheet, *Science*, 313, p. 1958–1960, doi:10.1126/science.1129007.

Chen, J. L., C. R. Wilson, B. D. Tapley, D. D. Blankenship, and E. R. Ivins (2007), Patagonia icefield melting observed by Gravity Recovery and Climate Experiment (GRACE), *Geophy. Res. Lett.*, 34, doi:10.1029/2007GL031871.

Chen, J. L., C. R. Wilson, D. D. Blankenship, and B. D. Tapley (2009) Accelerated Antarctic ice loss from satellite gravity measurements, *Nature Geoscience* 2, p. 859–862, doi:10.1038/NGEO694.

Chen, J. L., C. R. Wilson, and Y. H. Zhou (2012), Seasonal excitation of polar motion, *J. Geodyn.*, 62, p. 8–15.

Chen, J. L., C. R. Wilson, and B. D. Tapley (2013,) Contribution of ice sheet and mountain glacier melt to recent sea level rise, *Nature Geoscience*, 6, p. 549–552, doi: 10.1038/NGEO1829.

Chen, J. L., C. R. Wilson, J. C. Ries, and B. D. Tapley (2013), Rapid ice melting drives Earth's pole to the east, *Geophys. Res. Lett.*, 40(11), p. 2625–2630.

Chen, J. L., J. Li, Z. Z. Zhang, S. N. Ni (2014), Long-term groundwater variations in northwest India from satellite gravity measurements, *Global and Planetary Change*, 116, p. 130–138, doi: 10.1016/j.gloplacha.2014.02.007.

Chen, J. L., C. R. Wilson, J. Li., and Z. Zhang (2015), Reducing leakage error in GRACE-observed long-term ice mass change: a case study in west Antarctica, *J. Geodesy*, 89, p. 925–940, doi: 10.1007/s00190–015–0824–2.

Cheng, M. and B. D. Tapley, (1999), Seasonal variations in low degree zonal harmonics of the Earth's gravity field from satellite laser ranging observations, *Journal of Geophysical Research*, 104, p. 2667–2682, doi:10.1029/1998JB900036.

Cheng, M. and B. D. Tapley (2004), Variations in the Earth's oblateness during the past 28 years, *J. of Geophys. Res.*, 109(B18), doi:10.1029/2004JB003028.

Cheng, M. K., and J. R. Ries (2012), Monthly estimates of C20 from 5 SLR satellites based on GRACE RL05 models, GRACE Technical Note 07, The GRACE Project, Center for Space Research, University of Texas at Austin (ftp://podaac.jpl.nasa.gov/allData/grace/docs/TN-07_C20_SLR.txt).

Collilieux, X., Z. Altamimi, J. Ray, T. van Dam, and X. Wu (2009), Effect of the satellite laser ranging network distribution on geocenter motion estimation, *J. Geophys. Res.*, 114, doi:10.1029/2008JB005727.

Cox, C. M. and B. F. Chao (2002) Detection of a large-scale mass redistribution in the terrestrial system since 1998, *Science*, 297, p. 831–833.

Dahle, C. et al. (2013), *GRACE Level-2 Processing Standards Document for Level-2 Product Release 0005, Scientific Technical Report STR12/02 – Data*, GFZ, Postdam.

Dickey J. O., S. L. Marcus, Olivier de Viron, and I. Fukumori (2002), Recent Earth oblateness variations: unraveling climate and postglacial rebound effects, *Science* 298, p. 1975–1977.

de Viron, O., V. Dehant, H. Goosse, and M. Crucifix (2002), Effect of global warming on the length-of-day, *Geophys. Res. Lett.*, 29(7), doi: 10.1029/2001GL013672

Dobslaw, H., R. Dill, A. Grötzsch, A. Brzeziński, and M. Thomas (2010), Seasonal polar motion excitation from numerical models of atmosphere, ocean, and continental hydrosphere, *J. Geophys. Res.*, 115.

Eubanks, T. M. (1993), Variations in the orientation of the Earth, in *Contributions of Space Geodesy to Geodynamics-Earth Dynamics*, Vol. 24 Geodyn. Ser., pp. 1–54, eds Smith, D. E. & Turcotte, D. L., AGU, Washington, DC.

Famiglietti, J. S., M. Lo, S. L. Ho, J. Bethune, K. J. Anderson, T. H. Syed, S. C. Swenson, C. R. de Linage, and M. Rodell (2011), Satellites measure recent rates of groundwater depletion in California's Central Valley, *Geophys. Res. Lett.*, 38, doi:10.1029/2010GL046442.

Famiglietti, J. S. and M. Rodell, (2013), Water in the balance, *Science*, 340, p. 1300–1301, doi:10.1126/science.1236460.

Farrell, W. (1972), Deformation of the Earth by surface loads, *Rev. Geophys.*, 10, doi: 10.1029/RG010i003p00761.

Fu, Y., J. T. Freymueller, and T. Jensen (2012), Seasonal hydrological loading in southern Alaska observed by GPS and GRACE, *Geophys. Res. Lett.*, 39, doi:10.1029/2012GL052453.

Fu, Y., and J. T. Freymueller (2012), Seasonal and long-term vertical deformation in the Nepal Himalaya constrained by GPS and GRACE measurements, *J. Geophys. Res.*, 117, doi:10.1029/2011JB008925.

Fu, Y., D. F. Argus, J. T. Freymueller, and M. B. Heflin (2013), Horizontal motion in elastic response to seasonal loading of rain water in the Amazon Basin and monsoon water in Southeast Asia observed by GPS as inferred from GRACE, *Geophys. Res. Lett.*, 40, p. 6048–6053, doi:10.1002/2013GL058093.

Gross, R. S., I. Fukumori, and D. Menemenlis,(2003), Atmospheric and oceanic excitation of the Earth's wobbles during 1980–2000, *J. Geophys. Res*, 108(B8), 2370, doi:10.1029/2002JB002143.

Gross, R. S., I. Fukumori, D. Menemenlis and P. Gegout (2004), Atmospheric and oceanic excitation of length-of-day variations during 1980–2000, *J. Geophys. Res*, 109, doi:10.1029/2003JB002432.

Gross, R. S., I. Fukumori, and D. Menemenlis (2005), Atmospheric and oceanic excitations on decadal-scale Earth orientation variations, *J. Geophys. Res.*, 110, doi:10.1029/2004JB003565.

Gross, R. S. (2007), Earth rotation variations – long period, in *Treatise on Geophysics*, Vol. 3: *Geodesy*, p. 239–294, ed. Herring, T. A., Elsevier.

Gunter, B. C., O. Didova, R. E. M. Riva, S. R. M. Ligtenberg, J. T. M. Lenaerts, M. A. King, M. R. van den Broeke, and T. Urban (2014), Empirical estimation of present-day Antarctic glacial isostatic adjustment and ice mass change, *The Cryosphere*, 8, p. 743–760, doi:10.5194/tc-8-743-2014.

Han, S.-C., C. K. Shum, M. Bevis, C. Ji, and C. Y. Kuo (2006), Crustal dilatation observed by GRACE After the 2004 Sumatra-Andaman Earthquake, *Science*, 313, p. 658–662, doi:10.1126/science.1128661.

Han, S.-C., J. Sauber, and F. Pollitz (2015), Coseismic compression/dilatation and viscoelastic uplift/subsidence following the 2012 Indian Ocean earthquakes quantified from satellite gravity observations, *Geophy. Res. Lett.*, 42, p. 3764–3772, doi:10.1002/2015GL063819.

Han, S.-C., J. Sauber, and F. Pollitz (2016), Postseismic gravity change after the 2006–2007 great earthquake doublet and constraints on the asthenosphere structure in the central Kuril Islands, *Geophy. Res. Lett.*, 43, p.3169–3177, doi:10.1002/2016GL068167.

Höpfner, J. (2001), Atmospheric, oceanic and hydrological contributions to seasonal variations in length of day, *J. Geod.*, 75, 137–150.

Huang, M., L. Zhu, H. Gong, and Y. Shao (2016), Close correlation between global air temperature change and polar motion during 1962–2013, *J. Geophys. Res.*, 121, p. 11,248–11,263, doi:10.1002/2016JD024842.

Ivins, E. R. and T. D. James (2005), Antarctic glacial isostatic adjustment: a new assessment, *Antarctic Science*, 17, p. 537–549, doi:10.1017/S0954102005002968.

James T. S. and E. R. Ivins (1997), Global geodetic signatures of the Antarctic ice sheet, *J. Geophys. Res.*, 102, p. 605–633.

Jekeli C. (1981), *Alternative Methods to Smooth the Earth's Gravity Field*, Department of Geodetic Science and Surveying, Ohio State University, Columbus, OH.

Jochmann H. and H. Greiner-Mai (1996), Climate variations and the earth's rotation, *J. Geodyn.*, 21(2), p. 161–176.

Lambeck, K. (1980), *The Earth's Variable Rotation: Geophysical Causes and Consequences*, Cambridge University Press, New York.

Landerer, F. W., and S. C. Swenson (2012), Accuracy of scaled GRACE terrestrial water storage estimates, *Water Resour. Res.*, 48, doi:10.1029/2011WR011453.

Leuliette, E. W. and L. Miller (2009), Closing the sea level rise budget with altimetry, Argo, and GRACE. *Geophy. Res. Lett.*, 36, p. 4608, doi:10.1029/2008GL036010.

Li, J., J. L. Chen, and Z. Zhang (2014), Seismologic applications of GRACE time-variable gravity measurements, *Earthquake Science*, 27, p. 229–245, doi:10.1007/s11589-014-0072-1.

Luthcke, S. B., J. H. Zwally, W. Abdalati, D. D. Rowlands, R. D. Ray, R. S. Nerem, F. Lemoine, J. J. McCarthy, and D. S. Chinn (2006), Recent Greenland ice mass loss by drainage system from satellite gravity observations, *Science*, 314, p. 1286–1289, doi:10.1126/science.1130776.

Luthcke, S. B., A. A. Arendt, D. D. Rowlands, J. J. McCarthy, and C. F. Larsen (2008), Recent glacier mass changes in the Gulf of Alaska region from GRACE mascon solutions, *Journal of Glaciology*, 54, p. 767–777, doi:10.3189/002214308787779933.

Luthcke, S. B., T. J. Sabaka, B. D. Loomis, A. A. Arendt, J. J. McCarthy, and J. Camp (2013), Antarctica, Greenland and Gulf of Alaska land-ice evolution from an iterated GRACE global mascon solution, *Journal of Glaciology*, 59, p. 613–631, doi:10.3189/2013JoG12J147.

Munk, W. H. and G. J. F. MacDonald (1960), *The rotation of the Earth: a geophysical discussion*, Cambridge University Press, Cambridge.

Nakada M. and J. Okuno (2003), Perturbations of the Earth's rotation and their implications for the present-day mass balance of both polar ice caps, *Geophys. J. Int.* 152, p. 124–138.

Pavlis, N. K., S. A. Holmes, S. C. Kenyon, and J. K. Factor (2012), The development and evaluation of the Earth Gravitational Model 2008 (EGM2008), *Journal of Geophysical Research (Solid Earth)*, 117 (B16), p. 4406, doi:10.1029/2011JB008916.

Peltier, W. R. (2004), Global glacial isostasy and the surface of the Ice-Age Earth: the ICE-5G (VM2) model and GRACE, *Annual Review of Earth and Planetary Sciences*, 32, p. 111–149, doi: 10.1146/annurev.earth.32.082503.144359.

Pearlman, M. R., J. J. Degnan, and J. M. Bosworth (2002), The International Laser Ranging Service, *Advances in Space Research*, 30(2), p. 135–143.

Räisänen J. (2003), CO2-Induced changes in atmospheric angular momentum in CMIP2 experiments, *J. Climate* 16 (1), p. 132–143.

Rodell, M., P. R. Houser, U. Jambor, J. Gottschalck, K. Mitchell, C.-J. Meng, K. Arsenault, B. Cosgrove, J. Radakovich, M. Bosilovich, J. K. Entin, J. P. Walker, D. Lohmann, and D. Toll (2004), The Global Land Data Assimilation System, *Bull. Amer. Meteor. Soc.*, 85(3), p. 381–394.

Rodell, M., I. Velicogna, and J. S. Famiglietti (2009) Satellite-based estimates of groundwater depletion in India, *Nature*, 460, p. 999–1002, doi:10.1038/nature08238.

Rosen, R. D. and W. J. Gutowski Jr., (1992), Response of zonal winds andatmospheric angular momentum to a doubling Of $C0_2$, J. *Climate*, 5(12), p. 1391–1404.

Roy, K., and W. R. Peltier (2011), GRACE era secular trends in Earth rotation parameters: A global scale impact of the global warming process?, *Geophys. Res. Lett.*, 38(10).

Save, H., S. Bettadpur, and B. D. Tapley (2016), High resolution CSR GRACE RL05 mascons, J. Geophys. Res. *Solid Earth*, 121, p. 7547–7569, doi:10.1002/2016JB013007.

Schindelegger, M., S. Böhm, J. Böhm, and H. Schuh (2013), Atmospheric effects on Earth rotation, In *Atmospheric Effects in Space Geodesy*, p. 181–231, Springer Berlin.

Seitz F. and H. Schuh (2010), Earth rotation. In G. Xu, editor, *Science of Geodesy*. I. *Advances and Future Directions*, p. 185–227, Springer Berlin.

Swenson, S., D. Chambers, and J. Wahr (2008), Estimating geocenter variations from a combination of GRACE and ocean model output, *Journal of Geophysical Research (Solid Earth)*, 113, doi:10.1029/2007JB005338.

Tamisiea, M. E., J. X. Mitrovica, and J. L. Davis (2007), GRACE gravity data constrain ancient ice geometries and continental dynamics over Laurentia, *Science*, 316, p. 881–883, doi:10.1126/science.1137157.

Tapley, B. D., S. Bettadpur, M. M. Watkins,and C. Reigber (2004), The Gravity Recovery and Climate Experiment: mission overview and early results, *Geophys. Res. Lett.*, 31 (9), 10.1029/2004GL019920.

Trupin A. S. (1993) Effects of polar ice on the Earth's rotation and gravitational potential, *Geophys. J. Int.* 113, p. 273–283.

Trupin, A. S., M. F. Meier, and J. M. Wahr (1992), Effect of melting glaciers on the Earth's rotation and gravitational field: 1965–1984, *Geophys. J. Int.* 108, p. 1–15.

van Dam, T. and J. Wahr (1987), Displacements of the Earth's surface due to atmospheric loading: effects on gravity and baseline measurements, *J. Geophys. Res.*, 92, doi: 10.1029/JB092iB02p01281.

van Dam, T., P. C. D. Milly, A. B. Shamakin, G. Blewitt, D. Lavalee, and K. M. Larson (2001), Crustal displacements due to continental water loading, *Geophys. Res. Lett.*, 28, p. 651–654.

Velicogna, I. and J. Wahr (2006), Measurements of time-variable gravity show mass loss in Antarctica. *Science*, 311, p. 1754–1756, doi:10.1126/science.1123785.

Velicogna, I. and J. Wahr, (2013), Time-variable gravity observations of ice sheet mass balance: precision and limitations of the GRACE satellite data, *Geophys. Res. Lett.*, 40, p. 3055–3063, doi:10.1002/grl.50527.

Wahr, J. M. (1982), The effects of the atmosphere and oceans on the Earth's wobble. I. Theory, *Geophys. J. R. Astr. Soc.*, 70, p. 349–372.

Wahr, J. M. (1983), The effects of the atmosphere and oceans on the Earth's wobble and on the seasonal variations in the length of day. II. Results, *Geophys. J. R. astr. Soc.*, 74, p. 451–487.

Wahr, J., M. Molenaar, and F. Bryan (1998), Time variability of the Earth's gravity field: Hydrological and oceanic effects and their possible detection using GRACE, *J. Geophys. Res.*, 103, p. 30205–30230, doi:10.1029/98JB02844.

Wahr, J., S. Swenson, V. Zlotnicki, and I. Velicogna (2004) Time-variable gravity from GRACE: First results, *Geophys. Res. Lett.*, 31, L11501, doi:10.1029/2004GL019779.

Wahr, J., S. Swenson, and I. Velicogna (2006), Accuracy of GRACE mass estimates, *Geophys. Res. Lett.*, 33: L6401, doi:10.1029/2005GL025305.

Wouters, B., D. Chambers, and E. J. O. Schrama, (2008), GRACE observes small-scale mass loss in Greenland, *Geophys. Res. Lett.*, 35: L20501, doi:10.1029/2008GL034816.

Wu, X., M. B. Heflin, E. R. Ivins, D. F. Argus, and F. H. Webb (2003), Large-scale global surface mass variations inferred from GPS measurements of load-induced deformation, *Geophys. Res. Lett.*, 30(14), p. 1742, doi:10.1029/2003GL017546.

Yamamoto, K., Y. Fukuda, and K. Doi (2011), Interpretation of GIA and ice-sheet mass trends over Antarctica using GRACE and ICESat data as a constraint to GIA models, *Tectonophysics*, 511, p. 69–78.

Yoder, C. F., J. G. Williams, J. O. Dickey, B. E. Schutz, R. J. Eanes, and B. D. Tapley (1983), Secular variation of earth's gravitational harmonic J2 coefficient from LAGEOS and nontidal acceleration of earth rotation, *Nature*, 303, p. 757–762.

Zotov L, C. Bizouard and C. K. Shum (2016), A possible interrelation between Earth rotation and climatic variability at decadal timescale, *Geodesy and Geodynamics*, 7 (3), p. 216–222.

PART III

Future Earth and the Earth's Fluid Environment

7
Future Earth and the Cryosphere

Ian Allison, Regine Hock, Matt A. King, and Andrew N. Mackintosh

7.1 Ice and Snow on Earth

The term "cryosphere" (deriving from the Greek word *kryos* for frost or icy cold) is used to describe collectively those portions of the Earth's surface where water is in a solid form. It includes sea, lake, and river ice, snow cover, glaciers, ice caps and ice sheets, and frozen ground (including permafrost). While the focus of this chapter is on geodetic issues and Future Earth, which primarily involves ice sheets and glaciers on land and their role in global mass redistribution and sea level, the cryosphere as a whole is an integral part of the global climate system with important links and feedbacks generated through its influence on surface energy and moisture fluxes, clouds, precipitation, hydrology, and atmospheric and oceanic circulation. The basic properties that determine these roles are the high albedo of snow and ice surfaces; the latent heat involved in phase changes of ice/water; the height–mass balance feedback, which results from the great elevation of ice sheets and resultant effects on temperature and precipitation; and the insulating effect of snow cover on land and of floating ice on fresh water or seawater. Other major factors include the water volume stored in ice sheets and glaciers, the greenhouse gases locked up in permafrost, and the delays in annual energy and water cycles due to seasonal snow and ice cover. Through these and associated feedback processes, the cryosphere plays a significant role in global climate. Cryospheric processes therefore need to be included explicitly and correctly in climate models to project future response to global change. Change in the cryosphere is also an important indicator of climate variability and change.

The cryosphere also has a wide range of natural hazards, with associated risks to human populations, infrastructure, and economic activity. Cryospheric hazards are often strongly linked to weather and climate conditions and, in many cases, these hazards are increasing as Earth warms. They range from localised and very short-term events such as snow avalanches; through iceberg and sea ice hazards, surging glaciers, glacier-related outburst floods; to mesoscale phenomena such as permafrost degradation including increasing erosion of permafrost coasts, or diminishing meltwater supply resulting from changes to the hydrological cycle of entire high-mountain drainage basins. The truly global scale hazard of eustatic sea level rise from shrinking glaciers and ice sheets, and its impacts on coastal environments, occurs at time scales of decades to centuries or longer and is discussed further in this chapter. The risks resulting from cryospheric hazards are also increasing as human population densities increase and as a consequence of the expansion of human activities into the cryospheric environment, such as new shipping routes in polar oceans and tourist activities in high mountains.

The surface area and volume (or sea level equivalent for ice on land) of the various cryospheric components on Earth are shown in Table 7.1. Changes in the extent of land-based glaciers and ice sheets influence the distribution of water mass on Earth – between solid ice in polar regions and high mountains, and liquid water distributed over the global ocean. The sea level equivalent (SLE) of a mass of ice is typically calculated as the mass, converted to a volume using a density of 1000 kg m^{-3}, and divided by the ocean surface area of 3.625×10^{14} m^2, although more accurate estimates include, among other processes, shoreline and grounding line migration, isostatic adjustments of land surfaces, and melt water storage in subglacial and proglacial lakes, aquifers and enclosed drainage basins. One millimetre of sea level rise is equivalent to 362.5 gigatonne (Gt) of ice melt. Much of the ice in the Antarctic and Greenland ice sheets is below current sea level, and their SLE is estimated by allowing sea water (with an assumed density of 1028 kg m^{-3}) to fill that volume of ice (with an assumed density of 917 kg m^{-3}) that is below sea level (Figure 7.1).

Table 7.1 *Surface Area and Volume of Components of the Cryosphere*

Ice on Land	Surface Area (10^3 km^2)	% of Global Land Surface	Sea Level Equivalent (metres)
Antarctic ice sheet	12,295	8.3	58.3
Greenland ice sheet	1,801	1.2	7.36
Glaciers including those in the Greenland and Antarctic periphery	706	0.5	0.31-0.53
Terrestrial permafrost	13,200-18,000	9–12	0.02–0.10
Seasonally frozen ground (NH)	48,100	33	not applicable
Seasonal snow cover (NH)	1,920-45,170	1.3–30.6	0.001–0.01
Lake and river ice (NH)	1,620	1.1	not applicable
Total[a]	**77,700-82,500**	**53–56**	**~66.1**
Ice in the Ocean	**Surface area (10^3 km^2)**	**% of Global Ocean Surface**	**Volume (10^3 km^3)**
Antarctic ice shelves	1,617	0.45	~380
Antarctic sea ice, Feb. (Sept.–Oct.)	2,900 (18,900)	0.8 (5.2)	3.4 (11.1)
Arctic sea ice, Sept. (Feb.–Mar.)	6,200 (14,100)	1.7 (3.9)	13.0 (16.5)
Sub-sea permafrost	~2,900	~0.8	not available
Total[b]	**19,200-26,500**	**5.3–7.3**	

Notes: [a]To allow for areas of permafrost and seasonally frozen ground that are also covered by seasonal snow, total area excludes seasonal snow cover.
[b]Summer and winter totals are assessed separately.
The data are derived from a compilation in the IPCC Fifth Assessment Report (Vaughan et al. 2013), and the sources of these estimates are given in that report. For glaciers, the area estimate is based on the latest version of the Randolph Glacier Inventory (RGI Consortium 2017; see Section 7.5).

For glaciers, some ice is currently below sea level (Loriaux and Casassa 2013), but this process is not included in the SLE calculations in Table 7.1. For glaciers, also, melt water could be retained in subglacial or proglacial lakes (Haeberli and Linsbauer 2013), and stored in endorheic (closed) basins (e.g., Neckel et al. 2014), and the actual contribution to sea-level rise will be a few percent less than this.

7.2 Past Climate and Sea Level Change Involving the Cryosphere

Earth's climate has fluctuated naturally over tens of millions of years and ice sheets have waxed and waned with these fluctuations. Large changes in sea level, driven predominantly by melting and freezing of land-based ice sheets in the Northern Hemisphere, have accompanied these climate oscillations. However, contributions from the Antarctic Ice Sheet, particularly during warm periods, have also been important. Widespread glaciation of Antarctica and a shift toward colder temperatures at the Eocene–Oligocene boundary (~34 million years ago) was one of the most fundamental reorganizations of the global climate system evidenced in the geologic record (Barrett 1996, DeConto and Pollard 2003). There were major variations in Antarctic ice sheet volume and extent during the early to mid-Miocene (~20 million years ago) (Gasson et al. 2016).

A record of 5 million years (Pliocene–Pleistocene) of climate change derived from ocean sediment cores is given by Lisiecki and Raymo (2005). West Antarctic ice sheet growth and collapse through this period have been modelled by Pollard and DeConto (2009), and the ANDRILL 1B sediment core provides a proximal record of West and East Antarctic ice sheet changes through much of this period (Naish et al. 2009).

Over the past 800,000 years, Earth's climate has varied with ice-age cycles of around 100,000 years' duration. Sea-level changes of more than 100 m, primarily driven by growth and decay of ice sheets in the Northern Hemisphere, have accompanied these climate oscillations. Prior to 1.3 million years ago, each ice-age cycle lasted about 41,000 years. These cycles are initiated by the Earth's orbital variations, which cause small changes in the amount and timing of radiation from the

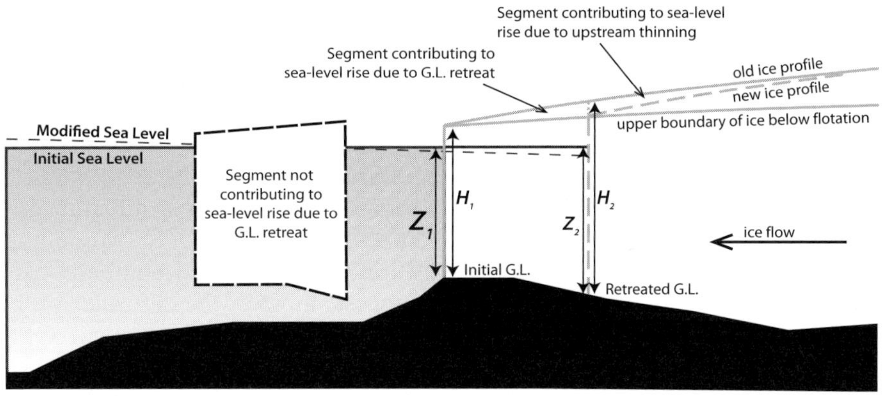

Figure 7.1 Schematic of sea-level change due to ice thinning and retreat of the grounding line (GL) for a marine ice sheet, one where the ice sheet is grounded below sea level. The grounding line is the boundary between the ice sheet resting on a solid bed and the floating ice shelves and glacier tongues. In practice, there is a zone, not a single line, where the ice is not freely floating. Gravimetric measurements are not sensitive to loss of ice mass below flotation, since this is replaced by ocean water of the same mass it displaces. Gravimetry, ice flux, and altimetric measurements of change each have different sensitivities and hence care is needed in their inter-comparison and the specification of what is being measured. Because of gravimetric and solid Earth effects, sea level falls within ~2000 km of a thinning ice sheet whilst the total ocean volume increases.

Sun. This small effect is amplified by a number of still poorly understood natural feedback processes, including the role of greenhouse gasses.

Societally relevant magnitudes, rates, and sources of sea-level rise from polar ice-sheet loss can be inferred from past Pleistocene warm periods (Dutton et al. 2015). During the most recent interglacial, the warm period around 125,000 years ago, before the last ice age, sea level was six to nine metres higher than at present (e.g., Dutton and Lambeck 2012, Kopp et al. 2013). During this interglacial, global air temperatures were around 2 °C higher than at present.

During the peak of the last ice age, about 21,000 years ago, a large amount of water was locked in terrestrial ice sheets, and large continental shelf areas were exposed as sea level fell about 125 m (averaged globally) below its present level (e.g., Bentley 1999, Lambeck et al. 2002, Lambeck et al. 2014, Stanford et al. 2010, Austermann et al. 2013). As global air temperatures increased after the last ice age, sea level rose until around 6,000 years ago, reclaiming continental shelves as ocean area. Over the last 6,000 years, the average rate of sea-level rise has been less than 0.1 m/century (Lambeck et al. 2014).

The global distribution of ice during the Last Glacial Maximum (LGM) and its subsequent melt history remain controversial, particularly relating to the LGM size of the Antarctic Ice Sheet. While local geological and ice sheet modelling suggests an Antarctic contribution of five to ten metres to post-LGM sea level (Mackintosh et al. 2011, Ivins et al. 2013, Gomez et al. 2013), reconstructions considering instead far-field sea level and/or geological constraints of Northern Hemisphere ice sheets indirectly suggest a larger Antarctic contribution of approximately ten to thirty metres (Argus et al. 2014, Clark and Tarasov 2014).

Several rapid increases in sea level occurred during this overall period of ice sheet retreat and sea level rise (Clark et al. 2009). During Meltwater Pulse 1a, the best-known of these events, which occurred between 14,310 and 14,650 calendar years ago, sea level increased at a rate of up to four metres per century (more than ten times the current rate of sea level rise) (Deschamps et al. 2012). The source of this meltwater flux is controversial (Liu et al. 2016). Initial meltwater 'finger-print' modelling based on the relative displacement of sea level at far-field sites suggests that Antarctica was the major source of this sea-level rise (Clark et al. 2002). In contrast, estimates of Antarctic ice sheet volume at the LGM and its subsequent loss based on geologically constrained ice sheet modelling (e.g., Bentley et al. 2010, Mackintosh et al. 2011, Whitehouse et al. 2012, Golledge et al. 2012) suggest that Antarctica contained insufficient excess ice to be responsible for this event. More recently, direct geological evidence that deposition of ice rafted debris occurred during Meltwater Pulse 1a has been discovered (Weber et al. 2014). Ice sheet modelling has recently been used to demonstrate a probable mechanism for rapid ice loss from both the North American and Antarctic Ice Sheets. Meltwater Pulse 1a may have occurred as a consequence of separation of the Laurentide and Cordilleran ice sheets in North America (Gregoire et al. 2012), as well as marine ice sheet instability in the Ross and Weddell Sea marine embayments in Antarctica (Golledge et al. 2014). The partitioning of meltwater release among these ice sheets remains controversial (Liu et al. 2016).

7.3 Ice, Solid Earth and Sea-Level Interactions

Redistribution of water mass on Earth with changes in ice extent alters the gravitational field and hence influences the distribution of sea-level change. In particular, ice mass loss from polar regions means that melt added to the ocean is preferentially redistributed towards equatorial regions. This redistribution of existing ocean water can cause a relative sea-level drawdown in excess of 1.5 mm/year in near-field cryospheric regions, and a corresponding amplification of sea-level rise in far-field equatorial regions (Riva et al. 2010, Slangen and Lenaerts 2016). This redistribution of existing ocean mass from cryospheric to non-cryospheric regions may be equivalent to up to 30% of the glacier and ice sheet contribution to sea-level rise. As a consequence, melting of the Greenland Ice Sheet results in a lowering of sea levels (in the absence of other changes) in northern Europe, and higher than average rise for much of the Pacific and south Atlantic. Melting West Antarctic ice results in higher than average rise for the north Pacific, north Atlantic and Indian oceans (Tamisiea and Mitrovica 2011). As a result of this effect, large-scale melting of far-field ice sheets provides a stronger mechanism for destabilising ice sheets in the opposite hemisphere than would otherwise occur (Konrad et al. 2015). By contrast, near the ice sheets that are losing mass, this effect can act to increase ice sheet stability (Figure 7.1; Gomez et al. 2010). Importantly, most modern ice sheet models do not yet include a full physical implementation of this effect (see Gomez et al. 2013, Konrad et al. 2015 for exceptions).

Cryospheric regions losing substantial ice mass experience elastic and viscoelastic rebound of the solid Earth as the crust is partly unloaded of its overburden pressure. The solid Earth's viscoelastic response to ice-ocean load changes is known as glacial isostatic adjustment (GIA), a process that generally persists for millennia after the load changes (Peltier et al. 2004), unlike the instantaneous elastic deformation. In terms of spatial characteristics of deformation, elastic deformation is concentrated more closely in relation to the load change compared with GIA. Surface deformation measurements are unable to separate purely elastic signal from GIA without the aid of separate models or observations.

Around much of the Greenland periphery, bedrock uplift rates, derived from GPS data, are in excess of ten millimetres per year over the 2007 to 2015 period, although subsidence is evident in the southwest because of Holocene readvance of the ice sheet (Bevis et al. 2012, Kahn 2016). Modelling of elastic deformation using observed contemporary ice mass changes suggests most of the deformation in this case is due to elastic effects. In Antarctica, GPS measurements of bedrock deformation have been used to identify substantial over-prediction of GIA and a requirement to improve GIA models (Bevis et al. 2009, Thomas et al. 2011). The spatial and temporal patterns of GPS uplift rates have also been used to show that the mantle viscosity below several glaciated regions, notably the Antarctic Peninsula, eastern Greenland, Patagonia, Alaska and Iceland, is much lower than considered in global GIA models (e.g., Nield et al. 2014, Khan et al. 2016, Richter et al. 2016). Upper mantle viscosities that are two to three orders of magnitude lower than global averages have been identified, meaning mantle relaxation times are measured in years rather than millennia in these locations. In such locations, present-day GIA is therefore largely a consequence of very recent loading changes as responses to millennial-scale load changes decayed long ago. There is a growing body of seismic and geodetic evidence that upper mantle viscosities may also be low over much of West Antarctica and eastern Greenland (Heeszel et al. 2016, Khan et al. 2016).

Seasonal and inter-annual signals are also evident in bedrock deformation and gravity field measurements, almost entirely due to purely elastic deformation (Bevis et al. 2012, Fu et al. 2011). Measurements of horizontal deformations are not yet as widely used as the vertical component of deformation, although they provide different information that can help validate GIA models or give insight into the relative location of ice mass change, provided contaminating effects such as plate tectonic signal can be removed (Wahr et al. 2013, King et al. 2016).

Accurate knowledge of GIA is required as it acts as an error in measurements of ice sheet volume, derived from altimetry, and ice mass change, derived from satellite gravimetry (see Section 7.4). This is particularly the case for the Antarctic Ice Sheet where uncertainties in ice sheet melt history propagate into possible systematic error in GIA models (King et al. 2012). GIA model error is compounded in some places by the application of an overly simple model of Earth rheology (including viscosity), where later variation in rheology is not considered. To address this, new 3D GIA models are being

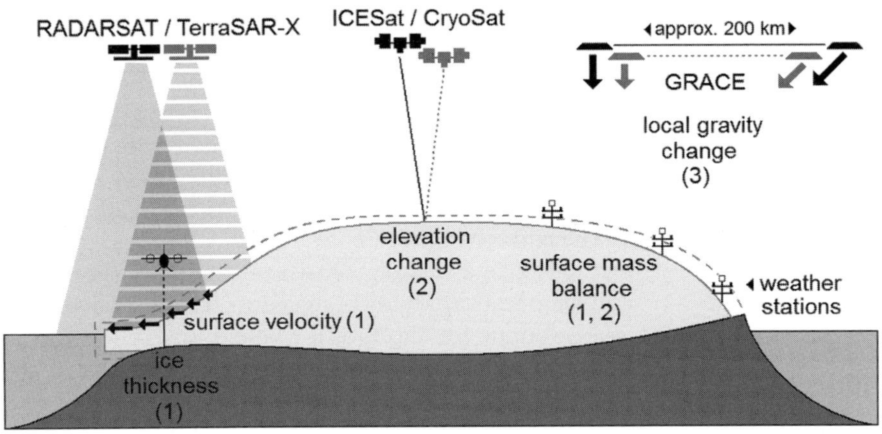

Figure 7.2 Methods of assessing an ice sheet mass balance: (1) the input–output method, (2) the volumetric method, and (3) the gravimetric method. Numbers in parentheses indicate data requirements of each method; the named satellites are those often used in ice sheet mass balance assessments. Surface mass balance is typically obtained from regional climate models that may be verified against in situ point measurements (adapted by permission from Elsevier (Allison et al. 2015), copyright 2015).

developed, although challenges remain in constraining their greater number of model parameters (van der Wal et al. 2015). The challenges involved in forward modelling GIA have resulted in various inverse solutions being computed; however, all forward and inverse solutions show important differences from each other and from independent GPS uplift estimates (Martín-Español et al. 2016a).

Long-term observation of polar motion, the oblateness of Earth's shape and the length of day provide constraints on both past, large-scale ice mass changes and lower mantle rheology (Sabadini et al. 1982). Geodetic measurements of the motion of Earth's rotation axis are only available since the beginning of the twentieth century, with a significant increase in precision through the advent of space geodetic techniques since the 1970s. These have allowed the identification of short-term changes in the direction of drift of Earth's rotational axis since the early 2000s (Chen et al. 2013, Adhikari and Ivins 2016), which have been largely attributed to changes in the rates of ice sheet mass loss.

7.4 Current Changes to Ice Sheets

Determining changes of the mass of ice in the Greenland and Antarctic ice sheets, and hence their contribution to sea level change, has long been a major challenge of polar and cryospheric science. But until relatively recent advances in measurement technology, the uncertainty in ice sheet mass balance estimates was greater than any net contribution to sea-level change. These advances, which have occurred over the last two decades or so, are largely based on remote sensing observations and numerical modelling. There are three basically independent measurement techniques, illustrated schematically in Figure 7.2.

The first technique, known as the mass budget or input–output method (Figure 7.2), estimates the difference between the net surface mass balance over the ice sheet (snow accumulation less melt and run-off, sublimation and wind transport of drift snow) and the ice discharge into the ocean at the ice sheet grounding zone, the region where the ice enters the ocean and contributes to sea-level rise, even though it is floating and may still be attached to the ice sheet. The net surface mass balance is usually derived from regional climate models (e.g., Lenaerts et al. 2012). Ice discharge is calculated using ice thickness data and the depth-averaged ice velocity. Surface velocity can be directly observed by satellite synthetic aperture radar (e.g., Rignot et al. 2011a) or by optical methods (e.g., Howat et al. 2011, Fahnestock et al. 2016) but needs to be corrected for the change in velocity with depth in grounded ice to estimate a depth-averaged velocity. Ice thickness can be obtained from airborne radar measurements along the grounding zone (e.g., Fretwell et al. 2013) or from satellite altimeter height measurements where the ice starts to float and is assumed to be in hydrostatic equilibrium. The input–output method can be applied to the ice sheet as a whole, or to individual drainage basins (Rignot et al. 2011b, subsequently revised and updated by Shepherd et al. 2012 and 2018).

The second technique, the volumetric or geodetic method, determines change in the ice sheet height from repeated airborne or space-borne measurements from a radar or laser altimeter (Figure 7.2). These height changes need to be converted to mass changes using an assumed density, which takes into account factors including surface mass balance and mean surface temperature, which affect the near-surface firn density (e.g., Sørensen et al. 2011).

The third approach, the gravimetry method (Figure 7.2), uses space-based observations of variations in the Earth's gravity field to assess changes in regional ice mass

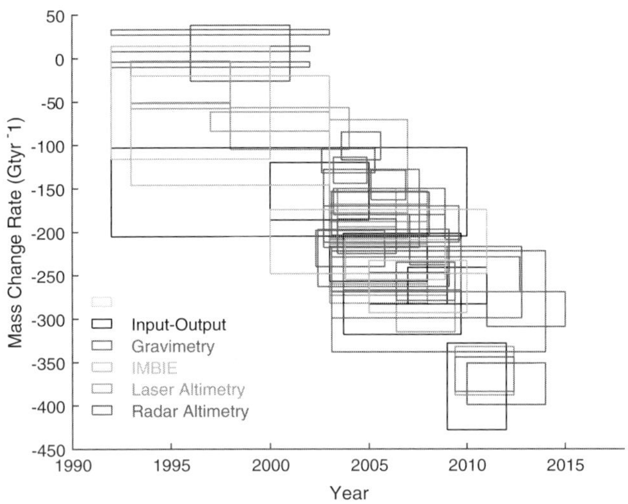

Figure 7.3 Summary of mass change estimates for the Greenland Ice Sheet. The individual studies compiled are given in Appendix 1. The coloured rectangles represent the time over which the measurements apply and the estimated range, given as (mean + uncertainty) and (mean – uncertainty) as reported in the original papers (adapted from Khan et al. 2015 © IOP Publishing. Reproduced with permission. All rights reserved). For the colour version, please refer to the plate section.

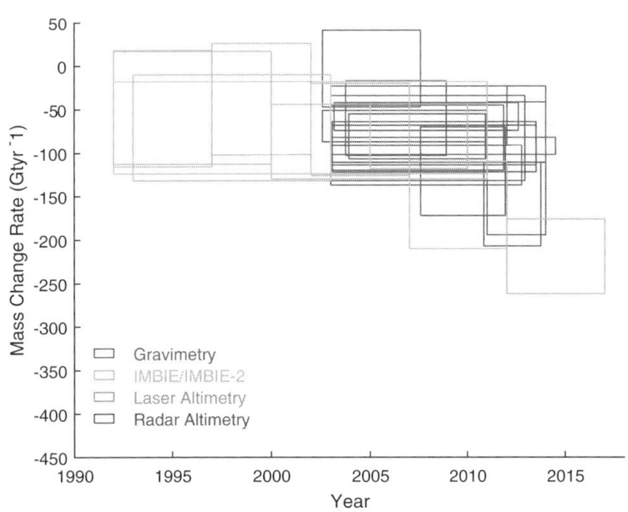

Figure 7.4 Summary of mass change estimates for the Antarctic Ice Sheet. The individual studies compiled are given in Appendix 1. The coloured rectangles represent the time over which the measurements apply and the estimated range, given as (mean + uncertainty) and (mean – uncertainty) as reported in the original papers (adapted from fig. 1, Scambos and Shuman, 2016 (copyright Cambridge University Press, reproduced with permission)). For the colour version, please refer to the plate section.

(e.g., Velicogna 2009, Chen et al. 2011, Schrama and Wouters 2011). This does not require knowledge of surface mass balance or firn density change, but it is sensitive to mass change signals from land and ocean hydrology, atmospheric mass exchange and solid Earth processes, particularly GIA (Section 7.3). Satellite gravity is limited to quite low spatial resolution (in the order of 300 km).

The mass change estimates produced by these different methods give a broadly consistent picture of the spatial and temporal variability of mass loss from the Greenland and Antarctic Ice Sheets as a whole. Large differences in the absolute values from the different methods in the earliest studies have now been substantially reduced with improved analysis and longer time series (Shepherd et al. 2012 and 2018, Hanna et al. 2013) as shown in Figures 7.3 and 7.4 for Greenland and Antarctica, respectively. There remain, however, systematic errors in the techniques, especially for Antarctica, where the distribution of GIA is still not well known and surface mass balance models require further development, which do not enable the uncertainty to be reduced below about forty gigatonne per year (or 0.11 mm SLE/year, 1-sigma).

The Ice-sheet Mass Balance Intercomparison Exercise (IMBIE) (Shepherd et al. 2012) applied these three mass balance methods to common geographical regions and time intervals for both ice sheets. Although the uncertainty associated with any single mass balance method was sometimes large, the overall certainty was considerably improved using a combination of methods. Over the period 1992 to 2011, IMBIE estimated a Greenland mass loss of 142 ± 49 Gt/year (0.39 ± 0.14 mm SLE/year) and an average Antarctic mass loss of 71 ± 53 Gt/year (0.20 ± 0.15 mm SLE/year) (Shepherd et al. 2012). IMBIE also derived estimates of mass change for both Greenland and Antarctica over shorter epochs, and these are shown in Figures 7.3 and 7.4. The IPCC Fifth Assessment Report (AR5) estimated both the Greenland and Antarctic mass balances from an average of studies, obtaining similar results to IMBIE for the same period. For Greenland, eighteen studies were used, giving an average loss of 125 ± 25 Gt/year (0.34 ± 0.07 mm SLE/year); and for Antarctica ten recent studies gave an average loss of 88 ± 35 Gt/year (0.24 ± 0.10 mm SLE/year) (Vaughan et al. 2013).

IMBIE-2 extended the Antarctic record to the period 1992-2017, using twice as many studies as the original intercomparison. This showed that the loss from Antarctica has increased to an average of 109 ± 56 Gt/year (0.30 ± 0.16 mm SLE/year) (Shepherd et al. 2018). The increased rate of loss has occurred from the Antarctic Peninsula, after 2000, and especially from the West Antarctic Ice Sheet where it trebled between the first and final five years of the 25-year study (from 53 ± 29 Gt yr^{-1} to 159 ± 26 Gt yr^{-1}). In both these regions the increased losses result from reductions in the thickness and extent of floating ice shelves due to ocean melt, which has led to acceleration and drawdown of marine-terminating glaciers.

The rate of mass loss from the Greenland Ice Sheet has also accelerated since 1992. Vaughan et al. (2013) reported that the average loss from 1992–2001 of 34 Gt/year [-6 to 74 Gt/year range at the 90% confidence interval], or 0.09 mm SLE/year, had increased to a loss of 215 [157 to 274] Gt/year (0.59 mm SLE/year) for 2002 to 2011. For 2014–2016 the rate of Greenland loss was ~190 Gt/year, somewhat less than the 2002–2016 mean of 269 Gt/year as estimated using GRACE gravimetry data (Velicogna et al. 2014, Tedesco et al. 2016, Harig and Simons 2016, Forsberg et al. 2016). But because of inter-annual variability, short-term rates may be less than the long-term average and it seems clear that Greenland ice mass loss is continuing.

The spatial pattern of change in Greenland also continues to evolve with some areas of past glacier acceleration now having slowed (Enderlin et al. 2014), and new areas of accelerated ice discharge (so-called dynamic ice mass loss) occurring, particularly in the northeast (Khan et al. 2014). Such strong temporal and spatial variability of change makes inter-comparison of different estimates particularly challenging. However, in part because of the increased coverage of measurements of ice sheet thickness, regional estimates of mass change from the three different techniques continue to converge and are in agreement in most areas within, often high, uncertainties (Xu et al. 2016). An extensive low-viscosity zone, as recently identified under eastern Greenland (Khan et al. 2016), could significantly change modelled GIA signal and hence GRACE gravimetry estimates of mass change in these regions.

These decadal-scale changes in the Greenland Ice Sheet have now been placed in the context of the entire twentieth century through recovery of digital elevation models from historic aerial photogrammetry. Using this approach, the net contribution of the ice sheet has been estimated over different periods and regions (Kjeldsen et al. 2015). They estimate Greenland-wide ice mass loss of 75.1 ± 29.4 Gt/year (1900–1983; 0.21 ± 0.08 mm SLE/year), 73.8 ± 40.5 Gt/year (1983–2003; 0.20 ± 0.11 mm SLE/year), and 186.4 ± 18.9 Gt/year (2003–2010; 0.51 ± 0.05 mm SLE/year). Over the twentieth century this equates to 25.0 ± 9.4 millimetres of SLE.

There have been significant recent advances in Antarctic GIA models, ice thickness measurements and regional climate models. These have been used in the most recent estimates of Antarctic mass change (e.g., Scambos and Shuman 2016, Shepherd et al. 2018) and, while earlier estimates such as AR5 remain in general agreement with these, Figure 7.4 shows only the more recent estimates for Antarctica. However, uncertainties are still large and disagreement remains as to the sign of the change in East Antarctica. A notable exception to estimates of overall Antarctic mass loss is the study of Zwally et al. (2015) based on radar and laser altimetry, which infers growth in East Antarctica exceeding losses in West Antarctica over the period 1992–2008. This result has been particularly controversial (Richter et al. 2016, Scambos and Shuman, 2015, Zwally et al. 2016a, b) and relies heavily on the calibration adopted for the laser altimeter.

Substantial differences between estimates remain when comparing individual drainage basins, or in the relative contribution of East and West Antarctica to the total, although important agreement is seen in some basins (e.g., Sutterley et al. 2014). There is also still ambiguity as to whether East Antarctica as a whole has been close to balance or in substantial positive mass balance since the 1990s (Zwally et al. 2015, King et al. 2012, Shepherd et al. 2018, Ivins et al. 2013). There is large inter-annual variability in total Antarctic snow accumulation and its distribution (e.g., Boening et al. 2012), making it difficult to infer any statistically significant trend in the overall Antarctic ice sheet mass balance over shorter time periods (e.g., Velicogna 2009, Wouters et al. 2013, Williams et al. 2014). As with Greenland, spatial and regional patterns of change are evident. Since 2010, substantial dynamic mass loss has emerged in the southern Antarctic Peninsula (Wouters et al. 2015, McMillan et al. 2014, Helm et al. 2014). The CryoSat-2 radar altimeter record has allowed much improved radar altimeter sampling in steep coastal regions when compared to earlier radar altimeters.

In a theoretical advance over the simple IMBIE and AR5 solution averaging, Martín-Español et al. (2016b) produced a statistical combination of various mass change datasets and model outputs, estimating ice-mass change and GIA. They found that over 2003–2012 Antarctica lost an average of 84 ± 22 Gt/year (0.24 ± 0.06 mm SLE/year), dominated by dynamic mass loss. East Antarctica was found to have increased in mass by 56 ± 18 Gt/year (-0.15 ± 0.05 mm SLE/year) while West Antarctica (-112 ± 10 Gt/year; 0.30 ± 0.03 mm SLE/year) and the Antarctic Peninsula (-28 ±7 Gt/year; 0.07 ± 0.02 mm SLE/year) lost mass.

Net mass loss from both Antarctica and Greenland is mostly from the lower-elevation ice sheet periphery, and particularly from large marine-terminating glaciers. In contrast, the higher-elevation ice sheet interiors are in steady state or slightly gaining mass. Between 2000 and 2008, about 50% of the ice loss from Greenland was from melt water runoff, and about 50% was due to dynamic mass losses (van den Broeke et al. 2009). In contrast, Antarctica experiences very little surface melt and nearly all ice is lost from Antarctica either by iceberg calving from the ice shelves and floating ice tongues that surround much of the coastline or by melt from the base of these (Depoorter et al. 2013, Rignot et al. 2013).

Most of the present Antarctic contribution to sea level rise results from the Amundsen Sea sector of West Antarctica, and there is evidence that this loss is increasing because of outlet glacier acceleration and increased iceberg calving and submarine melting. Enhanced submarine melting at ice–ocean interfaces has been shown to contribute to Antarctic ice loss (Holland et al. 2008, Pritchard et al. 2012). Rignot et al. (2014) used ERS1 and ERS2 satellite radar interferometry to show that between 1992 and 2011 the major glaciers in this region (Pine Island, Thwaites, Smith and Kohler) all underwent significant grounding line retreat of between ten kilometres and twenty-five kilometres. They noted that there were no major bed obstacles upstream of the 2011 grounding lines that would prevent the glaciers from further retreat and drawdown of the inland ice basin via the marine ice sheet instability mechanism (see Section 7.6). Using a numerical model, Joughin et al. (2014) showed that the Thwaites Glacier retreat rate was compatible with ocean melt of the floating tongue, and that the glacier may have already started the early stage of unstable collapse.

While most Antarctic loss is occurring from the West Antarctic Ice Sheet and the Antarctic Peninsula, new studies also raise concerns about what is happening in parts of East Antarctica, and particularly Totten Glacier. The Totten Glacier is also thinning significantly and these changes have been linked to coastal polynya variability (Khazendar et al. 2013). Large parts of East Antarctica connected to the Totten system, holding the equivalent of 3.5 m of SLE, may also be subject to a potential "marine ice sheet instability" (Greenbaum et al. 2015) with evidence of bi-stable conditions in the past (Aitken et al. 2016). Rintoul et al. (2016) observed that warm ocean water was entering the cavity beneath the Totten Glacier through a newly discovered deep channel, and that the heat transported into the cavity was sufficient to cause the large basal melt rates that have been observed.

7.5 Current Changes to Glaciers

Glaciers and ice caps cover an area of 706,000 km^2 and include approximately 215,000 individual glaciers according to the Randolph Glacier Inventory (RGI v6.0, RGI Consortium 2017), the first globally complete glacier inventory. This includes all glaciers other than ice sheets in Greenland and Antarctica but includes those in their periphery. Their total volume, derived from simple scaling relationships and ice-dynamical considerations, expressed as sea-level equivalent (SLE), is likely to range between 0.31 and 0.53 m (Huss and Farinotti 2012, Marzeion et al. 2012, Grinsted 2013, Radic et al. 2014).

Observations have revealed world-wide glacier thinning and retreat that is coincident with atmospheric warming, and that has direct socio-environmental consequences. Although only making up less than 1% of the Earth's global ice volume, these glaciers are major current contributors to sea-level change. Meier (1984) was the first to recognize their potentially significant sea-level impact, postulating that glaciers may have accounted for a third to half of the observed sea level rise during the eighty years prior to 1975. Glaciers also modulate streamflow, affecting the seasonality, magnitude, physical and biogeochemical properties and sediment content of glacier-fed rivers, with potential impacts for ocean circulation and coastal marine ecosystems. In addition, glacier decline can lead to natural hazards, such as outburst floods from proglacial lakes, or impact the recreational value of glacierized environments and thus tourism.

Glacier front observations indicate massive retreats of glaciers since the end of the Little Ice Age, which ended roughly between the middle of the nineteenth century and the beginning of the twentieth century depending on the region (Zemp et al. 2015). Direct measurements of mass balance, the change in the mass of a glacier or part of the glacier, can be made by measuring the difference between water-mass input to the glacier (accumulation) and mass loss (ablation) over a stated period. The first such long-term direct mass balance measurements are from Claridenfirn in Switzerland (Gabbi et al. 2014), where winter and summer balances at two locations have been measured uninterruptedly since 1914. However, systematic assessments of glacier-wide balances did not start until 1946/47, when detailed in situ measurements began on Storglaciaeren in Sweden (Holmlund et al. 2005), and soon thereafter on other glaciers, further spurred by the International Geophysical Year (IGY) in 1957/58. Nevertheless annual balances from only approximately one hundred glaciers on average since 1985 have been reported to the World Glacier Monitoring Service (WGMS; Zemp et al. 2009). Only fewer than forty glaciers have been measured continuously for more than forty years.

Another technique for determining changes in glacier mass balance is the so-called geodetic method, which involves repeated mapping of glacier surface elevations to estimate the volume change. The conversion of volume change to mass balance requires information on the density of the mass lost or gained. Mass balances derived from geodetic methods are available for roughly 450 individual glaciers (Cogley 2009, Zemp et al. 2015). Zemp et al. (2015) provide the so-far-most-comprehensive analysis of the main glaciological datasets compiled by the World Glacier Monitoring Service (WGMS), including observations of direct and geodetic mass balances and glacier lengths. Results reveal a clear picture of globally consistent centennial glacier retreat and mass loss with strong acceleration during the last three decades.

It is obvious that extrapolating these relatively few direct or geodetic measurements to all glaciers around the world to estimate their contribution to sea level is challenging. Measurements are not only scarce, but also geographically biased towards smaller and land-terminating glaciers. The global glacier estimates in the Fourth Assessment Report (AR4) of the International Panel on Climate Change (Lemke et al. 2007) were based on the consensus estimate by Kaser et al. (2006), who merged three independent assessments, which were obtained from this statistically small sample of observations by arithmetic (Cogley 2005) or area-weighted averaging (Ohmura 2004, Dyurgerov and Meier 2005) and spatial extrapolation (Cogley 2005). Results indicated that the global mass budget was close to balance around 1970, but net mass loss strongly accelerated thereafter in response to post-1970 global warming. For the period 1991–2004 an average mass loss rate of 0.77 ± 0.22 mm SLE/year was estimated for all glaciers, including those in the periphery of Greenland and Antarctica, which is more than twice the estimated rate for 1961–1990. Including a larger set of glaciologically and geodetically derived balances, Cogley (2009) calculated a considerably increased estimate of 1.1–1.4 mm SLE/year for the period 2001–2005.

In the last decade, new remote sensing technologies and missions, such as NASA's Ice, Cloud, and Land Elevation Satellite (IceSAT) and the Gravity Recovery and Climate Experiment (GRACE), have revolutionized our ability to assess mass changes, especially of large and remote glaciers in polar regions where direct observations are logistically challenging and therefore sparse. Gardner et al. (2013) took advantage of these new data, and computed the global glacier mass balance (including the glaciers in the periphery of Greenland and Antarctica) for the period October 2003 to October 2009, combining existing regional-scale and new estimates based on different methods. If several estimates from different methods were available for a region, the estimate derived from the method that was deemed most reliable was chosen. Extrapolation of glaciological observations was performed in smaller glacier regions, where observations are abundant and GRACE estimates have large uncertainties. Estimates derived from GRACE data and IceSAT data were used in remote regions with large ice cover but few direct observations. Their analysis yields a consensus estimate of 0.77 ± 0.08 mm SLE/year (-0.35 m w.e./year), accounting for 29 ± 13% of the observed sea level rise. Thus Earth's mountain glaciers and ice caps contributed almost as much as both ice sheets together during this period (0.79 ± 0.13 mm SLE/year; Shepherd et al. 2012). More than 80% of the ice lost was from glaciers in Alaska, the Canadian Arctic, the periphery of the Greenland Ice Sheet, the Southern Andes and the high mountains of Asia. Extending the global glacier estimates to cover the period 2002–2014, Reager et al. (2016) found a slightly smaller glacier contribution (0.65 mm SLE/year), while the contribution from both ice sheets has risen considerably (1.26 ± 0.08 mm SLE/year), but their glacier estimate excludes the contribution from glaciers in the Antarctic and Greenland periphery, and therefore is not directly comparable.

The analysis of Gardner et al. (2013) suggests that global estimates, based exclusively on extrapolation of the glaciological measurements, overestimate global mass loss substantially, most likely because of insufficient data in the heavily glacierized polar regions and a possible bias of the measured glaciers towards too-negative values. This may also explain why their sea-level contribution estimate for 2003–2009 is only roughly half of the recent estimate by Zemp et al. (2015) for the slightly longer period 2000–2010. The estimate of Zemp et al. is based on simple averaging of the WGMS geographically biased direct observations, and thus should be treated with caution. Despite these discrepancies with respect to absolute amounts, there is general consensus that glacier mass loss is globally coherent and has accelerated during the last few decades in response to global warming.

The IPCC Fifth Assessment Report (Vaughan et al. 2013) notes that almost all glaciers worldwide had continued to shrink since AR4. The AR5 estimate of total mass loss from all glaciers in the world, excluding those on the periphery of the ice sheets, was 0.62 ± 0.37 mm SLE/year) for the period 1971–2009. This included increasing rates of mass loss of 0.76 ± 0.37 mm SLE/year for the period 1993–2009 and 0.83 ± 0.37 mm SLE/year between 2005 and 2009. Vaughan et al. (2013) also noted that glacier extents are out of balance with the current climate, and that glaciers will continue to shrink in the future even without further temperature increase.

Marzeion et al. (2015) demonstrate that negative glacier mass balance since 1991 is predominantly due to anthropogenic climate warming. The anthropogenic fraction of global glacier loss was estimated to have increased from 25 ± 35% during the period 1851–2010 to 69 ± 24% during 1991–2010. Roe at al. (2017) show that the global pattern of glacier retreat is also virtually certain to be a consequence of climate changes. Departures from this global trend of negative mass balance and glacier retreat occurred in some regions, but were generally short-lived or locally restricted (e.g., Kapnick et al. 2014, Mackintosh et al. 2017).

New satellite missions providing observational capabilities with unprecedented spatial and temporal resolution will allow us to reduce the uncertainties in assessing current global glacier changes. Combining different methods, and thus exploiting the different strengths of each individual method in different regions, will be the key to accurate global-scale glacier mass-change assessments.

7.6 Projected Future Changes to the Cryospheric Contribution to Sea Level Rise

Anthropogenic warming of Future Earth is projected to increase global mean surface temperatures by 0.3 to 4.8 °C above pre-industrial values by the end of the twenty-first century and higher in centuries beyond that. More than 90% of the net energy increase in the Earth system since 1971 is stored in the ocean, mostly in the upper hundreds of metres (e.g., Domingues et al. 2008). Projections of future increases in ocean temperature from climate models are used to calculate the contribution to sea level rise, and its global distribution, from future thermal expansion of the ocean (Cazenave and Palanisamy, Chapter 11, this volume). Ocean warming will also result in enhanced melt of ice shelves and floating glacier tongues that fringe the Antarctic and Greenland Ice Sheets.

The contribution that large ice sheets will make to sea-level rise under such warming scenarios is difficult to quantify. The IPCC AR5 projected the Greenland contribution to sea-level rise by 2100 of 0.07 [0.02 to 0.13] m for RCP2.6 (a "Representative Concentration Pathway", defined by AR5 as giving a total radiative forcing from anthropogenic greenhouse gases of 2.6 W/m^2 by 2100) and 0.12 [0.05 to 0.23] m for RCP8.5. Antarctica, however, will contribute a relative sea-level reduction by 2100 of 0.00–0.02 m for RCP2.6, and a relative reduction of 0.01–0.07 m for RCP8.5 (Church et al. 2013). But these projections used extrapolation within a statistical framework and informed judgement, not deterministic modelling, to estimate the future dynamical response of marine-terminating glaciers and marine-based sectors of the Antarctic Ice Sheet.

In the *SeaRISE* project (Sea-Level Response to Ice Sheet Evolution; Bindschadler et al. 2013) ten ice sheet models of varying complexity and resolution, but with the same boundary conditions and driven by uniform climate forcing, were compared. Although there was a large range in projected sea-level contributions, Greenland was more sensitive to changes in atmospheric temperature and precipitation, while Antarctica was most sensitive to increased ice-shelf basal melting. For RCP8.5, the average sea-level contributions over the next century were 0.223 m (with a 0.62-m range) from Greenland, and 0.081 m (0.14-m range) from Antarctica. At 500 years in the future the mean contribution to sea level rise from Greenland was up to 0.37 m, and from Antarctica it was up to 3.35 m.

The dynamic response of a marine-terminating ice sheet is critical if a "marine ice sheet instability" (MISI) is triggered. First proposed more than forty years ago (e.g., Hughes 1973, Mercer 1978, Thomas 1979), MISI applies to a marine ice sheet where the bedrock slopes downward into the interior (a "reverse slope"). Ice shelves and their feeder glaciers are tightly coupled, with ice shelves "buttressing" the seaward flow of the grounded ice. The demise of ice shelves could lead to increased ice discharge at the grounding line, as well as grounding line retreat, as shown in Figure 7.5. MISI has been responsible for ice sheet changes in the recent geological past, causing rates of thinning similar to those observed in rapidly changing parts of Antarctica today (Jones et al. 2015).

As discussed previously, the rapid change observed in West Antarctica to the Pine Island, Thwaites, Smith and Kohler Glaciers (Rignot et al. 2014) has been attributed to early-stage onset of the marine ice sheet instability. Model studies of this system indicate that retreat may already be irreversible (e.g., Favier et al. 2014, Joughin et al. 2014).

Improved understanding of ice sheet dynamic processes (e.g., Schoof 2007, 2011), has led to improved models, which can better reproduce the observed retreat and speed-up outlet glaciers within the Greenland Ice Sheet (Nick et al. 2009, Vieli and Nick 2011). Prediction of decadal to centennial timescale ice sheet change is best achieved by using high-resolution models incorporating grounding line migration, full Navier–Stokes stress configuration, data integration and proper consideration of statistics (e.g., Pattyn et al. 2013, Ritz et al. 2015). But despite improved modelling techniques, marine ice sheet models aiming at predicting future mass loss from ice sheets still suffer from uncertainties with respect to grounding line migration (Pattyn and Durand 2013, Fogwill et al. 2014), and there are still considerable differences between different modelling

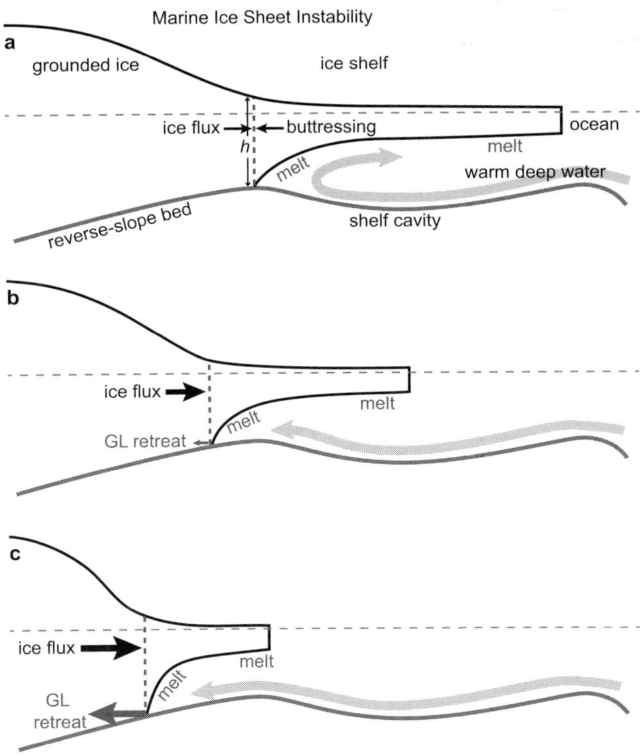

Figure 7.5 Schematic representation of the Marine Ice Sheet Instability (MISI) showing progressive ice retreat into a subglacial basin. Advection of warm circumpolar deep water (CDW) into the shelf cavity (grey arrow) causes basal melt under a stable marine terminating ice-sheet margin with a buttressing ice shelf (a). The flux of ice discharged into the ocean is strongly dependent on the grounding-line thickness h. With increased basal melt the ice shelf thins, reducing the buttressing, and the grounding line retreats onto the reverse-sloping bedrock (b). With increasing h from grounding-line retreat there is an increase in ice flow across the grounding line and a positive runaway feedback is triggered until the bed slope changes (adapted from DeConto and Pollard, 2016, with permission from Macmillan Publishers Ltd: Nature, copyright 2016).

methodologies (e.g., Pattyn and Durand 2013, Cornford et al. 2015, Golledge et al. 2015). Hydrofracturing of ice shelves and ice cliff failure are mechanisms that may lead to considerably increased rates of future sea-level rise (Pollard et al. 2015, DeConto and Pollard 2016).

The three post-AR5 projections for the whole Antarctic Ice Sheet compared in Table 7.2 are based on IPCC RCP scenarios. They have both similarities and differences. Ritz et al. (2015) and Golledge et al. (2015) provide fairly similar results in spite of their very different approaches. Deconto and Pollard (2016) show much more ice loss, but this is due to their inclusion of new and uncertain processes. Both Golledge et al. (2015) and Deconto and Pollard (2016) show limited ice loss under RCP2.6. However, the present-day climate in both studies is not perfectly reproduced, making it challenging to simulate future ice sheet response to low-emission climate change scenarios, where the scale of the forcing is similar to the uncertainty in present-day climate simulations.

Predicting centennial- to millennial-scale response of the ice sheets requires getting the physics right (e.g., DeConto and Pollard 2016). Aside from uncertainties regarding the ice-cliff failure mechanism, there are still basic problems in the integration of ocean and ice sheet models, and in considering the full impacts and feedbacks resulting from changing sub-ice shelf melt rates, tidal movement of floating ice, sea ice growth etc. The model of DeConto and Pollard (2016) suggests that, in addition to significant West Antarctic deglaciation, deglaciation within East Antarctic's Wilkes and Aurora basins could commence before 2100.

On the Greenland Ice Sheet, the significant and increasing surface melt will make a major contribution to future ice loss, but relatively narrow glacial troughs will limit dynamic ice flux to the ocean (Morlighem et al. 2014). In Antarctica, wide, deep troughs in both West (e.g., Bingham et al. 2012) and East Antarctica (Aitken et al. 2016, Mengel and Levermann 2014) and large coastal extents of floating ice will make ice–ocean interaction and the dynamic response more important.

Glaciers around the world other than the ice sheets will also continue to retreat into the future and contribute to sea-level rise. The strongly negative global mass balance indicates that glaciers are generally out of equilibrium with current climate. As the glaciers will seek to achieve a new equilibrium (balanced mass budget) they will retreat to higher (typically colder) areas. Hence there is a commitment for future mass loss that is yet to be realized, even if the climate does not change further. Several studies have quantified this committed net mass loss by evaluating accumulation area ratios (AAR), i.e. the ratios of the accumulation area to the total glacier area, which are closely related to total glacier mass balance for land-terminating glaciers. Bahr et al. (2009) found that global glacier volume needs to reduce by 27 ± 5% for AARs to reach equilibrium values and thus for the global glacier mass budget to be balanced. Mernild et al. (2013) found 38 ± 16% committed volume loss translating into 163 ± 69 mm SLE/year, if the 1971–2010 climate remains unaltered. While the AAR method yields an indication of a lower bound, future mass changes are more accurately assessed by process-based mass-balance models.

Global glaciers and ice sheets are typically modelled separately, since the types of models suitable for these

Table 7.2 *Projections of the Antarctic Contribution to Sea Level Rise over the Next Century and beyond from Different Recent Modelling Studies*

Model study	Ice sheet model strengths/limitations	Sea level contribution
Golledge et al. 2015	Geologically constrained spin-up Hybrid shallow ice/shallow shelf model Simplified climate forcing; medium level of parameterisation	Additional 40 cm by 2100 AD >8 m by 2500 AD (RCP 8.5)
Ritz et al. 2015	Hybrid shallow ice/shallow shelf model Constrained by observations Comprehensive ensemble/treatment of uncertainty.	Additional 30 cm by 2100 AD, 72 cm by 2200 (A1B scenario, similar to RCP6).
Deconto and Pollard 2016	Geologically constrained; hybrid shallow ice/shallow shelf model; includes two new processes, both highly uncertain	Additional **1 m by 2100 AD, 15** m by 2500 AD (RCP 8.5)

two sets of ice masses are distinctly different. In contrast to the abundance of models for ice sheets (e.g., Bindschadler et al. 2013), only relatively few models capable of modelling glaciers on a global scale have been reported in the literature, a total of seven models in the last ten years (see Radic and Hock (2014) for review; Huss and Hock 2015).

Models aiming to project global-scale glacier mass balances typically fall into three categories: (1) models based on simple extrapolations (Meier et al. 2007); (2) models based on mass-balance sensitivities, defined as the expected change in mass balance of a glacier in response to an instantaneous change in a climatic variable, typically air temperature or precipitation (e.g., Hock et al. 2009, Slangen et al. 2012) and (3) process-based models computing all major mass-balance components forced by transient climate projections.

The first type of model relies on extrapolation of recent glacier behaviour to the future without considering the physical underlying processes and climate scenarios. For example, Meier et al. (2007) investigated two simple scenarios: a) the mass loss rate in 2006 and b) constant acceleration, projecting a sea-level rise due to glacier wastage of between 0.1 and 0.25 m by the end of the twenty-first century. Although the approach is highly simplistic, results suggest that glaciers remain an important contributor to sea-level rise in the twenty-first century.

A sensitivity approach for global-scale glacier mass-balance projections was first applied by Gregory and Oerlemans (1998), further refined by Van de Wal and Wild (2001) and also used in the global glacier projections in the IPCC Third (TAR) and Fourth (AR4) Assessment Reports (IPCC 2001, 2007). Mass-balance sensitivities are derived from data or detailed modelling on selected well-observed glaciers, and then multiplied by the projected changes in the climate variable considered to compute future changes in mass balance. In TAR and AR4, sensitivities were assumed to decline as glaciers retreat. Slangen et al. (2012) computed global mass changes from sensitivities to temperature and precipitation. The latter were parameterised as a function of local precipitation and further differentiated between summer and non-summer months. The sensitivity approach does not account for any initial mass imbalance, which needs to be added to the projected mass change. Other shortcomings include large uncertainties with respect to determining how sensitivities vary spatially and how they may change in the future.

Global glacier projections in AR5 (Church et al. 2013) are based primarily on process-based modelling. Raper and Braithwaite (2006) were the first to use a process-based model driven by transient climate scenarios. They used a degree-day model and regressions of mass balance gradients against annual precipitation and summer temperature from a gridded climatology on a $1^\circ \times 1^\circ$ grid as well as climate data from two Global Climate Models (GCMs). Hirabayashi et al. (2010) developed a global-scale gridded model based on a degree-day approach in order to provide melt water input to a global hydrological model. The glacier area of each $0.5^\circ \times 0.5^\circ$ grid cell was treated as one large glacier, but sub-grid elevation bands were considered in the calculations. Radic and Hock (2011) for the first time modelled each glacier individually using the, at that time highly incomplete, World Glacier Inventory (Cogley, 2009). They applied a simple elevation-dependent temperature index model to the ~120,000 inventoried glaciers and upscaled results to the remaining glaciers in the world. Radic et al. (2014) updated their projections using the AR5 climate scenarios (IPCC 2013) based on the Climate Model Intercomparison Project (CMIP5) database (Taylor et al. 2012) and the globally complete Randolph Glacier Inventory (RGI Consortium 2017; Pfeffer et al. 2014).

Marzeion et al. (2012) applied a similar approach to reconstruct the mass changes in the past and project future glacier mass evolution, but they computed each glacier's melt solely from the air temperature at the elevation of the glacier tongue instead of performing the calculations for each elevation band. Giesen and Oerlemans (2013) took a different approach and applied a simplified surface energy balance model to a small subset of eighty-nine glaciers with mass balance observations, and statistically upscaled the projections to all other glaciers.

Most models rely on simple scaling methods, such as volume-area or length scaling (Bahr et al. 1997) to account for the glacier's dynamic adjustments in response to mass change. Area lost or gained is typically removed or added at the lowest elevation band, but surface elevation across the glacier is kept constant. Thus, the surface mass-balance feedback to glacier retreat/advance is considered, but not the feedback due to glacier thinning or thickening, which tends to be positive (i.e., amplifying the response) in contrast to the negative retreat/advance feedback. Scaling methods also assume that area adjustments in response to mass change are instantaneous. Marzeion et al. (2012) used volume-length scaling but introduced a relaxation time scale to account for delayed glacier area and length response. Raper and Braithwaite (2006) used a simple geometric model developed by Raper et al. (2000) that adjusts glacier area across the entire elevation range assuming idealized triangular-shaped area-altitude distributions with maximum area at the mean altitude between minimum and maximum elevation. As the minimum elevation changes, because of retreat/advance, so does the area-altitude distribution.

None of the process-based global-scale models described includes frontal ablation, i.e., the mass loss at marine-terminating glacier fronts through calving and submarine melt, thus systematically underestimating mass loss. Roughly 30% of the total glacierized area drains through marine-terminating calving fronts (Gardner et al. 2013) indicating a potentially large source of uncertainty. The simplicity of these global models was necessitated by the lack of data required by more sophisticated approaches. Taking advantage of the first global-scale ice thickness data set based on the RGI (Huss and Farinotti 2012), Huss and Hock (2015) overcome many of the shortcomings of the previous model generation. Their global model is the first to include, albeit simplistically, frontal ablation revealing that it makes up roughly 10% of total ablation globally. Regional differences are large, with the highest fraction exceeding 40% found for the glaciers in the Antarctic periphery. Converting the net mass losses into sea-level equivalent, the model also accounts for the effect that grounded ice of tidewater glaciers already displaces ocean water. The global glacier sea-level contribution is reduced by ~11% compared to the results when this effect is not accounted for as in all previous global modelling studies. Instead of scaling, Huss and Hock (2015) used the retreat parameterisation of Huss et al. (2010), which is based on observations that ice thickness change due to glacier volume loss generally increases exponentially towards the glacier front. Calculated volume change due to surface balance is converted into an elevation change for each elevation band according to an empirical function that relates normalized thickness change to normalized elevation range.

The five global-scale models published in the last ten years that include the glaciers in the Antarctic and Greenland periphery (Radic and Hock 2011, updated in Radic et al. 2014, Marzeion et al. 2012, Slangen et al. 2012, Giesen and Oerlemans 2013, Huss and Hock 2015) project multi-model mean global glacier mass losses between 15% and 55% over the twenty-first century, corresponding to 77 to 242 mm SLE. The models were forced by eight to fifteen Global Circulation Models (GCMs) and different emission scenarios, including the Representative Concentration Pathways used in AR5 (RCP2.6, RCP4.5, RCP6.0, RCP8.5 with numbers indicating the radiative forcing from anthropogenic greenhouse gases in watts per square metre by the year 2100), and the scenarios from the Special Report on Emissions Scenarios (SRES) used in TAR and AR4. These five models are summarized in Marzeion et al. (2016).

Results, even for identical climate forcing, vary considerably between studies, in particular on regional scales. Each study shows a large spread of glacier projections, when forced by different GCMs, indicating that the uncertainties in the climate projections, especially in precipitation, are a leading cause for uncertainties in projected global glacier evolution. Other uncertainties include inaccuracies in input data, simplified model physics and different model calibration procedures. Model calibration is typically done using the sparse glaciological data sets available through WGMS or other sources; hence the models suffer from large uncertainties due to the lack of datasets with sufficient geographic spread suitable for calibration. New data sets especially from satellite missions will greatly improve the capabilities of the global models to constrain model parameters.

Despite the large spread of results among the few global-scale studies it is robust to conclude that we can expect substantial global-scale glacier mass losses by the end of this century. A glacier model intercomparison project (GlacierMIP) is currently underway as a Targeted Activity of the World Climate Research Programme, Climate and Cryosphere (CliC) project, aiming to understand the differences in recent global projections and work towards improved future generation global-scale glacier models.

7.7 Techniques, Systems and Networks to Assess Cryospheric Change on Future Earth

The rapid changes in ice sheets and glaciers witnessed over the satellite era, and especially since the 1990s, demonstrate the value of long-term monitoring of these systems with high spatial and temporal sampling. Further, the great value of independent measurement techniques has been demonstrated. The future will likely introduce more dramatic and larger-scale cryospheric change, and ongoing and improved satellite missions are required. Understanding of regional, basin or glacier-scale change will allow new insights into the processes driving the change, and future intercomparison exercises will seek to improve intermeasurement agreement at these higher spatial scales. For example, IMBIE-2 now provides an updated assessment of Antarctic mass balance by combining altimetry, gravimetry and input—output method measurements (Shepherd et al. 2018).

New optical-based velocity measurements have already added extra sampling capability while methodological advances and data archaeology mean measurements now extend back to the 1970s for Landsat (Rosenau et al. 2015) or earlier for declassified spy imagery or aerial photography. The Sentinel-1 mission of the European Space Agency uses two satellites to deliver improved repeat-times of C-band Synthetic Aperture Radar data. Both GRACE Follow-On (launched in May 2018) and ICESat-2 (scheduled for later in 2018) will provide continuation of earlier missions.

With improved measurement resolution will come the need to provide background models with sufficient accuracy at these resolutions. This is particularly challenging for GIA and surface mass balance models. Modelled ice sheet surface mass balance, constrained by in situ observations, has an uncertainty of the order of 100 Gt/year (Lenaerts et al. 2012) and updated estimates can change by tens of gigatonnes per year. Improved understanding of the processes, and models, of firn densification is also required for converting elevation change to mass change. For GIA models, new understanding of Holocene ice changes is particularly important in West Antarctica, where low-viscosity mantle may predominate. Sustained, long-term GPS measurements are required to validate such models, or provide insights into mantle rheology in conjunction with seismological measurements. However, the absence of rock outcrops in the interior of East Antarctica and Greenland means that new approaches are needed to measure bedrock motion there. Improvements are needed to both model physics and boundary conditions.

Large regions of Antarctic bedrock remain unmapped, or mapped only at very low spatial resolution. While progress is being made onshore to fill large data gaps, progress offshore is slow, particularly in East Antarctica – offshore topography exerts first-order control of the flow of warm water across the continental shelf and gives insights into ice flow and retreat history during glacial periods. Geothermal heat flux remains poorly mapped under Greenland and Antarctica, especially for data based on in situ measurements. Satellite-based estimates for Antarctica show large differences and fewer than ten in-situ measurements exist. More complete inclusion of GIA is required within ice sheet models in order to capture its effects on ice sheet dynamics, while more sophisticated treatment of ice–ocean interactions is also required.

To quantify more accurately what changes are currently occurring to mountain glaciers, and what will happen in the future, ongoing improvements will be needed to satellite-based glacier inventories and monitoring of glacier volume/mass change. These must be supported by ground-based glacier mass balance measurements and development of schemes that are specific to regions and different types and sizes of glaciers to scale surface area to ice volume. Better knowledge of the thickness distribution of the world's glaciers is crucial to reduce the uncertainties in their future modelling. Better understanding of the processes controlling calving of tidewater glaciers is also needed.

7.8 Concluding Remarks

The cryospheric contributions to sea level rise are committed and will occur well beyond 2100, even without further temperature increase. Mass loss of mountain glaciers and Greenland glaciers will continue for decades. The contribution from ice sheets is already greater than that from glaciers and there is a possibility of Antarctic marine ice sheet instability that could lead to fast and high additional sea level rise.

Sea-level rise from melting glaciers and ice sheets will exert one of the major impacts on human societies on Earth in the future. It will be a major threat to low-lying coastal regions worldwide, including some of the most densely populated and agriculturally productive regions on Earth such as the Asian mega-deltas, with a total population of more than 250 million people. Rising sea level will directly impact coastal regions through inundation and shoreline erosion. These will be driven by tides and storm surges and will result in an increasing frequency of extreme flood events. Globally, sea-level rise will displace millions of people and the resultant induced migration and infrastructure adaption will likely cost hundreds of billions of dollars by the year 2100. Sea-level rise from a shrinking cryosphere in a warming world is slow but relentless, and it will continue for centuries to millennia. In addition, accurate knowledge about current and future glacier shrinkage is important to quantify the effects on freshwater availability in the world's glacier-fed catchments.

Acknowledgements

Data sets shown in Figures 7.3 and 7.4 were assembled with the assistance of Luke Trusel, Ted Scambos, Christopher Shuman and Abbas Khan. RH received funding from the National Science Foundation (grants 1543432 and 1603815) and NASA (#80NSSC17K0566) MAK is a recipient of an Australian Research Council Future Fellowship (project number FT110100207). His work was partially supported by Australian Research Council Special Research Initiative for Antarctic Gateway Partnership (Project ID SR140300001). ANM is partially supported by the NZ MBIE Past Antarctic Climates programme. IA, RH and ANM contributed this chapter as representatives of the IUGG International Association of Cryospheric Sciences (IACS), as well as their home institutions, and MAK as a representative of the International Association of Geodesy (IAG).

References

Adhikari, S. and E. R. Ivins (2016). Climate-driven polar motion: 2003–2015. *Science Advances*, 2(4), p.e1501693.

Aitken, A. R. A., J. L. Roberts, T. D. van Ommen, D. A. Young, N. R. Golledge, J. S. Greenbaum, D. D. Blankenship, and M. J. Siegert (2016). Repeated large-scale retreat and advance of Totten Glacier indicated by inland bed erosion. *Nature*, 533(7603), 385–389.

Allison, I., W. Colgan, M. King, and F. Paul (2015). Ice sheets, glaciers and sea-level. In: Haeberli, W. and C. Whiteman (eds.): *Snow and Ice-Related Hazards, Risks and Disasters*, pp. 713–747, Elsevier, Amsterdam.

Andersen, M. L., L. Stenseng, H. Skourup, W. Colgan, S. A. Khan, S. S. Kristensen, S. B. Andersen, J. E. Box, A. P. Ahlstrøm, X. Fettweis, and R. Forsberg (2015). Basin-scale partitioning of Greenland ice sheet mass balance components (2007–2011). *Earth and Planetary Science Letters*, 409, 89–95.

Arendt, A., A. Bliss, T. Bolch, J. Cogley, A. Gardner, J.-O. Hagen et al. (2015). Randolph glacier inventory – a dataset of global glacier outlines: Version 5.0. GLIMS Technical Report.

Argus, D. F., W. R. Peltier, R. Drummond, and A. W. Moore (2014). The Antarctica component of postglacial rebound model ICE-6G_C (VM5a) based on GPS positioning, exposure age dating of ice thicknesses, and relative sea level histories. *Geophysical Journal International*, 198(1), 537–563, doi:10.1093/gji/ggu140.

Austermann, J., J. X. Mitrovica, K. Latychev, and G. A. Milne (2013). Barbados-based estimate of ice volume at Last Glacial Maximum affected by subducted plate. *Nat. Geosci.*, 6(7), 553–557. http://dx.doi.org/10.1038/ngeo1859.

Bahr, D. B., M. Dyurgerov, and M. F. Meier (2009). Sea-level rise from glaciers and ice caps: a lower bound. *Geophysical Research Letters*, 36(3).

Bahr, D. B., M. F. Meier, and S. D. Peckham (1997). The physical basis of glacier volume-area scaling. *J. Geophys. Res.*, 102(B9), 20355–20362.

Barletta, V. R., L. S. Sørensen, and R. Forsberg (2013). Scatter of mass changes estimates at basin scale for Greenland and Antarctica. *The Cryosphere*, 7(5), 1411–1432.

Barrett, P. J. (1996). Antarctic paleoenvironment through Cenozoic times – a review. *Terr. Antarct.* 3, 103–119.

Bentley, M. J. (1999). Volume of Antarctic ice at the Last Glacial Maximum, and its impact on global sea level change. *Quaternary Science Reviews*, 18(14), 1569–1595.

Bentley, M. J., C. J. Fogwill, A. M. Le Brocq, A. L. Hubbard, D. E. Sugden, T. J. Dunai, and S. P. Freeman (2010). Deglacial history of the West Antarctic Ice Sheet in the Weddell Sea embayment: constraints on past ice volume change. *Geology*, 38(5), 411–414.

Bevis, M., J. Wahr, S. Khan, F. Madsen, A. Brown, and M. Willis et al. (2012). Bedrock displacements in Greenland manifest ice mass variations, climate cycles and climate change. *Proc. Nat. Acad. Sci. USA*, 109(30), 11944–11948. http://dx.doi.org/10.1073/pnas.1204664109.

Bevis, M., E. Kendrick, R. Smalley, I. Dalziel, D. Caccamise, I. Sasgen, M. Helsen, F. W. Taylor, H. Zhou, A. Brown, D. Raleigh, M. Willis, T. Wilson, and S. Konfal (2009). Geodetic measurements of vertical crustal velocity in West Antarctica and the implications for ice mass balance. *Geochemistry Geophysics Geosystems*, 10, Q10005, doi:10.1029/2009gc002642.

Bindschadler, R. A., S. Nowicki, A. Abe-Ouchi, A. Aschwanden, H. Choi, J. Fastook, (2013). Ice-sheet model sensitivities to environmental forcing and their use in projecting future sea level (the SeaRISE project). *Journal of Glaciology*, 59(214), 195–224.

Bingham, R. G., F. Ferraccioli, E. C. King, R. D. Larter, H. D. Pritchard, A. M. Smith, and D. G. Vaughan (2012). Inland thinning of West Antarctic Ice Sheet steered along subglacial rifts. *Nature*, 487(7408), 468–471.

Boening, C., M. Lebsock, F. Landerer, and G. Stephens (2012). Snowfall-driven mass change on the East Antarctic ice sheet. *Geophys. Res. Lett.*, 39(21), L21501. http://dx.doi.org/10.1029/ 2012gl053316.

Chen, J. L., C. R. Wilson, and B. D. Tapley (2006). Satellite gravity measurements confirm accelerated melting of Greenland ice sheet. *Science*, 313, 1958–60.

Chen, J. L., C. R. Wilson, and B. D. Tapley (2011). Interannual variability of Greenland ice losses from satellite gravimetry. *J. Geophys. Res.*, 116, B07406.

Chen, J. L., C. R. Wilson, J. C. Ries, and B. D. Tapley (2013). Rapid ice melting drives Earth's pole to the east. *Geophysical Research Letters*, 40(11), 2625–2630.

Church, J. A., P. U. Clark, A. Cazenave, J. M. Gregory, S. Jevrejeva, A. Levermann et al. (2013). Sea level change. In: Stocker, T. F., D. Qin, G.-K. Plattner, M. Tignor, S. K. Allen, J. Boschung, A. Nauels, Y. Xia, V. Bex, and P. M. Midgley (eds.): *Climate Change 2013: The Physical Science Basis, Contribution of Working Group I to the Fifth Assessment Report of the Intergovernmental Panel on Climate Change*. Cambridge University Press, Cambridge.

Clark, P. U., A. S. Dyke, J. D. Shakun, A. E. Carlson, J. Clark, B. Wohlfarth, J. X. Mitrovica, S. W. Hostetler, and A. M. McCabe (2009). The last glacial maximum. *Science*, 325(5941), 710–714.

Clark, P. U., J. X. Mitrovica, G. A. Milne, and M. E. Tamisiea (2002). Sea-level fingerprinting as a direct test for the source of global meltwater pulse IA. *Science*, 295(5564), 2438–2441.

Clark, P. U. and L. Tarasov (2014). Closing the sea level budget at the Last Glacial Maximum. *Proceedings of the National Academy of Sciences*, 111(45), 15861–15862, doi:10.1073/pnas.1418970111.

Cogley, J. G. (2005). *Mass and Energy Balances of Glaciers and Ice Sheets*. In: *Encyclopedia of Hydrological Sciences*. Wiley Online Library, John Wiley & Sons.

Cogley, J. G. (2009). Geodetic and direct mass-balance measurements: comparison and joint analysis. *Annals of Glaciology*, 50(50), 96–100.

Cornford, S. L., D. F. Martin, A. J. Payne, E. G. Ng, A. M. Le Brocq, R. M. Gladstone, T. L. Edwards, S. R. Shannon, C. Agosta, M. R. Van Den Broeke, and H. H. Hellmer (2015). Century-scale simulations of the response of the West Antarctic Ice Sheet to a warming climate. *The Cryosphere*, 9, 1579–1600.

DeConto, R. M. and D. Pollard (2003). A coupled climate–ice sheet modeling approach to the early Cenozoic history of the Antarctic ice sheet. *Palaeogeography, Palaeoclimatology, Palaeoecology*, 198(1), 39–52.

DeConto, R. M. and D. Pollard (2016). Contribution of Antarctica to past and future sea-level rise. *Nature*, *531*(7596), 591–597.

Depoorter, M. A., J. L. Bamber, J. A. Griggs, J. Lenaerts, S. Ligtenberg, M. van den Broeke, and G. Moholdt (2013). Calving fluxes and melt rates of Antarctic ice shelves. *Nature*, 502, 89–92.

Deschamps, P., N. Durand, E. Bard, B. Hamelin, G. Camoin, A. L. Thomas, G. M. Henderson, J. I. Okuno, and Y. Yokoyama (2012). Ice-sheet collapse and sea-level rise at the Bolling warming 14,600 years ago. *Nature*, 483(7391), 559–564.

Domingues, C. M., J. A. Church, N. J. White, P. J. Gleckler, S. E. Wijffels, P. M. Barker, and J. R. Dunn (2008). Improved estimates of upper-ocean warming and multidecadal sea-level rise. *Nature*, 453, 1090–1093. http://dx.doi.org/10.1038/nature07080.

Dutton, A., A. E. Carlson, A. J. Long, G. A. Milne, P. U. Clark, R. DeConto, B. P. Horton, S. Rahmstorf, and M. E. Raymo (2015). Sea-level rise due to polar ice-sheet mass loss during past warm periods. *Science*, 349(6244), aaa4019.

Dutton, A. and K. Lambeck (2012). Ice volume and sea level during the last interglacial. *Science*, 337(6091), 216–219.

Dyurgerov, M. B. and M. F. Meier (2005). *Glaciers and the changing earth system: a 2004 snapshot*. Institute of Arctic and Alpine Research, University of Colorado, Boulder.

Enderlin, E. M., I. M. Howat, S. Jeong, M.-J. Noh, J. H. van Angelen, and M. R. van den Broeke (2014). An improved mass budget for the Greenland ice sheet. *Geophysical Research Letters*, 41(3), 2013GL059010, doi:10.1002/ 2013gl059010.

Ewert H., A. Groh and R. Dietrich (2012). Volume and mass changes of the Greenland ice sheet inferred from ICESat and GRACE. *J. Geodyn.*, 59–60, 111–123.

Fahnestock, M., T. Scambos, T. Moon, A. Gardner, T. Haran, and M. Klinger (2016). Rapid large-area mapping of ice flow using Landsat 8. *Remote Sensing of Environment*, 185, 84–94.

Favier, L., G. Durand, S. L. Cornford, G. H. Gudmundsson, O. Gagliardini, F. Gillet-Chaulet, T. Zwinger, A. J. Payne, and A. M. Le Brocq (2014). Retreat of Pine Island Glacier controlled by marine ice-sheet instability. *Nature Climate Change*, 4(2), 117–121.

Fretwell, P. T., H. D. Pritchard, D. G. Vaughan, J. L. Bamber, N. E. Barrand, R. Bell et al. (2013). Bedmap2: improved ice bed, surface and thickness datasets for Antarctica. *The Cryosphere*, 7, 375–393.

Fogwill, C. J., C. S. Turney, K. J. Meissner, N. R. Golledge, P. Spence, J. L. Roberts, M. H. England, R. T. Jones, and L. Carter (2014). Testing the sensitivity of the East Antarctic ice sheet to southern ocean dynamics: past changes and future implications. *Journal of Quaternary Science*, 29(1), 91–98.

Forsberg, R., L. Sørensen, and S. Simonsen (2016). Greenland and Antarctica ice sheet mass changes and effects on global sea level. *Surv. Geophys.*, doi:10.1007/s10712-016-9398-7.

Fu, Y., J. T. Freymueller, and T. Jensen (2012). Seasonal hydrological loading in southern Alaska observed by GPS and GRACE. *Geophysical Research Letters*, 39(15), L15310, doi:10.1029/2012gl052453.

Gabbi, J., M. Carenzo, F. Pellicciotti, A. Bauder, and M. Funk (2014). A comparison of empirical and physically based glacier surface melt models for long-term simulations of glacier response. *Journal of Glaciology*, 60(224), 1140–1154.

Gardner, A. S., G. Moholdt, J. G. Cogley, B. Wouters, A. A. Arendt, J. Wahr et al. (2013). A reconciled estimate of glacier contributions to sea level rise: 2003 to 2009. *Science* 340(6134), 852–857, doi:10.1126/science.1234532.

Gasson, E., R. M. DeConto, D. Pollard, and R. H. Levy (2016). Dynamic Antarctic ice sheet during the early to mid-Miocene. *Proceedings of the National Academy of Sciences*, 113(13), 3459–3464.

Giesen, R. H. and J. Oerlemans (2013). Climate-model induced differences in the 21st century global and regional glacier contributions to sea-level rise. *Climate Dynamics*, 41(11–12), 3283–3300, doi: 10.1007/s00382-013-1743-7.

Golledge, N. R., C. J. Fogwill, A. N. Mackintosh and K. M. Buckley (2012). Dynamics of the last glacial maximum Antarctic ice-sheet and its response to ocean forcing. *Proceedings of the National Academy of Sciences*, 109(40), 16052–16056.

Golledge, N. R., L. Menviel, L. Carter, C. J. Fogwill, M. H. England, G. Cortese, and R. H. Levy (2014). Antarctic contribution to meltwater pulse 1A from reduced southern Ocean overturning. *Nature Communications*, 5.

Golledge, N. R., D. E. Kowalewski, T. R. Naish, R. H. Levy, C. J. Fogwill, and E. G. Gasson (2015). The multi-millennial Antarctic commitment to future sea-level rise. *Nature*, 526(7573), 421–425.

Gomez, N., J. X. Mitrovica, P. Huybers and P. U. Clark (2010). Sea level as a stabilizing factor for marine-ice-sheet grounding lines. *Nature Geosci.*, 3(12), 850–853, doi:10.1038/ngeo1012.

Gomez, N., D. Pollard, and J. X. Mitrovica (2013). A 3-D coupled ice sheet – sea level model applied to Antarctica through the last 40 ky. *Earth and Planetary Science Letters*, 384, 88–99, doi:10.1016/j.epsl.2013.09.042.

Greenbaum, J. S., D. D. Blankenship, D. A. Young, T. G. Richter, J. L. Roberts, A. R. A. Aitken, B. Legresy, D. M. Schroeder, R. C. Warner, T. D. van Ommen, and M. J. Siegert (2015). Ocean access to a cavity beneath Totten Glacier in East Antarctica. *Nature Geoscience*, 8(4), 294–298.

Gregoire L. J., A. J. Payne, and P. J. Valdes (2012). Deglacial rapid sea level rises caused by ice-sheet saddle collapses. *Nature*, 487, 219–222, doi:10.1038/nature11257.

Gregory, J. M., and J. Oerlemans (1998). Simulated future sea-level rise due to glacier melt based on regionally and seasonally resolved temperature changes. *Nature*, 391(6666), 474–476.

Grinsted, A. (2013). An estimate of global glacier volume. *The Cryosphere*, 7(1), 141–151, doi:10.5194/tc-7-141-2013.

Haeberli, W. and A. Linsbauer (2013). Global glacier volumes and sea level – small but systematic effects of ice below the surface of the ocean and of new local lakes on land. Brief communication. *The Cryosphere*, 7, 817–821.

Hanna, E. H., F. J. Navarro, F. Pattyn, C. M. Domingues, X. Fettweis, E. R. Ivins, R. J. Nicholls, C. Ritz, B. Smith, S. Tulaczyk, P. L. Whitehouse, and H. J. Zwally (2013). Ice-sheet mass balance and climate change. *Nature*, 498, 51–59. http://dx.doi.org/10.1038/nature12238.

Harig, C. and F. J. Simons (2015). Accelerated West Antarctic ice mass loss continues to outpace East Antarctic gains. *Earth and Planetary Science Letters*, 415, 134–141.

Harig, C. and F. J. Simons (2016). Ice mass loss in Greenland, the Gulf of Alaska, and the Canadian Archipelago: seasonal cycles and decadal trends. *Geophysical Research Letters*, 43(7), 3150–3159, doi:10.1002/2016GL067759.

Heeszel, D. S., D. A. Wiens, S. Anandakrishnan, R. C. Aster, I. W. D. Dalziel, A. D. Huerta, A. A. Nyblade, T. J. Wilson, and J. P. Winberry (2016). Upper mantle structure of central and West Antarctica from array analysis of Rayleigh wave phase velocities. *Journal of Geophysical Research: Solid Earth*, doi:10.1002/2015JB012616.

Helm, V., A. Humbert, and H. Miller (2014). Elevation and elevation change of Greenland and Antarctica derived from CryoSat-2. *The Cryosphere*, 8(4), 1539–1559.

Hirabayashi, Y., P. Doll, and S. Kanae (2010). Global-scale modeling of glacier mass balances for water resources assessments: Glacier mass changes between 1948 and 2006. *Journal of Hydrology*, 390(3–4), 245–256, doi:10.1016/j.jhydrol.2010.07.001.

Hock, R., M. de Woul, V. Radic, and M. Dyurgerov (2009). Mountain glaciers and ice caps around Antarctica make a large sea-level rise contribution. *Geophysical Research Letters*, 36, L07501 doi:10.1029/2008gl037020.

Holland, P. R., A. Jenkins, and D. M. Holland (2008). The response of ice shelf basal melting to variations in ocean temperature. *J. Clim.*, 21(11), 2558–2572.

Holmlund, P., P. Jansson, and R. Pettersson (2005). A re-analysis of the 58 year mass-balance record of Storglaciaren, Sweden. *Annals of Glaciology*, 42(1), 389–394, doi:10.3189/172756405781812547.

Howat, I. M., Y. Ahn, I. Joughin, M. R. van den Broeke, J. T. M. Lenaerts, and B. Smith (2011). Mass balance of Greenland's three largest outlet glaciers, 2000–2010. *Geophys. Res. Lett.*, 38, L12501. http://dx.doi.org/10.1029/2011gl047565.

Hughes, T. 1973. Is the West Antarctic ice sheet disintegrating? *Journal of Geophysical Research*, 78, 7884–7910.

Hurkmans, R. T. W. L., J. L. Bamber, C. H. Davis, I. R. Joughin, K. S. Khvorostovsky, B. S. Smith, and N. Schoen (2014). Time-evolving mass loss of the Greenland ice sheet from satellite altimetry. *The Cryosphere*, 8, 1725–40.

Huss, M. and D. Farinotti (2012). Distributed ice thickness and volume of all glaciers around the globe. *Journal of Geophysical Research: Earth Surface*, 117(F4), F04010 doi:10.1029/2012jf002523.

Huss, M. and R. Hock (2015). A new model for global glacier change and sea-level rise. *Frontiers in Earth Science*, 3, 54.

Huss, M., G. Jouvet, D. Farinotti, and A. Bauder (2010). Future high-mountain hydrology: a new parameterization of glacier retreat. *Hydrology and Earth System Sciences*, 14(5), 815–829.

IPCC (2001). Climate Change 2001: The Scientific Basis. Contribution of Working Group I to the Third Assessment Report of the Intergovernmental Panel on Climate Change. Houghton, J. T., Y. Ding, D. J. Griggs, M. Noguer, P. J. van der Linden, X. Dai, K. Maskell, and C. A. Johnson (eds.). Cambridge University Press, Cambridge.

IPCC (2007). Climate Change 2007: The Physical Basis. Contribution of Working Group 1 to the Fourth Assessment Report of the Intergovernmental Panel on Climate Change. Solomon, S., D. Qin, M. Manning, Z. Chen, M. Marquis, K. B. Averyt, M. Tignor, and H. L. Miller (eds.). Cambridge University Press, Cambridge.

IPCC (2013). Climate Change 2013: The Physical Basis. Contribution of Working Group I to the Fifth Assessment Report of the Intergovernmental Panel on Climate Change. Stocker, T. F., D. Qin, G.-K. Plattner, M. Tignor, S. K. Allen, J. Boschung, A. Nauels, Y. Xia, V. Bex, and P. M. Midgley (eds.). Cambridge University Press, Cambridge.

Ivins, E. R., T. S. James, J. Wahr, E. J. O. Schrama, F. Landerer, and K. Simon (2013). Antarctic contribution to sea-level rise observed by GRACE with improved GIA correction. *J. Geophys. Res.*, 118. http://dx.doi.org/10.1002/jgrb.50208.

Johannessen, O. M., K. Khvorostovsky, M. W. Miles, and L. P. Bobylev (2005). Recent ice-sheet growth in the interior of Greenland. *Science*, 310, 1013–1016.

Jones, R. S., A. N. Mackintosh, K. P. Norton, N. R. Golledge, C. J. Fogwill, P. W. Kubik, M. Christl, and S. L. Greenwood (2015). Rapid Holocene thinning of an East Antarctic outlet glacier driven by marine ice sheet instability. *Nature Communications*, 6.

Joughin, I., B. E. Smith, and B. Medley (2014). Marine ice sheet collapse potentially under way for the Thwaites Glacier Basin, West Antarctica. *Science*, 344(6185), 735–738.

Kapnick, S. B., T. L. Delworth, M. Ashfaq, S. Malyshev, and P. C. D. Milly (2014). Snowfall less sensitive to warming in Karakoram than in Himalayas due to a unique seasonal cycle. *Nature Geoscience*, 7, 834–840, doi:10.1038/ngeo2269.

Kaser, G., J. G. Cogley, M. B. Dyurgerov, M. F. Meier, and A. Ohmura (2006). Mass balance of glaciers and ice caps: consensus estimates for 1961–2004. *Geophysical Research Letters*, 33(19).

Khan, S. A., K. H. Kjaer, M. Bevis, J. L. Bamber, J. Wahr, K. K. Kjeldsen, A. A. Bjork, N. J. Korsgaard, L. A. Stearns, M. R. van den Broeke, L. Liu, N. K. Larsen, and I. S. Muresan (2014). Sustained mass loss of the northeast Greenland ice sheet triggered by regional warming. *Nature Clim. Change*, 4(4), 292–299, doi:10.1038/nclimate2161.

Khan, S. A., A. Aschwanden, A. Bjørk, J. Wahr, K. K. Kjeldsen, and K. H. Kjær (2015). Greenland ice sheet mass balance: a review. *Reports on Progress in Physics*, 78(4): 046801.

Khan, S. A., I. Sasgen, M. Bevis, T. van Dam, J. L. Bamber, J. Wahr, M. Willis, K. H. Kjær, B. Wouters, V. Helm, B. Csatho, K. Fleming, A. A. Bjørk, A. Aschwanden, P. Knudsen, and P. K. Munneke (2016). Geodetic measurements reveal similarities between post–Last Glacial Maximum and present-day mass loss from the Greenland ice sheet. *Science Advances*, 2(9), doi:10.1126/sciadv.1600931.

Khazendar, A., M. P. Schodlok, I. Fenty, S. R. M. Ligtenberg, E. Rignot, and M. R. Van den Broeke (2013). Observed thinning of Totten Glacier is linked to coastal polynya variability. *Nature Communications*, 4.

Kjeldsen, K. K., N. J. Korsgaard, A. A. Bjørk, S. A. Khan, S. Funder, N. K. Larsen, J. L. Bamber, W. Colgan, M. van den Broeke, M. L. Siggaard-Andersen, and C. Nuth (2015). Spatial and temporal distribution of mass loss from the Greenland ice sheet since AD 1900. *Nature*, 528(7582), 396–400.

King, M. A., R. J. Bingham, P. Moore, P. L. Whitehouse, M. J. Bentley, and G. A. Milne (2012). Lower satellite-gravimetry estimates of Antarctic sea-level contribution. *Nature*, 491(7425), 586–589.

King, M. A., P. L. Whitehouse, and W. van der Wal (2016). Incomplete separability of Antarctic plate rotation from glacial isostatic adjustment deformation within geodetic observations. *Geophysical Journal International*, 204(1), 324–330, doi:10.1093/gji/ggv461.

Konrad, H., I. Sasgen, D. Pollard, and V. Klemann (2015). Potential of the solid-Earth response for limiting long-term West Antarctic Ice Sheet retreat in a warming climate. *Earth and Planetary Science Letters*, 432, 254–264, doi: http://dx.doi.org/10.1016/j.epsl.2015.10.008.

Konrad, H., L. Gilbert, S. Cornford, A. Payne, A. Hogg, A. Muir, and A. Shepherd (2016). Uneven onset and pace of ice-dynamical imbalance in the Amundsen Sea Embayment, West Antarctica. *Geophysical Research Letters*, doi:10.1002/2016GL070733.

Kopp, R. E., F. J. Simons, J. X. Mitrovica, A. C. Maloof, and M. Oppenheimer (2013). A probabilistic assessment of sea level variations within the last interglacial stage. *Geophys. J. Int.*, 193, 711–716.

Krabill, W., E. Hanna, P. Huybrechts, W. Abdalati, J. Cappelen, B. Csatho, E. Frederick, S. Manizade, C. Martin, J. Sonntag, and R. Swift (2004). Greenland ice sheet: increased coastal thinning. *Geophysical Research Letters*, 31, L24402.

Lambeck, K., Y. Yokoyama, and T. Purcell (2002). Into and out of the last glacial maximum: sea-level change during oxygen isotope stages 3 and 2. *Quat. Sci. Rev.*, 21(1), 343–360.

Lambeck, K., H. Rouby, A. Purcell, Y. Sun, and M. Sambridge (2014). Sea level and global ice volumes from the

Last Glacial Maximum to the Holocene. *Proceedings of the National Academy of Sciences*, 111(43), 15296–15303.

Lemke, P., J. Ren, R. B. Alley, I. Allison, J. Carrasco, G. Flato, Y. Fujii, G. Kaser, P. Mote, R. H. Thomas, and T. Zhang (2007). Observations: changes in snow, ice and frozen ground. In: Solomon, S., D. Qin, M. Manning, Z. Chen, M. Marquis, K. B. Averyt, M. Tignor, and H. L. Miller (eds.), *Climate Change 2007: The Physical Basis. Contribution of Working Group 1 to the Fourth Assessment Report of the Intergovernmental Panel on Climate Change*, pp. 337–383 Cambridge University Press, Cambridge.

Lenaerts, J. T. M., M. R. van den Broeke, W. J. van de Berg, E. van Meijgaard, and P. Kuipers Munneke (2012). A new, high resolution surface mass balance map of Antarctica (1979–2010) based on regional climate modeling. *Geophys. Res. Lett.*, 39(1–5), L04501.

Lisiecki, L. E. and M. E. Raymo (2005). A Pliocene-Pleistocene stack of 57 globally distributed benthic $\delta^{18}O$ records. *Paleoceanography*, 20(1).

Liu, J., G. A. Milne, R. E. Kopp, P. U. Clark, and I. Shennan (2016). Sea-level constraints on the amplitude and source distribution of Meltwater Pulse 1A. *Nature Geoscience*, 9(2), 130–134.

Loriaux, T. and G. Casassa (2013). Evolution of glacial lakes from the Northern Patagonian Icefield and terrestrial water storage in a sea-level rise context. *Global Planet. Change*, 102, 33–40.

Luthcke, S. B., H. J. Zwally, W. Abdalati, D. D. Rowlands, R. D. Ray, R. S. Nerem, F. G. Lemoine, J. J. McCarthy, and D. S. Chinn (2006). Recent Greenland ice mass loss by drainage system from satellite gravity observations. *Science*, 314, 1286–1289.

Luthcke, S. B., T. J. Sabaka, B. D. Loomis, A. A. Arendt, J. J. McCarthy, and J. Camp (2013). Antarctica, Greenland and Gulf of Alaska land-ice evolution from an iterated GRACE global mascon solution. *Journal of Glaciology*, 59(216), 613–631.

Mackintosh, A., N. Golledge, E. Domack, R. Dunbar, A. Leventer, D. White, D. Pollard, R. DeConto, D. Fink, D. Zwartz, and D. Gore (2011). Retreat of the East Antarctic ice sheet during the last glacial termination. *Nature Geoscience*, 4(3), 195–202.

Mackintosh, A., B. Anderson, A. Lorrey, J. Renwick, P. Frei, and S. Dean (2017). Regional cooling caused recent New Zealand glacier advances in a period of global warming. *Nature Communications*, doi:10.1038/ncomms1420.

Martín Español, A., A. Zammit Mangion, P. J. Clarke, T. Flament, V. Helm, M. A. King, S. B. Luthcke, E. Petrie, F. Rémy, N. Schön, and B. Wouters (2016a). Spatial and temporal Antarctic ice sheet mass trends, glacio isostatic adjustment, and surface processes from a joint inversion of satellite altimeter, gravity, and GPS data. *Journal of Geophysical Research: Earth Surface*, 121(2), 182–200, doi:10.1002/2015JF003550.

Martín-Español, A., M. A. King, A. Zammit-Mangion, S. B. Andrews, P. Moore, and J. L. Bamber (2016b). An assessment of forward and inverse GIA solutions for Antarctica. *Journal of Geophysical Research: Solid Earth*, 121(9), 6947–6965, doi:10.1002/2016jb013154.

Marzeion, B., J. G. Cogley, K. Richter, and D. Parkes (2014). Attribution of global glacier mass loss to anthropogenic and natural causes. *Science*, 345, 919–921, doi:10.1126/science.1254702.

Marzeion, B., N. Champollion, W. Haeberli, K. Langley, P. Leclercq, and F. Paul (2017). Observation-based estimates of global glacier mass change and its contribution to sea-level change. *Surveys in Geophysics*, 1–26.

Marzeion, B., A. Jarosch, and M. Hofer (2012). Past and future sea-level change from the surface mass balance of glaciers. *The Cryosphere*, 6(6), 1295–1322.

Marzeion, B., P. W. Leclercq, J. G. Cogley, and A. H. Jarosch (2015). Brief communication: global reconstructions of glacier mass change during the 20th century are consistent. *The Cryosphere*, 9(6), 2399–2404.

McMillan, M., A. Shepherd, A. Sundal, K. Briggs, A. Muir, A. Ridout, A. Hogg, and D. Wingham (2014). Increased ice losses from Antarctica detected by CryoSat-2. *Geophysical Research Letters*, 41(11), 3899–3905.

Meier, M. F. (1984). Contribution of small glaciers to global sea level. *Science*, 226(4681), 1418–1421, doi: 10.1126/science.226.4681.1418.

Meier, M. F., M. B. Dyurgerov, U. K. Rick, S. O'Neel, W. T. Pfeffer, R. S. Anderson et al. (2007). Glaciers dominate eustatic sea-level rise in the 21st century. *Science*, 317(5841), 1064–1067.

Mengel, M. and A. Levermann (2014). Ice plug prevents irreversible discharge from East Antarctica. *Nat. Clim. Change*, 4(6), 451–455.

Mercer, J. (1978). West Antarctic ice sheet and CO_2 greenhouse effect: a threat of disaster. *Nature*, 271, 321–325.

Mernild, S. H., W. H. Lipscomb, D. B. Bahr, V. Radic, and M. Zemp (2013). Global glacier changes: a revised assessment of committed mass losses and sampling uncertainties. *The Cryosphere* 7(5), 1565–1577, doi: 10.5194/tc-7-1565-2013.

Morlighem, M., E. Rignot, J. Mouginot, H. Seroussi and E. Larour (2014). Deeply incised submarine glacial valleys beneath the Greenland ice sheet. *Nature Geoscience*, 7(6).

Naish, T., R. Powell, R. Levy, G. Wilson, R. Scherer, F. Talarico, L. Krissek, F. Niessen, M. Pompilio, T. Wilson, and L. Carter (2009). Obliquity-paced Pliocene West Antarctic ice sheet oscillations. *Nature*, 458(7236), 322–328.

Neckel, N., J. Kropacek, T. Bolch, and V. Hochschild (2014). Glacier mass changes on the Tibetan Plateau 2003–2009 derived from ICESat laser altimetry measurements. *Environ. Res. Lett.*, 9. http://dx.doi.org/10.1088/1748-9326/9/1/014009.

Nick, F. M., A. Vieli, I. M. Howat, and I. Joughin (2009). Large-scale changes in Greenland outlet glacier dynamics triggered at the terminus. *Nat. Geosci.*, 2, 110–114.

Nield, G. A., V. R. Barletta, A. Bordoni, M. A. King, P. L. Whitehouse, P. J. Clarke, E. Domack, T. A. Scambos, and E. Berthier (2014). Rapid bedrock uplift in the

Antarctic peninsula explained by viscoelastic response to recent ice unloading. *Earth and Planetary Science Letters*, 397, 32–41.

Ohmura, A. (2004). Cryosphere during the twentieth century. *The State of the Planet: Frontiers and Challenges in Geophysics*, American Geophysical Union, Geophysical Monograph 150, 239–257.

Pattyn, F., L. Perichon, G. Durand, L. Favier, O. Gagliardini, R. Hindmarsh et al. (2013). Grounding-line migration in plan-view marine ice models: results of the ice2sea MISMIP3d intercomparison. *J. Glaciol.*, 59, 410–422.

Pattyn, F. and G. Durand (2013). Why marine ice sheet model predictions may diverge in estimating future sea level rise. *Geophysical Research Letters*, 40(16), 4316–4320.

Peltier, W. (2004). Global glacial isostasy and the surface of the ice-age Earth: The ICE-5G (VM2) model and GRACE. *Annu. Rev. Earth Planet. Sci.*, 32, 111.

Pfeffer, W. T., A. Arendt, A. Bliss, T. Bolch, J. Cogley, A. Gardner, J. Hagen, R. Hock, G. Kaser, C. Kienholz, E. Miles, G. Moholdt, N. Mölg, F. Paul, V. Radic, P. Rastner, B. Raup, J. Rich, and M. J. Sharp (2014). The Randolph Glacier Inventory: a globally complete inventory of glaciers. *J Glaciol.*, 60(221), 537–551.

Pollard, D. and R. M. DeConto (2009). Modelling West Antarctic ice sheet growth and collapse through the past five million years. *Nature*, 458(7236), 329–332.

Pollard, D., R. M. DeConto, and R. B. Alley (2015). Potential Antarctic ice sheet retreat driven by hydrofracturing and ice cliff failure. *Earth and Planetary Science Letters*, 412, 112–121.

Pritchard, H. D., S. R. M. Ligtenberg, H. A. Fricker, D. G. Vaughan, M. R. van den Broeke and L. Padman (2012). Antarctic ice loss driven by ice-shelf melt. *Nature*, 484, 502–505.

Radic, V., A. Bliss, A. C. Beedlow, R. Hock, E. Miles, and J. G. Cogley (2014). Regional and global projections of twenty-first century glacier mass changes in response to climate scenarios from global climate models. *Climate Dynamics*, 42(1–2), 37–58, doi:10.1007/s00382-013-1719-7.

Radic, V. and R. Hock (2011). Regionally differentiated contribution of mountain glaciers and ice caps to future sea-level rise. *Nature Geoscience*, 4(2), 91–94, doi:10.1038/Ngeo1052.

Radic, V. and R. Hock (2014). Glaciers in the Earth's hydrological cycle: assessments of glacier mass and runoff changes on global and regional scales. *Surveys in Geophysics*, 35(3), 813–837, doi:10.1007/s10712-013-9262-y.

Ramillien, G., A. Lombard, A. Cazenave, E. R. Ivins, M. Llubes, F. Remy, and R. Biancale (2006). Interannual variations of the mass balance of the Antarctica and Greenland ice sheets from GRACE. *Glob. Planet. Change*, 53, 198–208.

Raper, S. C., and R. Braithwaite (2006). Low sea level rise projections from mountain glaciers and icecaps under global warming. *Nature*, 439(7074), 311–313, doi:10.1038/nature04448.

Raper, S. C. B., O. Brown, and R. J. Braithwaite (2000). A geometric glacier model for sea-level change calculations. *Journal of Glaciology*, 46(154), 357–368, doi:10.3189/172756500781833034.

Reager, J. T., A. S. Gardner, J. S. Famiglietti, D. N. Wiese, A. Eicker, and M. H. Lo (2016). A decade of sea level rise slowed by climate-driven hydrology. *Science*, 351(6274), 699–703, doi:10.1126/science.aad8386.

RGI Consortium (2017). *Randolph Glacier Inventory – a Dataset of Global Glacier Outlines*: Version 6.0: *Technical Report, Global Land Ice Measurements from Space*, Colorado. Digital Media, doi: https://doi.org/10.7265/N5-RGI-60.

Richter, A., E. Ivins, H. Lange, L. Mendoza, L. Schröder, J. L. Hormaechea, G. Casassa, E. Marderwald, M. Fritsche, R. Perdomo, and M. Horwath (2016). Crustal deformation across the Southern Patagonian Icefield observed by GNSS. *Earth and Planetary Science Letters*, 452, 206–215.

Richter, A., M. Horwath, and R. Dietrich (2016). Comment on 'Mass gains of the Antarctic ice sheet exceed losses' by H. J. Zwally and others (2015). *Journal of Glaciology*, 62(233), 604–606, doi:10.1017/jog.2016.60.

Rignot, E., J. Mouginot, and B. Scheuchl (2011a). Ice flow of the Antarctic ice sheet. *Science* 333, 1427–1430.

Rignot, E., I. Velicogna, M. R. van den Broeke, A. Monaghan, and J. Lenaerts (2011b). Acceleration of the contribution of the Greenland and Antarctic ice sheets to sea level rise. *Geophys. Res. Lett.*, 38, L05503.

Rignot, E., S. Jacobs, J. Mouginot, and B. Scheuchl (2013). Ice shelf melting around Antarctica. *Science*, 341(6143), 266–270. http://dx.doi.org/10.1126/science.1235798.

Rignot, E., J. Mouginot, M. Morlighem, H. Seroussi, and B. Scheuchl (2014). Widespread, rapid grounding line retreat of Pine Island, Thwaites, Smith, and Kohler glaciers, West Antarctica, from 1992 to 2011. *Geophysical Research Letters*, 41(10), 3502–3509.

Rintoul, S. R., A. Silvano, B. Pena-Molino, E. van Wijk, M. Rosenberg, J. S. Greenbaum, and D. D. Blankenship (2016). Ocean heat drives rapid basal melt of the Totten Ice Shelf. *Science Advances*, 2(12), p. e1601610.

Ritz, C., T. L. Edwards, G. Durand, A. J. Payne, V. Peyaud, and R. C. Hindmarsh (2015). Potential sea-level rise from Antarctic ice-sheet instability constrained by observations. *Nature*, 528(7580), 115–118.

Riva, R., J. Bamber, D. Lavallée, and B. Wouters (2010). Sea-level fingerprint of continental water and ice mass change from GRACE. *Geophys. Res. Lett.*, 37, L19605. http://dx.doi.org/10.1029/ 2010GL044770.

Roe, G., M. Baker, and F. Herla (2017). Centennial glacier retreat as categorical evidence of regional climate change. *Nature Geoscience*,10, 95, doi:10.1038/NGEO2863.

Rosenau, R., M. Scheinert, and R. Dietrich (2015). A processing system to monitor Greenland outlet glacier velocity variations at decadal and seasonal time scales utilizing the Landsat imagery. *Remote Sensing of Environment*, 169, 1–19. https://doi.org/10.1016/j.rse.2015.07.012.

Sabadini, R., D. A. Yuen, and E. Boschi (1982). Polar wandering and the forced responses of a rotating, multi-layered, viscoelastic planet. *Journal of Geophysical Research: Solid Earth*, 87(B4), 2885–2903, doi:10.1029/JB087iB04p02885.

Sasgen, I., M. van den Broeke, J. L. Bamber, E. Rignot, L. S. Sørensen, B. Wouters, Z. Martinec, I. Velicogna, and S. B. Simonsen (2012). Timing and origin of recent regional ice-mass loss in Greenland. *Earth Planet. Sci. Lett.*, 333–334, 293–303.

Sasgen, I., H. Konrad, E. R. Ivins, M. R. Van den Broeke, J. L. Bamber, Z. Martinec, and V. Klemann (2013). Antarctic ice-mass balance 2003 to 2012: regional reanalysis of GRACE satellite gravimetry measurements with improved estimate of glacial-isostatic adjustment based on GPS uplift rates. *The Cryosphere*, 7, 1499–1512.

Scambos, T. and C. Shuman (2016). Comment on 'Mass gains of the Antarctic ice sheet exceed losses' by H. J. Zwally and others (2015). *Journal of Glaciology*, 62(233), 599–603, doi:10.1017/jog.2016.59.

Schoof, C. (2007). Ice sheet grounding line dynamics: Steady states, stability, and hysteresis. *Journal of Geophysical Research: Earth Surface*, 112(F3).

Schoof, C. (2011). Marine ice sheet dynamics. Part 2. A Stokes flow contact problem. *Journal of Fluid Mechanics*, 679, 122–155.

Schrama, E. J. O. and B. Wouters (2011). Revisiting Greenland ice sheet mass loss observed by GRACE. *J. Geophys. Res.*, 116, B02407. http://dx.doi.org/10.1029/2009JB006847.

Schrama, E. J., B. Wouters, and R. Rietbroek (2014). A mascon approach to assess ice sheet and glacier mass balances and their uncertainties from GRACE data. *Journal of Geophysical Research: Solid Earth*, 119(7), 6048–6066.

Shepherd, A., E. R. Ivins, A. Geruo, V. R. Barletta, M. J. Bentley, S. Bettadpur et al. (2012). A reconciled estimate of ice-sheet mass balance. *Science*, 338(6111), 1183–1189.

Shepherd, A., E. Ivins, E. Rignot, B. Smith, M. van den Broeke, I. Velicogna et al. (2018). Mass balance of the Antarctic Ice Sheet from 1992 to 2017. *Nature, 556*, 219–222.

Siemes, C., P. Ditmar, R. E. M. Riva, D. C. Slobbe, X. L. Liu, and H. H. Farahani (2013). Estimation of mass change trends in the Earth's system on the basis of GRACE satellite data, with application to Greenland. *J. Geod.*, 87, 69–87.

Slangen, A. B. A., C. A. Katsman, R. S. W. van de Wal, L. L. A. Vermeersen, and R. E. M. Riva (2012). Towards regional projections of twenty-first century sea-level change based on IPCC SRES scenarios. *Climate Dynamics*, 38(5–6), 1191–1209, doi:10.1007/s00382-011-1057-6.

Slangen, A. B. and J. T. Lenaerts (2016). The sea level response to ice sheet freshwater forcing in the Community Earth System Model. *Environmental Research Letters*, 11(10), 104002.

Slobbe D. C., P. Ditmar, and R. C. Lindenbergh (2009). Estimating the rates of mass change, ice volume change and snow volume change in Greenland from ICESat and GRACE data. *Geophys. J. Int.*, 176, 95–106.

Sørensen, L. S., S. B. Simonsen, K. Nielsen, P. Lucas-Picher, G. Spada, G. Adalgeirsdottir, R. Forsberg and C. S. Hvidberg (2011). Mass balance of the Greenland ice sheet (2003–2008) from ICESat data – the impact of interpolation, sampling and firn density. *The Cryosphere*, 5, 173–186.

Stanford, J. D., R. Hemingway, E. J. Rohling, P. G. Challenor, M. Medina-Elizalde, and A. J. Lester (2010). Sea-level probability for the last deglaciation: a statistical analysis of far-field records. *Global Planet. Change*, 79(3), 193–203. http://dx.doi.org/10.1016/j.gloplacha.2010.11.002.

Sutterley, T. C., I. Velicogna, E. Rignot, J. Mouginot, T. Flament, M. R. van den Broeke, J. M. van Wessem, and C. H. Reijmer (2014). Mass loss of the Amundsen Sea embayment of West Antarctica from four independent techniques. *Geophysical Research Letters*, 41(23), 8421–8428, doi:10.1002/2014GL061940.

Tamisiea, M. E. and J. X. Mitrovica (2011). The moving boundaries of sea level change: understanding the origins of geographic variability. *Oceanography*, 24(2), 24–39, doi:10.5670/ oceanog.2011.25.

Taylor, K. E., R. J. Stouffer, and G. A. Meehl (2012). An Overview of CMIP5 and the experiment design. *Bulletin of the American Meteorological Society*, 93(4), 485–498, doi:10.1175/Bams-D-11-00094.1.

Tedesco, M., J. E. Box, J. Cappelen, R. S. Fausto, X. Fettweis, T. Mote, C. J. P. P. Smeets, D. van As, I. Velicogna, R. S. W. van de Wal, and J. Wahr (2016). Greenland ice sheet [in Arctic Report Card 2016], http://www.arctic.noaa.gov/Report-Card.

Thomas, I. D., M. A. King, M. J. Bentley, P. L. Whitehouse, N. T. Penna, S. D. P. Williams, R. E. M. Riva, D. A. Lavallee, P. J. Clarke, E. C. King, R. C. A. Hindmarsh, and H. Koivula (2011). Widespread low rates of Antarctic glacial isostatic adjustment revealed by GPS observations. *Geophysical Research Letters*, 38, L22302, doi:10.1029/2011GL049277.

Thomas, R. H. (1979). The dynamics of marine ice sheets. *Journal of Glaciology*, 24, 167–177.

Thomas, R. H., E. Frederick, W. Krabill, S. Manizade, and C. Martin (2006). Progressive increase in ice loss from Greenland. *Geophys. Res. Lett.*, 33, L10503.

van den Broeke, M., J. Bamber, J. Ettema, E. Rignot, E. Schrama, W. van de Berg et al. (2009). Partitioning recent Greenland mass loss. Science, 326. http://dx.doi.org/10.1126 /science.1178176.

van de Wal, W., P. L. Whitehouse, and E. J. O. Schrama (2015). Effect of GIA models with 3D composite mantle viscosity on GRACE mass balance estimates for Antarctica. *Earth and Planetary Science Letters*, 414, 134–143, doi:10.1016/j.epsl.2015.01.001.

van de Wal, R.S.W., and Wild, M. (2001). Modelling the response of glaciers to climate change by applying volume-area scaling in combination with a high resolution GCM. *Climate Dynamics*, 18(3–4), 359–366.

Vaughan, D.G., J.C. Comiso, I. Allison, J. Carrasco, G. Kaser, R. Kwok, P. Mote, T. Murray, F. Paul, J. Ren, E. Rignot,

O. Solomina, K. Steffen, and T. Zhang (2013). Observations: Cryosphere. In: *Climate Change 2013: The Physical Science Basis. Contribution of Working Group I to the Fifth Assessment Report of the Intergovernmental Panel on Climate Change* [Stocker, T.F., D. Qin, G.-K. Plattner, M. Tignor, S.K. Allen, J. Boschung, A. Nauels, Y. Xia, V. Bex, and P.M. Midgley (eds.). Cambridge and New York: Cambridge University Press, pp. 317–382, doi:10.1017/CBO9781107415324.012.

Velicogna, I., 2009. Increasing rates of ice mass loss from the Greenland and Antarctic ice sheets revealed by GRACE. *Geophys. Res. Lett.*, 36, L19503.

Velicogna, I. and J. Wahr (2006). Acceleration of Greenland ice mass loss in spring 2004. *Nature*, 443, 328–331.

Velicogna, I. and J. Wahr (2013). Time-variable gravity observations of ice sheet mass balance: precision and limitations of the GRACE satellite data. *Geophysical Research Letters*, 40(12), 3055–3063.

Velicogna, I., T. C. Sutterley, and M. R. van den Broeke (2014). Regional acceleration in ice mass loss from Greenland and Antarctica using GRACE time-variable gravity data. *Geophysical Research Letters*, 41(22), 8130–8137, doi:10.1002/2014GL061052.

Vieli, A. and F. M. Nick (2011). Understanding and modelling rapid dynamic changes of tidewater outlet glaciers: issues and implications. *Surv. Geophys.*, 32, 437–458.

Wahr, J., S. A. Khan, T. van Dam, L. Liu, J. H. van Angelen, M. R. van den Broeke, and C. M. Meertens (2013). The use of GPS horizontals for loading studies, with applications to northern California and southeast Greenland. *Journal of Geophysical Research: Solid Earth*, doi:10.1002/jgrb.50104.

Weber, M. E., P. U., Clark, G. Kuhn, A. Timmermann, D. Sprenk, R. Gladstone, X. Zhang, G. Lohmann, L. Menviel, M.O. Chikamoto, and T., Friedrich (2014). Millennial-scale variability in Antarctic ice-sheet discharge during the last deglaciation. *Nature*, 510(7503), 134.

Whitehouse, P. L., M. J. Bentley, G. A. Milne, M. A. King, and I. D. Thomas (2012). A new glacial isostatic adjustment model for Antarctica: calibrated and tested using observations of relative sea-level change and present-day uplift rates. *Geophys. J. Int.*, 190, 1464–1482.

Williams, S. D., P. Moore, M. A. King, and P. L. Whitehouse (2014). Revisiting GRACE Antarctic ice mass trends and accelerations considering autocorrelation. *Earth and Planetary Science Letters*, 385, 12–21.

Wouters, B., D. Chambers, and E. J. O. Schrama (2008). GRACE observes small-scale mass loss in Greenland. *Geophys. Res. Lett.*, 35, L20501.

Wouters, B., J. L. Bamber, M. R. van den Broeke, J. T. M. Lenaerts, and I. Sasgen (2013). Limits in detecting acceleration of ice sheet mass loss due to climate variability. *Nat. Geosci.*, 6(8), 613–616. http://dx.doi.org/10.1038/ngeo1874.

Wouters, B., A. Martin-Español, V. Helm, T. Flament, J. M. van Wessem, S. R. M. Ligtenberg, M. R. van den Broeke, and J. L. Bamber (2015). Dynamic thinning of glaciers on the Southern Antarctic Peninsula. *Science*, 348(6237), 899–903, doi:10.1126/science.aaa5727.

Xu, Z., E. J. O. Schrama, W. van de Wal, M. van den Broeke, and E. M. Enderlin (2016). Improved GRACE regional mass balance estimates of the Greenland ice sheet cross-validated with the input–output method. *The Cryosphere*, 10(2), 895–912, doi:10.5194/tc-10-895-2016.

Zemp, M., H. Frey, I. Gartner-Roer, S. U. Nussbaumer, M. Hoelzle, F. Paul et al. (2015). Historically unprecedented global glacier decline in the early 21st century. *Journal of Glaciology*, 61(228), 745–762, doi:10.3189/2015JoG15J017.

Zemp, M., M. Hoelzle, and W. Haeberli (2009). Six decades of glacier mass-balance observations: a review of the worldwide monitoring network. *Annals of Glaciology*, 50(50), 101–111.

Zwally, H. J., M. B. Giovinetto, J. Li, H. G. Cornejo, M. A. Beckley, A. C. Brenner, J. L. Saba, and D. Yi (2005). Mass changes of the Greenland and Antarctic ice sheets and shelves and contributions to sea-level rise: 1992–2002. *J. Glaciol.*, 51, 509–527.

Zwally, H. J., J. Li, A. C. Brenner, M. Beckley, H. G. Cornejo, J. Dimarzio, M. B. Giovinetto, T. A. Neumann, J. Robbins, J. L. Saba, D. Yi, and W. Wang (2011). Greenland ice sheet mass balance: distribution of increased mass loss with climate warming; 2003–07 versus 1992–2002. *J. Glaciol.*, 57, 88–102.

Zwally, H. J., J. Li, J. W. Robbins, J. L. Saba, D. Yi, and A. C. Brenner (2015). Mass gains of the Antarctic ice sheet exceed losses. *Journal of Glaciology*, 61(230), 1019–1036.

Zwally, H. J., J. Li, J. W. Robbins, J. L. Saba, D. Yi, and A. C. Brenner (2016a). Response to Comment by T. Scambos and C. Shuman (2016) on 'Mass gains of the Antarctic ice sheet exceed losses' by H. J. Zwally and others (2015). *Journal of Glaciology*, 62(235), 990–992, doi:10.1017/jog.2016.91.

Zwally, H. J., J. Li, J. W. Robbins, J. L. Saba, D. Yi, and A. C. Brenner (2016b). Response to Comment by A. Richter, M. Horwarth, R. Dietrich (2016) on 'Mass gains of the Antarctic ice sheet exceed losses' by H. J. Zwally and others (2015). *Journal of Glaciology*, 62(235): 993–995, doi:10.1017/jog.2016.92.

Appendix 1

Studies Included in Greenland Ice Sheet Mass Balance Estimates (Figure 7.3)

Input–output method	van den Broeke et al. (2009)	Rignot et al. (2011b)
	Sasgen et al. (2012)	Enderlin et al. (2014)
	Andersen et al. (2015)	
Radar altimetry method	Zwally et al. (2005)	Johannessen et al. (2005)
	Zwally et al. (2011)	Hurkmans et al. (2014)
Laser altimetry method	Krabill et al. (2004)	Thomas et al. (2006)
	Slobbe et al. (2009)	Zwally et al. (2011)
	Sørensen et al. (2011)	Ewert et al. (2012)
	Sasgen et al. (2012)	Khan et al. (2014)
	Hurkmans et al. (2014)	
Gravimetry method	Chen et al. (2006)	Luthke et al. (2006)
	Velicogna and Wahr (2006)	Ramillien et al. (2006)
	Wouters et al. (2008)	Slobbe et al. (2009)
	Velicogna (2009)	Ewert et al. (2012)
	Sasgen et al. (2012)	Siemes et al. (2013)
	Barletta et al. (2013)	Velicogna and Wahr (2013)
	Wouters et al. (2013)	Khan et al (2014)
IMBIE (combined methods)	Shepherd et al. (2012)	

Studies included in Antarctic Ice Sheet mass balance estimates (Figure 7.4)

*Radar altim*etry	Helm et al. (2014)	McMillan et al. (2014)
Gravimetry method	King et al. (2012)	Barletta et al. (2013)
	Ivins et al. (2013)	Luthcke et al. (2013)
	Sasgen et al. (2013)	Velicogna and Wahr (2013)
	Schrama et al. (2014)	Velicogna et al. (2014)
	Williams et al. (2014)	Harig and Simons (2015)
IMBIE (combined methods)	Shepherd et al. (2012)	
	Shepherd et al. (2018)	

8
Geographical Research and Future Earth

Michael E. Meadows

The question is not so much as whether, but how and to what degree, geographers might make their voices heard as society confronts accelerating environmental change (Castree, 2015: 4)

8.1 Introduction

Geography has a long and distinguished history as a scholarly discipline. In essence, geography is the subject that attempts to explore how environments emerge by natural processes; how societies produce, organize, use, and misuse environments and how societies themselves are influenced by the environments in which they are located. In straddling the natural and social sciences and humanities, geography aims to integrate the study of both natural and human realms and their interactions, focusing on space, places and regions, addressing and questioning both short-term and longer-term processes and their resultant patterns. In this sense, geography as a science is critical to understanding the processes, patterns and trajectories of our future Earth.

In pondering the potential role of the discipline in the context of the Future Earth platform it becomes evident that, while geography as an integrative discipline obviously lies at the very core of understanding human–environment relations, there has been a trend in academic departments towards divergence and a separation of 'human' from 'physical geography'. For example, Goudie (2016) notes that most physical geographers now publish their peer-reviewed research in periodicals other than those that may be regarded as mainstream geographical journals. The trend has manifested in other ways too, including the splitting of many university departments, with physical geographers joining their earth science cousins and the human geographers teaming up with, among others, anthropologists and sociologists. So, is the integrating capacity of geography a 'historical relic' and 'more rhetoric than reality' (Sharpe, 2009: 123), or is it, as argued by Lave et al. (2014), a possible source of renewed scholarly vigor?

Future Earth, and the array of human-environmental problems that have inspired it, challenges geographers to begin to rediscover the importance of the multifarious and complex connections between nature and society and recognize that a more fundamental (and useful) understanding of landscapes – and how their diverse components respond to, say, global climate change – *requires* explanations that are rooted in both the physical and the human (social, economic, political and cultural) domains. This is essentially what Lave et al. (2014) refer to as 'critical physical geography'. However, this should be more than human and physical geographers simply collaborating on problems of common interest but rather a fundamental recognition of the concept that, indeed, 'socio-biophysical landscapes are as much the product of unequal power relations, histories of colonialism, and racial and gender disparities as they are of hydrology, ecology, and climate change' (Lave et al., 2014: 2). No doubt Castree (2015a, b) would strongly concur that there must be more to global change science than simply incorporating more social science into the geosciences. Future Earth provides a platform for a broad spectrum of geographers to do so as we transition into what is increasingly being referred to as the Anthropocene. The term itself may be contested but as a concept it has proved a rallying point for the kind of integrated research that Future Earth requires. Certainly the time appears to be right for a more fundamental kind of engagement – in commenting on the debate around the formal designation of the Anthropocene as a stratigraphic entity, Cook et al. (2015: 231) reflect that 'if humans have become a rival to Nature, the epic nomenclature of the great forces – the eras, periods, and epochs of geological time – have finally been reconciled with the social'.

8.2 Geography and Global Environmental Change

In a recent essay revisiting the integration of physical and human geography, Goudie (2016) outlines seven key areas that offer intellectually challenging and potentially productive research opportunities, *viz*. natural hazards, global environmental change, earth system science, human impact, the Anthropocene, environmental history and the study and appreciation of landscapes. A number of these issues are very closely related and all, except perhaps the latter, fall under the conceptual umbrella of Future Earth. At the heart of these themes are the nature and scale of human impact on earth system processes, as well as its inverse – the impact of environment on people. These issues have long been a concern for geographers, both physical and human, who recognized the human–environment nexus as a stage on which they could ply their trade, albeit that by no means all were convinced of its value as a conceptual base (see Goudie, 1986). Nevertheless, numerous geographers have engaged profoundly in research and teaching with a focus on human impact. Physical geographers especially have made contributions to global environmental change and what is broadly termed earth system science, while – increasingly of late – they have turned to the Anthropocene as a lens through which to view their subject, the concept itself spawned out of a realization of the sheer magnitude of human impact on the global environment. In addition, the involvement of geographers in hazard and risk research (as well as practice, see Montz and Tobin, 2011) and GIS and remote sensing indicates that geographers have already had considerable intellectual influence on the spectrum of issues related to the theme of Future Earth. I reflect here on some of these contributions in making the case for the considerable potential of Future Earth to engage geographers from across the breadth of the discipline.

8.3 Human Impact on the Environment: Past, Present and 'Post-Holocene'

The classic 'Man's Role in Changing the Face of the Earth' (Thomas, 1955) summarized what was then known of the scale of human influence in transforming landscapes. Since then there have been many global reviews (see especially Turner et al., 1990, Meyer, 1996, Govorushko, 2016). The history of this kind of approach goes back much further and the legacy of George Perkins Marsh has been regularly lauded (most recently in Goudie, 2016). There is more to this approach than description and simplistic explanations of how people alter physical environmental conditions and induce undesirable, unintended consequences. Geographers following in the tradition of Carl Sauer (Leighly, 1976) contributed both physical and human geographical perspectives to an understanding of the dynamic and reciprocal role of people in the landscape. In South Africa; by way of example, Talbot's (1944) treatise on the impact of agriculture on soil degradation in the wheatlands to the north of Cape Town illustrates the complex and changing role of politics and economics in both initiating accelerated erosion and in implementing interventions to control it (see Meadows, 2003, 2013). While the human impact approach was – and is – favoured by physical geographers, the work of William Beinart (see, for example, Beinart, 1984) clearly demonstrates the way in which geography, political ideology, economics and culture all express themselves in landscapes at various temporal and spatial scales and that a linear, positivist approach can never achieve an adequate explanation. Human impact studies are also inextricably linked with environmental history and perhaps this in part underlies why there is a vibrant tradition among some biogeographers for Quaternary environmental reconstruction.

Clearly the amplification of the scale of human impact arising from exponential population growth accompanied by unprecedented technological development, industrialization and globalization has taken its toll on the global environment. The term 'Anthropocene' was introduced to the world of science at the turn of the new millennium (Crutzen and Stoermer, 2000) and has become a focus for geographers. While the idea that we are in a 'post-Holocene' world is not without its critics, it is abundantly clear that the Anthropocene is here to stay, 'propelled in a dizzying array of directions as if filling a linguistic vacuum' as Revkin (2016: www.anthropocenemagazine.org/anthropocenejourney/) puts it.

The term was introduced with the intention of emphasising the increasing impact of humans on the environment to the point at which they now dominate other 'natural' processes. Thus, anthropogenic impacts are considered both sufficiently potent and global in their extent to warrant the delimitation of a new post-Holocene epoch. This leads to the idea that we have created a 'geology of mankind' (Crutzen, 2002). There is no doubt that many geographers have embraced the concept, albeit not without a degree of circumspection. Geomorphologists in particular have proved especially willing participants in the debate. Brown et al. (2013) suggested that there is a 'geomorphological case' for the Anthropocene, while Lewin and Macklin (2014) assessed the possible value for geomorphologists of a formally defined Anthropocene. Goudie and Viles (2016) provide a comprehensive argument for the logic of geomorphological research viewed through an Anthropocene lens, while Meadows and Lin (2016) offer a range of case studies that offer ample evidence of the importance of geomorphology to society in the same context. A special issue of the journal *Earth Surface Processes* has emerged recently to consider the issue in depth, although the editorial piece (Brown et al., 2017) concludes that it is probably wiser for geomorphologists to stay out of the stratigraphical debate and rather continue to investigate 'processes and landform development, so providing the underpinning science for the study of this time of critical geological transition' (p. xx). This is a relatively conservative view in that it calls on geomorphologists essentially to 'stick to geomorphology' and this would certainly not satisfy Castree, whose 2015b essay argues for a more radical approach to geographical engagement with the Anthropocene concept and global change science in general.

Of course there are geographers who appear to be more comfortable adopting an approach to understanding the impacts of – for example, climate change – in ways that go beyond the purely scientific. Shackleton et al. (2015) present a fine example of this in advocating climate change adaptation research that focuses on issues of social justice, while Connelly-Boutin and Smit (2016) emphasise the global change research paradigms that have implications for the issue of food insecurity, which remains a blight in much of the developing world.

8.4 Hazards, Risks and Disasters

It is broadly accepted that the frequency and magnitude of extreme hydroclimatic events have increased in recent years and that this trend will continue because of increasing population, in terms of both numbers and density, as a result of accelerating urbanization against a background of climate change. Disasters emanating from these extreme events, together with those caused by earthquakes (most notably the catastrophic 9.0 magnitude event that devastated eastern Japan in March 2011), have prompted global initiatives, including the Sendai Framework for Disaster Risk Reduction 2015–2030 (United Nations, 2015). The stated goal of the framework is 'the substantial reduction of disaster risk and losses in lives, livelihoods and health and in the economic, physical, social, cultural and environmental assets of persons, businesses, communities and countries' (p. 12). It is immediately obvious that, if such a goal is to be anything like close to being achieved by 2030, then a multidimensional, integrated approach is needed that not only encompasses robust scientific assessments but takes account of institutional, political, health, psychological, educational, legal and economic factors, among others. Of course, this all goes way beyond the remit of academic geography; no single discipline could possibly hope to incorporate such a wide array of elements, but working across disciplinary boundaries would seem to be a promising first action step.

Indeed, geographers do work routinely and willingly across such boundaries, as is illustrated in several recently published case studies from East Asia. Taiwan, an island that is exposed to the influence of both frequent high-rainfall events and a high degree of tectonic activity, is a test case for the study of physical processes and human responses to environmental hazards. As Meadows and Lin (2016: p. 3) note: 'A combination of earthquakes – the island sits on the northwestern margin of the Philippine tectonic plate – steep topography, high relief, frequent summer and early autumn typhoons together with high urban population densities is a concoction that all too often can induce catastrophe.' The combination induces frequent landslides, river channel and coastal sediment and morphology dynamics that demand a high level of detailed technical competence coupled with a deep understanding of the social, political, economic and cultural circumstances if future catastrophes are to be prevented. Lin (2016), Jen et al. (2016), Shen (2016) and Su (2016) offer various perspectives on these hazards but all make the point strongly that integrating knowledge of the physical environment with that of society is essential to the reduction of disaster risk in Taiwan. Geographical contributions have also been prominent in reflecting on responses to the underwater earthquake that devastated eastern Japan in March 2011. Yoshino (2014) considers various everyday life and health aspects of the tsunami that was triggered by that event, while geographers are prominent among the list of authors and co-authors in Karan and Sugunuma's (2016) edited volume on the aftermath of the Tohoku disaster and explore issues such as the socioeconomic scope of response, recovery and reconstruction efforts as well some of the more physical environmental dimensions.

It is obviously inadequate to dwell only on past disasters, even if these can be used as a basis for learning about the kinds of responses necessary to avert crisis in the future. As Singh (www.southasiadisasters.net) points out, geographers need also to get involved at the science-policy interface, recognize the importance of multi-stakeholder approaches, help to develop and disseminate better participatory spatial technologies and promote public education to yield meaningful disaster risk reduction results – all of which are highly valuable but arguably less comfortable places for academic geographers to find themselves.

8.5 Remote Sensing and GIS

Both remote sensing and GIS are key tools in geography and, of course, in many other sciences. While the use of spatial technology does not in itself constitute a theoretical advance towards resolving the challenges facing humankind, it can be instrumental in facilitating a much more fundamental understanding of earth system processes and the relationships between them at various scales, in both the present day and the recent past. The extensive availability of satellite imagery at increasingly high levels of spatial resolution, much of which is open access and multitemporal, is an incredibly valuable resource. Remote sensing has typically been widely applied to measurement and monitoring of various aspects of the physical environment, but it is equally potent in observing the spatial manifestation of human activities, such as patterns of urbanization. Qian et al. (2016) provide a dramatic example, based on satellite data and official land use maps, of the remarkably rapid growth of the southern Chinese city of Shenzhen. As in the other mega-cities in the region, including Hong Kong and Guangzhou, construction land has expanded at unprecedented rates (up to 16.9% annually in Shenzhen). Such data would be difficult to obtain and interpret without the sequential imagery that satellite technologies afford.

Other forms of terrestrial or near-terrestrial technologies are now widely used for geospatial applications, such as unmanned aerial vehicles (Williams, 2016) and mobile phones (Deville et al., 2014), which reveal their enormous potential. In coupling the remotely sensed data with the analytical power of GIS, geographers, along with other scientists, have an extraordinary opportunity for integration across the physical and human domains, surely something that is essential to a fuller engagement with Future Earth. Geographers have consistently demonstrated the means by which satellite-derived information can be used to deepen our understanding of the connections between environment and society, for example through studying the relationships between remotely sensed land use change and other kinds of socio-economic data (see, for example, Lambin et al., 2013). As technology and data availability develop further, this kind of work is surely set to expand.

8.6 Steps to Sustainability?

The three identified research themes of Future Earth are designed to make progress in understanding a) the dynamic planet, b) global sustainable development and c) transformations to sustainability aimed towards the lofty goal of people thriving in a sustainable and equitable world (www.futureearth.org). These research goals also need to be seen in the context of the United Nations 17 Sustainable Development Goals (SDGs, www.un.org/sustainabledevelopment/sustainable-development-goals/; Griggs et al., 2013). The case has been outlined that geographers are in a position to participate in relevant aspects of earth systems science and, more broadly, global change science, as well as in making important intellectual inputs to sustainable development debates. The 'transformations to sustainability' theme is founded on the need to develop strategies and mechanisms that actually deliver sustainable solutions, and therefore goes beyond the scientific research that underpins Future Earth into the realms of education and advocacy. Even the highest-quality scientific work in this field can fail to translate into action if it is not appropriately communicated and converted into policy and practice across multiple sectors, understanding that actions on the part of individuals, communities, organisations, corporations and governments is one thing; changing behavior is another altogether. It is in this context that the International Geographical Union (IGU) initiative 'International Year of Global Understanding' (IYGU: www.global-understanding.info/) can perhaps play a role, and an emerging partnership with Future Earth would seem expedient in relation to the transformations research theme in particular.

The kernel of IYGU is best explained through the concept that everyday actions (by everybody) ultimately influence global change and that everyday decisions are related to and, indeed, depend on lifestyle. In essence: 'The International Year of Global Understanding (IYGU) addresses the ways in which we inhabit an increasingly globalized world. How do we transform nature? How will we build new social and political relationships for the emerging global reality? Societies and cultures determine the ways we live with and shape nature. They also influence how we perceive the global consequences of our everyday behavior. You can't change the world if you don't understand what your daily actions mean for the world as a whole' (IYGU brochure: www.global-understanding.info/). The International Council for Science (ICSU), the International Social Science Council (ISSC) and the International Council for Philosophy and the Humanities (CIPSH) jointly proclaimed 2016 as the International Year of

Global Understanding and, while its effectiveness as a transformational initiative remains to be seen, its underlying philosophy is contiguous with that of Future Earth.

A series of open access books ('briefs') is being published by IYGU (see, for example, Robinson et al., 2016), designed to promote a local to global perspective in respect of a diverse range of social, environmental, political and even cultural contexts. Each of these is an attempt to stimulate teaching and scientific research and at the same time raising public awareness. The series of books is intended to focus attention on issues of everyday life and sustainability in the light of the UN SDGs. The Regional Action Centers of IYGU (www.global-understanding.info/promotional-material/) are now established on every continent and there are clear opportunities for – and advantages to – working alongside Future Earth's own regional offices and fostering collaboration through its Open Network (www.futureearth.org/join-future-earth-open-network). IYGU is also consistent with the idea of 'Knowledge–Action Centers' (www.futureearth.org).

The key message from IYGU is one that the Future Earth platform should certainly embrace and help to disseminate – that global sustainability is impossible without local sustainability and that thoughts and actions by individuals that may seem disconnected in space and time are often fundamentally linked.

8.7 The Future of Future Earth and Geography?

Clearly, holistic, integrated approaches, characteristic of both 'traditional' and modern geographic research, offer important pointers towards the achievement of a more sustainable planet and they are central to the success of the Future Earth platform. Can geographers deliver that? The discipline historically was rooted in such integrated thinking and, while there has been divergence and specialization for many (Goudie, 2016), the preceding narrative suggests that some geographers have engaged – and will continue – to engage in interdisciplinary research endeavours to considerable effect. But is this enough? Castree (2015a, b) has offered a deeply thoughtful analysis of the place of geography in relation to current global change debates and is rather cautious, to say the least, regarding its role. He argues that the picture of geography as a holistic discipline 'that reaches the parts that others cannot' (Castree, 2015a: 2) is overplayed and not in a particularly good position either to redirect the discourse fundamentally around concepts such as the Anthropocene, or even to join the geoscientists 'on the march' in global change science. Castree's critique of the global change arena suggests that it fosters a hegemonic scientific worldview that 'refuses to fundamentally question the current political, economic and cultural order of things' (2016a: 7) and 'brings us out from under the shadow of these longstanding narratives' (2015a: 1). Geographers are not often regarded as influential in politics in the same way that, for example, economists are – but opportunities are surely emerging to inject 'new blood' into global change science debates and into policy.

While Castree's view is that only a handful of geographers will join the kind of 'thought collective' that could really make a difference to Future Earth's trajectory, there are some conceptual gaps that are perhaps more suited to their attention. Indeed, a global change research agenda that focuses on the physical environment and sees society (especially its less tangible and easily measurable attributes) as an 'add-on' is likely to fail. While that is certainly not the stated intention of Future Earth, the 'scientific-analytic' narrative employed is often couched in terms such as factors, drivers, causes, effects, feedbacks etc. that evolved from systems thinking and there is a danger that elements of society become yet another 'component' to be modeled. As Rickards (2015: 280) notes, the Anthropocene is a 'deeply cultural phenomenon' and there is a need for appropriate theory in resolving its manifestations that the geoscientists are unlikely to be much interested in. Harnessing the philosophical and critical tradition of human geography therefore seems an entirely logical and necessary intervention, even if Castree might argue that it is likely not radical enough.

There are still other challenges for geographers to confront. The Anthropocene, a widely accepted label irrespective of whether or not geologists are prepared to afford it the formal title of epoch, is an important underlying theme in Future Earth. But, as Cook and Balayannis (2015: 270) note, it is a concept 'steeped in fear' because of the adoption of the 'great acceleration' imagery of human impact and the kind of inexorable, directional escalation of problems that may imbue a sense of hopelessness. 'As a group, through shape, title, color, and font, these images co-produce an unavoidable sense of doom' (Cook and Balayannis, 2015: 275). While highlighting the trends of human impact on Earth is designed to educate and awaken interest and action, it can invoke fear and a sense of being overwhelmed that then has the consequence of inducing what might be called 'green fatigue' and apathy or, worse still, opposition to intended response. This is exacerbated by the very 'global' nature of environmental change as it can be seen to disempower, or at least marginalize, the ordinary person so that behaviors that might otherwise promote sustainability are eschewed. The kind of bottom-up, local-to-global philosophy (Werlen, 2015) that is espoused by IYGU represents an alternative approach that could begin to address essentially this kind of problem.

This begs the question 'What sort of geographical education for the Anthropocene?' (Pawson, 2015); Pawson, in answering it, highlights the imperative of addressing issues relating to the human–nature nexus at a scale that is relevant to individual or community action, i.e., 'countering grand narratives with local encounters' (p. 312). He goes on to recommend education – not only for geographers of course – that offers alternatives that are hopeful rather than laden by gloom. Active learning and community-based learning are two of the possible strategies that are advanced to achieve this end. Renegotiating and co-designing the curriculum in partnerships with, for example, the emerging generation of 'Anthropoceneans' (Head, 2015) seem desirable. After all, they have to live with the consequences of our decisions and actions.

Castree's (2015a) view is that the 'problematic' of the relationship between nature and society is currently more important than ever in higher education. It is undoubtedly true that not all – perhaps even not many – geographers will heed Castree's call for a more radical approach and campaign for the new political, economic and cultural order that he argues is necessary to address the current crisis fundamentally, but it still has an important ongoing place in the spectrum of natural and social science research. Indeed, contemporary geography may be heterogeneous but it has consistently demonstrated its effectiveness as a platform for synthesis and should, therefore, prove to be a highly valued source of intellectual reinforcement for Future Earth. In its turn, Future Earth perhaps presents a suitable unifying vehicle for the discipline itself.

References

Beinart, W. (1984). Soil erosion, conservationism and ideas about development: a southern African exploration, 1900–1960. *J. Southn. Afr. Stud.* **11**(1), 52–83.

Brown, A. G., Tooth, S., Chiverrell, R. C., Rose, J., Thomas, D. S. G., Wainwright, J., Bullard, J. E., Thorndycraft, V. R., Aalto, R., and Downs, P. (2013). The Anthropocene: is there a geomorphological case? *Earth Surf. Process Landf.* **38**, 431–434. doi: 10.1002/esp.3368.

Brown, A. G., Tooth, S., Bullard, J. E., Thomas, D. S. G., Chiverell, R. C., Plater, A. J., Murton, J., Thorndycraft, V. R., Tarolli, P., Rose, J., Wainwright, J., Downs, P., and Aalto, R. (2017). The geomorphology of the Anthropocene: emergence, status and implications. *Earth Surf. Process Landf.* doi: 10.1002/esp.3943.

Castree, N. (2015a). Geography and global change science: relationships necessary, absent, and possible. *Geogr. Res.* **53**(1), 1–15. doi: 10.1111/1745–5871.12100.

Castree, N. (2015b). Geographers and the discourse of an earth transformed: influencing the intellectual weather or changing the intellectual climate? *Geogr. Res.* **53**(1), 244–254. doi: 10.1111/1745–5871.12125.

Connelly-Boutin, L., and Smit, B. (2016). Climate change, food security, and livelihoods in sub-Saharan Africa. *Reg. Environ. Change.* **16**, 385–399. doi: 10.1007/s10113-015–0761-x.

Cook, B. R., and Babayannis, A. (2015). Co-producing (a fearful) Anthropocene. *Geogr. Res.* **53**(3), 270–279. doi: 10.1111/1745–5871.12126.

Cook, B. R., Rickards, L. A., and Rutherford, I. (2015). Geographies of the Anthropocene. *Geogr. Res.* **53**(3), 231–243. doi: 10.1111/1745–5871.12127.

Crutzen, P. J. (2002). Geology of mankind. *Nature*, **413**, 23.

Crutzen, P. J., and Stoermer, E. F. (2000). The 'Anthropocene'. *IGBP Newsletter* **41**, 17–18.

Deville, P., Linard, C., Martin, S., Gilbert, M., Stevens, F. R., Gaughan, A. E., Blondel, V. D., and Tatem, A. J. (2014). Dynamic population mapping using mobile phone data. *Proc. Natl. Acad. Sci.* **111**(45), 15888–15893. doi: 10.1073/pnas.1408439111.

Goudie, A. S. (1986). The integration of human and physical geography. *Trans. Inst. Br. Geogr.* **11**(4), 454–458.

Goudie, A. S. (2016). The integration of human and physical geography revisited. *Canadian Geographer*. doi: 10.1111/cag.12315.

Goudie, A. S., and Viles, H. (2016). *Geomorphology and the Anthropocene*. Cambridge University Press.

Govorushko, S. (2016). *Human Impact on the Environment: An Illustrated Atlas*. Springer.

Griggs, D., Stafford-Smith, M., Gaffney, O., Rockström, J., Öhman, M. C., Shyamsundar, P., Steffen, W., Glaser, G., Kanie, N., and Noble-Hamilton, I. (2014). Policy: sustainable development goals for people and planet. *Nature* **495**, 305–307. doi:10.1038/495305a.

Head, L. (2015). The Anthropoceneans. *Geogr. Res.* **53**(3), 313–320. doi: 10.1111/1745–5871.12124.

Jen, C. H., Lei, H. F., Chyi, S. J., and Hsueh, M. Y. (2016). The impact of Typhoon Morakot in 2009 on landslides, debris flows and population in the Chishan River catchment, Taiwan. In: M. E. Meadows, and J.-C. Lin (Eds.). *Geomorphology and Society*. Springer, 59–82.

Karan, P. P. and Suganuma, U. (Eds.) (2016). *Japan After 3/11: Global Perspectives on the Earthquake, Tsunami and Fukushima Meltdown*. Kentucky University Press.

Lambin, E. F., Gibbs, H. K., Ferreira, L., Grau, R., Mayaux, P., Meyfroidt, P., Morton, D. C., Rudel, T. K., Gaspari, I., and Munger, J. (2013). Estimating the world's potentially available cropland using a bottom-up approach. *Glob. Env. Ch.* **23**, 892–901. doi: 10.1016/j.gloenvcha.2013.05.005.

Lave, R., Wilson, M. W., Barron, E. S., Biermann, C., Carey, M. A., Duvall, C. S., Johnson, L., Lane, K. M., McClintock, N., Munroe, D., Pain, R., Proctor, J., Rhoads, B. L., Robertson, M. M., Rossi, J., Sayre, N. F., Simon, G., Tadaki, M., and van Dyke, C. (2014). Intervention: critical physical geography. *Canadian Geographer* **58**(1), 1–10. doi: 10.1111/cag.12061.

Leighly, J. (1976). Carl Ortwin Sauer, 1889–1975. *Geographers Biobibliographical Studies* **2**, 99–108.

Lewin, J., and Macklin, M. G. (2014). Marking time in geomorphology: should we try to formalise an Anthropocene definition? *Earth Surf. Process Landf.* **39**, 133–137. doi: 10.1002/esp.3484

Lin, J.-C. (2016). Geomorphological responses in a dynamic environment: how landforms interact with human activities in Japan. In: M. E. Meadows, and J.-C. Lin (Eds.). *Geomorphology and Society*. Springer, 43–58.

Meadows, M. E. (2003). Soil erosion in the Swartland, Western Cape Province, South Africa: implications of past and present policy and practice. *Env. Sci. Policy* **6**, 17–28. doi: 10.1016/S1462–9011(02)00122–3.

Meadows, M. E. (2013). William John Talbot. *Geographers' Biobibliographical Studies* **32**, 124–136.

Meadows, M. E., and Lin, J.-C. (Eds.). (2016). *Geomorphology and Society*. Springer.

Meyer, W. B. (1996). *Human Impact on the Earth*. Cambridge University Press.

Montz, B. E., and Tobin, G. A. (2011). Natural hazards: an evolving tradition in applied geography. *Applied Geography* **31**(1), 1–4. doi: 10.1016/j.apgeog.2010.06.005.

Pawson, E. (2015). What sort of geographical education for the Anthropocene? *Geogr. Res.* **53**(1), 306–312. doi: 10.1111/1745–5871.12122.

Qian, J., Peng, Y., Luo. C., Wu, C., and Du, Q. (2016). Urban land expansion and sustainable land policy in Shenzhen: a case study of China's rapid urbanization. *Sustainability* **8**. doi:10.3390/su8010016.

Rickards, L. A. (2015). Metaphor and the Anthropocene: presenting humans as a geological force. *Geogr. Res.* **53**(3), 280–287. doi: 10.1111/1745–5871.12128.

Robinson, J., Scott., A. J., and Talyor, P. J. (2016). *Working, Housing: Urbanizing: The International Year of Global Understanding – IYGU*. Springer.

Shackleton, S., Ziervogel, G., Sallu, S., Gill, T., and Tschakert, P. (2015). Why is socially-just climate change adaptation so challenging? A review of barriers identified from empirical cases. *WIREs Clim. Change*, 2015. doi 10.1002/wcc.335.

Sharpe, C. (2009). Past-president's address: is geography (the discipline) sustainable without geography (the subject). *Canadian Geographer* **53**(2), 123–138. doi: 10.1111/j.1541–0064.2009.00249.x.

Shen, S.-M. (2016). Delineation of historical fluvial territories and the implications for flood mitigation, with reference to four selected reaches in Taiwan. In: M. E. Meadows, and J.-C. Lin (Eds.). *Geomorphology and Society*. Springer, 83–102.

Su, S.-J. (2016). The political ecology of land subsidence: a case study of the solar energy-farming scheme, Pingtung County, Taiwan. In: M. E. Meadows, and J.-C. Lin (Eds.). *Geomorphology and Society*. Springer, 103–124.

Talbot, W. J. (1944). *Swartland and Sandveld*. Oxford University Press.

Thomas, W. L. (1955). *Man's Role in Changing the Face of the Earth*. Chicago University Press.

Turner II, B. L., Clark, W. C., Kates, R. W., Richards, J. F., Matthews, J. T., and W. B. Meyer (Eds.) (1990). *The Earth as Transformed by Human Action: Global and Regional Changes in the Biosphere over the Past 300 Years*. Cambridge University Press.

United Nations (2015). *Sendai Framework for Disaster Risk Reduction 2015–2030*. United Nations.

Werlen, B. W. (2015). From local to global sustainability: transdisciplinary integrated research in the digital age. In: B. W. Werlen (Ed.). *Global Sustainability*. Springer: 3–16.

Williams, R. (2016) The sky is the limit: reconstructing physical geography from an aerial perspective. *Journal of Geography in Higher Education*, **41**(1), 134–146. doi: 10.1080/03098265.2016.1241986.

Yoshino, M. (2014). Impact of tsunamis on human life and society. *Global Environmental Research* **18**, 3–8.

9

Water Security: Integrating Lessons Learned for Water Quality, Quantity and Sustainability

Elaine M. Faustman, Heidi Foth, Luis Schiesari, Julio Alejandro Navoni, Silvia Berlanga de Moraes Barros, Gisela de Aragão Umbuzeiro, Salmaan H. Inayat-Hussain, and Marissa N. Smith

9.1 Introduction

9.1.1 Defining Water Security

The United Nations definition for water security is the "capacity of a population to safeguard sustainable access to adequate quantities of, and acceptable quality water for, sustaining livelihoods, human well-being, and socio-economic development, for ensuring protection against water-borne pollution and water-related disasters, and for preserving ecosystems in a climate of peace and political stability" (UNU, 2013). In describing water security and its interface with Future Earth for this chapter, we will evaluate the use of the World Health Organization (WHO) risk assessment framework tool to explain how human health and ecological health are characterized and evaluated within this context. We will examine three case studies and illustrate methods for assessing water quality and human health well-being, ecosystem functioning and some aspects of food production. The first case study is on arsenic in water and soil in South America and risks of arsenic associated health impacts. The second case study is on pesticide use in sensitive areas in Brazil's Amazonian agricultural frontier and risk of loss of biodiversity. The third case study is on nitrate use in agriculture and will feature integrated approaches used by Germany. The case studies were chosen to illustrate global issues of water security and each addresses the questions in the template.

9.1.2 Case Study Template

- What is (are) the risk question(s)?
- What tools were used?
- What was observed?
- What are the key lessons?

Two workshops on water security have taken place. The first workshop was in Natal, Brazil, on October 2, 2016 as a satellite to the Congress of Toxicology for Developing Countries and the second occurred as a continuing education session at the International Congress of Toxicology in Merida, Mexico, in October 2016. This emphasis on Latin America arose from a report entitled "Climate Change and Scarcity and Quality Considerations for Water Security" published by the Inter-American Network of Academies of Sciences in 2015, and José Galizia Tundisi, a co-author of this report, opened the session in Natal, Brazil. Michael Clegg, Professor Emeritus at the University of California Irvine and Vice-President for External Relations ICSU, who spoke at both courses, provided a context from the International Council for Science (ICSU) and Future Earth.

Future Earth supplied an excellent venue for our discussions on water security issues around the globe. During the discussions, our multi-disciplinary scientists rose up to the challenge of defining approaches for thriving within a sustainable and equitable world. From these opening remarks, our health-based participants defined a risk-assessment-based framework for their case studies.

Approaches to address issues in water security for Future Earth have occurred within different regional sectors. However, overall the United Nations defines sustainable development goals and water security is a key component (UNICEF, 2015; UNU, 2013). They identify seven out of the 17 sustaibable development goals (listed in the following) that relate to various aspects of water security. For example, in goal 2, to achieve food security, there must be water security and especially water quality and sustainability because water must be available to support food production. Goal 6 specifically cites the need for sustainable water

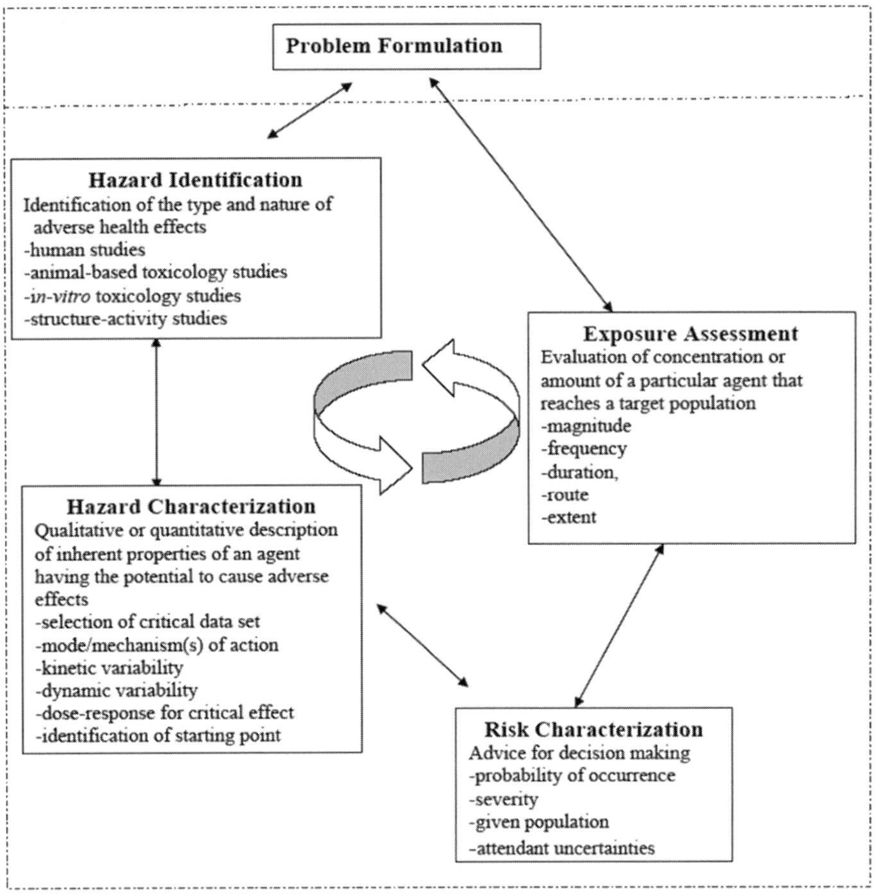

Figure 9.1 The key steps in a risk assessment (WHO, 2009).

and sanitation. As resilient infrastructure is built and sustainable industrialization occurs (goal 9), improved systems for promoting water security will be available. Goals 12 and 13 focus on sustainability and include food and water production. New sources of water in the face of local population growth and changing climate demographics are also critical to water security characterization. Historically, droughts and floods have affected our water supply for the past ten thousand years. Goals 14 and 15 promote the idea that conserved and healthy water and terrestrial ecosystems are critical to food and water security. There is an intimate interrelationship of water quality, food security and human and ecosystem health.

Seven Example Sustainable Development Goals: relevant for Future Earth and water security (WS):

- Goal 2. End hunger, achieve food security and improved nutrition and promote sustainable agriculture.
- Goal 6. Ensure availability and sustainable management of water and sanitation for all.
- Goal 9. Build resilient infrastructure, promote inclusive and sustainable industrialization and foster innovation.
- Goal 12. Ensure sustainable consumption and production patterns.
- Goal 13. Take urgent action to combat climate change and its impacts.
- Goal 14. Conserve and sustainably use the oceans, seas and marine resources for sustainable development.
- Goal 15. Protect, restore and promote sustainable use of terrestrial ecosystems, sustainably manage forests, combat desertification and halt and reverse land degradation and halt biodiversity loss.

Using risk assessment for water security:

In risk assessment, a standard framework for analysis is used and Figure 9.1 shows five key steps in risk assessment from WHO (WHO, 2009). Initially, the problem or risk that we desire to estimate is identified. This initial step for risk assessment (RA) is the "problem formulation" stage, which sets out the risk question as the object for the RA. In the examples in this chapter, the problem is related to contamination of water, compromising water security. This translates to asking a risk question aimed at understanding the human and ecological health implications of compromised water security issues related to the problem identified in the case study. Evaluation from such assessments is most often a combination of qualitative (describing the identified risks) and quantitative risk assessment steps of exposure assessment, hazard identification and risk characterization. Hazard identification is the second step, when multiple tools can be used to determine the potential health risks. This usually includes evaluation of the properties and reactivity of the chemical; during this step the chemical potency and structures are compared to other known chemicals that are toxic (i.e., compared to other carcinogens, mutagens, metals etc.). Additional scientific evidence regarding affected populations, including how the chemical or agent affects health over time or how it targets sensitive populations such as children, is discussed in the characterization steps.

Overall, *risk assessment* is defined as the estimation of the association between an exposure to an agent/process and the incidence of some adverse outcome, whereas *risk management* is the public policy developed to deal with hazards identified through risk assessment and includes the process of evaluating alternative regulatory options and selecting among them. This chapter will distinguish between risk assessment and risk management as it presents a human-health-centered view and frequently cites WHO documents on drinking water standards. However, we are placing these discussions in the larger concept of water security and sustainability, where risk management often encompasses all the steps from identification to mitigation of risks (Beer, 2003). Both approaches concur that risk assessment is critical to risk management whether the results of risk assessment drive the risk management process (human health fields) or whether risk assessment is included in the risk management process (environmental fields).

The case studies in this chapter exemplify the need to examine the seven sustainability issues from Beer (2003), listed in the following:

1. Determine concerns by using risk assessment techniques for various scenarios.
2. Identify the consequences by systematically identifying hazards.
3. Undertake calculations by using appropriate models.
4. Evaluate certainties and uncertainties involved in the calculations of vulnerability and exposure to hazards.
5. Compare with criteria to assess the need for future actions.
6. Determine and act on options to control, mitigate and adapt to the risk.
7. Communicate the results to those who need to know.

Arsenic Case Study

What is (are) the risk question(s)?

High levels of arsenic exposure are problematic throughout Latin America. In Argentina alone, it is estimated that between 1.2 and 3 million people are exposed to high concentrations of arsenic in drinking water (McClintock et al., 2012). High exposures have also been observed in Chile, Mexico, Peru, Bolivia and Ecuador (McClintock et al., 2012). Much of these exposures is suspected to be caused by water contamination.

Arsenic is a naturally occurring metal in many regions. Exposure to arsenic is concerning because it is associated with both cancer and non-cancer health effects. Inorganic arsenic is catalogued as a human carcinogen on the basis of epidemiological evidence (IARC, 1987). Non-cancer health effects include associations with hypertension and diabetes. Additionally, chronic endemic regional hydroarsenicism can occur in populations with long-term, low-level exposure from the prenatal period across the life course. The risk question for this case study is, Are Latin Americans living in regions with high levels of naturally occurring arsenic at increased risk of adverse health effects due to arsenic contaminated water?

What are the exposures and risks for Latin American populations living in regions with high naturally occurring arsenic?

There are many forms of arsenic, including the organic forms, such as arsenobetaine (AB), arsenocoline (AC), monomethylarsononic acid (MMA), dimethylarsinic acid (DMA), trimethylarsine oxide (TMAO), tetramethylarsonium (TMA^+), arsenosugars and arsenolipids and inorganic forms, such as As III, As V, AsH_3, and AsGa. The form of arsenic is important as it can influence bioavailability and toxicity. Arsenic forms, such as the sugar conjugates, occur naturally in seafood and will not be considered in this case but they can interfere with the total arsenic analysis. Some regions in Latin America have high concentrations of naturally occurring arsenic in the soil. Arsenic mobility from the soil to

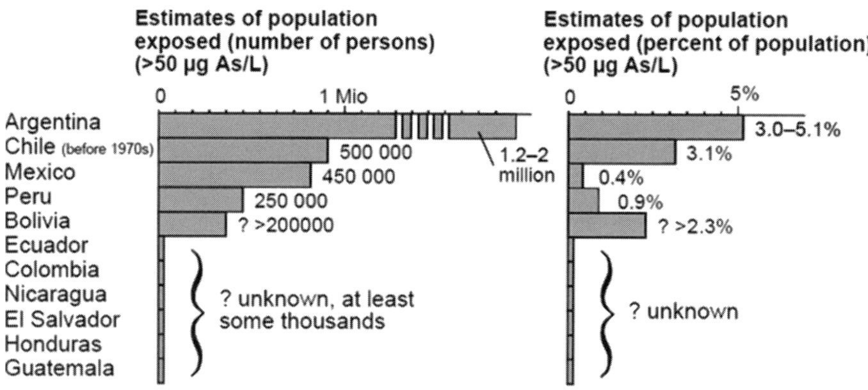

Figure 9.2 Latin American populations with estimated chronic exposure to arsenic from drinking water greater than 50 microgram/L. Populations in countries listed as having unknown arsenic exposure may still have high arsenic exposure, as this figure highlights those populations exceeding 50 microgram/L and the WHO recommends a safe limit of 10 microgram/L (McClintock et al., 2012).

water reservoirs depends on the chemical composition of the environment, including the pH. In many cases, however, arsenic readily enters the water supply. The World Health Organization provides the provisional guidance value of 0.01 mg/L for arsenic in drinking water, but estimates that in areas with naturally occurring sources of arsenic, concentrations can reach 12 mg/L (WHO, 2011).

What tools were used?

Arsenic exposure can be assessed through multiple biomarkers including hair, nails and urine concentrations of total and speciated arsenic. Environmental markers include arsenic content of soil, water and household dust. Measuring biomarkers for arsenic can be particularly helpful because it allows for the identification of various arsenic species. Because arsenic speciation can be indicative of exposure routes, this information can be easily fed into risk assessment and mitigation efforts. Biomarkers can be used to assess susceptibility factors (such as genetic polymorphisms), exposure and early biological effects. For this study, questionnaire (survey) data from sampling interviews with study participants were used to identify their behavior in relation to water use and thus the sources of arsenic from water.

Another tool employed in this case study was the measurement of arsenic in drinking water. In order to relate arsenic in drinking water to public health risks, it is important to consider the frequency of contact. In order to determine whether the potential for a health risk is present from the arsenic contamination of water, the average daily dose of arsenic can be estimated by using the concentration of arsenic in drinking water, ingestion rate and frequency, body weight and average lifespan. The reference dose is the daily dose that is considered safe for exposure across the lifetime. By comparing the average daily dose to the reference dose, we can identify exposures that exceed safe doses. This tool is particularly useful for chronic exposures, such as those related to water contamination. Acute exposures, such as those highlighted in other case examples in this chapter (e.g., pesticide application), can have different biological effects and require the use of more frequent sampling to assess exposure over time accurately.

What was observed:

- The case study showed the magnitude of arsenic exposure in Latin America.
- Global estimates have identified millions of people at risk; less is known about Latin America but populations are estimated to be higher (McClintock et al., 2012). Example findings are seen in Figure 9.2.

Key lessons:

- Arsenic exposure through contaminated drinking water is extensive throughout Latin America.
- There are insufficient biomonitoring programs to assess both the extensiveness of the exposure and associated risks adequately.
- Environmental assessments of arsenic in soil and water in Latin America are scarce.
- Arsenic exposure occurs from multiple routes.
- Addressing arsenic contamination requires input from interdisciplinary teams of experts including water engineers, health specialists, exposure scientists and epidemiologists to identify populations at risk, calculate an exposure and develop successful reduction strategies including controlling arsenic concentrations in water and promoting behaviors that prevent and reduce exposure.

Pesticide Use and Biodiversity Case Study

What is (are) the risk question(s)?

This case study focused on the impacts of pesticide use on biodiversity in native habitats recently converted to agriculture, known as agricultural frontiers (Schiesari et al. 2013). Agricultural frontiers are currently concentrated in tropical areas where biodiversity may be particularly susceptible, not only because there are more species but because the ones that are there are also proportionally more sensitive, vulnerable or endemic species. In many cases, aquatic organisms drive pesticide regulations because of their heightened sensitivity (Laetz et al., 2009). Further, agricultural frontiers are remote and therefore farmers have limited access to education in general, and governmental rural extension services that orient farmers to which particular pesticides to use, when, how, and in what quantities. Because pesticides are not only toxic to humans but also to other, nontarget aquatic and terrestrial organisms, the risk question for this case study is, What risks to aquatic and terrestrial biodiversity from pesticide use patterns in agricultural frontiers can be identified?

What tools were used?

Pesticide use patterns in three production scenarios in the Amazon Basin were characterized by interviews and analysis of detailed farm pesticide application records. These tools were selected because they allowed for the identification of both the volume and the frequency of pesticide application. While environmental sampling for pesticides can be a worthwhile investment, the seasonal nature of pesticide application requires multiple samples per year, which can be costly. Interviews and application records were a cost-effective tool for understanding the episodic and dynamic applications of pesticides. The first was a cross-sectional study investigating 220 smallholder fruit and vegetable producers (2005), the second a cross-sectional study investigating one large-scale, pilot sugarcane producer (2009/2010) and the third a longitudinal study documenting the large-scale conversion of pastureland into soybean plantation, and further intensification, in a high-profile agroindustrial farm (2003–2008). Pesticide application practices were compared with legal recommendations (e.g., minimum and maximum doses, frequency of applications) to identify inappropriate and over-application scenarios. Pesticides were classified on the basis of their toxicity and environmental classes as established by the Brazilian Ministries of Health, Environment and Agriculture and obligatorily presented in pesticide labels and leaflets. Because active ingredients varied from year to year, to assess the temporal trends in total pesticide toxicity on biodiversity the mass of each active ingredient applied per hectare was first converted into toxic units for model aquatic invertebrates (i.e., *Daphnia*), fish, algae and rats and summed.

What was observed?

- With poor education and no technical support, 71% and 96% of all recorded pesticide applications in small farms were outside technical recommendations in terms of dosage and frequency of application. Under-dosing was more frequent than over-dosing, but application frequencies were fivefold greater than recommended. Moreover, in several instances the pesticide formulations that were employed were not even recommended for the pests they were intended to control. This created a scenario of low agronomical efficacy and unnecessarily high exposures to humans and the environment.
- With more technical and financial resources, and aiming at more restrictive international markets, large-scale producers followed more closely technical recommendations and even voluntarily replaced pesticides considered more harmful to human and environmental health. Indeed, over time the mammalian (i.e., rat) toxicity of pesticides appears to have declined even though the number of active ingredients and dosage increased. Interestingly, however, toxicity to aquatic model organisms, such as invertebrates, fish and algae, increased up to 135 times (Figure 9.3).

Key lessons:

- Pesticide use is a highly technically elaborate class of land management practices. Not only are there dozens or even hundreds of pesticide formulations available in the market for pest and weed control of any given crop, but pesticide use also changes dynamically over time, creating ever-changing scenarios of exposure. Approaches to study pesticide impact must include longitudinal assessments to capture the ranges of exposures.
- Lack of technical knowledge of control management operating procedures and appropriate pesticide application protocols was evident in smallholder producers, who had no access to technical support and had a very deficient understanding of the complex recommendations provided in product labels and leaflets (Waichman et al., 2007).
- Because there was no correlation in the toxicity of active ingredients to terrestrial and freshwater model species, selection of pesticides more protective to human health (as indicated by toxicity to the rat) does not guarantee protection of freshwater biodiversity. Such observation reinforces the importance of

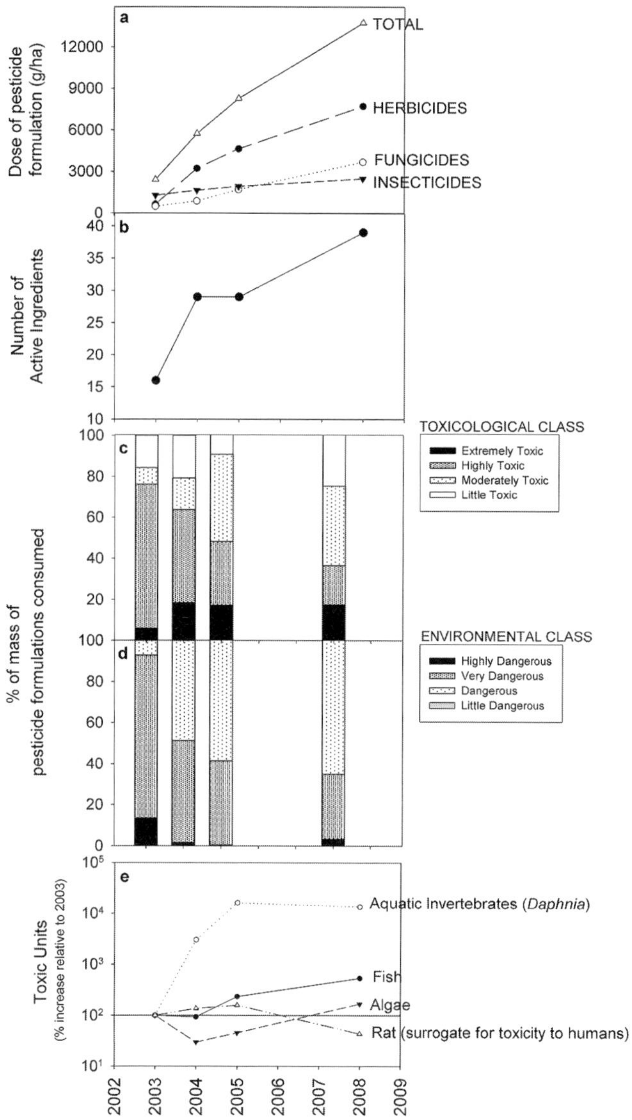

Figure 9.3 Temporal trends in the application of pesticides in a high-profile agroindustrial farm in the Amazonian 'arc of deforestation' over the first years of conversion of pastureland into a soy bean plantation. (a) Doses (mass per hectare) and (b) diversity of active ingredients over time. There was a gradual substitution of pesticide formulation towards less hazardous formulas for human health, as indicated by their (c) Toxicological Class (as defined by the Ministry of Health), and their (d) Environmental Class (as defined by the Ministry of the Environment), (e) The result of increased doses of presumably less hazardous compounds is that toxicity per unit area decreased for mammals but increased from 1.7 to 135 times for aquatic organisms (not log scale) (L. Schiesari et al., 2013).

preserving sufficiently wide and adequately vegetated buffer zones along streams to reduce exposure.

- Reducing pesticide risk to human health and biodiversity, particularly in remote frontier regions, depends on responsibility sharing by multiple stakeholders including intergovernmental organizations, national and local governments, NGOs, the pesticide industry, pesticide retailers, farmer organizations, large and small stakeholders, agricultural marketing organizations, consumers and academia (Schiesari et al., 2013).
- Focusing on increasing education and regulation, including regulation laws, policies and penalties for pesticide use, but also rewarding the voluntary adoption of better land management practices, could reduce use without impacting efficacy and have "co-solution" benefits for workers, residents and biodiversity.
- Agricultural frontiers are sensitive landscape interfaces that require special attention because of the unique proximity of diverse ecosystems and agricultural activities.

Nitrate Case Study

What is (are) the risk question(s)?

This is a case study on nitrate contamination of drinking water that provides an integrated approach to protecting water supplies in Germany using the European Water Framework Directives. Nitrogen is produced by industry, by automobiles and through agriculture. The nitrogen cycle transports nitrogen from pollution sources to water bodies, where it can remain or be evaporated to contaminate the air. Nitrate fertilizers are used in agriculture to increase productivity. In fact, fertilizer input and world population growth share similar trends (Figure 9.4). This is problematic because nitrates in drinking water can endanger ecological and human health. Nitrates in the environment can cause eutrophication and acidification, ultimately reducing biodiversity. Exposure to humans is of concern because nitrosamines are suspected carcinogens (WHO and IARC, 2007, vol. 89). Thus the risk question for this case study is, How do we provide high-quality water and still have sustainable agriculture in Europe?

What tools were used?

Risk assessment was used to identify sources of exposure and to establish health-protective nitrate

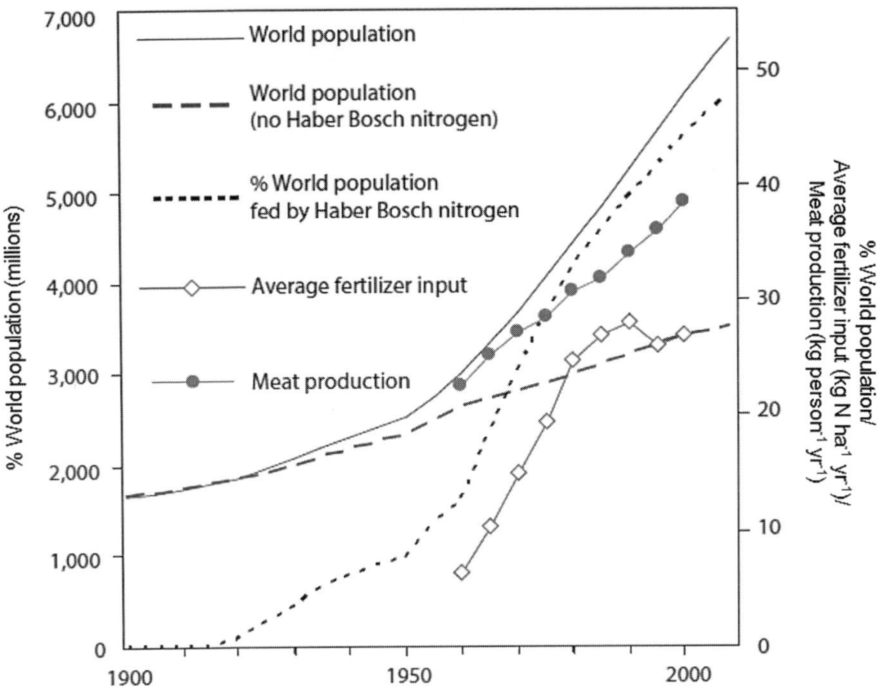

Figure 9.4 Global fertilizer use over time has increased as the world population and meat production have increased. The parallel lines indicate increase in percentage of the population fed (Erisman et al., 2008, p. 637).

standards. A key element of this case study is the identification of unique sources of nitrate contamination. Source identification was needed because it allowed for the linkage with agricultural and fertilizer contributions and ultimately risk management strategies to control nitrate sources (Figure 9.4). The water framework directive (WFD) set attainment goals for European water bodies, including rivers, lakes, groundwater bodies and coastal waters. Originally, the WFD set the goal of attaining "good status" for all of Europe's water bodies by 2015. Good status is achieved by obtaining good ecological and chemical standards. To obtain good chemical standards, water bodies must not have concentrations of pollutants that exceed environmental quality standards (Annex V, European Commission 2014).

What was observed:

- Most nitrogen pollution is from nitrate, ammonia and nitrous oxide released from agricultural activities.
- In 2009, 48% of German terrestrial ecosystems showed signs of eutrophication (Schaap et al., 2014).
- High nitrate concentrations (>50 mg/L) prevent 27% of German groundwater bodies from reaching "good status" based on the WFD criteria.
- Nitrogen in surface water has declined since 1983, largely because of reductions in point sources.

However, groundwater contamination has not decreased and in 2011 constituted the largest source of nitrogen.

Key lessons:

- Managing nitrogen contamination requires an integrated approach to consider environmental, energy and agricultural policies; thus interdisciplinary teams including stakeholders, academics and independent scientists are important. Stakeholder involvement is frequently at the center of risk assessment and management paradigms as it is critical to identifying hazards and sources of exposure and developing mitigation techniques that will work. Academic interdisciplinary teams can supply both specific expertise as well as the broader context. Working with independent scientists without funding from vested interest groups provides a clear advantage for analyzing the strengths and weaknesses of mitigation strategies.
- It is necessary to consider both environmental and public health protection in managing nitrogen in water bodies.
- Sustainable agriculture requires management and reduction of nitrogen outputs.
- Informing and involving stakeholders and the general public is important to developing feasible approaches for sustainable management.

9.2 Future Directions

In this chapter we have chosen examples to show how human and ecological impacts are incorporated in water security issues. In reviewing these methods it is important to realize that many cases, especially if done on a contaminant-specific basis, require site- and chemical-specific methods. More recently there has been a recognition by the scientists who work in these fields and in the interface with Future Earth that more integrative tools are needed that allow the risk manager and other data users to assess broader environmental databases, access information about impacts across boundaries and incorporate data with an emphasis on functionality and services (especially ecosystem services). Towards this end we have identified three new considerations that exemplify this need. The first two examples are from the field of environmental metagenomics. Using this approach, microbial DNA from environmental sources (such as water and soil) is sequenced and matched to known and suspected microbes through bioinformatic pipelines and standardized ontologies.

The first example for integrating information from multiple disciplines highlights the new cross-disciplinary bioinformatics platform and dashboards that facilitate the analysis, integration and sharing of data. MG-RAST, a cloud-based metagenomics analysis server and web framework, is one example of a bioinfomatic tool (Meyer et al., 2008). Not only can large datasets be uploaded and shared internationally, but the dashboard also allows users to identify others with related datasets to facilitate collaboration. Data sharing is used as an incentive for faster bioinformatic analysis. For example, the more willing researchers are to share their data, the faster their bioinformatic analyses will be completed.

The second example is from also from the metagenomics world. Microbial signatures for the oceanic metagenome are available for analysis from samples collected across the globe (Rusch et al., 2007). Access to these databases (available through MG-RAST) proved useful for looking at regional trends and anthropomorphic impacts on microbial populations (Port et al., 2012). Assessment of regional and international examples of antibiotic resistance signals in oceans was facilitated by these tools and availability of reference databases (Port et al., 2014, Wallace et al. 2017). These assessments were further facilitated by linkage with decision analytic tools allowing decision managers to put these results into useful public health metrics.

The third example of integrated assessments of health impacts are those facilitated by the development of novel bioassays that use aquatic organisms to predict water quality impacts. Effects on these model organisms (such as mortality, reproductive health and metabolic function) have been used successfully to predict degradation or improvements in water quality in the field that are relevant for human and other ecosystem members yet do not directly assess health impacts in the organisms to be protected (Umbuzeiro et al., 2017).

9.3 Conclusions

In this chapter, the authors examine water security within the vision of Future Earth to ensure that humans thrive in a "sustainable and equitable world" (ICSU, 2014). By focusing on one component, water security, we were able to link examples from the field to the set of relevant sustainability goals. Seven of the WHO development goals were directly applicable for this assessment of water security. Another focus of this chapter was to examine how health information was incorporated into three case studies for water security and how this information was utilized via a risk assessment context.

Lessons across the examples emphasized numerous shared lessons:

1. The importance of the cross-disciplinary research – did public health scientists not only had to talk with epidemiologists (studying population health) but also with toxicologists (human and ecological health scientists) and engineers and geoscientists (arsenic in soils).
2. The need for integrated tools that can link field assessment with regional and global impact prediction.
3. The importance of cross-disciplinary boards and committees to build assessment and confidence across our areas of specific training.
4. Success requires more trans-disciplinary educational opportunities and sharing of case studies to achieve our goals.
5. The need for integrated approaches for evaluating water security.
6. The importance of the risk assessment/risk management context.
7. New systems-based tools and big database-informed methods are available.
8. Agricultural examples provide an integrated context for identifying sustainable risk management options.

While these lessons learned may have previously been shared by other disciplines, the inclusion of human health in Future Earth discussions is novel and critical. The risk assessment framework allows for a standard approach that can be used in the geoscience and human health fields. Data can be collected through surveys, environmental measurements or biomarker assessments to characterize exposure. Biomarkers are particularly useful for integrating these concepts as they may include markers of exposure, susceptibility and early biological effects.

Acknowledgements

Special thanks to ICSU, who made these workshops possible, and to the International Union of Toxicology, who provided the venue for these two Water Security Workshops. Special thanks should also go to the Brazilian Society of Toxicology and to the Mexican Society of Toxicology, who provided the regional organization of these special congresses. The participants in these courses provided many additional case studies too numerous to share within this context; however, the authors hope that we were able to provide a forum for their voices in these important global issues.

References

Azevedo, L. B., van Zelm, R., Leuven, R. S., Hendriks, A. J. and Huijbregts, M. A. (2015). Combined ecological risks of nitrogen and phosphorus in European freshwaters. *Environmental Pollution*, **200**, 85–92.

Beer, T. (2003). Environmental risk and sustainability. In *Risk Science and Sustainability*. Springer, Dordrecht pp. 39–61.

German Advisory Council on the Environment (2015) *NITROGEN: Strategies for resolving an urgent environmental problem*. German Advisory Council on the Environment (SRU).

German Advisory Council on the Environment. (2015). *Biodiversity and human health need ambitious reduction commitments: A German perspective on the new draft NEC Directive with regards to ammonia and nitrogen oxides.* Partial translation of the Special Report "Nitrogen: Strategies for resolving an urgent environmental problem". German Advisory Council on the Environment (SRU).

Erisman, J. W., Sutton, M. A., Galloway, J., Klimont, Z. and Winiwarter, W. (2008). How a century of ammonia synthesis changed the world. *Nature Geoscience*, **1**(10), 636–639.

European Commission. (2014). Water Framework Directive. Official Journal of the European Union, Commission Directive 2014/101/EU, Amending Directive 2000/60/EC. http://eur-lex.europa.eu/legal-content/EN/TXT/?uri=celex:32014L0101.

European Environment Agency. (2012). *European waters-current status and future challenges*, Synthesis. ISSN 1725–9177, EEA Report No 9/2012.

Faustman, E. M. and Omenn, G. S. (2013) Risk assessment. In *Casarett and Doull's Toxicology: The Basic Science of Poisons*, 8th Edition, C. Klaassen (ed.), New York: McGraw-Hill.

Halpern, B. S., Walbridge, S., Selkoe, K. A., Kappel, C. V., Micheli, F., D'agrosa, C., Bruno, J. F., Casey, K. S., Ebert, C., Fox, H. E. and Fujita, R. (2008). A global map of human impact on marine ecosystems. *Science*, **319**(5865), 948–952.

IANAS. (2015). *Urban Water Challenges in the Americas.* A perspective from the Academies of Sciences. Inter-American Network of Academies of Sciences.

IARC. (1987). *IARC Monographs on the Evaluation of Carcinogenic Risks to Humans: Supplement 7*. The Agency. International Agency for Research on Cancer.

ICSU. (2014). *Future Earth 2025 Vision, Future Earth Research for Global Sustainability, Paris, France*. International Council for Science.

Navoni, J. A., De Pietri, D., Garcia, S. and Villaamil Lepori, E. C. (2012). Riesgo sanitario de la población vulnerable expuesta al arsénico en la provincia de Buenos Aires, Argentina. *Revista Panamericana de Salud Pública*, **31**(1), 1–8.

Laetz, C. A., Baldwin, D. H., Collier, T. K., Hebert, V., Stark, J. D. and Scholz, N. L. (2009). The synergistic toxicity of pesticide mixtures: implications for risk assessment and the conservation of endangered Pacific salmon. *Environmental Health Perspectives*, **117**(3), 348.

López-Doval, J. C., Montagner, C. C., de Alburquerque, A. F., Moschini-Carlos, V., Umbuzeiro, G. and Pompêo, M. (2017). Nutrients, emerging pollutants and pesticides in a tropical urban reservoir: Spatial distributions and risk assessment. *Science of The Total Environment*, **575**, 1307–1324.

McClintock, T. R., Chen, Y., Bundschuh, J., Oliver, J. T., Navoni, J., Olmos, V., Lepori, E. V., Ahsan, H. and Parvez, F. (2012). Arsenic exposure in Latin America: Biomarkers, risk assessments and related health effects. *Science of the Total Environment*, **429**, 76–91.

Meyer, F., Paarmann, D., D'Souza, M., Olson, R., Glass, E. M., Kubal, M., Paczian, T., Rodriguez, A., Stevens, R., Wilke, A. and Wilkening, J. (2008). The metagenomics RAST server–a public resource for the automatic phylogenetic and functional analysis of metagenomes. *BMC bioinformatics*, **9**(1), 386.

Port, J. A., Cullen, A. C., Wallace, J. C., Smith, M. N. and Faustman, E. M. (2014). Metagenomic frameworks for monitoring antibiotic resistance in aquatic environments. *Environmental Health Perspectives (Online)*, **122**(3), 222.

Port, J. A., Wallace, J. C., Griffith, W. C. and Faustman, E. M. (2012). Metagenomic profiling of microbial composition and antibiotic resistance determinants in Puget Sound. *PLoS One*, **7**(10), e48000.

Rusch, D. B., Halpern, A. L., Sutton, G., Heidelberg, K. B., Williamson, S., Yooseph, S., Wu, D., Eisen, J. A., Hoffman,

J. M., Remington, K. and Beeson, K. (2007). The Sorcerer II global ocean sampling expedition: Northwest Atlantic through eastern tropical Pacific. *PLoS Biol*, **5**(3), e77.

Schaap, M., Wichink Kruit, R., Hendriks, C., Kranenburg, R., Segers, A., Builtjes, P., Banzhaf, S. and Scheuschner, T. (2014). *Ermittlung und Bewertung der Einträge von versauernden und eutrophierenden Luftschadstoffen in terrestrische Ökosysteme*. Abschlussbericht zum F&E-Vorhaben: Umweltbundesamt, Dessau-Roßlau.

Schiesari, L., Waichman, A., Brock, T., Adams, C. and Grillitsch, B. (2013). Pesticide use and biodiversity conservation in the Amazonian agricultural frontier. *Phil. Trans. R. Soc. B*, **368**(1619), 20120378.

Umbuzeiro, G. D. A., Heringa, M. and Zeiger, E. (2017). In vitro genotoxicity testing: Significance and use in environmental monitoring, in *In vitro Environmental Toxicology - Concepts, Application and Assessment*, G. Reifferscheid and S. Buchinger, Editors. Springer International: Cham, Switzerland. p. 59–80.

UNICEF (2015). *Progress on Sanitation and Drinking Water 2015 Update and MDG Assessments*. UNICEF and World Health Organization: Geneva.

UNU. (2013). *Water Security and the Global Water Agenda-A UN Analytical Brief*, UNESCAP, United Nations University, INWEH: Ontario.

Waichman, A. V., Eve, E. and da Silva Nina, N. C. (2007). Do farmers understand the information displayed on pesticide product labels? A key question to reduce pesticides exposure and risk of poisoning in the Brazilian Amazon. *Crop Protection*, 26(4), 576–583.

Wallace, J. C., Port, J. A., Smith, M. N. and Faustman, E. M. (2017). FARME DB: A functional antibiotic resistance element database. *Database*, (1).

WHO and Inter-Organization Programme for the Sound Management of Chemicals. (2009). *Principles for modelling dose-response for the risk assessment of chemicals* (Vol. 239). World Health Organization: Geneva.

WHO and IARC. (2007). *Smokeless tobacco and some tobacco-specific N-Nitrosamines*. Vol 89. World Health Organization and International Agency for Research on Cancer: Geneva.

WHO and UNICEF. (2015). *Progress on sanitation and drinking water–2015 update and MDG assessment*. World Health Organization: Geneva.

WHO. (2000). *Joint IPCS/OECD Project on the Harmonization of Chemical Hazard/Risk Assessment Terminology (Harmonization Project)*. World Health Organization: Geneva.

WHO. (2011). *Guidelines for Drinking-water Quality*, 4th Edition. World Health Organization: Geneva.

10
Decadal Coupled Ocean–Atmosphere Interaction in North Atlantic and Global Warming Hiatus

Jianping Li, Cheng Sun, and Ruiqiang Ding

10.1 Introduction

It is well known that Earth's climate is warming, which has major global implications for human well-being. The change of global mean surface temperature (GMT) has been a focus of a great amount of research since it is a simple integral measure for a climate system and its change. Therefore, the study of hemispheric- or global-mean temperature change is a hot topic with significant importance for understanding the past and current climate change and its future predictions. Human activities are giving rise to increases in heat-trapping greenhouse gases, mainly carbon dioxide from fossil fuel burning, so GMT warming is expected. Indeed, the GMT has increased by about 0.8°C over the past century, and the rate of GMT increase has been mostly upward (Figure 10.1). Nevertheless, it is apparent there are strong variabilities in GMT from year to year, decade to decade and multidecade to multidecade as well. The GMT increased at a rate of 0.12°C per decade during the period 1951–2012, but the rate of GMT increase for 1998–2012 had dramatically slowed to 0.05°C per decade (Stocker et al. 2013). Explaining this slowdown in the rate of warming, referred to as "hiatus" or "pause" of global warming, brought great challenges to climate scientists (Li et al. 2013a; Fyfe et al. 2016). It was inconsistent with expectations that the GMT would rise at similar or even greater rates than it had in the second half of the twentieth century as the greenhouse gas concentration in the atmosphere still continued to rise. The greenhouse gas effect obviously fails to explain the slowdown in the rate of global warming.

Several possible mechanisms have been proposed to explain the recent slowdown in the rate of warming, mainly including three aspects: changes in incoming energy at the surface, redistribution of the energy within the climate system and a role of internal natural climate variabilities as well. Recent studies have suggested that possible candidates for the reduction in the surface radiative forcing are the cooling effect of small-scale volcanic eruptions (Haywood et al. 2014), decline in stratospheric water content (Solomon et al. 2010) and minimum phase of the cycle of total solar irradiance (Lean and Rind 2008). Alternatively, the energy can be redistributed within the climate system, leading to a cooling at the Earth surface. Oceans are thought to be the most likely candidates to store the energy below the surface because of changes in ocean circulation. Measurements of heat content in the deeper ocean have suggested that the heat uptake by the deeper ocean increased remarkably during the period of global warming slowdown (Balmaseda et al. 2013; Watanabe et al. 2013; Chen and Tung 2014). This indicates that despite the pause in surface warming, the climate system continued to absorb a substantial amount of heat during the past decade. Correspondingly, the surface sea temperature (SST) anomaly for the warming pause period showed a spatial pattern with substantial SST cooling over the tropical eastern Pacific, in accordance with the cool phase of the El Niño–Southern Oscillation (ENSO, Kosaka and Xie 2013; Trenberth and Fasullo 2013) or the Interdecadal Pacific Oscillation (IPO, Meehl et al. 2011; England et al. 2014). The ENSO is the dominant mode of ocean–atmosphere variability on an interannual time scale in the tropical Pacific, with SST in the central and eastern equatorial Pacific cycling between above- (El Niño) and below-average (La Niña), while the IPO shows an ENSO-like pattern of variability operating at decadal and interdecadal time scales.

Third, internal natural climate variabilities can also slow down GMT warming over short terms or decades. For instance, Li et al. (2013a) and Sun et al. (2015a) demonstrated that the North Atlantic Oscillation (NAO) leads to multidecadal variability in Northern Hemispheric mean surface air temperature (NHT) by

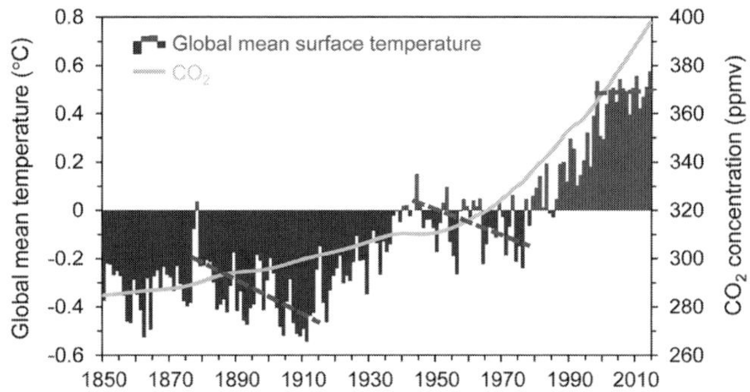

Figure 10.1 The GMT (bars) and global atmospheric CO_2 concentration (purple) from 1850 to 2014. The GMT data are from the HadCRUT4 dataset (Morice et al. 2012), relative to the base period 1961–1990. The global atmospheric CO_2 concentration is from the suite of atmospheric concentration observations and emissions estimates for greenhouse gases (Meinshausen et al., 2011). The green dashed lines show three linear trends for the periods of 1877–1913, 1944–1976 and 1998–2014, respectively. For the color version, please refer to the plate section.

15–20 years through decadal coupled ocean–atmosphere interaction in the North Atlantic, and decadal weakening of the NAO intensity since the mid-1980s acts to produce the recent warming hiatus.

Besides the mechanism-based explanations mentioned previously, some recent studies have raised questions about biases in the datasets used to identify the warming pause over the 1998–2012 period, and suggested that the recent warming hiatus was possibly an artifact of cold bias in the SST measurements by buoys (Karl et al. 2015; Hausfather et al. 2017). This has challenged the very existence of the recent hiatus of global warming. However, these studies cannot deny the existence of global warming hiatuses in past periods (e.g., 1944 to 1976) as discussed later.

Although much attention has been focused on the recent hiatus of global warming, we should note that there are two more hiatus intervals with much lower rates of temperature increase in the past one and one-half centuries. The first was from about 1877 to 1913, and the second was from 1944 to 1976 (Figure 10.1). This fact indicates that the recent warming pause is essentially a manifestation of the decadal to multidecadal variability of the GMT (Li et al. 2013a). Our understanding of the natural decadal climate variability so far is still very limited. Currently, the biases in the model representation of natural decadal climate variability are still very large (Stocker et al. 2013) and the decadal prediction skill is also very low (Newman 2013). Meanwhile, most previous studies on decadal climate variability only emphasized the role of the ocean (Schlesinger and Ramankutty 1994; Zhang et al. 2007; England et al. 2014), while the role of atmospheric variability is much less often addressed. Therefore, further investigations of the mechanisms influencing the decadal to multidecadal climate variability are required, particularly on the role of atmospheric variability and its coupling with the ocean. Li et al. (2013a) and Sun et al. (2015a) emphasized the important role of the NAO in multidecadal climate variability as well as the influences of NAO decadal weakening on the three global warming hiatuses in the past one and one-half centuries.

The term NAO, originally defined by Sir Gilbert Walker, is used to describe a large-scale see-saw in sea level pressure (SLP) between the Icelandic low and Azores high (Walker and Bliss 1932). Using the empirical orthogonal function (EOF) analysis of SLP or geopotential height, the NAO is identified as a dominant mode of atmospheric circulation variability in the North Atlantic region throughout the year, accounting for about one-third of the interannual variability. A negative (positive) NAO phase indicates weakening (strengthening) of both the Icelandic low and Azores high, which results in a decreased (increased) SLP gradient across the North Atlantic region and further reduces (enhances) the strength of the westerlies (Li and Wang, 2003). Although the NAO is generally defined from SLP data (Hurrell 1995), it is significantly associated with other meteorological factors in the North Atlantic sector, including the position and speed of the westerlies, the heat and moisture transport between the Atlantic and its adjacent continents, and the intensity and frequency of winter storm tracks.

The NAO has broad time scales ranging from intraseasonal to interdecadal (Feldstein 2000; Li 2005). On interdecadal time scales, the NAO shows a noticeable upward trend over the second half of the twentieth century (Hurrell 1995; Thompson et al. 2000). This upward trend has important impacts on regional and hemispheric climates. It explains much of the observed warming trend over Eurasia and North America (Hurrell 1995; Thompson et al. 2000), and furthermore, it significantly contributes to the warming of the Northern Hemisphere extratropical mean surface temperature (Hurrell 1996). This positive trend in the NAO is also linked to the interdecadal variations of Asian summer and winter monsoons (Gong et al. 2001). In the past two decades, the NAO has shown a decreasing trend (Cohen and Barlow 2005). In particular, the winters of 2009/2010 and 2010/2011 were characterized by strong negative NAO phases (Taws et al. 2011), and the negative NAO is suggested as one of the main factors causing the extremely cold temperatures in the Northern

Hemisphere (Wang et al. 2010; Sun and Li 2012). These studies indicate that the NAO fluctuations on decadal to multidecadal time scales are considerable, and closely linked with the Northern Hemisphere climate. Previous studies have also suggested that the NAO variability significantly contributes to the oceanic variability in the North Atlantic, such as the North Atlantic tripole pattern (NAT), and the strength of the ocean circulation (Visbeck et al., 2003 and references therein).

On decadal to multidecadal time scales, the North Atlantic SST variability is dominated by the Atlantic Multidecadal Oscillation (AMO). The AMO can be obtained as the leading EOF mode of annual mean SST over the North Atlantic region. It features a basin-wide homogeneous pattern with larger amplitudes in the subpolar regions. Thus, the AMO index can be easily defined as the area-weighted SST average over the North Atlantic basin (Enfield et al. 2001; Sutton and Hodson 2005; Ting et al. 2009). The AMO is thought to be closely linked to variations in the strength of the Atlantic Meridional Overturning Circulation (AMOC) (Delworth and Mann 2000; Knight et al. 2005). There is no consensus on the origins of the AMO, but many studies have suggested that internal climate variability plays an important role in multidecadal North Atlantic SST fluctuations (Ting et al. 2009; Zhang et al. 2013).

The persistent basin-scale SST anomalies associated with the AMO have significant impacts on the climate of the surrounding land masses. The AMO has been linked to rainfall variability in the United States (Enfield et al. 2001), and during the warm (cold) AMO phase, most of the United States receives less (more) than normal rainfall. The warm (cold) phase of the AMO can weaken (strengthen) the North Atlantic subtropical high and result in decreased (increased) moisture flow into the United States. Meanwhile, positive correlations have been found between the AMO index and rainfall amounts in northwest Europe and the Sahel (Sutton and Hodson 2005; Zhang and Delworth 2006). With respect to land surface air temperature (LSAT) variations, previous studies have linked a warm AMO phase to warmer-than-normal LSAT over North America and western Europe (Sutton and Hodson 2005; Ting et al. 2009). Hurricane activity over the North Atlantic basin has also been related to the AMO (Zhang and Delworth 2006). This is due not only to the SST anomalies but also to changes in vertical shear in the troposphere over the main hurricane development region.

The AMO can exert remote influences on climate variability outside the North Atlantic region. Both observations and numerical simulations have suggested that the AMO has a direct warming/cooling effect on the surface air temperature averaged over the entire Northern Hemisphere and that the multidecadal variability in the NHT can be largely explained by the AMO (Knight et al. 2005; Li et al. 2013a). The AMO has a remote impact on the Siberian warm season rainfall through a Rossby wave train across the northern Eurasian continent (Sun et al. 2015b), and this wave train pattern is clearly observed at both low and upper levels, which can be referred to as the Atlantic–Siberia teleconnection pattern. During summer, a positive correlation has been found between the strength of the Indian summer monsoon and the AMO (Zhang and Delworth, 2006). Several mechanisms have been put forward to explain this teleconnection, including the meridional shift of the Intertropical Convergence Zone and a Rossby wave response in the troposphere induced by the Atlantic SST anomalies (Kucharski et al. 2006, 2009). Meanwhile, the East Asian summer monsoon is also likely to be stronger during the warm AMO phase (Lu et al. 2006). During the cold season, the AMO also shows some connection with East Asian climate. There is some observational and modeling evidence that the warm AMO phase is associated with a weaker East Asian winter monsoon and warmer-than-normal LSAT over most of China (Li and Bates 2007). Recently, Sun et al. (2016) proposed an Africa–Asia multidecadal teleconnection pattern to elucidate the associated dynamical mechanism and pathway linking the AMO to multidecadal variability of East Asian winter monsoon.

A 50- to 70-year quasi-periodic cycle in the NAO has been reported in previous studies based on other NAO reconstructions (Cook et al. 1998; Li, 2005; Olsen et al. 2012; Dieppois et al. 2013). Meanwhile, previous studies using observation and reconstruction records have also suggested that the AMO index spectrum shows significant peaks near periods of around 50–70 years (Schlesinger and Ramankutty 1994; Delworth and Mann 2000). The time scale of AMO is similar to that found in the NAO index, implying a potential linkage between the AMO and NAO multidecadal variability. How are they related and what is the underlying mechanism? Sun et al. (2015a) have proposed a possible mechanism from the decadal coupled ocean–atmosphere interaction to address these questions. Considering that the AMO and NAO are the dominant modes of variability of the ocean and atmosphere, respectively, in the North Atlantic region, answering these questions is important for a better understanding of the North Atlantic ocean–atmosphere interaction on decadal or longer time scales.

10.2 Data and Methodology

10.2.1 Data

Global ocean and atmosphere data are relatively short and there is still large uncertainty in the reanalysis products regarding the ocean long-term variability, e.g., variations in the AMOC. Climate models provide a useful approach to analyze multidecadal variability. A long-term pre-industrial control simulation of version 4 of the Community Climate System Model (CCSM4) is used. CCSM4 is a fully coupled model of the Earth's physical climate system, and a general description is given in Gent et al. (2011). Several recent studies have used the CCSM4 pre-industrial control simulations to investigate the Atlantic multidecadal variability and found that the model performs reasonably well in capturing the variability (Danabasoglu et al. 2012; Zanchettin et al. 2013). The control simulation was integrated over thirteen hundred model years, and a three-hundred–year segment between model years 800 and 1100 is used in the present analysis. The reason for this choice is to avoid the initial adjustment period because of the model spin-up process (Danabasoglu et al. 2012). The other reason is that the multidecadal oscillation of the NAO in thethree-hundred-year segment is regular and very similar to the reconstruction, ensuring our analysis is relevant to the real world. The CCSM4 simulation data used for the analysis include variables at the ocean–atmosphere interface (e.g., SLP and SST) and variables in the deep ocean (e.g., sea water temperature and meridional overturning circulation trajectory).

10.2.2 Statistical Method

The Principal Oscillation Pattern (POP) analysis (von Storch et al. 1988) was applied to investigate the dominant SST patterns of the oscillatory multidecadal variability. In contrast to the EOF analysis, which considers only the spatial co-variability in a dataset, the POP method considers the full spatiotemporal structure. POPs are in general complex, with a real part (P_R) and imaginary part (P_I). The corresponding complex coefficient time series satisfy the standard damped harmonic oscillator equation, so that the evolution of the system in the POP space spanned by the real and imaginary part can be interpreted as the following oscillatory sequence of the spatial pattern: $\cdots P_R \to -P_I \to -P_R \to P_I \to P_R \cdots$. Prior to the POP analysis, band-pass filtering (Butterworth band-pass filter) was applied to annual mean SST anomalies in order to highlight the multidecadal variability. An EOF analysis was then performed on the filtered data, and only the leading 10 EOF modes were retained in the POP analysis. Note that the SST data were area-weighted prior to the analysis and no normalization was applied. Similar approaches to POP analysis have been adopted in other studies (e.g., Park and Latif 2010).

10.2.3 Atmospheric General Circulation Model (AGCM)

We perform idealized numerical experiments to investigate the atmospheric circulation response to a given SST pattern. All the experiments have been designed and implemented on the Simplified Parameterizations, primitivE-Equation DYnamics (SPEEDY) model developed at the Abdus Salam International Centre for Theoretical Physics (ICTP), Italy (Kucharski et al. 2013). SPEEDY is an intermediate complexity AGCM based on a spectral primitive-equation dynamical core in sigma coordinates, with semi-implicit treatment of gravity waves. The resolution used here is triangular 30 (T30, about 3.75° × 3.75° horizontal resolution), with eight vertical levels. The model is computationally inexpensive, because of the simplified parameterizations, as well as the relatively low horizontal and vertical resolutions. Nevertheless, the simulation performance of the SPEEDY model compares very well with that of more complex AGCMs with respect to extratropical large-scale variability (Kucharski et al. 2013 and references therein). The control run was integrated for fifteen years and forced with the climatological SST, and the last ten years of the integrations were used to provide the basic annual mean state. The sensitivity experiment was integrated for twelve years and forced with the given SST anomaly pattern imposed on the climatological SST. A ten-member ensemble mean was constructed from the last ten years of the integrations to reduce the uncertainties arising from different initial conditions. The atmospheric response to the SST forcing is defined as the difference between the ensemble means for the sensitivity and control runs.

10.3 The Decadal-Scale Coupled Ocean–Atmosphere Mode: The NAT-NAO-AMOC-AMO (NNAA) Mode

10.3.1 Dominant Modes of SSTA Multidecadal Variability

In order to help understand the physical mechanisms involved in the relationship between NAO and AMO over the fifty- to seventy-year scale, we firstly applied

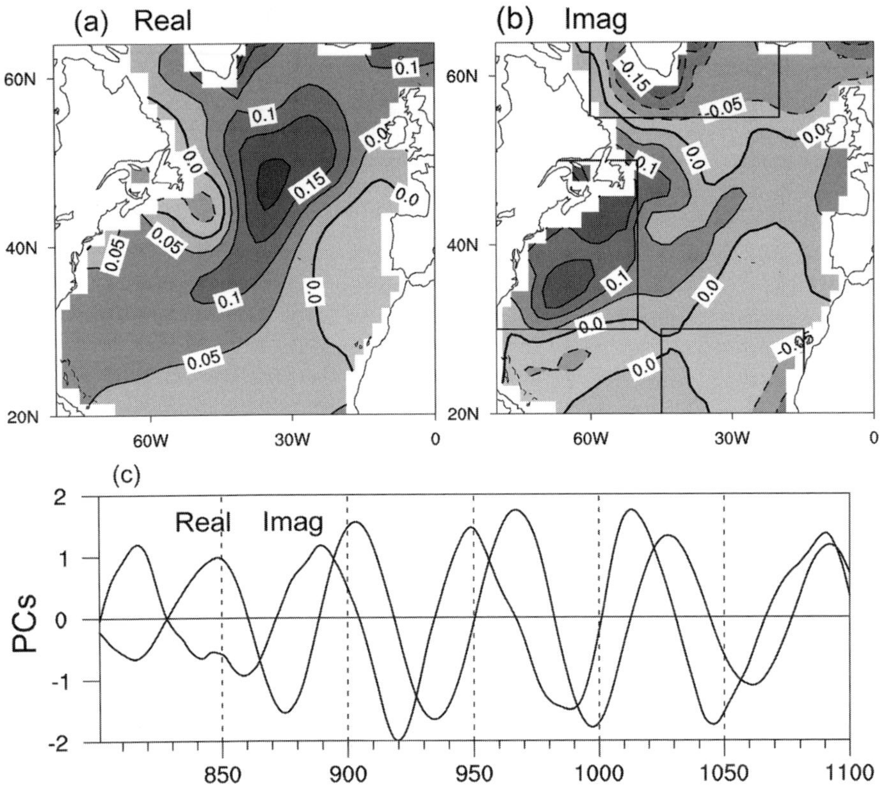

Figure 10.2 The first leading POP mode of band-pass (50–70 years) filtered annual SST anomalies over the North Atlantic Ocean in the CCSM4 simulations: (a) real-part pattern of POP1; (b) imaginary-part pattern of POP1; and (c) their corresponding PCs. The POP patterns are shown as the SST anomaly (in K) regressions onto the normalized PCs. The boxes marked in (b) indicate the regions used to define the NAT index (from Sun et al. (2015a)).

the POP analysis to detect the dominant SST patterns in association with the North Atlantic quasi-periodic multidecadal variability in CCSM4 simulations, and a fifty- to seventy-year band-pass filtering was performed on the annual mean North Atlantic SST data. Note that using a broader filter (e.g., a thirty- to ninety-year band-pass filtering) gives qualitatively similar results. The leading POP mode (Figure 10.2), accounting for about 60% of the filtered variability, is oscillatory, and characterized by a period of sixty years. This indicates that the fifty- to seventy-year quasi-periodic variability is mainly explained by POP1. The real- and imaginary-part PCs (Figure10.2c) show, as expected from the definition of POPs, a phase shift of about 90° during most of the analyzed period. The real-part pattern of POP1 (Figure 10.2a) features positive SST anomalies covering most of the North Atlantic region, similar to the basin-wide uniform SST pattern associated with the AMO (Enfield et al. 2001; Knight et al. 2005). The imaginary part (Figure 10.2b) exhibits a strong meridional gradient of SST across the North Atlantic, with cooling in the subpolar North Atlantic and warming in mid-latitude North Atlantic regions. Moreover, the imaginary part is reminiscent of the NAT (Visbeck et al. 2003, and references therein; Wu et al. 2009; Deser et al. 2010; Li et al. 2013b; Sun et al. 2015a). The lead–lag correlation between the AMO and NAT supports the oscillatory sequence of the real and imaginary parts of the POP, which is as follows: $\cdots NAT^+ \rightarrow AMO^+ \rightarrow NAT^- \rightarrow AMO^- \rightarrow NAT^+ \cdots$, where the superscripts + and – respectively denote the positive and negative phases. We hypothesize three possible mechanisms involved in the oscillatory sequence of the NAT and AMO. First, the NAT has a direct effect on the NAO; second, the NAO exerts some forcing effects on the AMO; and third, the AMO in turn provides some negative feedback on the NAT. These positive and negative feedback mechanisms may act successively, leading to the quasi-periodic multidecadal variability. In such a way, a decadal-scale coupled ocean–atmosphere mode, the NAT–NAO–AMOC–AMO (NNAA) mode, is formed.

10.3.2 The Direct Effect of the NAT on the NAO

We examined the SLP field over the North Atlantic region correlated with the NAT index in the CCSM4 simulations (Figure 10.3a). The SLP pattern in association with the NAT variability features a meridional

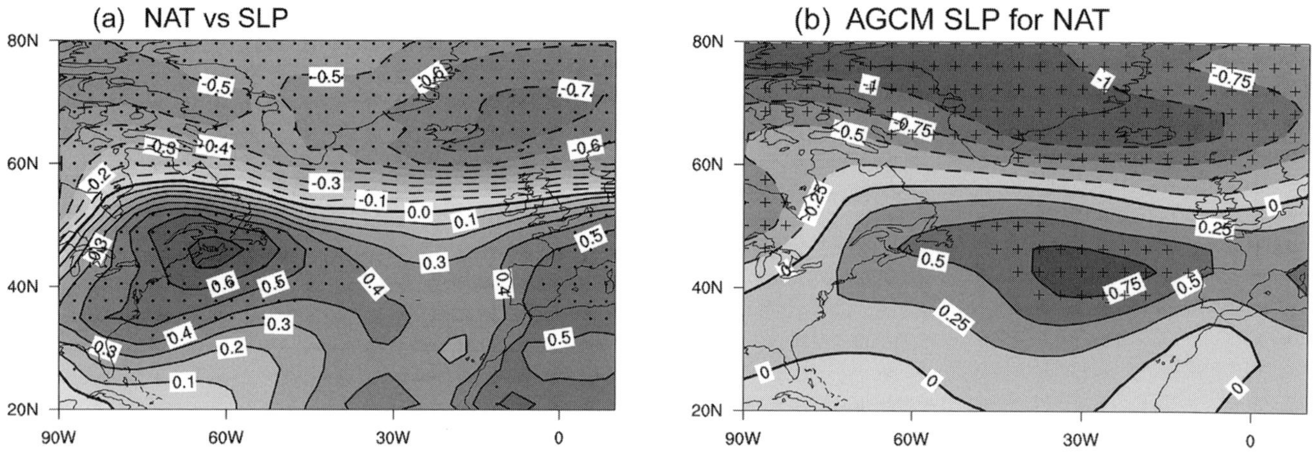

Figure 10.3 (a) Correlation maps of the SLP over the North Atlantic with the NAT index based on 21-yr running mean data in the CCSM4 simulation. Dots in (a) indicate that correlations are significant at the 95% confidence level. (b) AGCM simulated SLP responses (hPa) from the SPEEDY model forced by NAT-type SST anomalies. Crosses in (b) indicate the regions where the results from the sensitivity simulations are significantly different from the control at the 95% confidence level (student's t-test) (from Sun et al. (2015a)).

dipole with significantly negative correlations around Greenland and significantly positive correlations in the region of 30°–50°N. This pattern strongly projects onto the NAO. Moreover, we use the speedy AGCM to examine the atmospheric circulation response to the NAT SSTA forcing, and the AGCM simulated response to the NAT SST anomalies is also NAO-like (Figure 10.3b). Previous observation and modeling studies have provided evidence that the NAT is in close relationship with the NAO and that there is a positive feedback between them (Wu and Gordon 2002; Czaja and Frankignoul 2002). The gradient SST anomalies associated with the NAT induce changes in the position and strength of the storm tracks through thermal forcing, and the transient eddy forcing associated with the storm tracks tends to excite and maintain the NAO-like response. Sun et al. (2015a) also provided an explanation for the physical processes responsible for the strong coherence between the NAO and NAT. A schematic of the direct effect of the NAT on the NAO is shown in Figure 10.4. The atmosphere acts to damp the gradient SST anomalies associated with the NAT. This leads to changes of the synoptic eddy growth rate and the storm track tends to shift northward. The associated eddy forcing tends to excite and maintain the NAO-like response.

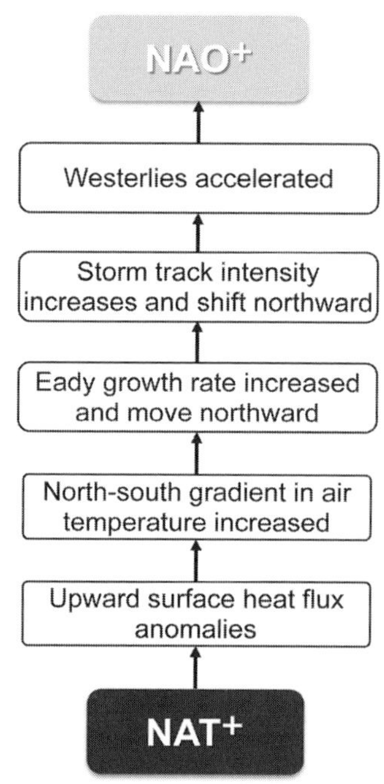

Figure 10.4 Schematic of the direct effect of the NAT on the NAO.

10.3.3 The NAO Forcing on the AMOC and AMO

The delayed effect of the NAO decadal variability on the AMO has been discussed extensively in numerical model studies (Visbeck et al. 1998; Delworth and Greatbatch 2000; Eden and Jung 2001). There is substantial modeling evidence that NAO-related surface turbulent heat flux, wind stress and freshwater flux over the North Atlantic Ocean can lead to multidecadal fluctuations of the AMOC, which in turn produce the SST signatures of the AMO. Observational studies have also demonstrated the NAO impact

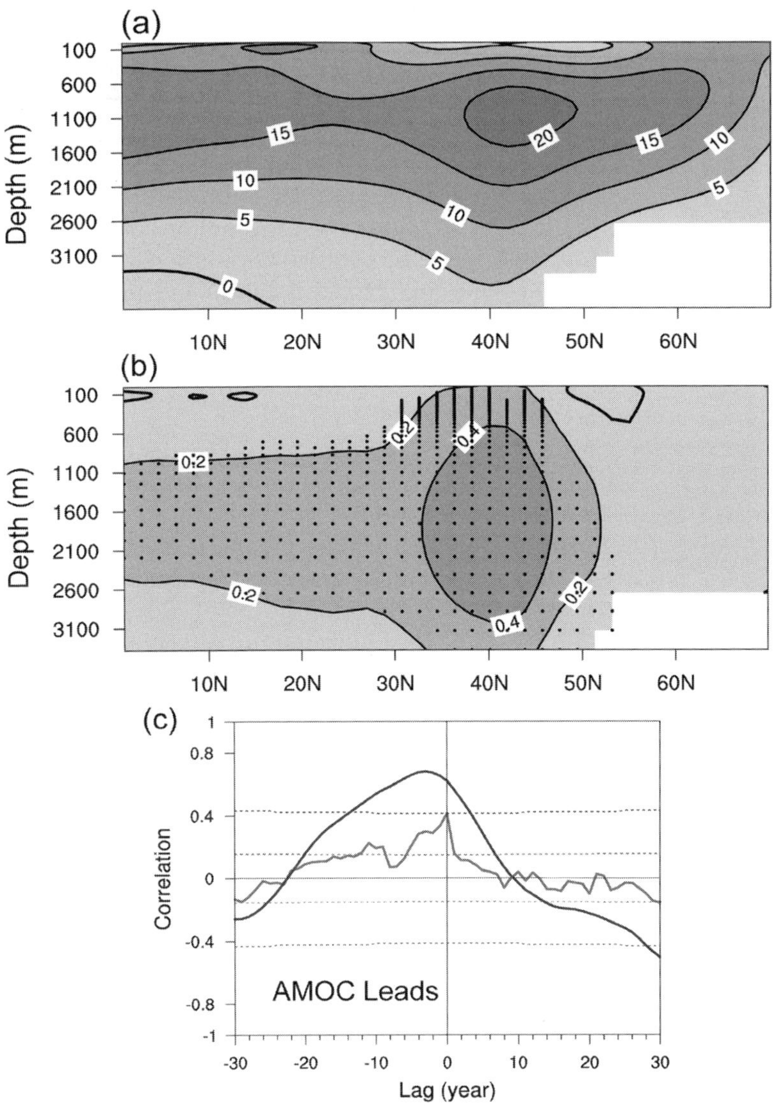

Figure 10.5 (a) Long-term mean AMOC trajectory (Sv, the unit of volume transport, 1 Sv = 10^6 m^3 s^{-1}) in the CCSM4 simulations. (b) Lagged regression of the annual mean AMOC trajectory (Sv) with respect to the normalized NAO index based on twenty-one-year running mean data, with the NAO leading by fifteen years. Dots indicate regressions significant at the 95% confidence level. (c) Cross-correlations of the simulated AMO index with the AMOC index based on original annual (red) and twenty-one-year running mean (blue) time series. Positive (negative) lags mean that the AMOC is lagging (leading). Dotted lines denote the 95% significance level (from Sun et al. (2015a)).

on the AMOC. The NAO directly affects the convection in the Labrador Sea, which then leads to changes in the AMOC (Latif et al. 2006a, b). Figure 10.5a shows the long-term mean trajectory of AMOC for the CCSM4 simulation. The simulated position of maximum overturning (about fifteen Sv) lies at the depth of one thousand to twelve hundred meters, and 40°–50°N, which is basically consistent with the hydrographic data. Figure 10.4b shows the lagged regression of the trajectory of AMOC onto the normalized NAO index when the NAO leads by fifteen years. The regression pattern shows significantly positive anomalies in the North Atlantic Ocean, and largely resembles the long-term mean structure of the AMOC. This indicates that the AMOC tends to strengthen (weaken) about one to two decades after the NAO phase turns positive (negative). The enhanced AMOC leads to stronger northward heat transport in the upper ocean, and, as a result, causes anomalously warm SST anomalies in the North Atlantic (Knight et al. 2005). An AMOC strength index is defined as the maximum of the annual mean AMOC trajectory at 40°N, similar to a previous study (Zhang 2008). The AMOC varies strongly in phase with the AMO over multidecadal time scales (Figure 10.5c) and is significantly correlated with the fifteen-year-ahead NAO (r = 0.23 for the original data and r = 0.54 for the decadally smoothed). Therefore, in accordance with previous studies (Latif et al. 2006a, b), the forcing effect of the NAO on the AMO

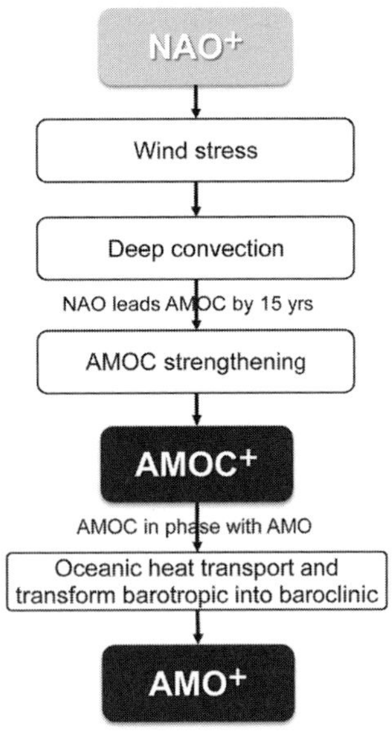

Figure 10.6 Schematic of the NAO forcing on the AMOC and AMO.

through AMOC changes may be a key mechanism of the North Atlantic multidecadal variability (Figure 10.6).

10.3.4 The Negative Feedback of the AMO on the NAT

Previous modeling studies have suggested that the key processes associated with the AMOC multidecadal fluctuations mainly consist of the advection induced by the overturning and the slow adjustment of the ocean circulation (Huang and Chou 1994; Greatbatch and Zhang 1995; Drbohlav and Jin 1998; Huang et al. 2014). As shown in Figure10.7a, the warm (cold) phase of the AMO is associated with a uniform surface warming (cooling) and a subsurface cooling (warming) in the North Atlantic Ocean. Since the warm phase of the AMO coincides with the strengthening of the AMOC (Figures 10.5c and 10.6), the warming upper ocean water continues to be advected northward to the downwelling region of the AMOC and then sinks, while the cooling subsurface ocean water continues to be moved southward. Figure 10.7b shows the regression of overturning heat transport (OHT). The regression map shows uniformly positive values in the North Atlantic Ocean, corresponding to an intensification of clockwise heat transport. Moreover, the OHT shows maximum values around 40°N. This may lead to a heat convergence (divergence) north (south) of these latitudes, inducing an ocean column warming (cooling) tendency over the entire upper ocean down to about two thousand meters. As a result, the enhanced AMOC is expected to be decelerated as a result of this developing positive meridional temperature gradient anomaly. Thus, the OHT associated with the enhanced AMOC acts as a negative feedback to prevent the AMOC from further strengthening by weakening the meridional temperature gradient after a delayed time of ocean adjustment.

The ocean temperature signal following the AMO is illustrated by a correlation of temperature with the normalized AMO index as a function of time lag (Figure 10.7c–e). The positive correlations are at first located in the upper North Atlantic and then propagate into the subpolar region, expanding downward; the negative correlations are shifted southward. This is consistent with the clockwise heat transport pathway of the OHT and further indicates that the ocean temperature is progressively modulated by the OHT associated with AMOC intensification. Previous studies have also shown a similar delayed response of ocean temperature to the preceding enhanced AMOC in both ocean (Greatbatch and Zhang 1995; Drbohlav and Jin 1998) and air–sea coupled (Ba et al. 2013) models. Moreover, the lagged correlation pattern at lag 18 is strikingly similar to the ocean temperature pattern associated with the NAT variability (Figure 10.7f), implying that the AMOC/AMO may provide a delayed effect on the NAT, and the NAT index is negatively correlated with the eighteen-year-ahead AMO index ($r = -0.27$ for the original data and $r = -0.52$ for the smoothed). To identify this delayed negative feedback further , Sun et al. (2015a) provided the physical explanation for the time delay. The time delay can be qualitatively interpreted as the time it takes for transport along the OHT pathway. A schematic of the negative feedback of the AMO on the NAT is shown in Figure 10.8. Therefore, the delayed negative feedback of the AMOC/AMO on the NAT/NAO through the oceanic advection process may be another important mechanism for the quasi-periodic multidecadal variability of the NAO.

10.3.5 A Mechanism of Multidecadal Variability in the North Atlantic

Figure 10.9 summarizes three key mechanisms involved in the fifty- to seventy-year quasi-periodic North Atlantic multidecadal variability. As mentioned previously, the positive NAT corresponds to the positive NAO phase in the atmosphere, and consequently, the positive NAO forces the enhancement of the AMOC, and leads to the AMO positive phase. The forcing effect is delayed by about fifteen years, possibly as a result of the large inertia associated with slow oceanic processes. The enhanced AMOC continues to affect the heat transport, and because of slow ocean adjustment, the North

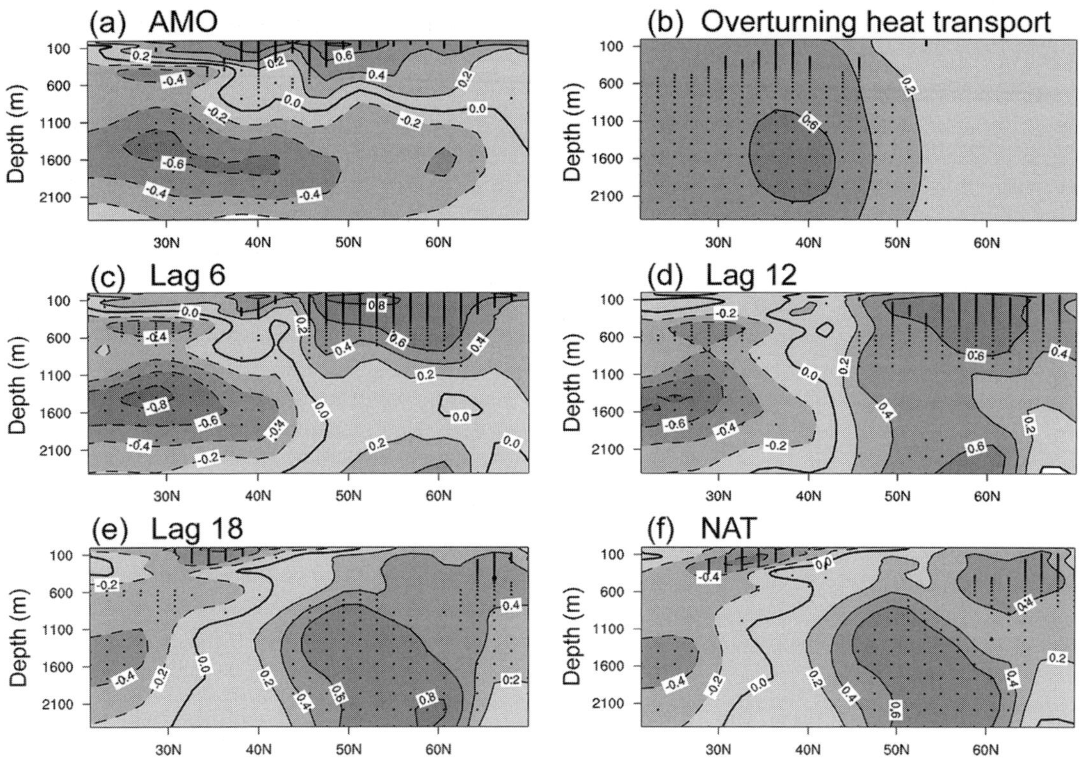

Figure 10.7 (a) Correlations of the North Atlantic zonal mean (75°7.5°W) ocean temperature with the AMO index in the CCSM4 simulations as a function of depth and latitude. (b) Regression of the OHT (10^{14} W, defined as $\rho C_p \int_{-H}^{z} \int_{Xe}^{Xw} \bar{v} T dxdz$, where ρ is density, C_p is the specific heat capacity, $\bar{v}$ is zonally averaged meridional velocity across the Atlantic basin, T is sea water temperature, z is ocean depth, H is the bottom of the ocean, and X_e and X_w are the eastern and western boundaries of the Atlantic basin) onto the normalized AMO index. (c)–(e) As in (a) but for lagged correlations. The lag (in years) is positive when AMO leads. (f) Correlations between the zonal mean ocean temperature and the imaginary-part PC of POP1 (NAT index). For convenience of comparison with (e), the sign of the NAT index is reversed. In (a)–(f), a 21-yr running mean was applied to the data prior to the correlation/regression analysis, and dots indicate correlations/regressions significant at the 95% confidence level (from Sun et al. (2015a)).

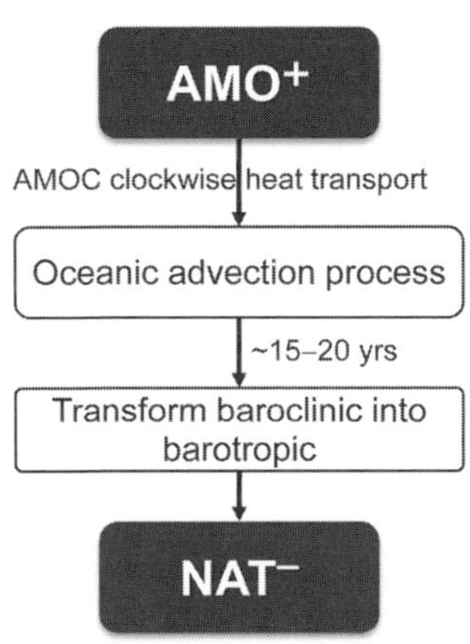

Figure 10.8 Schematic of the negative feedback of the AMO on the NAT.

Atlantic Ocean shows a delayed response (after about fifteen to twenty years) to the preceding enhanced AMOC with an SST pattern that resembles the NAT negative phase. The NAT negative phase coincides with the NAO negative phase in the atmosphere, and thus the cycle proceeds, but in the opposite sense. This mechanism testifies to the NAT–NAO–AMOC–AMO (NNAA) decadal coupled mode in the North Atlantic.

10.4 The Impact of the NNAA Decadal Coupled Mode on the Recent Warming Hiatus in the NHT

The NNAA decadal coupled mode has important influences on global and regional climate. An important example is its influence on the global warming

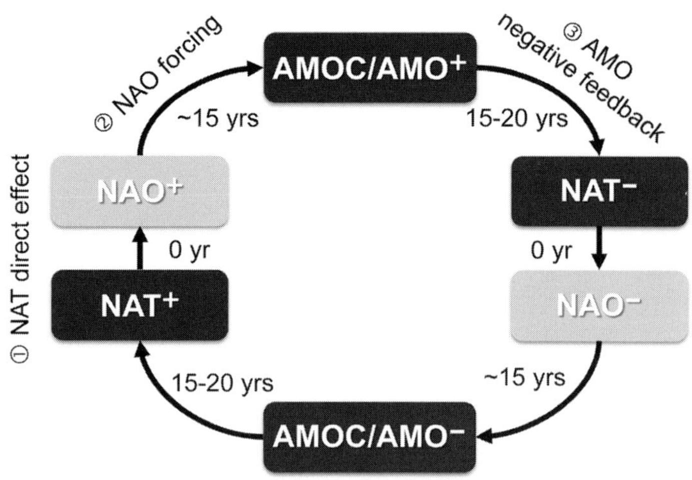

Figure 10.9 Summary of the mechanisms controlling the fifty- to seventy-year quasi-periodic cycle of the NNAA decadal coupled mode in the North Atlantic based on the analyses with the CCSM4 simulations. The positive NAT leads to the positive NAO. The positive NAO forces the enhancement of the AMOC, and leads to the AMO positive phase. The forcing effect is delayed by about about fifteen years, possibly because of the large inertia associated with slow oceanic processes. The enhanced AMOC continues to affect the heat transport, and because of slow ocean adjustment, the North Atlantic Ocean shows a delayed response (after about fifteen to twenty years) to the preceding enhanced AMOC with an SST pattern that resembles the negative NAT phase. The negative NAT phase coincides with the negative NAO phase in the atmosphere, and thus the cycle proceeds, but in the opposite sense. Blue (black) text indicates oceanic (atmospheric) phenomena (redrawn; based on figure 8 in Sun et al. (2015a)).

hiatus (Li et al. 2013a). The twentieth-century Northern Hemisphere mean surface temperature (NHT) is characterized by a multidecadal warming–cooling–warming pattern followed by a flat trend since about 2000 (i.e., the recent global warming hiatus). Li et al. (2013a) demonstrated that the multidcadal variability in NHT including the recent warming hiatus is tied to the NNAA decadal coupled mode and the NAO is implicated as a useful predictor of NHT multidecadal variability. The NAO signal leads the observed AMO and detrended NHT by about sixteen years because of the large thermal inertia associated with slow oceanic processes. The NAO signal in NHT can be defined as the damped response of the Northern Hemisphere climate system to the NAO variability, and by using the Hasselmann climate model (Hasselmann 1976), the detrended NHT response to the NAO can be simulated. Thus, the multidecadal variability in NHT can be explained and modeled very well by using the NAO signal (Figure 10.10). The results show that the warming hiatus in the period of 1944–1976 can be explained by the NAO multidecadal weakening from the mid-1920s to the late 1960s and the recent warming hiatus is attributed to NAO decadal weakening since the mid-1990s. In fact, the warming hiatus from 1877 to 1913 can also be explained by the NAO multidecadal weakening from the early 1860s to the late 1890s. An NAO-based linear model was therefore established to predict the NHT (Li et al. 2013a), which gives an excellent hindcast for NHT in 1971–2011 with the recent flat trend well predicted (Figure 10.10). The NHT in 2012–2027 is predicted to fall slightly over the next decades, because of the recent NAO decadal weakening that temporarily offsets the anthropogenically induced global warming.

10.5 Conclusion

A decadal coupled ocean–atmosphere interaction theory associated with the NNAA coupled mode is proposed to explain the fifty- to seventy-year quasi-periodic multidecadal variability over the North Atlantic region. This ocean–atmosphere interaction mode has important impacts on the global and regional climates. For example, the NNAA decadal coupled mode may lead to the NHT multidecadal variability and can cause the global warming hiatus, and the NAO can be implicated as a useful predictor for decadal variation of the NHT.

The present study has implications for predicting the regional and hemispheric climates over interdecadal time scales. Assuming that the approximate sixty-year cycle in the NNAA decadal coupled mode continues, we should be entering a negative phase of the NAO. This may correspond to more negative NAO events during the next decade. Considering that the negative NAO usually leads to cooling over most of the Northern Hemisphere continent (Hurrell 1995, 1996; Li and Wang 2003; Li et al. 2013a), the anthropogenic warming effect may be partially offset. Moreover, the NNAA decadal coupled mode also has implications for interpretation of proxy data records.

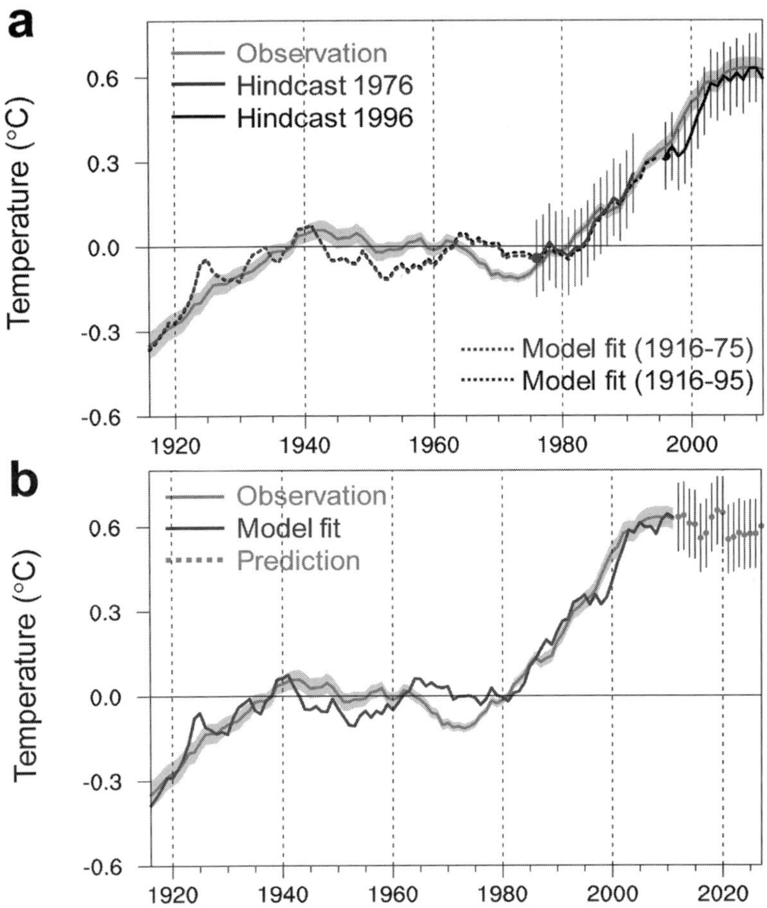

Figure 10.10 The observed, modeled, hindcast and predicted NHT. (a) The 11-year running mean of the observed NHT (red) from 1916 to 2011, and two sets of 16-year hindcast NHT starting at 1976 and 1996, respectively, using the NAO-based linear model. The solid circle represents the starting year of the hindcast and the fitted NHT before each hindcast is shown as a dotted line in the same color as the corresponding hindcast. Vertical bars denote the 95% confidence intervals for the linear model. (b) The 11-year running mean of the observed NHT (red) from 1916 to 2011, the model fit of NHT (blue), and the predicted NHT in 2012–2027. The pink shaded areas in (a) and (b) show the 2-sigma uncertainty ranges of the NHT series for the HadCRUT4 dataset. The description of the NAO-based linear model is presented in Li et al (2013a) (from Li et al., 2013a). For the color version, please refer to the plate section.

Acknowledgements

This paper was jointly supported by the MOST Key Project (2016YFA0601801) and SOA International Cooperation Program on Global Change and Air-Sea Interaction (GASI-IPOVAI-03, GASI-IPO-VAI-06).

References

Ba, J., N. S. Keenlyside, W. Park, M. Latif, E. Hawkins, and H. Ding (2013). A mechanism for Atlantic multidecadal variability in the Kiel Climate Model. *Clim. Dyn.*, 41, 2133–2144.

Balmaseda, M. A., K. E., Trenberth, and E. Källén (2013). Distinctive climate signals in reanalysis of global ocean heat content, *Geophys. Res. Lett.*, 40, doi:10.1002/grl.50382.

Chen, X. Y., and K. K. Tung (2014). Varying planetary heat sink led to global-warming slowdown and acceleration. *Science* 345(6199), 897–903.

Cohen, J., and M. Barlow. (2005). The NAO, the AO, and global warming: How closely related? *J. Clim.*, 18, 4498–4513.

Cook, E. R., R. D. D'Arrigo, and K. R. Briffa. (1998). A reconstruction of the North Atlantic Oscillation using tree-ring chronologies from North America and Europe. *Holocene*, 8, 9–17.

Czaja, A., and C. Frankignoul. (2002). Observed impact of Atlantic SST anomalies on the North Atlantic oscillation. *J. Clim.*, 15, 606–623.

Danabasoglu, G., S. G. Yeager, Y.-O. Kwon, J. J. Tribbia, A. S. Phillips, and J. W. Hurrell. (2012). Variability of the Atlantic Meridional Overturning Circulation in CCSM4. *J. Clim.*, 25, 5153–5172.

Delworth, T. L., and R. J. Greatbatch. (2000). Multidecadal thermohaline circulation variability driven by atmospheric surface flux forcing. *J. Clim.*, 13, 1481–1495.

Delworth, T. L., and M. E. Mann. (2000). Observed and simulated multidecadal variability in the Northern Hemisphere. *Clim. Dyn.*, 16, 661–676.

Deser, C., M. A. Alexander, S.-P. Xie, and A. S. Phillips. (2010). Sea surface temperature variability: Patterns and mechanisms. *Annual Review of Marine Science*, 2, 115–143.

Dieppois, B., A. Durand, M. Fournier, and N. Massei. (2013). Links between multidecadal and interdecadal climatic oscillations in the North Atlantic and regional climate variability of northern France and England since the 17th century. *J. Geophys. Res. Atmos.*, 118, 4359–4372.

Drbohlav, J., and F. F. Jin. (1998). Interdecadal variability in a zonally averaged ocean model: An adjustment oscillator. *J. Phys. Oceanogr.*, 28, 1252–1270.

Eden, C., and T. Jung. (2001). North Atlantic interdecadal variability: Oceanic response to the North Atlantic oscillation (1865–1997). *J. Clim.*, 14, 676–691.

Enfield, D. B., A. M. Mestas-Nunez, and P. J. Trimble. (2001). The Atlantic multidecadal oscillation and its relation to rainfall and river flows in the continental US. *Geophys. Res. Lett.*, 28, 2077–2080.

England, M. H., S. McGregor, P. Spence, G. A. Meehl, A. Timmermann, W. Cai, A. S. Gupta, M. J. McPhaden, A. Purich, and A. Santoso. (2014). Recent intensification of wind-driven circulation in the Pacific and the ongoing warming hiatus, *Nature Climate Change*, doi:10.1038/nclimate2106.

Feldstein, S. B. (2000). The timescale, power spectra, and climate noise properties of teleconnection patterns. *J. Clim.*, 13, 4430–4440.

Fyfe, J. C., G. A. Meehl, M. H. England et al. (2016). Making sense of the early-2000s warming slowdown. *Nat. Clim. Chang.*, 6, 224–228.

Gent, P. R. et al. (2011). The Community Climate System Model Version 4. *J. Clim.*, 24, 4973–4991.

Gong, D. Y., S. W. Wang, and J. H. Zhu. (2001). East Asian winter monsoon and Arctic Oscillation. *Geophys. Res. Lett.*, 28, 2073–2076.

Greatbatch, R. J., and S. Zhang. (1995). An Interdecadal Oscillation in an Idealized Ocean-Basin Forced by Constant Heat-Flux. *J. Clim.*, 8, 81–91.

Hasselmann, K. (1976). Stochastic climate models. 1. Theory. *Tellus*, 28, 473–485.

Haywood J. M., Jones A., and Jones G. S. (2014). The impact of volcanic eruptions in the period 2000–2013 on global mean temperature trends evaluated in the HadGEM2-ES climate model. *Atmos. Sci. Lett.*, 15 (2), 92–96, doi:10.1002/asl2.471.

Hausfather, Z., K. Cowtan, D. C. Clarke, P. Jacobs, M. Richardson, and R. Rohde. (2017). Assessing recent warming using instrumentally homogeneous sea surface temperature records. *Science Advances*, 3, e1601207.

Huang, B., J. Zhu, and H. Yang. (2014). Mechanisms of Atlantic Meridional Overturning Circulation (AMOC) variability in a coupled ocean-atmosphere GCM. *Adv. Atmos. Sci.*, 31, 241–251.

Huang, R. X., and R. L. Chou. (1994). Parameter sensitivity study of the saline circulation. *Clim. Dyn.*, 9, 391–409.

Hurrell, J. W. (1995). Decadal Trends in the North-Atlantic Oscillation – Regional Temperatures and Precipitation. *Science*, 269, 676–679.

Hurrell, J. W. (1996). Influence of variations in extratropical wintertime teleconnections on Northern Hemisphere temperature. *Geophys. Res. Lett.*, 23, 665–668.

Karl, T. R. et al. (2015). Possible artifacts of data biases in the recent global surface warming hiatus. *Science*, 348, 1469–1472, doi:10.1126/science.aaa5632.

Knight, J. R., R. J. Allan, C. K. Folland, M. Vellinga, and M. E. Mann. (2005). A signature of persistent natural thermohaline circulation cycles in observed climate. *Geophys. Res. Lett.*, 32, L20708, doi:10.1029/2005GL024233.

Kosaka, Y., and XIE, S.-P. (2013). Recent global-warming hiatus tied to equatorial Pacific surface cooling. *Nature*, 501: 403–407.

Kucharski, F., F. Molteni, and J. H. Yoo. (2006). SST forcing of decadal Indian Monsoon rainfall variability. *Geophys. Res. Lett.*, 33, L03709, doi:10.1029/2005GL025371.

Kucharski, F., F. Molteni, M. P. King, R. Farneti, I. S. Kang, and L. Feudale. (2013). On the Need of Intermediate Complexity General Circulation Models A "SPEEDY" Example. *Bull. Am. Meteorol. Soc.*, 94, 25–30.

Kucharski, F., A. Bracco, J. H. Yoo, A. M. Tompkins, L. Feudale, P. Ruti, and A. Dell'Aquila. (2009) A Gill-Matsuno-type mechanism explains the tropical Atlantic influence on African and Indian monsoon rainfall. *Q. J. Roy. Meteorol. Soc.*, 135, 569–579.

Latif, M., M. Collins, H. Pohlmann, and N. Keenlyside. (2006a). A review of predictability studies of Atlantic sector climate on decadal time scales. *J. Clim.*, 19, 5971–5987.

Latif, M. et al. (2006b). Is the thermohaline circulation changing? *J. Clim.*, 19, 4631–4637.

Lean, J. L., and D. H. Rind. (2008). How natural and anthropogenic influences alter global and regional surface temperatures: 1889 to 2006. *Geophys. Res. Lett.*, 35, L18701, doi:10.1029/2008GL034864.

Li, J. P., and J. X. L. Wang. (2003). A new North Atlantic Oscillation index and its variability. *Adv. Atmos. Sci.*, 20, 661–676.

Li, J. P. (2005). Coupled air-sea oscillations and climate variations in China. In: D. Qin (Eds.), *Climate and Environmental Evolution in China*, Vol. 1. Beijing: China Meteorological Press, pp. 324–333.

Li, J. P., C. Sun, and F. F. Jin. (2013a). A: NAO implicated as a predictor of Northern Hemisphere mean temperature multidecadal variability. *Geophys. Res. Lett.*, 40, 5497–5502.

Li, J. P., R. C. Ren, Y. Qi, F. Wang, R. Lu, P. Zhang, Z. Jiang, W. Duan, F. Yu, and Y. Yang. (2013b). B: Progress in air–land–sea interactions in Asia and their role in global and Asian climate change. *Chinese J. Atmos. Sci.*, 37(2), 518–538.

Li, S., and G. T. Bates. (2007). Influence of the Atlantic Multidecadal Oscillation on the winter climate of East China. *Adv. Atmos. Sci.*, 24, 126–135.

Lu, R. Y., B. W. Dong, and H. Ding. (2006). Impact of the Atlantic Multidecadal Oscillation on the Asian summer monsoon. *Geophys. Res. Lett.*, 33.

Meehl, G. A., J. M. Arblaster, J. T. Fasullo, A. Hu and K. E. Trenberth. (2011). Model-based evidence of deep-ocean heat uptake during surface-temperature hiatus periods. *Nature Climate Change*, 1, 360–364, doi: 10.1038/NCLIMATE1229

Meinshausen, M., Smith, S. J., Calvin, K. et al. (2011).The RCP greenhouse gas concentrations and their extensions from 1765 to 2300, *Climatic Change*, 109, 213. doi:10.1007/s10584-011-0156-z.

Morice, C. P., J. J. Kennedy, N. A. Rayner, and P. D. Jones. (2012). Quantifying uncertainties in global and regional temperature change using an ensemble of observational estimates: The HadCRUT4 data set, *J. Geophys. Res.*, 117, D08101, doi:10.1029/2011JD017187.

Newman, M. (2013). An empirical benchmark for decadal forecasts of global surface temperature anomalies. *J. Climate*, 26, 5260–5269.

Olsen, J., N. J. Anderson, and M. F. Knudsen. (2012). Variability of the North Atlantic Oscillation over the past 5,200 years. *Nat. Geosci.*, 5, 808–812.

Park, W., and M. Latif. (2010). Pacific and Atlantic multidecadal variability in the Kiel Climate Model. *Geophys. Res. Lett.*, 37, L24701, doi:10.1029/2006GL027655.

Schlesinger, M. E., and N. Ramankutty. (1994). An oscillation in the global climate system of period 65–70 years. *Nature*, 367, 723–726.

Solomon, S. et al. (2010). Contributions of stratospheric water vapor to decadal changes in the rate of global warming. *Science*, 327, 1219–1223.

Stocker, T. F. et al. (2013). Climate change 2013: The physical science basis - Summary for policymakers, in Climate Change 2013: The Physical Science Basis. Contribution of Working Group 1 to the Fifth Assessment Report of the Intergovernmental Panel on Climate Change, edited by T. F. Stocker et al., Cambridge and New York: Cambridge University Press, pp. 2–29.

Sun, C., and J. P. Li. (2012). Analysis of anomalously low surface air temperature in Northern Hemisphere during 2009/2010 winter. *Climatic Environ. Res.*, 17, 259–273.

Sun, C., J. P. Li, and F.-F. Jin. (2015). A: A delayed oscillator model for the quasi-periodic multidecadal variability of the NAO. *Clim. Dyn.*, 45, doi:10.1007/s00382-014-2459-z.

Sun, C., J. P. Li, and S. Zhao. (2015). B: Remote influence of Atlantic multidecadal variability on Siberian warm season precipitation. *Sci Rep*, 5, 16853.

Sun, C., J. Li, R. Q. Ding, and Z. Jin. (2016). Cold season Africa–Asia multidecadal teleconnection pattern and its relation to the Atlantic multidecadal variability. *Clim Dyn, Online*, doi:10.1007/s00382-016-3309-y.

Sutton, R. T., and D. L. R. Hodson. (2005). Atlantic Ocean forcing of North American and European summer climate. *Science*, 309, 115–118.

Taws, S. L., R. Marsh, N. C. Wells, and J. Hirschi. (2011). Re-emerging ocean temperature anomalies in late-2010 associated with a repeat negative NAO. *Geophys. Res. Lett.*, 38, L20601, doi:10.1029/2011GL048978

Thompson, D. W. J., J. M. Wallace, and G. C. Hegerl. (2000). Annular modes in the extratropical circulation. Part II: Trends. *J. Clim.*, 13, 1018–1036.

Ting, M. F., Y. Kushnir, R. Seager, and C. H. Li. (2009). Forced and internal twentieth-century SST trends in the North Atlantic. *J. Clim.*, 22, 1469–1481.

Torrence, C., and G. P. Compo. (1998). A practical guide to wavelet analysis. *Bull. Am. Meteorol. Soc.*, 79, 61–78.

Trenberth, K. E., and J. T. Fasullo.(2013). An apparent hiatus in global warming? *Earth's Future*, 1, 19–32, doi:10.1002/2013EF000165.

Visbeck, M., H. Cullen, G. Krahmann, and N. Naik. (1998). An ocean model's response to North Atlantic Oscillation-like wind forcing. *Geophys. Res. Lett.*, 25, 4521–4524.

Visbeck, M., E. P. Chassignet, R. Curry, T. Delworth, B. Dickson, and G. Krahmann. (2003). The ocean's response to North Atlantic Oscillation variability, in *The North Atlantic Oscillation: Climatic Significance and Environmental Impact*, Geophys. Monogr. Ser., vol. 134, pp. 113–145, , Washington, DC: AGU.

Von Storch, H., T. Bruns, I. Fischerbruns, and K. Hasselmann. (1988). Principal oscillation pattern-analysis of the 30-day to 60-day oscillation in General-Circulation Model Equatorial Troposphere. *J. Geophys. Res. Atmos.*, 93, 11022–11036.

Walker, G. T., and E. W. Bliss. (1932). World Weather V, *Mem. Quart. J. Roy. Meteor. Soc.*, 4, 53–84.

Wang, C., H. Liu, and S.-K. Lee. (2010). The record-breaking cold temperatures during the winter of 2009/2010 in the Northern Hemisphere. *Atmos. Sci. Lett.*, 11, 161–168.

Watanabe, M., Y. Kamae, M. Yoshimori, A. Oka, M. Sato, M. Ishii, T. Mochizuki, and M. Kimoto. (2013). Strengthening of ocean heat uptake efficiency associated with the recent climate hiatus. *Geophys. Res. Lett.*, 40, 3175–3179, doi: 10.1002/grl.50541.

Wu, P. L., and C. Gordon. (2002). Oceanic influence on North Atlantic climate variability. *J. Clim..*, 15, 1911–1925.

Wu, Z. W., B. Wang, J. P. Li, and F.-F. Jin. (2009). An empirical seasonal prediction model of the east Asian summer monsoon using ENSO and NAO. *J. Geophys. Res.*, 114, D18120, doi:10.1029/2009JD011733.

Zanchettin, D., O. Bothe, W. Müller, J. Bader, and J. H. Jungclaus. (2013). Different flavors of the Atlantic Multidecadal Variability. *Clim. Dyn.*, 42, doi:10.1007/s00382-013-1669-0.

Zhang, R. (2008). Coherent surface-subsurface fingerprint of the Atlantic meridional overturning circulation. *Geophys. Res. Lett.*, 35, L20705, doi:10.1029/2008GL035463.

Zhang, R., and T. L. Delworth. (2006). Impact of Atlantic multidecadal oscillations on India/Sahel rainfall and Atlantic hurricanes. *Geophys. Res. Lett.*, 33, L17712, doi:10.1029/2006GL026267.

Zhang, R., T. L. Delworth, and I. M. Held. (2007). Can the Atlantic Ocean drive the observed multidecadal variability in Northern Hemisphere mean temperature? *Geophys. Res. Lett.*, 34, L02709, doi:10.1029/2006GL028683.

Zhang, R., and Coauthors. (2013). Have aerosols caused the Observed Atlantic Multidecadal Variability? *J. Atmos. Sci.*, 70, 1135–1144.

11
Sea Level Rise and Future Earth

Anny Cazenave and Hindumathi Palanisamy

11.1 Introduction

Sea level variations spread over a very broad spectrum of spatio-temporal scales. On geological time scales (~5–100 million years [Myr]) large amplitude (>100 m) sea level changes depend primarily on tectonic processes (e.g., large-scale change in the shape of ocean basins associated with seafloor spreading and mid-ocean ridge expansion) as well as growth and decay of polar ice sheets. On 10 000- to 100 000-year time scales, glacial/interglacial cycles driven by changes of the Earth's orbit and obliquity also cause ~100-m-amplitude sea level variations. On shorter time scales, modest (on the order of a few meters or less) sea level changes occur at interannual, multidecadal, centennial and millenary time scales, in response to natural (external) climate forcing factors (change in solar irradiance and volcanic eruptions), as well as natural variability inside the climate system, related for example to coupled atmosphere–ocean perturbations such as El Niño–Southern Oscillation (ENSO), North Atlantic Oscillation (NAO), or Pacific Decadal Oscillation (PDO). At time scales of 1-year or less, sea level variations are caused by seasonal forcing, atmospheric loading, ocean tides as well as storm surges.

Influence of humans on climate through green house gas (GHG) emissions is noticeable since the beginning of the twentieth century. It is now well established that the Earth's climate is warming and that the main cause is the accumulation of GHG inside the atmosphere from anthropogenic fossil fuel combustion and land use change (mostly deforestation) (Intergovernmental Panel on Climate Change/IPCC 5th Assessment Report/AR5, Stocker et al., 2013). Global warming has already had several visible consequences, in particular increase of the Earth's mean surface temperature (Stocker et al. 2013) and ocean heat content (Levitus et al., 2012, Rhein et al., 2013), melting of sea ice (Vaughan et al., 2013), melting of glaciers (Gardner et al., 2013, Marzeion et al., 2015) and loss of ice mass from the Greenland and Antarctica ice sheets (Rignot et al., 2011, Shepherd et al., 2012, Vaughan et al., 2013). On average over the last 50 years, 93% of heat excess accumulated in the climate system because of GHG emissions has been stored in the ocean, the remaining 7% being used to warm the atmosphere and continents and melt sea and land ice (von Schuckmann et al., 2016). Because of ocean warming and land ice melt, sea level is rising.

In this chapter we focus on present-day sea level changes induced by anthropogenic forcing, natural systems and unforced natural variability. Time scales considered here range from a few years to several decades (seasonal and sub-seasonal variations are not addressed). We first present sea level observations, then discuss the various contributions to the global mean sea level rise and superimposed regional variability. We also examine closure of the sea level budget, and briefly discuss projections of future sea level.

Definition: Sea level is the height of the sea surface expressed either in a geocentric reference frame (absolute sea level) or with respect to the moving Earth's crust (relative sea level). Absolute sea level variations result from changes in the volume of water filling ocean basins (either due to water density or mass changes) while relative sea level variations designate sea surface height changes with respect to the ground (accounting thus for both 'absolute' sea level changes *plus ground motions).*

11.2 Observations of Recent-Past and Present-Day Sea Level Variations

11.2.1 Twentieth Century

Historical (i.e., since the beginning of the industrial era) sea level variations are derived from direct in situ measurements by tide gauges, instruments located along continental coastlines and islands. Prior to the beginning of the twentieth century, data are very sparse and based only on a few long records at sites located in western Europe. The largest tide gauge database of monthly and annual mean sea level records is the Permanent Service for Mean Sea Level (PSMSL, www.psmsl.org/; Woodworth and Player, 2003), which contains data for the twentieth century from ~2000 sites maintained by about 200 nations. The records are somewhat inhomogeneous in terms of data length and quality. For long-term sea level studies, only ~10% of this data set is usable, because of data gaps and limited tide gauge distribution in the past. Tide gauges measure sea level relatively to the ground. In active tectonic and volcanic regions, or in areas subject to strong ground subsidence due to natural causes (e.g., sediment loading in river deltas) or human activities (ground water pumping and oil/gas extraction), tide gauge data are directly affected by the corresponding ground motions (Woppelmann and Marcos, 2016). Post glacial rebound, the visco-elastic response of the Earth crust to last deglaciation, also called Glacial Isostatic Adjustment (GIA), is another process that gives rise to vertical land movements (Peltier, 2004, Tamisiea, 2011).

After the ~130-m sea level rise associated with the last deglaciation that started about 20 000 years ago, geological, geochemical and archeological observations indicate that the mean sea level remained almost stable during the last two to three millennia, then started to rise about two centuries ago (Lambeck et al., 2010, Kemp et al., 2011). A large number of studies have analysed tide gauge data from the PSMSL to provide an historical mean sea level time series. Different processing methodologies and corrections to the data have been applied by the investigators. Among recent works, let's quote those by Jevrejeva et al. (2006, 2014a), Woppelmann et al. (2009), Church and White (2011), Ray and Douglas (2011) and Hay et al. (2015). These studies report a mean sea level rise in the range 1.2–1.9 mm/yr since the early twentieth century. This large spread of estimates results from many factors, e.g., uneven and restricted distribution of tide gauges, treatment of data gaps in the records, geographical averaging procedure, impact of vertical crustal motions and methods to correct them, account or not of fingerprint constraints etc. Hamlington and Thomson (2015) showed that tide gauge selection has profound impact on historical sea level rate estimates. Figure 11.1 shows the historical mean sea level time series since 1900 based on different tide gauge reconstructions (data from Church and White, 2011, Ray and Douglas, 2011, Jevrejeva et al., 2014a and Hay et al., 2015). The altimetry-based global mean sea level is also shown (see Section 11.2.2).

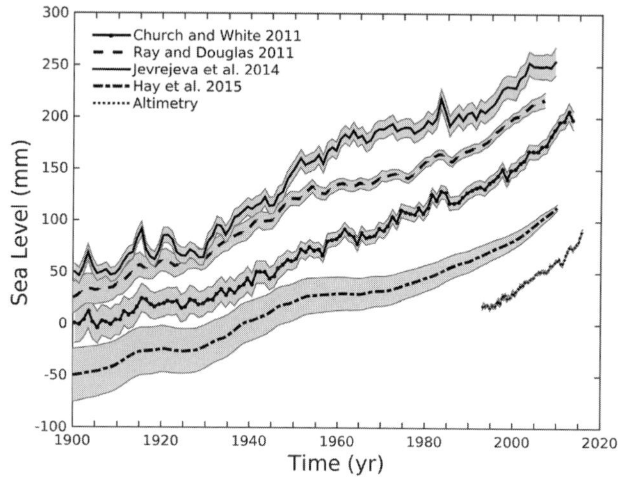

Figure 11.1 Twentieth-century mean sea level curves from different tide gauge reconstructions. From top to bottom: Church and White (2011), Ray and Douglas (2011), Jevrejeva et al. (2014a), Hay et al., (2015). Shaded areas represent data uncertainties.The last bottom curve is the altimetry-based global mean sea level since 1993. Arbitrary vertical translations are applied to the sea level curves for clarity.

11.2.2 Satellite Altimetry Era

Since the early 1990s, satellite altimetry has become the main tool for precisely and continuously measuring sea level with quasi-global coverage and a few days' revisit time (called the 'orbital cycle') (Nerem et al., 2010). Compared to tide gauges that provide sea level relative to the ground, satellite altimetry measures 'absolute' sea level variations. The concept of the satellite altimetry measurement is simple: the onboard radar altimeter transmits microwave radiation towards the sea surface that partly reflects back to the satellite. Measurement of the round-trip travel time provides the height of the satellite above the instantaneous sea surface (called 'range'). The quantity of interest in oceanography is the sea surface height above a reference fixed surface (typically a conventional reference ellipsoid or a mean sea surface). It is obtained by the difference between the height of the satellite above the reference (deduced from precise orbitography) and the range measurement. The estimated sea surface height needs to be corrected for various factors because of atmospheric (ionospheric and tropospheric) delays, as well as instrumental drifts and bias between successive altimetry missions. Other corrections due to geophysical effects, such as the sea state bias, solid Earth tides, pole and ocean tides are also applied. High-precision satellite altimetry started with the launch in 1991 of ERS-1 by the European Space Agency (ESA) and in 1992 of the joint NASA (National Aeronautics and Space Administration)–CNES (Centre National d'Etudes Spatiales) satellite TOPEX/Poseidon (T/P). Since then, several high-precision altimetry missions have followed: Jason-1 (2001), Jason-2 (2008) and Jason-3 (2016), the successors of T/P with similar orbital characteristics. ESA also developed ERS-2 (1995), Envisat (2002), CryoSat (2010) and Sentinel-3A (2016). Cryosat and Sentinel-3A use new technology, i.e., Synthetic Aperure Radar (SAR) altimetry. Sentinel-3A contributes to the COPERNICUS operational programme of the European Union. SARAL/Altika (2013), a joint Indian-French mission, operates in the Ka-band, allowing smaller radar footprint on ground than other missions (T/P, the Jason series, ERS and Envisat being equipped with Ku-band radars). A wide-swath interferometric altimetry mission (SWOT – Surface Waters Ocean Topography) able to study the mesoscale ocean circulation is currently developed by NASA and CNES for a launch in 2021. Altimeter satellites have also been put in orbit by the Chinese Space Agency (HY2A and HY2B). The T/P and Jason series have an orbital cycle of ten days but a large spacing between satellite tracks (~ 300 km at equator). They cover the 66°S–66°N latitude domain. The orbital cycle of the ESA missions and SARAL/AltiKa is 35 days, but the spacing between tracks is shorter and the latitudinal coverage goes up to 82°, allowing a large portion of the Arctic Ocean to be monitored. During the past ~25 years, sea surface height has been routinely measured by at least two altimeter satellites operating simultaneously.

The precision of an individual sea surface height measurement has now reached the 1–2 cm level (Ablain et al., 2009), allowing a precision on the global mean rate of rise of ~0.5–0.8 mm/yr (Ablain et al., 2015, 2017). This trend uncertainty was reduced by a factor of about 2 owing to recent efforts conducted in the context of the ESA Climate Change Initiative (CCI) project with complete reprocessing of all altimetry missions, and improvement of orbits and geophysical corrections and reduction of instrumental drifts and bias between missions (Ablain et al., 2015, 2017).

The temporal evolution of the global mean sea level (GMSL) from satellite altimetry since early 1993 is presented in Figure 11.2. It shows an almost quadratic increase, except for two temporary anomalies associated with ENSO events, in particular the 1997–1998 El Niño and the 2011 La Niña. Over this ~25 year-long period, the GMSL has been rising at a mean rate of ~3 mm/yr, a value significantly higher than the mean

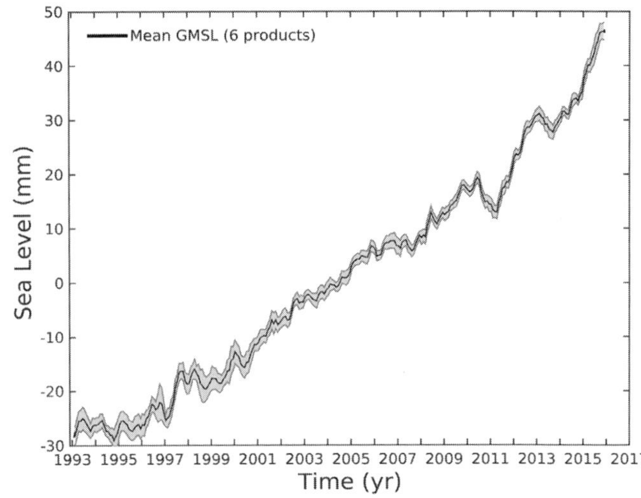

Figure 11.2 Global mean sea level time series from satellite altimetry from January 1993 to December 2015 (average time series based on altimetry data provided by six different processing centers; details on these data in Dieng et al., 2017). The shaded area represents the data uncertainty (estimated from the dispersion of the six GMSL time series around the mean).

rate recorded by tide gauges over the twentieth century. The altimetry record is now sufficiently long to conclude with some confidence that sea level rise is accelerating (Church et al., 2013).

The quasi-global coverage of the ocean domain by altimeter satellites has allowed measurements of regional sea level variations. A map of spatial linear trend patterns over the period January 1993 to September 2016 is presented in Figure 11.3. It shows that sea level is not rising uniformly. In some regions (e.g., Western Pacific, North Atlantic, Southern Ocean), rates of sea level rise are faster by a factor up to 3 than the global mean rate.

11.3 Causes of Present-Day Sea Level Rise

The physical processes causing global mean sea level rise and regional changes are not identical, although they are related. In terms of global mean, sea level change resuts from ocean thermal expansion and ocean mass changes due to melting and discharge of ice from the Greenland and Antarctica ice sheets, melting of glaciers, and changes in land water storage and atmospheric water content. As the oceans warm in response to anthropogenic forcing, sea waters expand, thus sea level rises. When mountain glaciers melt in response to increasing air temperature, sea level rises because of fresh water mass input into the oceans. Similarly, ice mass loss from

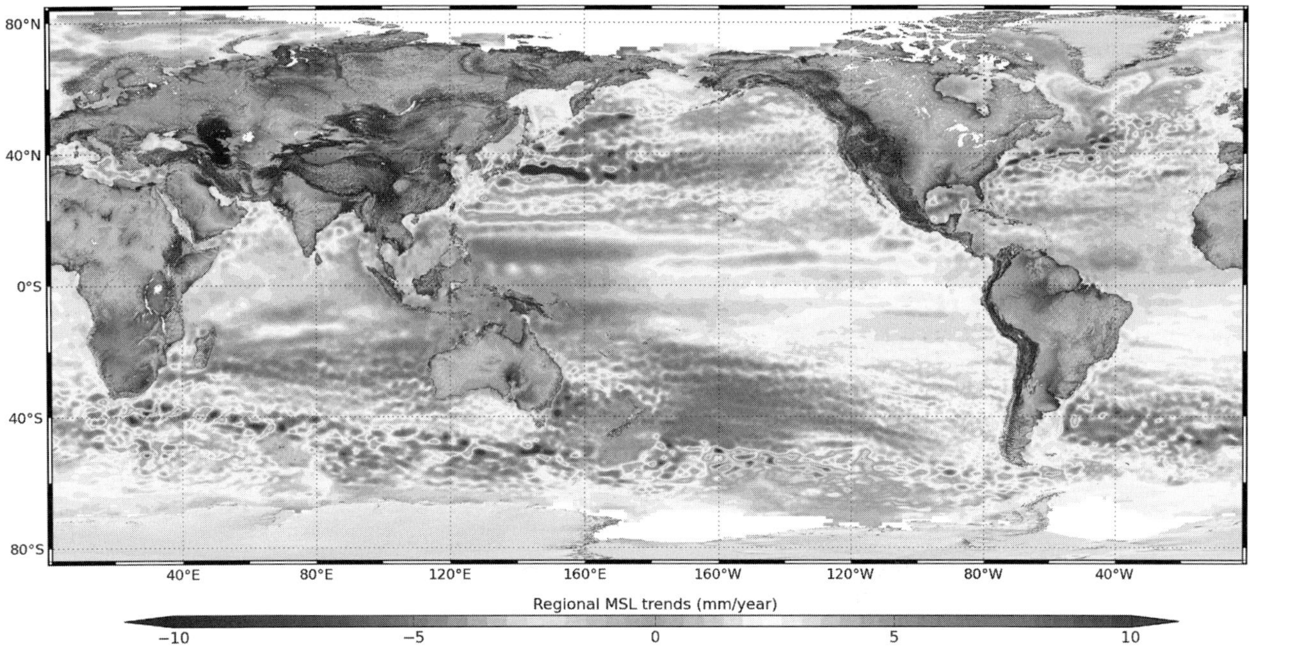

Figure 11.3 Spatial trend patterns in sea level (mm/yr) observed by satellite altimetry over the period January 1993 to September 2016. Sea level data from AVISO (www.aviso.altimetry.fr). For the color version, please refer to the plate section.

the ice sheets causes sea level rise. Modification of the land hydrological cycle and associated land water storage due to climate variability and direct anthropogenic forcing also leads to sea level change.

At regional scale, geographical patterns of sea level change result from the superposition of 'fingerprints' caused by different processes: changes in sea water density due to changes in temperature and salinity (phenomena called 'steric' effects), atmospheric loading, and solid Earth's deformations and gravitational changes in response to mass redistributions caused by land ice melt and land water storage changes (called 'static' factors; Stammer et al., 2013). The land ice melt-related static factor has two components: one associated with last deglaciation, the so-called GIA effect and the other with on-going land ice melt. The static factors – mostly known from modelling – give rise to complex regional patterns in sea level change (Bamber and Riva, 2010, Tamisiea, 2011, Spada et al., 2017): sea level drop in the immediate vicinity of the melting bodies but sea level rise in the far field (e.g., along the North East coast of America and in the tropics). The GIA effect depends on Earth's mantle viscosity and deglaciation history (Peltier, 2004, Lambeck et al. 2010) while the response of the solid Earth to on-going land ice melt essentially depends on lithosphere elasticity as well as amount & location of ice mass loss.

GMSL change as a function of time t is usually expressed by the sea level budget equation:

$$\mathrm{GMSL}(t) = \mathrm{GMSL}_{\mathrm{steric}}(t) + \mathrm{GMSL}_{\mathrm{oceanmass}}(t) \quad (11.1)$$

where $\mathrm{GMSL}_{\mathrm{steric}}(t)$ refers to the global steric sea level change (i.e., the contributions of ocean thermal expansion and salinity changes) and $\mathrm{GMSL}_{\mathrm{oceanmass}}$ (t) refers to the change in mass of the oceans.

Because of water mass conservation in the climate system, the ocean mass term (i.e., $\mathrm{GMSL}_{\mathrm{oceanmass}}$ (t), also noted $\mathrm{M}_{\mathrm{ocean}}$ (t)) is such that:

$$\mathrm{M}_{\mathrm{ocean}}(t) + \mathrm{M}_{\mathrm{Glaciers}}(t) + \mathrm{M}_{\mathrm{Greenland}}(t) + \mathrm{M}_{\mathrm{Antarct.}}(t) + \mathrm{M}_{\mathrm{LWS}}(t) + \mathrm{M}_{\mathrm{Atm}}(t) + \mathrm{M}_{\mathrm{Snow}}(t) + \text{missing mass terms} = 0 \quad (11.2)$$

where $\mathrm{M}_{\mathrm{Glaciers}}(t)$, $\mathrm{M}_{\mathrm{Greenland}}(t)$, $\mathrm{M}_{\mathrm{Antarct.}}(t)$, $\mathrm{M}_{\mathrm{LWS}}(t)$, $\mathrm{M}_{\mathrm{Atm}}(t)$ and $\mathrm{M}_{\mathrm{Snow}}(t)$ represent temporal changes in mass of glaciers, Greenland and Antarctica ice sheets, land water storage (LWS), atmospheric water vapor and snow mass changes. Missing terms could include for example permafrost melting.

At regional scales, the regional sea level (RSL) budget takes the form:

$$\mathrm{RSL}\ (t) = \mathrm{RSL}_{\mathrm{steric}}(t) + \mathrm{RSL}_{\mathrm{oceanmass}}(t) + \text{atmospheric loading} + \text{static terms} \quad (11.3)$$

where $\mathrm{RSL}_{\mathrm{steric}}(t)$ and $\mathrm{RSL}_{\mathrm{oceanmass}}$ (t) are regional steric sea level and ocean mass variations. As noted above, the 'static terms' consist of GIA and fingerprints of present-day ice-water mass redistributions (Milne et al., 2009, Tamisiea and Mitrovica, 2011).

11.3.1 Steric Sea Level

Steric sea level variations result from temperature and salinity-related density changes of water between two depth levels Z_1 and Z_2. Sea water density is a non-linear function of temperature T and salinity S. However, it is common to consider a first-order expansion of the inverse of density around a reference T_0/S_0 profile, usually estimated from a climatology. Hence, the total steric effect H can be separated into two terms: a (temperature-related) thermosteric component H_T, and a (salinity-related) halosteric component H_S, so that:

$$H = H_T + H_S = \int \alpha(T - T_0)dz + \int \beta(S - S_0)dz \quad (11.4)$$

where z is depth (integration between Z_1 and Z_2). α and β are coefficients of thermal and haline expansion (note that a temperature/salinity increase causes increase/decrease in steric sea level). In terms of global mean, H_S is almost zero to first-order (Bindoff et al., 2007). However in practice, as coverage of observations is never fully global, the halosteric term needs to be considered (provided that salinity data exist).

Since the middle of the 20th century, in situ ocean temperature (and to a lesser extent salinity) data have been collected by various devices, e.g., Expandable Bathy Thermographers (XBTs) from ships, buoys and moorings (Abraham et al., 2013) and since the beginning of the 2000s by automatic profiling floats from the Argo system (Roemmich et al., 2012). During the 1990s, temperature measurements of the upper ocean were collected down to ~700 m depth along commercial shipping routes. Although the coverage was improved compared to the previous decades, large regions remain un-sampled, in particular in the southern hemisphere and in the Arctic. Besides, salinity measurements were very few. Full deployment of the Argo floats was effective around 2005. At the time of writing, ~3900 floats provide systematic temperature and salinity data, with quasi-global coverage. Starting at 1000 m depth, Argo now extends T/S measurements down to 2000 m.

These T/S data sets have been processed by different groups to provide gridded time series of in situ ocean temperature and salinity, useful to estimate ocean heat content and steric sea level changes. Analyses of data showed that ocean heat content and ocean thermal expansion have significantly increased with time, in particular since 1970, although not linearly (e.g., Domingues et al., 2008, Levitus et al., 2012, Johnson et al., 2010, Lyman and Johnson, 2014). A steep increase was observed in thermal expansion during the 1993–2003 decade (e.g., Bindoff et al., 2007), but since about 2003, thermal expansion has increased less rapidly (Lymann et al., 2010, von Schuckmann and Le Traon, 2011). Although very sparse, the few available deep ocean temperature measurements (below ~2000 m) indicate that the deep ocean has also warmed in the recent decades, although less than the upper ocean (e.g., Purkey and Johnson, 2010).

For the period 1993–2010, the IPCC AR5 estimated the total steric contribution (upper and deep ocean) to 1.1 ± 0.3 mm/yr (Church et al., 2013). An update of the steric contribution by Dieng et al. (2017) based on an ensemble mean of several data sets provides a similar value for 1993–2015.

11.3.2 Land Ice Contribution to Sea Level

The main land ice contributions to sea level comes from glacier melting and ice mass loss from the Greenland and Antarctica ice sheets. Snow is another contribution but studies have shown that even though the snow pack extent is decreasing in response to global warming, corresponding contribution to sea level is small (Biancamaria et al., 2011). It is not considered here.

11.3.2.1 Glaciers

Being very sensitive to global warming, mountain glaciers and small ice caps have retreated worldwide during the 20th century, with significant acceleration since the early 1990s. Volume and mass balance studies of a significant number of glaciers, based on various in situ and remote sensing observing methods and modelling, have been proposed, and estimates of the contribution of glacier melt to sea level rise have been deduced (e.g., Kaser et al., 2006, Cogley, 2009, Jacob et al., 2012, Leclerq et al., 2011, Gardner et al., 2013, Marzeion et al., 2015, 2017). Because of high year-to-year variability in glacier melting, in situ-based mass balances are generally provided as 5-year averages, a strong limitation when the period of investigation is short. Remote sensing estimates based on the GRACE (Gravity Recovery and Climate Experiment) gravity mission suffer from errors due to unrelated signals (e.g., due to land hydrology) difficult to separate from the glacier signal because of insufficient GRACE spatial resolution (Velicogna and Wahr, 2013).

For the period 1993–2010, glaciers have contributed to 0.76 ± 0.4 mm/yr sea level rise (Church et al., 2013). Recent updates from Cogley (2009), Leclercq et al.

(2011) and Marzeion et al. (2017) leads to a similar contribution over 1993–2015.

11.3.2.2 Ice Sheets

If totally melted, Greenland and West Antarctica would raise sea level by about 7 m and 5 m respectively. Thus even a small amount of ice mass loss from the ice sheets would produce substantial sea level rise, with adverse societal and economical impacts on vulnerable low-lying coastal regions.

Since the early 1990s, different remote sensing observations (airborne, satellite laser and radar altimetry, Synthetic Aperture Radar Interferometry – InSAR, and since 2002, space gravimetry from GRACE) provide important observations of the mass balance of the ice sheets (Steffen et al., 2010, Church et al., 2013). These data indicate that Greenland and West Antarctica are loosing mass at an accelerated rate (Rignot et al., 2011). Several tens of articles have been published in the recent years on that topic, leading to important dispersion between the various mass balance estimates. An extended list of references on the Greenland and Antarctica mass balances based on these 3 techniques and their combination is provided in the IPCC AR5 (Church et al., 2013) as well as in Dieng et al. (2015c; Supplementary Information) (see also chapter by Allison et al., this volume). The papers by Zwally et al. (2011), Shepherd et al. (2012) and Hanna et al. (2013) provide an overview of these estimates and discuss the difference obtained when using different observing systems, considering different time spans or applying different processing methodologies. For the GRACE era (starting in 2002), most recent estimates includes those from Velicogna et al. (2014), Schrama et al. (2014), and Forsberg et al. (2017). Note that over the ice sheets, GRACE measurements need to be corrected for the GIA contribution. It is a small effect for Greenland, unlike for Antarctica. Because the GIA over Antarctica is highly model-dependent, it significantly impacts the GRACE-based Antarctica mass balance results and largely explain differences between published estimates.

All studies show clear acceleration of ice mass loss over the last decade for both ice sheets.

It is now well established that the recent ice mass loss partly results from accelerated glacier flow along some coastal margins of the ice sheets and further iceberg discharge into the surrounding ocean (e.g., Rignot et al., 2011, Vaughan et al., 2013, Allison et al., this volume). This results from short-term dynamical instabilities occurring in regions where coastal glaciers are grounded below sea level (this is especially the case in West Antarctica). Thinning and subsequent break-up of floating ice tongues or ice shelves that buttressed the glaciers result in rapid grounding line retreat and accelerated glacier flow (Pritchard et al., 2012, Hanna et al., 2013). Several recent observations have also shown that warming of subsurface ocean waters trigger these dynamical instabilities (e.g., Alley et al., 2016, Allison et al., this volume).

On average over the 1993–2010 time span, the two ice sheets have contributed 0.6 mm/yr to the global sea level rise (Church et al., 2013). Over the 1993–2015 time span, the updated estimate by Dieng et al. (2017) based on data from the Ice sheet Mass Balance Intercomparison Exercise (IMBIE) (http://imbie.org/imbie-2012; Shepherd et al., 2012) data until December 2003, completed by the ESA CCI products for the ice sheet mass balance (www.esa-icesheets-cci.org) indicates a total ice sheet contribution of 0.9 ± 0.2 mm/yr over 1993–2015.

11.3.3 Land Water Storage

Change in land water storage due to natural climate variability has little contributed to the GMSL long-term trend during the past few decades (Milly et al., 2010). On the other hand, at interannual time scale, in particular during ENSO events, fluctuating water storage on land causes temporary anomalies of the global mean sea level, of several millimeters as shown by the studies by Llovel et al. (2011), Boening et al. (2012), Cazenave et al. (2014). These studies showed that during El Niño or La Niña events, the GMSL presents positive or negative anomalies of one- to two-year duration due to excess or deficit of precipitation over the tropical Pacific (causing increase or decrease of ocean mass). Because the atmospheric reservoir cannot contain an important amount of water mass on a multi-year time scale, change in ocean mass is nearly compensated by change in land water storage in tropical river basins (e.g., Fasullo et al., 2013) (note that the atmospheric contribution is not totally negligible during ENSO events; it accounts by about 10% of the GMSL change, Dieng et al., 2015b).

Human activities (i.e., anthropogenic changes in the amount of water stored in soils, reservoirs and aquifers as a result of dam building along rivers, underground water mining, irrigation, urbanization, deforestation etc.) represent another potential contribution to sea level change. The dominant contributions are from dam building – causing sea level drop – and ground water depletion, causing sea level rise. For the last few decades, ground water pumping has considerably increased while the building of dams has decreased (Chao et al., 2008, Konikow, 2011, Wada

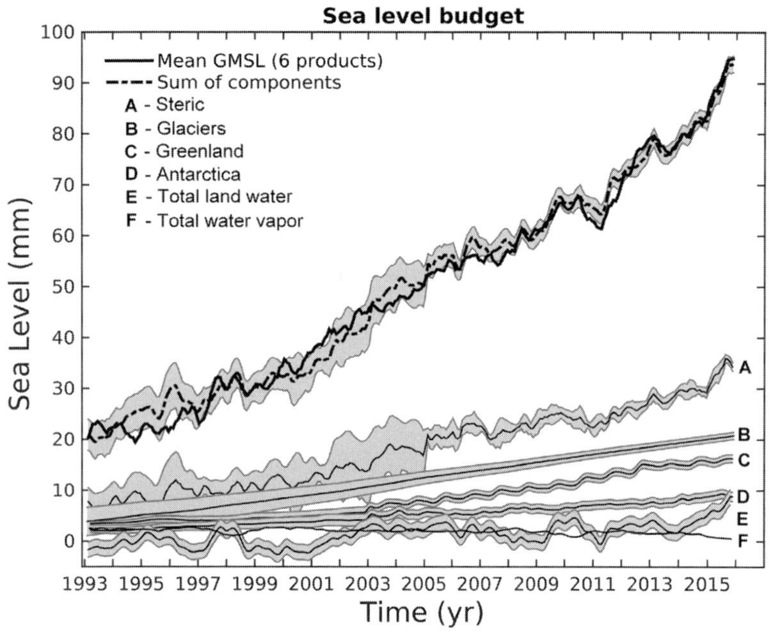

Figure 11.4 Sea level budget over the period January 1993 to December 2015. GMSL corrected for the TOPEX A instrumental drift (solid curve); Sum of components (dashed curve); Steric sea level (A); Glaciers contribution (B); Greenland contribution (C); Antarctica contribution (D); Net land water storage contribution (E); Atmospheric water vapor contribution (F). Shaded areas represent data uncertainties (redrawn from Dieng et al., 2017).

et al., 2012, 2016, 2017), so that there is a net slightly positive effect on the global mean sea level (less than 0.5 mm/yr). However, its exact contribution remains highly uncertain (e.g., Dieng et al., 2015c, Reager et al., 2016). The recent study by Wada et al. (2016) estimates to 0.12 ± 0.04 mm/yr the net terrestrial water contribution to the GMSL during 1993–2010.

11.4 Global Mean Sea Level Budget (Altimetry Era)

A large number of studies have tried to close the sea level budget, i.e., compare the observed rate of sea level rise with the sum of contributions estimated independently. Over the altimetry era, publications are numerous and regularly updated. Studies by Leuliette and Willis (2011), Chen et al. (2013) and Chambers et al. (2017) count among the most recent (see also Church et al., 2013 in IPCC AR5). In all cases, the sea level budget was considered as closed within quoted uncertainties. Dieng et al. (2017) revisited the GMSL budget over 1993–2015 using a large number of different data sets for all terms of the sea level budget and correcting for the instrumental drift of the TOPEX A altimeter that affected the first 6 years (1993–1998) of the altimetry record. Accounting for this correction and considering data ensemble means for the GMSL and components led to closure of the sea level budget within 0.2 mm/yr and to a slightly reduced rate of the GMSL rise for the whole altimetry era (3.0 mm/yr compared to 3.4 mm/yr). Another important implication is the larger rate of GMSL rise (by ~0.8 mm/yr) during the second half of the altimetry period compared to the first half (see also Watson et al., 2015). Figure 11.4 shows for the period January 1993 to December 2015 the evolution of the GMSL (corrected for the TOPEX A instrumental drift) and individual contributions, as well as the sum of components (details on the data sets can be found in Dieng et al., 2017).

As of 2002, data from the GRACE mission allow direct estimate of the ocean mass change without relying on individual mass components (i.e., glaciers, ice sheets and land waters). Since 2005, the Argo project provides ocean temperature and salinity data down to 2000 m, with quasi-global coverage. The GRACE and Argo data are thus extremely useful to study the sea level budget as of 2003/2005, providing upper bounds of missing contributions, such as the deep ocean below 2000m (Llovel et al., 2014, Dieng et al., 2015a). The GRACE-based ocean mass also allows studying the water mass budget by comparing it with the sum of other mass contributions (glaciers, ice sheets, land water storage) determined independently. However, the GIA correction to apply to the GRACE data over the oceans remains an important source of uncertainty. Its contribution is of the same order of magnitude as the ocean mass signal and slightly different fron one model to another (e.g., Chambers et al., 2010).

11.5 Regional Variability in Sea Level

Over the altimetry era, the regional variability in sea level trends is mainly due to large-scale changes in temperature and salinity-related density structure of the oceans, in response to forcing factors (e.g., heat and fresh water exchange at the sea–air interface and wind stress) and their interaction with the ocean circulation (Stammer et al., 2013). Some studies have shown that increase in wind stress and associated deepening of the thermocline explain well the strong sea level trends observed in the western tropical Pacific during the altimetry era (e.g., Timmermann et al, 2010, Merrifield and Maltrud, 2011, Palanisamy et al., 2015a). Figure 11.5 compares spatial trend patterns in altimetry-based and steric sea level over the 1993–2012 time span over the western tropical Pacific (global mean trends removed in both cases). Observed and steric regional trends show very good agreement. While in most regions, regional changes in sea level trends mostly result from ocean temperature change, in a few others, change in salinity can be also important.

Fingerprints of 'static' effects related to GIA and present-day land ice melt estimated from modelling (Tamisiea and Mitrovica, 2011, Spada, 2017) are almost undetectable in current observations because of remaining regional trend errors in both altimetry observations and steric estimates.

11.6 Human-Induced versus Internal Climate Variability Influence on Sea Level Change

This issue, often called 'detection and attribution', refers to the discrimination in a physical signal between changes due to natural (internal) variability of the climate system and changes due to anthropogenic forcing.

Detection and attribution studies applied to global mean sea level observations and steric sea level (Gleckler et al., 2012, Marcos and Amores, 2014, Becker et al., 2014, Slangen et al., 2014a, Marcos et al., 2017) suggest that anthropogenic forcing plays a substantial role in the observed rise. Marcos and Amores (2014) concluded that 87% of the observed steric trend in the upper 700 m since 1970 is induced by human activities. Slangen et al. (2014) provided evidence that the GMSL

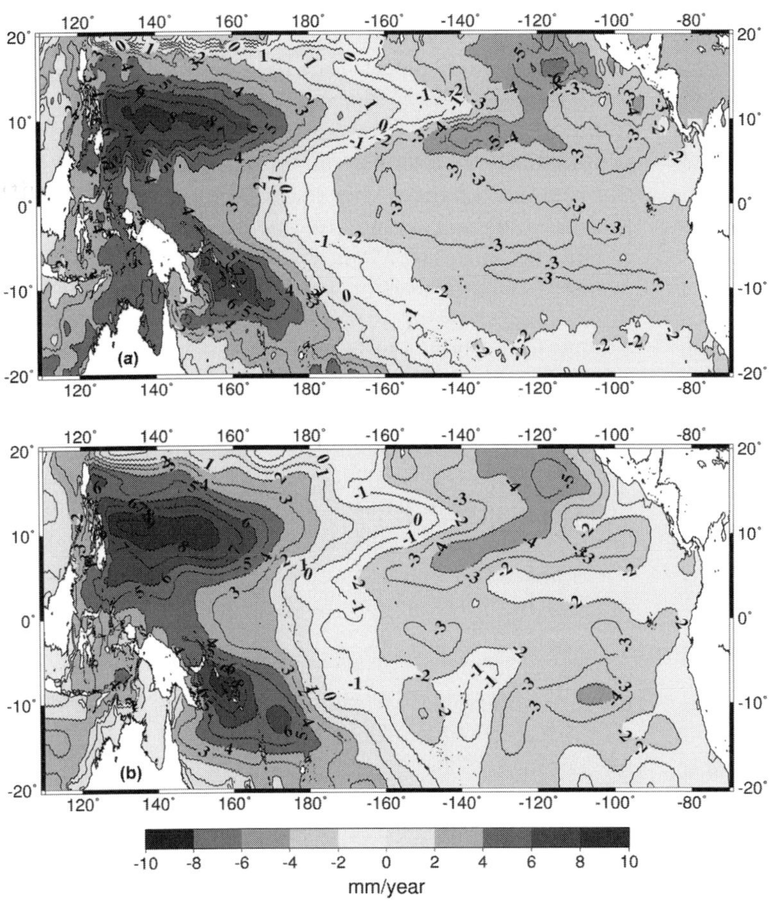

Figure 11.5 Spatial trend patterns in sea level (in millimeters per year) over January 1993 to December 2012 in the western tropical Pacific (global mean trends removed). Upper panel: observed by satellite altimetry (data from AVISO; www.aviso.altimetry.fr); Lower panel: steric sea level estimated from ocean temperature and salinity. Level curves at 1 mm/yr-interval (data from Ishii and Kimoto (2009); redrawn from Palanisamy et al. (2015a)).

has accelerated in the recent decades and that anthropogenic contribution accounts for more than 70% since 1970, a result confirmed by the Marcos et al. (2017) study showing that the human fingerprint dominates the steric and glaciers components, hence GMSL rise, over the second half of the twentieth century.

Direct observations from satellite altimetry and past sea level reconstructions based on statistical approaches and ocean reanalyses indicate that trend patterns in thermal expansion are not stationary but fluctuate in both space and time in response to natural/internal oscillation perturbations of the climate system such as ENSO, NAO and PDO. While several studies highlight the role of natural climate modes to explain the spatial trend patterns of the altimetry era (see Meyssignac et al. 2012, Zhang and Church, 2012, Stammer et al. 2013, Palanisamy et al., 2015b), others suggest that the fingerprint of anthropogenic forcing is already detectable (e.g., Hamlington et al. 2014). However, there is no clear consensus yet. Except in the southern ocean, where sea level changes may already result from human-induced forcing, it is likely that in other regions, the anthropogenic signal will not emerge from the unforced variability before several decades (Bilbao et al., 2015).

11.7 Future Sea Level Changes

IPCC AR5 projections of future sea level rise based on ensemble means of process-based climate models indicate that at the end of the twenty-first century, global mean sea level should be on average higher than today in the range 40–75 cm, depending on the radiative forcing scenario. Accounting for model dispersion leads to a larger uncertainty range, from ~25 cm to 95 cm (Church et al., 2013). Ocean warming, glacier melting and ice sheet mass loss may contribute by ~45%, 26% and 23%, the remaining being attributed to land water storage change. This proportion is roughly the same for all scenarios. Compared to IPCC AR4 (Meehl et al., 2007), the ice sheet contribution was revised upward because of accounts of the recently discovered dynamical instabilities in some coastal regions of Greenland and West Antarctica (Vaughan et al., 2013). As a result, IPCC AR5 projected a global mean sea level elevation of ~50–100 cm for 2086–2100 with respect to 1986–2000 for the RCP (Representative Concentration Pathway) 8.5 scenario. This value was about a factor of 2 higher than IPCC AR4 projections for a nearly similar scenario. Since the IPCC AR5 publication in 2013, it has been suggested that these projections could be underestimated (e.g., Clark et al., 2015, Ritz et al., 2015). For example, accounting for a new mechanism for Antarctica ice sheet instabilities (as yet unobserved), a recent study by De Conto and Pollard (2016) proposes an Antarctic contribution of ~1 m by 2100 (with respect to year 1950) leading to a GMSL rise >1.5 m for the RCP 8.5 scenario, and up to 15 m around 2500. Several other recent studies based on 'empirical' and probabilistic projections also suggest higher sea level rise projections than IPCC AR5 (e.g., Jevrejeva et al., 2014b, Kopp et al., 2014, Horton et al., 2014, Jackson and Jevrejeva, 2016). While future global mean sea level elevation is related to future trajectories of GHG emissions, it also depends on potential runaway of some components of the climate system such as the ice sheets. Above a given warming threshold, parts of the ice sheets may undergo irreversible melting leading to several meters of sea level rise on time scales from centuries to millennia (Robinson et al., 2012, Collins et al., 2013).

In addition, as it is already the case today, future sea level rise will not be uniform, amplification with respect to the global mean being expected in several oceanic regions as a result of steric and static factors (Church et al., 2013, Slangen et al., 2014b, 2017, Carson et al., 2016). Moreover, at local scale, a number of factors (see later discussion) may also give rise to sea level changes. The quantity of interest for coastal populations is the total sea level rise relative to the ground, i.e., the climate-related global mean rise plus the regional changes due to climate and non-climate factors, plus local changes due to short-scale ocean dynamics processes and finally vertical ground motions. Withdrawal of water from aquifers can produce dramatic relative sea level rise, especially in coastal megacities. During the second-half of the twentieth century, Tokyo, Shanghai, Bangkok and Jakarta subsided by several meters because of ground water withdrawal (Nicholls, 2010). Ground subsidence, hence relative sea level rise, is also widely observed in highly populated deltaic areas, in particular in Asia (e.g., Ericson et al., 2006).

Compared to extreme events, sea level rise is a slow process, but it has long-term consequences (Levermann et al., 2013, Clark et al., 2016). Because of the long-term responses of the oceans and the ice sheets to GHG-induced atmospheric warming, Clark et al. (2016) show that sea level will continue to rise well beyond the next few centuries. They project several meters or even tens of meters of elevation on a multi-millenial time scale. Exact rate and magnitude of long-term sea level rise will depend on near-term policy decisions about anthropogenic GHG emissions.

Adverse effects of sea level rise in coastal areas (inundation, shoreline erosion, increased flooding during storm surges, salinization of wetlands and aquifers etc.) are generally considered as a major threat, considering that coastal zones are the most densely populated and economically active areas of the world. They concentrate important infrastructures such as harbors and industries. It is now well established that at local scale, the coastal response to sea level rise depends on several non–linearly related factors such as coastal morphology, near-shore bathymetry, sediment supply from rivers and changing waves and currents (Fitzgerald et al., 2008, Cazenave and Le Cozannet, 2014, Passeri et al. 2015). While for the twentieth century, coastal changes mostly resulted from multiple drivers (Nicholls, 2010, Passeri et al., 2015), expected sea level rise by the end of the twenty-first century and beyond will make sea level rise the dominant forcing factor of future coastal system changes (Nicholls et al., 2007, Nicholls, 2010).

11.8 Current Challenges from a Geophysical Perspective

Sea level is one of the most important climate indicators as it integrates changes of several components of the climate sytem in response to anthropogenic global warming and natural forcing factors (including internal variability of the ocean–atmosphere system). Because of the high thermal inertia of the oceans and long life time of emitted carbon dioxide by fossil fuel combustion, sea level will continue to rise by a substantial amount in the coming decades and centuries (Church et al., 2013, Kopp et al., 2014), and even millennia (Levermann et al., 2013, Clark et al., 2016). In the long term, there is no doubt that a number of low-lying, highly populated coastal zones will suffer the impacts of sea level rise (Nicholls and Cazenave, 2010). In order to be able to develop adaptation strategies, populations and stakeholders should have at their disposal as precise as possible local projections of future sea level rise under different warming scenarios, accounting for all factors at work from global to local scales. The climate and other models used for the projections still suffer imperfection. Their improvement is conditioned by the availability of various types of observations used for model validation. Observations of sea level and its components are thus essential, allowing detection of new phenomena (e.g., ice sheet dynamical instabilities in the late 1990s) and better process understanding, hence improved modelling and projections.

Another important challenge is to measure sea level in coastal areas precisely to understand and model coastal impacts of rising sea levels better. There are reasons to suspect that coastal sea level changes are not just an extension of open ocean variations but may differ at the coast because of the presence of shelf currents, trapped Kelvin waves, baroclinic instabilities etc. Some other processes (e.g., waves, river runoff in river deltas) only occur at the coast and may also impact coastal sea level changes. However, satellite altimetry provides good estimates of sea level variations in open oceans up to 10–20 km from the coast only. Within the coastal band, classical nadir radar altimetry performs poorly because of perturbations from the land (Cipollini et al., 2017). To retrieve sea level variations in the coastal zone, specific post-processing of altimetry data is required, including "retracking" of radar echoes (waveforms). New altimetry technology such as Ka-band altimetry (e.g., on SARAL/Altika) or Synthetic Aperture Radar altimetry (e.g., on Cryosat and Sentinel 3) is another promising option for constructing sea level time series in coastal areas (Cipollini et al., 2017).

To summarize, from the observational and modelling points of view, what is crucially needed includes:

- Long-term, sustained monitoring of sea level and components from space-based and in situ observing systems (e.g., altimeter satellite constellation, GRACE-type space gravimetry, Argo)
- Reduction of bias and measurement uncertainties in order to meet the Global Climate Observing System (GCOS) requirements for long-term climate monitoring (GCOS, 2011)
- Estimates of missing contributions to sea level changes (e.g., contribution of the deep ocean below 2000 m) through various approaches (e.g., the sea level budget approach)
- Uncertainty reduction of the GIA factor for more accurate estimation of the ocean mass and Antarctica mass balance from space gravimetry (more coordinated work on GIA modelling is highly recommended)
- Improved sea level projections, with emphasis on coastal sea level projections, accounting for all climate- and non-climate-related processes at work from global to local scales
- Global data sets of coastal sea level variations in the coastal zones
- Multidisciplinary studies of coastal zone evolution under multiple forcing agents, with focus on the exact role of sea level rise in shoreline retreat and erosion

11.9 Conclusion

Measuring sea level rise and understanding its causes have considerably improved in the recent years, essentially because new in situ and remote sensing observations have become available. Sea level is presently rising at a mean rate of ~3 mm/yr and will continue in the future decades and centuries because of expected continuing global warming, thereby affecting a large number of world coastal regions. In this chapter we provided an overview of recent advances in sea level studies. We discussed observations of present-day sea level variations at global and regional scales, and of the components (ocean thermal expansion, land ice melt, land water storage changes). We also presented attempts to close the sea level budget based on comparisons between observed sea level change and sum of contributions. After a brief outlook of current estimates of sea level projections by the end of the twenty-first century and beyond, as well as associated coastal impacts of sea level rise, we listed a number of remaining issues related to gaps in observations, process understanding and modelling, that need be addressed to anticipate future changes better.

References

Ablain, M., Cazenave, A., Valladeau, G., and Guinehut S. (2009). A new assessment of the error budget of global mean sea level rate estimated by satellite altimetry over 1993–2008. *Ocean Science* 5(2), 193–201.

Ablain M. et al. (2015). Improved sea level record over the satellite altimetry era (1993–2010) from the Climate Change Initiative Project, *Ocean Sciences*, 11, 2029–2071.

Ablain M., Legeais, J. F., Prandi, P., Fenoglio-Marc, L., Marcos, M., Benveniste, J. and Cazenave, A. (2017). Altimetry-based sea level, global and regional, *Surveys in Geophysics*, 38, 7–31, doi:10.1007/s10712–016–9389–8.

Abraham, J. P., Baringer, M., Bindoff, N. L., Boyer, T., Cheng, L. J., Church, J. A., Conroy, J. L., Domingues, C. M., Fasullo, J. T., Gilson, J., Goni, G., Good, S. A., Gorman, J. M., Gouretski, V., Ishii, M., Johnson, G. C., Kizu, S., Lyman, J. M., Macdonald, A. M., Minkowycz, W. J., Moffitt, S. E., Palmer, M. D., Piola, A. R., Reseghetti, F., Schuckmann, K., Trenberth, K. E., Velicogna, I., and Willis, J. K. (2013). A review of global ocean temperature observations: Implications for ocean heat content estimates and climate change, *Rev. Geophys.*, 51, 450–483, doi:10.1002/rog.20022.

Alley K. E. et al. (2016). Impacts of warm water on Antarctic ice shelf stability through basal channel formation, *Nature Geosciences*, 9, 290–294.

Bamber J. and Riva R. (2010). The sea level fingerprint of recent ice mass fluxes, *Cryosphere*, 4, 621–627.

Becker M., Karpytchev M., and Lennartz-Sassinek S. (2014). Long-term sea level trends: Natural or anthropogenic? *Geophys. Res. Lett.*, 5571–5580, doi: 10.1002/2014GL061027.

Biancamaria, S., Cazenave, A., Mognard, N., LLovel, W., and Frappart, F. (2011). Satellite-based high latitudes snow volume trend, variability and contribution to sea level over 1989/2006, *Global and Planet. Change*, 75, 99–107, doi:10.1016/j.gloplacha2010.10.011.

Bilbao, R. A. F., Gregory, J. M., and Bouttes, N. (2015). Analysis of the regional patter of sea level change due to ocean dynamics and density change for 1993–2099 in observations and CMIP5 AOGCMs, *Clim Dyn.*, 45, 2647–2666.

Bindoff, N., Willebrand, J., Artale, V., Cazenave, A., Gregory, J., Gulev, S., Hanawa, K., Le Quéré, C., Levitus, S., Nojiri, Y., Shum, C. K., Talley, L., Unnikrishnan, A. (2007). Observations: Oceanic Climate and Sea Level. In: Climate change 2007: The physical Science Basis. Contribution of Working Group I to the Fourth Assessment report of the Intergouvernmental Panel on Climate Change [Solomon S., D. Qin, M. Manning, Z. Chen, M. Marquis, K. B. Averyt, M. Tignor and H. L. Miller (eds.)]. Cambridge University Press, Cambridge, UK, and New York, USA.

Boening, C., Willis, J. K., Landerer, F. W., and Nerem, R. S. (2012). The 2011 La Nina: So strong, the oceans fell, *Geophys. Res. Lett.*, 39, L19602, doi:10.1029/2012GL053055.

Carson, M., Kohl, A., Stammer, D., Slangen, A. B., Katsman, C. S., van de Wahl, R. S., Church J. and White, N. J. (2016). Coastal sea level changes, observed and projected during the 20th and 21st century, *Climatic Change*, 134, 269–281.

Cazenave, A., Dieng, H., Meyssignac, B., von Schuckmann, K., Decharme, B., and Berthier, E. (2014). The rate of sea level rise, *Nature Climate Change*, 4, doi:10.1038/NCLIMATE2159.

Cazenave, A., and Le Cozannet, G. (2014). Sea level rise and coastal impacts, *Earth's Future*, 2(2), 15–34, doi:10.1002/2013EF000188.

Chambers, D. P., Wahr, J. M., Tamisiea, M., and Nerem, R. S. (2010). Ocean mass from GRACE and glacial isostatic adjustment. *Journal of Geophysical Research*, 115, B11415.

Chambers, D. P., Cazenave, A., Champollion, N., Dieng, H. B., Llovel, W., Forsberg, R., von Schuckmann, K., and Wada, Y. (2017). Evaluation of the Global Mean Sea Level Budget between 1993 and 2014. *Surv. Geophys*, 38, 309–327, doi:10.1007/s10712-016-9381-3.

Chao, BF, Wu, Y. H., and Li, Y. S. (2008.) Impact of artificial reservoir water impoundment on global sea level. *Science* 320:212–14, doi:10.1126/science.115458.

Chen J. L., Wilson C. R., and Tapley B. D. (2013). Contribution of ice sheet and mountain glacier melt to recent sea level rise, *Nature Geoscience*, 6, 549–552.

Church, J. A., and White N. J. (2011). Sea-Level Rise from the Late 19th to the Early 21st Century, *Surveys in Geophysics*, *32*, 585–602, doi:10.1007/s10712-011-9119-1.

Church, J. A., Clark, P. U., Cazenave, A., Gregory, J. M., Jevrejeva, S., Levermann, A., Merrifield, M. A., Milne, G. A., Nerem, R. S., Nunn, P. D., Payne, A. J., Pfeffer, W. T., Stammer, D., and Unnikrishnan A. S. (2013). Sea Level Change. In: *Climate Change 2013: The Physical Science Basis. Contribution of Working Group I to the Fifth Assessment Report of the Intergovernmental Panel on Climate Change*, edited by T. F. Stocker, D. Qin, G.-K. Plattner, M. Tignor, S. K. Allen, J. Boschung, A. Nauels, Y. Xia, V. Bex and P. M. Midgley. Cambridge University Press, Cambridge and New York.

Cipollini P., Calafat, F. M., Jevrejeva, S., Melet, A., and Prandi, P. (2017). Monitoring sea level in the coastal zone with satellite altimetry and tide gauges, *Surveys in Geophysics*, 38, 33–57.

Clark P. U. et al. (2015). Recent progress in understanding and projecting regional and global mean sea level, *Curr. Clim. Change*, doi/10.1007/s40641–015-0024–4.

Clark P. U. et al. (2016). Consequences of twenty-first-century policy for multi millenial climate and sea level change, *Nature Climate Change*, 6, 360–369.

Cogley J. C. (2009). Geodetic and direct mass balance measurements: Comparison and joint analysis, *Annals of Glaciology*, 2009(50), 96–100.

Collins, M., Knutti, R., Arblaster, J., Dufresne, J.-L., Fichefet, T. Friedlingstein, P., Gao, X. W. J., Gutowski, Johns, T., Krinner, G., Shongwe, M., Tebaldi, C., Weaver A. J., and Wehner M. (2013): Long-Term Climate Change: Projections, Commitments and Irreversibility. In: *Climate Change 2013. The Physical Science Basis. Contribution of Working Group I to the Fifth Assessment Report of the Intergovernmental Panel on Climate Change*, edited by T. F. Stocker, D. Qin, G.-K. Plattner, M. Tignor, S. K. Allen, J. Boschung, A. Nauels, Y. Xia, V. Bex, and P. M. Midgley. Cambridge University Press, Cambridge and New York.

De Conto, R. M., and Pollard D. (2016). Contribution of Antarctica to past and future sea-level rise, *Nature*, 531, 591.

Dieng, H. B., Palanisamy, H., Cazenave, A., Meyssignac, B. and von Schuckmann, K. (2015a). The sea level budget since 2003: Inference on the deep ocean heat content. *Surv. Geophys.*, 36, 209–229, doi:10.1007/s10712-015-9314-6.

Dieng, H. B., Cazenave, A., von Shuckmann, K., Ablain, M., and Meyssignac, B. (2015b). Sea level budget over 2005–2013: Missing contributions and data errors. *Ocean Science*, 11, 789–802, doi:10.5194/os-11-789-2015.

Dieng, H., Champollion, N., Cazenave, A., Wada, Y., Schrama, E. and Meyssignac, B. (2015c), Total land water storage change over 2003–2013 estimated from a global mass budget approach, *Environmental Research Letters*, 10, 124010, doi:10.1088/1748-9326/10/12/124010.

Dieng, H., Cazenave A., Meyssignac B. and Ablain M. (2017). New estimate of the current rate of sea level rise from a sea level budget approach, *Geophys. Res. Lett.*, 44, doi:10.1002/2017GL073308.

Domingues C. et al. (2008). Improved estimates of upper ocean warming and multidecadal sea level rise. *Nature* 453:1090–1093, doi:10.1038/nature07080.

Ericson, J. P. (2006). Effective sea-level rise and deltas: Causes of change and human dimension implications, *Global and Planetary Change*, 50(1–2), 63–82.

Fasullo, J. T., Boening, C., Landerer, F. W., and Nerem, R. S. (2013). Australia's unique influence on global mean sea level in 2010–2011, *Geophys. Res. Lett.*, 40(16), 4368–4373, doi:10.1002/grl.50834.

Fitzgerald, D. M. et al. (2008). Coastal impacts of sea level rise, *Annual Rev. Earth Planet. Sci.*, 36, 601–647.

Forsberg R., Sorensen, L., and Simonsen S. (2017), Greenland and Antarctica ice sheet mass changes and effects on global sea level, *Surveys in Geophysics*, 38, 89–103.

Gardner, A. S. et al. (2013). A reconciled estimate of glacier contributions to sea level rise, 2003–2009, *Science*, 340, 852–857.

GCOS, 2011. Systematic Observation Requirements for Satellite-Based Data Products for Climate (2011 Update) – Supplemental details to the satellite-based component of the "Implementation Plan for the Global Observing System for Climate in Support of the UNFCCC (2010 Update)". GCOS-154 (WMO, December 2011).

Gleckler P. J.,Santer, B. D., Domingues, C., Pierce, D., W., Barnett, T. P., Church, J. A., Taylor, K. E., AchutaRao, K. M., Boyer, T. P., and Caldwell, P. M. (2012). Human-induced global ocean warming on multidecadal time scales, *Nature Climate Change*, doi:10.1038/NCLIMATE1553.

Hamlington B. D., Strassburg, M. W., Leben, R. R., Han, W., Nerem, R. S. and Kim, K. Y. (2014). Uncovering an anthropogenic sea-level rise signal in the Pacific Ocean, *Nature Climate Change*, doi:10.1038/NCLIMATE2307.

Hamlington, B. D., and Thompson, P. R. (2015), Considerations for estimating the 20th century trend in global mean sea level. *Geophys. Res. Lett.*, 42, 4102–4109, doi: 10.1002/2015GL064177.

Hanna, E. et al. (2013), Ice-sheet mass balance and climate change, *Nature*, 498, 51–59, doi:10.1038/nature12238.

Hay, C. C. et al. (2015). Probabilistic reanalysis of twentieth-century sea level rise, *Nature*, 517(7535), 481.

Horton, B. et al. (2014). Expert assessment of sea level rise by AD 2100 and AD 2300, *Quat Sci Rev*., 84, 1–6.

Ishii, M. and Kimoto, M. (2009). Reevaluation of historical ocean heat content variations with time-varying XBT and MBT depth bias corrections. *J. Oceanogr. 65*(3), 287–299, doi:10.1007/s10872-009-0027-7.

Jackson, L. P., and Jevrejeva, S. (2016). A probabilistic approach to 21st century regional sea level projections using RCP and high-end scenarios, Global Planet. *Change*, 146, 179–189.

Jacob, T., Wahr, J., Pfeffer, W. T., and Svenson S. (2012). Recent contribution of glaciers and ice caps to sea level rise, *Nature*, 482, 514–518.

Jevrejeva, S., Grinsted, A., Moore, J. C., and Holgate, S. (2006). Non linear trends and multiyear cycles in sea level records, *J. Geophys. Res.*, C09012, doi:1.1029/2005JC003229, 2006.

Jevrejeva, S. et al. (2014a). Trends and acceleration in global and regional sea levels since 1807, Global Planet. *Change*, 113, 11–22.

Jevrejeva, S. et al. (2014b). Upper limit for sea level projections by 2100, *Environ. Res. Lett.*, 9(10), 1–9.

Johnson G. C., Palmer, M. D., Smith, D. M., and Willis, J. K. (2010). Robust warming of the global upper ocean, *Nature*, 465, 334–337, doi:10.1038/nature09043.

Kaser, G, Cogley, J. G., Dyurgerov, M. B., Meier, M. F., and Ohmura, A. (2006). Mass balance of glaciers and ice caps: Consensus estimates for 1961–2004, *Geophys. Res. Lett*, 33, L19501, doi:10.1029/2006GL027511.

Kemp, A. C., Horton, B., Donnelly, J. P., Mann, M. E., Vermeer, M., and Rahmstorf S. (2011). Climate related sea level variations over the past two millennia, *PNAS*, doi/10.1073/pnas.1015619108.

Konikow, L. F. (2011). Contribution of global groundwater depletion since 1900 to sea level rise, *Geophys. Res. Lett.*, 38, L17401, doi:10.1029/2011GL048604.

Kopp, R. E., et al. (2014). Probabilistic 21st and 22nd century sea level projections at a global network of tide gauge sites, *Earth's Future*, 2, 383–406.

Lambeck, K. et al. (2010). Paleoenvironmental Records, Geophysical Modelling and Reconstruction of Sea Level Trends and Variability on Centennial and Longer Time Scales, In: *Understanding Sea Level Rise and Variability*, edited by J. A. Church et al. Wiley-Blackwell, Oxford, UK.

Leclercq, P. W., Oerlemans, J., Cogley, J. G. (2011). Estimating the glacier contribution to sea-level rise for the period 1800–2005. *Surv Geophys* 32:519–535.

Leuliette, E. W., and Willis, J. K. (2011). Balancing the sea level budget. *Oceanography*, 24, 122–129.

Levermann, A., Clark, P. U., Marzeion, B., Milne, G. A., Pollard, D., Radic, V., and Robinson A. (2013). The multimillenial sea level commitment of global warming, *PNAS*, 110(3), 13745–13750.

Levitus, S., Antonov, J. I., Boyer, T. P., Baranova, O. K., Garcia, H. E., Locarnini, R. A., Mishonov, A. V., Reagan, J. R., Seidov, D., Yarosh, E. S., and Zweng, M. M. (2012). World ocean heat content and thermosteric sea level change (0–2000 m), 1955–2010. *Geophys Res Lett*, 39, L10603, doi:10.1019/2012GL051106.

Llovel, W., Becker, M., Cazenave, A., Jevrejeva, S., Alkama, R., Decharme, B., Douville, H., Ablain, M., and Beckley, B. (2011). Terrestrial waters and sea level variations on interannual time scale, *Global Planet. Change*, 75, 76–82, doi:10.1016/j.gloplacha.2010.10.008.

Llovel, W., Willis, J. K., Landerer, F. and Fukumori, I. (2014). Deep ocean contribution to sea level and energy budget not detectable over the past decade *Nat. Clim. Change*, 4 1031–5.

Lyman, J. M., Godd, S. A., Gouretski, V. V., Ishii, M., Johnson, G. C., Palmer, M. D., Smith, D. M., and Willis J. K. (2010), Robust warming of the global upper ocean, *Nature*, 465:334–337, doi:10.1038/nature09043.

Lyman, J. M. and Johnson, G. C. (2014). Estimating global ocean heat content changes in the upper 1800 m since 1950 and the influence of climatology choice, *Journal of Climate*, 1945–1957, doi:10.1175/JCLIM-D-12-00752.1.

Marcos, M. and Amores, A. (2014). Quantifying anthropogenic and natural contributions to thermosteric sea level rise, *Geophys. Res. Lett.*, doi: 10.1002/2014GL059766.

Marcos, M., Marzeion, B., Dangendorf, S., Slangen, A. B., Palanisamy H., and Fenoglio-Marc L., (2017). Internal variability versus anthropogenic forcing on sea level and its components, *Surveys in Geophysics*, 38, 329–347.

Marzeion, B., Leclercq, P. W., Cogley, J. G., and Jarosch, A. H. (2015), Brief communication: Global reconstructions of glacier mass change during the 20th century are consistent. *Cryosphere,* 9, 2399–2404, doi:10.5194/tc-9-2399-2015.

Marzeion, B., Champollion, N., Haeberli, W., Langley, K., Leclercq, P., and Paul, F. (2017). Observation-based estimates of global glacier mass change and its contribution to sea level change, *Surveys in Geophysics*, 38, 105–129.

Merrifield, M. A., and Maltrud, M. E. (2011). Regional sea level trends due to Pacific trade wind intensification, *Geophys. Res. Lett.*, 38, L21605, doi:10.1029/2011GL049576.

Meyssignac, B., Salas-Melia, D., Becker, M., Llovel, W. and Cazenave, A. (2012). Spatial trend patterns in observed sea level: Internal variability and/or anthropogenic signature? *Climate of the Past*, 8, 787–802, doi:10.5194/cp-8-787-2012.

Meehl, G. A. et al. (2007). *Global climate projections, in Climate Change 2007: The Physical Science Basis. Contribution of Working Group I to the Fourth Assessment Report of the Intergovernmental Panel on Climate Change,* edited by S. Solomon et al., Cambridge University Press, Cambridge.

Milne, G. A., Gehrels, W. R., Hughes, C. W., and Tamisiea, M. E. (2009). Identifying the causes of sea-level change, *Nature Geoscience*, 2(7), doi:10.1038/ngeo544.

Milly, P. C. D., Cazenave, A., Famiglietti, J. S., Gornitz, V., Laval, K., Lettenmaier, D. P., Sahagian, D. L., Wahr, J. M., and Wilson C. R. (2010). *Terrestrial Water-Storage Contributions to Sea-Level Rise and Variability, in Understanding Sea-Level Rise and Variability*, edited by J. A. Church, P. L. Woodworth, T. Aarup, and W. S. Wilson, Wiley-Blackwell, Oxford, UK. pp. 226–255, doi:10.1002/9781444323276.ch8.

Nerem, R. S., Chambers, D. P., Choe, C., and Mitchum G. T. (2010). Estimating mean sea level change from the TOPEX and Jason altimeter missions. *Mar. Geodesy,* 33(1), 435–446.

Nicholls, R. J. et al. (2007), *Coastal systems and low-lying areas, in Climate Change 2007: Impacts, Adaptation and Vulnerability.* Fourth Assessment Report of the Intergovernmental Panel on Climate Change (IPCC, 2007), edited by M. L. Parry et al., Cambridge University Press, Cambridge. pp. 315–356.

Nicholls, R. J. (2010). Impacts of and Responses to Sea Level Rise, *In Understanding Sea Level Rise and Variability*, edited by J. A. Church, P. L. Woodworth, T. Aarup, and W. S. Wilson, Wiley-Blackwell, Oxford, UK.

Nicholls, R. J. and Cazenave, A. (2010). Sea level change and the impacts in coastal zones. *Science* 328, 1517–1520, doi:10.1126/science.1185782.

Palanisamy, H. Cazenave, A., Delcroix, T. and Meyssignac, B. (2015a). Spatial trend patterns in Pacific Ocean sea level during the altimetry era: The contribution of thermocline depth change and internal climate variability, *Ocean Dynamics*, doi:10.1007/s10236–014–0805–7.

Palanisamy, H., Meyssignac, B., Cazenave, A., and Delcroix, T. (2015b). Is the anthropogenic sea level fingerprint already detectable in the Pacific Ocean? *Environmental Research Letters*, 10, 124010, doi:10.1088/1748-9326/10/12/124010.

Passeri, D. L. et al. (2015). The dynamic effects of sea level rise on low-gradient coastal landscapes: A review, *Earth's Future*, 3(6), 159–181.

Pritchard, H. D., Ligtenberg, S. R. M., Fricker, H. A., Vaughan, D. G., van den Broeke, M. R., and Padman L. (2012). Antarctic ice-sheet loss driven by basal melting of ice shelves. *Nature*, 484, 502–505.

Ray, R. D., and Douglas, B. C. (2011). Experiments in reconstructing twentieth-century sea levels. *Prog. Oceanogr.*, 91, 496–515, doi:10.1016/j.pocean.2011.07.021.

Reager, J. T., Gardner, A. S., Famiglietti, J. S., Wiese, D. N., Eicker, A., and Lo M.-H. (2016). A decade of sea level rise slowed by climate-driven hydrology, *Science*, 351, 699–703.

Rhein, M., Rintoul, S. R., Aoki, S., Campos, E., Chambers, D., Feely, R. A., Gulev, S., Johnson, G. C., Josey, S. A., Kostianoy, A., Mauritzen, C., Roemmich, D., Talley L. D., and Wang F. (2013). Observations: Oceans. In*: Climate Change 2013: The Physical Science Basis. Contribution of Working Group I to the Fifth Assessment Report of the Intergovernmental Panel on Climate Change,* edited by T. F. Stocker, D. Qin, G.-K. Plattner, M. Tignor, S. K. Allen, J. Boschung, A. Nauels, Y. Xia, V. Bex and P. M. Midgley. Cambridge University Press, Cambridge and New York.

Roemmich, D., Gould, W. J., and Gilson J. (2012). 135 years of global ocean warming between the Challenger expedition and the Argo Programme, *Nature Climate Change*, 2(6), 425–428, doi:10.1038/nclimate1461.

Peltier, W. R. (2004). Global glacial isostasy and the surface of the ice-age Earth: The ICE-5G (VM2) model and GRACE. *Annual Review of Earth and Planetary Sciences* 32: 111–149.

Purkey, S. and Johnson, G. C. (2010). Warming of global abyssal and deep southern ocean waters between the 1990s and 2000s: Contributions to global heat and sea level rise budget. *J Clim*, 23:6336–6351.

Rignot, E., Velicogna, I., van den Broeke, M. R., Monaghan, A., and Lenaerts J. (2011). Acceleration of the contribution of the Greenland and Antarctic ice sheets to sea level rise, *Geophysical Research Letters*, *38*, doi:10.1029/2011gl046583.

Ritz, C. et al. (2015). Potential sea-level rise from Antarctic ice-sheet instability constrained by observations, *Nature*, 528, 115.

Robinson, A. et al. (2012). Multistability and critical thresholds of the Greenland ice sheet, *Nature Climate Change*, 1–4, doi:10.1038/NCLIMATE 1449.

Schrama, E J O, Wouters, B and Rietbroek, R. (2014). A mascon approach to assess ice shett and glacier mass balance and their uncertainties from GRACE data *J. Geophys. Res. Solid Earth*, 119, 6048–6066.

Shepherd, A. et al. (2012), A Reconciled Estimate of Ice-Sheet Mass Balance, *Science*, 338(6111), doi:10.1126/science.1228102.

Slangen, A. B., Church, J. A., Zhang, X. and Monselesan, D. (2014a). Detection and attribution of global mean thermosteric sea level change, *Geophys. Res. Lett.*, doi: 10.1002/2014GL061356.

Slangen, A. B. A., Carson, M., Katsman, C. A., van de Wal, R. S. W., Kohl, A., Vermeersen, L. L. A., and Stammer D. (2014b). Projecting twenty-first century regional sea-level changes. *Clim. Change*, 124, 317–332, doi:10.1007/s10584-014-1080-9.

Slangen, A. B. A., Aldoff, F., Jevrejeva, S., Leclercq, P. W., Marzeion, B., Wada, Y. and Winkelmann, R. (2017). A review of recent updates of sea level projections at global and regional scales, *Surveys in Geophysics,* 38, 385–406.

Steffen, K. et al. (2010). Cryospheric Contributions to Sea Level Rise and Variability. In: *Understanding Sea Level Rise and Variability*, edited by J. Church, P. L. Woodworth, T. Aarup, and S. Wilson. Blackwell, Oxford, UK.

Stocker, T. F. et al. (2013), *Climate Change 2013: The Physical Science Basis. Contribution of Working Group I to the Fifth Assessment Report of the Intergovernmental Panel on Climate Change*, edited by Stocker, T. F. et al. Cambridge University Press, Cambridge and New York.

Spada, G. (2017). Glacial isostatic adjustment and contemporary sea level rise: An overview, *Surveys in Geophysics*, 38, 153–185

Stammer, D., A. Cazenave, R. M. Ponte, and M. E. Tamisiea (2013). Causes for contemporary regional sea level changes, *Annual Review of Marine Science,* edited by C. A. Carlson and S. J. Giovannoni, 5, 21–46, doi:10.1146/annurev-marine-121211-172406.

Tamisiea, M. E. (2011). Ongoing glacial isostatic contributions to observations of sea level change, *Geophysical Journal International*, 186(3), 1036–1044, doi:10.1111/j.1365-246X.2011.05116.x.

Tamisiea, M. E. and Mitrovica, J. X. (2011). The moving boundaries of sea level change: Understanding the origins of geographic variability. *Oceanography,* 24(2), 24–39.

Timmermann, A., McGregor, S., and Jin F.-F. (2010). Wind effects on past and future regional sea level trends in the southern Indo-Pacific, *Journal of Climate*, 23(16), doi:10.1175/2010jcli3519.1.

Vaughan, D. G., Comiso, J. C., Allison, I., Carrasco, J., Kaser, G., Kwok, R., Mote, P., Murray, T., Paul, F., Ren, J., Rignot, E., Solomina, O., Steffen, K., and Zhang, T. (2013): Observations: Cryosphere. In: *Climate Change 2013: The Physical Science Basis. Contribution of Working Group I to the Fifth Assessment Report of the Intergovernmental Panel on Climate Change,* edited by T. F. Stocker, D. Qin, G.-K. Plattner, M. Tignor, S. K. Allen, J. Boschung, A. Nauels, Y. Xia, V. Bex, and P. M. Midgley. Cambridge University Press, Cambridge and New York. pp. 317–382, doi:10.1017/CBO9781107415324.012.

Velicogna, I., Sutterley, T. C., and van den Broeke, M. R. (2014). Regional acceleration in ice mass loss from Greenland and Antarctica using grace time variable gravity data Geophys. *Res. Lett*. 41, 8130–8137.

Velicogna, I., and Wahr, J. (2013). Time-variable gravity observations of ice sheet mass balance: Precision and limitations of the GRACE satellite data, *Geophysical Research Letters*, 40(12), 3055–3063, doi:10.1002/grl.50527.

Von Schuckmann, K., and Le Traon, P. Y. (2011), How well can we derive Global Ocean Indicators from Argo data? *Ocean Science*, 7(6), 783–791, doi:10.5194/os-7-783-2011.

von Schukmann, K. et al. (2016). Earth's energy imbalance: An imperative for monitoring, *Nature Climate Change*, 26, 138–144.

Wada, Y., van Beek, L. P. H., Weiland, F. C. S., Chao, B. F., Wu, Y. H., and Bierkens, M. F. P. (2012). Past and future contribution of global groundwater depletion to sea-level rise. *Geophysical Research Letters*, 39, L09402.

Watson, C. S., White, N. J., Church, J. A., King, M. A., Burgette, R. J., and Legresy, B. (2015), Unabated global mean sea level over the satellite altimeter era, *Nature Climate Change*, doi:10.1038/NCLIMATE2635.

Wada, Y., Lo, M.-H., Yeh, P. J. F., Reager, J. T., Famiglietti, J. S., Wu, R.-J., and Tseng, Y.-H. (2016). Fate of water pumped from underground and contributions to sea level rise, *Nature Climate Change*, doi:10.1038/NCLIMATE3001.

Wada, Y., Reager, J. T., Chao, B. F., Wang, J., Lo, M. H., Song, C., Li, Y., and Gardner, A. S. (2017). Recent changes in land water storage and its contribution to sea level variations, *Surveys in Geophysics*, 38, 131–151.

Woppelmann, G., Letetrel, C., Santamaria, A., Bouin, M. N., Collilieux, X., Altamimi, Z., Williams, S. D. P., and Miguez, B. M. (2009). Rates of sea-level change over the past century in a geocentric reference frame. *Geophys. Res. Lett*., 36, doi:10.1029/2009gl038720.

Wöppelmann, G., and Marcos, M. (2016). Vertical land motion as a key to understanding sea level change and variability, *Rev. Geophys*., 54, doi:10.1002/2015RG000502.

Woodworth, P. L., and Player, R. (2003). The permanent service for mean sea level: An update to the 21st century, *Journal of Coastal Research*, 19(2), 287–295.

Zhang, X., and Church, J. A. (2012). Sea level trends, interannual and decadal variability in the Pacific Ocean, *Geophys. Res. Lett*., *39*, doi:10.1029/2012GL053240.

Zwally, H. J., and Giovinetto, M. B. (2011). Overview and Assessment of Antarctic Ice-Sheet Mass Balance Estimates: 1992–2009, *Surveys in Geophysics*, *32*(4–5), 351–376, doi:10.1007/s10712-011-9123-5.

12
Ocean Circulation: Knowns and Unknowns

Harry L. Bryden and Lawrence A. Mysak

12.1 Introduction

The ocean and the atmosphere are the main dynamic components of the Earth's climate system. These components interact on various spatial scales on a daily, seasonal and interannual basis and, in particular, exchange vast amounts of heat, freshwater and CO_2. In addition, the ocean and atmosphere transport poleward substantial amounts of these quantities on daily to decadal and longer time scales.

The Sun's incoming short-wave radiation that heats the Earth is concentrated in tropical latitudes, whereas the outgoing long-wave (black body) radiation that cools the planet is more uniform over the globe. For the latter distribution to occur, a poleward transport of heat by the combined ocean and atmosphere is required. This combined (total) transport reaches a peak of almost 6 PW in mid-latitudes, and the ocean contributes 40% of the maximum total heat transport at 25°N (Figure 12.1, Bryden and Imawaki, 2001). To account for the global distribution of the air–sea exchange of freshwater, characterized by subtropical zones of net evaporation (evaporation minus precipitation) and subpolar regions of net precipitation, the poleward atmospheric transport of water in each hemisphere is required (Wijffels et al., 1992); this transport peaks at nearly 0.6 Sv (Figure 12.2, dashed curve). To maintain a global water balance, the ocean returns freshwater (low-salinity water) from the wet subpolar regions to the dry subtropics in almost equal amounts to those in the atmosphere (Figure 12.2, solid curve). Similarly, for the global balance of natural cCO_2, the CO_2 is largely taken up by the ocean from the atmosphere in the North Atlantic Ocean (Gruber et al., 2009). Some of this CO_2 is then outgassed back to the atmosphere in the Southern Ocean, being transported there from the North Atlantic at a rate of about 0.5 Pg C yr^{-1} (Figure 12.3). Additional CO_2 is transported by various currents to the equatorial Pacific, where it outgasses and returns to the North Atlantic via the atmosphere.

Numerous ocean currents contribute to the preceding transports that provide the near- global balances for heat, freshwater and CO_2. Any small residuals in the exchanges between the ocean and atmosphere lead to an accumulation (or decrease) of heat, freshwater and CO_2 in the ocean. The distribution of these first two changes in the ocean then modifies the ocean stratification and hence the large-scale circulation of the ocean over decadal and longer time scales.

The key to unraveling climate change, past and future, is to understand how the ocean and atmospheric circulations and ocean–atmosphere exchanges are affected by subtle changes in the Earth's radiation budget. In this chapter we l examine two climatic situations: (1) the past glacial climate, where subtle changes in the pattern of solar insolation due to the Earth's orbital cycles caused the ice ages, which are characterized by continental ice sheets of vast scale; and (2) future climate change, where increasing anthropogenic CO_2 and other greenhouse gases in the atmosphere result in an increase in net incoming radiation of the order of 1 W m^{-2} that is predicted to lead to global warming of 2°C or more (IPCC, 2013). We try here to summarize what is known and unknown about the ocean circulation and how the circulation is affected by changes in incoming solar radiation.

A global survey of ocean circulation was made for the first time during the 1990s under the internationally coordinated World Ocean Circulation Experiment (WOCE). Moored time series measurements were made to quantify the transport of (1) the upper western boundary currents (Gulf Stream, Kuroshio Current, Agulhas Current, Brazil Current and East Australian Current) in each ocean basin and (2) the deep western boundary currents that carry bottom waters from their

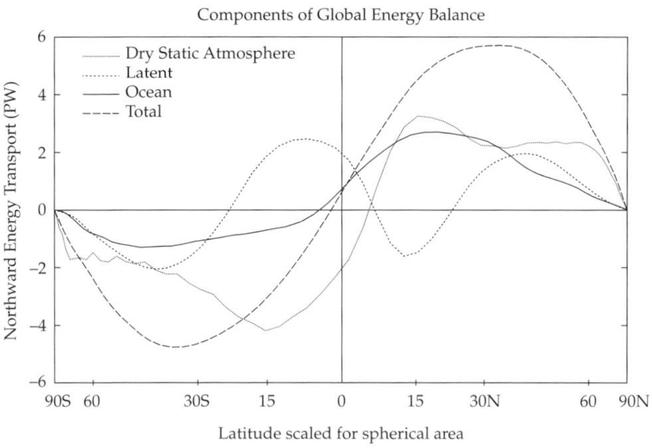

Figure 12.1 Components of atmosphere and ocean energy transport required to balance the net radiative heating/cooling of the Earth. Standard atmospheric energy transport is here divided into the dry static atmospheric energy transport and the latent heat transport because the latent heat transport is fundamentally a joint atmosphere–ocean process as the atmospheric water vapour transport is balanced by an opposing oceanic freshwater transport (see Figure 12.2). The ocean heat transport is determined by integrating over the oceans the spatial distribution of atmosphere–surface heat exchange calculated by subtracting the atmospheric energy transport divergence from the radiative heating at the top of the atmosphere (adapted from figure 6.1.3 of Bryden and Imawaki, 2001).

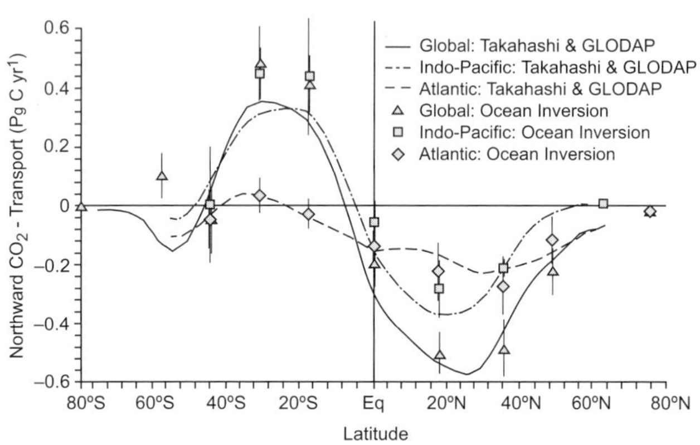

Figure 12.3 Northward transport of carbon dioxide for the global ocean and for the individual ocean basins. Both estimates from inverse flux estimates (symbols) and those inferred from the pCO_2-based flux estimates (lines) are shown (Figure 7 of Gruber et al., 2009).

source regions in the northern North Atlantic and Southern Ocean to the rest of the global ocean. Full-depth, high-quality hydrographic sections measured a complete suite of water properties including temperature, salinity, oxygen, nutrients, carbon system components and tracers in each ocean basin. For many

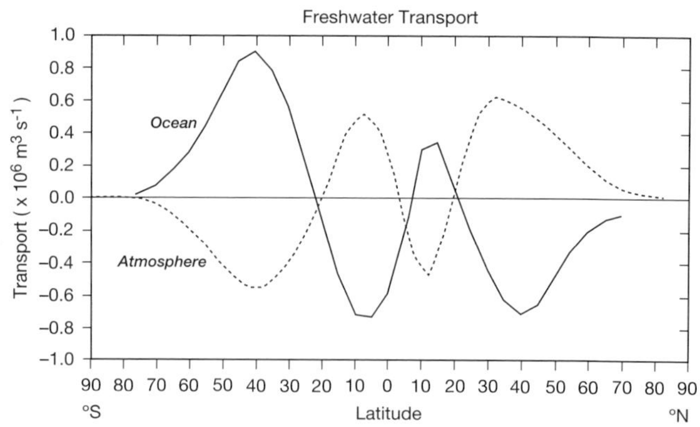

Figure 12.2 Northward transport of freshwater as a function of latitude in the ocean (solid line) and the atmosphere (dashed line) (redrawn figure 3 from Wijffels et al., 1992).

basins, this was the first time such observations were carried out. As a result, we can now quantify the strength and structure of the present ocean circulation and describe how and where water masses are formed and how they circulate. The progress made during WOCE is described in the thirty-three contributions to a summary book edited by Siedler et al. (2001).

Also during WOCE, satellite altimetry measurements of sea surface height commenced globally on a ten-day repeat cycle. As a consequence, we now have some understanding of oceanic variability on time scales of weeks to decades over regional to global spatial scales. About ten years ago, repeated sampling of temperature, salinity (and recently oxygen) from the surface to two-thousand-meter depth began with the deployment of three thousand profiling floats under the Argo programme. Thus the variability in the circulation of the upper two thousand meters in the global ocean can now be quantified on a continuing basis. We have moved since 1990 from an ocean circulation that was largely not sampled over vast regions to an ocean that has been comprehensively measured once during WOCE and that is now being routinely monitored from the surface to two-thousand-meter depth on a ten-day time scale. Two recent books (Talley et al., 2011; Wunsch, 2015) describe our post-WOCE understanding of the regional, basin-scale and global ocean circulation and its temporal variability. In a sense, we can conclude that the present ocean circulation is reasonably well described everywhere, and we are currently quantifying its temporal variability over time scales from twenty days to several decades.

There remain some parts of the ocean circulation that are not well understood, particularly in terms of several complex physical processes such as the flow over sills connecting two basins and the entrainment of the dense waters as they cascade down into the next basin. Other complex processes that need further elucidation include how and where mixing occurs in the open ocean, and how much of it is isopycnal and how much is diapycnal. We also need to identify more precisely the locations where deep water is formed in late winter and quantify how much deep water formation occurs in a severe winter. Another challenge is to understand how sea ice melts: from above, from below or from the edges where waves and currents chip away. The preceding processes all represent major research opportunities that need further investigation to enable researchers to include their effects in models of ocean circulation and in coupled atmosphere–ocean models used for climate projections one hundred years into the future. The inclusion of such processes may well be critical for accurate future projections, which are far removed from the present ocean circulation with which the models are initialised.

Traditionally, ocean scientists try to separate ocean circulation into wind-driven and thermohaline-driven components. From the distribution of the surface wind stress, one can derive (1) upwelling patterns in coastal and equatorial regions as well as in the Southern Ocean, (2) mid-ocean Sverdrup flows in each ocean basin and (3) poleward-flowing western boundary currents. There is general agreement that the wind-driven circulation is well understood in terms of patterns and magnitude. However, there remain substantial uncertainties as to how the observed circulation strengths relate to the magnitude of the wind forcing, and especially as to how the circulation changes over decadal time scales as the winds change (Beal and Elipot, 2016).

In terms of the thermohaline circulation (THC), we understand how deep water forms in late winter during severe buoyancy loss events, but we have a poor knowledge of how much deep water is formed each winter and of what sets the strength of the THC globally and in each ocean. This of course is a key issue in explaining how the THC will change under different radiative regimes, and how the ocean will then interact with the atmosphere over decadal to centennial time scales. Thus, it is generally accepted that the greatest need is to understand the thermohaline-driven ocean circulation in order to predict future climates.

In this chapter, we first discuss what we know and do not know about the ocean circulation during the ice ages, especially the thermohaline component in the Atlantic Ocean. Of special interest here is the role of the stability of the Atlantic meridional overturning circulation (AMOC) in explaining the abrupt climate changes that recurred on millennial time scales during the last glacial period (ca 100 to 10 kyr BP). These changes are known as Dansgaard-Oeschger (D-O) warm events and Heinrich cold events, and they were first observed in Greenland ice cores and North Atlantic sediment records (Figure 12.4). Secondly, we describe what we know about the evolution of the deep Atlantic Ocean circulation and water masses during the transition from the Last Glacial Maximum (at 21 kyr BP) to the present. Next, we examine how we expect the ocean circulation and ocean–atmosphere exchanges to evolve in the future under increasing atmospheric CO_2 and other greenhouse gases; we also try to identify the key unknowns in predicting future climate change.

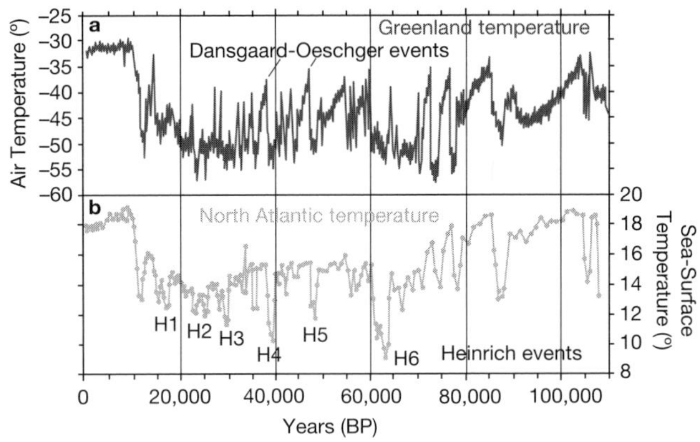

Figure 12.4 Climatic and oceanic variations in and around the North Atlantic Ocean during the past 110,000 years, as revealed in Greenland ice cores and North Atlantic sediment cores obtained off the Iberian Margin. Time progresses from right to left. (a) The Greenland air temperature, based on isotope thermometry, shows abrupt warming periods called Dansgaard-Oeschger events. (b) The sea surface temperature in the North Atlantic shows episodes of drastic cooling called Heinrich events. This record is based on biomolecular thermometry with long-chain (thirty-seven-carbon) organic molecules called alkenones measured in the sediment cores collected by E. Bard and colleagues (Bard, 2002). For the color version, please refer to the plate section.

12.2 Abrupt Climate Changes during the Last Glacial and Related AMOC Changes

The aforementioned abrupt and large-amplitude climate changes that occurred during the last glacial were discovered more than two decades ago (e.g., Dansgaard et al., 1993; Heinrich, 1988). Of the two types of notable changes seen in Figure 12.4, the long-lasting cold Heinrich events were most likely triggered by massive iceberg discharges from the Laurentian Ice Sheet into the North Atlantic (Figure 12.5; Bond and Lotti 1995). On the other hand, the origin and cause of the shorter-time-scale warm D-O events, which lasted about a thousand years and often recurred in groups, remain to be understood (Bender, 2013). However, it is now generally agreed that the stability of the AMOC and the switch of this circulation between different modes (patterns) play a central role (e.g., Ganopolski and Rahmstorf, 2001; Rahmstorf, 2002; Clement and Peterson, 2008; Kageyama et al., 2010; Xiao Zhang et al., 2014).

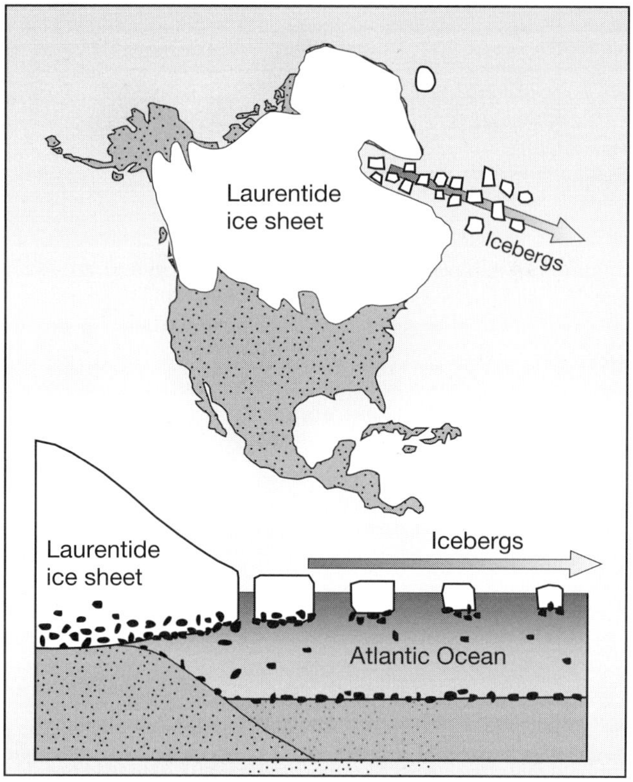

Figure 12.5 Iceberg surges into the North Atlantic during a Heinrich event (Bard, 2002).

A typical D-O event starts with an abrupt warming of order 10°C over Greenland that lasts a few decades; this is followed by a plateau period of slow cooling over several centuries (an "interstadial"), which then ends with a fairly rapid drop of temperature back to the cold background state (a "stadial"). Warming and cooling periods associated with a D-O event were not confined to Greenland and the North Atlantic; they also occurred in the Southern Hemisphere (Figure 12.6) and on other continents (Voelker et al., 2002). Also, in the case of a cold Heinrich event followed by a warm D-O event (e.g., see the pairs H5 and D-O12, H4 and D-O8, and H3 and D-O4 in Figure 12.6), a bipolar seesaw pattern is clearly seen between Greenland temperature (NGRIP curve in Figure 12.6) and the temperature in the Antarctic (top three curves in Figure 12.6): prior to a sudden warming in the north, there was a slow warming in Antarctica that lasted several thousand years. During such times the CO_2 concentration in the atmosphere increased about 20 ppm above a background level of ca 200 ppm because of increased ventilation in the Southern Ocean, which allowed the escape of this greenhouse gas from this region (Bender, 2013). The bottom curve in Figure 12.6 shows that during most of the interstadials in Greenland, the atmospheric methane (CH_4) concentration (as measured in Greenland ice cores) also increased. Since CH_4 originates from wetlands, it follows that the high-latitude warming events also affected continental areas at lower latitudes (Bender, 2013).

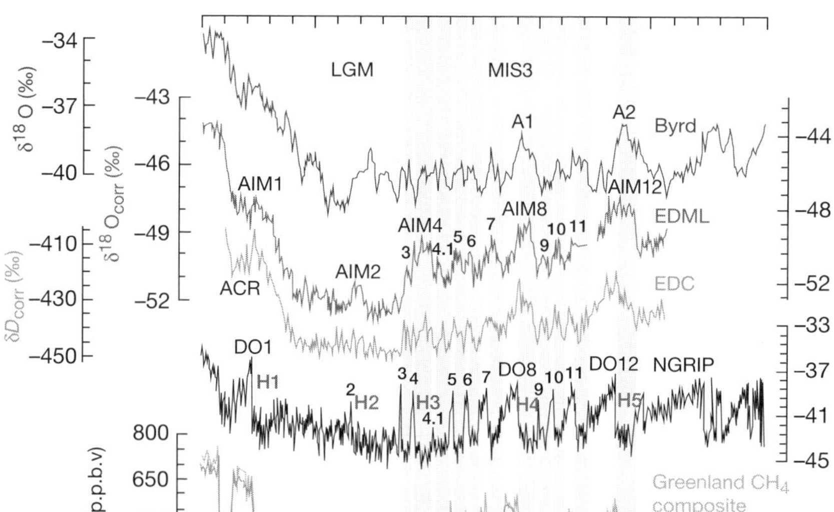

Figure 12.6 Climate records covering the period from 10 to 60 kyr; original data from EPICA Community Members. From top to bottom: the $\delta^{18}O$ or δD of ice (both indicating paleotemperature) from the Antarctic ice cores Byrd, EDML (EPICA Dronning Maud Land) and EDC (EPICA Dome C); the $\delta^{18}O$ of the NGRIP (North Greenland) ice core; and the CH_4 concentrations as measured in Greenland ice cores. In the EDML plot, AIM indicates Antarctic Isotope Maximum. In the NGRIP plot, the unlabeled numbers or the numbers labeled with DO are Dansgaard-Oeschger warm events, and the H numbers are the various Heinrich cold events (Clement and Peterson, 2008).

Probably the first explanation of the D-O events involved the concept of THC bi-stability (Broecker et al., 1985): North Atlantic Deep Water (NADW) formation is strong during an interstadial (warm) period and is shut off during a cold stadial, with some outside forcing causing a switch between the two states. However, this theory is based on Stommel's (1961) two-state (THC "on" or THC "off") box model of the THC (see box 2 in Rahmstorf, 2002). Such a model for the glacial THC is not supported by recent North Atlantic sediment data collected during the latter part of the glacial period. From eight time-slice reconstructions of the ocean circulation during the past 30 kyr, Sarnthein et al. (1994) showed that NADW formation occurred during both stadials and interstadials, and was only shut down during Heinrich events.

A second idea proposed to explain D-O events involves internal oscillations in the volume transport of the AMOC; this is known as the salt oscillator mechanism (Broecker et al., 1990). While challenged by some investigators (e.g., Rahmstorf, 1996), this concept has been explored with simple box models (Birchfield et al., 1994) and coupled General Circulation Models (Peltier and Vettoretti, 2014). In the latter study, the impact of a variable diapycnal diffusivity on the AMOC in the glacial version of the NCAR Community Earth System Model 1 (CESM1) is investigated. Peltier and Vettoretti show that a Heinrich event serves as a "kick" to induce millennial internal salt oscillations of relaxation oscillator form in the AMOC.

To discuss the third and possibly most plausible mechanism for explaining D-O events, it is first useful to provide a schematic picture of the three glacial THC modes or patterns that likely occurred at different times in the North Atlantic (Figure 12.7). These modes have been compiled from time slices of paleoclimate data (Sarnthein et al., 1994; Alley and Clark, 1999). From top to bottom, these have been labelled the interstadial (warm), stadial (cold) and Heinrich (off) modes because of their occurrence during interstadials, stadials and Heinrich events marked in Figure 12.6. In the warm mode, NADW formed in the Nordic Seas (as happens today), while in the cold mode, it formed in the North Atlantic in the subpolar region south of Iceland. In the Heinrich mode, NADW formation essentially ceased and Antarctic Bottom Water (AABW) filled the whole Atlantic basin. It is particularly important to note that there was a large southward latitudinal shift in the convection (sinking) region in transition from the warm mode to the cold mode.

In the pioneering paper of Ganopolski and Rahmstorf (2001) (hereafter GR 2001), simulations with the CLIMBER-2 climate model of intermediate complexity were used to argue that the cold stadial mode (Figure 12.7, middle panel) was likely the normal

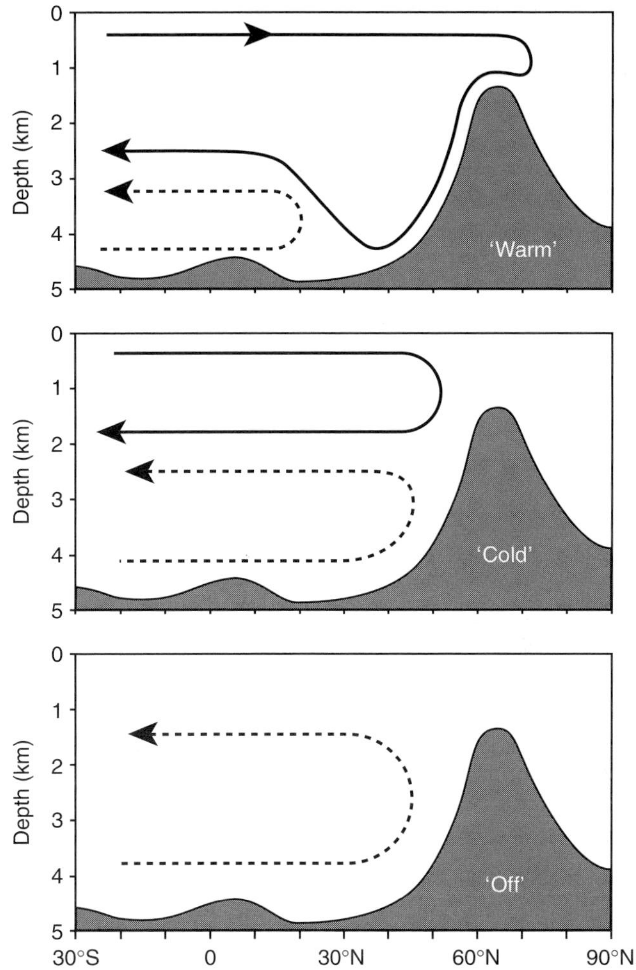

Figure 12.7 Schematic configuration of the three modes of ocean circulation (the AMOC) that prevailed during different times of the last glacial period. Shown is a section along the Atlantic; the rise in the topography in the north represents the shallow sill between Greenland and Scotland. North Atlantic overturning is represented by the clockwise circulation in the upper two panels; the Antarctic Bottom water circulation is represented by the anticlockwise flow pattern in all three panels (from Rahmstorf, 2002).

pattern of the AMOC during the glacial period. In this case NADW was formed south of Iceland in response to a freshwater forcing of order 0.03 Sv. Then, to produce the warm phase of a D-O event, the freshwater forcing was slowly reduced to negative values; when it reached -0.03 Sv the AMOC shifted rapidly to the strong warm interstadial (Figure 12.7, top), with NADW formation occurring in the Nordic Seas (GR 2001, Figure 12.5). An increase in the freshwater forcing then forced the AMOC back to the stadial mode, thus completing a hysteresis loop (Rahmstorf, 2002; Rahmstorf et al., 2005). By applying a fifteen-hundred-year cycle of freshwater forcing with an amplitude of 0.03 Sv, a sequence of D-O events was simulated by

GR 2001 similar to what is seen in Figures 12.4 and 12.6. GR 2001 also could model the bipolar seesaw. However, it is not clear what would cause this fifteen-hundred-year cycle, although weak variations in solar variability on this time scale have been proposed (Van Geel et al. 1999; Bond 2001), but contested by others (e.g., Bard and Frank, 2006). Nevertheless, because of the sudden latitudinal switch in the convection sites in this mechanism, the concept of stochastic resonance has been invoked as an explanation for D-O events (Ganopolski and Rahmstorf, 2002; Rahmstorf and Alley, 2002). In addition, to help elucidate the reason for the switching between modes, the sensitivity of the AMOC in the North Atlantic to various freshwater perturbations during the late part of the last glacial has been investigated (e.g., Van Meerbeeck et al., 2011).

In contrast to the preceding mechanisms involving the stability and/or changes in the AMOC, tropical processes have also been suggested as playing a role in triggering D-O events (for a review, see Clement and Peterson, 2008). In response to large tropical SST changes, there will generally be large shifts in the atmospheric planetary wave pattern, which in turn could induce notable changes in Greenland temperatures via anomalies in atmospheric heat transports.

One of the puzzling features of D-O events is that there were relatively few of them in the first half of the last glacial (see Figure 12.4). During this period the ice sheets were still growing in response to Milankovitch global cooling. Building on this observation, Wang and Mysak (2006) used the McGill Paleoclimate Model (a climate model of intermediate complexity) to show that a moderate global cooling together with white noise freshwater forcing forces a fairly strong AMOC into an unstable state when a threshold temperature T_1 is reached. This is a Hopf bifurcation point in the model. At this point, the AMOC in the model begins to cycle between the strong mode, with NADW formed by intense heat loss to the atmosphere, and a weak mode, which is maintained by sea ice brine rejection. These cycles in the AMOC produce the warming and cooling associated with D-O events. However, when the background climate reaches the second critical temperature T_2 ($<T_1$), another threshold in the model is reached, and the AMOC evolves into the weak mode (see figure 5 in Wang and Mysak, 2006). Thus under this scenario, D-O events would have occurred only during the latter stages of the last glacial, in agreement with the observations.

Finally, we mention that a number of investigators have proposed that even small changes in the Northern Hemisphere ice sheets could have induced abrupt climate shifts during the last glacial (e.g., Xu Zhang et al., 2014). Such ice-sheet changes could produce variations in ice-sheet calving and subsequent changes in the AMOC, which created a continuum of glacial millennial-scale climate variability (Menviel et al., 2014). A similar mechanism involving ice-sheet calving was also investigated some time ago with an early version of the McGill Paleoclimate Model (Wang and Mysak, 2001).

From the preceding discussion, it is evident that there is probably not a single trigger or mechanism that explains all the observed D-O events. It is conceivable that some of the mechanisms outlined work in tandem (Rahmstorf, 2002), and the challenge for climate modellers and paleoceanographers is to determine the appropriate mix of mechanisms for each D-O event.

In contrast to the preceding situation, the origin of the long-lasting (1–2 kyr) cold Heinrich events can be traced to the discharge of large quantities of ice from the Laurentian Ice Sheet into the North Atlantic (Figure 12.5) every 7–10 kyr (Figures 12.4 and 12.6). Each event is characterized by a distinct layer of ice-rafted debris that was deposited in the North Atlantic. From the chemical and mineral properties of this debris, the provenance of the discharged ice was found to be mainly the Hudson Strait (Hemming, 2004). However, recent modelling studies with the Marshall and Clarke (1997a, b) 3-D dynamic-thermodynamic ice-sheet model suggest that the more northerly regions around Lancaster Strait and Baffin Island may also be an important source for the ice-sheet discharges that initiate a Heinrich event (Papa et al., 2006).

Several investigators have shown that the massive ice-sheet discharges tended to occur when the Laurentian Ice Sheet (LIS) became very high, causing excessive bottom pressure and hence basal melting, which led up to 10% of the LIS to slide down toward the coastal regions during each Heinrich event (Calov et al., 2002; Papa et al., 2006). The resulting freshwater flux of order 0.1 Sv was sufficient to shut down NADW formation for more than 1 kyr (Sarnthein et al., 1994; GR 2001) and hence create the glacial off mode shown in Figure 12.7. Further, during this very cold stadial, (i) the sea ice in winter expanded dramatically to the south, which insulated Greenland from the warm ocean; (ii) the sea level rose several meters and (iii) warming occurred in the South Atlantic due to the reduced northward transport of heat by the AMOC (the bipolar seesaw) (Bender, 2013). Also, because of changes in the global atmospheric circulation, during the peak of a Heinrich event precipitation was elevated in the western United States and in southern Brazil (Bender, 2013). It has been suggested that the sea level rise during a Heinrich event and the concurrent subsurface warming in the Nordic and Labrador Seas served as two positive feedbacks that could have created further destabilization of the LIS and hence set the stage for future Heinrich events (Flückiger et al., 2006).

Heinrich events end after ca 1–2 kyr when the iceberg meltwater is flushed out over the North Atlantic and NADW production resumes; then the changes seen in the tropics and the Southern Ocean are reversed (Bender, 2013). From Figures 12.4 and 12.6 it is evident that a large D-O event generally occurred after a Heinrich event, and that the subsequent D-O events (interstadials) become progressively weaker until the next Heinrich event. This sequence of D-O events bracketed by a pair of Heinrich events is sometimes referred to as a Bond cycle (Rahmstorf, 2002). However, it is a mystery as to why such cycles only occurred between about 50 and 25 kyr BP.

12.3 Deglaciation and the Evolution of the Deep Atlantic Ocean Circulation since the Last Glacial Maximum

By the time of the Last Glacial Maximum (LGM) at 21 kyr BP, regular D-O and Heinrich events had all but ceased (except for one pair after the LGM; see Figure 12.6). During the LGM, ocean proxy data suggest that the AMOC was weaker and shallower than at present (Lynch-Stieglitz et al., 2007). According to $^{231}Pa/^{230}Th$ values from a subtropical Atlantic sediment core, the strength of the AMOC was reduced by up to 30% during the LGM (McManus et al., 2004). This was also the time when the atmospheric CO_2 dropped to its glacial minimum of ca 180 ppm, and the atmospheric conditions were cold, windy and dry. The mean global temperature was nearly 6°C below that at pre-industrial times (e.g., Schneider von Deimling et al., 2006). The simulation of the weaker AMOC during the LGM with different climate models has proven to be a challenge (e.g., Kageyama et al., 2013). For example, Oka et al. (2012) have shown that there is a thermal threshold that has to be crossed in order to reproduce an LGM-type AMOC in a coupled climate model: only for sufficient cooling and extensive sea ice cover in the northern North Atlantic is the deep convection suppressed enough to produce a shallow AMOC.

Another important feature of the LGM Atlantic was the existence of strong stratification due to the saltier-than-today Antarctic Bottom Water (AABW) (e.g., Curry and Oppo, 2005). This produced a salinity barrier that prevented dense waters from the Northern Hemisphere from sinking to great depths. Also, the LGM Atlantic had a much lower vertical mixing rate than present-day (Adkins, 2013), and the AABW filled most of the Atlantic basin. As mentioned, the North Atlantic produced a water mass that penetrated only to mid-depths and extended southward only to the tropics in the Northern Hemisphere. This shallow water mass

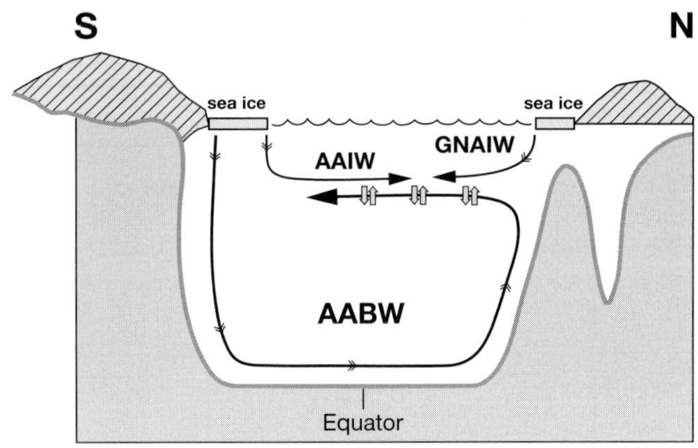

Figure 12.8 Schematic configuration of the AMOC during the Last Glacial Maximum . AABW: Antarctic Bottom Water; AAIW: Antarctic Intermediate Water; GNAIW: Glacial North Atlantic Intermediate Water (modified from a sketch by J. Marson, personal communication, 2015).

(the upper limb of the AMOC) is now referred to as Glacial North Atlantic Intermediate Water (GNAIW) (Duplessy et al., 1988). A schematic configuration of the AMOC during the LGM is shown in Figure 12.8.

Deglaciation started around 21 kyr BP, being initiated by summer insolation increases in high northern latitudes (about 30 W m^{-2} between 24 and 12 kyr ago (Rahmstorf, 2002)); at the same time CO_2 was released to the atmosphere from the oceans and land, which amplified and globalized the warming (Ruddiman, 2008). At first a fairly rapid warming was seen in the Antarctic (Figure 12.6, top three curves), but it was delayed in the north by a Heinrich event (H1 in NGRIP curve in Figure 12.6), which would have produced a weakened AMOC in the north. Following H1, in keeping with the evolution of earlier Heinrich events, the warm DO1 event occurred; it is also known as the Bølling-Allerød (BA) warm period; presumably the AMOC was then in the interstadial mode (Figure 12.7). Simultaneous with this warming, a small cooling in the Southern Hemisphere (SH) occurred; it is known as the Antarctic cold reversal (marked as ACR in Figure 12.6); this is consistent with the bipolar seesaw. A second freshwater-induced cold period occurred in the north approximately between 12.9 and 11.7 kyr, an event known as the Younger Dryas (YD) (Ruddiman, 2008). Because of the magnitude of the meltwater flux at the time, NADW formation probably ceased or was very weak (Fanning and Weaver, 1997; Manabe and Stouffer, 1997; McManus et al., 2004). The end of the YD was followed by a moderate warming leading to the Holocene, the fairly stable climate of the past 11.7 kyr.

There have been many modelling and data studies of the changes in the Atlantic since the LGM in response to

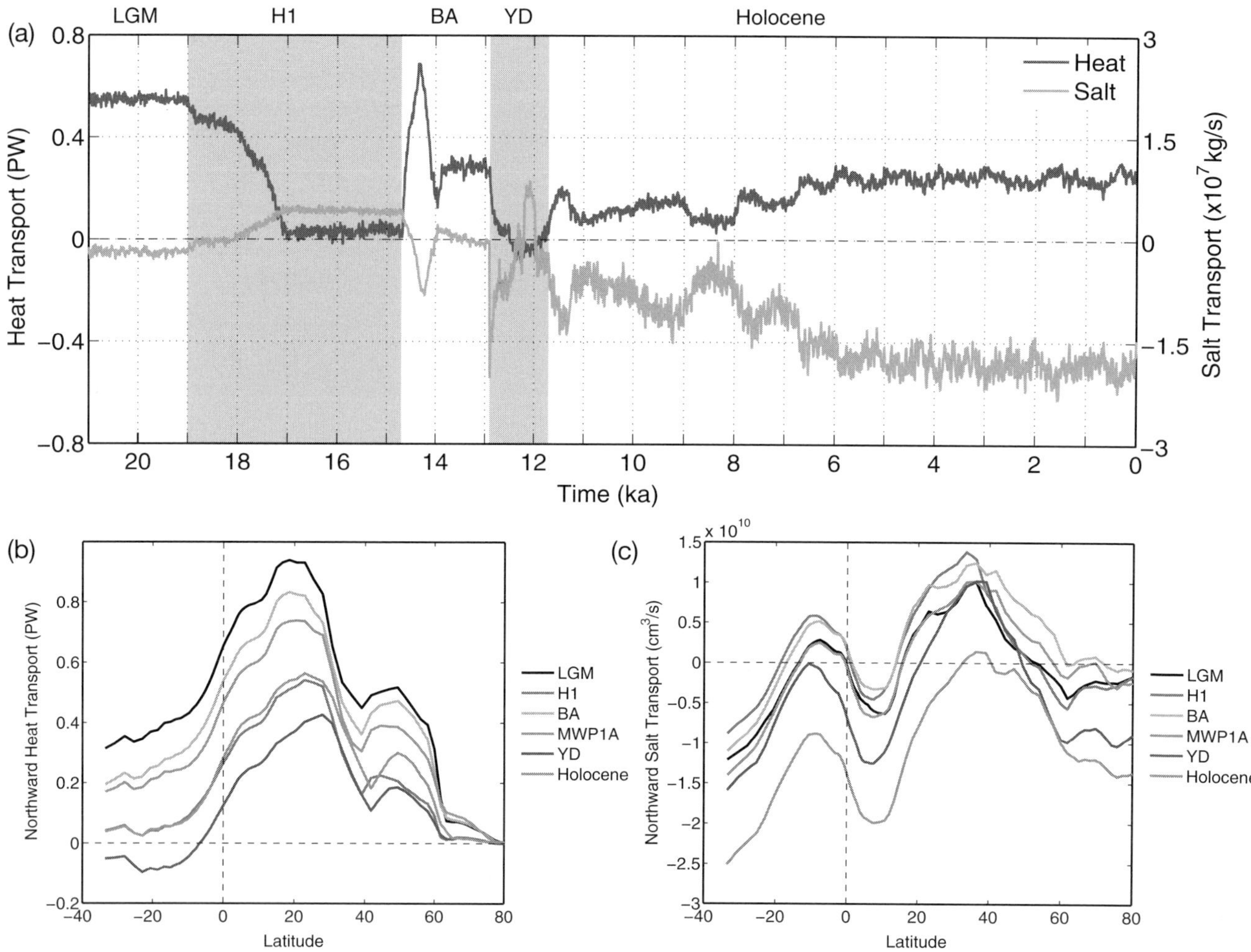

Figure 12.9 (a) Simulated meridional heat (purple) and salt (green) transport in the South Atlantic between 30°S and the equator. Positive values denote northward transport. The main climatic events during the last deglaciation are also marked: Last Glacial Maximum (LGM), Heinrich event 1 (H1), Bølling-Allerød (BA), Younger Dryas (YD) and the Holocene. (b) Northward transport of heat (PW) and (c) northward transport of salt ($cm^3\ s^{-1}$) averaged for each key period when freshwater was injected into the ocean (from Marson et al., 2014). For the color version, please refer to the plate section.

the various climatic forces in operation over the past 21 kyr: glacial meltwater, decreasing ice sheets, rising sea level, increasing solar radiation in the NH and increasing greenhouse gases (GHGs) (see Marson et al., 2014). However, most of these investigations have tended to focus on specific climatic events of the last 21 kyr, such as H1, BA and the YD. In contrast, Marson et al. (2014) investigated the evolution of the deep Atlantic from the LGM to the present day using the output of a transient climate simulation of the last 22 kyr carried out with the NCAR-CCSM3; this simulation is known as the TraCE-21K run (He, 2011). The model data from TraCE-21K have previously been used to study the importance of each forcing (orbital, GHG concentration and ice sheet configuration) to the global temperature evolution from the LGM to the BA (He et al., 2013). Also, Liu et al. (2009) used this run to investigate the BA warming, in a study called the DGL-A experiment. The study by Marson et al. (2014) extends these studies to the present day and focuses on the structural changes in the Atlantic deep ocean. Marson et al. (2014) showed that the present-day circulation of the South Atlantic Ocean was established approximately at the onset of the Holocene (about 11.7 kyr ago). The spreading of the NADW from its GNAIW roots to the South Atlantic was only possible after the deglacial freshening of the AABW eroded the deep salinity barrier that was present during the LGM.

Not surprisingly, in response to the meltwater forcing used in TraCE-21K (see figure 1 in Marson et al., 2014) and the subsequent changes in the AMOC, there were significant changes in the modelled meridional heat and

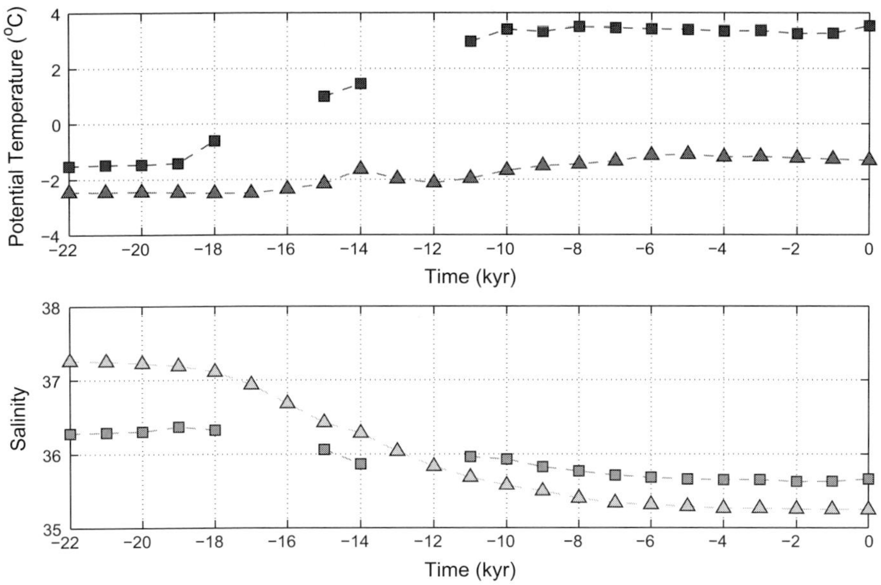

Figure 12.10 Evolution of potential temperature (top) and salinity (bottom) for the water types AABW (triangles) and GNAIW-NADW (squares) since the LGM. To obtain each AABW point at a given time, an average over the five coldest simulated temperatures (and corresponding salinities) between 60°S and 70°S and between 1500 and 5000 m was performed. Similarly, for each GNAIW point, an average over the five lowest salinities and corresponding lowest temperatures between 40°N and 80°N and between 400 and 1500 m was performed. For each NADW point, an average over the five highest salinities (where temperatures were less than 3.5°C) between 40°N and 80°N and between 1500 and 5000 m was performed (from Marson et al., 2015).

salt transports in the South Atlantic during the past 21 kyr. Figure 12.9a shows that the heat transport in this region is generally northward throughout this period, except for the cold H1 and YD events, when the transport was near zero or southward. At these times there was strong cooling in the NH (Figure 12.4). The large peak at approximately 14.7 kyr occurs just after the H1 meltwater forcing ceased, and it is associated with the BA warming. The northern ocean heat transport (NOHT) as a function of latitude for each climatic period is shown in Figure 12.9b. A notable feature here is the continual northward shift in the NOHT maximum as one moves through the past 21 kyr: this maximum moves from 20°N for the LGM and H1 events to around 30°N for the Holocene.

The northward ocean salt transport (NOST) in the South Atlantic is generally out of phase with the NOHT (Figure 12.9a). The NOST as a function of latitude for each climatic period is shown in Figure 12.9c. We note the existence of two maxima for each period, one at around 35°N, and one in the South Atlantic at around 10°S. From the equator to about 20°N, the salt transport is southward for all periods, and northward between 20°N and about 50°N for all periods except the Holocene. For the latter period, the transport is nearly always southward between 20°N and 80°N north.

As a continuation of Marson et al. (2014), Marson et al. (2015) investigated the evolution of the physical properties and geographical distribution of the AABW and NADW since the LGM also using the data fromTraCE-21K. They employed the mixing triangle technique (Mamayev, 1975; figure 1 in Marson et al., 2015) to determine the thermohaline variability and spatial distribution of the AABW and NADW over the past 21 kyr. In addition, the variability and distribution of the Central Water (CW) and Antarctic Intermediate Water (AAIW) were explored. Each water mass was tracked in the model data by using their extreme thermohaline characteristics:

(i) Maximum temperature for CW;
(ii) salinity minimum for AAIW;
(iii) temperature minimum at depth (below two thousand meters) for AABW;
(iv) salinity maximum at depth (below two thousand meters) for NADW, when this salinity is greater than the AABW's salinity.

Figure 12.10 (triangles) shows the evolution of the AABW thermohaline properties (potential temperature and salinity) every 1 kyr. The LGM AABW temperature (upper panel) starts at -2.5°C and ends up at -1.4°C, and the gradual warming is almost monotonic except for a small peak at 14 kyr, coincident with the meltwater pulse 1A (MWP-1A) (Weaver et al., 2003). More dramatic is the relatively large drop in AABW salinity, from 37.3 to 35.2 during the past 21 kyr (Figure 12.10, lower panel). This is the result of the massive discharge of freshwater into the ocean from ice-sheet melting.

Figure 12.10 also shows the evolution of the properties of the NADW and its precursor, the GNAIW (squares). LGM GNAIW was almost as cold as AABW at that time, but warmed relatively quickly between 19 kyr and 11 kyr during its transition to

NADW. The points missing during 17–16 and 13–12 kyr in the GNAIW/NADW curves are due to the fact that the mixed layer in the model at these times was shallower than 150 m and hence the deep water formation regions in the model were indeterminate (for further details, see Marson et al., 2015). Figure 12.10 (bottom panel) also shows that GNAIW is saltier than NADW, although it is relatively fresh compared to the AABW up to 14 kyr.

Marson et al. (2015) also showed that the simulated AABW filled the whole Atlantic basin until 15 kyr (see their figure 3); then its distribution was reduced by the high production of GNAIW at the onset of the BA and the appearance of NADW at about 11 kyr (see their figure 4). The GNAIW was a cold water mass, fresher than AABW, and shallower than NADW; it could not spread southward beyond the equator until the end of H1. GNAIW progressively weakened during H1, and was finally replaced with NADW after 11 kyr.

A schematic configuration of the AMOC for the present day is shown in Figure 12.11. Upon comparing this with Figure 12.8, we note that because of the reduced AABW density today, the NADW sinks deeper and expands farther southward. Also, because of the more northward position of the sea ice edge in the North Atlantic at present, the site of NADW formation is farther north.

In the last two sections, we have presented evidence and possible mechanisms for significant and relatively sudden shifts in the patterns of the AMOC during the past glacial and deglacial periods. Remarkably, these changes have occurred under slow changes in the radiative forcing, and thus it is a puzzle as to why the circulation changes have occurred so suddenly. Are these changes just due to LIS instabilities/calving and subsequent freshwater discharges that dramatically increased the stability (stratification) of the surface waters of the northern North Atlantic and hence reduced or shut down deep water formation there? And if so, what caused the timing of the ice-sheet instabilities/calving – is the latter related to internal rheological processes in the ice sheet? Is it due to crossing a height threshold during ice-sheet growth? Or is it due to under-ice sea water seepage caused by rising sea levels from earlier large-scale ice-sheet calving?

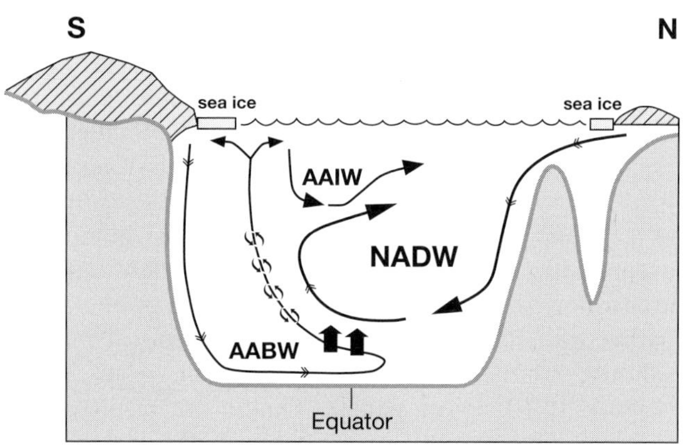

Figure 12.11 Schematic configuration of the modern AMOC. AABW: Antarctic Bottom Water; AAIW: Antarctic Intermediate Water; NADW: North Atlantic Deep Water (modified from a sketch by J. Marson, personal communication, 2015).

Today, one wonders whether the aforementioned ice-sheet instabilities/calving are likely to occur in the future. Of course now there is only the Greenland ice sheet (GIS) in North America and no LIS. While the GIS is now rapidly melting and actively calving, and consequently could provide a total freshwater input of more than ten thousand cubic kilometers during the period from 1995 to 2025 (Bamber et al., 2012), this input is only equivalent to a freshwater forcing of about 0.01 Sv over thirty years. Such a forcing is unlikely to lead to a complete shutdown of the AMOC, only to its slowdown, as is currently happening (Rahmstorf et al., 2015). According to past model studies, a freshwater input of about 0.1 Sv is required for a switch of the AMOC to the "off" mode.

Perhaps of greater concern today is the ocean impact of the continuing collapse of the Antarctic ice sheet due in part to a warming ocean under the ice shelves (DeConoto and Pollard, 2016). While this is estimated to lead to a sea-level rise of about one meter by 2100, the related Southern Hemisphere freshwater input is not likely to affect the strength of the AMOC significantly over this time frame (Stouffer et al., 2007).

12.4 Changes in Ocean Circulation Associated with Global Warming

As CO_2 accumulates in the atmosphere, surface air and ocean temperatures are rising because of the greenhouse effect whereby CO_2 in the atmosphere reduces the outgoing long-wave radiation and the resulting imbalance with incoming solar radiation leads to warming of the Earth. How much temperatures will rise is controversial because of feedbacks within the climate system, one of which is how the ocean circulation will change. Our best estimate for how much the global radiation budget is out of balance derives from observing the increase in ocean heat content, which is estimated to be about 0.7 W m^{-2} averaged over the global ocean (Rhein, et al., 2013).

To understand how this radiation imbalance will affect the atmosphere and ocean circulations, we use coupled ocean–atmosphere climate models with prescribed greenhouse forcing to project changes in ocean circulation over a few decades and compare the projected changes to observed changes.

The models predict warming almost everywhere with enhanced warming over the polar regions. Because a warmer atmosphere holds more water vapour, the hydrological cycle is enhanced, with increased evaporation from the oceans and increased precipitation (Collins et al., 2013). Ocean observations generally confirm these trends: polar regions do appear to have greater warming (Levitus et al., 2012); and the subtropical zones of high evaporation are becoming saltier, indicating enhanced evaporation, and the subpolar ocean regions of net precipitation are becoming fresher indicating enhanced precipitation (Durack et al., 2012; Levang and Schmitt, 2015). The vertical structure of the observed warming is less well represented in the models. About 65% of the observed ocean warming occurs in the upper seven hundred meters, but there is substantial warming down to two-thousand-meter depth and the deep waters below three thousand meters are also warming (Levitus et al., 2012, Purkey and Johnson (2010). Models generally show ocean warming more concentrated in the upper ocean, which suggests warming from the surface downward (Collins et al., 2013). The observed warming of the deep waters implies there are ventilation pathways that connect the intermediate and deep waters to distant surface warming (Figure 12.12). There is much to be learned about the ventilation processes by assessing the combined salinity and temperature changes in intermediate and deep waters and relating them to surface salinity and temperature changes (Zika et al., 2012).

In subpolar regions, particularly in the northern Atlantic, the combination of surface warming and freshening leads to a decrease in surface density, which makes the region more stratified (Capotondi et al., 2012). For the northern North Atlantic where much of the present-day deep water forms, increasing stratification suggests that deep water formation might decrease. This is manifest in coupled climate model projections where the AMOC slows down (Cubasch and Meehl, 2001; Collins et al., 2013) because there is less southward flow of NADW through the Atlantic over the twenty-first century. Indeed, in the observations there was a 20% decrease in the AMOC over the first ten years of monitoring (Smeed et al., 2014). But there are arguments that this decline was preceded by an equally strong increase in the AMOC during the preceding decade (Jackson et al., 2016). This is an example where model projections match observed changes, but still much uncertainty remains whether the observed changes are real climate trends or just the result of natural decadal variability.

To summarize what we know about the THC changes, the observations and model simulations are consistent: the ocean is warming, more so in polar regions; the saltier subtropical regions are getting saltier while fresher subpolar regions are getting fresher; and the warming and freshening subpolar regions are becoming less dense at the surface, implying less deep water formation in the northern North Atlantic. The latter is reflected in a declining southward deep water flow in the AMOC in both observations and model simulations.

For the wind-driven circulation, atmospheric observations indicate a poleward shift in the boundaries of the Hadley circulation, which enlarges the tropical zones, and a poleward shift and strengthening of the mid-latitude westerlies (Hartmann et al., 2013). As a result, the subtropical ocean gyres should expand poleward. Observational evidence (Rhein et al., 2013) for this poleward shift is strongest in the South Pacific, where the East Australian Current now penetrates into the Tasman Sea, and there has been a 2° latitude southward shift in the subtropical front. There is also evidence for poleward movement of the Brazil Current and Confluence in the South Atlantic, and the North Pacific subtropical gyre has expanded northward. Southern Ocean fronts also moved southward at a rate of 1° latitude over forty years and the Southern Ocean winds increased by about 10%. Models predict that the Antarctic Circumpolar Current (ACC) strengthens (Capotondi et al., 2012), but the observed baroclinic transport of the ACC has not increased (Chidichimo et al., 2014). In the North Atlantic any shift in the boundary of the subtropical gyre is too subtle to be observed above the interannual variability associated with the North Atlantic Oscillation. Long-term monitoring of the flow through the Florida Straits shows no trend in the transport of the Gulf Stream over the last fifty years (Meinen et al, 2010).

For changes over time scales of a few decades, coupled model projections are being carefully compared with observed variability in ocean circulation every six or seven years through publication of assessment reports under the auspices of the Intergovernmental Panel on Climate Change (e.g., Bindoff et al., 2007; Rhein, et al., 2013). With only about twenty-five years of observed ocean variability over global scales, it is difficult to separate long-term changes or trends from the normal seasonal to interannual variability in ocean circulation. At decadal time scales, model projections

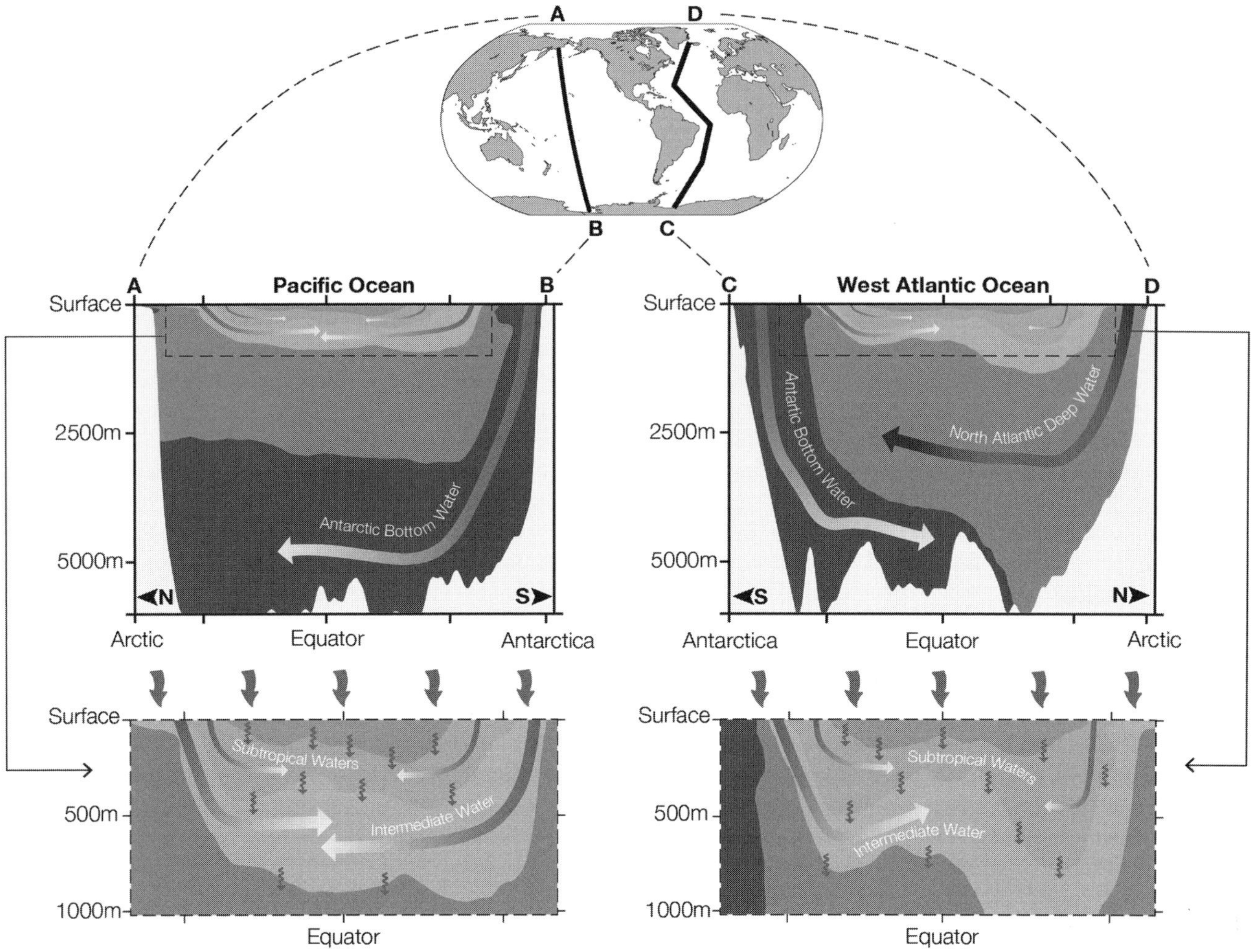

Figure 12.12 Ocean heat uptake pathways. Cold Antarctic Bottom Water (dark blue) sinks around Antarctica then spreads northward along the ocean floor into the central Pacific (upper left panel: red arrows fading to white indicate stronger warming of the bottom water most recently in contact with the ocean surface) and western Atlantic Oceans (upper right panel), as well as the Indian Ocean (not shown). Less cold, hence lighter, North Atlantic Deep Water (lighter blue) sinks in the northern North Atlantic Ocean (upper right panel: red and blue arrow in the deep water indicates decadal warming and cooling), then spreads south above the Antarctic Bottom Water. Similarly, in the upper ocean, cool Intermediate Waters (cyan) sink in sub-polar regions (red arrows fading to white indicating warming with time), before spreading toward the equator under warmer Subtropical Waters (green), which in turn sink (red arrows fading to white indicate stronger warming of the intermediate and subtropical waters most recently in contact with the surface) and spread toward the equator under tropical waters, the warmest and lightest (orange) in all three oceans. Excess heat or cold entering at the ocean surface (top curvy red arrows) also mixes slowly downward (sub-surface wavy red arrows). (Figure 1 in Rhein, Rintoul et al., 2013). For the color version, please refer to the plate section.

are being used to suggest circulation features that may be changing, such as the poleward shift and strengthening of the Southern Hemisphere westerlies and the ACC; but observed changes have not yet emerged to be significant above the background seasonal to interannual variability.

For changes over time scales of a century, there is effectively no observational record for changes in ocean circulation. There are suggestions of a fifty- to seventy-year cycle in Atlantic SST that has earned the label of Atlantic Multi-Decadal Oscillation (AMO), which is linked to supposed changes in the AMOC (Delworth and Mann, 2000). However, the accuracy of ocean observations becomes questionable for periods before World War II. Coupled climate models can and do project multi-decadal ocean circulation changes in a variety of forms (similar to the AMO–AMOC link); however, they cannot in any way be validated.

As indicated earlier, models do not represent many key small-scale ocean processes (such as mixing, flow over sills, deep water formation) in a realistic manner, and over century time scales small differences in representing these processes can accumulate, potentially leading to markedly different ocean circulations. As a result, model projections of ocean circulation changes over the hundred-year time scale have not been seriously analysed, with an exception being changes in the strength and structure of the AMOC. Over the next century, most coupled models project a slowdown but not a stoppage in the AMOC: there is a downward trend in the AMOC in the models that exceeds the interannual variability after about thirty years from the initialised state (Roberts et al., 2014). The accepted reason for the slowdown is that the warming and freshening of the northern Atlantic make the surface regions where deep water formation now occurs less dense and hence less capable of deep water formation. There is still concern that the models do not well represent the deep water formation process so it is possible that the sites of deep water formation will move poleward and still produce deep water of sufficient amounts to fuel the lower limb of the AMOC.

A potentially more serious issue for present-day coupled climate models is that there is no evidence in any of the models for a switch from one ocean circulation pattern to another substantially different one, as apparently happened regularly during the latter half of the last glacial period. It has been argued that present climate models are too stable in the sense that they cannot be run under paleo forcing scenarios to produce replicas of the paleo-circulation modes and the switches between them (Valdes, 2011). The models may be lulling us into a sense of security that only small changes in present-day ocean circulation will occur under CO_2 forcing.

The reasons for such stability may be subtle. For example, recent studies have suggested that the ocean freshwater transport by the AMOC determines whether the AMOC is bi-stable, that is, whether it can have two different circulation states for similar forcing (Dijkstra, 2007). If the AMOC transports freshwater southward out of the Atlantic, then a freshwater pulse in the north that shuts down deep convection and hence the AMOC cannot be exported out of the Atlantic without the AMOC; thus the AMOC shutdown persists until slow diffusive processes can restore the deep water formation. Most of the present coupled climate models have an AMOC that transports freshwater northward through the Atlantic, in contrast to the observed AMOC freshwater transport that is southward (Hawkins et al., 2011). Hence, it is argued that the models are too stable because the initialised ocean circulation has the wrong freshwater transport. Newer models with higher resolution are trying to repair this initialisation, and some models do now suggest that the ocean circulation and AMOC can suddenly shift into another state for gradual changes in freshwater forcing (Weaver et al., 2012).

Finally, we have little idea of what sets the strength and spatial extent of the THC and associated ocean stratification. How can we forecast the future circulation and stratification when we do not know why they have their present structure and strength? Some recent models are linking the THC and ocean stratification by suggesting that the wind stress over the Southern Ocean sets the amount of upwelling of North Atlantic deep water in the southern circumpolar region (Nikurushin and Vallis, 2012). The difference between the amount of deep water formed and the amount of wind-driven upwelling then changes the stratification throughout the global ocean. Other traditional models have argued that the rate of ocean mixing defines the amount of upwelling and strength of the stratification (Wunsch and Ferrari, 2005), in which case the representation of mixing in the climate models would be critical for accurate projections of century-scale changes in ocean circulation.

12.5 Summary

The pattern, structure and magnitude of the present ocean circulation are well known as the result of new global-scale, top-to-bottom measurements made during WOCE and their subsequent analysis. Unknowns are the size of water mass formation and the relation of critical processes to the strength and structure of the circulation over long time scales. These unknowns are critical for understanding past and future ocean circulations, when the radiative forcings are different.

The past ocean circulation is largely known from sediment core records of tracers, from which we have interpreted the spatial and temporal variability of ocean temperature in terms of differences from the present ocean circulation. There were clear shifts in different epochs and much work has been done to try to understand the causes. The major unknown is the past distribution of salinity (there is presently no tracer from which salinity can be reliably inferred), which is usually deemed critical for causing shifts in ocean circulation in the past.

To forecast the future ocean circulation, we are using coupled climate models and observations to document and understand changes from the present circulation over time scales of decades. Models and observations largely agree on patterns of warming, freshening and salinification, and on the changes in ocean circulation caused by changing winds. We are confident that discrepancies between model projections and observations over these time scales will be identified and will teach us more about ocean circulation processes. On the century time scale, where there is no observational record, model projections remain suspect because critical processes such as mixing and deep water formation are not (nor likely will be) well represented in numerical coupled climate models and the errors in their representation are likely to accumulate over long time scales. Finally, shifts between different ocean circulations, as observed in the paleo record, have not been found in coupled climate models as yet. What form the ocean circulation will take one hundred years from now is largely unknown.

Acknowledgements

The authors thank R. B. Alley, A. Ganopolski and D. M. Holland for valuable discussions on paleoclimate and ice sheet–ocean interactions, and J. Boeckh, G. McCarthy and D. Neelin for providing helpful comments on a first draft of this chapter.

References

Adkins, J. F. (2013). The role of deep ocean circulation in setting glacial climates. *Paleoceanography*, **28**, 539–561.

Alley, R. B., and Clark, P. U. (1999). The deglaciation of the Northern Hemisphere: a global perspective. *Annu. Rev. Earth Planet. Sci.*, **27**, 149–182.

Bamber, J. et al. (2012). Recent large increases in freshwater fluxes from Greenland into the North Atlantic. *Geophys. Res. Lett.*, **39**, L19501, doi: 10.1029/2012GL052552.

Bard. E. (2002). Climate shock: abrupt climate changes over millennial time scales. *Physics Today*, **55** (December), 32–38.

Bard, E., and Frank, M. (2006). Climate change and solar variability: what's new under the sun? Earth planet. *Sci. Lett.*, **248**, 1–14.

Beal, L. M., and Elipot, S. (2016). Decadal changes in the Agulhas Current. *Nature*, **540**, 570–573.

Bender, M. L. (2013). *Paleoclimate*. Princeton University Press, Princeton, NJ.

Bindoff, N. L., Willebrand, J., Artale, V., Cazenave, A., Gregory, J., Gulev, S., Hanawa, K., Le Quéré, C., Levitus, S., Nojiri, Y., Shum, S. K., Talley, L. D. and Unnikrishnan, A. (2007). Observations: Oceanic Climate Change and Sea Level. In *Climate Change 2007: The Physical Science Basis*. Contribution of Working Group I to the Fourth Assessment Report of the Intergovernmental Panel on Climate Change, edited by Solomon, S., Qin, D., Manning, M., Chen, Z., Marquis, M., Avery, K. B., Tignor, M. and Miller, H. L. Cambridge University Press, Cambridge.

Birchfield, G. E., Wang, H., and Rich, J. J. (1994). Century/millennium internal climate variability: an ocean-atmosphere-continental icesheet model. *J. Geophys. Res.*, **99**, 12459–12470.

Bond, G. C., and Lotti, C. (1995). Iceberg discharges into the North Atlantic on millennial time scales during the last glacial. *Science*, **267**, 1005–1010.

Bond, G. et al. (2001). Persistent solar influence on North Atlantic climate during the Holocene. *Science*, **294**, 2130–2136.

Broecker, W. S., Peteet, D. M., and Rind, D. (1985). Does the ocean-atmosphere system have more than one stable mode of operation? *Nature*, **315**, 21–26.

Broecker, W. S. et al. (1990). A salt oscillator in the glacial North Atlantic? 1. *The concept. Paleoceanography*, **5**, 469–477.

Bryden, H. L., and Imawaki, S. (2001). Ocean heat transport. In *Ocean Circulation and Climate*, edited by Siedler, G., Church, J., and Gould, J. Academic Press, Amsterdam, 455–474.

Calov, R., Ganopolski, A., Petoukhov, V., and Claussen, M. (2002). Large-scale instabilities of the Laurentide Ice Sheet simulated in a fully coupled climate-system model. *Geophys. Res. Lett.*, **29** (24), 69.1–69.4, doi:10.1029/2002GL016078.

Capotondi, A., Alexander, M. A., Bond, N. A., Curchitser, E. N., and Scott, J. D. (2012). Enhanced upper ocean stratification with climate change in the CMIP3 models, *J. Geophys. Res.*, **117**, C04031, doi:10.1029/2011JC007409.

Chidichimo, M. P., Donohue, K. A., Watts, D. R., and Tracey, K. L. (2014). Baroclinic Transport Time Series of the Antarctic Circumpolar Current Measured in Drake Passage, *J. Phys. Oceanogr.*, **44**: 1829–1853, doi:10.1175/JPO-D-13-071.1.

Clement, A. C., and Peterson, L. C. (2008). Mechanisms of abrupt climate change of the last glacial period. *Rev. Geophys.*, **46**, RG4002, doi:10.1029/2006RG000204.

Collins, M., Knutti, R., Arblaster, J., Dufresne, J.-L., Fichefet, T., Friedlingstein, P., Gao, X., Gutowski, W. J., Johns, T., Krinner, G., Shongwe, M., Tebaldi, C., Weaver, A. J. and Wehner, M. (2013): Long-term Climate Change: Projections, Commitments and Irreversibility. In *Climate Change 2013: The Physical Science Basis*. Contribution of Working Group I to the Fifth Assessment Report of the Intergovernmental Panel on Climate Change, edited by Stocker, T. F., Qin, D., Plattner, G.-K., Tignor, M., Allen, S. K., Boschung, J., Nauels, A., Xia, Y., Bex V. and Midgley, P. M. Cambridge University Press, Cambridge.

Cubasch, U., and Meehl, G. A. (2001). Projections for future climate change. *Climate Change 2001: The Scientific Basis*, edited by J. T. Houghton et al. Cambridge University Press, Cambridge, 525–582.

Curry, W. B., and Oppo, D. W. (2005). Glacial water mass geometry and the distribution of $\delta^{13}C$ of ΣCO_2 in the western Atlantic Ocean. *Paleoceanography*, **20**, PA1017, doi:10.1029/2004PA001021.

Dansgaard, W. et al. (1993). Evidence for general instability of past climate from a 250-kyr ice-core record. *Nature*, **364**, 218–220.

DeConto, R. M., and Pollard, D. (2016). Contribution of Antarctica to past and future sea-level rise. *Nature*, **531**, 591–597.

Delworth, T. L., and Mann, M. E. (2000). Observed and Simulated Multidecadal Variability in the Northern Hemisphere, *Clim. Dyn.*, **16**, 661–676.

Dijkstra, H. A. (2007). Characterization of the multiple equilibria regime in a global ocean model. *Tellus, A*, **59** (5), 695–705, doi:10.1111/J.1600-0870.2007.00267.x

Duplessy, J. et al. (1988). Deepwater source variations during the last climatic cycle and their impact on the global deepwater circulation. *Paleoceanography*, **3**, 343–360.

Durack, P. J., Wijffels, S. E., and Matear, R. J. (2012). Ocean salinities reveal strong global water cycle intensification during 1950 to 2000. *Science*, **336**, 455–458.

Fanning, A. F., and Weaver, A. J. (1997). Temporal-geographical meltwater influences on the North Atlantic conveyor: implications for the Younger Dryas. *Paleoceanography*, **12**, 307–320.

EPICA Community Members (2006). One-to-one coupling of glacial climate variability in Greenland and Antarctica. *Nature*, **444**, 195–198

Flückiger, J., Knutti, R., and White, W. C. (2006). Oceanic processes as potential trigger and amplifying mechanisms for Heinrich events. *Paleoceanography*, **21**, PA2014, doi:10.1029/2005PA001204.

Ganopolski, A., and Rahmstorf, S. (2001). Rapid changes of glacial climate simulated in a coupled climate model. *Nature*, **409**, 153–158.

Ganopolski, A., and Rahmstorf, S. (2002). Abrupt glacial climate changes due to stochastic resonance. *Phys. Rev. Lett.*, **88**, 038501-1-038501-4.

Gruber, N., Gloor, M., Mikaloff Fletcher, S. E., Doney, S. C., Dutkiewicz, S., Follows, M. J., Gerber, M., Jacobson, A. R., Joos, F., Lindsay, K., Menemenlis, D., Mouchet, A., Muller, S. A., Sarmiento, J. L. and Takahashi, T. (2009). Oceanic sources, sinks, and transport of atmospheric CO_2. *Global Biogeochem. Cycles*, **23**, GB1005, doi:10.1029/2008GB003349.

Hartmann, D. L., Klein Tank, A. M. G., Rusticucci, M., Alexander, L. V., Brönnimann, S., Charabi, Y., Dentener, F. J., Dlugokencky, E. J., Easterling D. R., Kaplan, A., Soden, B. J., Thorne, P. W., Wild, M. and Zhai, P. M. (2013). Observations: Atmosphere and Surface. In *Climate Change 2013: The Physical Science Basis*. Contribution of Working Group I to the Fifth Assessment Report of the Intergovernmental Panel on Climate Change, edited by Stocker, T. F., Qin, D., Plattner, G.-K., Tignor, M., Allen, S. K., Boschung, J., Nauels, A., Xia, Y., Bex V. and Midgley, P. M. Cambridge University Press, Cambridge.

Hawkins, E., Smith, R., Allison, L., Gregory, J., Woollings, T., Pohlmann, H., and de Cuevas, B. (2011). Bistability of the Atlantic overturning circulation in a global climate model and links to ocean freshwater transport. *Geophys. Res. Lett.*, **38**, L16699.

He, F. (2011). Simulating transient climatic evolution of the last deglacial cycle with CCSM3. Unpublished Ph.D. thesis, University of Wisconsin, Madison, WI 53706.

He, F. et al. (2013). Northern Hemisphere forcing of Southern Hemisphere climate during the last deglaciation. *Nature*, **494**, 81–85.

Heinrich, H. (1988). Origin and consequences of cyclic ice rafting in the northeast Atlantic Ocean during the past 130,000 years. *Quat. Res.*, **29**, 143–152.

Hemming, S. R. (2004). Heinrich events: massive late Pleistocene detritus layers of the North Atlantic and their global climate imprint. *Rev. Geophys.*, **42** (1), RG1005, doi:10.1029/2003RG000128.

IPCC (2013). Summary for Policymakers. In *Climate Change 2013: The Physical Science Basis. Contribution of Working Group I to the Fifth Assessment Report of the Intergovernmental Panel on Climate Change*, edited by Stocker, T. F., Qin, D., Plattner, G.-K., Tignor, M., Allen, S. K., Boschung, J., Nauels, A., Xia, Y., Bex V., and Midgley, P. M. Cambridge University Press, Cambridge.

Jackson, L. C., Peterson, K. A., Roberts, C. D., and Wood, R. A. (2016). Recent slowing of Atlantic overturning circulation as a recovery from earlier strengthening, *Nature Geoscience*, **9**, 518–522, doi:10.1038/ngeo2715.

Kageyama, M., Paul, A., Roche, D. M., and Van Meerbeeck, C. J. (2010). Modelling glacial climatic millennial-scale variability related to changes in the Atlantic overturning circulation: a review. *Quat. Sci. Rev.*, **29**, 2931–2956.

Kageyama, M. et al. (2013). Climatic impacts of fresh water hosing under Last Glacial Maximum conditions: a multimodel study. *Clim. Past*, **9**, 935–953.

Levang, S. J., and Schmitt, R. W. (2015). Centennial changes of the global water cycle in CMIP5 Models, *J. Phys. Oceanography*, **45**, 6489–6502.

Levitus, S., Antonov, J. I., Boyer, T. P., Baranova, O. K., Garcia, H. E., Locarnini, R. A., Mishonov, A. V., Reagan, J. R., Seidov, D., Yarosh, E. S., and Zweng, M. M. (2012). World ocean heat content and thermosteric sea level change (0–2000m) 1955–2010. *Geophys. Res. Lett.*, **39**, L10603.

Liu, Z. et al. (2009). Transient simulation of the last deglaciation with a new mechanism for Bølling-Allerød warming. *Science*, **325**, 310–314.

Lynch-Stieglitz, J. et al. (2007). Atlantic meridional overturning circulation during the Last Glacial Maximum. *Science*, **316**, 66–69.

Mamayev, O. I. (1975). *Temperature-Salinity Analysis of World Ocean Waters*. Elsevier, New York.

Manabe, S., and Stouffer, R. (1997). Coupled ocean-atmosphere model response to freshwater input: comparison to Younger Dryas event. *Paleoceanography*, 12, 321–336.

Marshall, S. J., and Clarke, G. K. C. (1997a). A continuum mixture model of ice stream thermomechanics in the Laurentide Ice Sheet 1. *Theory. J. Geophys. Res.*, **102** (B9), 20599–20613.

Marshall, S. J., and Clarke, G. K. C. (1997b). A continuum mixture model of ice stream thermomechanics in the Laurentide Ice Sheet 2. Application to the Hudson Strait ice stream. *J. Geophys. Res.*, **102** (B9), 20615–20637.

Marson, J. M., Wainer, I., Mata, M. M. and Liu, Z. (2014). The impacts of deglacial meltwater forcing on the South Atlantic Ocean deep circulation since the Last Glacial Maximum. *Clim. Past*, **10**, 1723–1734.

Marson, J. M., Mysak, L. A., Mata, M. M., and Wainer, I. (2015). Evolution of the deep Atlantic water masses since the last glacial maximum based on a transient run of NCAR CCSM3. *Clim. Dyn.*, doi:10.1007/s00382-015-2876-7.

McManus, J. F. et al. (2004). Collapse and rapid resumption of Atlantic meridional circulation linked to deglacial climate changes. *Nature*, **428**, 834–837.

Meinen, C. S., Baringer, M. O., and Garcia, R. F. (2010). Florida Current transport variability: an analysis of annual and longer-period signals. *Deep-Sea Res. Pt. I*, **57**, 835–846.

Menviel, L. et al. (2014). Atlantic-Pacific seesaw and its role in outgassing CO_2 during Heinrich events. *Paleoceanography*, **29**, 58–70.

Nikurashin, M., and Vallis, G. (2012). A theory of the interhemispheric meridional overturning circulation and associated stratification. *J. Phys. Oceanogr.*, **42**, 1652–1667.

Oka, A., Hasumi, H., and Abe-Ouchi, A. (2012). The thermal threshold of the Atlantic meridional overturning circulation and its control by wind stress forcing during glacial climate. *Geophys. Res. Lett.*, **39**, L09709, doi:10.1029/2012GL051421.

Papa, B. D., Mysak, L. A., and Wang, Z. (2006). Intermittent ice sheet discharge events in northeastern North America during the last glacial period. *Clim. Dyn.*, **26**, 201–216.

Peltier, W. R., and Vettoretti, G. (2014). Dansgaard-Oeschger oscillations predicted in a comprehensive model of glacial climate: a "kicked" salt oscillator in the Atlantic. *Geophys. Res. Lett.*, **41**, doi:10.1002/2014GL061413.

Purkey, S. G., and Johnson, G. C. (2010). Warming of global abyssal and deep Southern Ocean waters between the 1990s and 2000s: contributions to global heat and sea level rise budgets. *J. Clim.*, **23**, 6336–6351.

Rahmstorf, S. (1996). On the freshwater forcing and transport of the Atlantic thermohaline circulation. *Clim. Dyn.*, **12**, 799–811.

Rahmstorf, S. (2002). Ocean circulation and climate during the past 120,000 years. *Nature*, **419**, 207–214.

Rahmstorf, S., and Alley, R. B. (2002). Stochastic resonance in glacial climate. *Eos*, **83**, 129–135.

Rahmstorf, S. et al. (2005). Thermohaline circulation hysteresis: A model intercomparison. *Geophys. Res. Lett.*, **32**, L23605, doi:10.1029/2005GL023655.

Rahmstorf, S. et al. (2015). Exceptional twentieth-century slowdown in Atlantic Ocean overturning circulation. *Nature Climate Change*, **5**, 475–480.

Rhein, M., Rintoul, S. R., Aoki, S., Campos, E., Chambers, D., Feely, R. A., Gulev, S., Johnson, G. C., Josey, S. A., Kostianoy, A., Mauritzen, C., Roemmich, D., Talley, L. D., and Wang, F. (2013). Observations: Ocean. In *Climate Change 2013: The Physical Science Basis*. Contribution of Working Group I to the Fifth Assessment Report of the Intergovernmental Panel on Climate Change, edited by Stocker, T. F., Qin, D., Plattner, G.-K., Tignor, M., Allen, S. K., Boschung, J., Nauels, A., Xia, Y., Bex V., and Midgley, P. M. Cambridge University Press, Cambridge.

Roberts, C. D., Jackson, L., and McNeall, D. (2014). Is the 2004–2012 reduction of the Atlantic meridional overturning circulation significant? *Geophys. Res. Lett.*, **41** (9), 3204–3210, doi: 10.1002/2014GL059473.

Ruddiman, W. R. (2008). *Earth's Climate: Past and Future*, 2nd edition. W. H. Freeman, New York.

Saenko, O. A., Fyfe, J. C., and England, M. H. (2005). On the response of the oceanic wind-driven circulation to atmospheric CO_2 increase. *Clim. Dyn.*, **25**, 415–426.

Sarnthein, M. et al. (1994). Changes in east Atlantic deepwater circulation over the last 30,000 years: eight time slice reconstructions. *Paleoceanography*, **9**, 209–267.

Schneider von Deimling, T., Ganopolski, A., Held, H., and Rahmstorf, S. (2006). How cold was the last glacial maximum? *Geophys. Res. Lett.*, **33**, L14709, doi:10.1029/2006GL026484.

Siedler, G., Church, J., and Gould, J. (editors) (2001). *Ocean Circulation and Climate*. Academic Press, Amsterdam.

Smeed, D. A., McCarthy, G., Cunningham, S. A., Frajka-Williams, E., Rayner, D., Johns, W. E., Meinen, C., Baringer, M. O., Moat, M. I., Duchez, A., and Bryden, H. L. (2014). Observed decline of the Atlantic Meridional Overturning Circulation 2004 to 2012, *Ocean Science*, **10**, 29–38.

Stommel, H. (1961). Thermohaline convection with two stable regimes of flow. *Tellus*, **13**, 224–230.

Stouffer, R. J., Seidov, D., and Haupt, B. J. (2007). Climate response to external sources of freshwater: North Atlantic versus the Southern Ocean. *J. Climate*, **20**, 436–448.

Talley, L. D., Pickard, G., and Emery W. J. (2011). *Descriptive Physical Oceanography: An Introduction*. Academic Press, Amsterdam.

Valdes, P. (2011). Built for stability. *Nature Geoscience*, **4**, 414–416, doi:10.1038/ngeo1200.

Van Geel, B. et al. (1999). The role of solar forcing upon climate change. *Quat. Sci. Rev.*, **18**, 331–338.

Van Meerbeeck, C. J., Roche, D. M., and Renssen, H. (2011). Assessing the sensitivity of the North Atlantic Ocean circulation to freshwater perturbation in various glacial climate states. *Clim. Dyn.*, **37**, 1909–1927.

Voelker, A. H. et al. (2002). Global distribution of centennial-scale records for marine isotope stage (MIS) 3. *Quat. Sci. Rev.*, **21**, 1185–1214.

Wang, Z., and Mysak, L. A. (2001). Ice sheet-thermohaline circulation interactions in a climate model of intermediate complexity. *J. Oceanography*, **57**, 481–494.

Wang, Z., and Mysak, L. A.(2006). Glacial abrupt climate changes and Dansgaard-Oeschger oscillations in a coupled climate model. *Paleoceanography*, **21**, PA2001, doi:10.1029/2005PA001238.

Weaver, A. J., Saenko, O. A., Clark, P. U., and Mitrovica, J. X. (2003). Meltwater pulse 1A from Antarctica as a trigger of the Bølling-Allerød warm interval. *Science*, **299**, 1709–1713.

Weaver, A. J. et al. 2012: Stability of the Atlantic meridional overturning circulation: A model intercomparison. *Geophys. Res. Lett.*, **39**, L20709.

Wijffels, S. E., Schmitt, R. W., Bryden, H. L., and Stigebrandt, A. (1992). Transport of freshwater by the oceans. *J. Phys. Oceanography*, **22**, 155–162.

Wunsch, C. 2015. *Modern Observational Physical Oceanography*, Princeton University Press, Princeton, NJ.

Wunsch, C., and Ferrari, R. (2004). Vertical mixing, energy and the general circulation of the ocean. *Annu. Rev. Fluid Mech.*, **36**, 281–314.

Zhang, Xiao, Prange, M., Merkel, U., and Schulz, M. (2014). Instability of the Atlantic overturning circulation during Marine Isotope Stage 3. *Geophys. Res. Lett.*, **41**, doi:10.1002/2014GRL060321.

Zhang, Xu, Lohmann, G., Knorr, G., and Xu, X. (2013). Different ocean states and transient characteristics in Last Glacial Maximum simulations and implications for deglaciation. *Clim. Past*, **9**, 2319–2333.

Zhang, Xu, Lohmann, G., Knorr, G., and Purcell, C. (2014). Abrupt glacial climate shifts controlled by ice sheet changes. *Nature*, **290**, 290–294.

Zika, J. D., England, M. H., and Sijp, W. P. (2012). The ocean circulation in thermohaline coordinates. *J. Phys. Oceanogr.*, **42**, 708–724, doi:10.1175/JPO-D-11-0139.1.

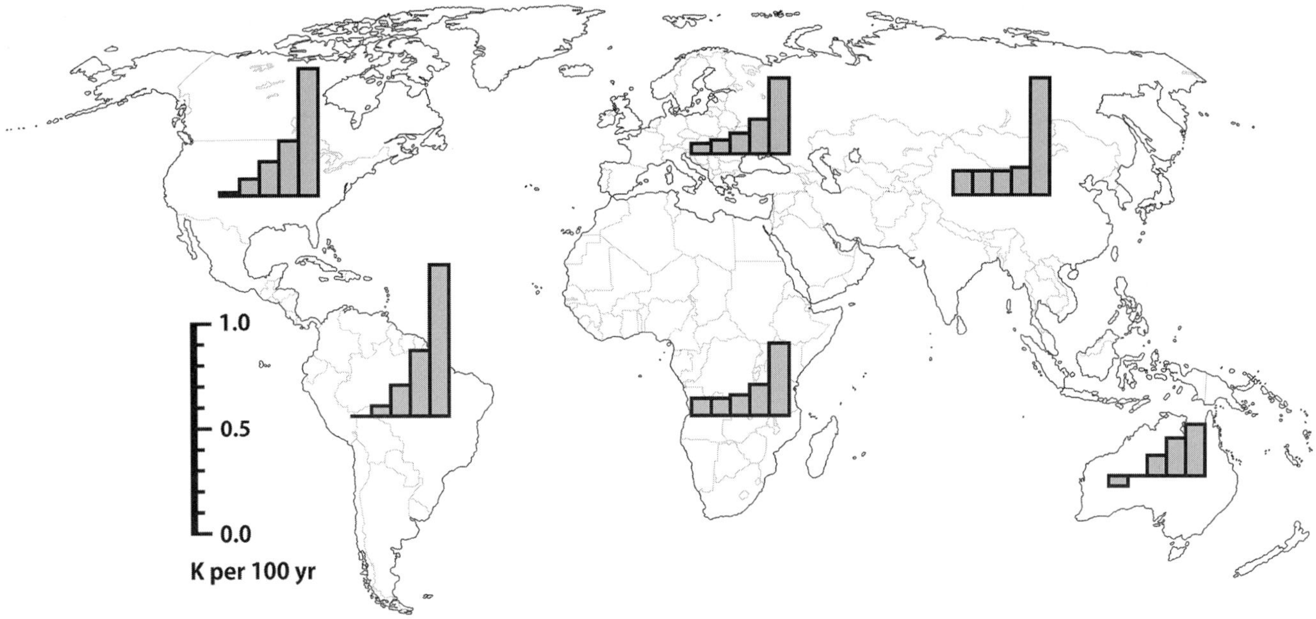

Figure 13.1 Reconstructions of surface temperature history from borehole temperature-depth profiles. Bar graphs show the magnitude of estimated surface warming during the last five centuries (from the sixteenth century on the left to the twentieth century on the right) (after Huang et al., 2000).

of the heat island effect due to urbanization (Taniguchi and Uemura, 2005). The effects of both of these two processes can have consequences on the underground water system. Subsurface temperatures in four Asian cities (Tokyo, Osaka, Bangkok, and Seoul) were compared, to evaluate the effects of subsurface warming due to urbanization and global warming and the relationship to the developmental stage of each city (Taniguchi et al., 2007). Mean surface warming in each city ranged between 1.8°C and 2.8°C as confirmed by air temperature changes during the last hundred years. The depth of the impact of surface warming measured as a departure from the regional geothermal gradient was found to be deepest at 140 m below the surface in Tokyo, followed by 80 m in Osaka, and 50 m in Seoul and Bangkok. This result shows that the effect of subsurface warming due to global warming and heat island was largest in Tokyo, followed by Osaka, Seoul, and Bangkok.

Subsurface warming means that there are increases in subsurface thermal storage in each Asian city. Taniguchi (2011), as mentioned, showed the change in subsurface thermal storage was greatest in Tokyo, followed by Osaka, Seoul, and Bangkok. Numerical analyses using one-dimensional heat conduction theory (Taniguchi et al., 2007) with different magnitudes and timings of the initiation of surface warming showed that the subsurface thermal storage was greater when the magnitude of surface warming was higher, and when the elapsed time from the start of surface warming due to urbanization was longer. This trend was confirmed by air temperature records from each study area during the last one hundred years (Taniguchi et al., 2007, Taniguchi, 2011).

Reconstructions of the surface warming history by use of subsurface temperatures have been made in rural, suburban, and urban areas of Bangkok (Yamano et al., 2009). The results show that the surface warming started earlier in the current urban area, followed by the current suburban area, and then the rural area. The magnitude of effects of subsurface warming was largest in urban followed by suburban and then rural areas. The heat island effect due to urbanization was preserved deeper if the timing of urbanization was earlier. Therefore, a record of the expansion of the city is preserved as heat island records in the subsurface environment.

13.4 Land-Ocean Interaction and Loads from Land to the Ocean in Asia

As shown in Section 13.2, land subsidence due to excessive groundwater pumping occurred in many Asia coastal cities, where alluvial depositions with clay layers were compacted after extraction of pore water from

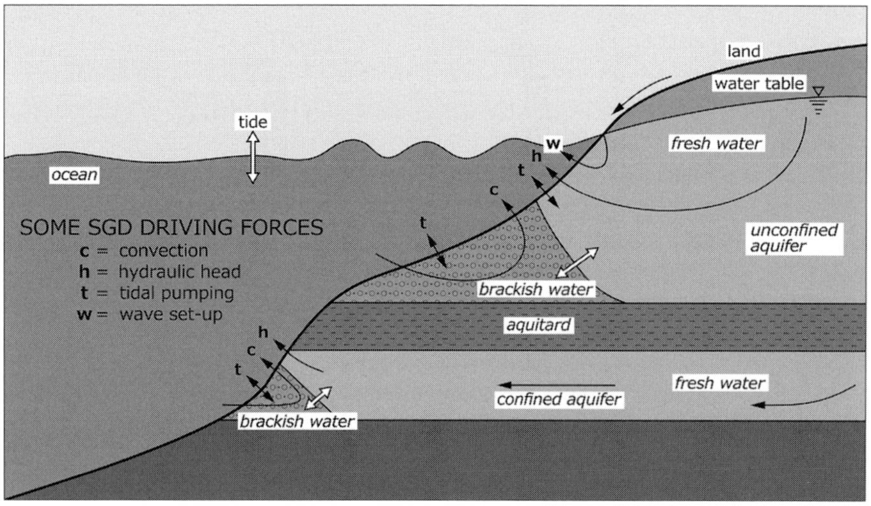

Figure 13.2 Submarine groundwater discharge in the coastal zone (after Taniguchi et al. 2002).

clay layers. Excessive groundwater use in coastal cities also caused another subsurface environmental problem in the form of saltwater intrusion in many Asian coastal cities. In general, water discharge from the land to the ocean occurs by river water and groundwater as submarine groundwater discharge (SGD; Figure 13.2); however, saltwater intrusion is in the opposite direction to the water movement, i.e., from the ocean to the land. Figure 13.3 shows a typical water exchange under the ground between land and ocean in Tokyo, Osaka, Bangkok, and Jakarta from the 1920s to 2000 (Taniguchi, 2011). As can be seen from Figure 13.3, net groundwater discharge from land to the ocean occurred through the period (1920s to 2000s) in Tokyo and Osaka. Although excessive groundwater pumping and land subsidence occurred during the 1960s and 1970s in Tokyo and Osaka, net water exchange underground was from land to the ocean.

even though excessive groundwater pumping and land subsidence during the 1960s and 1970s. On the other hand, groundwater discharge (land to the ocean) occurred in Bangkok until the 1980s; however, net groundwater exchange between land and ocean changed to saltwater intrusion (i.e., from the ocean to the land) from the 1980s. This happened also in Jakarta during the 1990s.

Nutrients and heavy metal contamination in each aquifer of Asian coastal cities were also evaluated. Compositions of nitrogen contamination of the groundwater in each city show that nitrate contamination dominated in Jakarta; on the other hand, ammonium contamination was dominant in Bangkok (Umezawa et al., 2009). Stable isotope ratios of N and C of groundwater can be used to indicate the

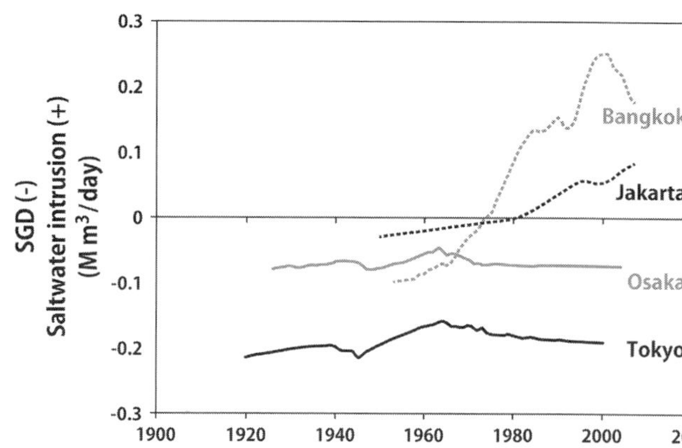

Figure 13.3 Net of groundwater exchange between land and ocean: positive shows saltwater intrusion and negative shows submarine groundwater discharge (after Taniguchi, 2011).

origin of the contamination in Manila, Bangkok, and Jakarta, such as domestic or farm origin, and denitrification processes in Bangkok (Umezawa, et al., 2009). Nitrite contamination was found in Jakarta and Manila, yet denitrification was found to occur in Bangkok even though loads of nitrogen are large. This is attributed to the natural conditions whether oxic or anoxic depending on the land slope (geomorphology), geology (volcanic or sediment), and hydroclimate conditions (Hosono et al., 2010).

Groundwater salinization was also found in Osaka, Bangkok, and Jakarta. The difference in marine alluvium volume (same as topographic gradient), natural recharge, and intensive pumping period controlled the degree of salinization. On the other hand, less terrestrial submarine groundwater discharge (SGD) but large material flux by total SGD was found in Asian coastal cities (Burnett et al., 2007). Spatial variations in SGD were estimated around each city, using topographic models and Rn measurements (Taniguchi, 2011).

We evaluated the relative vulnerability risk of the subsurface environment in Asian cities based on the accumulation of contaminants in the subsurface environment and that transport of pollutants. For instance, relatively higher risks of nitrate are found in Jakarta and Manila, and arsenic pollution was found in other cities, depending upon the redox state (Hosono, 2010). The pollution accumulation and transport were controlled by natural factors such as topography, climate, and geology as well as human impacts such as pumping rate and pollution load. Transport of material loads from surface to subsurface can be regulated. The transport of pollutants in Tokyo, Osaka, and Seoul increased at an early stage of urbanization; however, it turned into a decrease after implementation of environmental regulation such as lead-free gasoline. On the other hand, accumulation of contaminants kept increasing in all cities in both developed and developing countries.

13.5 Groundwater Depletion in Asia

Groundwater depletion has occurred rapidly and on a global scale (Gleeson et al., 2010), because of increases in groundwater abstraction due mainly to agricultural production. Groundwater abstraction of 312 km^3/year in 1960 increased to 734 km^3/year in 2010, and the groundwater depletion of 126 km^3/year accelerated to 283 km^3/year in 2010 (Wada et al., 2010), which is equivalent to a global decrease of 1.5 mm/year in the water column (one-sixth of global river discharge).

This groundwater depletion is particularly serious in Asia. According to the estimates of continental-scale groundwater depletion (Taylor et al., 2013), Asia had the highest degree of groundwater depletion using either flux-based or volume-based methods (Table 13.1) to measure it. The groundwater depletion in Asia measured by a flux-based method was 150 km^3/year, which is equivalent to 0.42 mm/year water of sea-level rise, and that by a volume-based method was 111 km^3/year, which is equivalent to 0.31 mm/year of sea-level rise. Groundwater depletion was largest in Asia at a continental scale, followed by North America, Africa, Europe, South America, and Australia.

The ecological footprint is an index to represent global sustainable development. Gleeson et al. (2013) showed the groundwater footprints (GFs) based on current groundwater use in aquifers (Figure 13.4). Groundwater footprint is defined as an area that is required for current use of groundwater. The groundwater is sustainable when the ratio of GF/Aa is less than 1, where Aa is the actual area of the aquifer. The global average of GF/Aa was estimated to be 3.5 (Gleeson et al., 2013); therefore, this is not a sustainable situation. In particular, at the upper Ganges, northern part of India, the GF/Aa was estimated to be more than 20, which was the largest in the world.

13.6 Transformation toward Sustainable Water Use in Asia

For sustainable water use in Asia, groundwater management is one of the keys for achieving sustainability. As a case study, Kumamoto city, Kyushu island, Japan, transformed toward sustainable water use. Kumamoto city with an area of 390 km^2 and population of 730,000 depends 100 percent on groundwater. Recent trends in the Kumamoto area during the last two decades showed decreases in groundwater level at the discharge area by

Table 13.1 *Estimates of Global and Continental Groundwater Depletion*

	Flux-based method		Volume-based method	
Region	Groundwater depletion ($km^3\ yr^{-1}$)	Sea-level rise ($mm\ yr^{-1}$)	Groundwater depletion ($km^3\ yr^{-1}$)	Sea-level rise ($mm\ yr^{-1}$)
World	204 ± 30	0.57 ± 0.09	145 ± 39	0.40 ± 0.11
Asia	150 ± 25	0.42 ± 0.07	111 ± 30	0.31 ± 0.08
Africa	5.0 ± 1.5	0.014 ± 0.004	5.5 ± 1.5	0.015 ± 0.004
North America	40 ± 10	0.11 ± 0.03	26 ± 7	0.07 ± 0.02
South America	1.5 ± 0.5	0.0042 ± 0.0014	0.9 ± 0.5	0.002 ± 0.001
Australia	0.5±0.2	0.0014±0.0006	0.4±0.2	0.001±0.0005
Europe	7±2	0.02±0.006	1.3±0.7	0.004±0.002

(After Taylor et al., 2013)

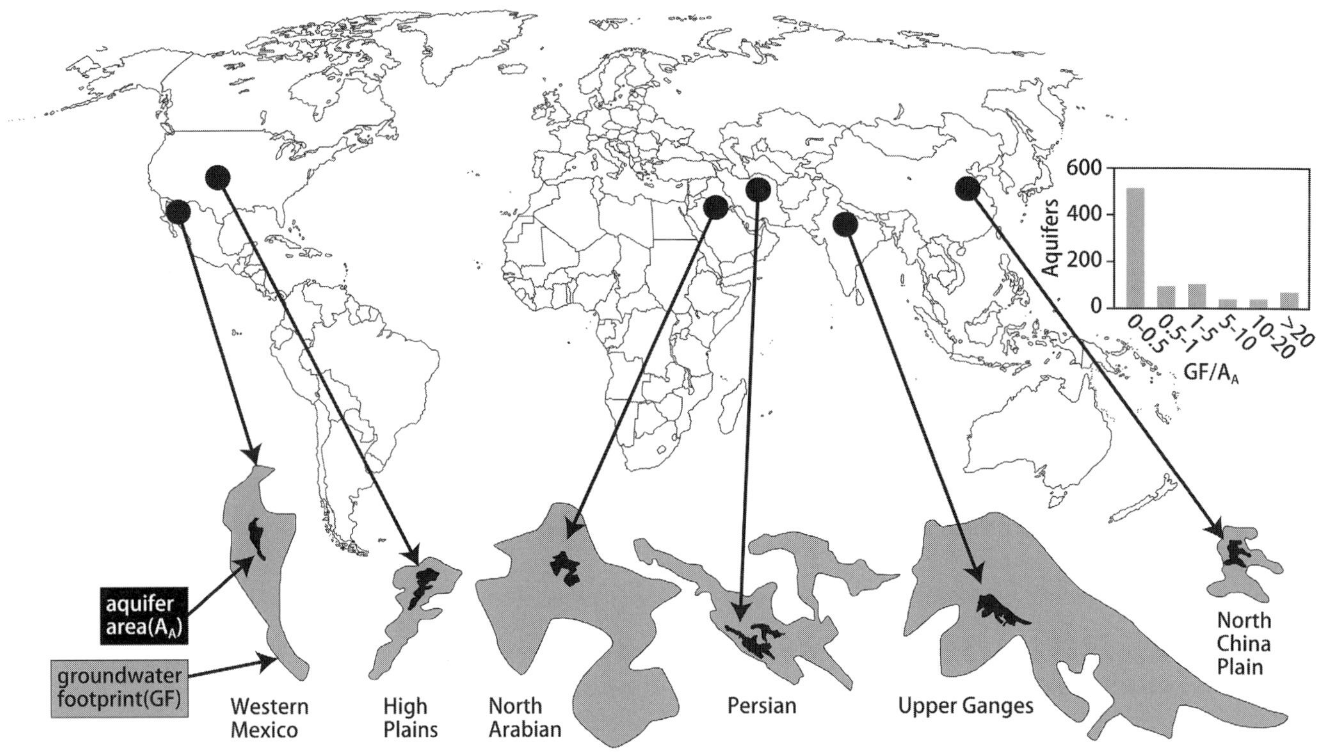

Figure 13.4 Global footprints of aquifers (after Gleeson et al., 2013).

0.5 m for fifteen years, a decrease in groundwater level at the recharge area by 5 m over twenty-five years, and decreases in spring water discharge at Lake Ezu, which is located in the groundwater discharge area, by 0.1 million ton over 10 years. All data showed groundwater depletion in Kumamoto. However, because the groundwater pumping rate also decreased by 40 million tons over 20 years, the reason for the groundwater depletion in Kumamoto was not an increase in groundwater pumping.

Integrated research on groundwater in Kumamoto included the analyses of long-term hydrometeorological data, land cover/use change data, and new hydrogeological data with a numerical simulation model. The results revealed that the decrease in groundwater recharge due to land use change was the main reason for their groundwater depletion (Shimada et al., 2012). This land use change was caused by the decrease in the area of paddy fields due to government regulation and urbanization. Irrigated water from rivers adjacent to the paddy fields allowed for groundwater recharge. The urbanized area in Kumamoto increased from 90 km^2 to 230 km^2 during the last thirty years. Impermeable layers due to urbanization decreased groundwater recharge. Therefore, the decrease of the area of paddy fields and the increase in urbanized areas in the groundwater recharge area eventually decreased the groundwater recharge rate.

To solve these problems, a wide-area administrative consortium on water management was established for not only Kumamoto city but also the surrounding seven cities within the groundwater recharge area of the Kumamoto groundwater system. Kumamoto city agreed to pay Kumamoto city's taxpayers' money to the farmers in the surrounding cities as a subsidy if they put water from the nearby Shira River in the unused paddy fields even if they did not grow rice. Kumamoto city paid 11,000 Japanese yen if farmers put water in the paddy fields for more than twenty-five days. They paid twice if the farmers put water in the field for more than eighty-five days.

This integrated wide-area administrative consortium worked very well, and groundwater recharge volume of 8,500,000 m^3 over an area of 220 ha/month in 2004 increased to 17,000,000 m^3 over an area of 500 ha/month in 2009. The spring water discharge from Lake Ezu and groundwater levels in Kumamoto have been increasing since 2006. This change of water management system toward sustainable use of water in Kumamoto is a successful example of transformation toward sustainability.

Another example of sustainability for water management in Asia is the water–energy–food nexus (Taniguchi et al., 2013). The water–energy–food nexus is one of the eight challenges and Knowledge Action Networks (KANs) in the Future Earth framework. Water, food, and energy are also listed in the seventeen Sustainable Development Goals (SDGs). Water is connected with energy and food as tradeoff and synergy. This tradeoff of resources may cause conflict between stakeholders, which can be prevented through knowledge acquired by transdisciplinary research including stakeholder analyses and participatory research. Production, transport, distribution, and consumption of the water–energy–food nexus are analyzed in the Asia-Pacific region by including self-sufficiency of water, energy, and food in each country and examining the diversity of each resource (Taniguchi et al., 2015) at a regional scale. Higher ratios of water and energy consumption for fisheries production were found in some Asian countries such as Japan, China, and Indonesia. Water footprints including water consumption for food production and water uses for energy production and energy footprints including energy consumption for water production and energy consumption for food production were also analyzed in the Asia–Pacific region as were food consumption for energy production (Taniguchi et al., 2018). These nexus analyses provide useful lessons for transformation to sustainability under various social systems with different regulations, laws, and lifestyles.

13.7 Conclusion

Groundwater usage in Asia accelerated during the Anthropocene because of Asian regional development producing increased demands for foods, industries, and changing individual lifestyles, as well as alternative water resource usage arising from accommodating climate change. Groundwater footprints in some aquifers in Asia exceeded a value of 1, which indicates unsustainable conditions. Decreases in groundwater storage and residence time due to excessive groundwater pumping cause a decrease of resilience for the society, and water security, in terms of self-sufficiency and diversity of the water resources, is diminished.

Interdisciplinary studies on groundwater in Asia show changes in the subsurface environment during the Anthropocene by combining geodesy using GRACE satellite data, using geothermal studies with reconstructions of surface temperature changes due to global warming and urbanization from borehole temperature-depth profiles, and using oceanography with groundwater discharge into the ocean. Those examples are parts of the themes of the dynamic planet and global development in the Future Earth framework because rapid changes in the economy and environment of Asia occurred during the Anthropocene, and thus Asia is a hot spot of global environment changes.

Case studies of transformation toward sustainable water use in Asia are shown in Kumamoto area as well as the water–energy–food nexus in the Asia-Pacific region. A wide-area administrative consortium on integrated water management of the Kumamoto aquifer system was effective with subsidence payments from Kumamoto city to the farmers outside Kumamoto city for water introduction to the unused paddy fields for groundwater recharge. Analyses of the water–energy–food nexus are useful for transformation toward sustainability as a result of a changing social system.

References

Burnett, W. C., Wattayakorn, G., Taniguchi, M., Dulaiova, H., Sojisuporn, P., Rungsupa, S., and Ishitobi, T. (2007). Groundwater-derived nutrient inputs to the Upper Gulf of Thailand. *Cont. Shelf Res.* **27**, 176–190.

Gleeson, T., VanderSteen, J., Sophocleous, M. A., Taniguchi, T., Alley, W. M., Allen, D. M., and Zhou, Y. (2010). Groundwater sustainability strategies. *Nature Geoscience*, 3, 378–379.

Gleeson, T., Wada, Y., Bierkens, M. F. P., and van Beek, L. P. H. (2013). Sustainable water balance of global aquifers revealed by groundwater footprint. *Nature*, 488, 197–200

Huang, S., Pollack, H. N., and Shen, Po-Yu. (2000). Temperature trends over the past five centuries reconstructed from borehole temperatures, *Nature*, 403, 756–758.

Hosono, T. (2010). The nitrate-arsenic boundary as an important concept in aquatic environmental studies. Taniguchi M., and Shiraiwa T., eds., *The Dilemma of the Boundary*. Springer.

Kagabu M, Shimada, J., Delinom, R., Tsujimura, M., and Taniguchi, M. (2011). Groundwater flow system under a rapidly urbanizing costal city determined by hydrogeochemistry, *J. Asian Earth Sci*,. 40, 226–239.

Knorr, W., Prentice, I. C., House, J. I., and Holland E. A. (2005). Long-term sensitivity of soil carbon turnover to warming, *Nature*, 433: 298–301.

Moss R. H. et al. (2010). The next generation of scenarios for climate change research and assessment, *Nature*, Vol 463 (11) February 2010, doi:10.1038/nature08823.

Shimada, J., Ichiyanagi, K., Kagabu, M., Saita, S., and Mori, K. (2012). Effect of artificial recharge using abandoned rice paddies for the sustainable groundwater management in Kumamoto, Japan. *World Environmental and Water Resources Congress 2012: Crossing Boundaries, Proceedings of the 2012 Congress.* 59–69.

Taniguchi, M., Burnett, W. C., Cable, J. E, and Turner, J. V. (2002). Investigation of submarine groundwater discharge, *Hydrol. Process.*, 16, 2115–2129.

Taniguchi, M., and Uemura, T. (2005). Effects of urbanization and groundwater flow on the subsurface temperature in Osaka, *Japan, Physics of Earth and Planetary Interior*, 152: 305–313.

Taniguchi, M., Uemura T., and Jago-on K. (2007). Combined effects of urbanization and global warming on subsurface temperature in four Asian cities, *Vadose Zone Journal*, 6: 591–596.

Taniguchi, M, Burnett, W. C, and Ness G. D. (2009). Integrated research on subsurface environments in Asian urban areas, *Sci. Total Environ.*, doi:10.1016/i.scitotenv.2009.02.002.

Taniguchi M. (2011). Groundwater and subsurface environmental assessments under the pressures of climate variability and human activities in Asia. Taniguchi, M. ed., *Groundwater and Subsurface Environments – Human Impacts in Asian Coastal Cities*. Springer, 3–18.

Taniguchi, M., Allen, D., and Gurdak, J. (2013). Optimizing the water-energy-food nexus in the Asia Pacific ring of fire, *EOS*, vol. 94, 435, doi:10.1002/2013EO470005.

Taniguchi, M., Masuhara, N., and Burnett, K. (2015): Water, energy, and food security in the Asia Pacific region. *Journal of Hydrology: Regional Studies*, doi:10.1016/j.ejrh.2015.11.005.

Taniguchi, M, Masuhara, N, and Teramoto, S. (2018). Trade-offs in the water-energy-food nexus in the urbanizing Asia Pacific region, *Water International* (in press).

Taylor, RG, Scanlon, B., Döll, P., Rodell, M., van Beek, R., Wada, Y., Longuevergne, L., Leblanc, M., Famiglietti, J. S., Edmunds, M., Konikow, L., Green, T. R., Chen, J., Taniguchi, M., Bierkens, M. F. P., MacDonald, A., Fan, Y., Maxwell, R. M., Yechieli, Y., Gurdak, J. J., Allen, D. M., Shamsudduha, M., Hiscock, K., Yeh, P. J.-F., Holman I., and Treidel, H. (2013). Groundwater and climate change. *Nature Climate Change*. DOI:10.1038/nclimate1744.

Umezawa, Y., Hosono, T., Onodera, S., Siringan, F., Buapeng, S., Delinom, R., Jago-on, K. A., Yoshimizu, C., Tayasu, I., Nagata, T., and Taniguchi, M. (2009). The characteristics of nitrate contamination in groundwater at developing Asian-Mega cities, estimated by nitrate $d^{15}N$ and $d^{18}O$ values. *Sci. Total Environ.* **407**, 3219–3231.

Wada, Y., van Beek, L. P. H., van Kempen, C. M., Reckman, J. W. T. M., Vasak, S. (2010) Global depletion of groundwater resources, *Geophysical Research Letters* 37 (20)

Yamamoto K, Fukuda Y, Nakaegawa T, and Nishijima J. (2009). Landwater variation in four major river basins of the Indochina peninsula as revealed by GRACE, *Earth Planets Space*, 59, 193–200.

Yamano, M., Goto, S., Miyakoshi, A., Hamamoto, H., Lubis, R. F., Monyrath, V., Kamioka, S., Huang, S., and Taniguchi, M. (2009). Study of the thermal environment evolution in urban areas based on underground temperature distributions, *Sci. Total Environ.*, 407(9), 3120–3128.

14

Africa's Broken Food Systems: Unravelling the Hidden Fortune under Climate Change

Richard Munang and Robert Mgendi

14.1 Background and Current Status of Affairs

In Africa, agriculture is not only a source of food, but also of livelihoods, employing up to 64% of the population (Calestos, 2011). Women produce up to 80% of the food (Jeanne & Andrew, 2002). This implies that performance of this sector has direct socio-economic ramifications. While this is the case, the sector's productivity is relatively low. Though being the major source of employment for an average 64%, it contributes a lowly 25%–34% to continental GDP. Agriculture productivity in Africa considerably lags other developing regions (*Africa Competitiveness Report*, 2015).

This low productivity scenario is exacerbated by high inefficiencies along the entire agro-value chain. These manifest as postharvest losses, degraded ecosystems resulting in lost potential food and overall productivity threatened by climate change. The 2015 Africa Competitiveness Report observes that Africa imports food staples valued at about US$25 billion annually, essentially because food production, supply and consumption systems are not functioning optimally. In 2011, the region's total food imports excluding fish were valued at US$35 billion.[1] Under status quo, food importation is projected to exceed $110 billion by 2025.[2]

On postharvest losses (PHLs), low value addition means Africa's average annual cereal grain losses are high (*Africa Competitiveness Report*, 2015). These are estimated at US$4 billion annually,[3] enough grain[4] to feed an extra 48 million people for a year (Formo et al., 2014). PHLs in sub-Saharan Africa average 30% of total production (*Africa Competitiveness Report*, 2015). In 2010, the estimated cumulative PHLs of cereals, roots and tubers, fruits and vegetables, meat, milk and fish in Africa were about 100 million tonnes with total value of US$48 billion.[5] When juxtaposed with Africa's US$35 billion food import bill in 2011, recovering these losses would essentially eliminate the need for imports without increasing production while injecting an extra US$35 billion to capitalize other sectors of the continent's economy.

On degraded ecosystems, sub-Saharan Africa's (SSA's) food loss due to agro-ecosystem degradation is estimated as high as 6.6 million tonnes of grain annually, enough to meet annual calorific needs of approximately 31 million people (Munang et al., 2015). Land and ecosystems degradation in sub-Saharan Africa is estimated at $68 billion annually.[6] It is noteworthy that healthy ecosystems are the foundation of long-term productivity underpinning food production through ecosystem goods and services such as water, soils and pollinators. For instance, insect pollination by bees is an ecosystem service that is necessary for 75%[7] of all crops used as human food. Increasing the quantity and variety of pollinating insects can increase crop yields by more than 20%.[8]

On climate change, the 2015 Africa Adaptation Gap Report[9] observes that for a below-2°C global warming scenario the agriculture sector will be hit by up to 40% yield declines, resulting in a 25%–90% increase in incidences of undernourishment, not to mention economic losses. According to this report by Schaeffer et al., (2015),[9] the globe is on track for a warming of around 3–3.5°C by 2100 with more thana 66% confidence level, implying impacts could be worse. Adapting to climate change is therefore an imperative to safeguard Africa's future food security. The costs will be no less than $50 billion annually by 2050 and the pace of international support does not reflect this continental urgency.

The 2015 *Africa Competitiveness Report* notes that the below par performance in agriculture undermines poverty reduction and inclusive growth in Africa. There is therefore an urgent imperative to optimize productivity of Africa's agriculture, by eliminating these inefficiencies.

14.2 Potentials for Agricultural Transformation

Africa has adequate potential to transform its agriculture and optimize productivity. Africa holds 65% of the world's arable land[10] and 10% of renewable internal freshwater resources.[11] With a growing middle class currently at an estimated 300 million people,[12] the African food market alone is projected to grow to US$150 billion by 2030. If harnessed, the entire agriculture and agro-business sector is projected to grow to be worth an estimated US$1 trillion by 2030,[13] thus enhancing agriculture's contribution to Africa's GDP. When optimized, growth in agriculture is at least two to four times more effective[14] in reducing poverty than growth in other sectors. Agricultural growth also stimulates productivity in other sectors, e.g. processing and transport, whose value chains link with the agro-chain, resulting in economy-wide impacts.

Furthermore, in Africa, a 10% increase in crop yields translates to approximately a 7% reduction in poverty. Neither the manufacturing nor services sector can achieve an equivalent impact.[15] This, coupled with its inclusivity, indicates its unique potential for enhancing inclusive economic growth for substantial poverty reduction.

Optimizing food production by embracing Ecosystems Based Adaptation approaches (EBA) that work with nature will restore ecosystems to reverse land degradation and hence enhance recovery of the $68 billion currently lost due to degradation, resulting in yield increases of up to 128%[16] and accompanying farmer income increases, improved nutrition, enhanced capacity of ecosystems and enhanced community climate resilience. Linking these ecological approaches to supply and demand side value chains will cut PHLs. Embracing off-grid clean energy for processing along this chain will cut emissions and cumulatively these downstream value chain activities will spur additional income and business opportunities along the entire agro-value chain, to create potentially as many as 17 million sustainable jobs annually[17] towards combating poverty. Thus optimizing the entire agro-value chain in a continuum should be the focus of transformative action in Africa's agriculture.

14.3 Solutions I – Key Players and Partnerships

Enhancing productivity of Africa's agriculture by focusing on optimizing the entire value chain in a continuum holds the key to a transformative sector capable of delivering environmentally sustainable, socially inclusive growth for the continent.

This vital truth has been acknowledged in pivotal policy declarations and development blueprints at the continent level led by the African Union (AU) and at the global level through the Sustainable Development Goals (SDGs).

Among these are the Maputo[18] and Malabo[19] Declarations and related Vision 2025 for Africa's agriculture, a commitment by the AU heads of state and government to end hunger, halve poverty in Africa by 2025 and reduce postharvest losses by 50%. Also included is the Comprehensive African Agriculture Development Programme (CAADP) implementation strategy,[20] which recognizes EBA approaches as among key strategies to enhance agro-productivity in Africa. Another is the AU Agenda 2063,[21] which recognizes agriculture as the means to achieve inclusive sustainable development in the continent. It also underscores the need for gender parity and enhancing women's agro-productivity through access to financing, for instance, where it calls for dedication of 30% of agricultural financing to women. It also underscores the need to achieve more than 50% clean energy, which will be vital in modernizing and transforming Africa's agriculture in a sustainable way.

In addition, the 2015 African Ministerial Conference on the Environment (AMCEN), through the Cairo Declaration,[22] underscored the importance for Africa of sustainably leveraging its natural capital and ecosystems for inclusive growth and poverty alleviation.

An overriding theme implied in these blueprints is the need for modernization and optimization of Africa's agriculture while ensuring that the productivity of ecosystems that underpin agricultural productivity are safeguarded for future generations.

At the global level, the Agenda 2030 and the SDGs align with these noble continental aims. Specifically, SDG 1 aims for poverty eradication. SDG 2 aims to end hunger, achieve food security and improved nutrition and promote sustainable agriculture, with targets to this effect to be achieved by 2030. SDG 13 calls for action to combat climate change. SDG 15 calls for sustainable management and restoration of ecosystems. SDG 5 aims to empower women including in agriculture, where they produce up to 80% of the food.

These development blueprints and policy declarations provide the first principles and theoretical solutions to transforming Africa's agriculture. Their implementation constitutes an impactful practical solution.

14.4 Solutions II – Climate Action as an Opportunity to Mend Africa's Broken Food Systems

Optimizing agriculture productivity cannot be considered in isolation from climate change and ecosystem degradation.

Linking farm EBA approaches to clean energy for processing and other post-farm-gate commercial value chains, especially efficient linkage to supply and demand markets, e.g. by leveraging at scale available Information and Communication Technology (ICT) based marketing tools such as the M-Farm[23] mobile application and access to agro-value chain financing in a continuum, can potentially ensure Africa actualizes sustainable agro-industrialization, simultaneously unlocking socio-economic benefits of food security and job and income opportunities along the entire agro-value chain while actualizing crucial climate and environmental action.

14.4.1 Basis to Leverage Opportunities in the Paris Deal to Actualize Agro-Industrialization

In light of the Paris Agreement, such an approach will satisfy provisions of the agreement, including Article 7 on adaptation, given that EBA is a climate adaptation technique; Article 2, the overarching purpose of the agreement on enhancing climate resilience and in the context of food security, esp. 2(1b) on low emissions, climate resilient food production; Article 4 on mitigation through both enhanced sinks (e.g., through EBA approaches like agro-forestry) and reduced sources (e.g., bridging the energy gap by tapping into Africa's significant renewable/clean energy potential, prioritizing clean energy powered processing, ICT based market linkages instead of paper processes). This approach will also satisfy Article 3 on Nationally Determined Contributions (NDCs), given that with negligible industrial emissions, a majority of Africa's NDC commitments target land-based actions, aiming to reduce land-based emissions. This can be achieved through reversing degradation (through EBA approaches such as re-forestation, Farmer Managed Natural Regeneration and agroforestry), enhancing carbon sequestration and sinks (through EBA approaches of conservation agriculture, including minimum tillage, Farmer Managed Natural Regeneration, agroforestry).

14.4.1.1 Fulfilling the Agenda 2030 and Paris Agreement through Sustainable Agro-Industrialization

Integrating EBA as mainstream techniques for on-farm production will contribute to fulfilling Article 7 on adaptation and SDG 13 on climate action. EBA will also boost food security with up to 128% yield increases hence contribute to SDG 2, while enhancing farmer incomes to combat poverty (SDG 1) and enhancing capacity of ecosystems to continue providing ecosystem goods and services that enable communities to adapt to climate change (SDG 13 and SDG 15). On mitigation, EBA approaches such as agro-forestry or farmer managed natural regeneration (FMNR) will enhance carbon sinks. For example, based on inference, one large-scale forest regeneration project of 25,000 Ha can ensure a country sequesters up to 15.6 million tons of CO_2 (Rohit, Brent, & John, 2006). In addition, conservation techniquessuch as zero-tillage will minimize sources, cumulatively contributing to Article 4. Furthermore, under Article 6(4) of the Paris Agreement, countries can qualify for Certified Emissions Reduction (CER) incentives.

On value addition and agro-industrialization, linking on-farm production to ICT based market and supply chain linkages that have a lower carbon print relative to traditional paper processes and physical travel and clean energy powered value addition will minimize greenhouse gas emission sources and thus comply with Article 4, and SDG13 (solar powered irrigation pump in place of diesel-powered pumps can sequester up to 3,000,000 tCO_2 equivalent by 2030)[24] while catalyzing agro-industrialization to create up to 17 million jobs along the entire value chain and catalyzing an agro-sector worth US$1 trillion by 2030.[25] This will contribute directly toward SDG 1 to combat poverty and SDG 2 on food security and catalyze SDGs 7and 8 that enhance high-quality jobs and structural transformation. EBA-driven agriculture is also known to result in food with more immune boosting compounds (Kirsten and Jens, 2001). On the other hand, clean energy options reduce pollution, all contributing to SDG 3 on good health, while also enhancing health of ecosystems, thus contributing to SDG 15 on reversing degradation.

Sustainably financing climate action: Simple EBA techniques applied judiciously will enhance Africa's recovery of the $68 billion lost annually to land degradation, which could be re-invested to voluntary adaptation finance as called for under Article 9(2) of the Paris Agreement. This is an innovative angle that will wean Africa from dependency on unpredictable international sources alone and fast track the establishment of a more balanced climate financing model, a composite combining both international and domestic sources. This is advocated for by the UNEP Inquiry into sustainable climate financing (UNEP, 2016).

Most importantly, this will demonstrate globally how the region prioritizes adaptation; hence it will provide an objective basis to lay claims under Article 9, for additional international support.

These EBA techniques have already proven their practicability.

For example, Mali has restored 500,000 hectares over a decade through using simple FMNR (Reij and Winterbottom, 2015). The same technique in Senegal[26] has regenerated indigenous trees on 40 000 hectares of cropland. This ecological technique has increased tree density on cropland from an average of four to thirty-three trees per hectares and improved soil fertility, crop yields, and wildlife, and reduced soil erosion, while in Ethiopia,[27] it has restored 2 700 hectares of barren mountain terrain.

Through application of a simple indigenous soil and water conservation technique known as Zai,* Niger rehabilitated up to 300 000 hectares of its crusted and barren soil(Godfray et al., 2010).

14.4.2 Tapping into COP21 Opportunities

By virtue of directly fulfilling adaptation and mitigation provisions of the Paris Agreement, EBA based agro-industry powered by clean energy as discussed provides Africa the *locus standi* to leverage the Paris Agreement towards optimizing agro-productivity and mending its broken food systems.

Specific provisions to be leveraged include especially Articles 9, 10 and 11, which mandate that developed countries support developing countries with both financial and non-financial means of implementation to meet their obligations under the deal. In this case, these could be directed at tapping into vast renewable energy potential to bridge the energy gap and power agro-industries, and to implement large-scale EBA-driven agriculture interventions. For instance, the Paris process secured commitments by a global alliance to mobilize up to $1 trillion to finance[28] renewable energy development. The Green Climate fund (para 59, section III) is another opportunity that can be unlocked through these alignments. These are good starting points.

14.4.3 The Ecosystems Based Adaptation for Food Security Assembly (EBAFOSA)– Delivery Pathway

Alignment to fulfilling provisions of the Paris deal is by no means enough. In practice, unlocking this potential will require broad participation – involving both state and non-state actors as advocated under section V of the Paris Agreement and building mutual partnerships to unlock the critical means of implementation as called for in SDG17. Actualizing this collective effort through country driven but continentally synergized and globally supported efforts is what African Ministers of Environment have committed to doing. This is demonstrated by the adoption of the Sixth Africa Ministerial Conference on the Environment (AMCEN) Cairo Declaration,[29] which recognized the pan-African Ecosystems Based Adaptation for Food Security Assembly[30] as a priority policy–action framework for bridging implementation gaps towards actualizing sustainable agro-industry in Africa, simultaneously ensuring socio-economic priorities of food security, job creation and macro-economic growth are achieved alongside taking climate action, hence actualizing multiple SDGs. This mandate was endorsed by ministers under AMCEN decision 1/SS6.[31]

The innovativeness of EBAFOSA is in its inclusiveness, which convenes cross-cutting global, regional and national stakeholders based in a country to foment mutual partnerships that leverage their respective strengths to bridge gaps in policy processes, financing, technology, techniques, skills, markets, commercialization and continental exchange of innovative practices towards a common aim: policy and ground actions to actualize sustainable agro-industrialization capable of simultaneously satisfying climate obligations hence leveraging opportunities and provisions under the COP21 deal while also ensuring socio-economic benefits of food security and contribution to macro-economic growth and creation of income and job opportunities, including high-quality industrial based jobs for structural transformation, poverty reduction and inclusive growth. EBAFOSA strengths are also in its country-drivenness to enhance ownership, and building on established and planned initiatives at both policy and ground actions by the multiple stakeholders to ensure harmonized, complementary progress for optimized outcomes. This is unlike implementing initiatives in siloes, where aggregate outcomes will be sub-optimal.

Example of EBAFOSA actions to bridge gaps:

Policy gaps: Within government, EBAFOSA is bridging policy gaps by breaking inter-ministerial siloes through convening requisite ministries of environment, agriculture, industry energy, finance etc., for collaborative, enabling policies for:

- agriculture policies to incorporate EBA-based on-farm production techniques of conservation agriculture: agro-forestry, farmer managed natural regeneration, etc., into mainstream

* a traditional land rehabilitation technology "invented" by farmers in Burkina Faso to rehabilitate degraded dry lands and to restore soil fertility to the benefit of farmers living in dry lands (https://www.cbd.int/doc/case-studies/tttc/tttc-00163-en.pdf).

agriculture practices to ensure simultaneous actualization of both food security and environmental and climate requirements of reforestation, enhancing carbon sinks, ecosystems preservation and climate adaptation.
- targeted clean energy investment in rural areas – decentralized transmission renewable plants to power rural agro-industry.
- ensuring of prioritized investment in rural transport, i.e. feeder roads in farm areas to facilitate efficient linkage of produce to markets, hence commercialization of sustainable agro-industry.
- trade/industry policies to create special economic zones aimed at attracting investors to set up clean energy based agro-industries in rural areas where farmers are.

Human capacity building gaps: EBAFOSA is targeting peer-to-peer learning among cross-cutting stakeholders across the forty national branches in Africa by coordinating through the national secretariat, transfer of innovative practices and lessons at both strategic policy and operational levels from exemplary EBAFOSA national branches for replication. This enhances skills improvement across the continent, which is a key step in engineering a structural transformation.

Institutional capacity building: EBAFOSA action is country driven with country institutions and personnel taking the lead in implementation, while EBAFOSA secretariat and partners coordinate sharing and transfer of innovative practices and provide technical back-stopping. This facilitates institutional capacity building through transfer of globally and continentally sourced applicable innovative practices and technologies.

Technology gaps: EBAFOSA is targeting south–south partnerships with technology leaders, e.g., Chinese companies on renewable energy, leading private sector in an EBAFOSA country (e.g., Kenya on decentralized renewable); exemplary and accessible innovative resilience EBA technologies in an EBAFOSA country, e.g., Zai in Burkina Faso, for transfer to demand areas in all the EBAFOSA national branches. Leveraging at scale application of available ICT based agro-market linkage tools, e.g., M-Farm in Kenya instead of paper processes and travelling, a cheaper option that also simultaneously cuts emissions.

14.5 Conclusion

A crucial way to enhance sustainability of climate actions in Africa is to link them directly to socio-economic priority areas of countries – food security, job creation, entrepreneurial opportunities, macro-economic growth, as opposed to climate actions siloed as environmental obligations. This is by targeting climate actions at catalytic sectors capable of simultaneously unlocking socio-economic development opportunities and meeting climate objectives. EBA-driven agriculture and its linkage to clean energy and other commercial value chains towards actualizing sustainable agro-industrialization have been argued as the right approach for Africa. Such an approach to climate actions beyond environmental obligation enhances participation of state and non-state actors including the private sector, academia, NGOs, CSOs, ground actors and individual citizens as called for under Section 5 of the Paris Agreement, who are critical to mobilizing means of implementation. This is the basis for mending Africa's broken food system under the changing climate.

The views expressed here are those of the authors and do not necessarily represent those of the institution with which they are affiliated.

Internet Resources

[1] Africa Progress Report (2014). *Grain Fish Money, Financing Africa's Green and Blue Revolutions.* Retrieved from www.afdb.org/fileadmin/uploads/afdb/Documents/Project-and-Operations/Africa_Progress_Report_2014.PDF

[2] Johm, K.B. et al (2016). *Feed Africa: Strategy for Agricultural Transformation in Africa 2016–2025.* African Development Bank. Retrieved from www.afdb.org/fileadmin/uploads/afdb/Documents/Generic-Documents/Feed_Africa-_Strategy_for_Agricultural_Transformation_in_Africa_2016-2025.pdf

[3] Nomathemba, M. et al (2010). *FAO/World Bank workshop on reducing post-harvest losses in grain supply chains in Africa.* Retrieved from www.fao.org/fileadmin/user_upload/ags/publications/FAO_WB_ph_web.pdf

[4] FAO (2013). *Food Wastage Footprints.* Retrieved from www.fao.org/fileadmin/templates/nr/sustainability_pathways/docs/Factsheet_FOOD-WASTAGE.pdf

[5] Nomathemba, M. et al (2010). *FAO/World Bank workshop on reducing post-harvest losses in grain supply chains in Africa.* Retrieved from www.fao.org/fileadmin/user_upload/ags/publications/FAO_WB_ph_web.pdf

[6] Montpellier Panel Report (2014). *No Ordinary Matter: Conserving, Restoring and Enhancing Africa's Soils.* Retrieved from http://ag4impact.org/wp-content/uploads/2014/12/MP_0106_Soil_Report_LR1.pdf

[7] http://www.fao.org/3/a-at523e.pdf

[8] http://presse.inra.fr/en/Press-releases/World-food-security-affected-by-a-shortage-of-pollinating-insects

[9] Schaeffer, M., Baarsch, F., Balo, G., de Bruin, K., Callard, R., Fallasch, F., Melkie, M. E., Verwey, L., Freitas,

S., de Marez, L., van Rooij, J., & Hare, B (2015). *Africa's Adaptation Gap 2 Technical Report. Bridging the Gap – Mobilising sources*. Retrieved from http://apps.unep.org/redirect.php?file=/publications/pmtdocuments/-Africa%E2%80%99s_Adaptation_Gap_2__.pdf

[10] UNESCO position paper retrieved from www.unesco.org/new/fileadmin/MULTIMEDIA/FIELD/Nairobi/sciencesclimatechange_01.pdf

[11] https://data.worldbank.org/indicator/ER.H2O.INTR.PC?view=map

[12] Mubila, M. and Aissa, M. B. (2011). *The Middle of the Pyramid: Dynamics of the Middle Class in Africa*. African Development Bank. Retrieved from www.afdb.org/fileadmin/uploads/afdb/Documents/Publications/The%20Middle%20of%20the%20Pyramid_The%20Middle%20of%20the%20Pyramid.pdf

[13] www.worldbank.org/en/news/press-release/2013/03/04/africas-food-markets-could-create-one-trillion-dollar-opportunity-2030

[14] http://belfercenter.ksg.harvard.edu/files/TheNewHarvest-rev.pdf?webSyncID=5d7735de-05a6-a77e-062e-e846db86ea0e&sessionGUID=a377cb71-e520–6936-ad5d-ceb69c7559d9

[15] Imhoff, D. (2015). *Will the World Bank Get Truly Climate Smart?* Bretton Woods project. Retrieved from www.brettonwoodsproject.org/2015/03/will-the-world-bank-get-truly-climate-smart/

[16] *Agroecology and the Right to Food*, Report presented at the 16th Session of the United Nations Human Rights Council [A/HRC/16/49], 8 March 2011. Available online at: www.srfood.org/images/stories/pdf/officialreports/20110308_a-hrc-16-49_agroecology_en.pdf

[17] Bafana, B. (2014). *Denting Youth Unemployment through Agriculture*. UN Africa Renewal. Retrieved from www.un.org/africarenewal/magazine/special-edition-agriculture-2014/denting-youth-unemployment-through-agriculture

[18] AU 2003 Maputo Declaration on Agriculture and Food Security. Retrieved from www.nepad.org/resource/au-2003-maputo-declaration-agriculture-and-food-security

[19] Malabo Declaration on Accelerated Agricultural Growth and Transformation for Shared Prosperity and Improved Livelihoods. Malabo, Equatorial Guinea. 26–27 June, 2014. Retrieved from https://au.int/sites/default/files/documents/31006-doc-malabo_declaration_2014_11_26-.pdf

[20] Implementing the Comprehensive Africa Agriculture Development Programme and Restoring Food Security in Africa "The Roadmap." Retrieved from www.nepad.org/download/file/fid/3606%20

[21] The African Union Commission Agenda 2063 Framework Document. The Africa We Want. A shared strategic framework for inclusive growth and sustainable development and a Global Strategy to Optimize the Use of Africa's Resources for the Benefit of All Africans. September, 2015. Retrieved from www.un.org/en/africa/osaa/pdf/au/agenda2063-framework.pdf

[22] Cairo Declaration on Managing Africa's Natural Capital for Sustainable Development and Poverty Eradication. Retrieved from www.un.org/en/africa/osaa/pdf/au/cap_naturalcapital_2015.pdf

[23] www.mfarm.co.ke/

[24] www.reeep.org/projects/solar-powered-irrigation-kenya-futurepump

[25] www.worldbank.org/en/news/press-release/2013/03/04/africas-food-markets-could-create-one-trillion-dollar-opportunity-2030

[26] Senegal Food and Livelihoods Enhancement Initiative Ensuring Food Security in a Changing Climate. World Vision. Retrieved from www.worldvision.com.au/docs/default-source/publications/climate-change/senegal-food-and-livelihoods-enhancement-initiative-project-overview.pdf?sfvrsn=8

[27] Ethiopia Climate Project Receives Africa's First Forestry Carbon Credits under the CDM. World Bank. Retrieved from www.worldbank.org/en/news/feature/2012/10/09/ethiopia-climate-project-receives-africa-s-first-forestry-carbon-credits

[28] www.fc-es.net/category/reporting-from-cop21-in-paris-2015/

[29] AMCEN sixth special session Cairo Declaration, 2016. Retrieved from www.unep.org/sites/default/files/amcen6/amcen_sixth_special_session_cairo_declaration_final.pdf

[30] www.ebafosa.org/

[31] 6th AMCEN Special Session Decision 1/SS6: Implementation of 2030 Agenda for Sustainable Development, the Sustainable Development Goals and African Union's Agenda 2063: A Contribution of the Natural Capital. Retrieved from www.unep.org/sites/default/files/amcen6/decision_1_implementing_sdgs_in_africa_final.pdf

References

Africa Competitiveness Report (2015). *World Economic Forum*. Geneva.

Calestos, J. (2011). *The New Harvest, Agricultural Innovation in Africa*. New York: Oxford University Press.

Formo, R. K., Jorstad, H., Nellemann, C., Mafuta, C., Munang, R., Andrews, J., & Hval, J. N., (2014). *Food Wasted, Food Lost – Food Security by Restoring Ecosystems and Reducing Food Loss*. United Nations Environment Programme and GRID-Arendal, Nairobi and Arendal.

Godfray, H., Beddington, J. R., Crute, I. R., Haddad, L., Lawrence, D., Muir, J. F., Pretty, J., Robinson, S., Thomas, S. M., & Toulmin, C (2010). Food security: the challenge of feeding 9 billion people. *Science*. Vol. 327, Issue 5967, pp. 812–818.

Jeanne, H. C., and Andrew, R. T. (2002). *The Rise of Women Entrepreneurs*. Westport, CT: Quorum Books.

Kirsten, B., and Jens, P. M. (2001). Organic agriculture: does it enhance or reduce the nutritional value of plant foods?

Journal of the Science of Food and Agriculture. Special Issue: Nutritional Enhancement of Plant Foods *(NEO-DIET)* Volume 81, Issue 9, pp. 924–931.

Munang R., Mgendi R., Alverson K., O'Brien-Onyeka M, Ochieng C., Molua E., Ottichilo W. Mugerwa K., Bunch R., Bogale A., Bekele W., Sudarkasa M., Badiane O., Ayuk E., Amis M., Tabot-Tabot E., Kinney K., Kwangwari C., Luganda P., Aidoo K., Mulangu F., & Kariuki J. (2015) *Ecosystem Based Adaptation (EBA) for Food Security in Africa – Towards a Comprehensive Strategic Framework to Upscale and Out-Scale EbA-Driven Agriculture in Africa.* United Nations Environment Programme (UNEP), Nairobi.

Reij, C., & Winterbottom, R. (2015). *Scaling up Re-Greening: Six Steps to Success. A Practical Approach to Forest and Landscape Restoration.* World Resources Institute.

Rohit, J., Brent, S., & John, K. (2006). *Status of Carbon Sequestration Projects in Africa: Potential Benefits and Challenges to Scaling Up.* World Agroforestry Centre, Nairobi.

United Nations Environment Programme (2015). *The Emissions Gap Report 2015.* United Nations Environment Programme (UNEP), Nairobi.

United Nations Environment Programme (2016). *The Financial System We Need: From Momentum to Transformation.* 2nd ed.

PART V

Future Earth and Urban Environments

15
Nutrition, Urban Environments, and Future Earth

Godwin D. Ndossi and Keto E. Mshigeni

15.1 Introduction

This Chapter will focus on the relationships of nutrition, urbanization and the Earth's future environments. It will highlight the dynamics of population change in general, provide global and regional trends, and relate these to concurrent changes in the environment and how it might affect future growth and sustainability. Nutrition is an important determinant of health and well-being. It is argued that people will migrate into, or emigrate out of, one place to another, in search of employment and better living conditions. The trend is that rural populations are migrating to urban areas, thus stretching the limited available resources, including infrastructure and services, particularly those in developing countries. Torrey (2010) has argued that people in urban areas in developing countries eat more food than their rural counterparts, and are more likely to experience higher rates of obesity as a result of lower levels of activity. Urban populations too tend to change their dietary habits, demanding more energy-intense foods such as meat and highly refined foods that are easy to prepare. Such foods impact the rural-urban food supply system, and are likely to affect the environment adversely. The authors attempt to relate current trends in urbanization and the way these might influence our thinking in the way we plan and manage our future cities. In addition, some aspects of food security and climate change will be covered under this chapter, but specific details on these topics will appear in their respective chapters.

15.2 Global Population Size and Projections

In this section we are concerned with global population size, composition in terms of gender and age, and spatial distribution (density), births and deaths. This is because these factors have implications for health and nutrition, economic growth, educational development, as well as other social dimensions of our lives. According to the United Nations report titled *World Population Prospects: The 2015 Revision* (United Nations, 2015a), the global population reached 7.3 billion in 2015. Males were 50.4 percent and females 49.6 percent. In 1950 the world population was 2.5 billion people. That means that compared to the current population, the global population increased by 4.8 billion people in 65 years. How has this increase impacted the lives of people in general?

In many parts of the world, in developing as well as developed countries, we have seen a trend whereby people tend to move from rural to urban areas in search of jobs, or simply better livelihood opportunities. Rural poverty in part caused by prolonged drought, has forced people to move from villages to towns. Dar es Salaam, Tanzania, is a city with a population of nearly 5 million residents, and it is projected to grow to 10 million by 2040. The city was planned for a much smaller population, and hence the influx of people into it has pushed its residents into zones prone to seasonal river flooding and inundation due to sea-level rise. This phenomenon is seen in many coastal urban centres around the world.

Urban centres become an attraction because they have amenities that are not available in rural areas. Hence they have had to accommodate larger populations than originally planned and the urban space and social services have had to be stretched beyond limits. Out of need for space and shelter, the immigrating populations have, at times, been forced to inhabit the urban fringes, thus leading to the birth of peri-urban environments.

The effect of this process is not always inconsequential: human activities in the peri-urban space have sometimes caused serious environmental damage, leading to peri-urban floods. The response has usually been crisis mitigation rather than well-thought-out comprehensive plans.

Table 15.1 *Population of the World and Major Areas, 2015, 2030, 2050 and 2100, According to the Medium-Variant Projection*

	Population (millions)			
Major area	2015	2030	2050	2100
World	7349	8501	9725	11213
Africa	1 186	1 679	2 478	4 387
Asia	4 393	4 923	5 267	4 889
Europe	738	734	707	646
Latin America and the Caribbean	634	721	784	721
Northern America	358	396	433	500
Oceania	39	47	57	71

Source: United Nations, Department of Economic and Social Affairs, Population Division (2015b). World Population Prospects: The 2015 Revision. New York: United Nations.

What are the projections for the future? In the next 14 years, the global population is projected to reach 8.5 billion; in 2030, it will reach 9.7 billion, and in 2050 it will reach 11.2 billion. In 2100 approximately 60 percent of the world population will be found in Asia, 16 percent in Africa, 10 percent in Europe, 9 percent in Latin America and the Caribbean and 5 percent in Northern America and Oceania (Table 15.1, Figure 15.1). It is notable that China and India alone account for approximately 37 percent of the global population, with a population of 1.4 and 1.3 billion, respectively (United Nations, 2015b).

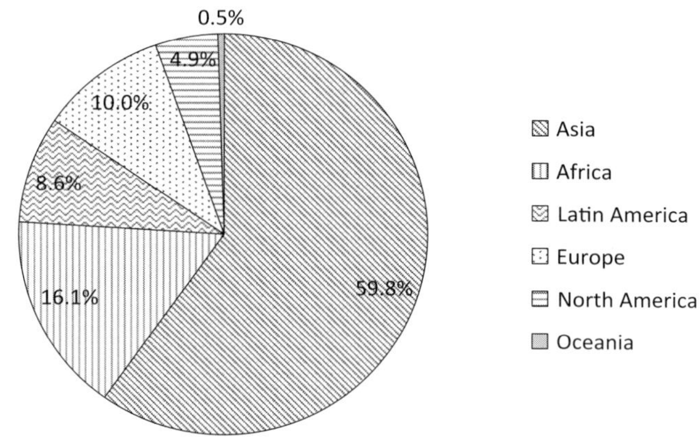

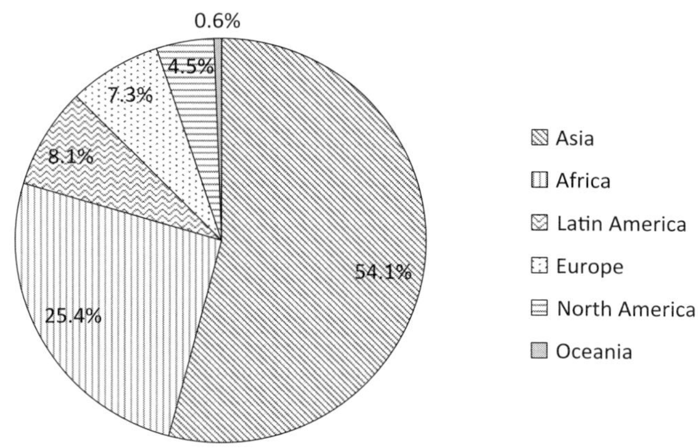

Figure 15.1 2015–2050 Population projection by major region
Source: United Nations. (2015b). World Population Prospects, the 2015 Revision. PowerPoint presentation by Gerhard K. Heilig, 7 September 2015 (www.gerhard-k-heilig.com).

15.2.1 What Are the Drivers of These Demographic Changes?

On the one hand we can consider fertility rates across different countries, and on the other, mortality rates. By considering the balance between these two processes (i.e., net-reproduction rate)[1] it has been shown that the population of Africa is almost doubling between generations, while those of Asia, Northern America and Europe are shrinking between generations by 3 percent, 10 percent and 24 percent, respectively (United Nations, 2015b). One can argue that in the foreseeable future the population of Africa and Asia will be increasing faster than in other regions, and that such population growth patterns will affect migration trends between rural and urban areas, within these two regions and across to other regions of the world.

If we consider populations by age structure, we note that 26 percent of the world population is children below the age of fifteen years, 62 percent are people aged fifteen to fifty-nine years, and 12 percent are people sixty years and older. Populations aged twenty to sixty-four years constitute the productive force of society, although it is also fair to point out that there is a sizeable proportion of people older than sixty-four years who are active in the labour force. Future population projections indicate that the proportion of people aged sixty years or more will more than double by 2050: Europe leading with 34 percent, Asia, Latin America and the Caribbean transforming to more than 25 percent and Africa to 9 percent (United Nations, 2015a). These projections show that the population in Europe is aging, and that in the next four decades Africa will have a largely young

[1] The Net-Reproduction Rate (NRR) is the average number of daughters that female members of a birth cohort would bear during their reproductive life span if they were subject throughout their lives to the observed age-specific fertility and mortality rates of the given time period. The NRR is expressed as the number of daughters per woman.

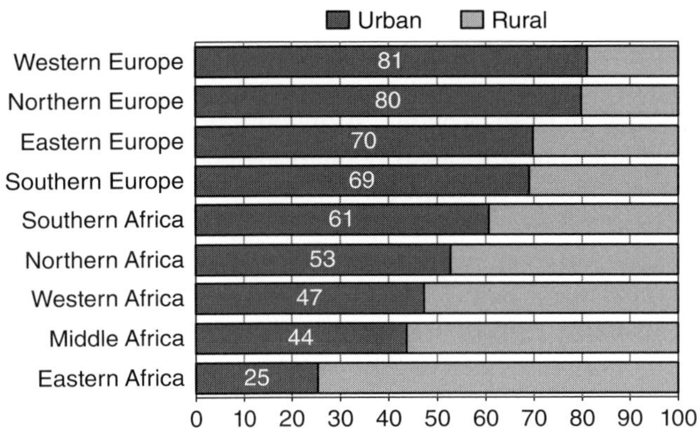

Figure 15.2 Africa – Europe urbanization, 2015 (%urban/rural) Source: United Nations (2015b). World Population Prospects, the 2015 Revision. PowerPoint presentation by Gerhard K. Heilig, 7 September, 2015 (www.gerhard-k-heilig.com).

population compared to the other geographical regions, thus pointing to potential challenges (for example, combating hunger and malnutrition, expanding school enrollment and provision of health services) and opportunities (e.g., increased labour force, expansion in service provision) related to a growing population.

Globally 54 percent of the population lives in urban areas and by 2050 the global population is projected to be 34 percent rural and 66 percent urban (United Nations, 2015c). North America, Latin America and the Caribbean and Europe are more urbanized compared to Africa and Asia, which are mostly rural. The trend is for all regions to urbanize further, thus posing sustainable development challenges(Figure 15.2).

As rural populations move to the urban areas, cities face the challenge of providing adequate services including food, health, housing and recreational facilities. All these impact the quality of life of urban dwellers. In sub-Saharan Africa it is projected that the percentage of people aged nine years and older will increase from 68 percent to 81 percent by 2050, translating to a higher caloric requirement compared to that of children aged eight years and below (Devlin and Bremner, 2012).

15.3 Food and Nutrition Security in Urban Environments

It is difficult to talk about nutrition without talking about food security. In this section we will attempt to highlight the importance of urban and peri-urban agriculture (UPA) as a source of food and income, particularly to the poor households in urban and peri-urban areas. We provide in Box 1 definitions of terminologies that are commonly used in this context.

Box 1 Key Definitions and Use of Terminology

- **Food security** exists when all people, at all times, have physical, social and economic access to sufficient, safe and nutritious food that meets their dietary needs and food preferences for an active and healthy life.
  Household food security is the application of this concept to the family level, with individuals within households as the focus of concern (FAO, 2009).
- **Food insecurity** exists when people do not have adequate physical, social or economic access to food as defined above (FAO, 2009).
- **Nutrition security** exists when food security is combined with a sanitary environment, adequate health services, and proper care and feeding practices to ensure a healthy life for all household members.
  This section refers to food and nutrition security throughout, because achieving nutrition security is imperative to reducing malnutrition (Shekar, 2006).
- **Undernourishment** measures aspects of food security and exists when energy intake is below the minimum dietary energy requirement, which is the amount of energy needed for light activity and a minimum acceptable weight for attained height (FAO, 2009). Although undernourishment is based on national level data, it may be used as a proxy for food consumption in contexts where regional or household level data are unavailable or unreliable. It varies by country and from year to year, depending on the gender and age structure of the population. Sometimesthe words “hunger” and “undernourishment” are used interchangeably.
- **Undernutrition** exists when insufficient food intake and repeated infections result in one or more of the following: underweight for age, short for age (stunted), thin for height (wasted), and functionally deficient in vitamins and/or minerals (micronutrient malnutrition).
- **Malnutrition** is a broad term that refers to all forms of poor nutrition. Malnutrition is caused by a complex array of factors including dietary inadequacy (deficiencies, excesses or imbalances in energy, protein and micronutrients), infections and socio-cultural factors. Malnutrition includes undernutrition as well as overnutrition (overweight and obesity) (Shekar, 2006).

We start from the premise that the world has adequate food to feed everyone if it were to be distributed equitably. However, this is not the case and so various population groups in different parts of the world in developing as well as developed countries, in rural as in urban areas, face food insecurity for various reasons. The most immediate reason is lack of, or lack of accessibility to, adequate food necessary to enable one to meet nutritional needs. In rural areas food insecurity could be due to a number of factors, including poor weather, leading to poor harvests, poverty, lack of land and breakdown of traditional family structure, and community safety networks, among others. As people move to urban centres in search of opportunities, they are faced with a number of problems: if they are not gainfully employed, they may not have decent housing or accommodation, and therefore live in squalid conditions or slums characterized by lack of potable water, electricity services and poor sanitation (Te Lintelo, Marshall and Bhupal, 2001). Under these conditions they become susceptible to infectious diseases due to poor sanitation and hygiene. Urban centres may have a wide range of food commodities that are not accessible to people living in slums because of income poverty as a result of unemployment. Furthermore, urban dwellers may also face food insecurity because they have no land to grow their own food, etc. As a result of these factors urban populations tend to have altered food consumption patterns (Fustos, 2011) that may predispose them to malnutrition and ill health.

In order to meet their food and income needs, urban and peri-urban dwellers engage in agricultural activities (Figures 15.3 and 15.4) as a coping mechanism. It is a practice that is observed in both developed and developing regions of the world, to varying extents (Egal, Valstar and Meershoek, 2001). It is estimated that 15–20 percent of the world's food is produced in urban areas, further highlighting the contribution of urban agriculture to global food requirements (Wurwarg, 2008).

In big cities people have come up with innovative ideas such as the "multilevel elevated urban farm"(for example, in Atlanta, Georgia, United States) (Card, 2011) or the "Elevated Highway Farm to Feed Chicago" (Gang, 2010) in order to maximize utility of available urban land. UPA involves a system of various agricultural activities for food and non-food production (Madaleno, 2001). These activities include vegetable gardening, fruit trees, horticultural crops, mushrooms, algae (e.g. *Spirulina*), ornamental plants, animals, poultry and fish. The list is long and depends on available land but also local regulations guiding the undertaking of such livelihoods.

In some cases the activities are carried out in open spaces (often illegally occupied), on roadsides, under power lines and along riverbanks. Because of its informality urban agriculture is often done under extremely difficult conditions (Drescher, 2004; Beach, 2013). These setbacks notwithstanding, the benefits of urban agriculture clearly stand out: a source of income for poor households (Beach, 2013), assured access to a variety of fresh and nutritious foods and improved nutritional status of children (Maxwell, Levin and Csete, 1998). One might also consider an aesthetic reason of making the city green to look beautiful! What is needed is a comprehensive multipronged approach to deal with the complex environmental, social, cultural and economic factors that impact the health and well-being of urban and peri-urban populations. If we are to achieve sustainable urban land use and development, then we must align policy, governance and practice to support the informal trade that is linked to the urban food system and the rural-urban supply chain.

15.4 Global Nutrition Situation

The global nutrition situation is characterized by high levels of malnutrition. "Malnutrition" is a term used to

Figure 15.3 Vegetable garden in urban Dar es Salaam, Tanzania (G. D. Ndossi). For the color version, please refer to the plate section.

Figure 15.4 Vegetable garden in Fuzhou, Fujian, China (G. D. Ndossi). For the color version, please refer to the plate section.

describe an abnormal physiological condition caused by deficiencies, excesses or imbalances in energy, protein and/or other nutrients (DFID, 2009). Malnutrition includes several forms of undernutrition (namely, stunting, wasting and underweight) as well as overnutrition (overweight and obesity). The direct cause(s) of malnutrition is/(are) related to food in terms of its adequacy and variety, but other factors such as water and sanitation, health, care, education, women's empowerment, access to resources, social-cultural factors, and others all have a role to play in its causation. The impact of malnutrition has health, economic and social consequences at both the individual and community levels. At the individual level good or adequate nutrition is necessary for achieving optimal physical and mental development during childhood. As such it has been linked to better school performance and higher salary or wage in later years. At population level good nutrition has been linked to increased economic productivity and growth (UN-SCN, 2015; Chase and Ngure, 2016; Hoddinott et al., 2013) and thus has implications for national economy.

15.5 Global Nutrition Initiatives

Table 15.2 shows some of the commitments and declarations made by world leaders in support of nutrition actions. It is important to note that a number of these commitments build on each other.

In September 2000 world leaders met at the United Nations headquarters in New York, where they adopted the UN Millennium Declaration. Nations committed to a new global partnership to reduce extreme poverty and setting out a series of time-bound targets – with a deadline of 2015 – which became known as the eight Millennium Development Goals (MDGs). Nearly all of the MDGs were relevant to nutrition. The set date for achieving the MDGs was 2015. As will be noted in the section that follows there has been considerable progress in attaining the MDGs and that it is on the strength of the momentum built that the Sustainable Development Goals (SDG) have been proposed.

At the UN Conference on Sustainable Development: Rio+20 held in 2012, the UN secretary general proposed an ambitious goal, the "Zero Hunger

Table 15.2 *Building a Global Commitment to Nutrition*

Year	Global commitment to nutrition
2011	The United Nations releases a political declaration on noncommunicable diseases (NCDs) as the outcome of a High-Level Meeting on the Prevention and Control of NCDs.
2012	At the World Health Assembly, national governments adopt a series of nutrition targets as part of the Comprehensive Implementation Plan on Maternal, Infant, and Young Child Nutrition.
2013	The governments of the United Kingdom and Brazil together with the Children's Investment Fund Foundation cohost a summit designed to raise commitment to actions to achieve the Global Targets on Maternal, Infant, and Young Child Nutrition. At the World Health Assembly, national governments adopt a series of targets on the prevention and control of NCDs, including nutrition-relevant targets.
2014	The United Nations holds a follow-up meeting to the 2011 High-Level Meeting on the Prevention and Control of NCDs to review progress. Countries make clear commitments to, by 2015, set national NCD targets for 2025 and establish process indicators taking into account the nine NCD targets.
2014	Governments come together at the Food and Agriculture Organization/World Health Organization International Conference on Nutrition (ICN2) and agree on a set of 10 commitments in the Rome Declaration on Nutrition and the accompanying Framework for Action.
2015	Countries assemble at the United Nations to adopt a new nutrition target as part of the Sustainable Development Goals to, by 2030, end all forms of malnutrition.
2016	The United Nations General Assembly declares a Decade of Action on Nutrition from 2016 to 2025. The Decade of Action would translate the ICN2 commitments into coherent and coordinated actions and initiatives by all national governments, both low and high income.
2016	Proposed date for the Nutrition for Growth (N4G) Summit in Rio de Janeiro, Brazil
2016	Japan's leadership on nutrition is growing in advance of the 2016 Group of 7 meeting and the lead-up to the 2020 Tokyo Olympics and Paralympics.

Source: International Food Policy Research Institute (IFPRI), 2016. *Global Nutrition Report 2016: From Promise to Impact: Ending Malnutrition by 2030*. Washington, DC.

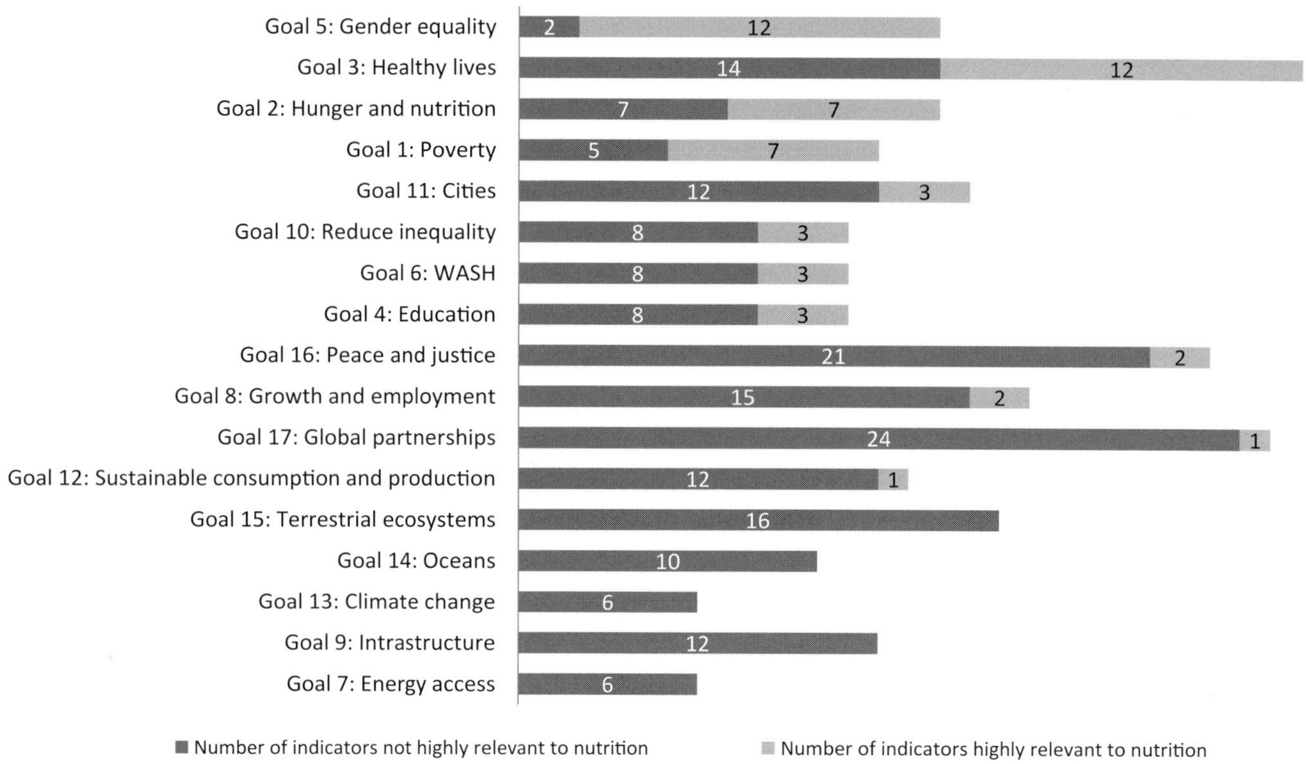

Figure 15.5 Number of indicators in each SDG that are highly relevant for nutrition.
Source: International Food Policy Research Institute (IFPRI), 2016.

Challenge," as one of the seventeen Sustainable Development Goals. The vision is to eliminate global hunger by 2025. In September 2015 the UN Sustainable Development (SD) Summit set the global agenda for SD until 2030. It has been estimated that at least twelve of the seventeen SDGs contain indicators relevant to nutrition (Figure 15.5).

15.6 What Is the Global Burden of Malnutrition?

Recent estimates indicate that globally, 868 million people, or 12.5 percent of the world's populations are undernourished and that 26 percent of children younger than age five are too short for their age(stunted). Furthermore, 2 billion people show signs of one or more micronutrient deficiencies and 1.4 billion people are overweight, of whom 500 million are obese (FAO, IFAD and WFP, 2013). Data from 159 countries show that 57 of them have severe levels of both undernutrition and overnutrition, which may co-exist in the same household or even individual (FAO and WHO, 2014).

If no action is taken, malnutrition can contribute to the global burden of disease and therefore health care costs, as well as loss in productivity. It has been estimated that lost productivity and direct healthcare costs could account for as much as 5 percent of global gross domestic product (GDP) equivalent to US$ 3.5 trillion per year (FAO, IFAD and WFP, 2013). Similar calculations for undernutrition and micronutrient deficiencies put the costs at 2–3 percent of the global GDP, equivalent to US$1.4–2.1 trillion per year. Some countries have committed to improving the nutrition situation of their people, and levels of malnutrition in those countries have been reduced; examples of these countries are Ghana, Peru, Brazil and the State of Maharashtra in India (IFPRI, 2016). According to the recent *Global Nutrition Report 2015*, most countries are making progress in reducing levels of malnutrition, even though they are not on course to meet the set global nutrition targets.Table 15.3 shows the relative improvement in the prevalence of selected nutrition indicators as of 2016.

Other highlights in the *Global Nutrition Report* are that the number of stunted children is declining in every

Table 15.3 *The Global State of Malnutrition*

Indicator	Number of individuals	Current prevalence (%)
Under-5 stunting	159 million in 2014 (255 million in 1990)	23.8 (39.6 in 1990)
Under-5 overweight	41 million in 2014 (31 million in 1990)	6.1 (4.8 in 1990)
Under-5 wasting	50 million in 2014	7.5
Under-5 severe wasting	16 million in 2014	2.4
Anemia in women ages 15–49 years (nonpregnant and pregnant)	533 million in 2011	29 for nonpregnant women in 2011 (33 in 1995) 38 for pregnant women in 2011 (43 in 1995)
Exclusive breastfeeding (under 6 months)	NA	39 in 2014
Low birth weight	20 million in 2014	15
Adult overweight (ages 18+)	1.9 billion in 2014	39
Adult obesity (ages 18+)	600 million in 2014	13
Adult diabetes (raised blood glucose) (ages 18+)	NA	9

Source: International Food Policy Research Institute. *Global Nutrition Report 2016: From Promise to Impact: Ending Malnutrition by 2030*. Washington, DC

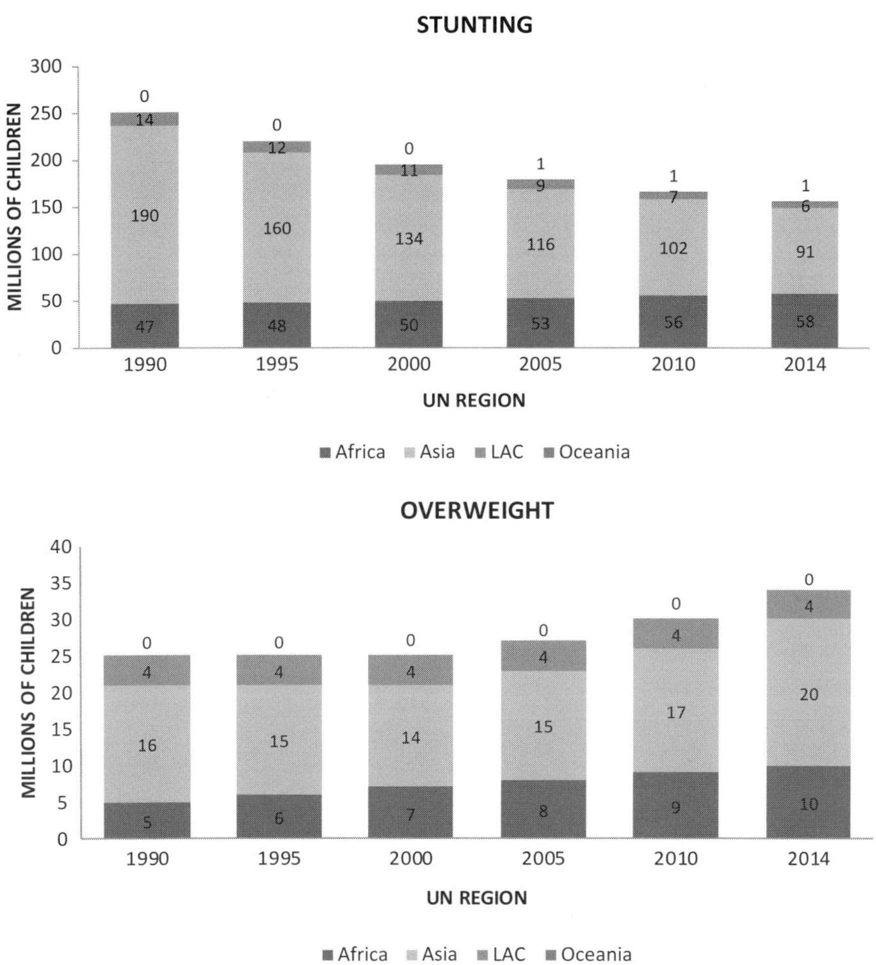

Figure 15.6 Number of children below age five affected by stunting and overweight by region, 1990–2014. Source: International Food Policy Research Institute (IFPRI), 2016.

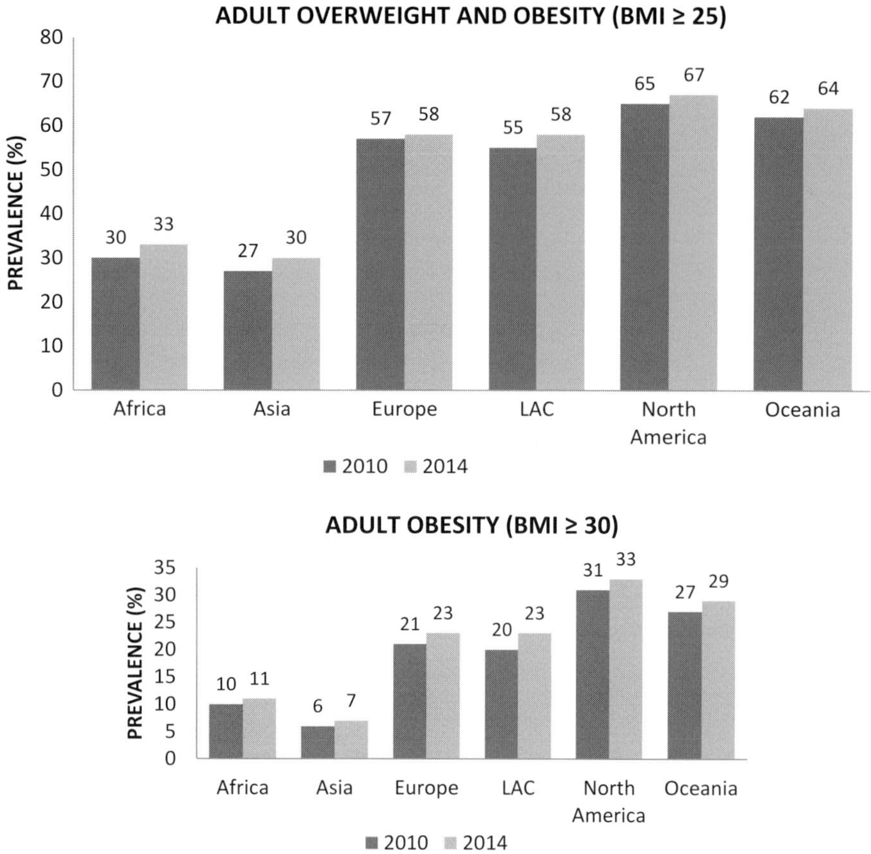

Figure 15.7 Adult overweight and obesity, adult obesity by UN region, 2010 and 2014.
Source: International Food Policy Research Institute (IFPRI), 2016.

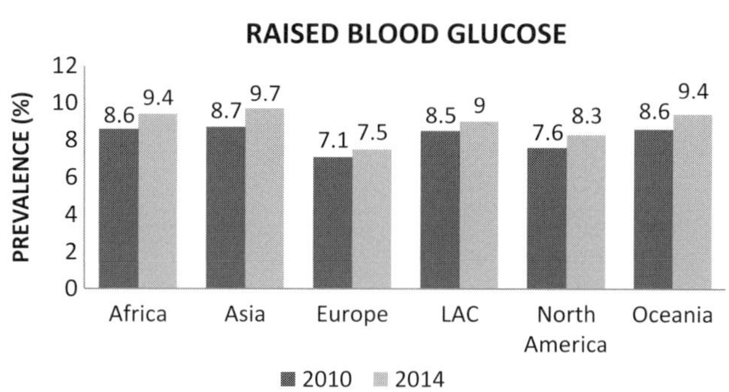

Figure 15.8 Adult diabetes by UN region, 2010 and 2014.
Source: International Food Policy Research Institute (IFPRI), 2016.

region except Africa; the number of overweight children less than age five is increasing, especially in Asia, and prevalence of adult overweight and obesity is increasing in all regions (Figure 15.7).

Furthermore, many countries are either on track or making progress on the World Health Assembly global target for exclusive breastfeeding of infants younger than six months, under-five stunting, wasting and overweight. Nearly all countries are off course to meet the set target for anaemia of women of reproductive age (fifteen to forty-nine years), adult overweight, and obesity and raised blood glucose (Figures 15.7 and 15.8).

15.7 Rural/Urban Differentials

Figure 15.9 shows the trend in urbanization in different regions of the world. The level and growth of urbanization differ considerably by region. But on the whole the trend of urbanization is growing for all regions. In 2014, 54 percent of the global population lived in urban centres and it is projected that by 2050, 66 percent of the global population will be urban (United Nations, 2015c). Data on child nutritional status show that urban children are better nourished and have lower prevalence of underweight compared to their rural counterparts (Smith, Ruel and Ndiaye, 2005; Van de Poel, O'Donnell and Van Doorslaer, 2007; UNICEF,

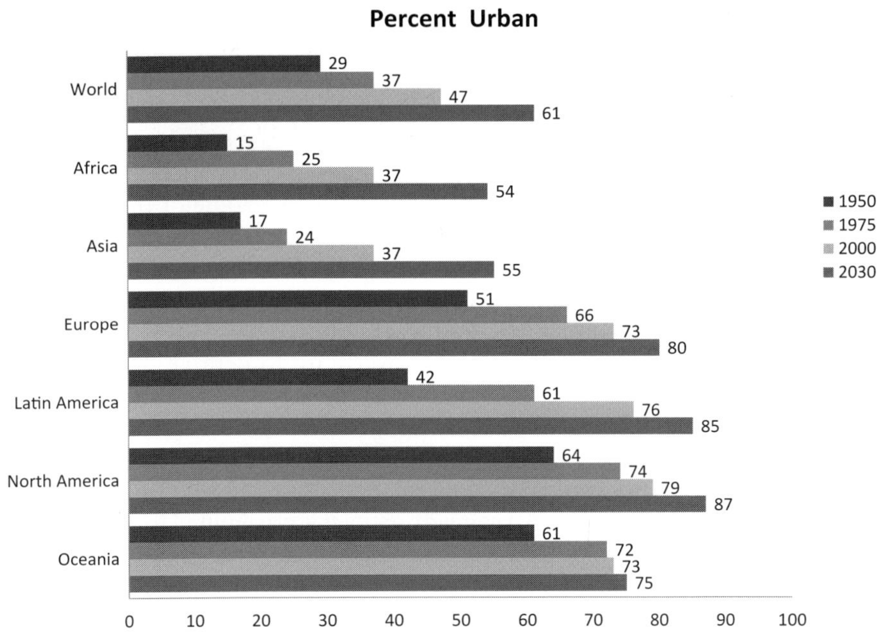

Figure 15.9 Population living in urban areas.
Source: United Nations (2004). *World Urbanization Prospects: The 2003 Revision* (2004).

2013). This is due to the fact that the social-economic determinants of child nutritional status, such as maternal education, health care services, water and sanitation, as well as availability of a more diverse diet, are more favourable to urban children than to rural children. All things considered, therefore, urban children are more likely to be better nourished.

Differences are also noticed among adults in rural versus urban environments with respect to prevalence of overweight and obesity. In general, the tendency is for urban populations to have higher prevalence of overweight and obesity than their rural counterparts. And this is true for both men and women. Urban populations tend to eat more food; they have a diversity of choices and hence the tendency for higher consumption of animal-source foods, fruits and vegetables. If income allows, urban populations will increase their intake of processed foods either out of choice or as a necessary strategy to cut down on food preparation time because of work commitments. In this way their fat, sugar and salt intake is likely to increase as is their risk to become overweight and obese. Data in this area are limited but the differential burden of malnutrition between rural and urban areas is apparent in many countries. In fact, in India the rural/urban divide for underweight has been cited among adults living in slum areas of eight cities in which the prevalence was 23 percent while the prevalence in rural areas in the same states was close to 40 percent (cited in FAO, IFAD WFP, 2013). It is argued that the labourious working conditions, weak education system and lack of healthcare services may have jointly contributed to the rural adult nutrition indicators' being worse than those in urban slum areas.

15.8 Environmental Effects of Urbanization

As we have seen earlier, urban centres will be preferred homes for future global populations. The trend has already been set where the number of cities with 10 million or more people (megacities) is on the increase. In 1975 there were four megacities and in 2014 the number had increased to twenty-eight (United Nations, 2014). It is expected that in the foreseeable future new megacities will grow out of small- to medium-size cities. Unless deliberate efforts are taken to plan for the influx of people into cities, we are likely to witness scarcity of supportive infrastructure for the burgeoning urban population. Service availability will be stretched to the limits. Large populations are also likely to exert pressure on the environment, thereby making it fragile and susceptible to the pangs of climate change. Greater reliance on crops for biofuels and dietary changes towards consumption of processed foods by urban populations will add another negative dimension to the impact of urbanization on the environment.

15.9 Urbanization and Climate Change

The world's urban centres are also the key focal points for industrial development activities. This is largely due to infrastructural attractions (e.g., reliable transport, availability of sources of power, a reliable water supply, good telecommunication network, availability of human capital and market). Since the labour force available for emerging industrial activities in developing nations is relatively cheap compared to that in industrialized nations, many of the world's industrial giants are increasingly setting new industrial plants in the South. Another factor attracting investors to urban centres of developing nations is lack of stringent anti-pollution policies in these countries. The Future Earth we want will not tolerate ecological sins of urban pollution. Stringent anti-pollution policies must be exercised globally to save our planet.

If greenhouse gases are considered to be environmental pollutants, then the key environmental pollutants from urban centres are carbon dioxide, sulphur dioxide, and other gases emitted by cars, by industrial plants, and by people and their domesticated animals (e.g., cattle emit appreciable quantities of methane); effluents from industrial processing plants, and sewage from people's residential dwellings; and various categories of solid wastes, which must be creatively and innovatively handled, e.g., processing them into novel value-added products, and creating new jobs for the people in the process. In the past, in many countries, industrial effluents and domestic sewage were simply drained into the sea, or nearby rivers; solid wastes were simply taken to dump sites and burnt, polluting the atmosphere (UNDP, 2004a), and during periods of excessive rains and flooding of rivers, these wastes floated around, becoming a serious health hazard.

15.10 Future Earth for Sustainable Development

We inherited from our ancestors a beautiful planet endowed with captivating birds, insects and other arthropods; with an amazing biodiversity of reptiles, fish, amphibians, invertebrates, vertebrates and other biota; and with myriad other creatures, big and small: ranging from bacteria to algae to vascular plants on land, in lakes, rivers and seas. These biota include the amazing myriad wild game that collectively attract millions of tourists who travel to take a glimpse of them in places such as the Ngorongoro and Serengeti in Tanzania, and elsewhere in Africa.

The countless land, freshwater and marine biota on our planet include mysterious species with variability in beauty, colour, natural products chemistry and ecology. Some bacteria and blue-green algae in our planet's ecosystems thrive in geothermal, boiling hot-spring water habitats.

The rich heritage of our planet's biota includes species that are edible and highly nutritious, and others that are medicinally very potent. Some of the biota are currently not being used for food in one country, but are perceived as a nutritious delicacy in another.

Sustainable development dictates that in whatever activities we do, we do not destroy the balance of nature. We should never dare to deprive future generations of opportunities also to witness this intriguing wealth of the Earth's biodiversity heritage.

But, alas! Some of the biota have already vanished forever! The innocent dodo bird of Mauritius (*Raphus cucullatus*), is already extinct: it was hunted to total extinction by the earlier explorers of the African continent.

Many forest tree lichens, especially those sensitive to pollution by elevated levels of atmospheric gases, are rapidly disappearing from forests of industrialized European countries, and some may be on the verge of extinction.

The release of excessive levels of carbon dioxide and other gases into the atmosphere by industrialized cities, and the subsequent discharge of acid rains into the sea are also threatening the existence of many calcified marine biota, including colourful corals, bivalve mollusks, and other calcareous biota, whose protective shells are dissolved by the ambient carbonic acid in seawater.

Man's agricultural activities in tropical regions often also disturb the Earth's balance of nature. Excessive leaching away of agricultural fertilizers and other mineral nutrients from agricultural fields and their accumulation in lakes, dams and rivers, have also caused an excessive growth of the water hyacinth hydrophyte (*Eichhornia crassipes*) in these water bodies.

Originally imported to Africa from Latin America via Europe as an indoor ornamental plant, *Eichhornia* subsequently fled domestication and got into rivers, where it multiplied profusely, because of the high level of nutrients in the freshwater bodies (Mshigeni, 1995). The hydrophyte multiplied to an extent that it is now conceived as an unwanted noxious weed, a weed to be eradicated since it interferes with local fishing and water navigation activities. In places, the entire river or dam water surface is totally covered with it. Indeed, water hyacinth is now a grave ecological problem in the Lake Victoria basin, in the Zambezi river ecosystem, and in the Akosombo Dam of Ghana.

Worth emphasizing here is the point that, in our various socioeconomic activities, we should refrain from disturbing the Earth's balance of nature, and we should protect all categories of our indigenous biota for posterity. And, when we encounter challenges like that of the water hyacinth cited here, we should be innovative and dream new approaches through which we can transform the challenges into socioeconomic opportunities.

Sustainable new uses for the water hyacinth biomass could include using it as an organic agro-fertilizer, or as a substrate for farming edible and highly nutritious mushrooms (Chang, et al., 1998; UNDP, 2004b; He, et al., 2016).

15.11 Conclusions

The world population is growing in all regions and the growth is more rapid in developing countries compared to developed countries. With growing population there is pressure on the environment as resources and services are being stretched to meet demand, particularly in urban centres. The population dynamics or movement between rural and urban environments, on the one hand, and the urban food systems and the rural-urban supply chain, on the other, impact the food and nutrition security of a given urban community. Hence, city authorities should plan and monitor urban development with a view to ensuring adequate availability and accessibility to food for all.

The global state of malnutrition indicates that many countries have serious levels of both undernutrition and overnutrition, the so-called double burden of malnutrition. It is known that malnutrition contributes to the global burden of disease as well as loss in productivity and therefore in order to attain sustainable development countries must improve the nutrition situation of their people. This challenge was taken up by world leaders at the UN Conference on Sustainable Development: Rio+20 held in 2012 where the seventeen SDGs were proposed. It has been estimated that at least twelve of the seventeen SDGs have indicators relevant to nutrition. This should be considered good news since it demonstrates that efforts addressing the SDGs have complementary outcomes. Another consideration is that although the SDGs are global, these need to be actualized within a local context taking into account the relative differences in the prevalence of malnutrition between specific communities, e.g., rural versus urban.

Finally, for global sustainability, human activity should not disturb the balance of nature, but rather deliberate efforts should made in preserving the Earth's biodiversity by adopting relevant supportive policies and practices.

References

Beach, M. (2013). *Urban Agriculture Increases Food Security for Poor People in Africa*. Washington, DC: Population Reference Bureau.

Black, R. E., Victora, C. G., Walker, S. P., Bhutta, Z. A., Christian, P., de Onis, M., Ezzati, M., Grantham-McGregor, S., Katz, J., Martorell, R., and Uauy, R. (2013). "Maternal and Child Undernutrition and Overweight in Low-Income and Middle-Income Countries." *Lancet* **382** (9890): 427–451.

Card, M. (2011). Proposed Multilevel Elevated Urban Farm in Atlanta. Team Elevate. www.teamelevateatlantageorgia.weebly.com (Accessed 12 May 2016).

Chang, S. T., Mshigeni, K. E., Pauli, G., and Tagwira, M. (1998). The impact of mushroom cultivation on the control of the water hyacinth weed problem in Africa, in, *A New Hope for Sustainable Development in Africa: Zero Emissions and Total Productivity of Raw Materials*. ed. K. E. Mshigeni et al. Windhoek: University of Namibia, pp. 169–175.

Chase C., and Ngure F. (2016). *Multisectoral Approach to Improving Nutrition: Water, Sanitation and Hygiene*. Washington, DC: Water and Sanitation Program World Bank Group.

Department for International Development (DFID). (2009). *The Neglected Crisis of Undernutrition: Evidence for Action*. London: DFID, pp. 71.

Devlin, K., and Bremner, J. (2012). *How Changing Age Structure and Urbanization Will Affect Food Security in Sub-Saharan Africa*. Washington, DC: Population Reference Bureau.

Drescher, A. W. (2004). Food for the Cities: Urban Agriculture in Developing Countries. Applied Geography of the Tropics and Subtropics (APT) Proc. IC on Urban Horticulture, R. Junge-Berberovic et al. (eds.). *ActaHort* **643**, 227–231.

Egal, F., Valstar, A., and Meershoek, S. (2001). *Urban Agriculture, Household Food Security and Nutrition in Southern Africa*. Rome: FAO.

FAO. (2009). *State of Food Insecurity in the World, 2009*. Rome: Food and Agriculture Organization of the United Nations.

FAO, IFAD and WFP. (2013). *The State of Food Insecurity in the World 2013. The Multiple Dimensions of Food Security*. Rome: FAO. www.fao.org/docrep/018/i3434e/i3434e.pdf

FAO, WFP and IFAD. (2012). *The State of Food Insecurity in the World 2012. Economic Growth is Necessary but not Sufficient to Accelerate Reduction of Hunger and Malnutrition*. Rome: FAO. www.fao.org/docrep/016/i3027e/i3027e.pdf.

FAO and WHO. (2014). Second International Conference on Nutrition. Rome, 19–21 November 2014. Conference Outcome Document: Rome Declaration on Nutrition.

FAO. Food for the Cities: Urban food security, nutrition and livelihoods. Rome: FAO.

Fustos, K. (2011). *Rising Global Food Prices Threaten to Increase Poverty*. Washington, DC: Population Reference Bureau.

Gang, J. (2010). An Elevated Highway Farm to Feed Chicago by Yuka Yoneda. www.inhabitat.com/feeder-an-elevated-highway-farm-to-feed-chicago.

He, H., Liu, M., Xiao, Z., and Zhu, Z. (2016). Test on mechanical processing of exotic invasive plants: water hyacinth for straw mushroom cultivation. *Edible and Medicinal Mushrooms*, **24** (3): 173–175.

Herforth, A., and Dufour, C. (2013). *Key Recommendations for Improving Nutrition through Agriculture: Establishing a Global Consensus SCN News*. Rome:FAO. www.unscn.org/files/Publications/SCN_News/SCNNEWS40_final_standard_res.pdf.

Hoddinott, J., Behrman, J. R., Maluccio, J. A., Melgar, P., Quisumbing, A. R., Ramirez-Zea, M., Stein, A. D., Yount, K. M., and Martorell R. (2013). Adult consequences of growth failure in early childhood. *Am J ClinNutr*; **98**: 1170–1178.

International Food Policy Research Institute (IFPRI). (2014). *Global Nutrition Report. Actions and Accountability to Accelerate the World's Progress on Nutrition*.Washington DC: IFPRI. http://dx.doi.org/10.2499/9780896295643.

International Food Policy Research Institute (IFPRI). (2016). *Global Nutrition Report 2016: From Promise to Impact: Ending Malnutrition by 2030*. Washington, DC: IFPRI.

Madaleno, I. M. (2001). *Cities of the Future: Urban Agriculture in the Third Millennium*. Food, Nutrition and Agriculture. Rome: FAO.

Maxwell, D., Levin, C., and Csete J. (1998) Does urban agriculture help prevent malnutrition? Evidence from Kampala. *Food Policy*, **23**(5): 411–424.

Mshigeni, K. E. (1995). The water hyacinth "weed" in Africa: a problem or an opportunity? *Discovery and Innovation*, **7**: 99–100.

Mucha, N., Blomberg, J. H., and Stockdale, K. (2016). *Nutrition Landscape Analysis for 14 Countries*. Geneva: Global Alliance for Improved Nutrition (GAIN).

Shekar, M. (2006). *Repositioning Nutrition as Central to Development*. Washington, DC: World Bank.

Smith, L. C., Ruel, M. T., and Ndiaye, A. (2005). Why is child malnutrition lower in urban than in rural areas? Evidence from 36 developing countries. *World Development*, **33**(8): 1285–1305.

Te Lintelo, D., Marshall, F., and Bhupal. D. S. (2001). Peri-urban agriculture in Delhi, India. *Food, Nutrition and Agriculture*, **29**:4–13.

Torrey, B. B. (2010) Population Dynamics and Future Food Requirements in Sub-Saharan Africa, in *The African Food System and Its Interaction with Human Health and Nutrition*, ed. PerPinstrup-Andersen. Ithaca, NY: Cornell University Press.

United Nations Children's Fund (UNICEF). (2013). Statistics by area/Child nutrition/Underweight disparities. Childinfo: Monitoring the situation of womenand children. www.childinfo.org/malnutrition.

United Nations. (2004). World Urbanization Prospects: The 2003 Revision (2004).

United Nations, Department of Economic and Social Affairs, Population Division. (2014). World Urbanization Prospects: The 2014 Revision, Highlights (ST/ESA/SER.A/352).

United Nations, Department of Economic and Social Affairs, Population Division. (2015a). World Population Prospects: The 2015 Revision, Key Findings and Advance Tables. Working Paper No. ESA/P/WP.241.

United Nations. (2015b). World Population Prospects, the 2015 Revision. PowerPoint presentation by Gerhard K. Heilig, 7 September, 2015. www.gerhard-k-heilig.com.

United Nations. (2015c). *World Urbanization Prospects: The 2014 Revision – Highlights*. New York: United Nations. http://esa.un.org/unpd/wup/Highlights/WUP2014-Highlights.pdf.

United Nations, Department of Economic and Social Affairs, Population Division (2015). World Urbanization Prospects: The 2014 Revision (ST/ESA/SER.A/366).

United Nations Development Programme. 2004a. *UJAMS Project, Namibia: Setting a Model for Africa's Cities*. Windhoek: University of Namibia.

United Nations Development Programme. 2004b. *Water Hyacinth, Africa: The Use of the Water Hyacinth Weed in Africa as a Potent Socioeconomic Resource*. Windhoek: University of Namibia.

UN-SCN. (2010). *Sixth Report on the World Nutrition Situation: Progress in Nutrition*. Geneva, Switzerland: United Nations, Standing Committee on Nutrition and International Food Policy Research Institute. www.unscn.org/files/Publications/RWNS6/report/SCN_report.pdf.

UN-SCN. (2015). *Nutrition and the Post-2015 Development Agenda: Seizing the Opportunity*. SCN News #41. Geneva, Switzerland: United Nations Standing Committee on Nutrition.

Van de Poel, E., O'Donnell, O., and Van Doorslaer, E. (2007). Are urban children really healthier? Evidence from 47 developing countries. *Social Science & Medicine*, **65**(10): 1986–2003.

Vira, B., Wildburger, C., and Mansourian, S. (eds.) (2015) *Forests, Trees and Landscapes for Food Security and Nutrition: A Global Assessment Report*. World Series, Volume 33. Vienna: IUFRO.

Von Grebmer, Saltzman, K., Birol, A., Wiesmann, E., Prasai, D., Yin, N., Yohannes, S., Menon, Y., Thompson, P., and Sonntag, A. (2014). *2014 Global Hunger Index: The Challenge of Hidden Hunger*. Bonn, Washington, DC, and Dublin: Welthungerhilfe, International Food Policy Research Institute and Concern Worldwide. http://dx.doi.org/10.2499/9780896299580.

Webb Girard, A., Self, J. L., McAuliffe, C., and Oludea, O., (2012). The effects of household food production strategies on the health and nutrition outcomes of women and young children: a systematic review. *Paediatric and PerinatalEpidemiology*, **26** (Suppl. 1): 205–222. http://dx.doi.org/10.1111/j.1365-3016.2012.01282.x.

Wurwarg, J. (2008) Urban Agriculture for Sustainable Development, Poverty Alleviation and Food Security. Paper by FAO's Food for the Cities Multidisciplinary Initiative Paper overview.

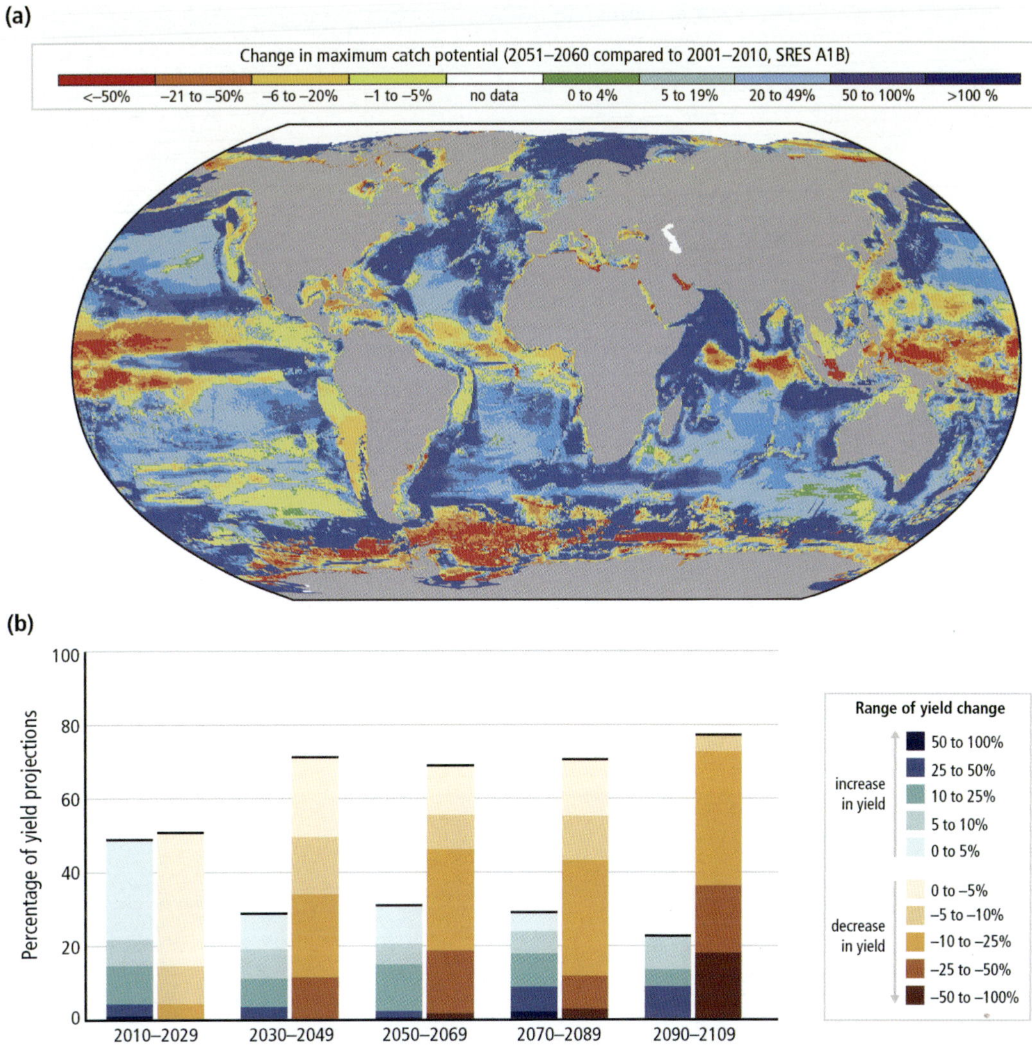

Figure 2.3 Climate change poses risks for food production: (a) change in maximum fish catch potential of ~1,000 species (2051–2060 compared to 2001–2010, SRES A1B, 2 °C warming); (b) summary of projected changes in crop yields (mostly wheat, maize, rice and soy) due to climate change over the twenty-first century. The figure combines 1,090 data points from crop model projections, covering different emission scenarios, tropical and temperate regions and adaptation and no-adaptation cases. Changes are relative to late twentieth-century levels (IPCC, 2014).

Figure 3.1 Photo of aurora from Greenland (Craig Heinselman).

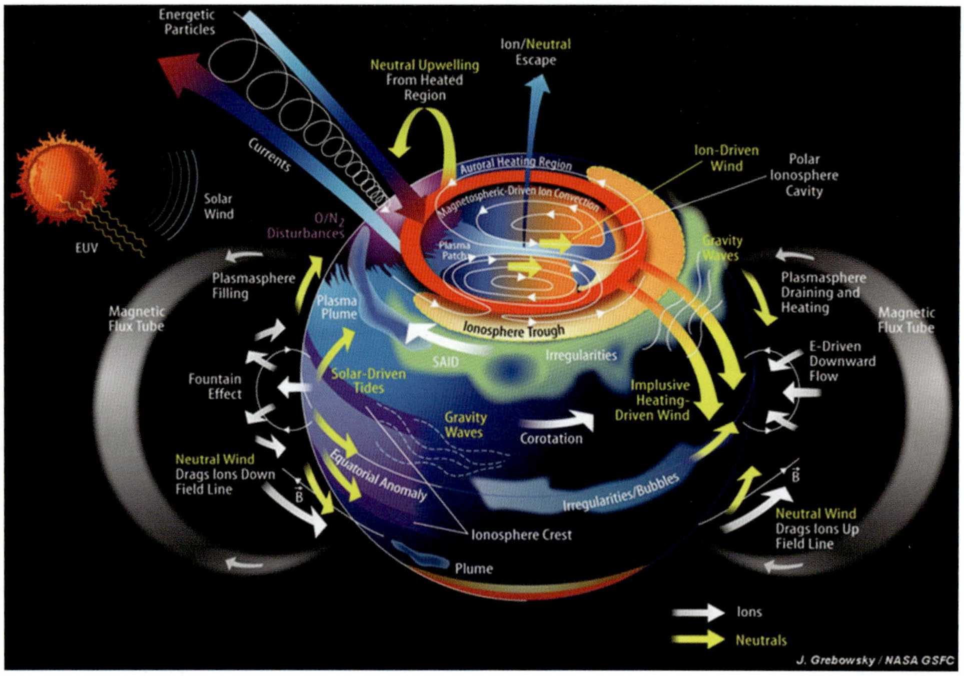

Figure 3.2 The magnetosphere comprising a number of different regions, each characterised by specific particle populations, particle drifts and magnetic fields (courtesy of J. Grebowsky).

Figure 4.9 An artist's impression of the change in observed atmospheric effects on Titan before, during and after equinox in 2009. The Titan globes also provide an impression of the detached haze layer that extends all around the moon (blue) (credit: NASA).

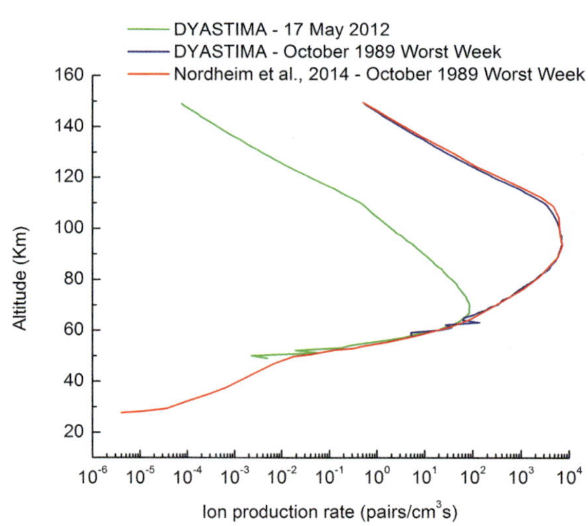

Figure 4.10 Solar energetic particles impacting the Venusian atmosphere can initiate space weather phenomena such as the ionization of the atmospheric molecules. The profile of the ionization rate varies significantly with the event. During conditions similar to those of the October 1989 and May 2012 events, the maximum of the ionization rate in the Venusian atmosphere is expected in relatively high altitudes and its intensity can be up to 105 times higher than the background (from Plainaki et al. 2016b).

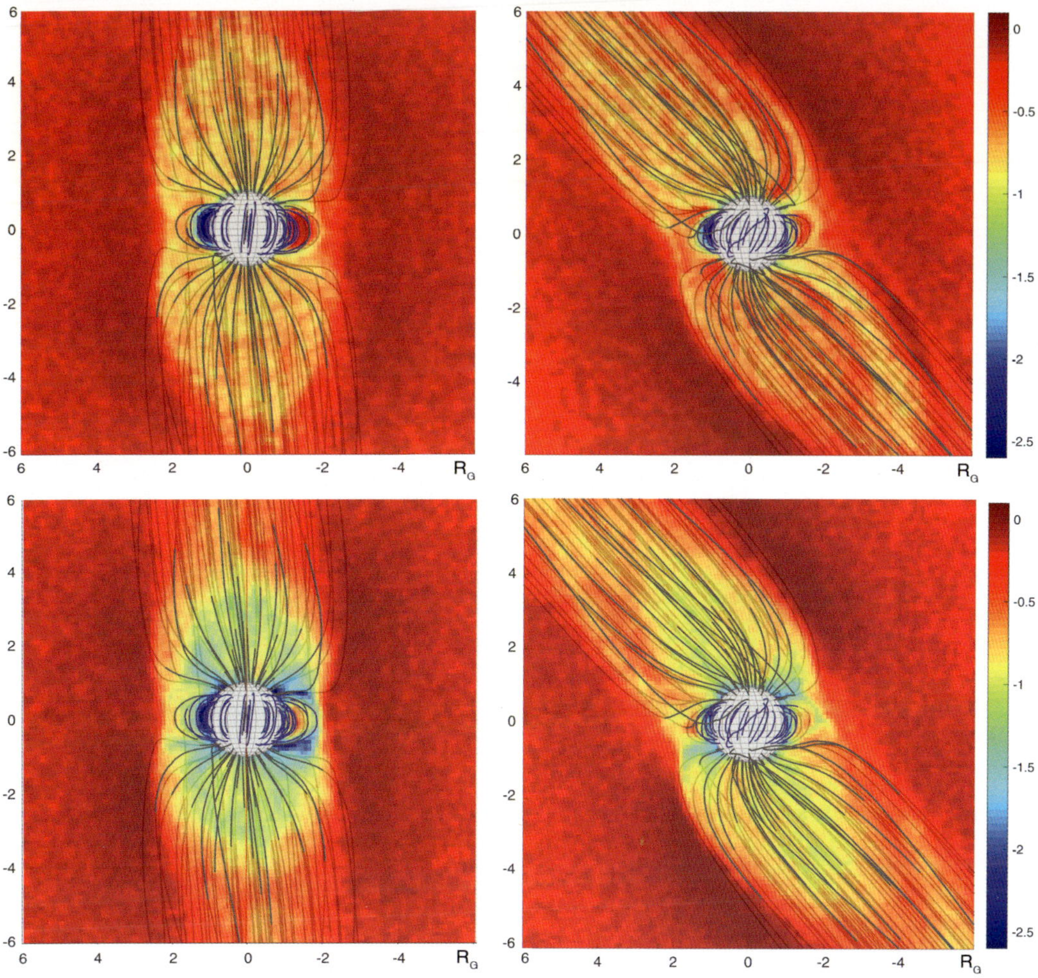

Figure 4.11 O^+ circulation in Ganymede's magnetosphere. The two upper panels refer to 10-keV O^+ and the two lower ones to 1-keV O^+. The panels on the left refer to orbital phases when Ganymede is close to the centre of Jupiter's plasma sheet (simulation consistent with the G8 Galileo flyby), whereas the panels on the right to those when Ganymede is above the plasma sheet (simulation consistent with the G2 Galileo flyby). The magnetic field configurations are given by the Jia et al. (2009) model (credit: Stefano Massetti).

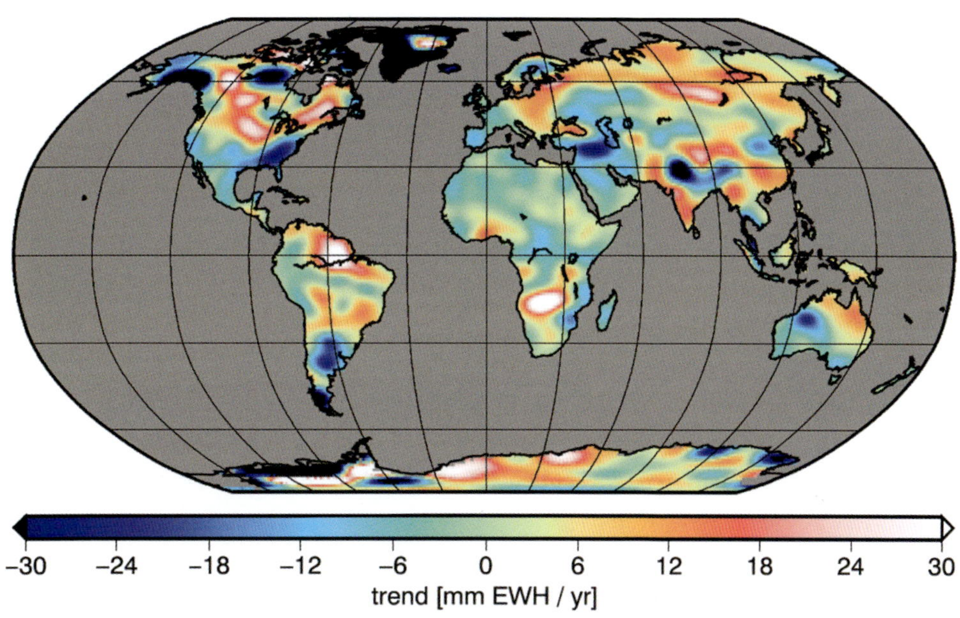

Figure 5.5 Average EWH trend for the period 2003–2012 from GRACE CSR-RL05 time series (after Eicker et al., 2016).

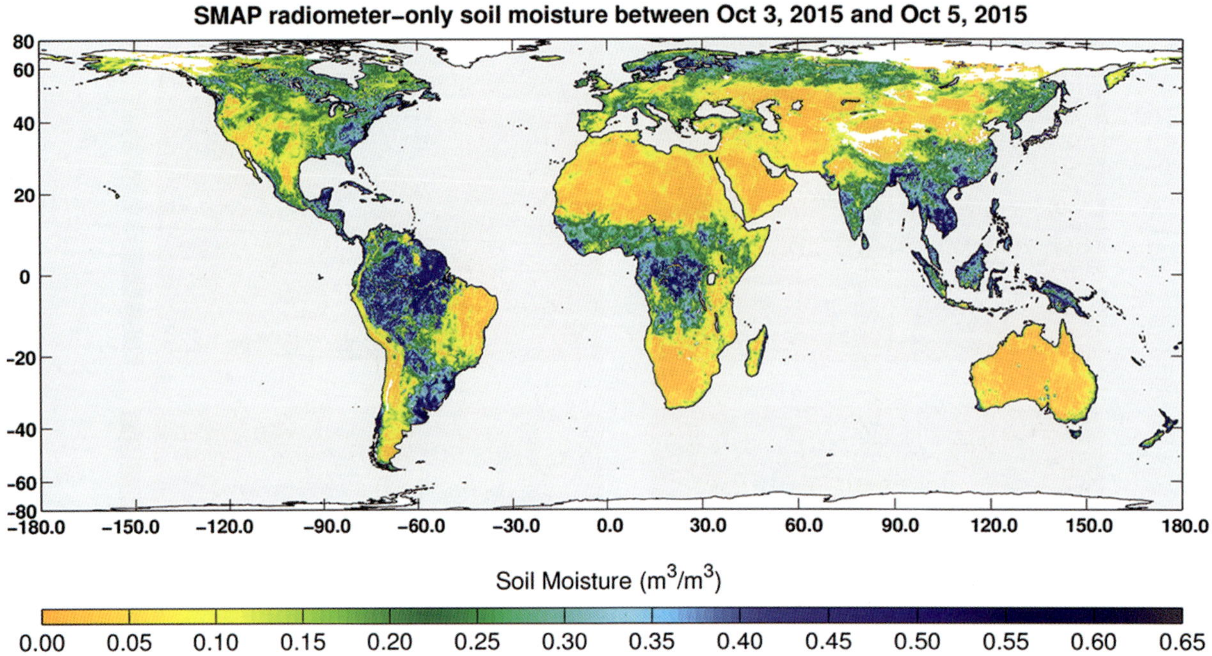

Figure 5.7 SMAP radiometer-only soil moisture between 3 and 5 October 2015 (courtesy NASA/JPL-Caltech, available at https://smap.jpl.nasa.gov/news/1250/smap-soil-moisture-data-available-at-nasa-national-snow-and-ice-data-center/).

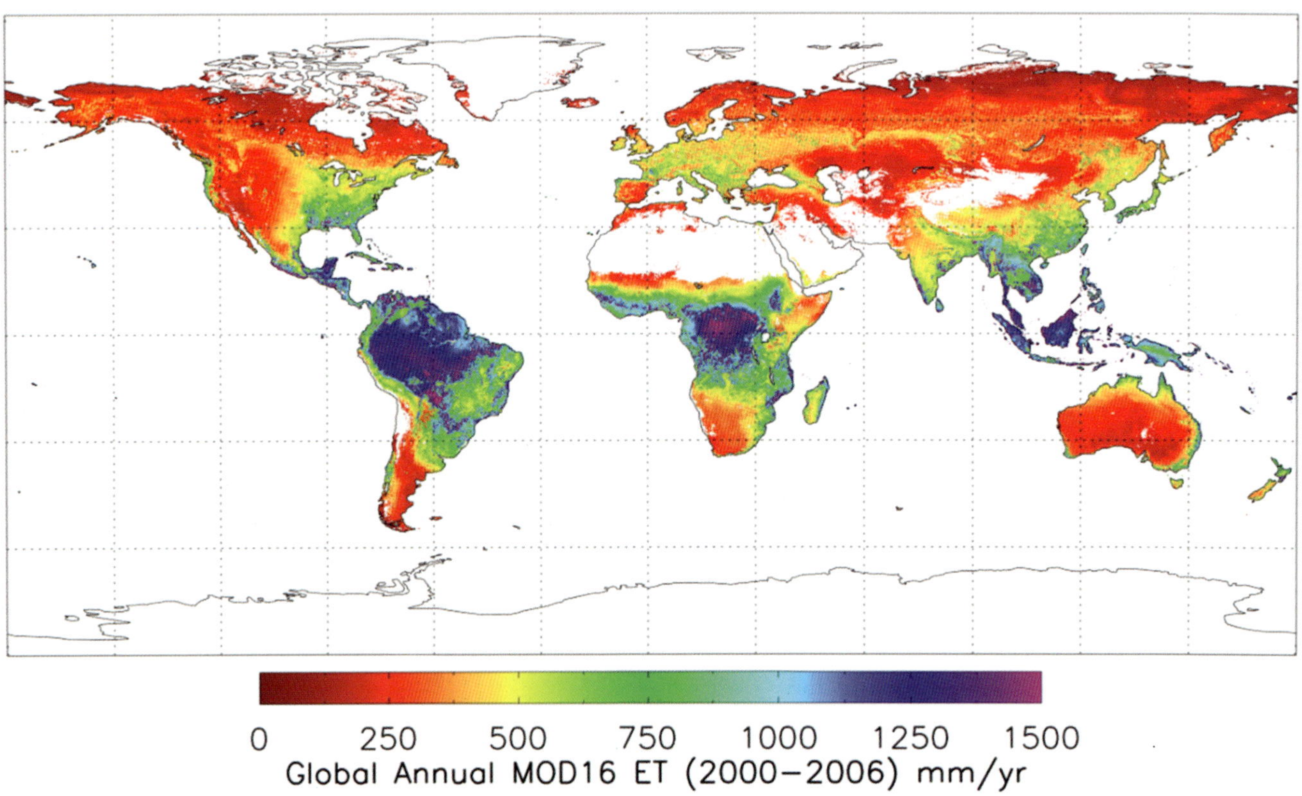

Figure 5.9 Global average ET over the 2000–2006 period (courtesy Numerical Terradynamic Simulation Group (NTSG), Univ. of Montana, MODIS Global Evapotranspiration Project (MOD16) www.ntsg.umt.edu/project/mod16.php).

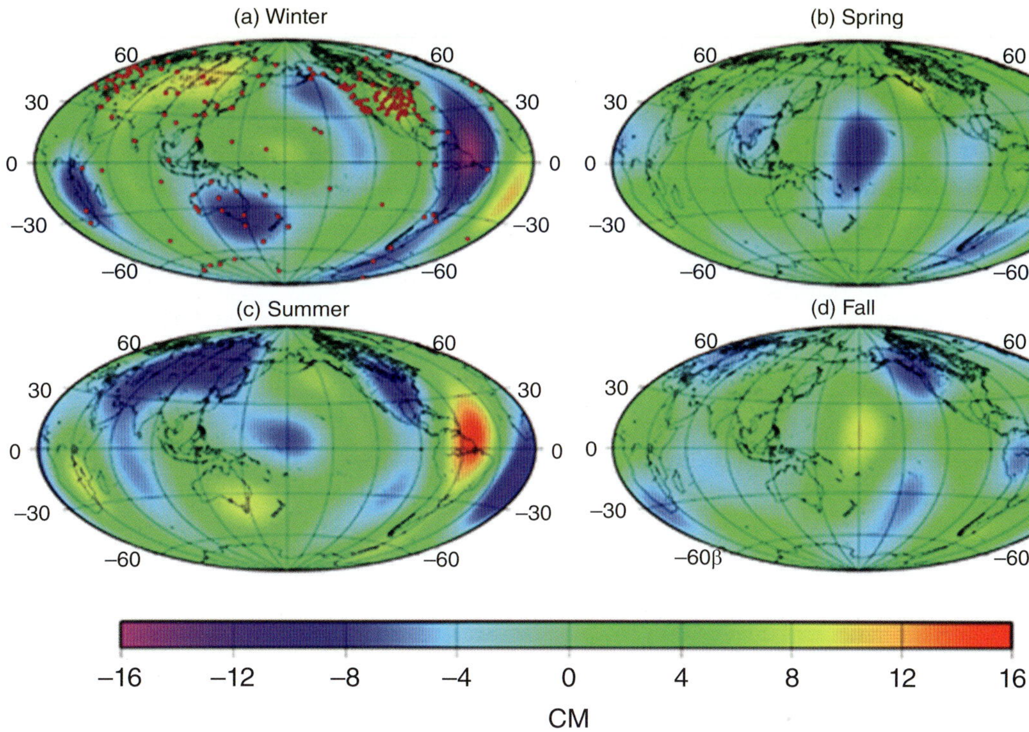

Figure 6.2 Inversion of three-dimensional displacements from two hundred global sites (red circles) over the continents plus estimates of surface displacements from ocean bottom pressure over the oceans. Inversion up to spherical harmonic degree and order 5, annual surface mass variation in centimeter equivalent water thickness. (a) Winter at mid-January, (b) spring at mid-April, (c) summer at mid-July, (d) fall at mid-October (from figure 1 of Wu et al., 2003).

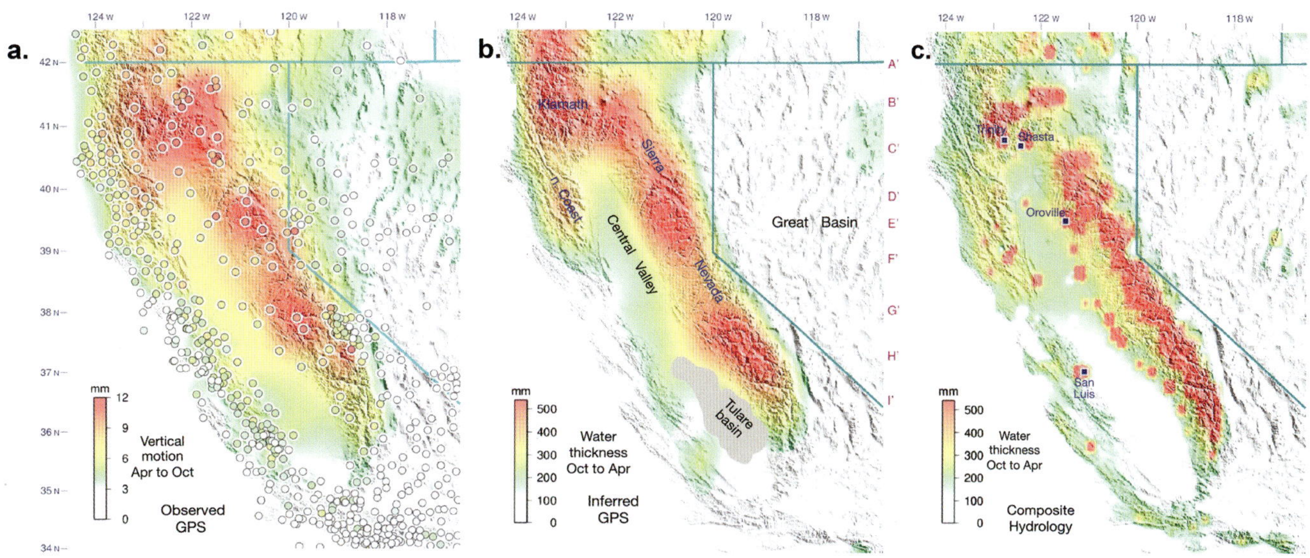

Figure 6.3 A comparison of surface uplift (a), inverted estimate of the annual change in water storage (b), and water storage change derived from a composite model of water storage (c) (composite image derived from figures 2 and 3 of Argus et al., 2014).

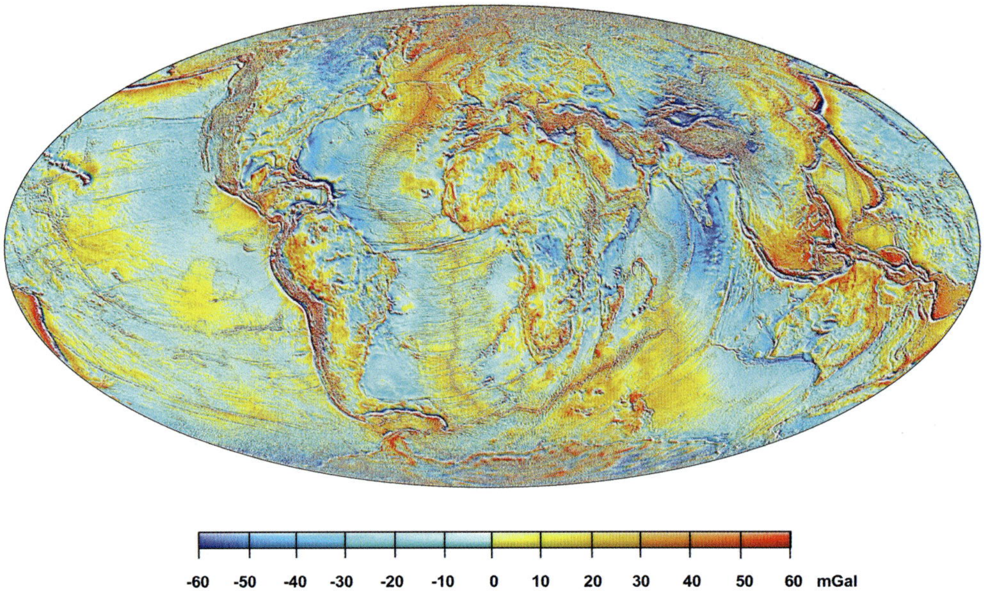

Figure 6.5 The static component of Earth's gravity field based on the EGM08 model (Pavlis et al., 2012). The figure shows gravity anomalies in mGal (modified from an original map in Cazenave and Chen, 2010).

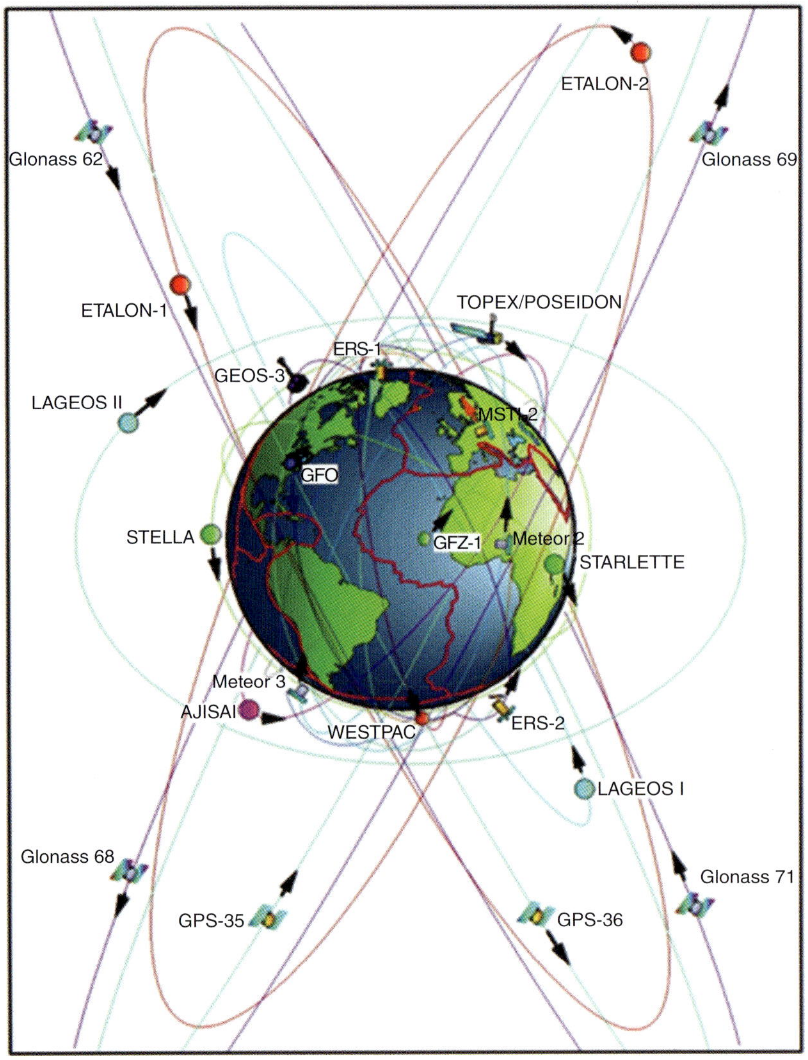

Figure 6.6 Constellation of satellites supported by SLR (source: NASA International Laser Ranging Service).

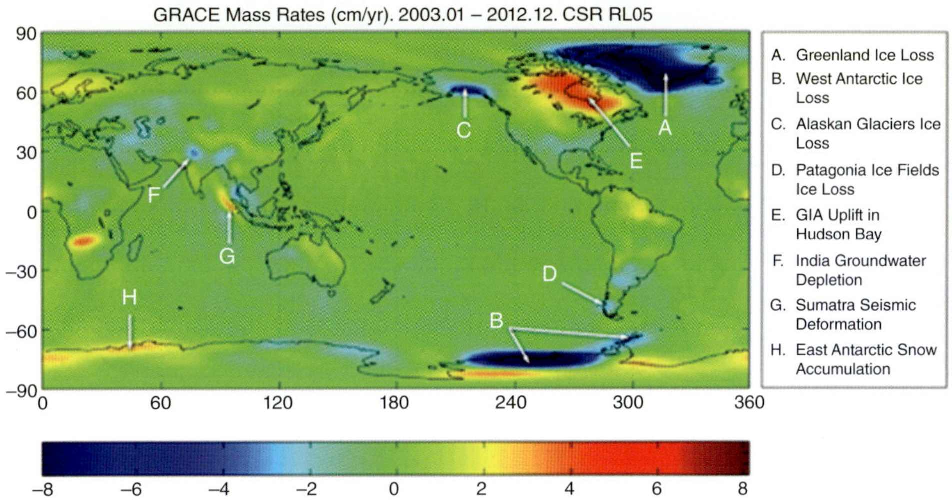

Figure 6.8 GRACE-observed long-term mass change rates (in centimeters per year of equivalent water height) over the ten-year period 2003–2012, based on the CSR RL05 monthly gravity solutions, with decorrelation filtering and 300-km Gaussian smoothing applied. Some long-term mass change signals from climate change and solid Earth geophysical processes are marked by white arrows and labeled on the right.

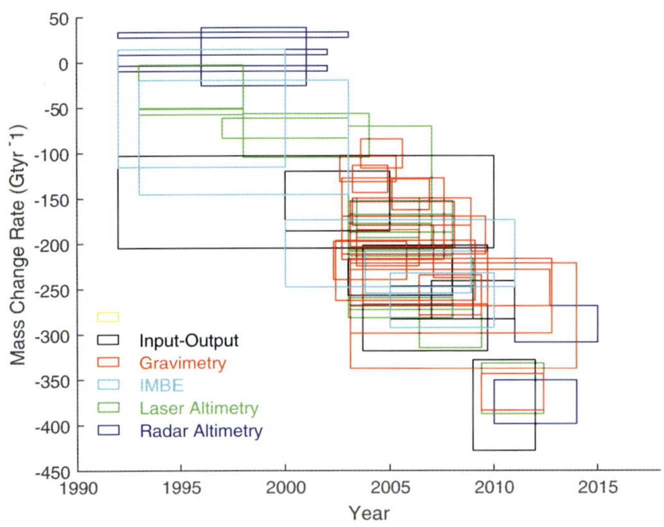

Figure 7.3 Summary of mass change estimates for the Greenland Ice Sheet. The individual studies compiled are given in Appendix 1. The coloured rectangles represent the time over which the measurements apply and the estimated range, given as (mean + uncertainty) and (mean – uncertainty) as reported in the original papers (adapted from Khan et al. 2015 © IOP Publishing. Reproduced with permission. All rights reserved).

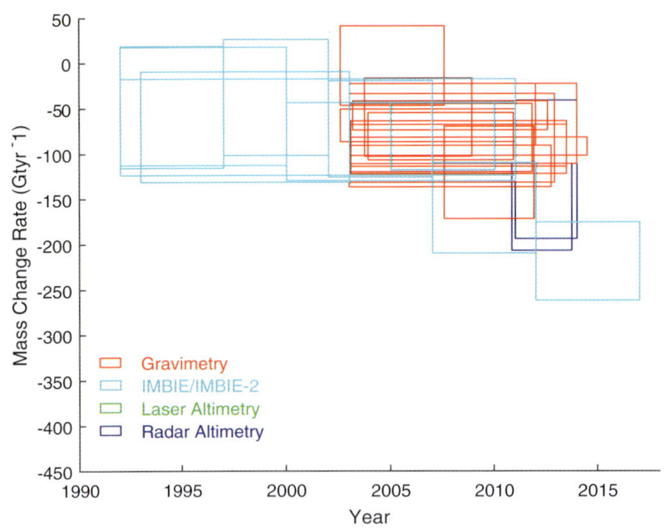

Figure 7.4 Summary of mass change estimates for the Antarctic Ice Sheet. The individual studies compiled are given in Appendix 1. The coloured rectangles represent the time over which the measurements apply and the estimated range, given as (mean + uncertainty) and (mean – uncertainty) as reported in the original papers (adapted from fig. 1, Scambos and Shuman, 2016 (copyright Cambridge University Press, reproduced with permission)).

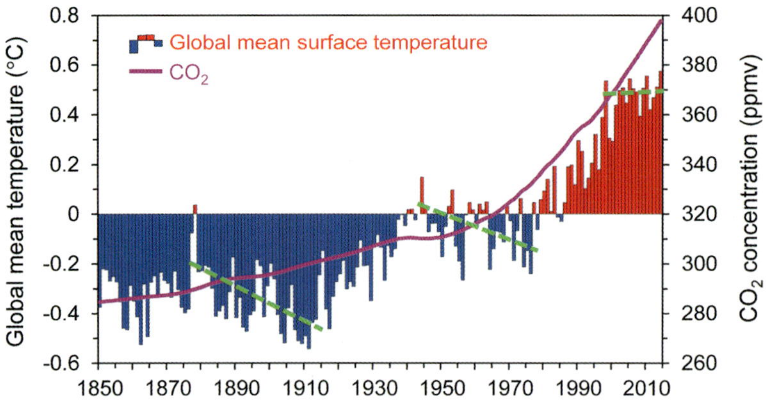

Figure 10.1 The GMT (bars) and global atmospheric CO_2 concentration (purple) from 1850 to 2014. The GMT data are from the HadCRUT4 dataset (Morice et al. 2012), relative to the base period 1961–1990. The global atmospheric CO_2 concentration is from the suite of atmospheric concentration observations and emissions estimates for greenhouse gases (Meinshausen et al., 2011). The green dashed lines show three linear trends for the periods of 1877–1913, 1944–1976 and 1998–2014, respectively.

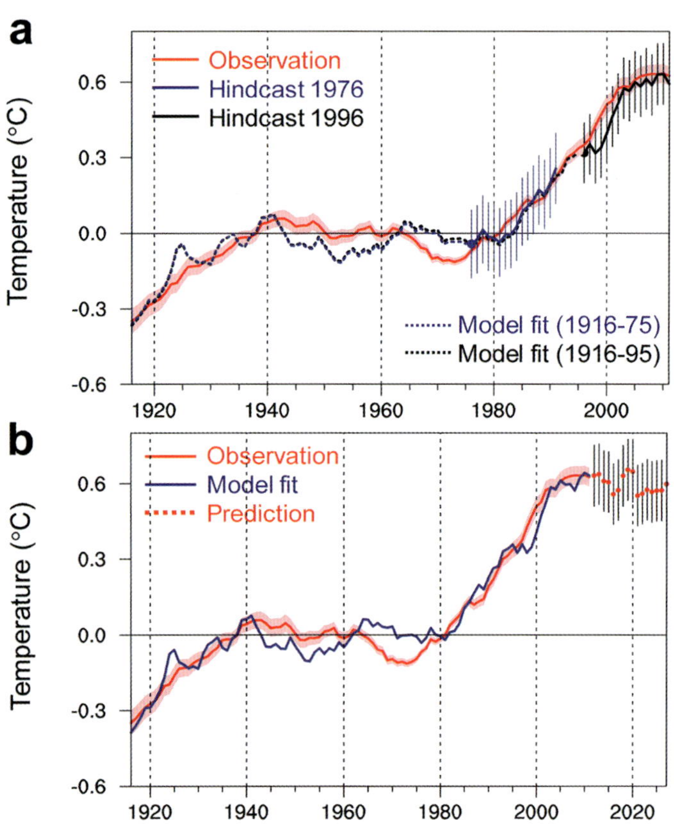

Figure 10.10 The observed, modeled, hindcast and predicted NHT. (a) The 11-year running mean of the observed NHT (red) from 1916 to 2011, and two sets of 16-year hindcast NHT starting at 1976 and 1996, respectively, using the NAO-based linear model. The solid circle represents the starting year of the hindcast and the fitted NHT before each hindcast is shown as a dotted line in the same color as the corresponding hindcast. Vertical bars denote the 95% confidence intervals for the linear model. (b) The 11-year running mean of the observed NHT (red) from 1916 to 2011, the model fit of NHT (blue), and the predicted NHT in 2012–2027. The pink shaded areas in (a) and (b) show the 2-sigma uncertainty ranges of the NHT series for the HadCRUT4 dataset. The description of the NAO-based linear model is presented in Li et al (2013a) (from Li et al., 2013a).

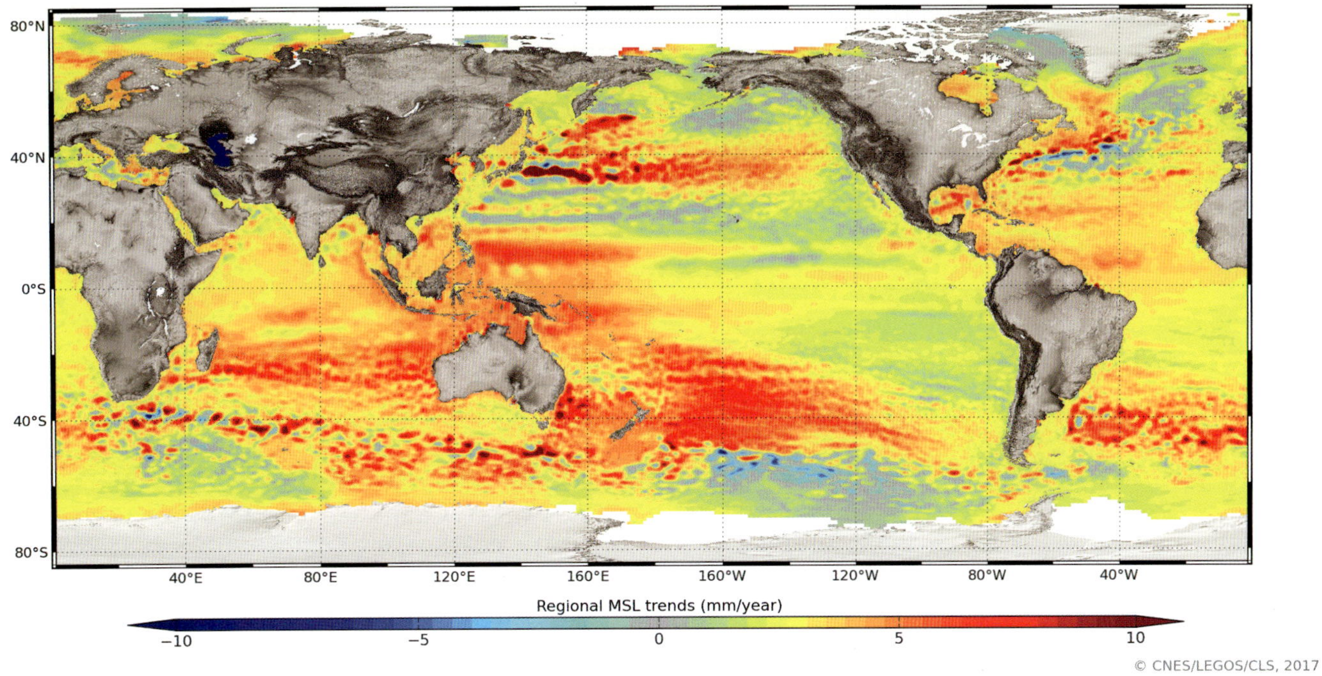

Figure 11.3 Spatial trend patterns in sea level (mm/yr) observed by satellite altimetry over the period January 1993 to September 2016. Sea level data from AVISO (www.aviso.altimetry.fr).

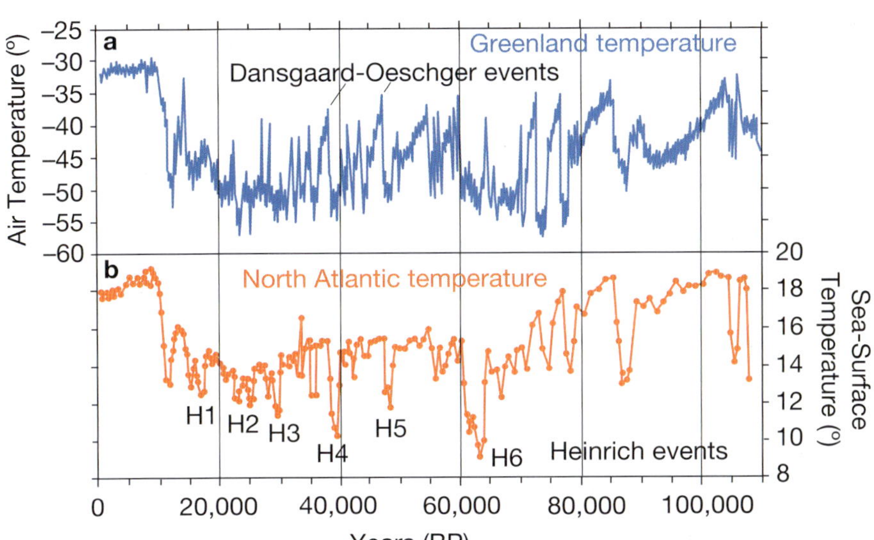

Figure 12.4 Climatic and oceanic variations in and around the North Atlantic Ocean during the past 110,000 years, as revealed in Greenland ice cores and North Atlantic sediment cores obtained off the Iberian Margin. Time progresses from right to left. (a) The Greenland air temperature, based on isotope thermometry, shows abrupt warming periods called Dansgaard-Oeschger events. (b) The sea surface temperature in the North Atlantic shows episodes of drastic cooling called Heinrich events. This record is based on biomolecular thermometry with long-chain (thirty-seven-carbon) organic molecules called alkenones measured in the sediment cores collected by E. Bard and colleagues (Bard, 2002).

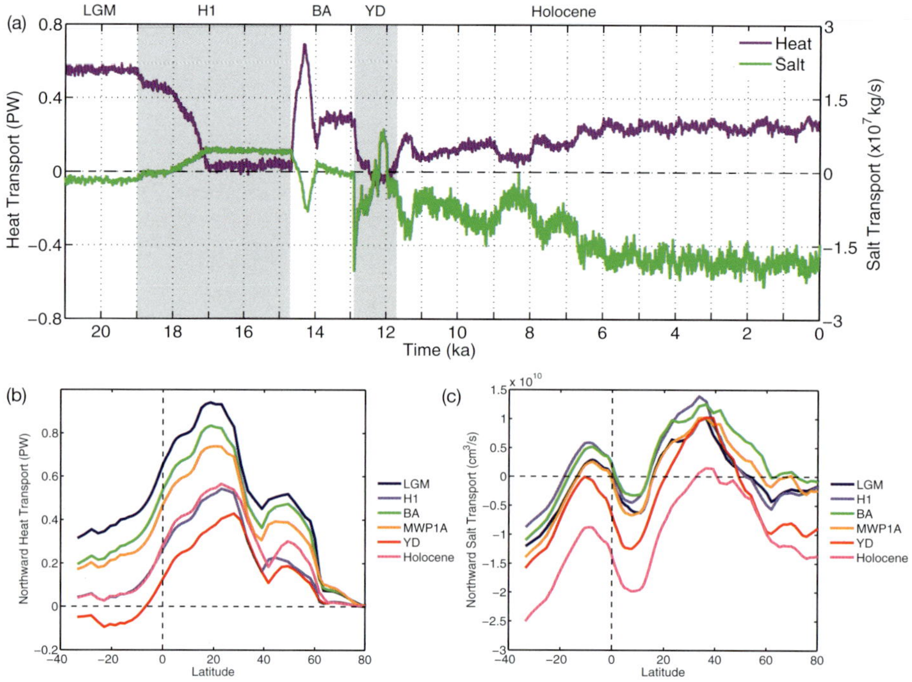

Figure 12.9 (a) Simulated meridional heat (purple) and salt (green) transport in the South Atlantic between 30°S and the equator. Positive values denote northward transport. The main climatic events during the last deglaciation are also marked: Last Glacial Maximum (LGM), Heinrich event 1 (H1), Bølling-Allerød (BA), Younger Dryas (YD) and the Holocene. (b) Northward transport of heat (PW) and (c) northward transport of salt ($cm^3\ s^{-1}$) averaged for each key period when freshwater was injected into the ocean (from Marson et al., 2014).

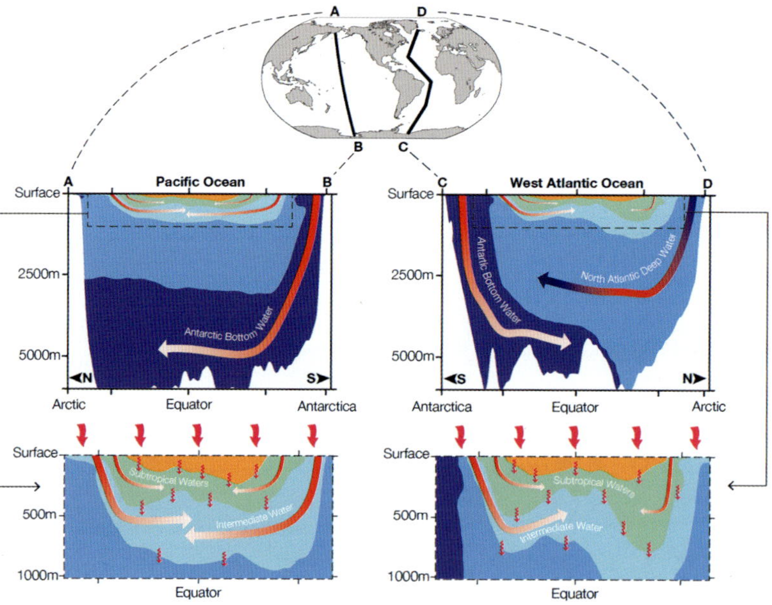

Figure 12.12 Ocean heat uptake pathways. Cold Antarctic Bottom Water (dark blue) sinks around Antarctica then spreads northward along the ocean floor into the central Pacific (upper left panel: red arrows fading to white indicate stronger warming of the bottom water most recently in contact with the ocean surface) and western Atlantic Oceans (upper right panel), as well as the Indian Ocean (not shown). Less cold, hence lighter, North Atlantic Deep Water (lighter blue) sinks in the northern North Atlantic Ocean (upper right panel: red and blue arrow in the deep water indicates decadal warming and cooling), then spreads south above the Antarctic Bottom Water. Similarly, in the upper ocean, cool Intermediate Waters (cyan) sink in sub-polar regions (red arrows fading to white indicating warming with time), before spreading toward the equator under warmer Subtropical Waters (green), which in turn sink (red arrows fading to white indicate stronger warming of the intermediate and subtropical waters most recently in contact with the surface) and spread toward the equator under tropical waters, the warmest and lightest (orange) in all three oceans. Excess heat or cold entering at the ocean surface (top curvy red arrows) also mixes slowly downward (sub-surface wavy red arrows). Figure is FAQ 3.1 (Figure 1 in Rhein, Rintoul et al., 2013).

Figure 15.3 Vegetable garden in urban Dar es Salaam, Tanzania (G. D. Ndossi).

Figure 15.4 Vegetable garden in Fuzhou, Fujian, China (G. D. Ndossi).

Figure 16.4 The trend towards smart phone–managed food systems: an unstaffed or 'peopleless' food and meal vending outlet in Taipei.

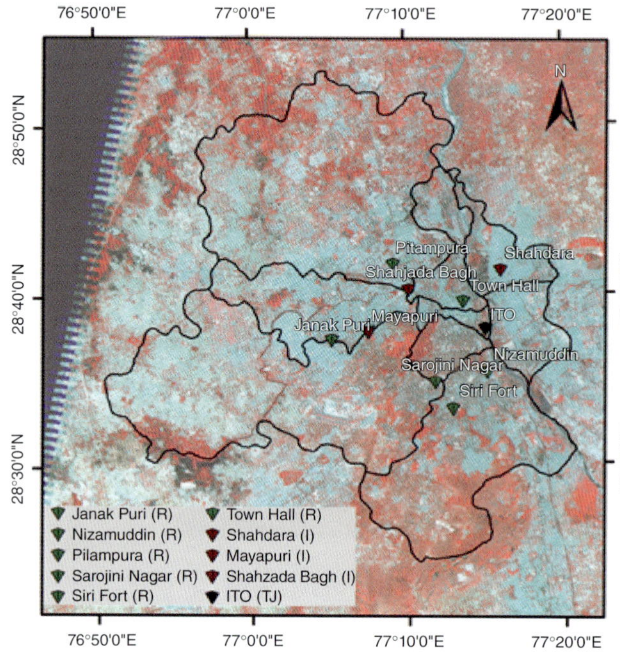

Figure 17.2 Location and nature of air quality monitoring stations in Delhi (background image is Landsat TM 5). The green symbol represents residential sites, red represents industrial sites and black represents traffic junction. (Source: CPCB, 2012; background image is taken from earthexplorer.usgs.gov.).

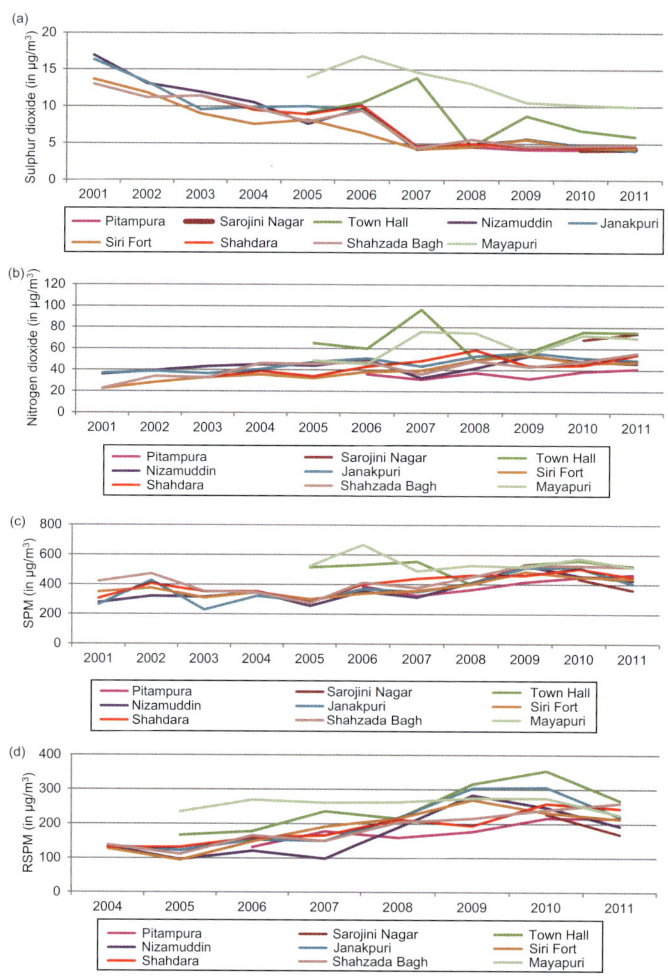

Figure 17.3 Annual trend of major pollutants in Delhi, 2001–2011. a. SO_2, b. NO_2, c. SPM and d. RSPM. The figure is prepared by the author using Central Pollution Control Board (CPCB) daily pollutant level data. Note: RSPM data not available until 2003; x-axis represents years and y-axis represent pollutant levels in $\mu g/m^3$.

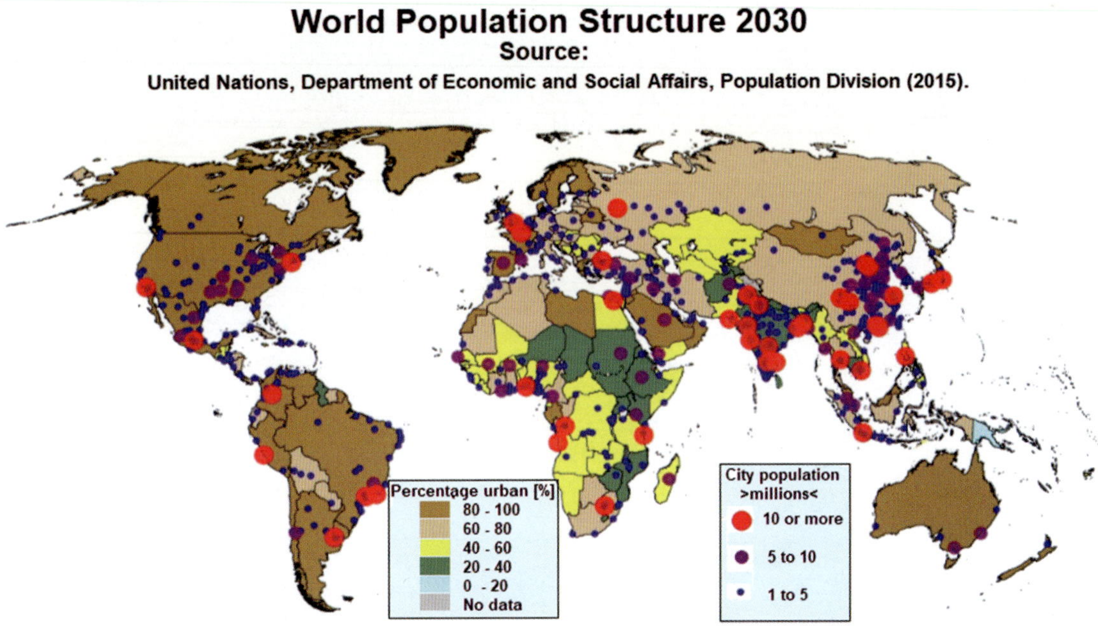

Figure 19.6 (b) Urbanization Structure [Alex][UN-Pop].

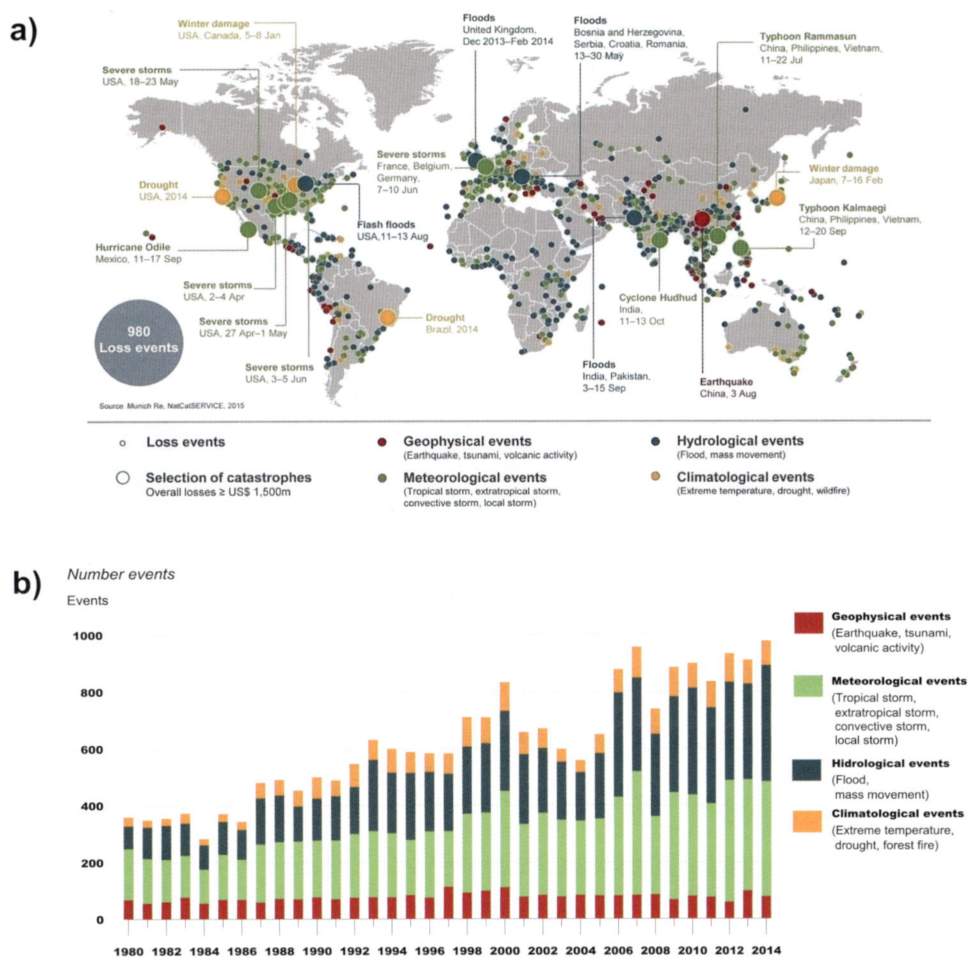

Figure 23.2 Natural disasters recorded in the NatCatSERVICE of Munich Re. (a) Worldwide major loss events in 2014 (from Munich Re, 2015). (b) Histogram of number of disasters worldwide between 1980 and 2014, divided into geophysical (earthquakes, tsunamis, volcanic eruptions), meteorological (storms), hydrological (floods, landslides) and climatological (drought, hot waves, forest wildfires) events (from Wirtz et al., 2014).

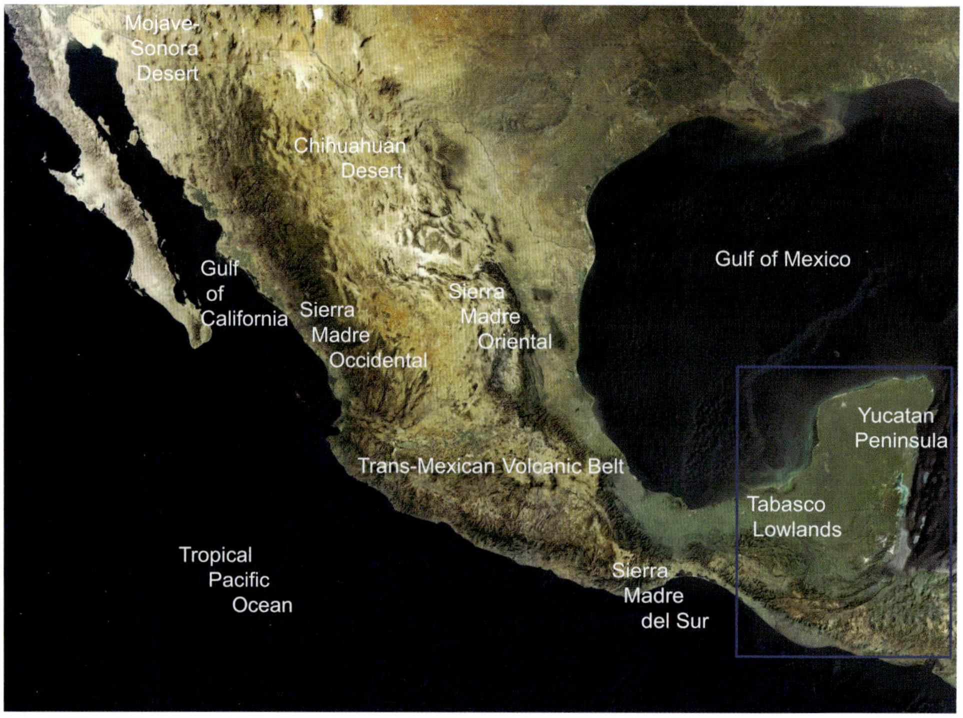

Figure 23.4 LANDSAT mosaic image of Mexico. Territory is characterized by elevated topography, with the volcanic sierras of the Sierra Madre Occidental and Trans Mexican volcanic belt and the thrust belt of the Sierra Madre Oriental and Sierra Madre del Sur. Area of southern Mexico marked in the rectangle is shown in Figure 23.8.

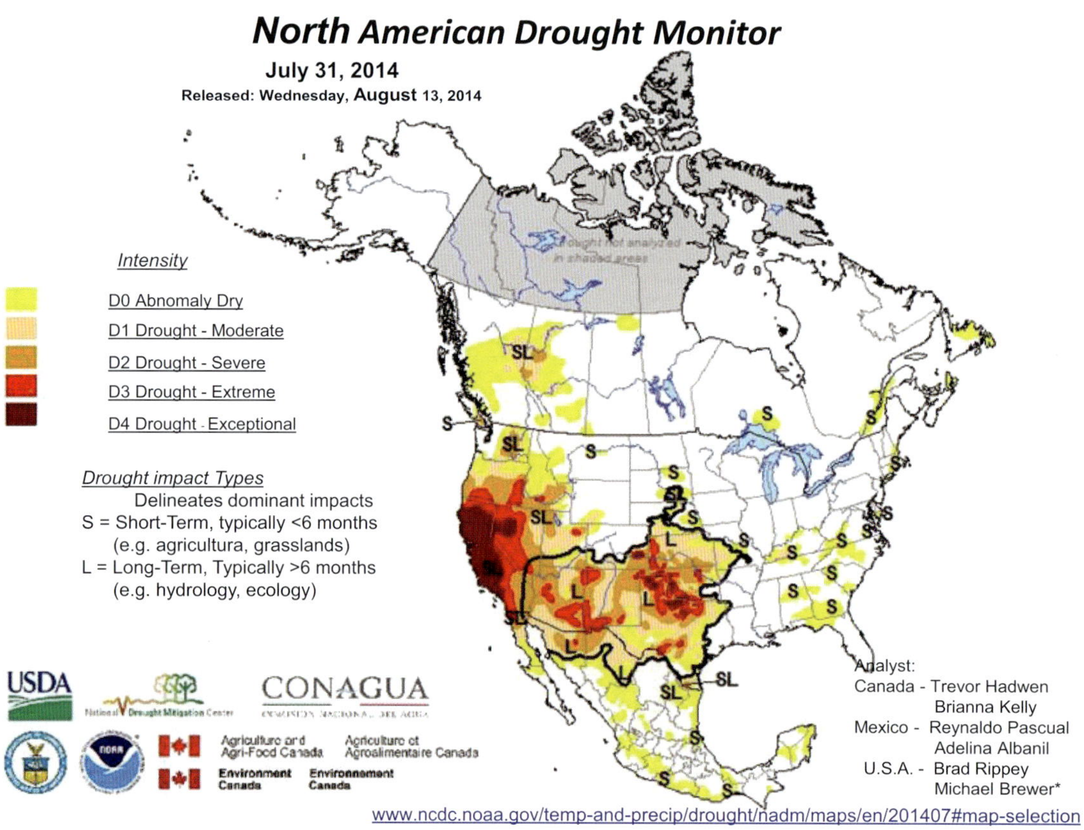

Figure 23.5 The North America Drought Monitor drought conditions and impact in North America for July 2014 (from North America Drought Monitor, August 2014).

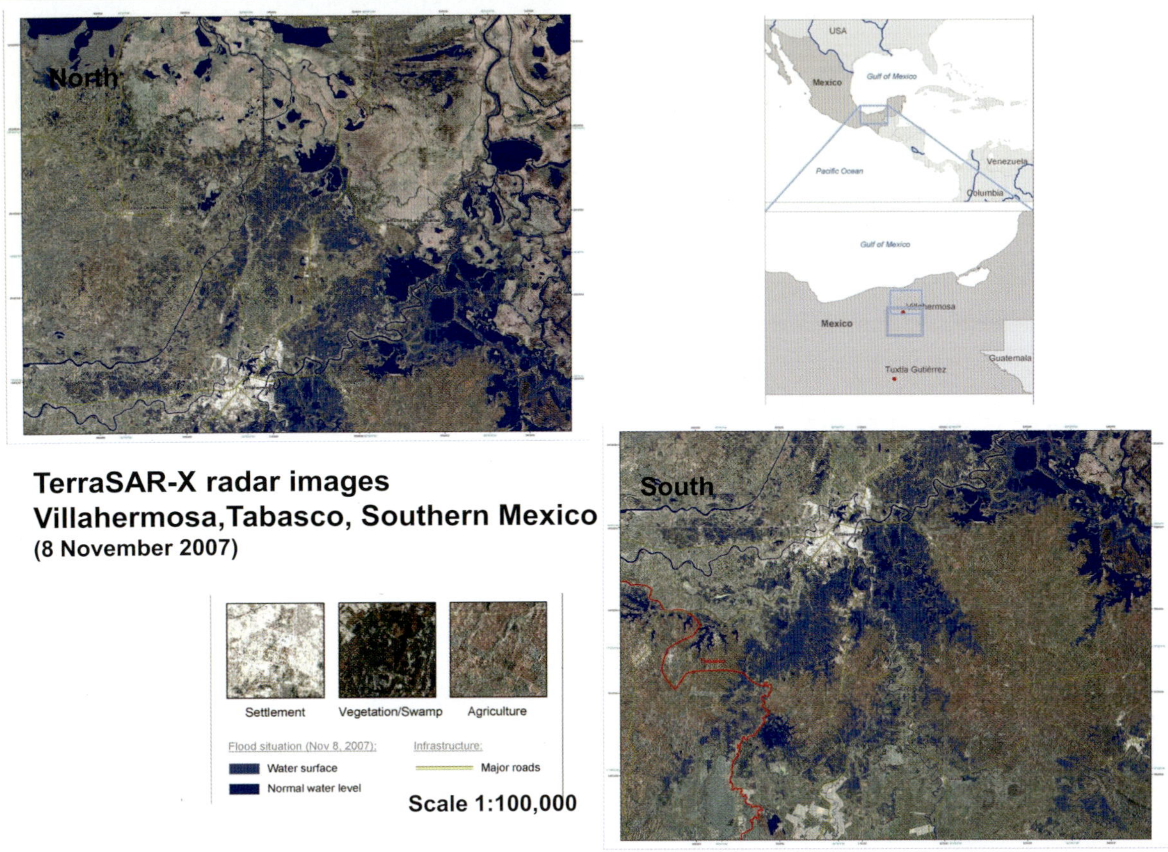

Figure 23.7 TerraSAR-X high-resolution radar satellite images of the Villahermosa region, Tabasco State, southern Mexico. Images of northern and southern Villahermosa area in scan mode taken on November 8, 2007 (SAR images from German Aerospace Center DLR, Center for Satellite Based Disaster Information ZKI flood maps).

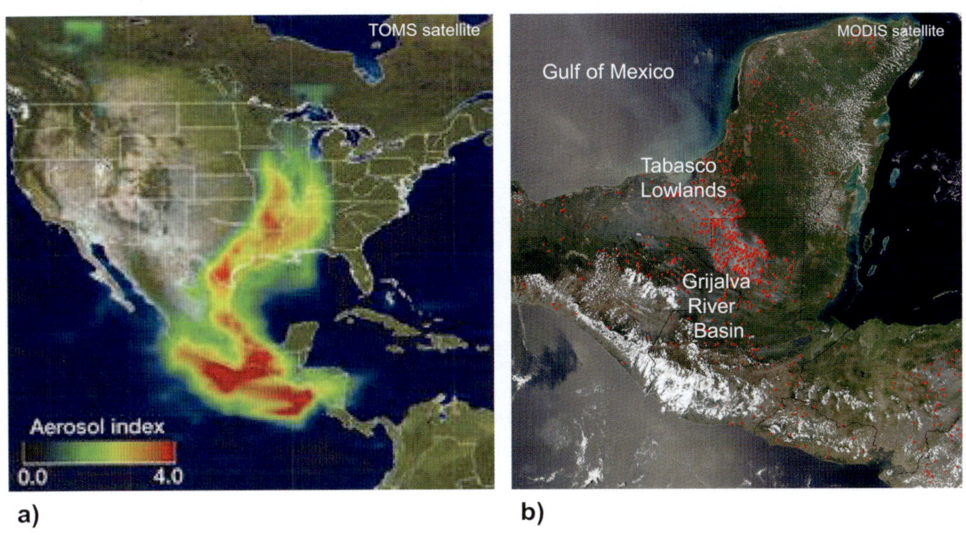

Figure 23.8 (a) Satellite TOMS image showing a smoke plume as it moved through North America (Kaufman et al., 2002). Plume raised from wildfires in southern Mexico (image on 15 May 1998; by J. Herman, NASA/GSFC; Scores of fires were burning in Mexico and Central America on 16 April 2003. (b) Satellite MODIS image (moderate resolution imaging spectroradiometer from Aqua satellite) of southern Mexico and northern Central America. Image shows location of wildfires, marked in red (image on 16 April 2003; by Jeff Schmaltz, MODIS Rapid Response Team, NASA/GSFC).

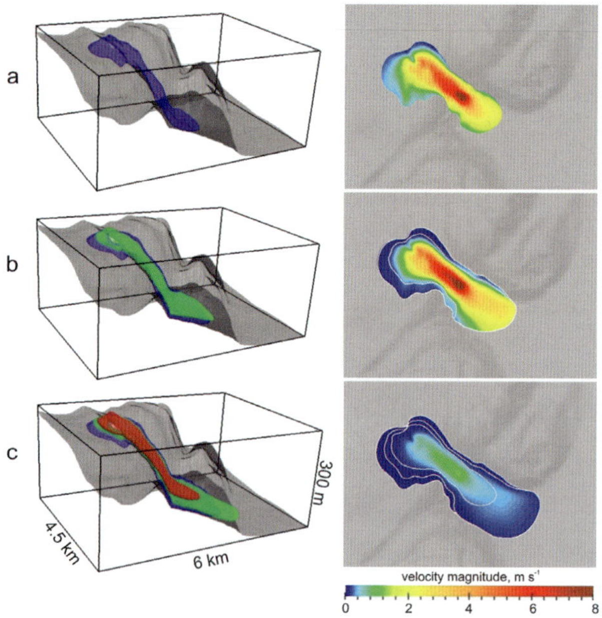

Figure 26.3 Model of the lava flow with compositional changes; snapshots at times 9000 s (a), 33000 s (b), and 63000 s (c). Blue-, green-, and red-coloured lavas are modelled fluids 1, 2, and 3, respectively (see Table 26.1 for their physical properties). Each time snapshot presents a 3-D view of the flow (left panels) and the velocity magnitude (right panels). The white curves in the right panels show the extent of lava flow at the end of the second and third episodes of lava effusion (after Tsepelev et al., 2016).

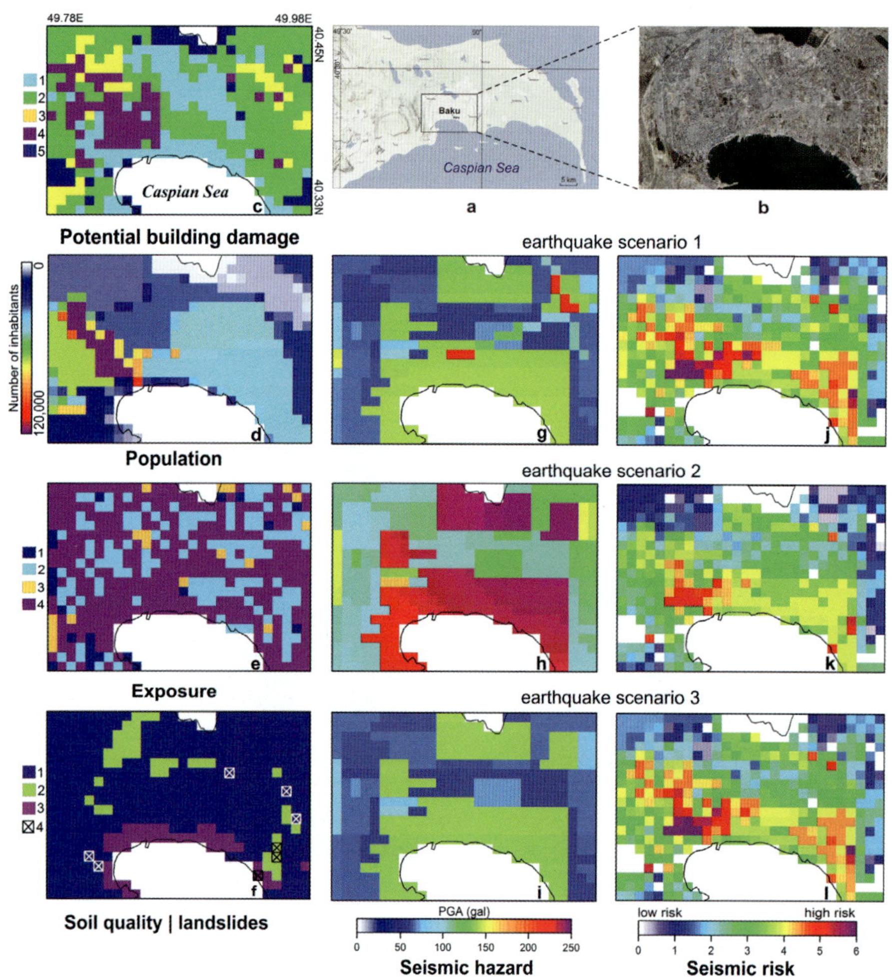

Figure 26.4 Earthquake risk for Baku, the capital city of Azerbaijan. (a) Topography of the Absheron peninsula and the location of Baku (the studied region is marked by a solid rectangle). (b) Satellite image of the studied region. (c) Distribution of building constructions: high (type 1), good (type 2), good to poor (type 3), and poor (type 4) quality. Construction-free areas are marked by 5. (d) Distribution of urban population. (e) Distribution of exposure. 1: low-value objects (park, lake, and other natural objects); 2: moderate-value objects (main road, square, and their intersections); 3: high-value objects (culture-valued objects); and 4: very-high-value objects (residential objects, hospital, school, commercial, governmental and state buildings, nation-valued object, important communication system). (f) Soil types. 1: Quaternary sand, clay, intrusions of limestone, and sandy clay; Upper Tertiary limestone and sandstone (minor amplification of shaking). 2: Quaternary sand, gravel-pebble, and limestone with clay intrusions (moderate amplification of shaking). 3: water-saturated sand, clay, and rubble (significant amplification of shaking). 4: landslide area. Computed surface peak ground acceleration (seismic hazard, g-i) and seismic risk (j-l) for three earthquake scenarios. Coastlines are marked by black curves (modified after Babayev et al., 2010).

16

Nutrition Science and Future Earth: Current Nutritional Policy Dilemmas

Mark L. Wahlqvist

16.1 Responsibilities

The over-riding responsibilities of nutrition science now and for the foreseeable future are those of food security and health for Earth's people and for planetary ecology. This means for the nutritional well-being of all things animate, but taking full account of the inanimate world and the cosmos in which we find ourselves (Figure 16.1). To meet these responsibilities is a formidable prospect. But we can take some encouragement from the food security heroes of the past, who understood their present and the limitations of what they decided to do. I nominate the examples of Tu Giay, the Vietnam peoples' doctor, who found new ecologically sound ways of ensuring food security in the face of conflict (Wahlqvist, 2015b), and Norman Borlaug, the architect of the Green Revolution that saved millions from starvation through plant breeding and fertiliser application (Quiñones et al., 1997; Briney and Writer, 2015). Borlaug, who laboured well into later life in his discipline, insisted in one-on-one discussion that there could not be another green revolution since, no matter what the plant potential, nutrient availability as fertiliser, notably phosphate, was finite. A broader perspective would add those who contributed the sociology and science of fertility control, family planning and population science to our ability to ensure food security (Seto et al., 2012); Margaret Sanger and Gregory Pincus stand out as the strategist and reproductive scientist team who led this revolution, which is still sadly short of its objectives in much of the world (Watkins, 1998; Lutz and Samir, 2010).What has failed us dismally until now is the misuse of fossil fuels as an energy source to replace our own efforts for transport and, particularly, for the entire food system (Wahlqvist and Lee, 2006; Wahlqvist, 2009b; Wahlqvist et al., 2009) (Figure 16.2).

16.2 Premise

The premise for bothering about an account of nutrition science in Earth's future is that, without the attention and commitment of the global science community, together with an informed and motivated citizenry, the future of the planet may well be one without us. Global warming of 1–2°C, which we had almost reached in 2016, will be difficult to tolerate and survive. Hansen et al. project catastrophic ice melts, sea level rises and superstorms within decades (Hansen et al., 2016). The Chinese Academy of Science projections for warming on the Tibetan plateau indicate that all of the rivers of Asia that emanate from there will have largely dried in twenty to thirty years, affecting more than 2 billion people (Kang et al., 2010; Yang et al., 2011; The Third People, 2015). The numbers of climate change refugees will be unprecedented. In the meantime, countries such as Australia are troubled by a few hundred refugees arriving in boats; Europe and the United States by millions. Anticipation, planning and mitigation of the climate change–associated food, nutrition and health crises are obligations that nutrition science, among the many sciences and technologies, must share. We now, since the industrial revolution of the eighteenth century, live in an epoch recognised as the Anthropocene or human-induced by many (Steffen, 2014; The Guardian, 2016). Nutrition science and scientists must quickly accommodate to this reality and work towards the prevention and mitigation of the adverse effects of climate change. That said, there is a cumulative knowledge base and wisdom from past involvement in poverty and hunger research, in natural disaster and conflict relief that can be marshalled as of now in anticipation of the difficulties ahead. A landmark, if somewhat neglected, example is of the nature of individual, family and community adaptation to restricted energy intake by Payne and Lipton (Payne and Lipton, 1994). It is also evident that

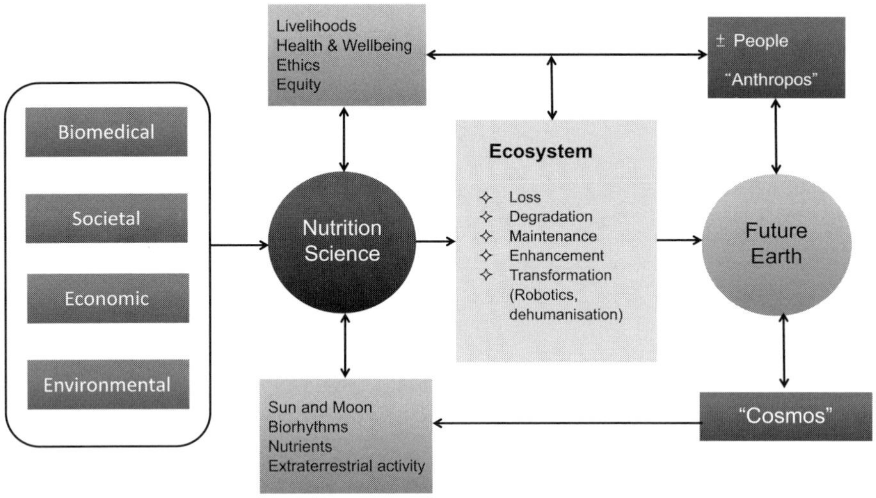

Figure 16.1 The pathways by which nutrition science may be developed and contribute to ecosystem status and Earth's future as it relates to people ('anthropos'), their livelihoods and well-being and to the 'cosmos' through nutrient provision and biorhythms.

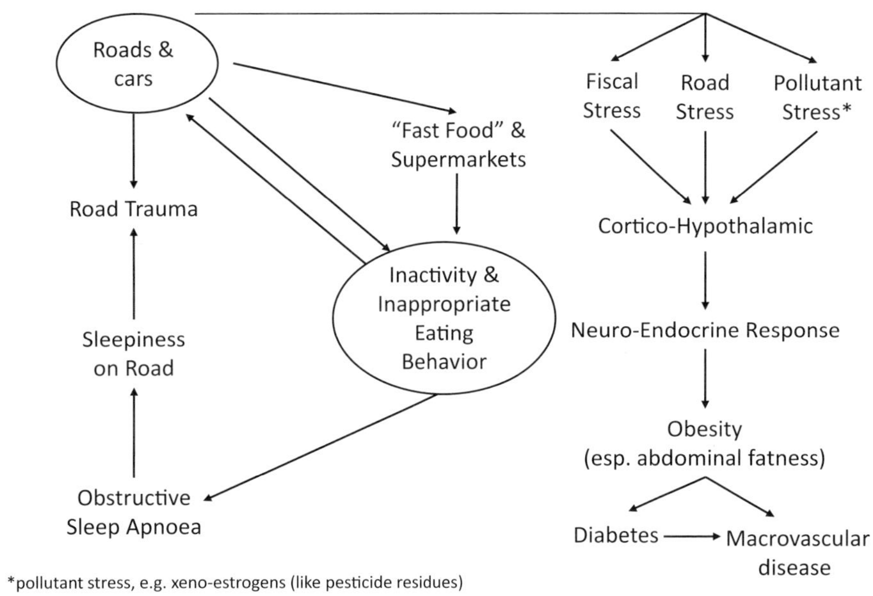

Figure 16.2 Fundamental and intermediate causes of ecologically and nutritionally-related disease (END): The example of eating, activity, cars and the road (based on Wahlqvist ML & Lee MS. *J Med Sci* 2006; 26(5):157–164).

increased exposure to fast food outlets by Taiwanese schoolgirls is associated with increased growth velocity, which is probably connected to the increased risk of pre-menopausal breast cancer among Chinese women (Chiang et al., 2011). These observations among others demonstrate what a questionable and fraught prioritisation by international organisations is that presently of 'stunting', as against adaptive and healthy shortness, without a contextual, food-based life course, intergenerational and clinical appreciation of its meaning (Payne and Lipton, 1994; Cannon, 2002; Samaras, 2011; World Health Organization, 2014). The question of optimal stature needs to take account of the impact of size on the food system, especially as global population numbers increase.

The health consequences of climate change mediated through the food system are now well rehearsed (Butler, 2009a; Butler, 2009b; Wahlqvist et al., 2009; Bulter, 2014). The food system does not meet the required expectations when it does not deliver a biodiverse diet, but this may be for reasons within and beyond the food system such as governance and climate change. Agriculture in the broadest sense, along with home gardens and seafood harvest, as well as family fish ponds and fishing are the ultimate sources of our food, with hunting largely a thing of the past, except for indigenous peoples. These facilities are all ecosystem dependent. Thus, merging the food and health systems as the planetary capacity to meet food needs is an overdue imperative (Wahlqvist et al., 2009; Dube et al., 2012).

16.3 New Nutrition Science

Towards the end of the twentieth century, population increase became asymptotic and the patterns of food consumption and health were changing rapidly by historical comparison, influenced by agribusiness, globalised trade, food technology, advertising, the media and socio-economic differentials (World Health Organization, 2013; FAO and IFA and WFP, 2014; World Health Organization, 2014). Across the globe, there was and is an inadequate consumption of nutritious food and limited physical activity. Where satiation has been possible, it has been increasingly met by ultraprocessed food lacking the properties required for optimal health (Monteiro, 2009; Monteiro et al., 2011; Wahlqvist, 2016b). Food-based dietary guidelines, which captured the current understanding of scientific food and health principles, ought to have allowed this evolving health scene to be checked and channelled in a more beneficial way (World Health Organization and Food and Agriculture Organization of the United Nations, 1995; Wahlqvist et al., 1999). However, they have not been successfully operationalised where they are most needed (Wahlqvist, 2009a). With rare exception, those being communities where there is collective commitment to healthy food systems as in EPODE in Europe and its counterparts in Australia, childhood obesity, alongside so-called stunting, in socioeconomically disadvantaged communities continue unabated (Borys, 2011; Pincock, 2011). It has been hoped that the advent of digital technology might better connect households and communities to network more successful approaches to nutritionally related health problems (Wahlqvist, 2009a; Smith et al., 2014; Jung et al., 2015; Kong et al., 2017).

While nutrition science is by necessity multi- and inter-disciplinary, the imperative for it was advanced by a working party of the International Union of Nutrition Sciences (IUNS), which reported in Giessen in 2005 (Beauman et al., 2005; Leitzmann and Cannon, 2005; Wahlqvist, 2005; Wahlqvist, 2008). Nutrition science was seen to have three dimensions, biomedical, societal and environmental. Subsequently, some of the working party proposed that the economic aspects of nutrition science should be more explicit than understood by 'societal' and so this was added as a fourth dimension. The basic concept has remained and been widely promulgated and implemented. It will be of increasing importance as the Earth undergoes major change in the foreseeable future.

16.4 Ecosystems and Their Ecological Creatures

There is increasing evidence that our health is highly dependent on our environment (Hartig, 2008; Mitchell and Popham, 2008). This environment needs to be conducive to home and shelter, viable food systems, education and provision of healthy workplaces, to enable recreation and socialisation, communication and sustainable transport and offer healthcare facilities.

The environment needs a measure of 'naturalness' for synchrony with our biology. Parks and gardens are, for example, predictors of better health outcomes and survival (Soderback et al., 2004; Austin et al., 2006; Ozer, 2007). Those who engage in gardening, probably for quite complex reasons, maintain better health (Hartig, 2008). In socioeconomically disadvantaged communities, home and school gardens contribute to the immediate nutritional needs of families and provide income and opportunities for education and livelihoods in general (Brown and Jameton, 2000; Galluzzi et al., 2010; Block et al., 2012; Williams and Dixon, 2013). The concept of 'nature deficit disorders' suggests a more fundamental understanding of who we are (Louv, 2005; Hale et al., 2011). Along similar lines, the Australian philosopher Glen Albrecht has coined the term 'topophilia' to describe "a love of place" and 'solastalgia' for "a form of psychic or existential stress caused by environmental loss" in situations where the landscape has been violated by coal mining in the Hunter Valley, a place otherwise renowned for its exceptional horticulture, including viticulture (Albrecht et al., 2007). With accelerating ecosystem loss, the question is, How much of this is tolerable in terms of how food and health systems relate (Gibson et al., 2015b; Gibson et al., 2015a)?

Wahlqvist has argued that we are 'ecological creatures', intimately connected with and dependent on our environment (Wahlqvist and Specht, 1998; Wahlqvist, 2014b; Wahlqvist, 2016a). Not to behave as we have evolved ecologically is to invite, even guarantee, that we will be dysfunctional. Perhaps we could evolve in ecological concert successfully, but the time frame is likely to be much longer than what we currently require and exceed by far (Wahlqvist, 1970).

16.5 Food and Health Systems

Surprisingly, there is a general failure in the health system to acknowledge that all bodily functions, systems and organs are nutritionally dependent (Wahlqvist and Lee, 2006). This under-appreciation of food–health relationships translates into an unnecessarily poor performance of the related food and health systems, especially across diverse ecological settings from rural to urban, by climate and by connectedness to each other for the resources required to function well (Gibson et al., 2015b; Gibson et al., 2015a). Health disorders can be defined from different perspectives, as can nutritional disorders – biomedical, societal, economic and environmental. Within each of these, there are selective perspectives that tend to change with time. For example, within the biomedical, reductionist approaches have driven a genetic and molecular diagnostic and therapeutic agenda, which is now changing as mitochondrial DNA (maternal in origin), non-coding RNA (ncRNA) and microbiomic genomes are evidently able to integrate vast amounts of environmental information (Wright and Bruford, 2011). This is over and above that keenly sought by way of single nucleotide polymorphisms (SNPs) in eukaryotic DNA as therapeutic management points. Even more fundamentally, our eukaryotic genome probably has environmentally derived endosymbiotic origins (Margulis, 1981; Wahlqvist, 2016a). In turn, this places a greater weight on the way human, societal and planetary biology interface.

The social determinants of disease, which inevitably include those reflective of the social role of food, are well established in WHO reports (Marmot et al., 2008). Similarly, health status is increasingly recognised as environmentally dependent. As ecological creatures, we merge with our environment through locomotion (exploration, hunting and gathering food, movement of many kinds, musculo-skeletal function in general, vehicular usage or non-usage), sensory inputs (sight, sound, smell, taste tactile inputs), hormonal factors (such as phyto-estrogens and endocrine disruptors from food, water and the atmosphere, whether naturally occurring or contaminants and pollutants), our microbiomes (skin, mouth, gut, respiratory, reproductive tract and breast), immune system (including the microbiomes, inhaled and ingested particles and chemicals, ecosystem dynamics and buffer zones) and natural rhythms (especially dependent on the Sun and Moon, but also, in less well-defined ways, the wider cosmos). Biorhythms are a way in which our synchrony with nature is expressed (Koukkari and Sothern, 2007). Thus, we are widely connected with the animate and inanimate world, a less discrete species than we think and our health patterns ones not so individual as we have presumed. Infectious disease is just one example of how the health of one affects the health of others; and the health of the environment affects all those to whom it relates. This has encouraged a reconceptualisation of health disorders as 'ecosystem health disorders' (Wahlqvist, 2014b).

Urbanisation (Seto and Ramankutty, 2016) has become the major form of human settlement with more than half the world's population now living in towns and cities (Cohen, 2003; Lutz and Samir, 2010; Seto et al., 2012; Samir and Wolfgang, 2014). It is driven by population growth and by the quest for facilities which must be shared for the practicality and cost of their provision. It has inevitably changed the need for planetary resources required for a built environment, particularly those of energy, food and water systems and for construction materials. Its rapidity and extent now present a food security and health challenge that will continue until the human population begins to decline, even if greater efficiencies in dwelling and living are realised (Gong et al., 2012; Gibson et al., 2015a; Seto and Ramankutty, 2016). The question is whether the crises arising from over-population can be averted in the available time frame, which is probably until about 2050 (Lutz and Samir, 2010). This is a global and not just a local or regional time frame, because the multi-systems failure, as it develops, is contributing to global as well as micro-climate change. For example, cities are warmer than their rural predecessors, they represent ecosystem loss and degradation, they require energy for function and liveability, they present demands for food and water distribution more remote from source and they require powered transport and communication systems (Cincotta et al., 2000). The most dramatic effects of rapid urbanisation can be expected through the loss of safe agricultural and potable water, already evident in every region of the world. The emerging loss of Asian river systems, with its dire consequences, has been identified at the beginning of this discourse. Prolonged drought in Australia in the early part of the twenty-first century contributed to food pricing and security abroad for its trading partners (IFPRI, 2008). Conversion of food to biofuel agriculture was also a major factor (Mitchell, 2008). No less consequential are the increased frequency and severity of storms and, depending on hemisphere, typhoons, hurricanes and cyclones. In Asia and the Pacific, food prices rise sharply and poverty is exacerbated when typhoons affect agricultural areas (von Braun and Bos, 2005; Tuyen le, 2009; Wahlqvist, 2014a; Koh et al., 2016; Numbeo, 2016). Not only are there short-term, but also long-term health consequences of these events. The solutions to the water shortage dilemma will be sought in more efficient digitally controlled delivery and use where possible, through more and more diversions and through desalination of sea-water (Wahlqvist, 2016). The latter will itself be demanding of energy and threaten ecosystems with its salt by-product. Insofar as storms and floods are concerned, the prevention resides firmly with the containment of global warming, principally and urgently with the need to abandon fossil fuels (Hansen et al., 2016). For the management of the events and their sequelae, who can live and harvest food where, as determined by fishing rights, land tenure and opportunities to migrate, are

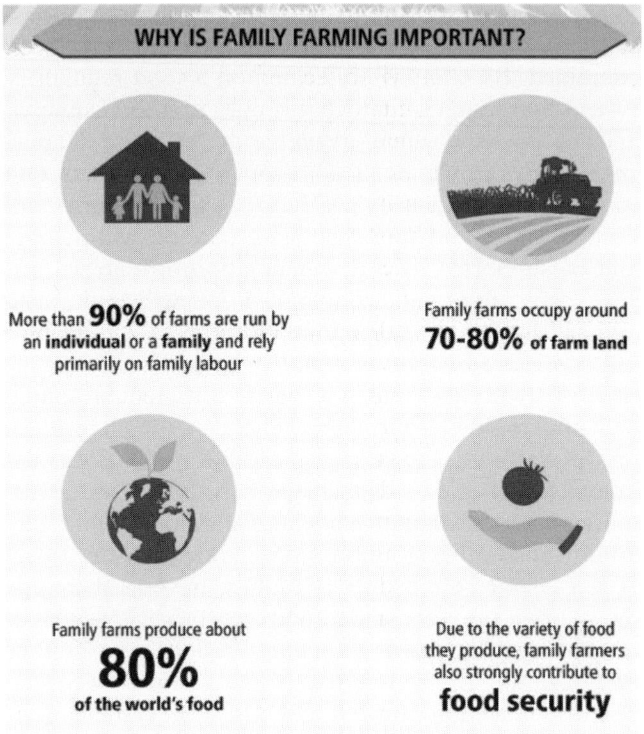

Figure 16.3 The value of family farming (from FAO, International Year of Family Farming 2014 www.gao.org/family-farming-2014/en/).

key. There are opportunities to maintain productive food systems within urban settings, and not only by trade between urban and rural households and communities (Gibson et al., 2015b; Gibson et al., 2015a). Planned rapid urbanisation in China has the potential to accommodate such possibilities (Gibson et al., 2015b; Gibson et al., 2015a). For an implementable socioeconomic and ecological framework, the work of Elinor Ostrom has appeal, although its potential where the need is greatest may be compromised by conflict and crisis (Ostrom, 2009).

The irony of urbanisation in so-called human development is that it is less conducive to ecological maintenance and enhancement, which are the ultimate basis of human well-being and health, because humans are ecological beings. Family farms, defined by FAO as those that 'rely primarily on family members for labor and management' are more ecologically valuable than agri-business (FAO, 2014; FAO, 2016; Shiva, 2016) (Figure 16.3). So while we may seek food production efficiencies through aggregation of land-holdings and mechanisation, these approaches need the caveat of genuine, not simply token, ecosystem retention if we are not to hasten our own demise in the process, not knowing the limits to ecosystem requirements for our nutritional well-being and health (Wahlqvist, 2014b; Wahlqvist, 2016a). Food and health systems and security go together (Wahlqvist et al., 2009; Wahlqvist and Kuo, 2009; Gibson et al., 2015a).

16.6 Nutritional Policy Dilemmas

In formulating food and nutrition policy for optimal health, we need to have our habitat uppermost in mind. When, where and to whom we are born expose us to a particular food system and culture over which we at first have little if any control, except that we impinge on it and our parents may alter it. The circumstances determine much of our own and the planet's future (Wahlqvist, 2016a). However, wherever we are, governance and resilience are known to have major consequences for our destiny, especially that to do with socio-economic status, food and water security, education and healthcare (Ogunrinade et al., 1999). There are opportunities and risks in the first thousand days from conception for our future as espoused in the SUN (Scaling Up Nutrition) program. But for generations before and after conception, any adverse trajectory is in powerful operation (United Nations, 2012). The capacity to manipulate it 'successfully' is limited, albeit progress has been made with the UN Systems MDGs (Millennium Development Goals) and more is anticipated with the SDGs (Sustainable Development Goals) of the same UN system (United Nations, 2016). These, in one way or another, relate to food and health and are directed towards a better, more peaceful, equitable, liveable and enduring world than presently obtains (International Food Policy Research Institute., 2014).

Although it might appear as the opposite extreme to nutrition policy for the socio-economically disadvantaged, the advent of a more technological, digital world provides new opportunities to address nutritionally related health at all stages of development. Increasingly, and without land-line communication, the remotest of people can access food and nutrition information, and food and health systems themselves, but what is found will be of varying quality and reliability (Lang and Heasman, 2004; Wang et al., 2006; Topol E., 2012). The opportunity to manage one's nutritional health with risk-minimal digital technology is also an evolving science with an uncertain future (Omar et al., 2005). It is tied up with the relative merits of a personal versus a public nutritional emphasis, by an increasing quest for short-term 'efficiency' in both arenas, and encouraged by so-called precision medicine, as though that should be superior. In reality, health and nutritional status is greater than the sum of its parts, so that approaches targeted to one or an other factor will generate peculiar side effects as do highly selective pharmaceuticals.

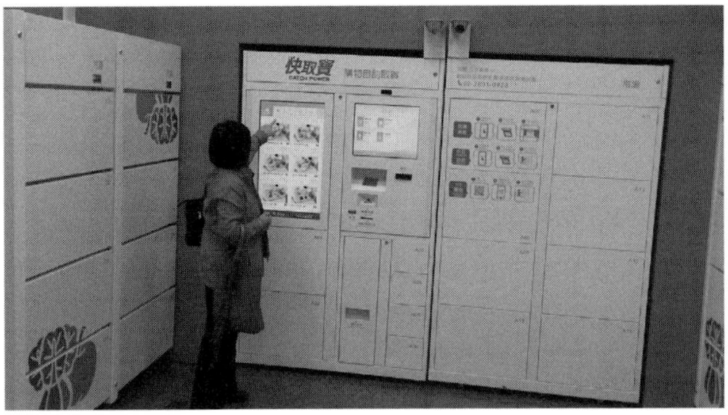

Figure 16.4 The trend towards smart phone–managed food systems: an unstaffed or 'peopleless' food and meal vending outlet in Taipei. For the color version, please refer to the plate section.

In looking to digital and robotic solutions of our future nutritional and health needs, not only must we consider their sustainability and environmental consequences, but also the role that people themselves play in determining the overall value of food. Food delivers health outcomes partly through its social role. Few landmark or festive events take place without food. Professor Hiroko Sho, with her extensive family knowledge of both Ryuku dynasties, underscores this point in regard to Okinawan longevity. Okinawans have a comparatively high number of food-based social events in any year (Sho, 2001).

While virtual technology may allow us to simulate and strategise various climate change and health scenarios, it in itself cannot conjure up food, nutrients or the physical system to produce them. At worst, it may delude people that all is well and disempower them from the effort required to anticipate and solve the problems ahead. The abilities to grow, harvest, store and cook food are just such skills that defy the virtual world although not perhaps to suggest better ways of executing these tasks. The evidence that both shopping and cooking are predictors of survival attest to this interpretation of virtual technology (Chang et al., 2012b; Chen et al., 2012). Interestingly, in informatics-rich Taiwan, one now finds side by side both the traditional fresh food market and peopleless vending shops for ready-made meals that may be instructed via a smart phone (Figure 16.4). What if the latter overtakes the former? Perhaps it is a more informed approach to food choice than that which now obtains in the West with mindless drive-through 'fast foods' taken with one hand, the other on the steering wheel! With such disengagement and lack of interest in food quality, it is no surprise that many have lost control of the food system as it pertains to their health and the planet on which they depend.

The world of additive manufacturing has been known to food processors for as long as ingredients have been combined. Its safety is predicated on a food regulatory system that requires that what is put into food is known and, as far as possible, available at the point of purchase; for example, by labelling or by a bar, QR or QQ code. Three-dimensional (3-D) food printers will become more and more the state of food technology, and be supported by a new generation of material science, especially nanotechnology. Food will look, taste and sell like natural or minimally processed food, but fall short of real food complexity by orders of magnitude. It will also have a vastly different resource demand, for better or worse. These major changes in the future food system have profound implications for human health and the planet (Berman, 2012; Tay, 2016).

Every set of dietary guidelines developed, and certainly those of WHO and FAO, known as the Food Based Dietary Guidelines (FBDGs), have as their primary advice to consume foods of biodiverse origin (World Health Organization and Food and Agriculture Organization of the United Nations, 1995). For the operationalisation of these FBDGs, food systems need to be supportive and households informed, empowered and mobilised to behave accordingly (Wahlqvist, 2009a). This has been achieved to varying extents between and within communities. The added challenge now is of rapidly declining biodiversity. Given the arguments that we are ecologically entwined obligates, how far can we safely go with ecosystem loss and how little food biodiversity can we afford?

There are serious threats to the successful implementation of the FBDGs as a way of mitigating the risks of climate change to health through dietary biodiversification.

The first is that much of the global food supply is now in the hands of those who prefer for organisational, professional, commercial, practical or even fatalistic reasons to manage the nutrition and health agenda as a reductionist 'nutrients problem' rather than a more integrative 'foods and food patterns problem'. This has steered the human diet away from one that is environmentally sensitive to one disorder-prone. The proliferation of ultraprocessed foods across all socio-economic gradients is particularly telling in this regard (Monteiro, 2009; Monteiro et al., 2011; Wahlqvist, 2016a). It is hoped that the current efforts of FAO and various NGOs, such as Food Tank (Shah, 2016) and the Nourish Network (Nourish Network, 2016), to reverse this trend gain traction.

Foods eaten in the smallest amounts nutritionally necessary is not only a valuable conservationist approach, but also one that reduces the risks of excess and toxicity. For example, less than one serving of a dairy food on most days is associated with a reduced risk of stroke in people of Chinese ancestry, for whom stroke is a major cause of morbidity and mortality (Huang et al., 2014a). At the same time, dairy becomes compatible with well-being for those who are not lactase persisters since symptoms are rarely seen above an ingestion of 25 g lactose (Lukito et al., 2015; Vandenplas, 2015; Wahlqvist, 2015a). In this way, many more people can benefit with less environmental consequence. Another potential benefit is that undigested lactose survives to the colon, where it would be fermented and contribute to microbiomic profile, epithelial integrity and the gut-distant organ functional relationships. Thus lactose might have functions akin to those of non-digestible oligosaccharides, which include anti-infection functions (Lukito et al., 2015; Wahlqvist, 2015a; Andreas et al., 2016). Similarly, tomatoes have spread globally from their origins in the Americas and become a favourite flavour and food item. Their lycopene bioavailability is greater from cooked than raw tomatoes. As tomato juice (equivalent to several tomatoes) perhaps more readily than the lesser amount in a salad, they are associated with hyperuricaemia and gout (Flynn et al., 2015), but uric acid might also be neuroprotective (Hong et al., 2015). Although relatively energy dense, a daily handful of mixed nuts has not been shown to increase the likelihood of overfatness, but consistently reduces the risk of cardiovascular disease and diabetes (van den Brandt and Schouten, 2015). This may be a general feature of seed crops such as legumes with major implications for reorienting agriculture in a more sustainable direction (Darmadi-Blackberry et al., 2004; Chang et al., 2012a; Foyer et al., 2016; Kouris-Blazos and Belski, 2016). Nevertheless, both favourable and unfavourable properties of all food crops must be taken into account; small amounts of diverse foods increase environmental stewardship and nutritional security for more people and present less risk of adverse effects.

Producing, buying and eating only what we need have the added value of food waste reduction (Lin et al., 2009; Huang, 2010) where some 30%–40% is wasted in most locations. In Taipei, household biomass collection most days of the week, with recycling either as fertiliser or as pig feed, has dramatically changed food habits and reduced waste disposal by more than 80% (Lin et al., 2009; Huang, 2010).

Soil and agricultural water contamination with industrial waste presents a major food safety problem in recently, if not previously, industrialised societies, reducing arable land. New threats are emerging with the mishandling of materials in the high-technology industries and with fracking for oil and gas (Peduzzi and Ruth, 2013). Some of these problems are slow to emerge. Environmentally contaminated foods, as with endocrine disruptors from the plastics, agricultural and household (e.g., fire retardants) chemicals industry (Damstra et al., 2002), are now associated with insulin resistance, diabetes, obesity and their sequelae (Wang et al., 2008).

Often packaging contributes more waste and contamination than the food or beverage it is meant to protect, transport or store. Plastic is the biggest culprit, not only compromising arable land, but each of the world's oceans. Microplastic accumulates in five major ocean vortices. It is increasingly difficult to find coastal or even deep sea marine life not contaminated with micro- or even nanoplastic. The added problem of these plastics is that they have surface binding sites for endocrine disruptors (ERs), so amplifying their toxicity. These tiny plastic particles have slow in vivo kinetics and cross into tissues including those of the central nervous system (Thompson et al., 2009; Seltenrich, 2015). It is estimated that if the use of plastic packaging is not stopped, its present ocean burden will double within ten years. This alone can be expected to change the present position of seafood as health protective radically to one with unacceptable risk. At the same time, fish stocks are in decline. It is trends like these that signal major changes in health patterns in the foreseeable feature. It may become necessary to accept risk–benefit ratios for fish consumption that are presently anathema to regulatory authorities, simply for reasons of survival, but with greater morbidity, especially where, as in North East (NE) Asia and the Pacific, fish is eaten several times a week. These are new risk communication scenarios for food and nutrition scientists. This constitutes a powerful incentive to find alternatives to current food packaging methods.

Yet another form of agricultural site contamination to constitute a grave threat is that of antibiotic resistance genes (ARGs) (Su et al., 2011). In China, multiply antibiotic-resistant (including vancomycin) gut microbes have been found in up to 45% of pig livestock (Zhu et al., 2013). In addition, polymyxin resistance has been found in Guangdong province (Liu et al., 2016). Excreta contaminate soil and workers' apparel. These genes can be rapidly transferred from the intestinal commensal microbial population to pathogens. Thus, there is now the increased risk that illnesses caused by such pathogens will be untreatable with any available antibiotic. The use of antibiotics as growth promotants has contributed to this situation and must be questionable given the risk it has presented of widespread resistance; additional meat production with antibiotics is only of the order of 2%–3% (Collignon et al., 2005). The health risk is not only to humans, but to the livestock themselves, whose loss of gut microbiome (with diversion of nutrients to the animal carcass) makes them immuno-compromised. In turn, the veterinary use of antibiotics needs to be demarcated from those prioritised for human therapeutics. Inappropriate farm practices contribute to food insecurity, which may be dire if a society depends heavily on few animal-derived foods, as with the pig, poultry, ruminants, fish or crustaceans. The ban on use of antibiotics as growth promotants in Sweden in 1986 has had no adverse clinical or commercial effects on the livestock industry and allowed a marked reduction in antibiotic usage (Wierup, 2001). This is yet another argument for a biodiverse diet.

Population demographics are undergoing marked change especially with mass migration in the face of conflict and rapid ageing, as much as one year for every three years in socio-economically advanced or advancing countries (such as China) (Lutz et al., 2008; Samir and Wolfgang, 2014). Whereas most nutrition science has heretofore focussed on maternal and child health, the nutritional determinants and outcomes of ageing are now receiving more attention. This has stimulated the field of nutritional economics on account of the burden and costs of chronic disease and disability (Lo et al., 2012a; Lo et al., 2012b; Lo et al., 2013; Lo et al., 2016a; Lo et al., 2016b), particularly metabolic disease such as diabetes, cardiovascular disease, musculoskeletal disease and neurodegenerative disease (Hsu et al., 2011; Lee et al., 2011a; Wahlqvist et al., 2012a; Wahlqvist et al., 2012b). Encouraging is that biodiverse diets are conducive not only to longer survival, but also to a decreased burden of disease (Lee et al., 2011b). Moreover, even with established disease, food diversity appears to mitigate the impact (Chen et al., 2011; Huang et al., 2014b; Huang et al., 2015) (Table 16.1).

In an effort to manage some of these and other food security difficulties, a group of mainly agricultural and environmental scientists issued the *Kavli Declaration* in the United Kingdom in 2012 (Godfray and Garnett, 2014; Poppy et al., 2014):

> By 2050, humanity will be unable to meet its needs for food through current agricultural practices. We must drastically transform the global agricultural system to deliver food security and net greenhouse gas absorption without losses of water availability and wild habitats. This can only be achieved by more resource-efficient agriculture. This will need to combine locally relevant crop and animal genetic improvement and resilient agronomic practices that harness local ecosystem services to minimize inputs and close nutrient loops while sequestering carbon. The success of these on-farm activities will depend on restoring degraded lands and safeguarding remaining natural habitats to ensure the continued provision of wider ecosystem services.

Similar calls have been made by the health sector through the Lancet Commission, making the point that climate change 'strikes at the heart of humanity' and needs to be framed as a health issue (Wang and Horton, 2015). This commission considers that the indirect effects of climate change on our present and future health are wide and include the effects of air pollution, disease vector proliferation, food and nutrition insecurity, loss of home and mental illness.

Table 16.1 *Food Diversity Is Important for a Number of Reasons*

- It encourages sustainability.
- It enhances food quality.
- It provides alternatives.
- *Homo sapiens* is omnivorous with a plant food orientation.
- The diet is more likely to be adequate in micronutrients and favourable bioactive components.
- It provides a dilutional effect on toxicants and contaminants (e.g., melamine, maleic acid, endocrine disruptors).
- It favours energy regulation.
- It predicts favourable health outcomes.
- It is a measure of household and personal food security.

> To facilitate action to address the threat of climate change, our 2015 Commission on Health and Climate Change provides nine recommendations for governments to consider. They include: scaling up financing for climate-resilient health systems worldwide; ensuring a rapid phase out of coal from the global energy mix; encouraging a transition to cities that support and promote healthy lifestyles for the individual and the planet; establishing the framework for a strong, predictable, and international carbon pricing mechanism; rapidly expanding access to renewable energy in low-income and middle-income countries; ensuring adequate local capacity and political support to develop low-carbon healthy energy choices; adopting mechanisms to facilitate collaboration between Ministries of Health and other government departments and empowering health professionals; agreeing and implementing an international agreement which supports countries in transitioning to a low carbon economy; and investing in climate change and public health research.

The editors of the *British Medical Journal* think likewise (McCoy et al., 2014).

16.8 Nutritional Equity

As in the general societal arena, equity is a pressing global issue, not only in terms of access to food, but also in regard to food quality (Marmot et al., 2008; Friel and Baker, 2009). Although basically dependent on a functionally food diverse system and a planet with sufficient biodiversity, social, cultural and economic factors largely determine food equity (Wahlqvist and Specht, 1998; Friel and Baker, 2009). It is one reason why food diversity is a measure of household food security in the UN System (Hoddinott and Yohannes, 2002; Arimond and Ruel, 2004; Swindale and Bilinsky, 2006). It is nevertheless remarkable that even impecunious people, perhaps if determined and networked, can find ways to maintain fruit and vegetable intakes, which are usually the principal contributors to food diversity scores (Lo et al., 2012a). Food biodiversity for all seems an intractable objective, and failure to achieve it gives rise to nutrient-specific approaches to food-related health problems (Friel and Baker, 2009; Wahlqvist et al., 2009). The nutrient approach seriously shortchanges what foods and food patterns can offer to sustainable health outcomes. However, food accessibility will be challenged by more climate change–related natural disasters, conflict and refugee status (Butler, 2009a; Butler, 2009b; Bulter, 2014; Wahlqvist, 2014a). Together with accelerated ecosystem loss, these challenges will dictate a more concerted effort to overcome the related nutritional inequities. The International Food Policy Research Institute (IFPRI) reports suggest that improved 'resilience' is a way to address this kind of nutritional dilemma (Fan et al., 2014). This presumably requires the comprehensive commitment of nutritional science, along with other instrumentalities, at its biomedical, societal, economic and environmental best – and leadership!

In the course of ecosystem deployment for a liveable and sustainable future, livelihoods need consideration. They comprise the ability to secure, in a sustainable and equitable way, food and water, shelter, clothing, warmth (and cooling with global warming), education and health care, along with some form of renewable energy driven transport and communication. We all, unless immature or indisposed, should be able to exert ourselves cooperatively to achieve these ends. Many now suffer the indignity of unemployment, as if we need to be beholden to someone else for our livelihood. Somehow, we need to replace employment with livelihood acquisition. Elinor Ostman's work, to which reference has been made, offers promise that we might achieve that. There is currently a growing trend toward cashless societies, as in Sweden. This might be a moment to replace some cash, at least, with transactions in 'Livelihood Units' (LUs). It would be difficult for this to embrace the components extant in the UNDP index (United Nations, 2016) since these are comparative between locations. LUs would be used to incentivise people where they are to engage in ecosystem enhancement and its connectedness with health, food and be mindful of a Future Earth (Figure 16.1). Such an approach would foster the emergence of a more ecologically relevant workforce, but not a stratification by employment status. It could factor in gender, age and various life-course needs. Its nutritional profile could be dynamic and tailored to community aspirations and inter-connectedness with other communities (Wahlqvist, 2009a).

16.9 Nutritional Ethics

As climate change progresses, the food system will be increasingly compromised but, optimistically, a mitigated eventuality. It is already clear that those who are ecologically advantaged now may be less affected. The global ethical question will, in the future, be more explicit, where presently it is mostly a question of land and fishing rights, knowledge and power. What and who in the future will determine who gets access to potable water and food (Solomons, 2002; Friel and Baker, 2009)?

16.10 Monitoring and Surveillance

How can we be confident that progress is being made with the building and restoration of the platform for an ecologically relevant nutrition science, supportive of human and planetary health? Ecosystem assessment tools are available through the work of the UN System (Wahlqvist, 2014b). Others that are primarily related to food and nutritional security are available (Arimond and Ruel, 2004; Lee et al., 2011b; Dupont, 2013; Peng et al., 2015). Advances could be made to simplify and network such indices in order that individuals, households and communities could gain feed-back and improve and accelerate the process in conjunction with the best available science and technology.

16.11 Conclusions

Collectively, we can optimise our own ecosystem, agree on livelihood provision, eat no more than we need and have a plant-based biodiverse diet; not waste food; use renewable energy, not fossil fuels; and ensure that all have access to education and health care (Stehfest, 2014; Tilman and Clark, 2014). These are ethical as well as knowledge-driven deductions for a food and nutrition future in which science and technology at large can play a constructive role.

References

Albrecht, G., Sartore, G. M., Connor, L., Higginbotham, N., Freeman, S., Kelly, B., Stain, H., Tonna, A. and Pollard, G. (2007). Solastalgia: the distress caused by environmental change. *Australas Psychiatry*, **15 Suppl 1**, S95–S958.

Andreas, N. J., Al-Khalidi, A., Jaiteh, M., Clarke, E., Hyde, M. J., Modi, N., Holmes, E., Kampmann, B. and Mehring Le Doare, K. (2016). Role of human milk oligosaccharides in Group B Streptococcus colonisation. *Clin Transl Immunology*, **5**(8), e99.

Arimond, M. and Ruel, M. T. (2004). Dietary diversity is associated with child nutritional status: evidence from 11 demographic and health surveys. *J Nutr*, **134**(10), 2579–2585.

Austin, E. N., Johnston, Y. A. M. and Morgan, L. L. (2006). Community gardening in a senior center: a therapeutic intervention to improve the health of older adults. *Therapeutic Recreation Journal*, **40**(1), 48–57.

Beauman, C., Cannon, G., Elmadfa, I., Glasauer, P., Hoffmann, I., Keller, M., Krawinkel, M., Lang, T., Leitzmann, C., Lotsch, B., Margetts, B. M., McMichael, A. J., Meyer-Abich, K., Oltersdorf, U., Pettoello-Mantovani, M., Sabate, J., Shetty, P., Soria, M., Spiekermann, U., Tudge, C., Vorster, H. H., Wahlqvist, M. and Zerilli-Marimo, M. (2005). The principles, definition and dimensions of the new nutrition science. *Public Health Nutr*, **8**(6a), 695–698.

Berman, B. (2012). 3-D printing: The new industrial revolution. *Business Horizons*, **55**, 155–162.

Block, K., Gibbs, L., Staiger, P. K., Gold, L., Johnson, B., Macfarlane, S., Long, C. and Townsend, M. (2012). Growing community: the impact of the Stephanie Alexander Kitchen Garden Program on the social and learning environment in primary schools. *Health Educ Behav*, **39**(4), 419–432.

Borys, M. (2011). EPODE International Network. Retrieved 25 April 2018, from http://epodeinternationalnetwork.com/

Briney, A. and Writer, C. (2015). Green Revolution: History and Overview of the Green Revolution. Retrieved 23 August 2016, from http://geography.about.com/od/globalproblemsandissues/a/greenrevolution.htm.

Brown, K. H. and Jameton, A. L. (2000). Public health implications or urban agriculture. *J Public Health Policy*, **21**(1), 20–39.

Bulter, C. D. (2014). *Climate Change and Global Health*. Oxfordshire: CABI.

Butler, C. D. (2009a). Food security in the Asia-Pacific: climate change, phosphorus, ozone and other environmental challenges. *Asia Pac J Clin Nutr*, **18**(4), 590–597.

Butler, C. D. (2009b). Food security in the Asia-Pacific: Malthus, limits and environmental challenges. *Asia Pac J Clin Nutr*, **18**(4), 577–584.

Cannon, G. (2002). Nutrition: the new world map. *Asia Pac J Clin Nutr*, **11** Suppl, S480-S497.

Chang, W. C., Wahlqvist, M. L., Chang, H. Y., Hsu, C. C., Lee, M. S., Wang, W. S. and Hsiung, C. A. (2012a). A bean-free diet increases the risk of all-cause mortality among Taiwanese women: the role of the metabolic syndrome. *Public Health Nutr*, **15**(4), 663–672.

Chang, Y. H., Chen, R. C., Wahlqvist, M. L. and Lee, M. S. (2012b). Frequent shopping by men and women increases survival in the older Taiwanese population. *J Epidemiol Community Health*, **66**(7), e20.

Chen, R. C., Chang, Y. H., Lee, M. S. and Wahlqvist, M. L. (2011). Dietary quality may enhance survival related to cognitive impairment in Taiwanese elderly. *Food Nutr Res*, **55**. http://foodandnutritionresearch.net/index.php/fnr/article/view/577/1121

Chen, R. C., Lee, M. S., Chang, Y. H. and Wahlqvist, M. L. (2012). Cooking frequency may enhance survival in Taiwanese elderly. *Public Health Nutr*, **15**(7), 1142–1149.

Chiang, P. H., Wahlqvist, M. L., Lee, M. S., Huang, L. Y., Chen, H. H. and Huang, S. T. (2011). Fast-food outlets and walkability in school neighbourhoods predict fatness in boys and height in girls: a Taiwanese population study. *Public Health Nutr*, **14**(9), 1601–1609.

Cincotta, R. P., Wisnewski, J. and Engelman, R. (2000). Human population in the biodiversity hotspots. *Nature*, **404**(6781), 990–992.

Cohen, J. E. (2003). Human population: the next half century. *Science*, **302**(5648), 1172–1175.

Collignon, P., Wegener, H. C., Braam, P. and Butler, C. D. (2005). The routine use of antibiotics to promote animal growth does little to benefit protein undernutrition in the developing world. *Clin Infect Dis*, **41**(7), 1007–1013.

Damstra, T., Barlow, S., Bergman, A., Kavlock, R. and Der Kraak, G. V. (2002). *Global assessment of the state-of-the-science of endocrine disruptors*. Geneva: World Health Organization.

Darmadi-Blackberry, I., Wahlqvist, M. L., Kouris-Blazos, A., Steen, B., Lukito, W., Horie, Y. and Horie, K. (2004). Legumes: the most important dietary predictor of survival in older people of different ethnicities. *Asia Pac J Clin Nutr*, **13**(2), 217–220.

Dube, L., Pingali, P. and Webb, P. (2012). Paths of convergence for agriculture, health, and wealth. *Proc Natl Acad Sci USA*, **109**(31), 12294–12301.

Economist Intelligence. (2013) Global Food Security Index 2013: An Annual Measure of the State of Global Food Security. Retrieved September 7, 2013, from http://foodsecurityindex.eiu.com/Index.

Fan, S., Pandya-Lorch, R. and Yosef, S. (2014). *Resilience for Food and Nutrition Security*. Washington, DC: International Food Policy Research Institute.

FAO. (2014). Family Farmers: Feeding the World, Caring for the Earth. Retrieved 6 September 2016, from www.fao.org/resources/infographics/infographics-details/en/c/270462/.

FAO. (2016). 2014 International Year of Family Farming. Retrieved 6 September 2016, from www.fao.org/family-farming-2014/en/.

FAO and IFA and WFP. (2014). The State of Food Insecurity in the World. Strengthening the Enabling Environment for Food Security and Nutrition. Retrieved 6 September 2016, from www.fao.org/3/a-i4030e.pdf.

Flynn, T. J., Cadzow, M., Dalbeth, N., Jones, P. B., Stamp, L. K., Hindmarsh, J. H., Todd, A. S., Walker, R. J., Topless, R. and Merriman, T. R. (2015). Positive association of tomato consumption with serum urate: support for tomato consumption as an anecdotal trigger of gout flares. *BMC Musculoskelet Disord*, **16**, 196.

Foyer, C. H., Lam, H.-M., Nguyen, H. T., Siddique, K. H. M., Varshney, R. K., Colmer, T. D., Cowling, W.,

Bramley, H., Mori, T. A., Hodgson, J. M., Cooper, J. W., Miller, A. J., Kunert, K., Vorster, J., Cullis, C., Ozga, J. A., Wahlqvist, M. L., Liang, Y., Shou, H., Shi, K., Yu, J., Fodor, N., Kaiser, B. N., Wong, F.-L., Valliyodan, B. and J., C. M. (2016). Neglecting legumes has compromised human health and sustainable food production. *Nature Plants*, **2**, 1–10.

Friel, S. and Baker, P. I. (2009). Equity, food security and health equity in the Asia Pacific region. *Asia Pac J Clin Nutr*, **18**(4), 620–632.

Galluzzi, G., Eyzaguirre, P. and Negri, V. (2010). Home gardens: neglected hotspots of agro-biodiversity and cultural diversity. *Biodiversity and Conservation*, **19**(13), 3635–3654.

Gibson, V., Zhu, Y. G., Ge, R. and Wahlqvist, M. L. (2015a). Food systems and life expectancy with rapid urbanisation in provincial China. *Asia Pac J Clin Nutr*, **24**(4), 731–743.

Gibson, V., Zhu, Y. G., Ge, R. and Wahlqvist, M. L. (2015b). Preferred ecosystem characteristics: their food and health relevance to China's rapid urbanisation. *Asia Pac J Clin Nutr*, **24**(4), 556–574.

Godfray, H. C. and Garnett, T. (2014). Food security and sustainable intensification. *Philos Trans R Soc Lond B Biol Sci*, **369**(1639), 20120273. doi: 10.1098/rstb.2012.0273.

Gong, P., Liang, S., Carlton, E. J., Jiang, Q., Wu, J., Wang, L. and Remais, J. V. (2012). Urbanisation and health in China. *Lancet*, **379**(9818), 843–852.

Hale, J., Knapp, C., Bardwell, L., Buchenau, M., Marshall, J., Sancar, F. and Litt, J. S. (2011). Connecting food environments and health through the relational nature of aesthetics: gaining insight through the community gardening experience. *Soc Sci Med*, **72**(11), 1853–1863.

The Guardian. (2016). The Anthropocene epoch: scientists declare dawn of human-influenced age. Retrieved 5 September 2016, from www.theguardian.com/environment/2016/aug/29/declare-anthropocene-epoch-experts-urge-geological-congress-human-impact-earth.

Hansen, J., Sato, M., Hearty, P., Ruedy, R., Kelley, M., Masson-Delmotte, V., Russell, G., Tselioudis, G., Cao, J., Rignot, E., Velicogna, I., Tormey, B., Donovan, B., Kandiano, E., von Schuckmann, K., Kharecha, P., Legrande, A. N., Bauer, M. and Lo, K. W. (2016). Ice melt, sea level rise and superstorms: evidence from paleoclimate data, climate modeling, and modern observations that 2°C global warming could be dangerous. *Atmos Chem Phys*, **16**, 3761–3812.

Hartig, T. (2008). Green space, psychological restoration, and health inequality. *Lancet*, **372**(9650), 1614–1615.

Hoddinott, J. and Yohannes, Y. (2002). *Dietary diversity as a food security indicator*. Washington, DC: International Food Policy Research Institute.

Hong, J. Y., Lan, T. Y., Tang, G. J., Tang, C. H., Chen, T. J. and Lin, H. Y. (2015). Gout and the risk of dementia: a nationwide population-based cohort study. *Arthritis Res Ther*, **17**, 139.

Hsu, C. C., Wahlqvist, M. L., Lee, M. S. and Tsai, H. N. (2011). Incidence of dementia is increased in type 2 diabetes and reduced by the use of sulfonylureas and metformin. *J Alzheimers Dis*, **24**(3), 485–493.

Huang, L. Y., Wahlqvist, M. L., Huang, Y. C. and Lee, M. S. (2014a). Optimal dairy intake is predicated on total, cardiovascular, and stroke mortalities in a Taiwanese cohort. *J Am Coll Nutr*, **33**(6), 426–436.

Huang, S. T. (2010). A recycling index for food and health security: urban Taipei. *Asia Pac J Clin Nutr*, **19**(3), 402–411.

Huang, Y. C., Wahlqvist, M. L., Kao, M. D., Wang, J. L. and Lee, M. S. (2015). Optimal dietary and plasma magnesium statuses depend on dietary quality for a reduction in the risk of all-cause mortality in older adults. *Nutrients*, **7**(7), 5664–5683.

Huang, Y. C., Wahlqvist, M. L. and Lee, M. S. (2014b). Appetite predicts mortality in free-living older adults in association with dietary diversity. A NAHSIT cohort study. *Appetite*, **83**, 89–96.

IFPRI. (2008). Rising Food Prices: What Should Be Done? Retrieved 6 September 2016, from www.ifpri.org/blog/rising-food-prices-what-should-be-done.

International Food Policy Research Institute. (2014). Global Nutrition Report 2014. Retrieved 1 September 2016, from http://ebrary.ifpri.org/utils/getfile/collection/p15738coll2/id/128484/filename/128695.pdf.

Jung, H. J., Lee, S. E., Kim, D., Noh, H., Song, S., Kang, M., Song, Y. J. and Paik, H. Y. (2015). Improvement in the technological feasibility of a web-based dietary survey system in local settings. *Asia Pac J Clin Nutr*, **24**(2), 308–315.

Kang, S., Xu, Y., You, Q., Flügel, W.-A., Pepin, N. and Yao, T. (2010). Review of climate and cryospheric change in the Tibetan Plateau. *Environ Res Lett*, **5**, 015101.

Koh, K.-L., Kelman, I., Kibugi, R. and Osorio, R.-L. E. (2016). *Adaptation to Climate Change: ASEAN and Comparative Experiences*. Singapore: World Scientific.

Kong, K., Zhang, L., Huang, L. and Tao, Y. (2017). Validity and practicability of smartphone-based photographic food records for estimating energy and nutrient intake. *Asia Pac J Clin Nutr*. **26**. doi: 10.6133/apjcn.042016.05.

Koukkari, W. L. and Sothern, R. B. (2007). *Introducing Biological Rhythms: A Primer on the Temporal Organization of Life, with Implications for Health, Society, Reproduction, and the Natural Environment. USA*: New York: Springer Science+Business Media.

Kouris-Blazos, A. and Belski, R. (2016). Health benefits of legumes and pulses with a focus on Australian sweet lupins. *Asia Pac J Clin Nutr*, **25**(1), 1–17.

Lang, T. and Heasman, M. (2004). *Food Wars. The Global Battle for Mouths, Minds and Markets*. London: Earthscan from Routledge.

Lee, M. S., Hsu, C. C., Wahlqvist, M. L., Tsai, H. N., Chang, Y. H. and Huang, Y. C. (2011a). Type 2 diabetes increases and metformin reduces total, colorectal, liver and pancreatic cancer incidences in Taiwanese:

a representative population prospective cohort study of 800,000 individuals. *BMC Cancer*, **11**, 20.

Lee, M. S., Huang, Y. C., Su, H. H., Lee, M. Z. and Wahlqvist, M. L. (2011b). A simple food quality index predicts mortality in elderly Taiwanese. *J Nutr Health Aging*, **15**(10), 815–821.

Leitzmann, C. and Cannon, G. (2005). Dimensions, domains and principles of the new nutrition science. *Public Health Nutr*, **8**(6a), 787–794.

Lin, A. Y., Huang, S. T. and Wahlqvist, M. L. (2009). Waste management to improve food safety and security for health advancement. *Asia Pac J Clin Nutr*, **18**(4), 538–545.

Liu, Y. Y., Wang, Y., Walsh, T. R., Yi, L. X., Zhang, R., Spencer, J., Doi, Y., Tian, G., Dong, B., Huang, X., Yu, L. F., Gu, D., Ren, H., Chen, X., Lv, L., He, D., Zhou, H., Liang, Z., Liu, J. H. and Shen, J. (2016). Emergence of plasmid-mediated colistin resistance mechanism MCR-1 in animals and human beings in China: a microbiological and molecular biological study. *Lancet Infect Dis*, **16**(2), 161–168.

Lo, Y. T., Chang, Y. H., Lee, M. S. and Wahlqvist, M. L. (2012a). Dietary diversity and food expenditure as indicators of food security in older Taiwanese. *Appetite*, **58**(1), 180–187.

Lo, Y. T., Chang, Y. H., Wahlqvist, M. L., Huang, H. B. and Lee, M. S. (2012b). Spending on vegetable and fruit consumption could reduce all-cause mortality among older adults. *Nutr J*, **11**, 113.

Lo, Y. T., Wahlqvist, M. L., Chang, Y. H., Kao, S. and Lee, M. S. (2013). Dietary diversity predicts type of medical expenditure in elders. *Am J Manag Care*, **19**(12), e415–e423.

Lo, Y. T., Wahlqvist, M. L., Chang, Y. H. and Lee, M. S. (2016a). Combined effects of chewing ability and dietary diversity on medical service use and expenditures. *J Am Geriatr Soc*, **64**(6), 1187–1194.

Lo, Y. T., Wahlqvist, M. L., Huang, Y. C. and Lee, M. S. (2016b). Elderly Taiwanese who spend more on fruits and vegetables and less on animal-derived foods use less medical services and incur lower medical costs. *Br J Nutr*, **115**(5), 823–833.

Louv, R. (2005). *Last Child in the Woods: Saving Our Kids from Nature Deficit Disorder*. Chapel Hill, NC: Algonquin Books of Chapel Hill.

Lukito, W., Malik, S. G., Surono, I. S. and Wahlqvist, M. L. (2015). From 'lactose intolerance' to 'lactose nutrition'. *Asia Pac J Clin Nutr*, **24 Suppl 1**, S1–S8.

Lutz, W. and Samir, K. C. (2010). Dimensions of global population projections: what do we know about future population trends and structures? *Philos Trans R Soc Lond B Biol Sci*, **365**(1554), 2779–2791.

Lutz, W., Sanderson, W. and Scherbov, S. (2008). The coming acceleration of global population ageing. *Nature*, **451**(7179), 716–719.

Margulis, L. (1981). *Symbiosis in Cell Evolution: Life and Its Environment on the Early Earth*. San Francisco: W. H. Freeman.

Marmot, M., Friel, S., Bell, R., Houweling, T. A. and Taylor, S. (2008). Closing the gap in a generation: health equity through action on the social determinants of health. *Lancet*, **372**(9650), 1661–1669.

McCoy, D., Montgomery, H., Arulkumaran, S. and Godlee, F. (2014). Climate change and human survival. *BMJ*, **348**, g2351.

Mitchell, D. (2008). A note on rising food prices. Retrieved 5 September 2016, from www.bio-based.eu/foodcrops/media/08-07ANoteonRisingFoodPrices.pdf.

Mitchell, R. and Popham, F. (2008). Effect of exposure to natural environment on health inequalities: an observational population study. *Lancet*, **372**(9650), 1655–1660.

Monteiro, C. A. (2009). Nutrition and health: the issue is not food, nor nutrients, so much as processing. *Public Health Nutr*, **12**(5), 729–731.

Monteiro, C. A., Levy, R. B., Claro, R. M., de Castro, I. R. and Cannon, G. (2011). Increasing consumption of ultra-processed foods and likely impact on human health: evidence from Brazil. *Public Health Nutr*, **14**(1), 5–13.

The Nourish Network. (2016). Retrieved 4 September 2016, from www.facebook.com/nourishnetwork/?fref=ts.

Numbeo. (2016). Food Prices in Japan. Retrieved 3 September 2016, from www.numbeo.com/food-prices/country_result.jsp?country=Japan.

Ogunrinade, A., Oniang'o, R. and May, J. (1999). *Not by Bread Alone: Food Security and Governance in Africa*. Witwatersrand, South Africa: Witwatersrand University Press.

Omar, A., Wahlqvist, M. L., Kouris-Blazos, A. and Vicziany, M. (2005). Wellness management through Web-based programmes. *J Telemed Telecare*, **11 Suppl 1**, 8–11.

Ostrom, E. (2009). A general framework for analyzing sustainability of social-ecological systems. *Science*, **325**(5939), 419–422.

Ozer, E. J. (2007). The effects of school gardens on students and schools: conceptualization and considerations for maximizing healthy development. *Health Educ Behav*, **34**(6), 846–863.

Payne, P. and Lipton, M. (1994). *How Third World Rural Households Adapt to Dietary Energy Stress*. Washington, DC: International Food Policy Research Institute.

Peduzzi, P. and Ruth, H. R. R. (2013). Gas fracking: can we safely squeeze the rocks? *Environmental Development*, **6**, 86–99.

Peng, C. J., Lee, M. S., Wahlqvist, M. L., Pan, W. H., Lee, W. C., Ling, C. and Guo, H. R. (2015). Needs-based food and nutrient security indices to monitor and modify the food supply and intakes: Taiwan, 1991–2010. *Food Policy*, **57**, 142–152.

Pincock, S. (2011). Boyd Swinburn: combating obesity at the community level. *Lancet*, **378**(9793), 761.

Poppy, G. M., Jepson, P. C., Pickett, J. A. and Birkett, M. A. (2014). Achieving food and environmental security: new approaches to close the gap. *Philos Trans R Soc Lond B Biol Sci*, **369**(1639), 20120272.

Quifiones, M. A., Borlaug, N. E. and Dowswell, C. R. (1997). A fertilizer-based green revolution for Africa. *Replenishing soil fertility in Africa: SSSA Publication Number 51*. R. J. Buresh, P. A. Sanchez and F. Calhoun. Wisconsin, USA: Soil Science Society of America, pp. 81–95.

Samaras, T. (2011). Reasons to be small. *World Nutrition*, **2**(3), 108–135.

Samir, K. C. and Lutz W. (2017). The human core of the shared socioeconomic pathways: Population scenarios by age, sex and level of education for all countries to 2100. *Global Environmental Change*, 1(42), 181–192.

Seltenrich, N. (2015). New link in the food chain? Marine plastic pollution and seafood safety. *Environmental Health Perspectives*, **123**(2), A35–A41.

Seto, K. C., Guneralp, B. and Hutyra, L. R. (2012). Global forecasts of urban expansion to 2030 and direct impacts on biodiversity and carbon pools. *Proc Natl Acad Sci USA*, **109**(40), 16083–16088.

Seto, K. C. and Ramankutty, N. (2016). Hidden linkages between urbanization and food systems. *Science*, **352** (6288), 943–945.

Shah, N. (2016). Open Agriculture Initiative: Is Digital Farming the Future of Food? Retrieved 23 August 2016, from http://foodtank.com/news/2016/05/open-agriculture-initiative-digital-farming.

Shiva, V. (2016). *Who Really Feeds the World? The Failures of Agribusiness and the Promise of Agroecology*. Berkeley CA: North Atlantic Books.

Sho, H. (2001). History and characteristics of Okinawan longevity food. *Asia Pac J Clin Nutr*, **10**(2), 159–164.

Smith, L. P., Hua, J., Seto, E., Du, S., Zang, J., Zou, S., Popkin, B. M. and Mendez, M. A. (2014). Development and validity of a 3-day smartphone assisted 24-hour recall to assess beverage consumption in a Chinese population: a randomized cross-over study. *Asia Pac J Clin Nutr*, **23**(4), 678–690.

Soderback, I., Soderstrom, M. and Schalander, E. (2004). Horticultural therapy: the 'healing garden'and gardening in rehabilitation measures at Danderyd Hospital Rehabilitation Clinic, Sweden. *Pediatr Rehabil*, **7**(4), 245–260.

Solomons, N. W. (2002). Ethical consequences for professionals from the globalization of food, nutrition and health. *Asia Pac J Clin Nutr*, **11 Suppl 3**, S653–s665.

Steffen, W. (2014). The anthropocene: a planet under pressure. *Climate Change and Global Health*. Oxfordshire: CABI, pp. 1–10.

Stehfest, E. (2014). Diet: Food choices for health and planet. *Nature*, **515**(7528), 501–502.

Su, H. C., Ying, G. G., Tao, R., Zhang, R. Q., Fogarty, L. R. and Kolpin, D. W. (2011). Occurrence of antibiotic resistance and characterization of resistance genes and integrons in Enterobacteriaceae isolated from integrated fish farms in South China. *J Environ Monit*, **13**(11), 3229–3236.

Swindale, A. and Bilinsky, P. (2006). *Household dietary diversity score (HDDS) for measurement of household food access: indicator guide*. Washington, DC: Food and Nutrition Technical Assistance Project, Academy for Educational Development.

Tay, N. (2016). Additive manufacturing (3D printing) in food. Retrieved 6 September 2016, from https://novoed.com/3d-opportunity/reports/259391.

The Third People. (2015). Tibetan plateau turning to desert, says Chinese Academy of Science. Retrieved 5 September 2016, from www.thethirdpole.net/2015/12/07/tibetan-plateau-turning-to-desert-says-chinese-academy-of-science/.

Thompson, R. C., Moore, C. J., vom Saal, F. S. and Swan, S. H. (2009). Plastics, the environment and human health: current consensus and future trends. *Philos Trans R Soc Lond B Biol Sci*, **364**(1526), 2153–2166.

Tilman, D. and Clark, M. (2014). Global diets link environmental sustainability and human health. *Nature*, **515**(7528), 518–522.

Topol E. (2012). *The Creative Destruction of Medicine: How the Digital Revolution Will Create Better Health Care*. New York: Basic Books.

Tuyen le, D. (2009). Food in health security in South East Asia. *Asia Pac J Clin Nutr*, **18**(4), 493–497.

United Nations. (2012). United Nations System Standing Committee on Nutrition: SUN (Scaling Up Nutrition) Movement. Retrieved 5 September 2016, from www.unscn.org/en/sun-scaling-up/.

United Nations. (2016). UN System Sustainable Development. Retrieved 1 September 2016, from https://sustainabledevelopment.un.org/.

United Nations. (2016). Human development data. Retrieved 4 September 2016, from hdr.undp.org/en/data.

van den Brandt, P. A. and Schouten, L. J. (2015). Relationship of tree nut, peanut and peanut butter intake with total and cause-specific mortality: a cohort study and meta-analysis. *Int J Epidemiol*, **44**(3), 1038–1049.

Vandenplas, Y. (2015). Lactose intolerance. *Asia Pac J Clin Nutr*, **24 Suppl 1**, S9–S13.

von Braun, J. and Bos, M. S. (2005). The changing economics and politics of rice: implications for food security, globalization, and environmental sustainability. *Rice is life: scientific perspectives for the 21st century*. K. Toriyama, K. L. Heong and B. Hardy. Manila: International Rice Research Institute, pp. 7–20.

Wahlqvist, M. L. (1970). *Arterial wall metabolism*. Adelaide: Adelaide University.

Wahlqvist, M. L. (2005). The new nutrition science: sustainability and development. *Public Health Nutr*, **8**(6a), 766–772.

Wahlqvist, M. L. (2008). New nutrition science in practice. *Asia Pac J Clin Nutr*, **17 Suppl 1**, 5–11.

Wahlqvist, M. L. (2009a). Connected Community and Household Food-Based Strategy (CCH-FBS): its importance for health, food safety, sustainability and security in diverse localities. *Ecol Food Nutr*, **48**(6), 457–481.

Wahlqvist, M. L. (2009b). Why food in health security (FIHS)? *Asia Pac J Clin Nutr*, **18**(4), 480–485.

Wahlqvist, M. L. (2014a). *Climate Change and Health in East Asia: A Food in Health Security Perspective. Climate*

Change and Global Health. C. D. Bulter (Editor). Oxfordshire: CABI, pp. 153–161.

Wahlqvist, M. L. (2014b). Ecosystem health disorders – changing perspectives in clinical medicine and nutrition. *Asia Pac J Clin Nutr*, **23**(1), 1–15.

Wahlqvist, M. L. (2015a). Lactose nutrition in lactase non-persisters. *Asia Pac J Clin Nutr*, **24 Suppl 1**, S21–S25.

Wahlqvist, M. L. (2015b). Tu Giay: The nutritionist victor of the Vietnam war *World Nutrition*, **6**(7–8), 613–622.

Wahlqvist, M. L. (2016a). Ecosystem dependence of healthy localities, food and people. *Ann Nutr Metab*, **69**, 75–78.

Wahlqvist, M. L. (2016b). Food structure is critical for optimal health. *Food Funct*, **7**(3), 1245–1250.

Wahlqvist, M. L. (2016). Future food. *Asia Pac J Clin Nutr*, **25**(4), 706–715.

Wahlqvist, M. L., Keatinge, J. D., Butler, C. D., Friel, S., McKay, J., Easdown, W., Kuo, K. N., Huang, C. J., Pan, W. H., Yang, R. Y., Lee, M. S., Chang, H. Y., Chiu, Y. W., Jaron, D., Krawinkel, M., Barlow, S., Walsh, G., Chiang, T. L., Huang, P. C. and Li, D. (2009). A Food in Health Security (FIHS) platform in the Asia-Pacific Region: the way forward. *Asia Pac J Clin Nutr*, **18**(4), 688–702.

Wahlqvist, M. L., Kouris-Blazos, A., Crotty, P., Worsley, T., Harvey, P., Tieru, H. and Cavalli-Sforza, L. T. (1999). *Development of Food-Based Dietary Guidelines for the Western Pacific Region*. Manila: World Health Organization.

Wahlqvist, M. L. and Kuo, K. N. (2009). Securing health through food systems: an initiative of the nutrition consortium of the National Health Research Institutes in Taiwan and Asia Pacific regional partners as a network. *Asia Pac J Clin Nutr*, **18**(4), 472–479.

Wahlqvist, M. L. and Lee, M. S. (2006). Nutrition in health care practice. *J Med Sci*, **26**(5), 157–164.

Wahlqvist, M. L., Lee, M. S., Chuang, S. Y., Hsu, C. C., Tsai, H. N., Yu, S. H. and Chang, H. Y. (2012a). Increased risk of affective disorders in type 2 diabetes is minimized by sulfonylurea and metformin combination: a population-based cohort study. *BMC Med*, **10**, 150.

Wahlqvist, M. L., Lee, M. S., Hsu, C. C., Chuang, S. Y., Lee, J. T. and Tsai, H. N. (2012b). Metformin-inclusive sulfonylurea therapy reduces the risk of Parkinson's disease occurring with Type 2 diabetes in a Taiwanese population cohort. *Parkinsonism Relat Disord*, **18**(6), 753–758.

Wahlqvist, M. L. and Specht, R. L. (1998). Food variety and biodiversity: Econutrition. *Asia Pac J Clin Nutr*, **7**(3/4), 314–319.

Wang, H. and Horton, R. (2015). Tackling climate change: the greatest opportunity for global health. *Lancet*, **386** (10006), 1798–1799.

Wang, N., Zhang, N. and Wang, M. (2006). Wireless sensors in agriculture and food industry – recent development and future perspective. *Computers and Electronics in Agriculture*, **50**(1), 1–14.

Wang, S. L., Tsai, P. C., Yang, C. Y. and Guo, Y. L. (2008). Increased risk of diabetes and polychlorinated biphenyls and dioxins: a 24-year follow-up study of the Yucheng cohort. *Diabetes Care*, **31**(8), 1574–1579.

Watkins, E. S. (1998). *On the Pill: A Social History of Oral Contraceptives, 1950–1970*. Baltimore, MD: The Johns Hopkins University Press.

Wierup, M. (2001). The Swedish experience of the 1986 year ban of antimicrobial growth promoters, with special reference to animal health, disease prevention, productivity, and usage of antimicrobials. *Microb Drug Resist*, **7**(2), 183–190.

Williams, D. R. and Dixon, P. S. (2013). Impact of garden-based learning on academic outcomes in schools synthesis of research between 1990 and 2010. *Review of Educational Research*, 0034654313475824.

World Health Organization (2014). *Childhood stunting: challenges and opportunities report*. Geneva: World Health Organization.

World Health Organization and Food and Agriculture Organization of the United Nations. (1995). Preparation and use of food-based dietary guidelines. Retrieved 6 September 2016, from www.who.int/nutrition/publications/nutrientrequirements/WHO_TRS_880/en/.

World Health Organization. (2013). Global nutrition policy review: what does it take to scale up nutrition action? Retrieved 6 September 2016, from http://apps.who.int/iris/bitstream/10665/84408/1/9789241505529_eng.pdf.

Wright, M. W. and Bruford, E. A. (2011). Naming 'junk': human non-protein coding RNA (ncRNA) gene nomenclature. *Hum Genomics*, **5**(2), 90–98.

Yang, K., Ye, B., Zhou, D., Wu, B., Foken, T., Qin, J. and Zhou, Z. (2011). Response of hydrological cycle to recent climate changes in the Tibetan plateau. *Climatic Change*, **109**(3), 517–534.

Zhu, Y. G., Johnson, T. A., Su, J. Q., Qiao, M., Guo, G. X., Stedtfeld, R. D., Hashsham, S. A. and Tiedje, J. M. (2013). Diverse and abundant antibiotic resistance genes in Chinese swine farms. *Proc Natl Acad Sci USA*, **110**(9), 3435–3440.

17

Air Pollution and Human Health Risk Reduction: The Case Study of Delhi Megacity, India

R. B. Singh and Aakriti Grover

17.1 Introduction

Health is determined by many factors such as the environmental surroundings, lifestyle, genetics and dietary habits. Human health and local climatic conditions have a complex relationship. As a result of unplanned, unregulated and haphazard urban growth in Delhi, the associated micro and meso scale climatic changes are imposing challenges to human health. As a result, urban areas, the prime centers of development, are experiencing a significant rise in the burden of diseases. This is largely in response to industrialization, vehicularization, motorization and energy consumption (Chen and Kan, 2008).

The atmospheric composition of gases has been altered by industrial and vehicular growth. There are rising concentrations of oxides of nitrogen, sulphur and carbon dioxide. Added to this, suspended particulate matter (SPM) and compounds such as benzene and ozone (Knowlton et al., 2004) are increasing in concentration. There are multiple causes of respiratory problems; a strong positive relationship has been established between respiratory illness and rising pollutant levels. Air pollutants have far-reaching damaging effect on human health including chronic heart and lung disease (Frumkin, 2002), asthma (Beggs and Bambricks, 2005; Magas et al., 2007), cancer (Bentham, 1992), minor eye, nose or skin irritations, allergy (Kim et al., 2015) and even death (Lvovsky, 1998; Fang et al., 2013; Kim et al., 2015). However, there are variations and patterns in the trend of air pollution and disease occurrence. Also, the levels of vulnerabilities vary, particularly with respect to gender and age, that are exposure and location dependent (Curtis et al., 2006).

It is reported that among the non-communicable diseases, cancer and cardiovascular diseases are leading causes of mortality in urban India, and exposure to poor air quality is one of the underlying reasons (World Health Organisation, 2016). It is alarming to note that diseases of the respiratory system are the third largest cause of all deaths in the country (Government of India, 2010). The number of deaths from diseases of the respiratory system in Delhi increased from 3,231 to 3,746 (2001–2010) (Government of NCT of Delhi, 2010).

The exposure to pollutants and immunity levels of individuals determine their state of health. Children and the elderly constitute the most vulnerable groups. Besides this, children have lower breathing zone, small airways, developing lungs and immature immune systems that make them much more vulnerable to the effect of pollutants (Central Pollution Control Board, 2008). The elderly are also more susceptible than other age groups. There are, however, morbidity and mortality differentials with respect to physical-biological-socio-economic status.

One conceptualization of the linkages between various aspects of health and environment can be described using system analysis (Gatzweiler et al., 2016). The level of exposure and susceptibility to health risk is dependent on individual physical and biological factors (Figure 17.1).

Furthermore, mitigating and aggravating factors produce differential impacts. Combinations of individual and physical environmental factors determine the severity of impact, which ranges from minor illness to death. The urban population is increasing at a fast rate, especially in developing countries, and these growing cities are expected to have higher levels of risk and exposure to health hazards in the future (Beer, 2001). Air pollution levels in Delhi were exceptionally high in November 2016 (post Diwali celebration) such that the concentration of harmful gases remained suspended for a couple of days and as a result schools were closed for three days. This chapter aims (1) to understand the trend of air pollution in Delhi and (2) to analyze the impacts of air pollution on human health.

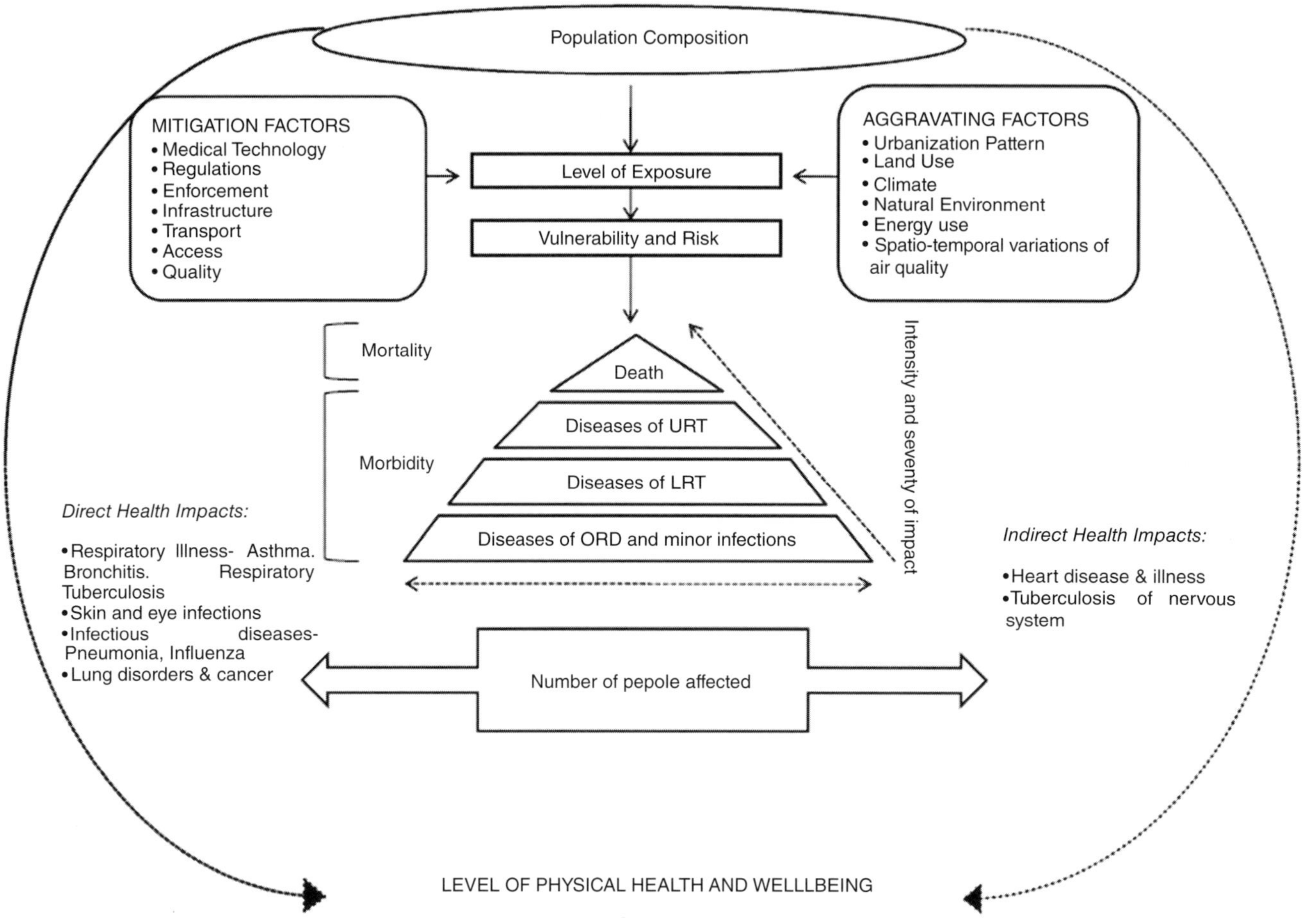

Figure 17.1 System analysis framework to conceptualize linkages between urban environment and level of physical health and well-being.

17.2 Study Area

The National Capital Territory (NCT) of Delhi is the national capital of India and is the political, judicial and administrative center of the country. The NCT of Delhi is located between 28° 24' 17" and 28° 53' 00" N and 76° 50' 24" and 77° 20' 37" E. With an area of 1,483 km^2, it encompasses a total population of 16.7 million, 97.5 per cent of which is urban (Directorate of Census Operations, 2011). Delhi has a density of 11,297 persons/km^2. For administrative purposes, Delhi is divided into nine districts and twenty-seven *tehsils*, or sub-districts. There are three statutory towns in the Delhi administrative city: the Municipal Corporation of Delhi (MCD), New Delhi Municipal Council (NDMC) and Delhi Cantonment Board (DCB).

Delhi experiences extreme climatic conditions due to its inland location. The perennial River Yamuna is the only major river in the city. Satellite data analyzed by the Forest Department of India indicate 176.20 km^2 of the area, which is 11.88 per cent of the total geographical area of the city, is under forest cover (Forest Survey of India, 2011). The total tree cover is 120 km^2 viz. 8.08 per cent of total area (Forest Survey of India, 2011).

The number of vehicles per thousand population was estimated to be 253 in 1999–2000 and rose to 436 in 2011–2012 (Planning Department of Delhi, 2013). Two-wheeled vehicles constitute the largest proportion (62.46 per cent) followed by cars and Jeeps (31.5 per cent). The road length in Delhi increased from 14,316 km (1980–1981) to 28,508 km (2000–2001) and reached 32,663 km in 2011 (Planning Department of Delhi, 2013).

17.3 Data and Methods

The present research relies on various secondary sources of data on air pollution and health impacts.

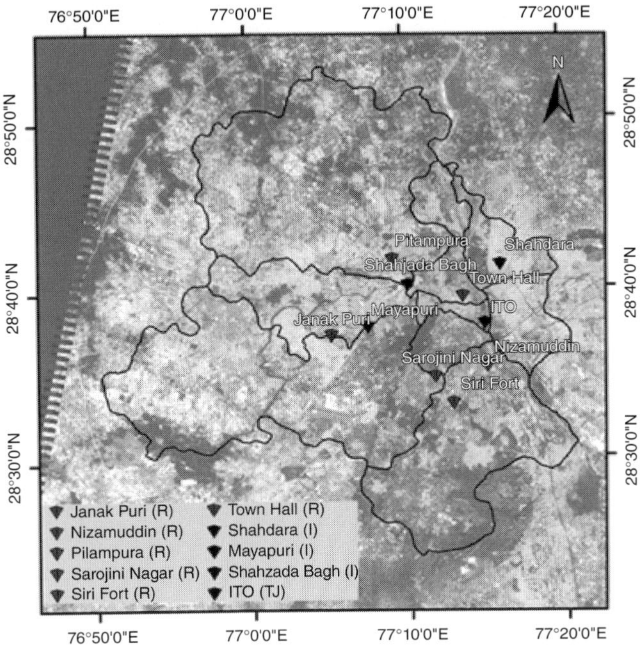

Figure 17.2 Location and nature of air quality monitoring stations in Delhi (background image is Landsat TM 5). The green symbol represents residential sites, red represents industrial sites and black represents traffic junction. (Source: CPCB, 2012; background image is taken from earthexplorer.usgs.gov.). For the color version, please refer to the plate section.

The Pollution Control Board (PCB) under the Ministry of Environment and Forests is responsible for collection of pollution data for all states and union territories of India. The pollution recording stations are representative of three categories on the basis of their function, that is, residential, industrial and traffic junctions. All the sites regularly monitor the SPM, respirable suspended particulate matter (RSPM), sulphur dioxide (SO_2) and nitrogen dioxide (NO_2) levels. SPM and RSPM are monitored every eight hours for twenty-four hours, whereas SO_2 and NO_2 are monitored every four hours. The CPCB provided the air pollutant data for nine stations, including JanakPuri, Nizamuddin, Siri Fort, Sarojini Nagar, Pitampura and Town Hall, which are residential sites, and Mayapuri, Shahdara and Shahjada Bagh, which represent industrial areas, have been used to analyze the status of air pollution in the city (Figure 17.2 and Table 17.1).

The spatio-temporal analysis of pollution data was carried out after averaging the daily pollution records to monthly and annual averages. Trends in levels of SPM, RSPM, SO_2 and NO_2 were analyzed (2001 to 2011) with respect to the annual and monthly fluctuations and with respect to the nature of the station.

Table 17.1 *Permissible Limits and Sources for Selected Pollutants*

Pollutants	Annual mean concentration range ($\mu g/m^3$)	24-Hour mean concentration range ($\mu g/m^3$)
SO_2	50	80
NO_2	40	80
RSPM	60	100
SPM	40	60

Source: Annual Mean Concentration Range given by National ambient air quality, 2009 guidelines adopted from CPCB, 2012.

To examine the impact of pollutants on human health, data on cause of mortality have been considered. There are two major limitations to establishing a link between air pollution and health. Firstly, one may argue that a disease may be due to many factors and not only air pollution. On the other hand, the disease or illness tends to be aggravated as pollution levels increase. Secondly, in the absence of respiratory illness data, the data on cause of mortality are used.

Among the direct causes, diseases of the respiratory system are those most likely to arise from air pollution and are divided into three parts: diseases of the upper respiratory tract (URT), diseases of the lower respiratory tract (LRT) and other respiratory system diseases (ORD) (Table 17.2). URT diseases have symptoms that involve symptoms such as runny and stuffy nose, sinusitis, sore throat, wet cough, dry cough, cold head, fever, burning or red eyes. The LRT symptoms include wheezing, phlegm, shortness of breath, chest discomfort or pain.

17.4 Results and Discussion

17.4.1 Trends in Air Pollution

Annual mean levels of SO_2 indicate that it has been mostly below the prescribed limit ($50 \mu g/m^3$) during the monitored period (Figure 17.3a). Power sector and thermal power plants are the main sources of SO_2; some researchers have also listed the transport sector as a source (Chelani and Devotta, 2007; Datta et al., 2010). Hence, the low concentrations of SO_2 may be due to the policies requiring the use of higher quality of diesel and compressed natural gas (CNG) in public vehicles. Goyal et al. (2006) also point out similar trends in SO_2. The levels of NO_2, on the other hand,

Table 17.2 *Classification of Causes of Death Due to Air Pollution*

	Diseases of the respiratory system		
Major classification	Diseases of the upper respiratory tract (URT)	Diseases of the lower respiratory tract (LRT)	Other respiratory system diseases (ORD)
Disease	Acute pharyngitis and acute tonsilitis	Acute Bronchitis and acute bronchiolitis	Influenza
	Acute laryngistis and trachetis	Bronchitis, chronic and unspecified emphysema	Pneumonia
	Acute upper respiratory infections	Asthma	Pleurisy
	Other diseases of URT	Other lower respiratory disorders	All other diseases of the respiratory system

Source: Compiled from Government of NCT of Delhi, 2001, 2004, 2005, 2006, 2007, 2008, 2009, 2010, 2011, 2012

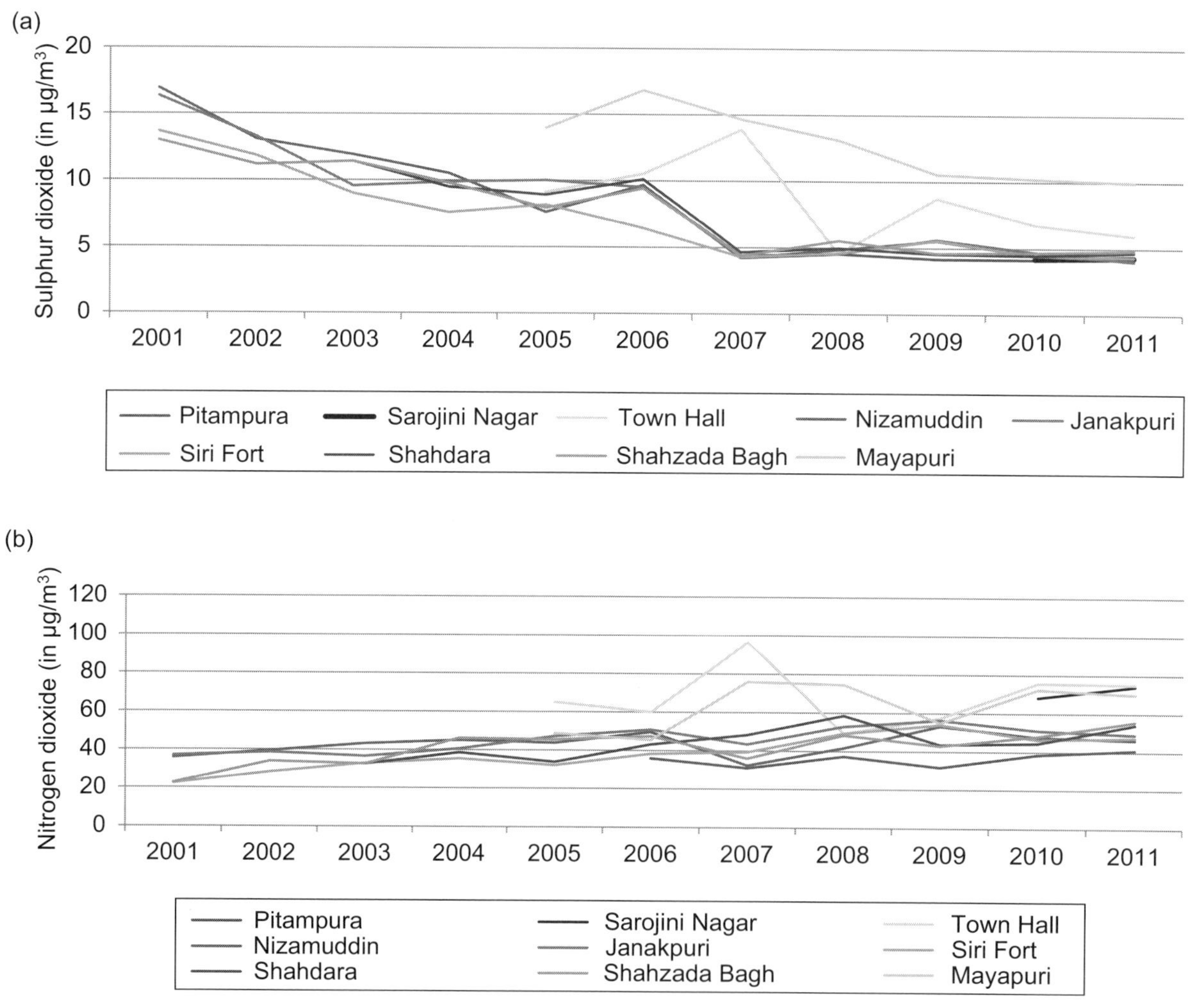

Figure 17.3 Annual trend of major pollutants in Delhi, 2001–2011. a. SO_2, b. NO_2, c. SPM and d. RSPM. The figure is prepared by the author using Central Pollution Control Board (CPCB) daily pollutant level data.
Note: RSPM data not available until 2003; x-axis represents years and y-axis represent pollutant levels in $\mu g/m^3$. For the color version, please refer to the plate section.

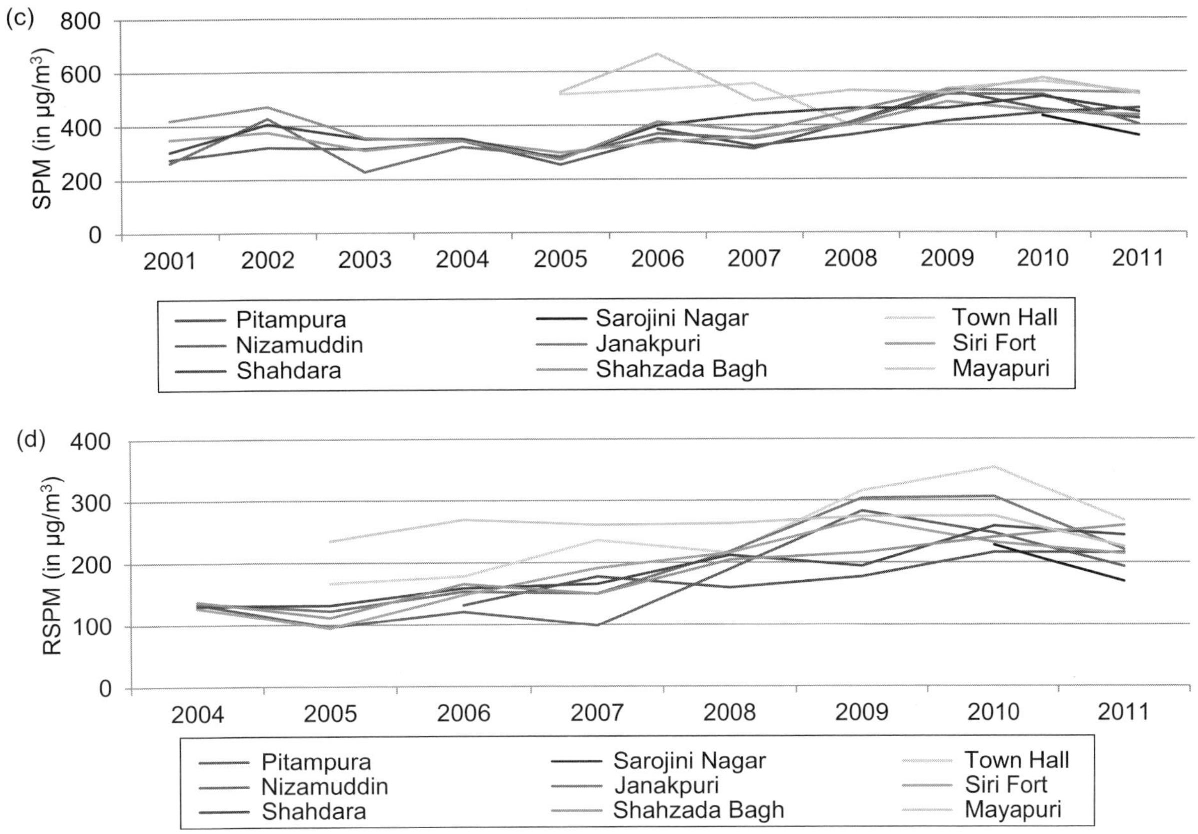

Figure 17.3 (*cont.*)

indicate an increasing trend and are well above the permissible limit of 40µg/m^3 (Figure 17.3b). A substantial rise in the vehicular fleet is considered the major cause for this increase. The rise of NO_2 is a threat to environmental and human health as the reactions of NO_X in the presence of sunlight produce ground level ozone that is dangerous to human health (Lo and Quattrochi, 2003).

The SPM and RSPM levels are seen to increase markedly during the monitored period (Figure 17.3c and d). Incomplete combustion from industries and vehicles is a major source of particulates of a range of sizes. Natural sources including dust from unpaved roads or roadsides combined with particular meteorological conditions add to the load. Other anthropogenic sources include agricultural fields, construction sites, refuse burning etc. These are a menace to human health and well-being including physical, social and mental conditions. In the process of rapid urbanization and the increased demand for intra- and inter-city movement, human activities produce large quantities of particles of complex structure.

The annual temporal analysis of SPM and RSPM demonstrates a rising trend (permissible limits for SPM and RSPM are 40 and 60µg/m^3, respectively). The particulates constitute a prominent share of the pollutant load of the city and, hence, may have profound effects on the lungs, respiratory tract and circulatory systems.

Because of variations in temperature, rainfall, humidity, wind direction and other climatic factors, the levels of pollutants in Delhi vary between seasons. There are variations in the level of increase of NO_2 between seasons. The winter months are associated with maximum concentrations, which gradually decrease in spring and summer to reach lowest values during the rainy season (Figure 17.4a).

RSPM is one of the most dangerous components for human health. The RSPM levels are markedly low in the rainy season (Figure 17.4b) because of the settling of minute particles under the influence of rainfall and high humidity. The highest RSPM levels are found in the winter, when they are very much above the prescribed limits (60µg/m^3). Similar findings are cited in

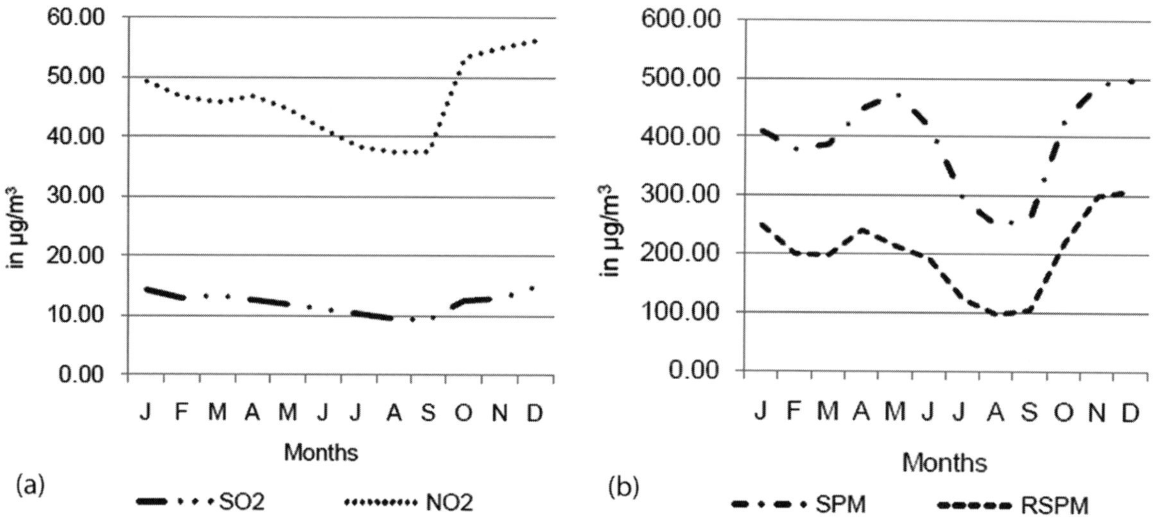

Figure 17.4 Monthly average of (a) SO_2, NO_2 (b) RSPM, SPM.

detailed research by CPCB (2006). The SPM levels are also high in all seasons except the rainy season. The contributing factors of high SPM are incomplete combustion in vehicles and construction works. The peak winter and summer seasons have maximum SPM (Figure 17.4b). In the former, the air is stable and there is less mixing in the atmosphere that contributes to high SPM levels. On the other hand, in the latter season, towards spring and summer, pollen and dryness of the air are responsible for the SPM content. Shandilya et al. (2006) indicate that the finer particles stick together or deposit on surfaces easily and rapidly in the rainy season, leading to cleaner air during monsoons.

17.4.2 Vehicular Growth as a Major Factor Influencing Air Quality of Delhi

While Delhi contains 1.4 per cent of India's population, it accounts for 7 per cent of total vehicular load of the country. According to Das and Parikh (2004) Delhi was the fourth most polluted city in the world, and the dominant contributor sector to air pollution was the transport sector, as illustrated by the strong correlation between population and GDP growth, motor vehicle numbers and pollution increase.

The total number of vehicles in Delhi increased from 1.92 million (1991) to 5.32 million (2010). Private motor cars have rapidly proliferated in the city, now a total of 1.46 million (2010), up from a mere 0.42 million (1991) (Figure 17.5). There is high dependency on private vehicles for transportation, which is clearly reflected by the increasing proportion of private cars to total vehicle composition. While in 1991, 22.2 per cent of the total vehicular load was of private cars, this had risen to 27.4 per cent in 2010.

17.4.3 Impact of Air Pollution on Human Health

The cause of deaths due to illness can be influenced by multiple factors including heredity, lifestyle, exposure level, occupation, dietary habits among others. Diseases may be acquired or inherited and may be aggravated by, for example, environmental conditions. In order to avoid these complexities, the immediate recorded cause of mortality is considered here as a base for establishing the degree of correlation with pollution and health. This takes into account the fact that the immediate surroundings have a profound impact on health in the sense of both developing a disease and aggravating the conditions that activate a hereditary illness.

Apart from disorders of the respiratory system, air pollution may cause harm to eyes, skin, lungs and other respiratory organs. In extreme cases it may lead to mortality from respiratory illness and cancer. Air pollution may also worsen the circulatory system, making it more vulnerable to failure.

SPM may cause pneumoconiosis, restrictive lung disease, asthma and cancer. Cropper et al. (1997), using the mortality data obtained from NDMC, established that there is a positive relationship between particulate pollution and daily non-traumatic deaths from respiratory and cardiovascular problems.

Table 17.3 *Total Deaths from Disease from URT, LRT and ORD in Delhi*

Cause of death	2001	2004	2005	2006	2007	2008	2009	2010	2011
Diseases of URT	49	243	48	9	43	17	21	499	653
Diseases of LRT	1,108	695	1,043	1,089	922	949	1,088	598	886
ORD	2,074	1,350	1,678	1,218	1,500	1,730	2,006	2,649	2,403

Source: Based on data from Government of NCT of Delhi, 2001, 2004, 2005, 2006, 2007, 2008, 2009, 2010, 2011

Table 17.4 *Area, Population and Mortality Details of Different Regions, Delhi*

	Area (2001) (per cent)*	Population (2001)*	Density (2001) (persons/km^2)*	Total institutional deaths (2001–2011)**
MCD	94	1,34,23,227 (97%)	9,607	4,16,094
NDMC	3	3,02,363 (2%)	7,074	2,02,529
DCB	3	1,24,917 (1%)	2,907	14,496
Urban	62	NA	13,957	NA
Rural	38	NA	1,692	NA

Sources: *Directorate of Economics and Statistics, 2014; **Directorate of Economics and Statistics, 2014. Compiled from Government of NCT of Delhi, 2001, 2004, 2005, 2006, 2007, 2008, 2009, 2010, 2011

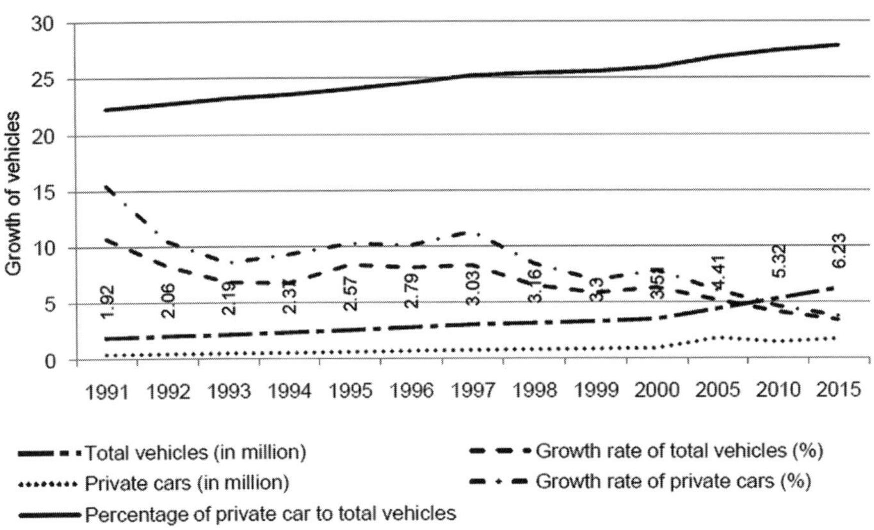

Figure 17.5 Growth of vehicles in Delhi, 1991–2015 (source: compiled from Firdaus and Ahmad, 2011; Transportation Department, 2005).

Similarly, Firdaus (2010) concluded that there is high correlation between SPM and health consequences. Using primary survey data, he found chronic respiratory symptoms and airway diseases among residents of high-pollution areas.

RSPM is released from industries, combustion of fossil fuels, vehicle exhausts and is known to cause respiratory illnesses such as chronic bronchitis and asthma, heart disease etc. Agarwal et al. (2006) suggest that SPM and RSPM have a direct bearing on the

Table 17.5 *Deaths from Respiratory Illness by Region in Delhi, 2001–2012*

	Direct health effects			
	Respiratory diseases			
Statutory towns and rural areas	Pneumonia	Influenza	Bronchitis and asthma	Whooping cough
MCD	6,441	2,588	5,330	300
NDMC	6,592	74	3,687	2
DCB	550	3	403	0
Rural	623	115	566	41

Total figures do not include data for 2002 and 2003.
Source: Based on data from Government of NCT of Delhi, 2001, 2004, 2005, 2006, 2007, 2008, 2009, 2010, 2011, 2012

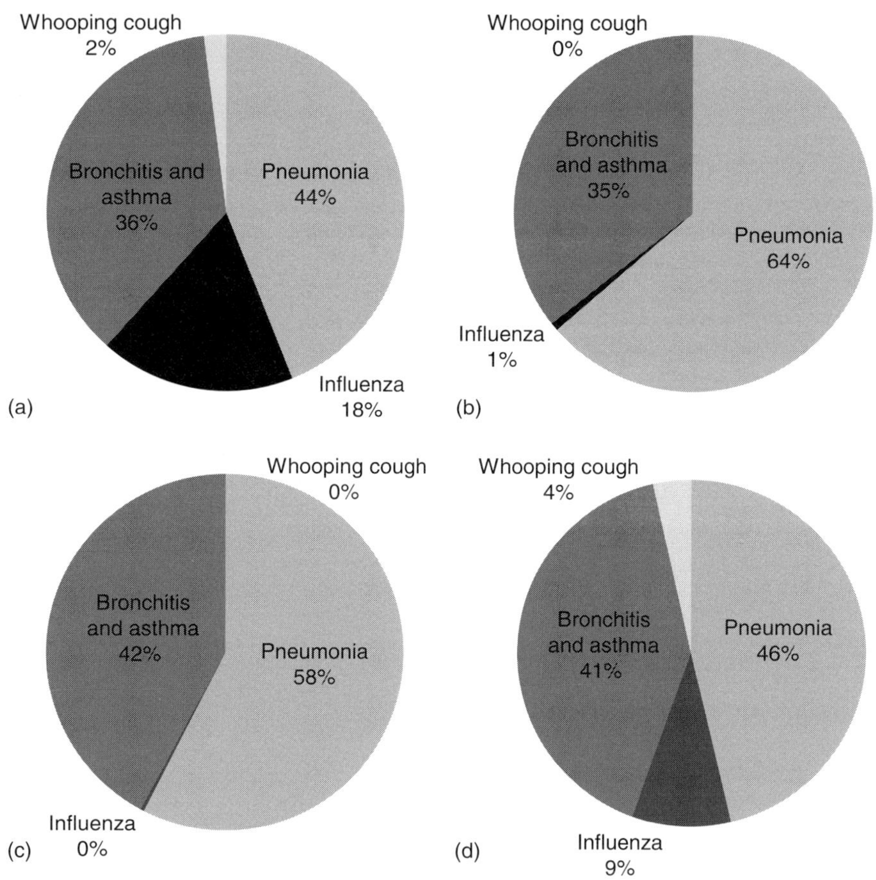

Figure 17.6 Regionwise direct and indirect causes of death, Delhi (in per cent) in (a) MCD, (b) NDMC, (c) DCB and (d) rural areas.
Note: does not include data for 2002 and 2003.
(Source: Based on data from Government of NCT of Delhi, 2001, 2004, 2005, 2006, 2007, 2008, 2009, 2010, 2011, 2012).

number of chronic obstructive airway/pulmonary disease (COPD) cases in Delhi. The extensive study by Rizwan et al. (2013) reflects on the associated morbidity due to air pollution. Dry cough, wheezing, breathlessness and chest discomfort are notable health problems in the city. In the same study, hypertension was also found to be positively correlated with RSPM, and Delhi is associated with significantly higher levels of chronic headache, eye irritation and skin irritation. Delhi also showed statistically significant increases in various types of lung function deficits. Oxides of nitrogen generated from power plants, electric utility boilers and

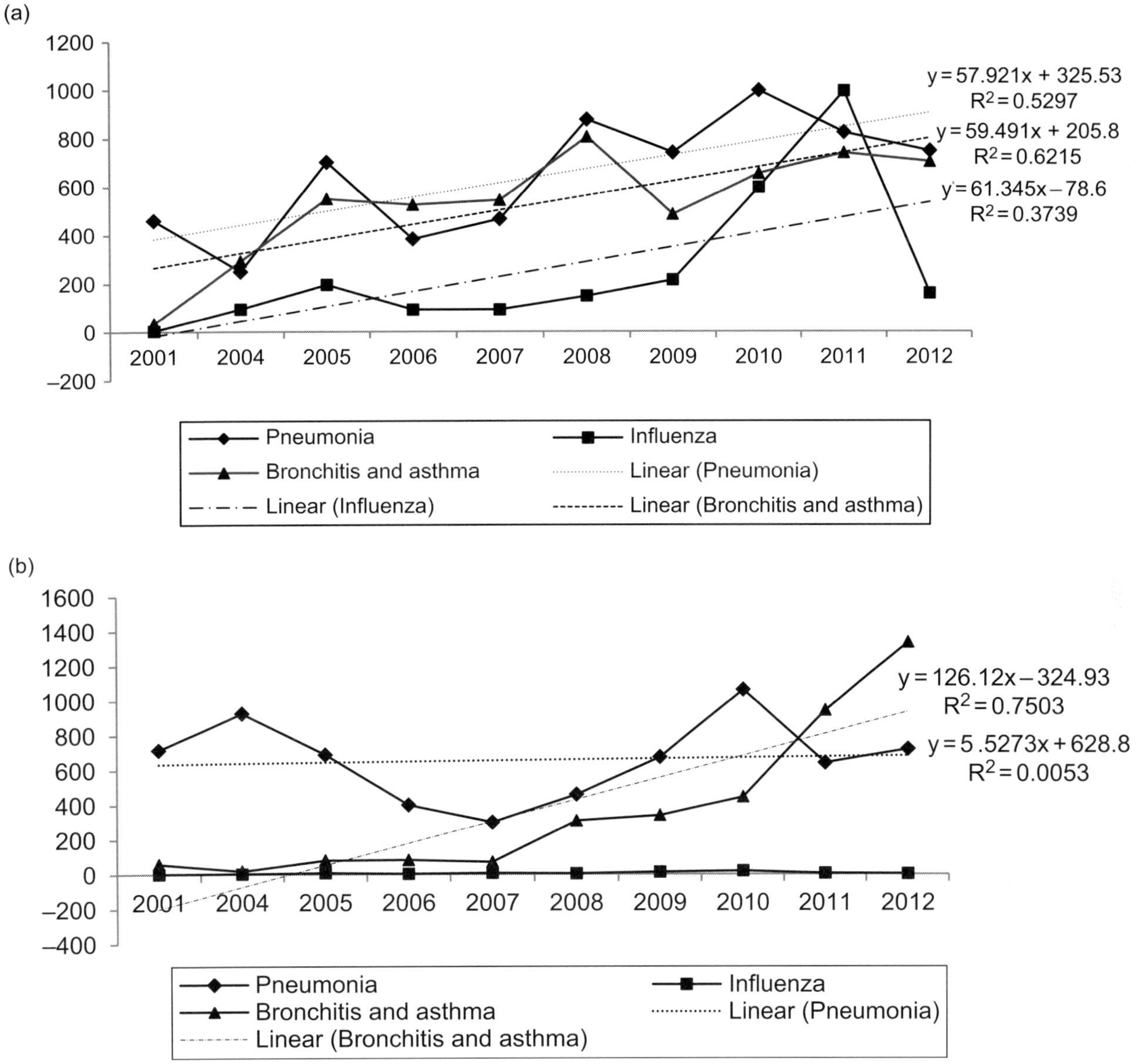

Figure 17.7 Trend analysis of deaths from respiratory diseases per region of Delhi (a) MCD, (b) NDMC, (c) DCB and (d) rural areas.
Note: does not include data for 2002 and 2003; x-axis represents number of deaths and y-axis represents year.
(Source: Based on data from Government of NCT of Delhi, 2001, 2004, 2005, 2006, 2007, 2008, 2009, 2010, 2011, 2012).

vehicle emission may cause airway resistance, chest tightness, lung and eye irritation and viral infections. Jayaraman and Nidhi (2008) correlated the daily count of patient data in the respiratory unit of one hospital and atmospheric levels of ozone, NO_2 and RSPM. The results reveal a significant impact of air pollution factors on all age groups for all variables. In another study, Nidhi and Jayaraman (2007) also state that gaseous pollutants such as SO_2 and NO_2, in spite of being mostly at low levels, have significant impact on human health.

The deaths from URT significantly increased from 9 in 2006 to 653 in 2011. On the other hand, LRT diseases have fluctuated with a minimum of 598 deaths in 2010 and a maximum of 1,108 deaths in 2001. ORD has always been the largest contributor of respiratory related mortality (Table 17.3).

(c)

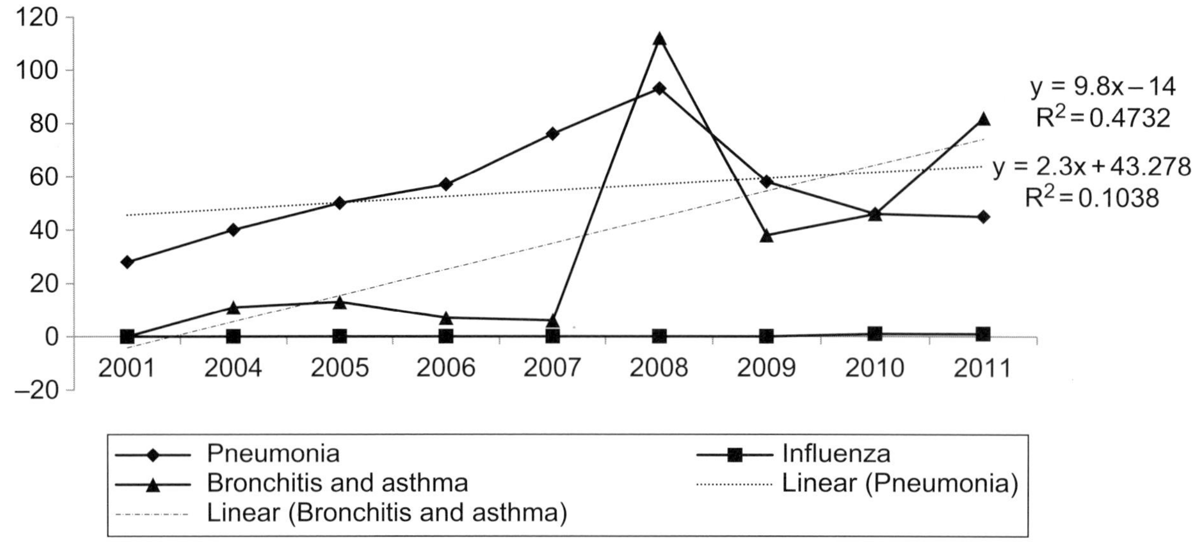

(d)

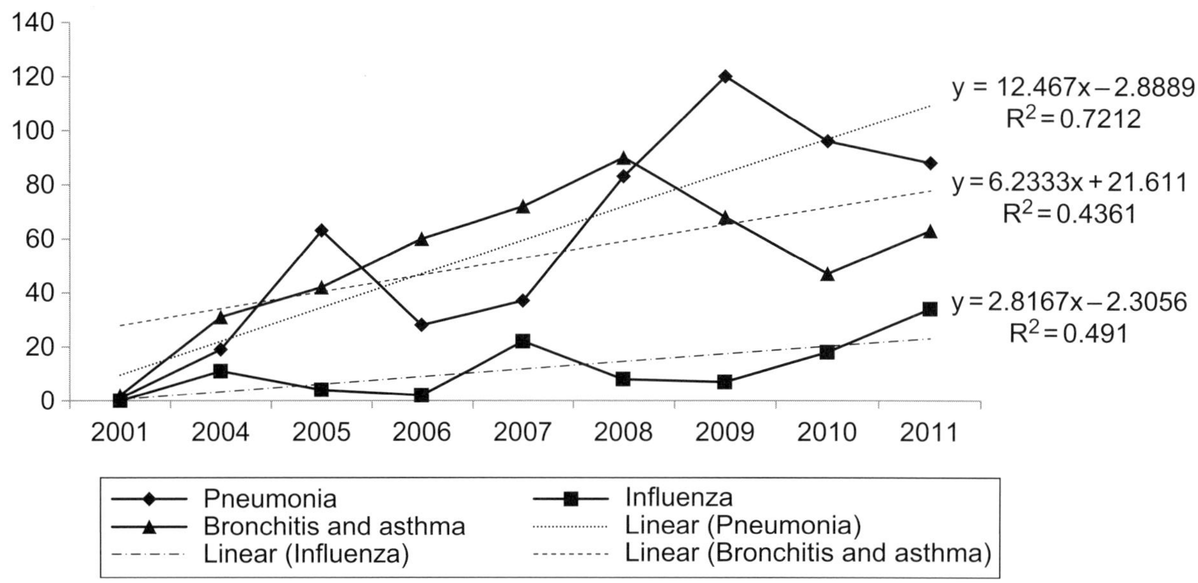

Figure 17.7 (*cont.*)

17.4.4 Spatial Analysis of Impact of Air Pollution on Human Health in Delhi

The urban area of Delhi is administered by three agencies, i.e., NDMC, MCD and DCB (Table 17.4). The MCD occupies both the maximum area (94 per cent) and maximum population (97 per cent in 2001) in Delhi and proportionally accounts for the highest number of deaths. The DCB is geographically the smallest region and counts for just 3 per cent of the area, 1 percent of the population and the lowest population density (2,907 persons/km^2). The spatial analysis of mortality from diseases caused due to air

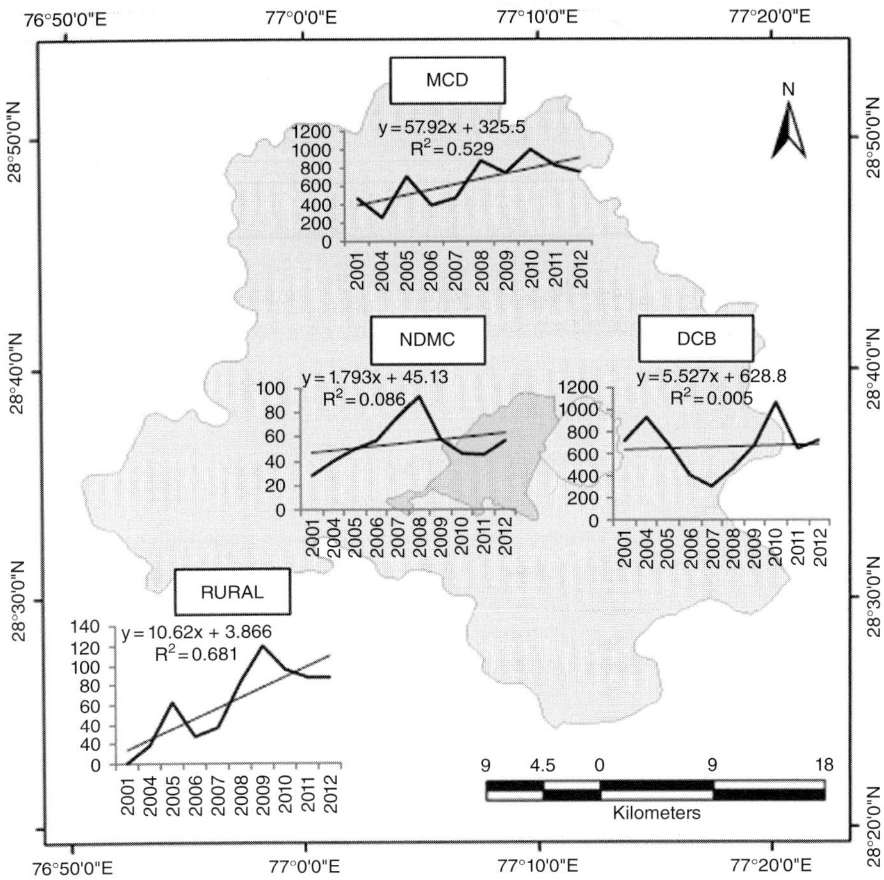

Figure 17.8 Spatio-temporal trend of mortality from pneumonia in regions of Delhi, 2001–2012.
Note: does not include data for 2002 and 2003; x-axis represents number of deaths and y-axis represent year.
(Source: Based on data from Government of NCT of Delhi, 2001, 2004, 2005, 2006, 2007, 2008, 2009, 2010, 2011, 2012).

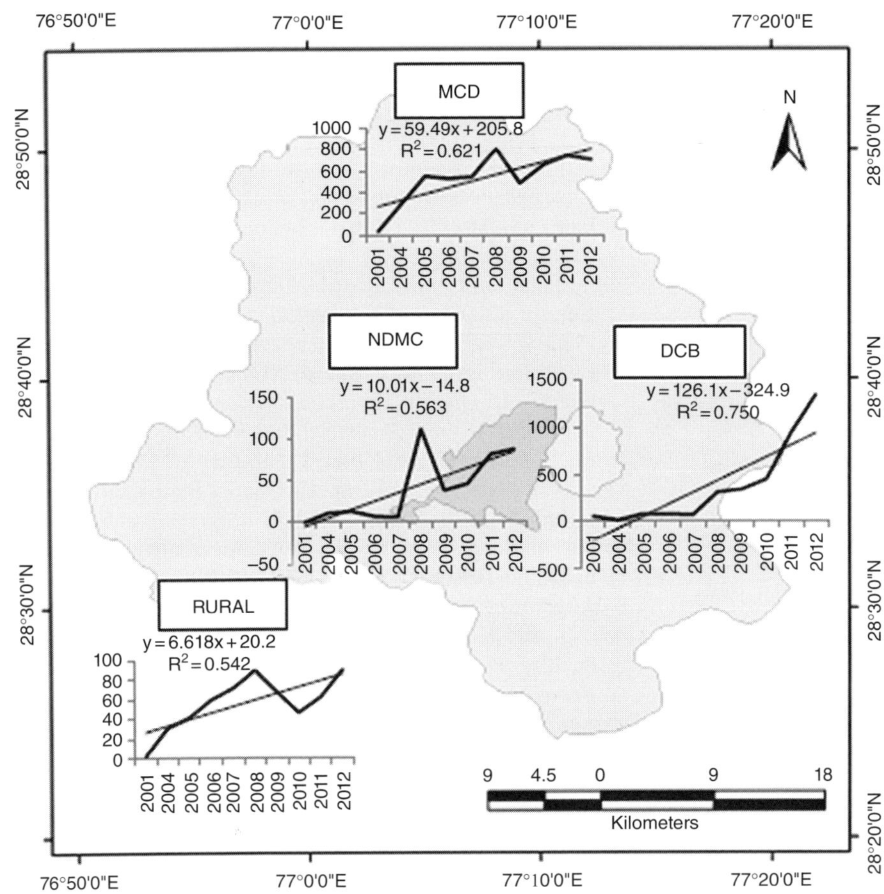

Figure 17.9 Spatio-temporal trend of mortality from bronchitis and asthma in regions of Delhi, 2001–2012.
Note: does not include data for 2002 and 2003; x-axis represents number of deaths and y-axis represents year.
(Source: Based on data from Government of NCT of Delhi, 2001, 2004, 2005, 2006, 2007, 2008, 2009, 2010, 2011, 2012).

pollution is done with respect to MCD, NDMC and DCB. Among the direct health effects, i.e., respiratory related mortality, maximum deaths are due to pneumonia and followed by bronchitis and asthma and influenza (Table 17.5).

Further, the regionwise trend analysis of four major diseases of the respiratory system from 2001 to 2012 reveals that bronchitis and asthma have observed maximum rise (Figure 17.6). The mortality from respiratory system diseases such as bronchitis and asthma and pneumonia has increased in MCD, NDMC and DCB since 2001 (Figure 17.7a,b,c).

The spatio-temporal analysis was done for three urban regions controlled by MCD, NDMC and DCB and for rural Delhi (Figures 17.8 and 17.9). The study reveals that deaths due to pneumonia, asthma and bronchitis have been increasing in the study area, however, there are spatial variations in the magnitude of change. In general, the deaths due to asthma and bronchitis have experienced a sharper increase than the deaths caused by pneumonia.

The second most important causes of death are bronchitis and asthma. The spatial analysis of mortality from bronchitis and asthma reveals an increase in all regions. NDMC was associated with the strongest levels of increase followed by MCD and DCB, suggesting the dire need for stricter and more effective policies to improve urban health.

17.5 Conclusion and Recommendations

The urban atmosphere of Delhi is deteriorating as a result of a rise in air pollution. The relationship between air pollution levels and various parameters associated with human health in Delhi suggests that PM exerts the strongest negative influence. Delhi has a large, increasing population and these rising health risks pose a substantial threat to the sustainable growth of the city.

For more sustainable growth of the national capital, there needs to be a focus on the entire complex of elements affecting physical human health and well-being, but it is clearly most important that pollutant levels be controlled. Green technology and cleaner fuels must be used.

There should be stringent implementation of the rules regarding the pollution norms of vehicles and industries (small, medium and large). Maintaining a clean environment in the city requires the establishment of a green belt. Cleaner fuels and very efficient diesel and petrol vehicles are required. In addition, the government should allocate additional funds for the development of heath facilities, including public hospitals. Furthermore, afforestation programmes, car pooling and use of public transport should be encouraged, as these can offer improvements in air quality and, accordingly, health.

Acknowledgements

We acknowledge the reviewers for valuable comments and suggestions for the improvement of this research paper. Thanks are also due to Dr. D. D. Basu from CSE; Dr. Chintan Chaudhary, Consultant at Safdarjung Hospital, Delhi for fruitful discussions. We are deeply grateful to Dr. Sanjeev Agarwal and Dr. Sangeeta from the Central Pollution Control Board for providing the pollution data for Delhi.

References

Agarwal, R., Jayaraman, G., and Anand, S., et al. (2006). Assessing respiratory morbidity through pollution status and meteorological conditions for Delhi. *Environmental Monitoring and Assessment*, **114**, 489–504.

Beer, T. (2001). Air quality as a meteorological hazard, *Natural Hazards*, **23**, 157–169.

Beggs, P. J. and Bambrick, H. J. (2005). Is the global rise of asthma an early impact of anthropogenic climate change? *Environmental Health Perspectives*, **113**, 915–919.

Bentham, G. (1992). Global climate change and human health. *GeoJournal*, **26**, 7–12.

Central Pollution Control Board. Ministry of Environment and Forests. (2008). *Epidemiological Study on Effect of Air Pollution on Human Health (Adults) in Delhi* (Delhi, India).

Central Pollution Control Board. Ministry of Environment. (2006). *Air Quality Trends and Action Plan for Control of Air Pollution from Seventeen Cities* (Delhi, India).

Central Pollution Control Board. Ministry of Environment. (2012). National Ambient Air Quality Status and Trends in India – 2010 (Delhi, India), 1–172.

Chelani, A. B. and Devotta, S. (2007). Air quality assessment in Delhi: Before and after CNG as fuel. *Environmental Monitoring and Assessment*, **125**, 257–263.

Chen, B. and Kan, H. (2008). Air pollution and population health: A global challenge. *Environmental Health and Preventive Medicine*, **13**, 94–101.

Cropper, M. L., Simon, N. B., and Alberni, A. et al. (1997). The health effects of air pollution in Delhi, India.

Policy Research Working Paper 1860, *The World Bank Development Research Group*, 1–40.

Curtis, L., Rea, W., and Smith-Willis, P. et al. (2006). Adverse health effects of outdoor air pollutants. *Environment International*, **32**, 815–830.

Das, A. and Parikh, J. (2004). Transport scenarios in two metropolitan cities in India: Delhi and Mumbai. *Energy Conversion and Management*, **45**, 2603–2625.

Datta, A., Saud, T., Goel, A. et al. (2010). Variation of SO_2 over Delhi. *J Atmos. Chem.*, **65**, 127–143.

Directorate of Census Operations. (2011). *Provisional Population Totals - NCT of Delhi.* Ministry of Home Affairs, Government of India.

Directorate of Economics and Statistics. (2012). Statistical abstract of Delhi. Government of NCT of Delhi, 1–194.

Fang, Y., Naik, V., Horowitz, L,W., and Mauzerall, D. L. (2013). Air pollution and associated human mortality: The role of air pollutant emissions, climate change and methane concentration increases from the preindustrial period to present. *Atmos Chem Phys.*, **13**, 1377–1394.

Firdaus, G. (2010). Urbanization and changing air quality in Delhi: A comparative analysis and strategy for its better management. *Asian Journal of Water, Environment and Pollution*, **8**, 23–31.

Firdaus, G. and Ahmad, A. (2011). Changing air quality in Delhi, India: Determinants, trends and policy implications. *Reg Environ Change*, **11**, 743–752.

Forest Survey of India. (2011). *State of Forest Report. Ministry of Environment and Forests*, (Dehradun, India).

Frumkin, H. (2002). Urban sprawl and public health. *Public Health Reports*, **117**, 201–217.

Gatzweiler, F. W., Zhu, Y-G., Diez Roux, A. V. et al. (2016). Advancing health and wellbeing in the changing urban environment: Implementing a system approach. Springer Singapore, 1–78.

Government of India, Ministry of Home Affairs. (2010). Report on medical certification of cause of death 2010, Office of Registrar General (India).

Government of NCT of Delhi. (2001). *Report on Medical Certification of Cause of Death (2001) Government of India*, Ministry of Home Affairs, Office of Registrar General, India.

Government of NCT of Delhi. (2004). *Report on Medical Certification of Cause of Death (2004) Government of India*, Ministry of Home Affairs, Office of Registrar General, India.

Government of NCT of Delhi. (2005). *Report on Medical Certification of Cause of Death (2005) Government of India*, Ministry of Home Affairs, Office of Registrar General, India.

Government of NCT of Delhi. (2006). *Report on Medical Certification of Cause of Death (2006) Government of India*, Ministry of Home Affairs, Office of Registrar General, India.

Government of NCT of Delhi. (2007). *Report on Medical Certification of Cause of Death (2007) Government of India*, Ministry of Home Affairs, Office of Registrar General, India.

Government of NCT of Delhi. (2008). *Report on Medical Certification of Cause of Death (2008) Government of India*, Ministry of Home Affairs, Office of Registrar General, India.

Government of NCT of Delhi. (2009). *Report on Medical Certification of Cause of Death (2009) Government of India*, Ministry of Home Affairs, Office of Registrar General, India.

Government of NCT of Delhi. (2010). *Report on Medical Certification of Cause of Death (2010) Government of India*, Ministry of Home Affairs, Office of Registrar General, India.

Government of NCT of Delhi. (2011). *Report on Medical Certification of Cause of Death (2011) Government of India*, Ministry of Home Affairs, Office of Registrar General, India.

Government of NCT of Delhi. (2012). *Report on Medical Certification of Cause of Death (2012) Government of India*, Ministry of Home Affairs, Office of Registrar General, India.

Goyal, S. K, Ghatge, S. V., Nema, P. et al. (2006). Understanding urban vehicular pollution problem vis-à-vis air quality: Case study of a megacity (Delhi, India). *Environmental Monitoring and Assessment*, **119**, 557–569.

Jayaraman, G. and Nidhi. (2008). Air pollution and associated respiratory morbidity in Delhi. *Health Care Manage. Sci.*, **11**, 132–138.

Kim, K-H., Kabir, E. and Kabir, S. (2015). A review on the human health impact of airborne particulate matter. *Environment International*, **74**, 136–143.

Knowlton, K., Rosenthal, J. E., Hogrefe, C. et al. (2004). Assessing ozone-related health impacts under a changing climate. *Environmental Health Perspectives*, **112**, 1557–1563.

Lo, C. P. and Quattrochi, D. A. (2003). Land-use and land-cover change, urban heat island phenomenon, and health implications: A remote sensing approach. *Photogramm. Eng. Remote Sens.*, **69**, 1053–1063.

Lvovsky, K. (1998). Economic costs of air pollution with special reference to India. *National Conference on Health and Environment*. Paper prepared for the National Conference on Health and Environment Delhi, India, July 7–9, 1998, 1–26.

Magas, O. K., Gunter, J. T., and Regens, J. L. (2007). Ambient air pollution and daily pediatric hospitalizations for asthma. *Env. Sci.Pollut. Res.*, **14**, 19–23.

Mohan, M. and Kandya, A. (2007). An analysis of the annual and seasonal trends of air quality index in Delhi. *Environmental Monitoring and Assessment*, **131**, 267–277.

Nidhiand Jayaraman, G. (2007). Air quality and respiratory health in Delhi, *Environ. Monit. Assess.*, **135,** 313–325.

Planning Department of Delhi. (2013). Economic survey of Delhi 2012–13, Department of Planning (India).

Rizwan, S. A., Nongkynrih, B., and Gupta, S. K. (2013). Air pollution in Delhi: Its magnitude and effects on health. *Indian J. Community Med.*, **38**, 4–8.

Shandilya, K. K., Khare, M., and Gupta, A. B. (2007). Suspended particulate matter in rural-industrial Satna and in urban-industrial South Delhi. *Environ. Monit. Assessment*, **128**, 431–445

Transportation Department. (2005). *Government of NCT*, Delhi.

World Health Organisation. (2016). Health as the pulse of the new urban agenda. United Nations Conference on Housing and Sustainable Urban Development, Quito–October 2016.

PART VI

Future Earth and Food Security

18

Targeting Research towards Achieving Food Security in an Era of Climate Change

Bruce M. Campbell, Dhanush Dinesh, and Sophia Huyer

18.1 Introduction

Agriculture is at the center of three of the greatest challenges facing humanity in the twenty-first century: firstly, the food security challenge, the components of which are changing as populations grow, get richer and remain unequal (Godfray et al., 2010); Secondly, the adaptation challenge, which encompasses the longer-term shifts in climate as well as climate variability, with agriculture being one of the sectors that will be hardest hit by climate change (Thornton et al., 2014); thirdly, the environmental challenge, with agriculture a major cause of the Earth system shifting beyond planetary boundaries (Steffen et al., 2015; Campbell et al., 2017). One component of this environmental challenge is reducing greenhouse gas (GHG) emissions from agriculture (Wollenberg et al., 2016).

The global agricultural community has responded by developing the concept of climate-smart agriculture (CSA), the objectives of which are to tackle food security, climate change adaptation and reducing emissions (Lipper et al., 2014). CSA interventions do not endeavour to achieve all three objectives in all contexts; the most relevant combination of objectives is determined in a context-specific manner (Lipper et al., 2014). There are now multiple initiatives around the globe that support CSA implementation and scaling up. At the global level, the Global Alliance for Climate-Smart Agriculture (GACSA) has more than 140 members including national governments, civil society organizations, private sector companies and research organizations. GACSA facilitates knowledge exchange for CSA implementation and promotes cross-regional cooperation. At the regional level, CSA alliances have been formed in Africa and North America (UN, 2014; NACSAA, 2015). At the sub-regional level, Southeast Asian countries under the auspices of the Association of Southeast Asian Nations have adopted regional guidelines for CSA (ASEAN, 2015) and West African countries have launched a West African CSA Alliance (ECOWAS, 2015). At the national level, many countries are supporting CSA, and case studies are being developed about CSA efforts in Costa Rica, France, Ireland, Malawi, Tanzania and Vietnam (GACSA, 2016). While these efforts represent growing commitment to CSA at multiple levels, we argue that the challenges are severe and urgent, and that much more research is required that works seamlessly on the boundary of research and action.

18.2 Food Security Challenge

Food security is itself a triple challenge – with a growing ‘triple burden’ of malnutrition (IFPRI, 2015). Just less than 1 billion people are chronically undernourished, some 2 billion lack micronutrients, while more than 2 billion persons over-consume. This must be set against population change with 2 billion more people expected by 2030, and richer households consuming more food and, particularly, more meat (Alexandratos & Bruinsma, 2012). For agriculture, and in the absence of consumption changes, some 60% more food will be required by 2050 by comparison to 2005. The sector will also have to provide employment to another 200 million smallholder farmers by 2050 (Campbell & Thornton, 2014).

Food security issues differ sharply amongst continents, amongst countries and within countries. While more than 25% of people are undernourished in Sub-Saharan Africa, about 65% are overweight in North America (FAO, 2013a; WHO, 2016). Rates of obesity and diabetes are rising faster in the Pacific than any other part of the world (Hawley et al., 2015). Though much has been done to increase food security in South East and East Asia, some 10% of the population remains insecure, while meat consumption is predicted to double from 2005 to 2050 (FAO 2013a; Alexandratos & Bruinsma, 2012). Of the developing country regions, Latin America has made the most advances in reducing food insecurity, but hotspots remain (e.g., in some countries in Central America).

18.3 Adaptation Challenge

Climate change will have, and is already having, profound impacts on agriculture. Climate change is often viewed as a problem of the future, but the recent Intergovernmental Panel on Climate Change Fifth Assessment Report (AR5) shows that impacts on agriculture are already being felt (Porter et al., 2014; Vermeulen 2014). The impacts are on all sub-sectors in agriculture (crops, livestock and fisheries) and all facets of food security (availability of food, access to food, utilization of food and stability of food supplies over time). But these impacts differ widely amongst regions and countries. Tropical countries that already have high food insecurity are particularly exposed to climate risks. Food security is expected to be at most risk from climate change in Central America, northeast Brazil, parts of the Andean region, Sub-Saharan Africa and South Asia.

While technological advances have resulted in yield increases, these would have been even greater in the absence of climate change. Maize, wheat and other major crops have experienced significant climate-associated losses of 40 million tonnes per year (from 1981 to 2002) (Lobell & Field, 2007). Globally wheat production is estimated to fall by 6% for each 1°C of further temperature increase, a fall equivalent to a quarter of current global trade (Asseng et al., 2015). Such global figures hide the strong spatial differences. For example, for maize, Thornton et al. (2010) suggested that yields would decrease by 9% in Uganda but increase by 18% in Kenya; and within these countries there would be further differences among agro-ecological zones. As climate change progresses it is highly likely that shifts from one crop to another, or shifts out of cropping, will have to take place. Rippke et al. (2016) suggest that during the twenty-first century all cropping systems in Sub-Saharan Africa will undergo transformational adaptation, but this will generally be limited to small areas for many crops; only for beans, maize and banana will more than 30% of the current area be affected. Climate change impacts are also likely to affect fisheries and livestock systems. In the tropics, yields from marine capture fisheries are projected to decrease by up to 40% by 2055, compared to 2005 levels (Cheung et al., 2010). While an average increase of 30%–70% in high-altitude regions (Cheung et al., 2010) may help make up for this decrease, small-scale fisheries in the tropics will be the hardest hit by such redistribution (Vermeulen, 2014). Considerably less work has been done on climate change impacts on livestock systems; known impacts include changes to quantity and quality of feeds, increased heat stress in animals, changes to water availability, changes to disease distribution, biodiversity losses, impacts on rural livelihoods and indirect impacts such as that on human health (Thornton et al., 2009). While the impacts described relate to long-term impacts of climate change, climate variability is also affecting food production. Ray et al. (2015) estimate that climate variability accounts for a third of the yield variability globally.

The nature of the adaptation challenge shows some differences amongst regions. In Asia, a particular problem will be sea level rise and saline intrusion in the large deltas crucial for rice production and regional food security. In arid and semi-arid regions of Africa, the Middle East and South Asia, a key focus issue will be climate variability and drought, as illustrated by the Syrian drought in the 2000s, the worst on instrumental record (Kelley et al., 2015).

18.4 Environmental Challenge

The planetary boundaries framework is useful for defining a safe operating space for humanity (Steffen et al., 2015). Agriculture is perhaps the most important sector in terms of driving the Earth system beyond safe levels (Campbell et al., 2017). Nine planetary boundaries are recognised. Of these, bio-geochemical flows (nitrogen and phosphorus cycles) and biosphere integrity have been transgressed, with agriculture's role in driving change exceeding 80%. Excessive nitrogen leads to soil and air pollution, increases GHG levels, drives biodiversity loss and pollutes coastal marine waters and watersheds (Swaney et al., 2012). Agriculture's share of total global anthropogenic N use (in organic and inorganic fertiliser) has been estimated at 86% (Galloway et al., 2008). Phosphorus is an important input into agriculture. Use in agriculture has led to eutrophication of freshwater and estuarine systems (Diaz & Rosenberg, 2008). A key driver of biosphere integrity is land cover change. Also important are habitat fragmentation and the spread of weeds. Agriculture has a major role to play in all these processes.

Three planetary boundaries are recorded as being in the zone of uncertainty (high risk): land-system change, freshwater use and climate change. Agricultural lands are one of the largest terrestrial biomes on the planet, occupying ca. 40% of land surface (Foley et al., 2005). It is estimated that at least 80% of deforestation is due to agriculture, while agriculture accounts for ca. 70% of global freshwater withdrawals. Agriculture is responsible for about a quarter of GHG emissions (directly in the form of N_2O and CH_4; and indirectly through driving deforestation and CO_2 release).

Other planetary boundaries are in the safe zone or have yet to be quantified, but even here agriculture can be important in driving change: ocean acidification, stratospheric ozone depletion, atmospheric aerosol loading and introduction of novel entities. Agriculture has contributed ca. 25% of the CO_2 driving ocean acidification. Given the historical emissions from chlorofluorocarbon, and their longevity, the current influence of agriculture on ozone depletion is low but will increase to become the dominant factor, through N_2O emissions. A significant global source of atmospheric aerosols is crop residue burning (van der Werf et al., 2010). Introduction of novel entities covers new types of engineered materials or organisms, and for agriculture would include use of pesticides and genetically modified organisms.

18.5 Responding to the Challenges: Climate-Smart Agriculture

There is a global movement on CSA. According to Lipper et al. (2014), CSA is designed to promote coordinated actions by farmers, researchers, the private sector, civil society and policymakers. The three objectives of CSA are to (1) to increase agricultural productivity sustainably to support equitable increases in incomes, food security and development; (2) to adapt and build resilience to climate change; and (3) to develop opportunities to reduce GHG emissions (FAO, 2013b). In CSA, the environmental challenge is more narrowly framed as a mitigation challenge, though other authors stress that CSA needs to have broader environmental objectives (Neufeldt et al., 2013). Similarly, while some focus on the productivity dimensions of CSA, others prefer a broader food security and food system framing.

There are many bio-physical research issues to be addressed in CSA, such as improved downscaling of climate models to assess the agricultural impacts of climate change, enhanced breeding to develop stress-tolerant varieties and breeds, measurement of GHG emissions in complex smallholder farming systems and understanding synergies and trade-offs between adaptation and mitigation in agricultural development. However, there is also a vast social science and economics agenda, as illustrated by Lipper et al. (2014), who state that the four main action areas for CSA are (1) building evidence, (2) increasing local institutional effectiveness, (3) fostering coherence between climate and agricultural policies and (4) linking climate and agricultural financing.

18.6 Major Research Thrusts to Tackle the Challenges

We identify six major areas of research that are needed to ensure food security in an era of climate change.

1. **Foresight.** There is plenty of research on the impacts of climate change on future food security, but we would question whether this is rapidly advancing our knowledge on the topic and leading society to address food security issues better (Campbell et al., 2016). One problem holding back such research is the relative lack of improvement in the skill of climate models for spatially explicit analysis of agricultural productivity (Ramirez-Villegas et al., 2013). A more promising approach to foresight is scenario building with stakeholders (e.g., Vervoort et al., 2014). Scenarios methodology is derived from systems science and aims to explore uncertainty and complexity in the decision-making context rather than providing a single forecast when such prediction is not possible. Stakeholders can be engaged from the start in the process, to foster co-learning amongst multiple stakeholders, including researchers. This can lead to rapid policy development, with researchers playing a supporting role to policy players (Mason-D'Croz et al., 2016). Choudhury et al. (2013) found that credible, legitimate and salient participatory scenarios with a focus on capacity building could be an effective tool to bridge disciplinary boundaries.

2. **Developing and extending stress-tolerant breeds/varieties and practices, with low emissions.** While this research is important, it is probably the area of work that is best represented in the current global research portfolio, with some significant successes to date. For example, more than thirty-three megatonnes of seed of drought-tolerant maize has been distributed in thirteen African countries, with the yields of the new varieties higher than those of currently available commercial maize varieties (Fisher et al., 2015). Breeding for stress tolerance is also relevant to livestock, and appropriate breeding strategies (within and between breeds) have been shown to increase resilience and productivity, while reducing emissions (de Haas et al., 2016). Successes have also been seen in the development of practices that increase tolerance to climatic stress. In India, the practice of laser land levelling (levelling the fields under irrigated rice–wheat systems) helps cope with increasing water stress and has been shown to increase yields, reduce water use and reduce GHGs (Aryal et al., 2015). It is conservatively estimated that this technique has now been extended to some 544,000 hectares in the state of Haryana (Gill, 2014). Much work is needed on GHG emissions in smallholder systems, given the

complexity of such systems and the limited data available (Rosenstock et al., 2016). And the synergies and trade-offs between adaptation and mitigation constitute another research gap.

3. **Improving and delivering seasonal forecasts and advisories.** High-yield systems are only feasible when the risks to farmers are reduced, because high yields require more resource inputs (through improved seeds, better fertiliser use and/or greater labour inputs). Many farmers, especially smallholders, are reluctant to adopt new practices given the risky environment they operate in, related to markets, weather, pest outbreaks etc. One tool to reduce risk is that of seasonal forecasts, through which farmers get a better idea of what the season holds for them, and can thus better plan their strategies for that season. An example of this kind of work is in Colombia where understanding of weather–rice yield relationships (Delerce et al., 2016) was linked to seasonal forecasts from the meteorological department and led to seasonal advisories to farmers. In the first season of the advisories, 170 Colombian rice farmers avoided losses by taking the advice not to plant in the first of the two annual growing seasons, thus avoiding economic losses estimated at USD 3.6 million (CCAFS, 2016). Research is needed on improving seasonal forecasts, linking forecasts to agricultural impacts and supplying farmers with advisories in the appropriate form.

4. **Building safety nets.** Because of climatic variability, there will be seasons when crops fail and/or when livestock mortality increases. It will be essential to build safety nets. Two areas of work are particularly exciting – index-based insurance and productive social safety nets. Index-based insurance involves developing indices that are used to decide when insurance payments are made, thus bypassing the need for lengthy insurance assessments. This both benefits smallholder farmers, who otherwise can be forced to sell their assets (invariably agricultural assets), as well as insurance providers through reduction of transaction costs. Use of mobile phones to buy insurance and to receive insurance payments is likely to revolutionalise these schemes. Productive social safety net programs involve building assets through a variety of interventions targeted to poorer, more vulnerable households. This could involve work enhancing ecological assets that mitigate extreme events (e.g., restoration of rangelands, reduction of erosion and paid labour on public environment projects), asset transfers (e.g., direct livestock provision) and seed fairs, combined with emergency payments when extreme events hit (Nelson, 2011).

5. **Targeting solutions – dealing with social differentiation.** Food security and adaptation challenges are likely to be most extreme for the poorest of the poor, more marginalised people and/or women. And yet some of the solutions mentioned previously (e.g., greater use of improved more costly seed) are likely to be out of the reach of such individuals. It is crucial to understand social differentiation and the likely impacts of interventions on inequality and inclusion relating to differences in access to agricultural inputs, information, technologies and finance, as well as participation in decision making at household and community levels (Huyer, 2016; Jost et al., 2015). Gender analysis of CSA practices found that the labour productivity of many women smallholders is constrained by lack of access to labour-saving technologies and the most basic of farm tools (Murray et al., 2016); while a study in Bangladesh found that lower levels of education and financial literacy for women put them at a disadvantage when making decisions on crop insurance (Clarke and Kumar, 2016).

6. **Barriers to technological uptake and policy change.** There are many agricultural technologies that have been developed but remain "on the shelf". Campbell and Thornton (2014) estimated that adoption of agricultural innovations amongst smallholder farmers was low over the past fifteen years, ranging between 0% and 15%. In many cases, the policy environment is not conducive to changes in agriculture and food systems, and many recognise a major science–policy gap (Campbell et al., 2016). As important as the technology research is, more so is the work on the barriers to uptake to support transformation in the sector. This would include work on local institutional effectiveness, work on the degree of coherence between agricultural and climate policies and work on the investments and financing needed to foster change. Examples include work on farmer-led innovation processes of research and development – local capacity to innovate and to influence formal agricultural research and development institutions was increased as a result (Waters-Bayer et al., 2015). Similarly, research in Vietnam on adoption of CSA technologies found they need to be flexible to take into account rapid changes in economic, social, gender and environmental contexts. Inclusive consultations with local authorities, enterprises and donors is required, while connecting into existing development plans and programmes (Tuan et al., 2016). de Pinto et al. (2018) built self-reflection into the research process, discussing how they facilitated an outcome whereby the Colombian government adopted their results in preparation of its Intended Nationally Determined Commitment (INDC) as part of the global climate negotiations. Their experience confirms the importance of stakeholder interaction, transparency and openness of processes, as well as willingness to break disciplinary and institutional barriers.

18.7 Research Approaches to Tackle the Challenges

To tackle the grand challenges and to make rapid progress on changing agricultural systems, it is our vision that much greater effort has to be directed to reducing the research–implementation divide. Key in this is engaging with change agents, so that research can feed directly into their actions. The change agents and the ultimate impacts are identified through embedding research into impact pathways, with associated theories of change. Vermeulen and Campbell (2015) identify ten principles for a new vision of an effective Agricultural Research for Development (AR4D) program:

1. **Navigate towards specific points of leverage**: Explore complexity and confront wicked problems, but – rather than becoming either reductionist or vast attempts to model complexity – use "best-bet" prioritisation to navigate towards a limited number of leverage points most likely to drive change.
2. **Allocate resources in three thirds – needs, research, capacity**: Invest a third of resources in working with next users, particularly key change agents, to build relationships and to define their needs from research; a third on research per se and a third on enhancing next users' capacity to improve the uptake of the research.
3. **Join in external processes**: Participate in the processes of next users (governments, organizations, businesses, inter-governmental and multi-stakeholder processes) as much as possible rather than creating new processes and events.
4. **Use research products to build scientific credibility**: Maintain scientific credibility through high-quality publications – recognizing that they are seldom direct pathways to impact but provide the legitimacy that gives the researcher or research agency a seat at the table in driving processes and decisions through expert input.
5. **Sustain co-learning throughout policy engagement and implementation**: Tailor and translate diverse public goods outputs (scientific results, databases, decision tools etc.) in co-learning processes at all stages of the policy and practice cycle.
6. **Tackle power and influence**: Actively address gender and other power differences within deliberative approaches. Ignoring these differences will inhibit the relevance of scientific inputs to policy and action. In most cases science is only one among many influences on policy and action, and scientific inputs are not given privilege on account of being more "objective" or "factual".
7. **Invest in, and monitor, capacity enhancement**: Support next users to enhance their capacities to ask better questions of science, achieve associated development outcomes and adapt to new knowledge. This is likely to be crucial in AR4D.
8. **Mainstream higher-level goals**: Ensure that higher-level goals of poverty reduction, gender equality, social inclusion, environmental sustainability and improved nutrition are considered at all stages. It is all too easy for a researcher to lose sight of the higher goals of the work when closely involved in the delivery of a specific project.
9. **Create mechanisms for internal learning**: Include processes to review the theory of change, re-align the strategy for impact and seize emerging opportunities.
10. **Communicate strategically and actively**: Linking the communications strategy tightly with the impact pathway is likely to be important. This can involve imaginative use of the full range of tools and approaches, from peer-reviewed publications through to social media and reality TV, to reach a wide range of next users. The key need is for communicators to be embedded in impact pathway processes.

18.8 Conclusions

The triple challenges of food security, adaptation and environmental sustainability are serious, and sectoral transformation is needed to respond to these challenges. More effort is needed on the technical solutions, which include stress-tolerant breeds/varieties and practices, with low emissions, but as important will be the research on the barriers to uptake of those solutions. We have too many solutions sitting on the shelf, which could generate impact if barriers to uptake are addressed. For farmers to take up new practices, there needs to be a conducive policy context, and the risks farmers face need to be mitigated. Foresight research, which embraces participatory scenario development, can help foster policy outcomes. For reducing risks, more effort can be given to improving and delivering seasonal forecasts and advisories, and to building safety nets. As it is more marginalised households that are particularly challenged by food insecurity and climate change, research should enable policymakers and development practitioners to target such groups. A key cross-cutting issue for sectoral transformation is the need for sustainable finance. Greater amounts of finance will be needed to support transformation; for example, it is estimated that the adaptation costs for Africa (across all sectors) could be USD 7–15 billion/year by 2020 (current levels are USD 1–2 billion/year, UNEP 2015). In this context, with 60% of the African labour force dependent on agriculture, there is a pressing need for the AR4D community to develop technical and institutional solutions into which adaptation finance can be channelled. So, we need a new approach to research, whereby researchers are much more engaged with stakeholders, especially change agents. This will involve co-learning, more focus on power and politics than technical solutions alone and giving more attention to capacity development with change agents and to strategic communication initiatives.

Acknowledgements

We acknowledge the CGIAR Fund Council, Australia (ACIAR); European Union, International Fund for Agricultural Development (IFAD); New Zealand, Netherlands, Switzerland, United Kingdom and Thailand for funding to the CGIAR Research Program on Climate Change, Agriculture and Food Security (CCAFS).

References

Alexandratos, N., & Bruinsma, J. (2012). World agriculture towards 2030/2050: The 2012 revision. *ESA Working Paper No. 12–03*. Rome: Food and Agriculture Organization of the United Nations, Rome, Italy. www.fao.org/docrep/016/ap106e/ap106e.pdf. Accessed 29.09.16.

ASEAN (2015). ASEAN Regional Guidelines for Promoting Climate Smart Agriculture (CSA) Practices:Endorsed by the 37th AMAF 10 September 2015, Makati City, Philippines. www.asean.org/storage/images/2015/October/ASEAN-Regional-Guidelines-on-Promoting-CSA-Practices/ASEAN%20Regional%20Guidelines%20on%20Promoting%20CSA%20Practices-endorsed%2037th%20AMAF.pdf. Accessed 28.09.16.

Aryal, J. P., Mehrotra, M. B., Jat, M. L., & Sidhu, H. S. (2015). Impacts of laser land leveling in rice–wheat systems of the north-western indo-gangetic plains of India. *Food Security*, *7*(3), 725–738. DOI: 10.1007/s12571-015-0460-y

Asseng, S., Ewert, F., Martre, P., Rötter, R. P., Lobell, D. B., & Cammarano, D. et al. (2015). Rising temperatures reduce global wheat production. *Nature Climate Change*, *5*(2), 143–147.

Campbell, B. M., Beare, D. J., Bennett, E. M., Hall-Spencer, J. Ingram, J. S. I., Jaramillo, F., Ortiz, R., Ramankutty, N., Sayer, J. A., & Shindell, D. (2017). Agriculture production as a major driver of the Earth System exceeding planetary boundaries. *Ecology and Society*, *22*(4).

Campbell B., & Thornton P. (2014). *How many farmers in 2030 and how many will adopt climate resilient innovations? CCAFS Info Note. CGIAR Research Program on Climate Change,* Agriculture and Food Security (CCAFS), Copenhagen.

Campbell, B. M., Vermeulen, S. J., Aggarwal, P. K., Corner-Dolloff, C., Girvetz, E., Loboguerrero, A. M., Ramirez-Villegas, J., Rosenstock, T., Sebastian, L., Thornton, P. K., & Wollenberg, E (2016). Reducing risks to food security from climate change. *Global Food Security*, **11**, 34–43.

CCAFS (2016). Cracking patterns in big data saves Colombian rice farmers' huge losses. Outcome Case. CGIAR Research Program on Climate Change, Agriculture and Food Security (CCAFS), Copenhagen.

Chaudhury, M., Vervoort, J., Kristjanson, P., Ericksen, P., & Ainslie, A. (2013). Participatory scenarios as a tool to link science and policy on food security under climate change in East Africa. *Regional Environmental Change*, *13*(2), 389–398.

Cheung, W., Lam, V., Sarmiento, J., Kearney, K., Watson, R., Zeller, D., & Pauly, D. (2010): Large scale redistribution of maximum fisheries catch in the global ocean under climate change. *Global Change Biology*, **16**(1), 24–35.

Clarke, D. J., & Kumar, N. (2016). Microinsurance decisions gendered evidence from rural Bangladesh. *Gender, Technology and Development*, **20**(2), 218–241.

de Haas, Y., Davis, S., Reisinger, A., Richards, M. B., Difford, G., & Lassen, J. (2016). Improved ruminant genetics: Implementation guidance for policymakers and investors. *Climate-Smart Agriculture Practice Brief*. CGIAR Research Program on Climate Change, Agriculture and Food Security (CCAFS). Copenhagen.

Delerce, S., Dorado, H., Grillon, A., Rebolledo, M. C., Prager, S. D., Patiño, V. H., Varón, G. G., & Jiménez, D. (2016). Assessing weather-yield relationships in rice at local scale using data mining approaches. *PloS One*, *11*(8), e0161620.

De Pinto, A. Loboguerrero, A. M., Londoño, M., Ovalle Sanabria, K., & Suarez Castaño, R. (2018). Informing climate policy through institutional collaboration: Reflections on the preparation of Colombia's nationally determined contribution. *Climate Policy*, **18**(5), 612–626.

Diaz, R. J., & Rosenberg, R. (2008). Spreading dead zones and consequences for marine ecosystems. *Science*, **321** (5891), 926–929. DOI: 10.1126/science.1156401.

ECOWAS (2015). *West Africa Climate-Smart Agriculture Declaration: High Level Forum of Climate-Smart Agriculture Stakeholders in West Africa*. www.ecowas-agriculture.org/sites/default/files/High%20Level%20Forum%20of%20CSA%20Stakeholders%20in%20WA_Final%20Declaration_eng.pdf. Accessed 28.09.16.

FAO (2013a). The state of food insecurity in the world 2013: The multiple dimensions of food security. Rome: FAO. www.fao.org/docrep/018/i3434e/i3434e.pdf. Accessed 6.11.13.

FAO (2013b). *Climate-Smart Agriculture Sourcebook*. Food and Agriculture Organization of the United Nations. Rome.

Fisher, M., Abate, T., Lunduka, R. W., Asnake, W., Alemayehu, Y., & Madulu, R. B. (2015). Drought tolerant maize for farmer adaptation to drought in sub-Saharan Africa: Determinants of adoption in eastern and southern Africa. *Climatic Change*, **133**(2), 283–299.

Foley, J. A., DeFries, R., Asner, G. P., Barford, C., Bonan, G., Carpenter, S. R., Chapin, F. S., Coe, M. T., Daily, G. C., Gibbs, H. K., & Helkowski, J. H. (2005). Global consequences of land use. *Science*, ***309***(5734), 570–574. DOI: 10.1126/science.1111772.

GACSA 2016. Webinars of the Enabling Environment Action Group. *Global Alliance for Climate-Smart Agriculture*. Rome. www.fao.org/gacsa/webinars/en/#c430423. Accessed 28.09.16.

Galloway, J. N., Townsend, A. R., Erisman, J. W., Bekunda, M., Cai, Z., Freney, J. R., Martinelli, L. A., Seitzinger, S. P., & Sutton, M. A. (2008). Transformation of the nitrogen cycle: Recent trends, questions, and potential solutions. *Science*, ***320***(5878), 889–892. DOI: 10.1126/science.1136674.

Gill, G. (2014). *An assessment of the impact of laser-assisted precision land levelling technology as a component of climate-smart agriculture in the state of Haryana, India*. CGIAR Research Program on Climate Change, Agriculture and Food Security (CCAFS), Copenhagen.

Godfray, H. C. J., Beddington, J. R., Crute, I. R., Haddad, L., Lawrence, D., Muir, J. F., Pretty, J., Robinson, S., Thomas, S. M., & Toulmin, C. (2010). Food security: the challenge of feeding 9 billion people. *Science*, **327**(5967), 812–818. DOI: 10.1126/science.1185383.

Hawley, N. L., & McGarvey, S. T. (2015). Obesity and diabetes in Pacific Islanders: The current burden and the need for urgent action. *Current Diabetes Reports*, ***15***(5), 1–10.

Huyer, S. 2016. Introduction: Gender, climate change and agriculture. *Gender, Technology and Development*, ***20***(2). DOI: 10.1007/s11892-015-0594-5.

IFPRI (2015). *Global Nutrition Report: Actions and Accountability to Advance Nutrition and Sustainable Developmen*. International Food Policy Research Institute, Washington DC.

Jost, C., Kyazze, F., Naab, J., Neelormi, S., Kinyangi, J., Zougmore, R., Aggarwal, P., Bhatta, G., Chaudhury, M., Tapio-Bistrom, M., Nelson, S., & Kristjanson, P. (2015). Understanding gender dimensions of agriculture and climate change in smallholder farming communities. *Climate and Development*, ***0***(0), 1–12. article. http://doi.org/10.1080/17565529.2015.1050978.

Kelley, C. P., Mohtadi, S., Cane, M. A., Seager, R., & Kushnir, Y. (2015). Climate change in the fertile crescent and implications of the recent Syrian drought. *Proceedings of the National Academy of Sciences*, ***112***(11), 3241–3246.

Lipper, L., Thornton, P., Campbell, B. M., Baedeker, T., Braimoh, A., Bwalya, M., Caron, P., Cattaneo, A., Garrity, D., Henry, K., Hottle, R., Jackson, L., Jarvis, A., Kossam, F., Mann, W., McCarthy, N., Meybeck, A., Neufeldt, H., Remington, T., Sen, P. T., Sessa, R., Shula, R., Tibu, A., & Torquebiau, E. F., (2014). Climate-smart agriculture for food security. *Nature Climate Change*, **4**(12), 1068–1072.

Lobell, D. B., & Field, C. B. (2007). Global scale climate–crop yield relationships and the impacts of recent warming. *Environmental Research Letters*, **2**(1), 014002.

Mason-D'Croz, D., Vervoort, J., Palazzo, A., Islam, S., Lord, S., Helfgott, A., Havlík, P., Peou, R., Sassen, M., Veeger, M., van Soesbergen, A., Arnell, A. P., Stuch, B., Arslan, A., & Lipper, L. (2016). Multi-factor, multi-state, multi-model scenarios: Exploring food and climate futures for Southeast Asia. *Environmental Modelling and Software*, **83**, 255–270.

Murray, U., Gebremedhin, Z., Brychkova, G., & Spillane, C. (2016). Smallholder farmer and climate smart agriculture: Technology and labour-productivity constraints among women smallholders in Malawi. *Gender, Technology and Development*, **20**(2), 117–148.

NACSAA (2015). North American Climate Smart Agriculture Alliance: Statement at COP21. www.sfldialogue.net/files/NACSAA_cop21_statement.pdf. Accessed 28.09.16.

Nelson, V. (2011). *Gender, Generations, Social Protection and Climate Change: A Thematic Review*. Overseas Development Institute (ODI), London.

Neufeldt, H., Jahn, M., Campbell, B. M., Beddington, J. R., DeClerck, F., De Pinto, A., Gulledge, J., Hellin, J., Herrero, M., Jarvis, A., LeZaks, D., Meinke, H., Rosenstock, T., Scholes, M., Scholes, R., Vermeulen, S., Wollenberg, E., & Zougmoré, R. (2013). Beyond climate-smart agriculture: Toward safe operating spaces for global food systems. *Agriculture & Food Security*, **2**, 12.

Porter, J. R., Xie, L., Challinor, A., Cochrane, K., Howden, M., Iqbal, M. M., Lobell, D., & Travasso, M. I. (2014). Food security and food production systems. In: *Climate Change 2014: Impacts, Adaptation, and Vulnerability*. Contribution of Working Group II to the Fifth Assessment Report of the Intergovernmental Panel on Climate Change. www.ipcc-wg2.gov.

Ramirez-Villegas, J., Challinor, A. J., Thornton, P. K., & Jarvis, A. (2013). Implications of regional improvement in global climate models for agricultural impact research. *Environ. Res. Lett.* **8**, 24018. http://dx.doi.org/10.1088/1748-9326/8/2/024018.

Ray, D. K., Gerber, J. S., MacDonald, G. K., & West, P. C. (2015). Climate variation explains a third of global crop yield variability. *Nature Communications*, **6**, 5989.

Rippke, U., Ramirez-Villegas, J., Jarvis, A., Vermeulen, S. J., Parker, L., Mer, F., Diekkrüger, B., Challinor, A. J., & Howden, M. (2016). Timescales of transformational climate change adaptation in sub-Saharan African agriculture. *Nature Climate Change*. **6**(6), 605.

Rosenstock, T. S., Rufino, M. C., Butterbach-Bahl, K., Wollenberg, E., & Richards, M. (eds.) 2016. *Methods for Measuring Greenhouse Gas Balances and Evaluating Mitigation Options in Smallholder Agriculture*. Springer Open, London.

Steffen, W., Richardson, K., Rockström, J., Cornell, S. E., Fetzer, I., Bennett, E. M., Biggs R. et al. (2015). Planetary boundaries: Guiding human development on a changing planet. *Science*, **347** (6223), 1259855. DOI:10.1126/science.1259855.

Swaney, D. P., Hong, B., Ti, C., Howarth, R. W., & Humborg, C. (2012). Net anthropogenic nitrogen inputs to watersheds and riverine N export to coastal waters:

A brief overview. *Current Opinion in Environmental Sustainability*, ***4***(2), 203–211.

Thornton, P. K., Van de Steeg, J., Notenbaert, A., & Herrero, M. (2009). The impacts of climate change on livestock and livestock systems in developing countries: A review of what we know and what we need to know. *Agricultural Systems*, ***101***(3), 113–127.

Thornton, P. K., Jones, P. G., Alagarswamy, G., Andresen, J., & Herrero, M. (2010). Adapting to climate change: agricultural system and household impacts in East Africa. *Agricultural Systems*, ***103***(2), 73–82.

Thornton, P. K., Ericksen, P. J., Herrero, M., & Challinor, A. J. (2014). Climate variability and vulnerability to climate change: a review. *Global Change Biology*, ***20***(11), 3313–3328.

Tuan, D. M., Elisabeth, S., & Hai, L. (2016). *Participatory identification of climate-smart agriculture priorities*. CGIAR Climate Change, Agriculture and Food Security Program, Copenhagen.

UN (2014). *Climate Summit 2014 – Agriculture Annex*. United Nations, New York.

UNEP (2015). *Africa's Adaptation Gap 2: Technical Report*. United Nations Environment Programme, Nairobi.

Van der Werf, G. R., Randerson, J. T., Giglio, L., Collatz, G. J., Mu, M., Kasibhatla, P. S., Morton, D. C., DeFries, R. S., Jin, Y., & van Leeuwen, T. T. (2010). Global fire emissions and the contribution of deforestation, savanna, forest, agricultural, and peat fires (1997–2009). *Atmospheric Chemistry and Physics*, ***10***(23), 11707–11735.

WHO (2016). Obesity and overweight. *Fact sheet No. 311*. World Health Organization. Geneva.: WHO. www.who.int/mediacentre/factsheets/fs311/en/. Accessed 3.10.2016.

Vermeulen, S. J. (2014). Climate change, food security and small-scale producers. *CCAFS Info Brief*. CGIAR Research Program on Climate Change, Agriculture and Food Security (CCAFS). Copenhagen.

Vermeulen, S., & Campbell, B. (2015). Ten principles for effective AR4D programs. *CCAFS Info Note*. CGIAR Research Program on Climate Change, Agriculture and Food Security (CCAFS), Copenhagen.

Vervoort, J. M., Thornton, P. K., Kristjanson, P., Förch, W., Ericksen, P. J., Kok, K., Ingram, J. S. I., Herrero, M., Palazzo, A., Helfgott, A. E., & Wilkinson, A. (2014). Challenges to scenario-guided adaptive action on food security under climate change. *Global Environmental Change*, ***28***, 383–394.

Waters-Bayer, A., Kristjanson, P., Wettasinha, C., van Veldhuizen, L., Quiroga, G., Swaans, K., & Douthwaite, B. (2015). Exploring the impact of farmer-led research supported by civil society organisations. *Agriculture & Food Security*, ***4***, 1–7. http://doi.org/10.1186/s40066-015-0023-7.

Wollenberg, E., Richards, M., Smith, P., Havlík, P., Obersteiner, M., Tubiello, F. N., Herold, M., Gerber, P., Carter, S., Reisinger, A., van Vuuren, D., Dickie, A., Neufeldt, H., Sander, B. O., Wassmann, R., Sommer, R., Amonette, J. E., Falcucci, A., Herrero, M., Opio, C., Roman-Cuesta, R., Stehfest, E., Westhoek, H., Ortiz-Monasterio, I., Sapkota, T., Rufino, M. C., Thornton, P. K., Verchot, L., West, P. C., Soussana, J. F., Baedeker, T., Sadler, M., Vermeulen, S. J., & Campbell, B. M. (2016). Reducing emissions from agriculture to meet the 2° C target. *Global Change Biology*. **22**(12), 3859–3864.

19

The Contribution of Food Engineering to Achieve Global Food Security

*Walter E. L. Spieß**

19.1 Introduction

19.1.1 The Environment and Its Impact on Food Production

The coming decades pose a multitude of problems to the global community. Prominent challenges to be faced are certainly the predicted environmental changes and the increasing demand for food given the increasing world population; however, new technological and social developments also have to be addressed by politics, science and commerce. The global climate will see rising environmental temperatures, droughts, floods and hurricanes: phenomena that not only endanger human lives and civilisatory entities but also impose severe stress on agricultural production (see Figure 19.1). The growing world population will produce new forms of communal living; of special interest in this context are the formation of large human agglomerates, i.e. megacities, mainly in Asia, South America and Africa.

The food engineering, or in more general terms the food science/technology, community as part of the international science community, is engaged in assuring an adequate food supply to the global population by safeguarding harvested products, processing raw materials into edible products and dealing with the transport, storage, distribution and final consumption of these products; under the pressure of limited conventional resources for the growing world community, food science/engineering is exploiting and developing new unconventional resources and products (as well as conventional products) to satisfy the hunger of its fellow people while respecting cultural and religious costumes and practices.

It is anticipated that with rising environmental temperatures in future years more areas around the globe experience droughts, especially in the Americas, Southern Africa, the Middle East and Australia that will reduce the availability of agricultural area. Poor weather conditions are responsible for confined harvests and also for severe post-harvest losses. Many of those losses can only be reduced to a limited extent by technical interventions especially in extreme weather situations.Besides the general climatic conditions the impact of rising carbon dioxide concentration in the atmosphere (Figure 19.2) on agricultural production has also to be considered together with the rise of the global average temperatures (atmosnews, CO_2 –Temperature).

A positive aspect of the rise of the carbon dioxide concentration is carbon dioxide fertilization; it has, however, to be considered that not all plants tolerate higher carbon dioxide concentrations and higher temperatures without a longer adaption period; the total effect of carbon dioxide fertilization is therefore considered as overall negative. According to various model scenarios it is mainly the Northern Hemisphere that will benefit from the expected climate change (see Figure 19.3).

19.1.2 The Availability of Resources

It is a generally accepted and well-based assumption that the available biomass will be sufficient to supply the growing world population in an adequate way with high-quality food at least to 2050, even under the adverse climatic effects, shown in Figure 19.4, in which the global impacts of climate change on crop productivity from simulations published in 1994 and 2010 are presented. (The 2010 study simulated changes in yields of eleven crops for the year 2050, averaged across three greenhouse emission scenarios and five General Circulation Models (World Development Reports, 2010).

The assumption that there will be sufficient food to supply the growing world population is based amongst others on an optimal utilization of the agricultural

* This Chapter is dedicated to the memory of Rose Spiess, who challenged Food Engineering to reduce Food Insecurity in Countries with a poor food supply.

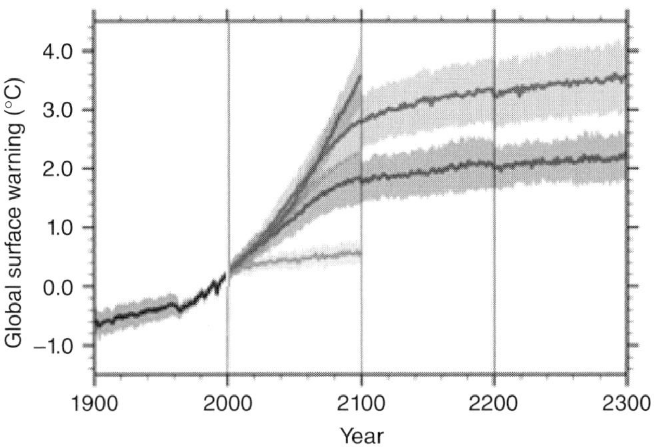

Figure 19.1 Predicted temperature rise according to various model scenarios.
Source: Contribution of Working Group I to the Fourth Assessment Report of the Intergovernmental Panel on Climate Change [Solomon, S., D. Qin, M. Manning, Z. Chen, M. Marquis, K. B. Averyt, M. Tignor and H. L. Miller (eds.)]. Cambridge University Press, Cambridge and New York, USA Global Climate Projections, www.ipcc.ch/ pdf/assessment-report/ar4/wg1/ar4-wg1-chapter10.pdf. Accessed 06.06.2017.

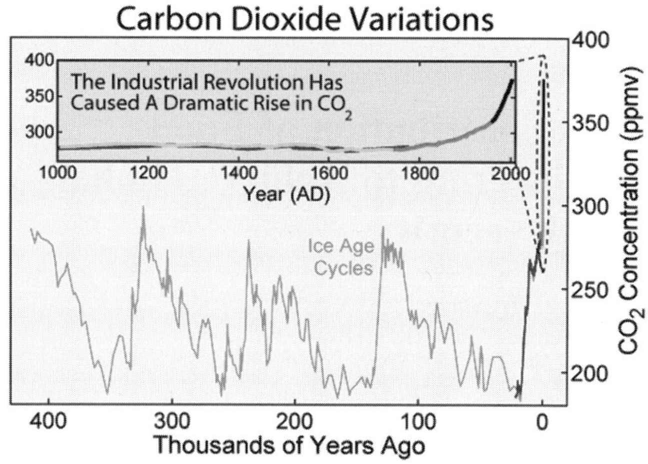

Figure 19.2 Carbon dioxide concentration in atmosphere. Sources: Scripps Institution of Oceanography at the University of California San Diego. Data from before the Keeling Curve: Lüthi, D., Le Floch, M., Bereiter, B., Blunier, T., Barnola, J.-M., Siegenthaler, U., Raynaud, D., Jouzel, J., Fischer, H., Kawamura, K. and Stocker, T. F. 2008. High-resolution carbon dioxide concentration record 650,000-800,000 years before present. *Nature*, 2008; 453: 379–382.

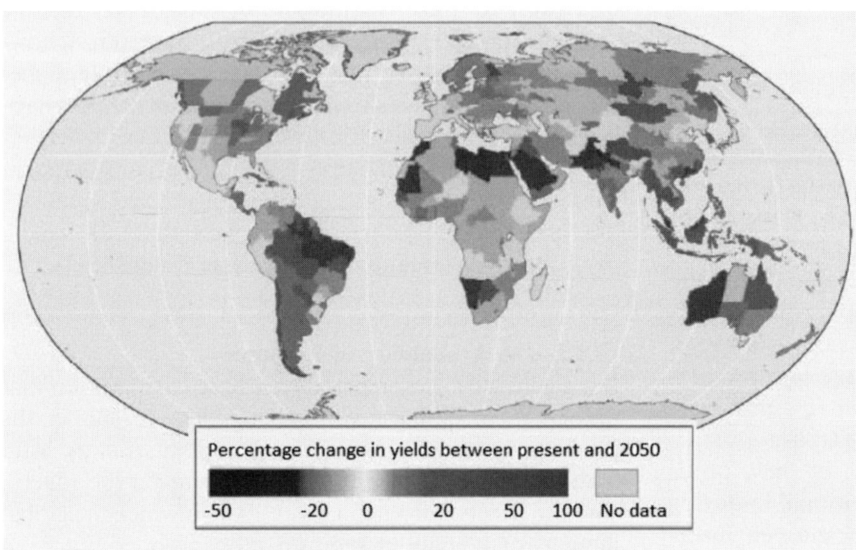

Figure 19.3 Global impacts of climate change on crop productivity from simulations.
Source: World Bank development report 2010: "Development and climate change" (World Bank, Washington, DC, 2010).

production and the exploitation of so-far under-/unutilized resources; of interest in this context are international statistical data that indicate that the population growth rates will decline in the years to come quite dramatically, whereas the cereal production growth rates will stabilize at a level of 1.2 percent (Figure 19.4).

The land area presently utilized for agricultural production is not sufficient to feed 2 to 3 billion more people. It will be necessary to extend and also upgrade the present share of farmland used for food production. Furthermore, higher productivity of the agricultural sector is necessary and requires amongst other factors fertilizing already highly stressed soils, improving irrigation systems and developing plant materials that are adapted to changing environmental conditions including less water. In producing the required amount of food, the global and local ecosystems are likely to be further compromised, negatively impacting biodiversity and hastening climate change.

In the last century all types of agricultural land increased in area and utilization (Figure 19.5). It is expected that by 2030 about 47 percent of the total land area will be used for agricultural production. To provide some relief of the pressure put on the agricultural production it will be important to reduce any type of losses to which food materials are exposed on their way from production to consumption.

Reserves that could be used for expanding agricultural land are mainly located in the Southern Hemisphere, so. for example, in Brazil and other South American countries and in southern Africa, in Mozambique amongst others. In Asia the reserves are limited: only China has the potential to expand its agricultural land if those areas can be sufficiently irrigated. In North America also the lack of water is the limiting factor. In Europe the land reserves are close to zero (see also Figure 19.5).

Despite the potential to expand the agricultural production in developing and transition countries it is expected that those countries are not in a position to provide sufficient food for their growing populations especially with regard to detrimental climatic conditions. The available and forecasted data (Table 19.1) indicate that industrial countries, especially in the Northern Hemisphere, will be the main suppliers of food. It is obvious that this is not only true for storable and conditionally storable foods, e.g., cereals in general,, pulses and roots/tubers, but also for highly perishable food materials, e.g., milk and dairy products, beef and veal, pig meat, and poultry meat.

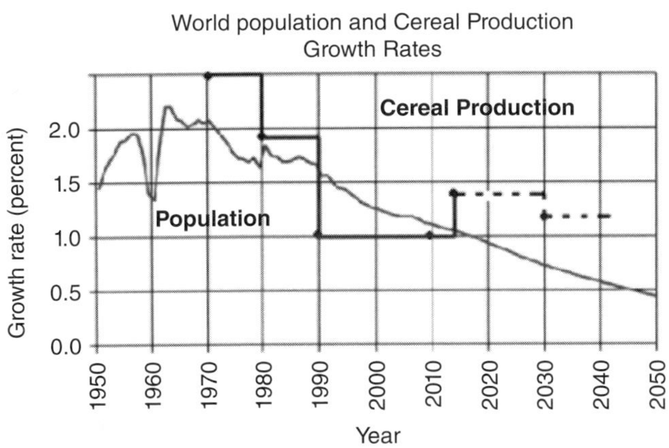

Figure 19.4 Population and cereal production growth rates.

A comparison of locations of major food production with the main population centers (Figure 19.6) to be supplied indicates that in the future major food production centers will be located with few exceptions in the Northern Hemisphere, e.g., in Canada, Scandinavia, Russia and Chile; whereas major population centers are grouped around 50° N latitude, production centers are, with regard to location, not congruent with the centers of demand, thus forcing food to cross country frontiers and social frontiers. Those supply systems are in the majority of the cases technically sophisticated and highly vulnerable. In practical terms it means that major amounts of food have to be shelf stable or even to be stabilized by adequate processing methods to pass through the food supply chain with a minimum of losses and to tolerate the frequently time-consuming transport, storage and distribution processes. In addition, the technological requirements demanded for stabilization and transport have to be affordable and meet social, cultural, religious and physiological requirements and needs. Special considerations have to be devoted in this relation to megacities that require new types of local distribution methods and methods of serving food; the high concentration of small size apartments with limited storage and food preparation facilities requires, e.g., frequent deliveries and intelligent composition of ready-to-eat/easy-to-prepare products.

With increasing global economic growth and income, the demand for meat increased and meat production grew disproportionally. However, the high growth rates especially in developing countries in the 1980s and 1990s have not been maintained at the start of the twenty-first century primarily because of a slowdown of the world economy. The production growth rates for meat are, however, still almost twice as high as the growth rates of cereals. The main consumers of meat

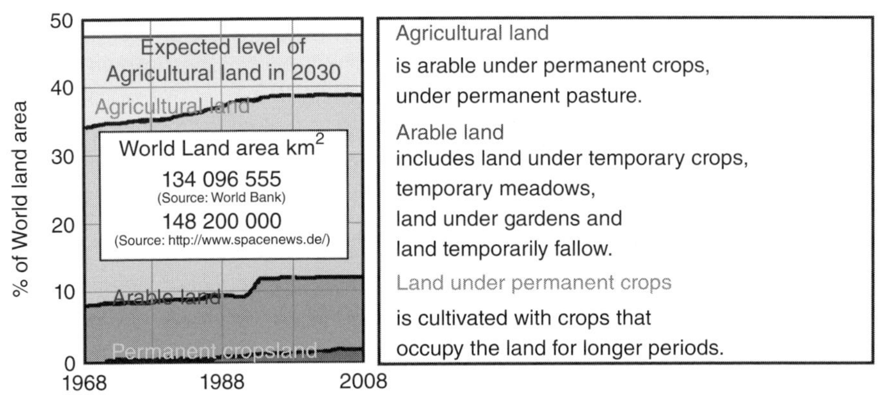

Figure 19.5 World land area and agricultural use.

Table 19.1 *Changes and Balance in the Commodity Composition of Food by Major Country Groups*

		Balances for selected commodities (million tons)								
		Developing countries			Industrial countries			Transition countries		
Produce	Year	Demand All uses	Production	Net trade	Demand All uses	Production	Net trade	Demand All uses	Production	Net trade
Wheat	2015	462	358	−104	159	263	104	110	114	4
	2030	566	425	−141	165	299	133	120	132	12
Rice (paddy)	2015	679	685	5	23	24	0.4	3	2	−1
	2030	771	778	7	24	23	−1	4	2	−2
Coarse grains	2015	628	539	−89	424	508	83	125	133	8
	2030	837	709	−128	471	585	115	139	154	15
Milk and dairy (fresh milk equivalent)	2015	375	346	−29	240	268	28	97	100	3
	2030	523	484	−39	251	286	36	99	104	5
Beef and veal	2015	42	41	−1	25	27	2	7	6	−1
	2030	56	55	−1	25	27	2	8	7	−1
Mutton and lamb	2015	42	41	−1						
	2030	17	15	−2	2	3	1	1	1	0
Pig meat	2015	70	69	−1	31	32	1	8	8	–
	2030	84	83	−1	32	33	1	8	9	1
Poultry meat	2015	61	59	−2	33	38	5	5	4	−1
	2030	97	94	−3	39	44	5	6	6	–
Roots and tubers	2030	808	818	69	100	88	−12	94	95	1
Pulses	2030	64	62	−2	13	13	1	3	4	1
Vegetable oils/seed	2030	152	156	4	53	52	−1.3	9	9	0.0

are the growing middle class in developing countries, especially in China and India, where consumers are turning away from plant-based traditional diets to higher meat consumption (Figure 19.7), a challenge for sociologists, nutritionists and food engineers.

In the case of cereal production it has to be considered that only about 50 percent of the production is destined for direct human consumption; a rather big share goes into animal feed and industrial uses; amongst others, about 10 percent of the cereal production is needed for biofuel production in order to relieve carbon dioxide emissions of the transport sector (Table 19.2). If the share that goes into the animal feed sector and other uses can be substituted by biological non-food materials such as algae and lignin containing materials that have been processed by fermentation or through pyrolysis the stress on the global food supply could be dramatically released. Promising steps are under way (Figure 19.14).

19.2 Intermediate Conclusion

There are sufficient resources available to meet the needs of the growing world population for more food and to produce materials that can be used as feed material and for other uses Such as the production of biofuel, which is important to reduce global greenhouse gas emissions. Water continues to be a limiting natural resource and should also be considered in the discussion around an adequate food supply

Food engineering/technology has made progress along with other technical disciplines with regard to improved processing; modern processes are more effective in the use of utilities, with the result of a better exploitation of available resources; product handling is more smooth and gentle with the result of better quality and process steps are interlinked to save space, time and transfer costs, as entirely new techniques allow for new products. The impact of these new improved techniques is, however, in the first instance directed towards quality improvements and product affordability.

Unfortunately improved modern, highly sophisticated technologies have not contributed substantially to a quantitative increase of the food supply. Mainly countries in the Southern Hemisphere, with a few exceptions, e.g., Australia and Argentina, did not participate in the general development of food science/technology with the result that in those countries major

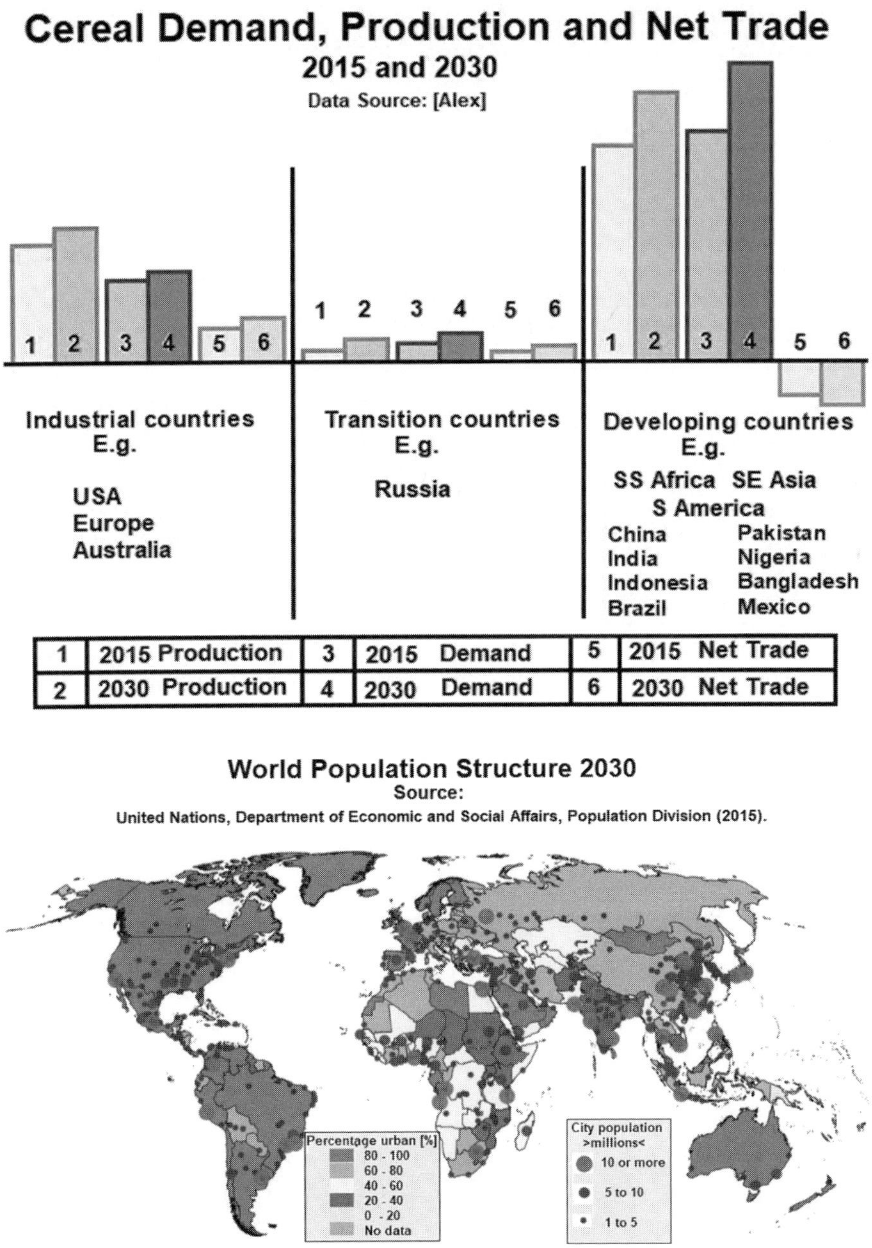

Figure 19.6 (a) Major Cereal Production areas and (b) Urbanization Structure (Alexandratos, N. and J. Bruinsma. 2012. *World agriculture towards 2030/2050: the 2012 revision.* ESA Working paper No. 12-03. Rome, FAO)(UN Department of Economic and Social Affairs, Population Division). For the color version of (b), please refer to the plate section.

segments of the population are undernourished (Figure 19.8). In order to achieve food security in disadvantaged areas, agricultural production has to increase with regard to quantity and quality; global food security, however, can only be achieved by a strategy that allows on a global basis for:

- reduction of losses;
- better utilization of conventional resources and opening of new unconventional resources;
- improvement of enzymatic processes to reclaim and process food waste and non-food biomaterials into food and feed;
- development of new storage, packaging and transportation technologies (required to maintain freshness of fruits/vegetables and to some extent of processed products over longer periods together with a reduction of the environmental stress, e.g., reduced greenhouse gas emission, by those processing elements);
- new types of distribution and retail outlets (digitalized shopping, to-door delivery, frequent deliveries in time because of minimal home storage facilities);
- new types of food preparation.

Most effective in providing an adequate food supply are certainly the reduction of losses and the substitution of

Table 19.2 *Cereal Consumption in the Period 2012 to 2030*

Year	Demand, MMT			Production, MMT	Demand per capita, kg/capita per year		
	Food	Feed	Other uses		Food	Feed	Other uses
1979–81	706	575	156	1442	159	129	35
1997–99	1003	657	204	1889	164	107	33
2015	1227	911	241	2387	168	124	33
2030	1406	1148	277	2839	169	138	33
Cereal consumption for biofuel production				156 *⟩ 190**⟩	*⟩ Maize **⟩ Wheat		

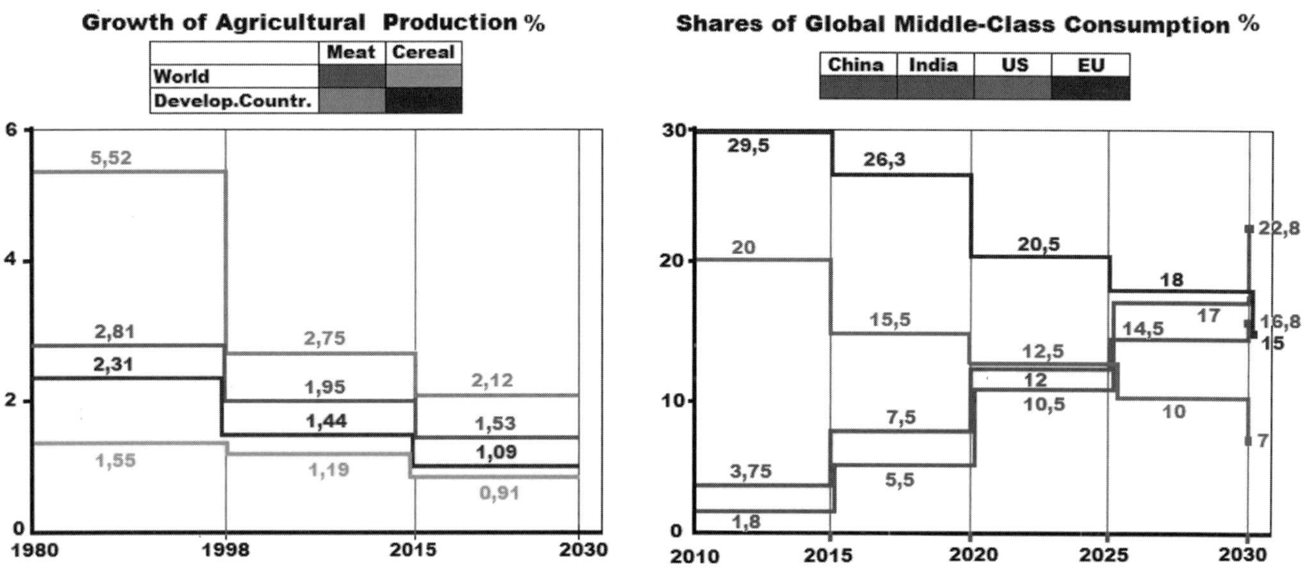

Figure 19.7 Growth of agricultural production versus development of middle-class consumption.
Source: Alexandratos, N. and J. Bruinsma. 2012. *World agriculture towards 2030/2050: the 2012 revision*. ESA Working paper No. 12-03. Rome, FAO.

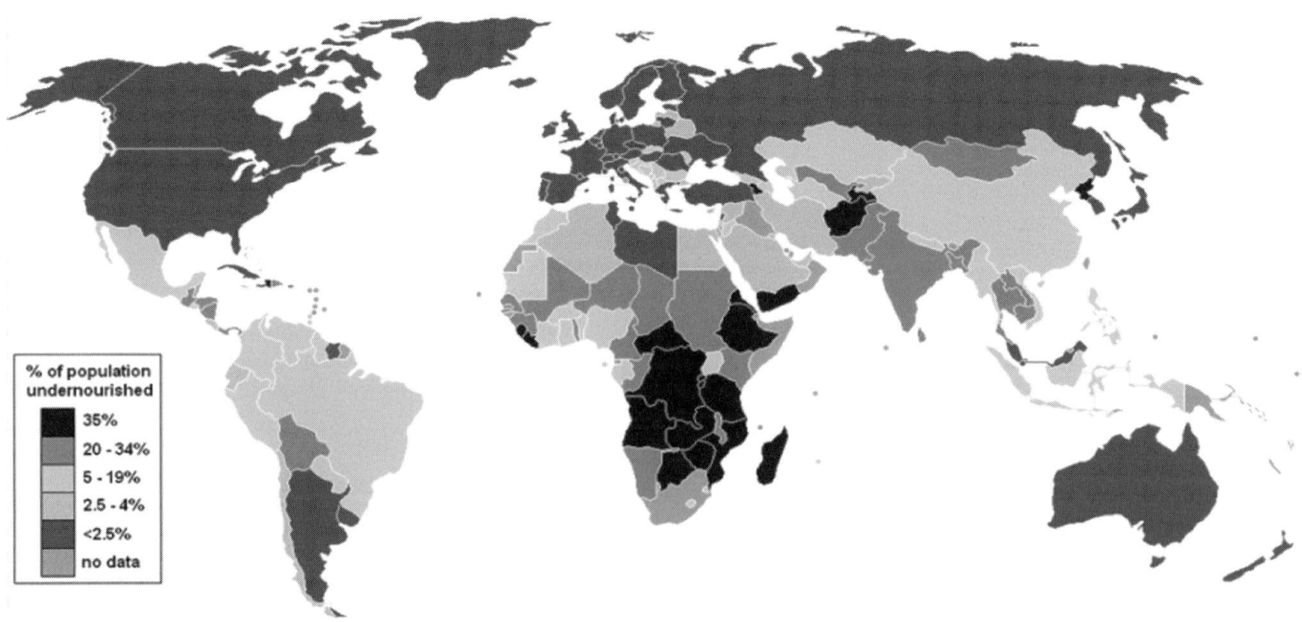

Figure 19.8 World hunger map 2014.
Source: www.wfp.org/content/hunger-map-2015, visited 14.11.2016

the share of raw produce that goes into feed production and that is used for other non-food purposes. The measures to be applied are in most cases variations on established methods that have to be improved and adopted to new challenges.

Both the quantitative and qualitative aspects have to be considered for new food production/generation. New materials and processing methods have to reflect cultural and religious traditions and respond to consumer demands, e.g., evolving lifestyles and/or new insight into nutritional-physiological processes. Of special interest to the consumer are:

- Food grown and processed in an ecological correct/safe, resource-saving way
- Food grown in the region for direct market supply
- Food that is adequate for specific physical activities and health situations, e.g., according to DNA analysis
- Food that recognizes the social, ideological and religious orientation of the consumer
- Food that is easy to prepare and eat

19.3 Measures to Increase and Stabilize the Global Food Supply

19.3.1 Food Losses

19.3.1.1 Quantification of Food Losses

Food losses undermine to a large extent an adequate food supply of major population segments in developing countries. Especially in North Africa, West and Central Asia and Sub-Saharan Africa but also in Latin America and South and South-East Asia the harvesting and post-harvesting losses are very high and above the global average. Reasons for this situation are manifold: lack of education, unfavorable weather conditions, political unrest and in many cases a lack of technical infrastructure and the absence of technical know-how. In the industrialized regions – North America; Oceania; Europe, incluing the Russian Federation; Japan; the Republic of Korea and China – the post-harvest losses are much lower mainly because of higher technical standards in handling harvested produce. In the latter group of countries lower losses also occur during processing/packing and distribution again because of advanced processing and distribution operations; the processing losses are mainly due to cleaning waste; losses during distribution can be attributed in the case of biological materials to respiration activities. In developing countries losses during processing/packing and distribution are again above average mainly because of deficient equipment and inadequate technical education. In developing countries with an insufficient food supply the food losses during the consumption phase are rather low in contrast to industrialized regions, where consumers tend to be less careful with the food that arrives at their homes (see also Table 19.3).

19.3.1.2 Technical Solutions to Reduce Food Losses

The technical toolbox to preserve the quality of harvested produce offers a large number of solutions, many of which are very effective but complex processes that demand elaborate processes and equipment such as irradiation and well-educated staff, solutions that are standard in industrialized countries. Drying, controlled atmosphere (CA) storage, freezing/chilling and with limitation heat processing are classical processes that can be put in place even under unfavorable conditions that may prevail in many developing countries; despite apparently simple process features they allow for very effective post-harvest and final processing. The equipment required can be constructed with easily available materials in many cases at low cost; in cases when major amounts have to be processed upgraded sophisticated high-tech versions of this equipment are demanded. In combination with efficient solar energy collectors rather primitive constructions can even be operated and equipped with electrical process aids at remote areas not connected to power networks.

The implementation of processes such as drying, CA storage, refrigeration and elimination of deteriorated materials in areas with exceptionally high food losses during harvest and post-harvest periods that amount to 20.9 percent (Latin America), 18.6 percent (North Africa together with West and Central Asia) and 25,2 percent (Sub-Saharan Africa) may help to improve the global food supply substantially. It will not be possible to eliminate losses entirely; figures around 7 to 8 percent, however, can be achieved.

19.3.1.2.1 Convective (Solar) Drying

One of the oldest techniques to stabilize harvested food materials is drying. It has been applied in all climatic zones and to all types of raw materials. Prominent examples of traditionally dried products are raisins, figs and apricots from Mediterranean climates; stock fish (dried cod) or pemmican ingredients in Northern Europe and America and *chuño* (freeze-dried potatoes) in the Andes region.

There is a large variety of drying equipment available, mostly adapted to specific conditions. In recent years solar dryers, basically upgraded traditional forms of flat bed dryers, have received specific interest. Besides the type of dryer presented in Figure 19.9, more sophisticated types of equipment are used especially in

Table 19.3 *Food Chain Losses of Valuable Resources*

Weight losses %	Food chain production steps	Factors helping to reduce losses: examples
10.9[1] 13.4[2] 8.7[3]	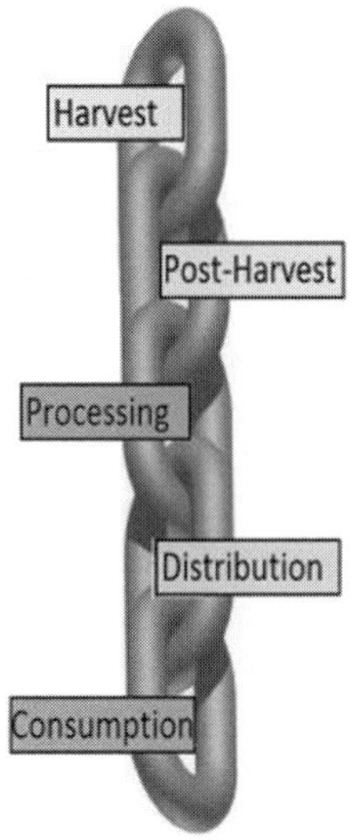 Harvest	Select varieties suited for optimal technical handling and marketing potential; utilize analytical methods to determine produce maturity; respond to market conditions and opportunities; minimize handling operations; if applicable allow for on-the-field curing
7.3[1] 12.7[2] 3.4[3]	Post-Harvest	Utilize integrated harvesting–preprocessing methods; utilize standardized containers for transport and storage; arrange for fast transport to processing; if applicable apply low temperatures/controlled atmospheres at intermediate storage
4.1[1] 6.3[2] 3.1[3]	Processing	Apply energy-/water-saving technologies; process/preprocess raw materials at the location of production to reduce transport volume; initiate infrastructure development and stimulate the generation of technical intelligence in developing countries
4.0[1] 5.6[2] 2.2[3]	Distribution	Utilize standardized transport containers including consumer units; avoid interruption of temperature-/atmosphere-controlled systems during bulk and retail storage
6.6[1] 12.6[2] 1.3[3]	Consumption	Organize purchasing activities according to consumption patterns; provide adequate storage facilities near the points of consumption; adjust portion size according to dietary requirements
Total losses*32.9[1] 36[2] 28[3]	General recommendations to reduce losses: Educate all food chain partners regarding sensitive food chain actions; organize cooperation over production levels; develop protocols for cooperation of all food chain partners; define interface conditions; standardize transport systems; use easy-to- clean/sanitize equipment	

*kcal Losses ~25 percent
[1]Mean value
[2]Maximum value
[3]Minimum value

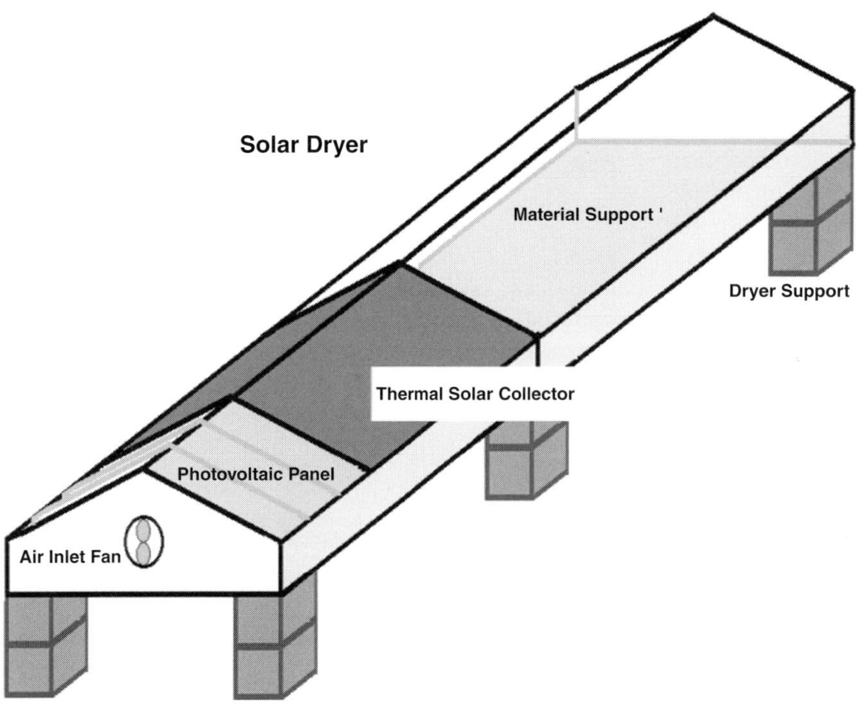

Figure 19.9 Solar dryer: simple solar dryer equipment may typically consist of a support tray covered with a transparent film. In the front part a solar collector (black stained flat material) is arranged, near the collector segment the material support is organized. Fresh air enters the collector segment, where it is heated; from there the air moves over the material segment. Improved versions are equipped with a fan that is powered by a photovoltaic panel. The dryer can be built with waste materials.

industrialized countries; under favorable conditions solar dryers can also be operated successfully in hot but also in moderate climates. Solar drying can be applied to almost all agricultural produce especially fruits and vegetables; because of mild processing conditions the products are in general of good or even superior quality. The preparation of the material to be dried is rather simple; it has to cleaned, cut into pieces and in the case of vegetables briefly dipped into hot water.

19.3.1.2.2 Osmotic Drying

The principle of osmosis is used to remove water from solid food material in various traditional preserving methods, e.g., salting of meat and fish and candying of fruit. The technique is simple: plant or animal tissue is immersed, whole or in pieces, into concentrated aqueous solution with an osmotic pressure higher than the tissue pressure. Difference of concentration between food and solution gives rise to a migration of water from the tissue into the solution and an impregnation of the tissue by the environmental solutes (Figure 19.10).

Osmotic treatment – also known as osmotic dehydration or dewatering impregnation soaking (DIS) – has wide potential application: fruits, vegetables, meat and fish can be osmotically treated previous to freezing, chilling, convective drying, microwave or high-pressure processing and other conventional and new processes; combined processes can be designed to obtain food products and food ingredients. Special emphasis has to be given to the reconditioning of the diluted osmotic solution. Modern membrane technology allows recycling processes: cheap-waste energy can be used for thermal concentration.

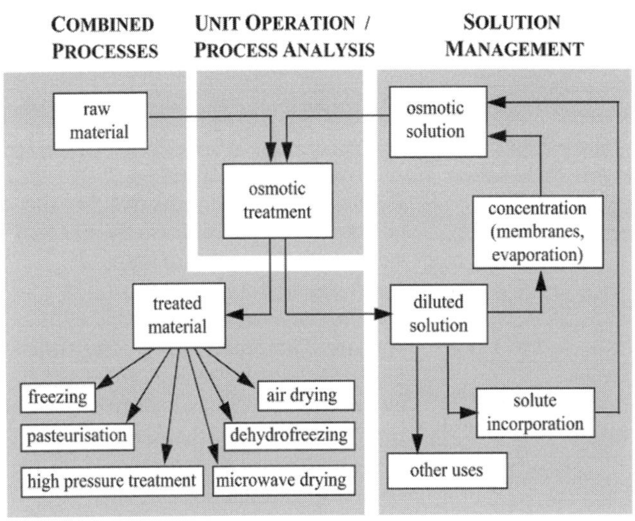

Figure 19.10 Osmotic processing: the equipment consists of a container for the osmotic solution into which the raw material is immersed. In straightforward designs the solution, e.g., sugar or salt solutions, is discarded when the concentration of solutes in the dried product is in equilibrium with the concentration of the solution; in advanced designs the osmotic solution is cleaned and concentrated either by thermal or by membrane processes. Spent solution in fruit processing is also used as an additive in other processes, e.g., marmalade production. Large amounts of discarded salt-containing solutions are considered an environmental hazard.

Currently, renewed attention is paid to this technique as a pre-step to further processing to improve the nutritional and sensory quality of food. Potential energy saving is also an essential advantage of this processing technique. In classical drying processes water is generally removed by evaporation or by sublimation, i.e., by high-energy-consuming methods, whereas by osmotic treatment up to 60 percent of the initial water can be removed from the solid food at moderate temperatures and without phase change. Under regular industrial production conditions, an energy cost reduction of 40 to 50 percent can be expected. The savings can be even higher if the process is applied to partial water removal of products to be frozen in a next processing step; in this case freezing loads and transport volumes are reduced in addition.

19.3.1.2.3 Controlled Atmosphere Storage (CA Storage)

Controlled atmosphere storage is an old technology that has been applied in rural areas; huge CA-storage facilities for, e.g., apples, are the modern sophisticated versions of earth-covered heaps (Figure 19.11).

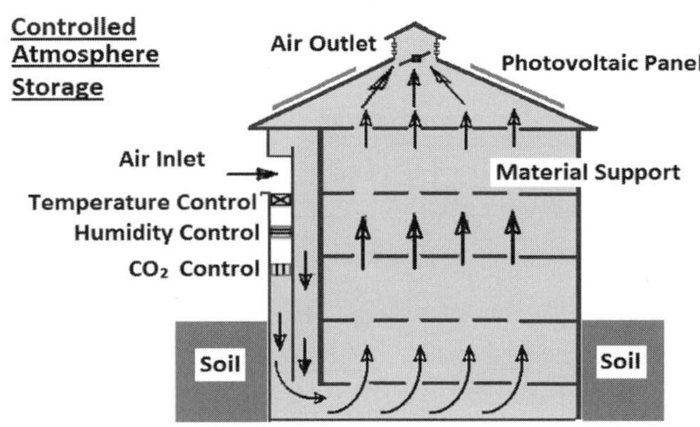

Figure 19.11 Controlled atmosphere (CA) storage facility: modern CA equipment consists of a gastight housing, as well as units to control the gas composition and the humidity of the atmosphere as well as the room temperature. In primitive storage facilities mainly the CO_2 level of the atmosphere is elevated. In more sophisticated facilities the ethylene produced by stored fruits is absorbed, the O_2 level (~2 percent) and the temperature (~2°C) are lowered and the CO_2 level (~2.5 percent) is elevated. Average storage time extension is about six months.

Especially suited for CA storage are climacteric fruits, e.g., apples, avocado, cherry, kiwi, nectarines and peaches. Non-climacteric produce, e.g., artichokes, asparagus, broccoli and cabbage, also may benefit from CA storage conditions because their respiration rates, i.e., metabolic processes, are lowered/slowing down and elevated CO_2 levels are unfavorable for bacterial growth.

19.3.1.2.4 Refrigeration

Refrigeration of agricultural produce immediately after harvesting helps to reduce post-harvest losses considerably; because of the simplicity of refrigeration systems the uninterrupted operation of such systems requires a minimum of technical infrastructure and know-how (see Figure 19.12). Most agricultural fruits and vegetable allow for shelf life extension by chilling immediately after harvest; climacteric fruits benefit especially from a reduction of environmental temperatures and a modification of the storage atmosphere. Tropical fruits and vegetables should, however, be treated with care because they may suffer in low temperatures <–4°C; temperatures slightly above 12°C are safe.

19.3.1.2.5 Elimination of Defective/Contaminated Material and Reduction of Mycotoxin Contamination

The high levels of aflatoxin B1 and related compounds in agricultural produce, particularly in peanuts, are a serious health threat in many tropical countries and are a major impediment to trade. In the case of small-hold farming, the commodity is frequently highly contaminated because of poor post-harvest processing and storage practices. Although aflatoxins have been recognized as a problem for more than fifty years, efforts to reduce aflatoxin contamination of peanuts have met with poor success because of the lack of suitable methods and equipment to reduce aflatoxins to acceptable levels. A newly developed promising method is to eliminate contaminated peanuts by sorting of material that has undergone a heat and de-skinning treatment at ~140°C to accentuate discolorations of toxin loaded peanuts (Figure 19.13). In small units, the sorting can be done manually, supported by visual sensing of discolored materials or entirely automatically in larger processing units; contaminated maize can be sorted by UV-light sensors.

In the case of small-sized production units photovoltaic panels can be used for electrical energy supply and solar-thermal panels for generating warm water, especially for cleaning/sanitation purposes and laboratory use. The basic process is designed in such a way that even farming communities can manufacture the necessary equipment and operate small processing units. Up-scaled sophisticated units are able to provide decontaminated material for larger segments of population.

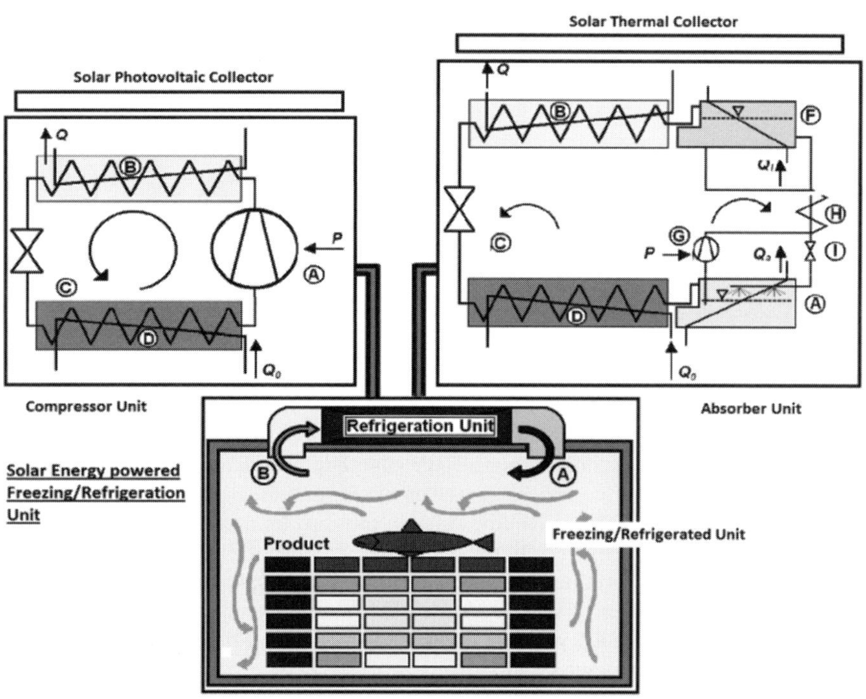

Figure 19.12 Basic processes for refrigeration operations: basically both technical principles do reduce environmental temperatures, compressor systems and absorber systems. Compressor units have to be powered; electrical energy is provided by photovoltaic panels; absorbers units can be operated by thermal energy to provide solar thermal collectors. Both energy systems can be operated as stationary units, e.g., freezing/chilling operations, cold stores refrigerators etc., or as non-stationary/transport units. In each case the refrigerated units have to be insulated adequately to reduce the energy requirements. For periods without sunshine energy storage systems have to be applied; in addition to expensive electrical batteries, thermal phase-change batteries are very appropriate.

19.3.2 Substituting Agricultural Produce Designated as Feed and Not Utilized for Food Production

Through

- Better utilization of conventional resources and opening of new unconventional resources
- Improvement of enzymatic processes to reclaim and process food waste and non-food biomaterials into food and feed

19.3.2.1 Availability of Produce Residues and Biomass

Agricultural societies utilize their crop production to stabilize food supply and create income to satisfy needs of daily life. Over the centuries the techniques to exploit crop production have been refined and improved in many ways. Traditional methods to exploit agricultural production have reached their limits and do not allow for any further exploitation of valuable materials. Considering, for example, the fact that cereals produce approximately one part of straw for one part of grains it becomes obvious that the residues that could be utilized to produce valuable byproducts present a potential for the production of valuable bioproducts (Table 19.4); those residues are presently used in a small amount directly for farm purposes; the main amount is composted, burned or otherwise wasted.

With the advent of modern technology this situation will change in a dramatic way; currently unutilized parts of the agricultural production will be, together with other biomaterials, important raw materials for the production of new types of food, feed, biofuel and other non-food industrial products (Table 19.5).

Table 19.4 *World Cereal Production*

Year	Cereal production, MMt	Straw production, MMt
2015	2387	~2200
2030	2839	~2700

Table 19.5 *Agricultural/Biomass Suited for Production of Bioproducts*

Produce	Residue	Possible conversion process	Product (examples)
Wheat, rice, barley, oat corn, grass	Straw, stalks, husks, cobs, bran	Combined: chemical and microbial	Single-cell protein, biofuel
Potato	Peel	Microbial	Biofuel
Soybean, other leguminoses	Various soy residues	Microbial	Single-cell protein
Sugar cane, sugar beet	Bagassse	Extraction; microbial	Biofuel, alcohol, sugar
Palm oil fruit	Skin, oilmeal	Extraction	Solid biofuel, fertilizer
Wood/cellulosic pulps	Sulphite/enzymatic liquor	Microbial	Single-cell protein
Starch from production	Starch milk etc.	Microbial	Biofuel
Grape/apple	Pomace	Extraction; microbial	Fruit oil, feed, biofuel
Orange, other citrus fruit	Peel	Extraction; microbial	Fruit oil, feed, biofuel
Jatropha (*Jatropha curcas*)	Fruit and skin	Extraction	Diesel oil, animal feed
Cattle, swine	Manure	Microbial	Methane

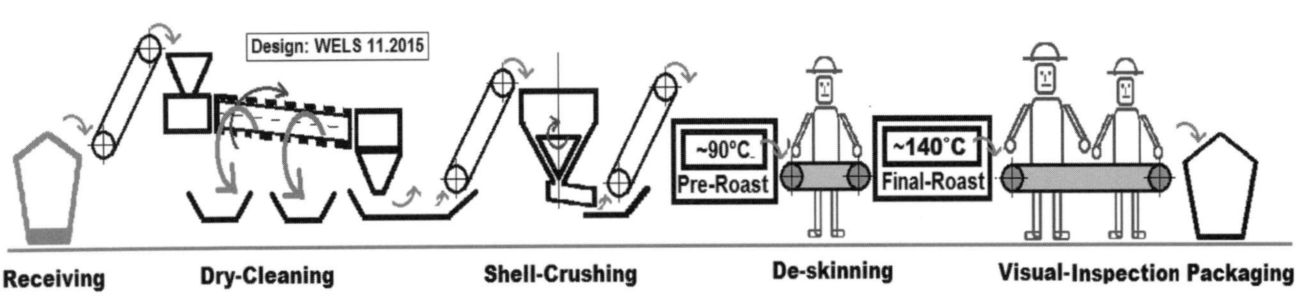

Figure 19.13 Layout of peanut decontamination. Major process steps: Receiving of raw produces, i.e., peanuts directly after harvest/ intermediate storage in almost harvested conditions; dry cleaning, possibly after a drying step, to remove soil, vines and debris in different sections to differentiate combustible materials from non-combustible materials; de-shelling (de-husking, de-hulling) to crush/remove the shells, stones and debris that have been carried along; after de-shelling the hulls/pods and sub-sized and shriveled material should be sorted out; roasting and de-skinning in two steps; pre-roasting at ~90°C followed by de-skinning then final roasting to accentuate discoloration at ~140°C; manual sorting (eliminating discolored peanuts).

Table 19.6 *Cellulose, Hemicelluloses and Lignin Content of Selected Biomass*

Produce	Cellulose %	Hemicellulose %	Lignin %
Wood	42–60	24 –40	18–30
Bagasse	42–50	29–42	16–21
Maize stalk	37–44	32–35	15–19
Wheat straw	35–39	22–24	18–25
Rice straw	(~35)	(~20)	~14,2
Rye straw	(~35)	(~20)	~13,5
Sorghum	(~35)	(~20)	~11
Kelp (*Dactylis glomerata* L *Festulolium braunii*)	~29	~19	~4

In order to utilize agricultural produce and biomass in industrial processes in an economic way it is important that the material be available in major amounts and over longer operational seasons; furthermore the utilization of biomass should not impede agricultural practices where the material has been used, e.g., as fertilizer soil conditioner or protection of soil against erosion.

19.3.2.2 Conversion of Biomass

The dry matter of agricultural crops, and of plant biomass at large, consists besides fats and proteins mainly of polysaccharides, cellulose and hemicelluloses and lignin (Table 19.6). Cellulose is easily decomposed into beta-glucose by cellulase that is produced by fungi and bacteria. Unfortunately only cotton is made up of pure cellulose; in other cellulose-containing biomass it forms complexes with lignin that prevent enzymatic degradation. To make the cellulose available for the enzymatic breakdown the lignin–cellulose complex has to be pretreated (Table 19.7) either by organic solvents, e.g., ethanol, or hydrolyzed, e.g., hydrochloric acid; the resulting reaction products can be further processed by cellulases into sugars that form the educt for single-cell or biofuel production. At the moment no entirely enzymatic saccharification process is available; processes that could offer advantages are under investigation and are expected to be ready for industrial application in about ten years (Figure 19.15).

Industrial single-cell and biofuel production require very elaborate and demanding processes; economic feasibility is only possible with larger processing units; e.g., the capacity of a single-cell protein plant is in general above 16 000 up to 40 000 Mt/a of yeast (Marx, 1989). Biofuel plants are in general much larger: the medium capacity of a bioethanol plant in the United States is about 407 million liter per year, whereas biodiesel plants are smaller: the medium capacity of a plant in the United States is 65 million liter per year (biomass).

Table 19.7 *Major Biomass Decomposition Methods*

Decomposition of biomass	
Pretreatment	**Operation**
Physical–mechanical	Milling, extrusion
Thermal	Pyrolysis, Steam explosion/high-pressure Steaming
Chemical or physicochemical	Alkali, acid treatment; solvent extraction, gas or oxidizing treatments
Biological	Microbial, enzymatic deconstruction

Table 19.8 *Processing of Biomass into Biofuel*

Production of biogas and ethanol from biomass pretreatment methods (see Figure 19.14)	
Substrate	**Yield (L t^{-1})**
Reed (untreated)	302 (Biogas VS (Volatile Solids))
Reed (steam explosion)	677 (Biogas VS)
Corn stover	222–342 (Ethanol)
Switchgrass	183 (Ethanol)
Sugar cane bagasse	241–469 (Ethanol)
Cellulosic materials (e.g., straw)	330 (Ethanol)
Rice straw	~200 (Ethanol)
Fermentative Production of Ethanol and Diesel Oil from Agricultural Sources	
Substrate	**Yield (L t^{-1})**
Maize	430
Wheat	350
Sugar beet	100
Rapeseed (Biodiesel)	370
Biomass (to Liquid) (Biodiesel)	500

Table top adopted from Lizasoain (2016); table bottom adopted from Vogelbusch (2017)

The main process step common to biofuel and single-cell protein production is the decomposition of the biomass into products that can be fermented (sugars) (see Table 19.8) or processed by, e.g., hydrocarbonisation (see Table 19.8) (Figure 19.14).

19.3.2.2.1 Production of Biofuel

"First-generation" biofuel is made from by enzymatic conversion from sugar or starch or by extraction from vegetable oil, and "second-generation" biofuels are

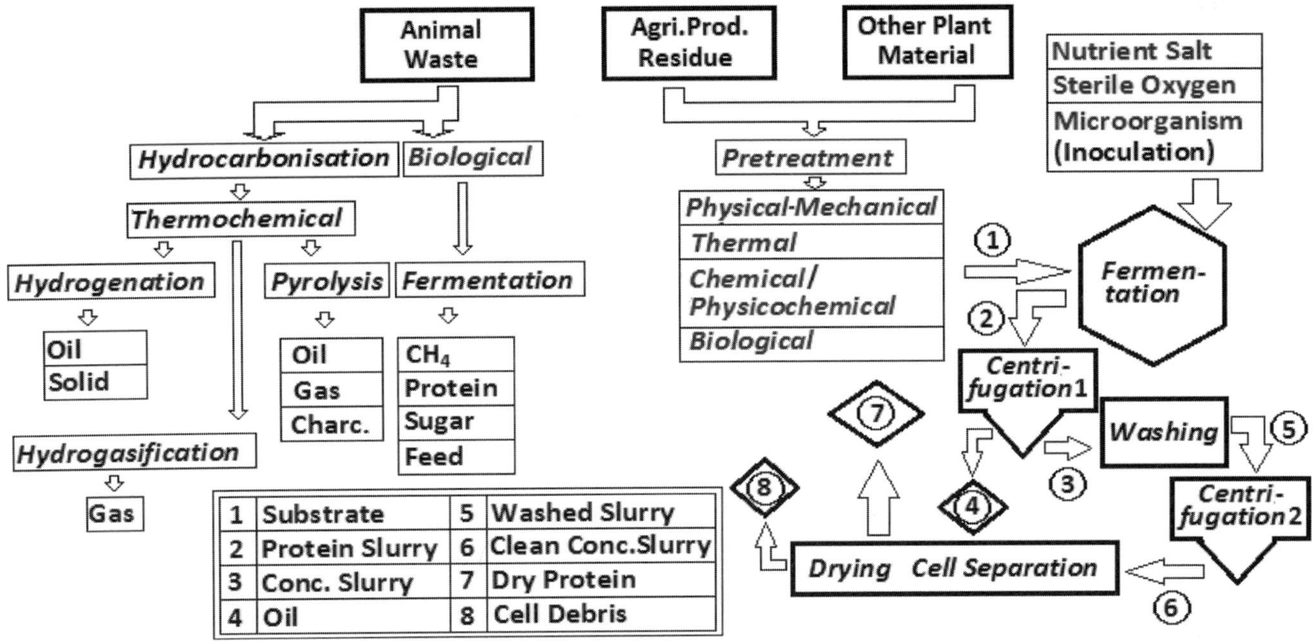

Figure 19.14 Pathways of biomass processing.

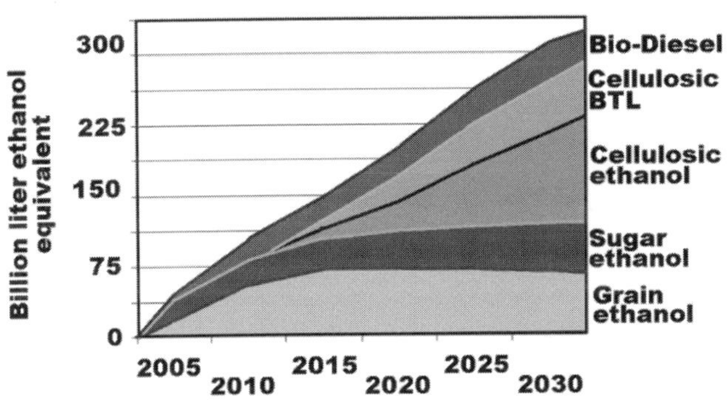

Figure 19.15 World biofuel production by type.
Source: World Biofuels Production Potential, Understanding the Challenges to Meeting the U.S. Renewable Fuel Standard, Office of Policy Analysis, Office of Policy and International Affairs, U. S. Department of Energy, Washington, DC 20585, September 15, 2008.

made from pretreated biomass. It is expected that by 2030 the second-generation biofuel will cover roughly 60 percent of the biofuel demand.

The world production of biofuels amounts presently to roughly 150 billion liter ethanol equivalents annually; the major part is bioethanol; biodiesel contributes about 10 percent (Figure 19.15).

Besides ethanol methanol and butanol as well as other related substances such as bioethers are obtained by fermentative processes from carbohydrate substrates. Biodiesel is mainly obtained from plant oils; an exception is the utilization of animal tallow, e.g., in Australia.

The conversion rates of biomass into biofuel depend very much on the type of biomass and its pretreatment, i.e., the accessibility of the cellulose and byproducts such as fats and proteins (Figure 19.14).

19.3.2.2.2 Production of Single-Cell Protein

The principles of the production of single-cell protein have been knownfor more than a century; already in the 1960s about 250 000 of yeast was produced as food and feed material; in 1970 the Soviet Union produced close to 1 MMT yeast, again as food and feed material.

The production comprises, as in biofuel production, several processing steps. In the first step the feed stock consisting of pretreated residues of various raw materials is inoculated for fermentation; yeasts, fungi, bacteria and algae are suited as inoculates (Table 19.9).

The yield of the biomass conversion depends – as in the case of biofuel production – on the type and pretreatment of the biomass utilized in the process (Table 19.10).

Fermentation as such is carried out as submerged, semisolid and/or solid-state fermentation. The downstream processes that follow the fermentation comprise the concentration and processing of the fully developed biomass in the broth, which amounts in the case of yeast to roughly 5 percent. The harvested material has to be amongst others decanted, centrifuged, washed and dried. In case of higher qualities suited for human consumption the protein has to be stripped from the cell by

Table 19.9 *Microorganisms Suited for Single-Protein Production*

Examples of Microorganisms			
Yeast	Fungi (*mycoprotein*)	Bacteria	Algae
Saccharomyces cerevisiae	*Aspergillus oryzae*	*Spirulina*	*Chlorella*
Pichia pastoris	*Fusarium venenatum*	*Rhodobacter capsulatus*	
Candida utilis utilis	*Sclerotium rolfsii*		
Torulopsis	*Polyporus*		
Geotrichum candidum	*Trichoderma*		
	Scytalidium acidophilum		

Table 19.10 *Biomass Conversion into Single-Cell Protein*

Biomass conversion		
Substrate	Microorganism	Yield dry cell material/ substrate (kg/kg)
Molasses	*Candida utilis/tropicalis*; *Saccharomyces* sp.	0.25–0.33
Citrus pulp	*Candida utilis*	0.18
Whey	*Candida krusel/fragilis/pseudo topicalis*; *Saccharomyces fragilis*; *Torula casei*	0.03
Spent sulphite liquor	*Candida utilis/tropicalis*	0.008
Potato-, fruit-,vegetable-wastewater	*Endomycopsis fibuliger; Candida utilis*	0.03–0.35
Cereal starch hydrolysates	*Candida utilis*	0.5–0.6
Wood hydrolysate	*Candida utilis*	0.3
Bagasse	*Candida utilis*	0.1–0.3
Corn stalk hydrolysate	*Candida utilis*	0.13
Methanol	*Candida* sp.; *Pseudomonas* sp.; *Methylomonas* sp.	0.25–0.5
Ethanol	*Candida* sp.; *Hansenula*; *Pichia*	0.6–0.7
n-Propanol; iso-propanol	*Candida* sp.	0.4
acetate; malate	*Candida* sp.	0.35
Methane	*Methanomonas* sp	0.3–1.4
n-Parfine	*Candida* sp.specifically *C. lipolytica tropicalis*; *Trichosporon japonica*; *Pseudomonas* sp.	1

chemical (acid), enzymatic (lytic enzymes), physical (freeze-thaw) or mechanical (milling) treatment (Table 19.11). High nucleic acid concentrations have to be reduced because of health reasons to acceptable levels by chemical and enzymatic treatments.

In a broader sense also the bodies of fungi and their mycelia are considered as single-cell protein sources; they have to be extracted and tailored for consumption.

Single-cell proteins can be used in its concentrated/dried state directly without treatment as animal feed. The

Table 19.11 *Protein Content of Single Cells*

Species	Protein content in dry matter %
Bacteria	50–85
Yeast	45–55
Fungi	30–45
Algae	45–65

extracted and purified material can be texturized or otherwise processed into palatable food; usually the raw material is spiced or supplemented by various food ingredients.

19.3.2.2.3 Production of Insect Protein

The utilization of low-grade biological materials/organic waste as feed for edible insects offers a further source of high-quality protein; in addition to amino acids insect protein contains valuable vitamins for human consumption (Table 19.13).

Insect feed can consist of organic wastes that cannot be utilized any further; above that the production can be organized in a very ecologically friendly manner, e.g., without the mission of greenhouse gases. The protein can be directly consumed or used as an addition to other products or as a protein source in feed-stock compositions. When it is used in food items intended for human consumption ethical and religious concerns have to be considered.

Table 19.12 *Conversion of Feed into Insect Protein*

Species	Comparable requirements in feed units to produce one unit of protein
Cricket	1
Cattle	6
Sheep	4
Pig/broiler chicken	2

19.3.3 Utilization of Processing Residues of Agricultural Produce into Added-Value Products

With the development of refined comminution, separation, extraction, drying, distillation etc. techniques many residues become – besides being used for SCP and biofuel production – valuable raw materials not only for the production of food or feed but also for the pharmaceutical and cosmetic industries; furthermore, new raw materials could be unlocked. Altogether the quantity of those products is for the time being rather limited; they are offering opportunities for expansion and interesting new features; e.g., jatropha and similar produce produced in larger amounts may release the pressure on the biomarket and create new sources of income for farmers in Africa and India; the same is true with a different emphasis for an expansion of aquatic resources (Table 19.13).

19.3.4 Development of New Storage, Packaging and Transportation Technologies - New Types of Distribution and Retail Outlets

Basically the entire agricultural production has to be stored, transported and distributed; finally a major share intended for human consumption has to pass retail outlets.

According to the World Bank (World Bank Indicator) the value of global agricultural production amounts to 3.192 trillion US dollars; the world trade in agricultural products is approaching in value 2000 billion US dollars; this means that major amounts of food

Table 19.13 *Utilization of Processing Residues*

Examples			
Produce	Process residue	Byproduct 1	Byproduct 2
Jatropha fruit		Diesel biooil	Press cake (feed)
Rice	Bran (10% of produce)	Rice-bran oil (15–18% of bran)	Defatted bran
Oil palm kernels		Palm kernel meal (animal feed)	Palm press fiber (solid fuel)
Grape	Grape pomace	Alcohol; animal feed; poly-phenol and antioxidant concentrates (grape skin/seed extracts)	Grape seed oil, grape seed hulls
Citrus fruits	Citrus pulp (50%–70% of produce)	Ensilaged pulp (animal feed)	Citrus seed oil, pectin
Algae, e.g. *Dunaliella salina; Chlorella* spp.	*Chlorella* paste	Chlorella powder (Animal feed)	Nutritional supplement, Dietetic oil
Phaeodactylum tricornutum	Edible oil	Oil derivates	
Palmaria	Food	Health food	Medical applications
Silk worm	Silk worm meal	Protein-rich food	Animal feed

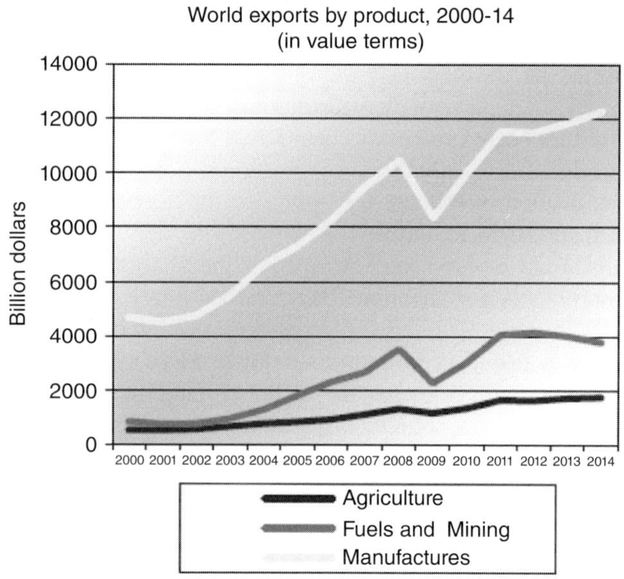

Figure 19.16 World exports by product.
Source: www.wto.org/english/res_e/statis_e/its2014_e/its14_merch_trade_product_e.htm

production have to cross country borders (for details see Table 19.1); to that 710 Mt fruit and 1011 Mt vegetable (IBIS) have to be added (Figure 19.16). Cereals, oil/oilseed, sugar and pulses, all products with a reduced water content – dry goods – are storable over certain periods under standardized conditions of environmental temperature and humidity and freedom from vermin; produce and products containing major amounts of water are susceptible to deterioration and require under trade conditions with special provisions; their cross-border trade is mainly regulated through the Agreement to Transport Perishables (ATP).

The handling of dry goods and perishables has undergone revolutionary improvements in the last years; on the technical side:

- Containerization, in the case of perishables bulk aseptic food processing
- Modular packaging
- High-bay warehouse–pallet rack system
- Data logging
- Bar-code systems

are essential elements. Indispensible for the effective management, monitoring and control of the modern food chain are also IT and GPS systems. The intelligent application of these elements allows providing better quality to the consumer and reducing losses of valuable produce/products above all.

The material flow of any type of goods from the production site to the consumer is organized/operated to a high degree by containerization. This technology is effectively utilized in the food industry, especially in handling perishables, e.g., fruits and vegetable but also other produce and products. Containers can be adjusted to producer/product and consumer needs with regard to size and environmental conditions, i.e., gas composition, humidity and temperature. Those conditions can be controlled and monitored by adequate data logging systems that even may ask for action in case of interruptions. The present generation of active and interactive sensors is available as mini- (one to two centimeters in length), medium-, and maxi-sized units; the latter are even capable of analyzing changing environmental conditions (Bauer and Spieß, 1981). In combination with global positioning systems (GPS) not only can container conditions be monitored; they can also be tracked.

It is expected that with progress in nanotechnology even small particulate foods can be equipped with sensors especially for quality control purposes.

Containerization of perishable food materials offers many possibilities for the food industry and at the very end to the consumer. The most attractive issue in this context is certainly to transport produce from agricultural production centers directly to centers of consumption. Harvested material can be trimmed or tailored according to consumer demands on the production site before being containerized under specific temperature regimes and conditions. As an example containerization at the smallest economic feasible level allows for preparation of small customized units at agricultural production sites, which then can be sent off to the interested consumers as part of larger units. Mini-containers can be realized as CA containers, and bundled with sister containers into standard palette size containers for transportation. Labeled with bar codes those and of course larger units can undergo transport and storage and be commissioned according to customer orders. In detail containers can be made of steel, wood and/or any type of plastic materials; they can be surface coated to absorb gases selectively, to reject surface contamination and/or to be easy to clean. Smaller containers that are designed to arrive at the point of consummation may carry messages regarding product handling, quality management and before or best-by dates.

Dominant methods of transportation of goods under a temperature regime are road transport and in special cases as air cargo. Despite the general trend to buy local products and products with a low environmental footprint, consumers are still prepared to buy seasonal produce from far away – in cases when it is offered at affordable prices – a situation that enables farmers in developing countries to upgrade and add value to their produce.

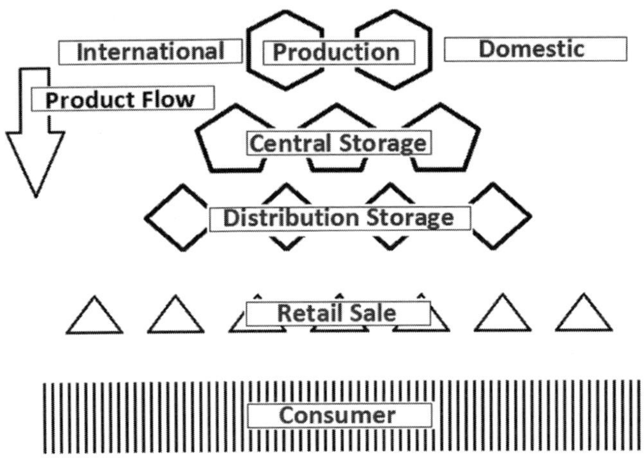

Figure 19.17 Distribution levels of perishables.

On their way from production/processing sites to the site of final consumption the transported materials in most cases have to go through central storage and distribution stores (see Figure 19.17). Containerization and the use of temperature controlled, computer supported high-bay storage facilities together with bar code systems and extremely efficient and very often completely automatic storage and distribution processes are possible; the process altogether allows for adequate short transportation/distribution processes. Containerization of smaller units also facilitates new methods of distribution such as digitalized shopping, to-door delivery and frequent in-time deliveries because of minimal home storage facilities (Spieß 1998).

A very special method of containerization is the containerization of liquids and pulps under aseptic conditions; starting with the aseptic handling of small-sized containers engineering efforts have succeeded in creating aseptic conditions in tanker vessels designed for the transport of orange juice and similar liquids. Available are epoxy-coated steel tank containers holding from 150 m^3 to 4000 m^3 as sterile storage facilities. For transport purposes sterile containerized bag systems of any size are also available. These developments allow the processing/semi finishing of highly perishable fruits and vegetables at the production site, reducing harvesting and transport costs, stretching final processing periods and generating income in agricultural areas.

19.3.5 New Types of Food Preparation

New ways of living especially in human agglomerates, i.e., megacities where new types of accommodation will be developed, will require new techniques of food preparation because cooking and food storage facilities, if they are available, will be miniaturized. With electronic communication systems special ways of making food available at the table will unfold. New types of cooking equipment in which blending, cooking, frying and baking are combined need little space and are very versatile; when combined with digital fabrication (fabbing, or 3D printing) systems the preparation of meals can be organized remotely. It also has to be expected that more consumers will decide to have their meals at restaurants or canteens, a fact that implies more industrialized cooking. In summary, engineered processes of food preparation will simplify food preparation for the consumer in the megacity, allowing also for a reduction of waste at the last step of the food chain.

19.4 Further General Developments and Projections to Stabilize Food Supply

19.4.1 Development of Food Engineering

The development of new technologies and processes to secure a sufficient food supply to the global community has not come to an end; new challenges will arise and will need responses. To respond in an adequate way the food engineering community has to maintain its efforts to update its global developing strategy. This can only be done by an intensive discussion of individual approaches and concepts within the community and a dialogue with all partners involved in structuring the food chain.

The request for a dialogue with all partners involved in the food chain is basically not a new one; it was, however, in most cases limited to agricultural, engineering, and nutritional science; meteorology, urban planning, architecture and consumer science were excluded – to mention some examples – in addition ethical issues of concern to food production and processing should be considered to a more pronounced degree.

Essential for the direct development of food engineering as a discipline is the understanding of changes – down to the molecular level – that food materials undergo during processing, storage and distribution and during home preparation. The understanding of interactions of the food components especially during processing that determine the sensory and physiological properties of the final product is the basis for the development of new processes and products.

Classical fields that need further development are

- Biotechnology, genomics, proteomics and related fields and the new biology
  - Integrated approaches have to be developed that combine (plant) meta-genomics with genomics of microorganisms and human genomics

- Processing techniques responding to nutritional requirements
  - Thermal processing with a minimum damage to critical components
  - Non-thermal processing application should be expanded into new areas
  - Mechanical processing/structuring should be applied to structure synthetic and natural proteins or carbohydrates into edible products
  - Mechanical separation techniques should be applied to remove less valuable/ harmful components, enrich valuable components and fractionate components with different physical properties
- Nanotechnology and new accelerated reaction kinetics should be explored in close partnership with all involved in production processes (from processing down to consumption and physiological fate) especially with regard to the impact of nano-sized components in food on human health (risk analysis). Important will also be the evaluation of the stability of nano-material in food matrices and during processing, storage and digestion. Of further interest is the biopersistence of nano-particles
- Improvement of sensory properties/fortification of new and unconventional food products/processes and elucidation of those innovations will be required to unlock consumer resistance
- Hygiene
  - Globalization with an increasing international trade of food materials requires that food products be free of any pathogens and quality-impairing substances especially, in the case of produce/ products intended for international trade
- Water
- Despite the fact that the amount of water used in food processing is almost negligible compared with the amount consumed for agricultural production, the use of water in processing is being considered as a key operation that is able to reduce the local environmental impact of food processing
- Food chemistry/ analytical issues/ enzyme technology
  - Enzyme technology can play a pivotal role in unlocking new resources and better utilization of conventional resources

19.4.2 The Educational Domain/Addressing Educational Tasks

In order to keep losses down and the quality up for food professionals, a thorough knowledge of chemistry is necessary to understand factors affecting the quality of foods; the same is true for an orientation in engineering; in this case besides a broad natural science background a profound education in relevant engineering subjects is required. Therefore, these subjects should have a prominent place in the curricula of all university studies on food science and technology; after all, food economy should not be neglected. Basic knowledge of food chemistry is also a requirement for anyone dealing with nutrition in general or nutritional physiology and related fields. In order to grasp labeling information on packaged food items educated customers should at least understand such terms as proteins, carbohydrates, lipids, mineral components, vitamins, and additives. Very delicate and to some extent problematic areas are the media (TV, radio, journals, newspapers etc.) dealing with food and nutrition. It would be too much to require that journalists dealing with those subjects should have a profound food science related background; however, it would help to allow for an objective coverage of food-related news, especially in critical situations. Altogether there is a great need for strengthening and improving the educational exposure to food science and technology. The food industry in highly industrialized as well as in developing countries is permanently looking for qualified personnel; therefore, increased emphasis should be put on attracting qualified students into food science and related areas.

The international food science community has a special obligation to disseminate food-related information especially into areas where food insecurity prevails. Information on unconventional food resources and edible materials not necessarily on the daily diet sheet but also on inedible produces may help to mitigate to some extent the lack of an adequate food supply and may help to prevent ingestion of toxic substances. It is important that the provision of food-related information be presented in a format understandable to the addressee, especially in areas with low literacy rates (see Figure 19.18).

19.4.3 Ethical/Religious Issues

Worldwide production and trade of food require that general ethical and religious rules and regulations are observed in food production and processing. Therefore, emphasis should be placed on ethical production. Food that is not in conformity with the consumers' religious and ethical fundamental attitudes will be rejected even in case of extreme need.

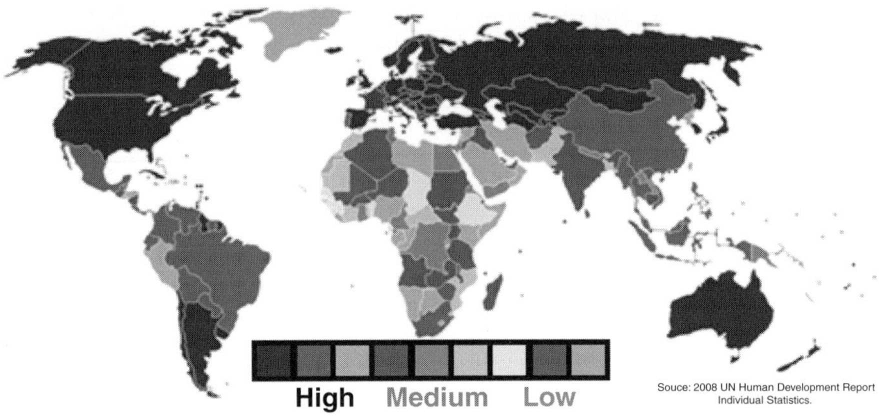

Figure 19.18 Global literacy rates.

19.4.4 The Industry and Government Cooperation Domain

Increasing emphasis on the formation of partnerships involving food science and technology is mandatory to address the complexity of issues involved in food production and processing and stabilization of the food chain. This includes industry–academia–government–consumer partnerships, which should also be extended to international partnerships. These partnerships also form the basis for the development of future projects. The contacts are also essential for global and domestic food trade, food safety and food security issues.

19.4.5 The Consumer Domain

Consumers should be involved at early stages in problem recognition and possible solutions. Furthermore, consumers should be educated to make the correct choices of food, appropriate to their health and social situation, and to minimize food losses through correct handling of food when it is under their control.

19.5 Conclusions

In the next decades the world population will grow to 8.5 billion or even more people. To supply adequate amounts of food to this clientele the conventional resources will not be sufficient, especially if the trend to more animal-based nutrition will continue.

Potential reserves to increase the available food supply are available and have to be unlocked; e.g., the large losses within the food chain and the exploitation of new resources have to be targeted in the first instance. Drying, refrigeration immediately after harvesting and CA storage are appropriate means to stabilize the quality and to extend the shelf life of animal-based products and plant produce. The technical tools for these processes are available; however, their application, especially in developing countries, requires a well-educated, trained and skilled labor force. The same is true for the exploitation of unconventional resources; those processes are technically highly demanding and are complex not only with regard to the technologies involved but also with regard to scale. These challenges require, besides technical and management abilities, strategic planning skills in order to make those operations profitable over longer periods.

The enzymatic breakdown of cellulose, hemicelluloses and lignin into fermentable sugars allows production of biofuel without affecting valuable cereal resources.

Also very important with regard to the stabilization of global food security is the involvement of the consumer in food chain operations; especially important is an understanding of the impact of specific dietary modes on the environment, e.g., the negative implications of the consumption of animal-based food, which requires major amounts of water for its production and emits greenhouse gases, and the rejection of food items produced from unconventional raw materials such as insects.

Considering the entire situation the prospects of feeding the growing world population can be viewed in a positive way; if the intellectual capacities of mankind are used in a constructive way, no supply shortfalls should occur and hamper population growth.

References

Alexander, P., Brown, C., Arneth, A., Dias, C., Finnigan, J., Moran, D. and Rounsevell, M. D. A. Could consumption of insects, cultured meat or imitation meat reduce global agricultural land use? In: *Global Food Security*, 22.04.2017. www.elsevier.com/locate/gfs.

Alexandratos N. and Bruinsma. J. 2012. World agriculture towards 2030/2050: the 2012 revision. ESA Working paper No. 12–03. Rome, FAO. www.fao.org/docrep/016/ap106e/ap106e.pdf. Accessed 06.06.2017.

Alfenore S. et al., Current status and future prospects of conversion of lignocellulosic resources to biofuels using yeasts and bacteria, *Process Biochemistry*, 2016; 51 (11): 1747–1756.

atmosnews. www2.ucar.edu/atmosnews/; https://www.nasa.gov/press-release/nasa-releases-detailed-global-climate-change-projections. Accessed 06.06.2017.

ATP. www.unece.org/trans/main/wp11/atp.html. Accessed 06.06.2017.

Bakker, R. R. C., Elbersen, H. W., Poppens, R. P. and Lesschen, J. P., Rice straw and wheat straw: potential feed stocks for the bio-based. *Economy*, June 2013. NL Agency - Ministry of Economic Affairs.

Bauer B. and Spieß W. E. L., Gerät zur (elektronischen-) Erfassung von Zeit und Temperatur, Gebrauchsmuster-anmeldung (utility patent application) P31 39 663.1, 06.10.1981. 1981.

Berkeleyearth. http://berkeleyearth.org/land-and-ocean-data// http://berkeleyearth.org/graphics/global-energy-and-emissions/#section-0–1. Accessed 06.06.2017.

Biomass. http://biomassmagazine.com/articles/9801/comparing-plant-capacities-of-u-s-ethanol-biodiesel-industries, Accessed 20.11.2016.

Ciepiela G. A., Godlewska A. and Jankowska J. The effect of seaweed *Ecklonia maxima* extract and mineral nitrogen on fodder grass chemical composition. *Environ Sci Pollut Res Int*. 2016; 23: 2301–2307. Published online 2015 Sep 26. doi: 10.1007/s11356–015–5417–3 PMCID: PMC4717178.

Detroy, R. W. and Hesseltine, C. W. Availability and utilization of agricultural and agro industrial wastes. *Process Biochemistry* Sept 1978; 13(9), http://handle.nal.usda.gov/10113/28215. Accessed 06.06.2017.

Feed, www.feedipedia.org/node/728. Accessed 10.09.2016.

Grüne-Oldenburg, www.gruene-oldenburg-land.de/ov-harpstedt/unsere-themen/energie/expand/380962/nc/1/dn/1/. Accessed 15.11.2016.

hdr.undp. http://hdr.undp.org/en/content/adult-literacy-rate-both-sexes-ages-15-and-older. Accessed 18.10.2015.

High Level Panel of Experts on Food Security and Nutrition. HLPE report 8, Food losses and waste in the context of sustainable food systems: A report by The High Level Panel of Experts on Food Security and Nutrition June 2014 www.fao.org/cfs/cfs-hlpe (Modified). Accessed 06.06.2017.

IBIS World Industry Report: *Global Fruit and Vegetables Processing*. September 2015.

Lizasoain J. Biogas production from reed biomass: Effect of pretreatment using different steam explosion, *Biomass and Bioenergy*, 2016; 95: 84–91.

Lüthi, D., Le Floch, M., Bereiter, B., Blunier, T., Barnola, J.-M., Siegenthaler, U., Raynaud, D., Jouzel, J., Fischer, H., Kawamura, K. and Stocker, T. F. 2008. High-resolution carbon dioxide concentration record 650,000-800,000 years before present. *Nature*, 2008; 453: 379–382.

Marx J. L. Editor. *A Revolution in Biotechnology*. ICSU Press, Cambridge University Press, 1989.

Meehl, G. A., Stocker, T. F., Collins, W. D., Friedlingstein, P., Gaye, A. T., Gregory, J. M., Kitoh, A., Knutti, R., Murphy, J. M., Noda, A., Raper, S. C. B., Watterson, I. G., Weaver A. J. and Zhao, Z.-C. 2007: Global Climate Projections. In: *Climate Change 2007: The Physical Science Basis.*

Contribution of Working Group I to the Fourth Assessment Report of the Intergovernmental Panel on Climate Change [Solomon, S., D. Qin, M. Manning, Z. Chen, M. Marquis, K. B. Averyt, M. Tignor and H. L. Miller (eds.)]. Cambridge University Press, Cambridge and New York, USA Global Climate Projections, www.ipcc.ch/pdf/assessment-report/ar4/wg1/ar4-wg1-chapter10.pdf. Accessed 06.06.2017.

Palmaria. https://en.wikipedia.org/wiki/Palmaria_palmata. Accessed 06.06.2017.

Rehm, H.-J. Industrielle Mikrobiologie. 2. völlig neu bearbeitete Auflage. Springer-Verlag, Berlin-Heidelberg-New York 1980. XX, 718 Seiten. mit 215 Abb. u. 89 Tab.

Single-cell protein. https://en.wikipedia.org/wiki/Single-cell_protein. Accessed 18.11.2016.

Spieß, W. E. L., Böhme, T., and Wolf, W. 1998. *Food Storage Stability*, edited by Taub I. A., and Singh R. P. CRC Press.

Spieß, W. E. L., Behsnilian, D. 2001. Introduction into osmotic treatment. *Journal of Food Engineering*, 2001; 49: 75–76.

Spieß, W. E. L. 2002. Freezing and chilling of food, lecture text book; Hohenheim (Germany) University 1976, Seoul (Republic of Korea) National University 2002.

Spieß, W. E. L. 2008. Improving the utilization of agricultural raw materials to reduce food insecurity, the modest contribution of food science and technology to fight food insecurity around the globe, PPP Presentation ICSU-IUFoST Symposium, IUFoST World Congress 2008.

Spieß, W. E. L. and Moy G. 2014. Reducing aflatoxin exposure in Sub-Saharan Africa by post-harvest sorting of peanuts, business plan submitted to IUFoST 23.07.2014.

The State of Food and Agriculture. Biofuels: prospects, risks and opportunities FAO, Rome, 2008, http://www.fao.org/docrep/011/i0100e/i0100e00.htm, Accessed 06.06.2017.

Thehindu.com. http://www.thehindu.com/multimedia/archive/02001/UN_GN1_2001757a.jpg. Accessed 19.10.2015.

United Nations, Department of Economic and Social Affairs, Population Division (2015); https://esa.un.org/unpd/wup/Maps/CityDistribution/CityPopulation/2030_City_Urban, Accessed 23.01.2018.

Vogelbusch. http://www,bioethanol,vogelbusch,com/en/faq,php; Accessed 06.06.2017.

World Food Program. https://www.wfp.org/content/hunger-map-2015, Accessed 14.11.2016.

World Bank Indicator: http://data.worldbank.org/indicator/NV.AGR.TOTL.CD. Accessed 06.06.2017.

World Biofuels Production Potential; Understanding the Challenges to Meeting the U.S. Renewable Fuel Standard; Office of Policy Analysis; Office of Policy and International Affairs; U. S. Department of Energy; Washington, DC 20585; September 15, 2008.

World Development Report 2010: Development and Climate Change. Washington, DC. © World Bank. https://openknowledge.worldbank.org/handle/10986/4387 License: CC BY 3.0 IGO." Creative Commons Attribution license (CC BY 3.0 IGO) Accessed 06.06.2017.

World Population to 2300. World Population Prospects, UN Department of Economic and Social Affairs Population Division https://esa.un.org/unpd/wpp/publications/files/key_findings_wpp_2015.pdf. Accessed 06.06.2017.

WTO-merch. https://www.wto.org/english/res_e/statis_e/its2014_e/its14_merch_trade_product_e.htm. Accessed 06.06.2017.

Verma, M., Godbout, S., Brar, S. K., Solomatnikova, O., Lemay, S. P. and Larouche, J. P. Biofuels production from biomass by thermochemical conversion technologies, *International Journal of Chemical Engineering*, 2012; Article ID 542426. http://dx.doi.org/10.1155/2012/542426. Accessed 06.06.2017.

20
Supply Chains and Future Earth

Albert McGill

20.1 Introduction

As an outcome of the UN Conference on Sustainable Development (Rio+20, 2012) the Future Earth international research programme was launched. It was intended to "provide critical knowledge required for societies to face the challenges posed by global environmental change and to identify opportunities for a transition to global sustainability" (Future Earth, 2013). In proposing the need for a step change in Earth system research, human activities are given as a major factor in the alteration of such systems at all levels, local, regional and global, leading to changes in the Earth's climate and the loss of biodiversity.

The threat to human prosperity could be balanced by incentives to develop opportunities to support sustainable development. Of the three broad research themes proposed (Dynamic Planet, Global Development, and Transformations towards Sustainability) global development is most closely dependent on supply chains and most closely aligned with global food security. In addressing the needs for 'sustainable, secure and fair stewardship of food, water, biodiversity, energy, materials and other ecosystem functions and services' this theme requires delivery systems that will be dependent on efficient supply chains.

The relevant focus of the Future Earth vision as it targets global development will be outlined. The fundamental aspects of supply chains will be described and those directly related to global food security will be expanded. The challenges posed to these supply chains by adverse aspects of climate change will be identified, especially when disasters occur.

20.2 Future Earth

The vision of Future Earth, as articulated in the ICSU 2025 statement, is "for people to thrive in a sustainable and equitable world" (Future Earth, 2014). Its ten-year research programme "will define opportunities to reduce risks and vulnerabilities, to enhance resilience and innovation, and show ways to implement transformations to prosperous and equitable futures" (Future Earth, 2013). The research has eight focal challenges, four of which (numbers 1, 4, 5 and 6) are most sensitive to the influence of supply chains:

1. Deliver water, energy and food for all, and manage the synergies and trade-offs among them, by understanding how these interactions are shaped by environmental, economic, social and political changes.
2. (4) Build healthy, resilient and productive cities by identifying and shaping innovations that combine better urban environments and lives with declining resource footprints, and provide efficient services and infrastructures that are robust to disasters.
3. (5) Promote sustainable rural futures to feed rising and more affluent populations amidst changes in biodiversity, resources and climate by analyzing alternative land uses, food systems and ecosystem options, and identifying institutional and governance needs.
4. (6) Improve human health by elucidating, and finding responses to, the complex interactions amongst environmental change, pollution, pathogens, disease vectors, ecosystem services and people's livelihoods, nutrition and well-being.

All of these challenges are informed by the basic principles outlined in the definition of food security, namely, availability, access, affordability and preference or choice (FAO, 1996). The means whereby goods and services move from their sources to their ultimate consumer is by the mechanism of supply chains.

20.3 Supply Chains

There are a number of definitions for supply chains ranging from the more simple (Google), "The sequence of processes involved in the production and distribution of a commodity", through the academic, "A supply chain is a network of organizations that work together to convert and move goods from the raw materials stage to the end customer. These organizations are linked together through physical, information and monetary flows" (Boyer & Verma, 2010), to the more comprehensive from Wikipedia, "A supply chain is a system of organizations, people, activities, information and resources involved in moving a product or service from supplier to customer. Supply chain activities involve the transformation of natural resources, raw materials and components into a finished product that is delivered to an end customer." For the vision of Future Earth to become a reality, the smooth operation of supply chains is critical.

Supply chains, and in particular those associated with the food industry (illustrated in Figure 20.1), are subject to risks, as has been discussed by Dani & Deep, 2010. Many of these risks are associated with increases in regulations regarding food safety and with greater consumer demands for fresher foods of known origin. Other risks affect the supply chain itself and include loss of power, flooding and transportation strikes.

It may be useful to consider the evolution of a supply chain network, and this has been set out in Figure 20.2. In its simplest form (a) the producer is the consumer and no chain is required. This situation applies to subsistence farmers but also to households that produce much of their food for personal consumption. It can also apply to some aspects of urban and peri-urban agriculture (Beer et al., 2016). As the distance increases the produce has to be transported or transmitted from the producer to the consumer (b), with increasing complexity as the processor (c), retailer (d) and wholesaler (e) are added to the system. Thus supply chains are, in reality, supply networks. An expanded version of the system is provided in Figure 20.2 (f), but this still omits the developing demand-based link between consumers and primary providers, for example, through farmers' markets, personal allotments in cities or planned urban agriculture projects. The growth of schemes in urban agriculture such as vertical farms and horticulture in

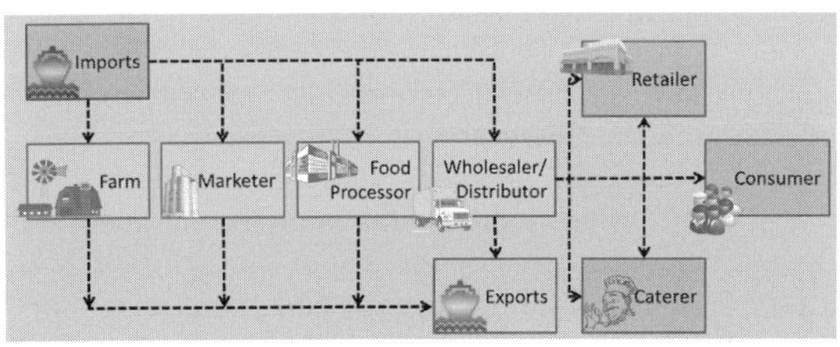

Figure 20.1 The food supply chain and trends affecting it.

public spaces can shorten the supply chain network, making it less vulnerable to adverse weather events (Figure 20.3).

There has been much debate on the effectiveness of urban agriculture projects. The United Nations Development Programme's review of urban food supply around the world has indicated that local production increases nutrition, enhances food security and creates employment opportunities for communities (Smit et al., 1996). Urban agriculture provides 60% of the vegetables and 90% of the eggs consumed by residents in Shanghai (Bhatt & Farah, 2009), 47% of the produce in Bulgaria (Premat, 2005), 60% of the vegetables produced in Cuba (Premat, 2005) and 90%–100% of the leafy vegetables for poor households in Harare (Mougeot, 2005). More recently (Garfield & Jacobs, 2017) reports of success in growing produce hydroponically in shipping containers in Brooklyn, New York, are an outcome of entrepreneurial activity by Kimbal Musk and Tobias Peggs through their company, Square Roots, which has developed through an extensive franchise operation. They claim the ability to produce, for example, fifty thousand mini-heads of lettuce per annum. A number of similar businesses are in competition, so there should be a line connecting consumers and producers in Figure 20.2 (f).

The consumers themselves require transport, storage and disposal systems both to access and to handle the acquired produce, an aspect that limits the availability of many products to different sectors of communities. For every link in the supply chain there exists a specific dedicated supply chain that both enables and limits its operations.

One of the most basic of Maslowian needs is the supply of food and water (Maslow, 1943). The supply chains that satisfy these needs are critical to the health and well-being of all peoples, and any factors that disrupt these chains will limit the future of all involved at each stage of the system from producers to consumers. The fragility of the food supply chain and how risks to its operations can be analyzed and addressed have been discussed by Dumke (2010). A food supply chain for the United Kingdom is illustrated in Figure 20.4 with a more commodity-related example given in Figure 20.5. The increasing complexity of supply chains as developed by the European Technology Platform, Food for Life (2005), is illustrated for the total food system in Figure 20.6.

Although large quantities of goods may be moved by sea and rapid deliveries of goods may be delivered by air freight, the most significant transport route for movements and final delivery of goods is by road. This includes the means whereby consumers access these goods from retailers and convey them to their homes.

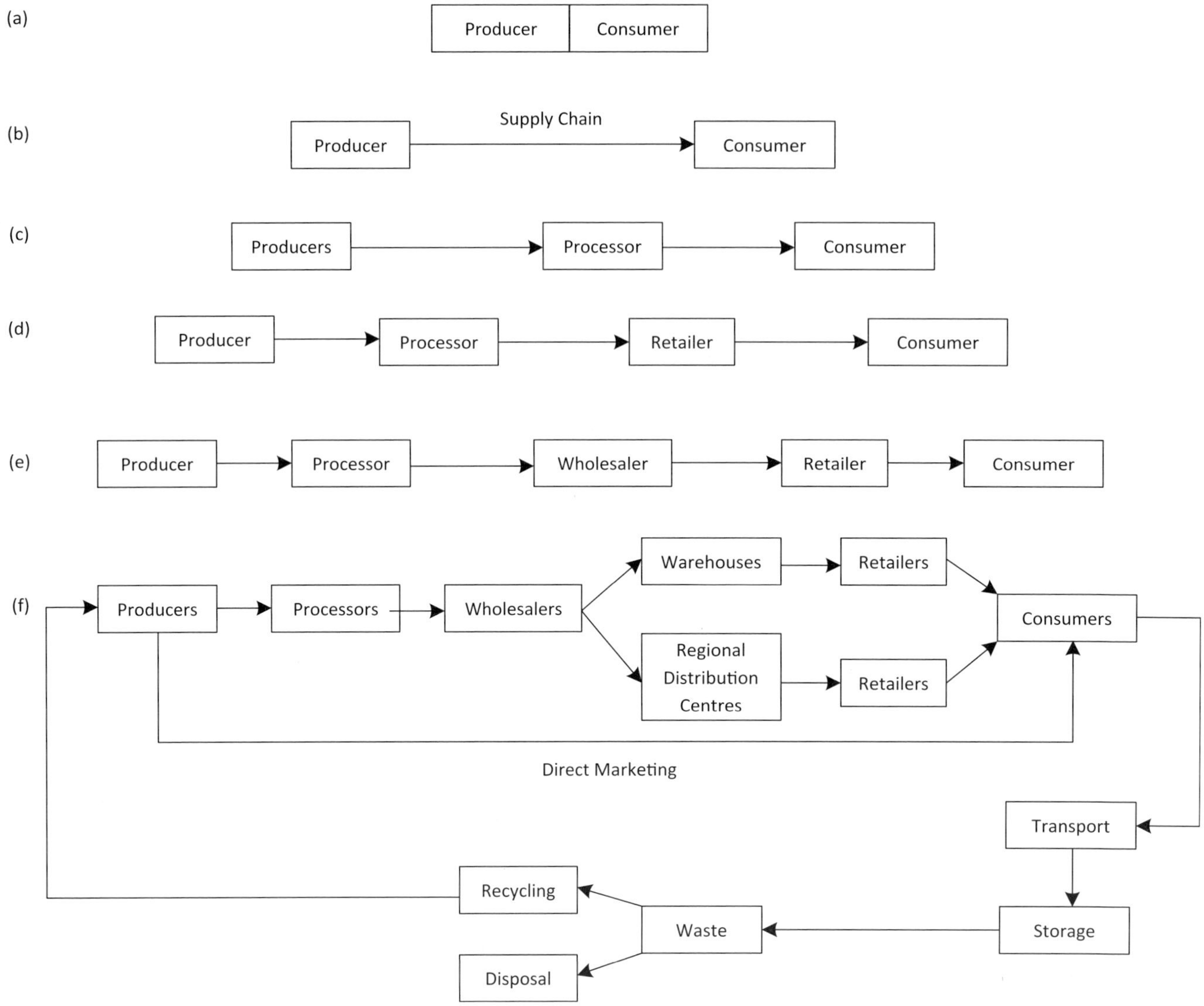

Figure 20.2 Development of supply chain networks.

Thus the major infrastructure component and the most variable is the road network in any environment. Any improvement in road structures or access or any factors that damage or interfere with these structures will enhance or limit the efficiency of most supply chain networks. Apart from war and associated conflicts, the major impact on supply chains (and the health and well-being of all peoples) is extreme weather, which may be exacerbated by climate change.

20.4 Climate Change

Climate change may be considered under three headings: incremental, severe and catastrophic.

Incremental changes are continuous and show predicted increases in overall global temperatures that may well average 2° C but can be as much as 8° C in specific locations. This will require a series of adaptations including a shift in the location of crop production to more temperate zones with major changes in the livelihoods of primary producers and a re-organization of supply chain routes and composition – a potential loss of availability in the vision of Future Earth.

Severe changes include increasing incidence of droughts, storms, floods, tidal inundation, bush and forest fires and land slippage. All of these are likely to interfere with and limit transport and communications, damaging the supply chain networks, but may well be limited to directly affected areas of the world. Although

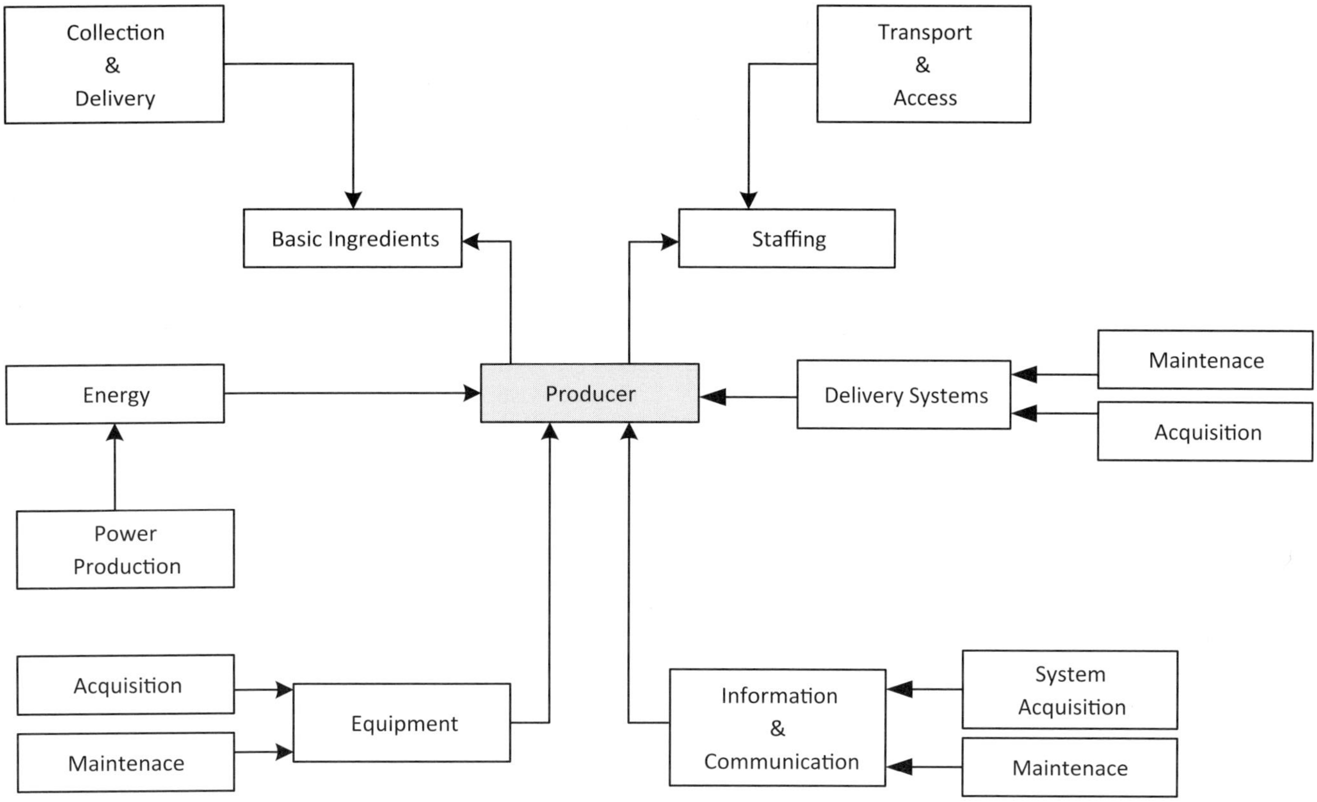

Figure 20.3 Supporting supply chain for producers and processors.

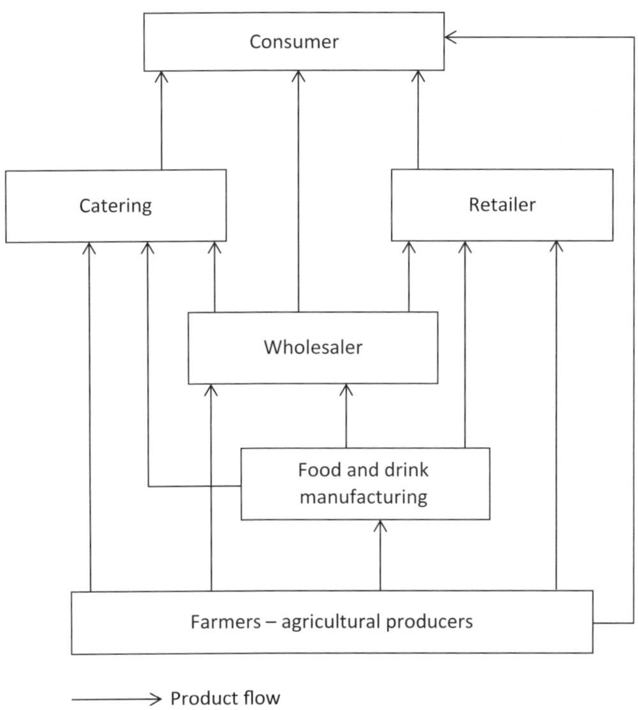

Figure 20.4 The UK food supply chain.

the impacts locally may be severe, the effects globally may be mitigated by the efficiency of international networks to re-route and reorganize the required distribution systems. However, they are most likely to affect the price of commodities, placing them outside the reach of those of more limited finances – a loss of affordability in the vision of Future Earth.

Catastrophic changes refer to events that have extensive and/or global impacts; earthquakes and volcanic eruptions may not be the result of climate change but cause similar levels of infrastructure disruption and loss of life, whereas typhoons, hurricanes, tornadoes, tsunamis, glacier melts and extensive landslides have similar impacts. Such events may extend across continents and frequently destroy infrastructure, especially road and rail networks critical to the success of supply chain operations – a loss of access to food in the vision of Future Earth. These events cause major crop and animal losses and loss of transport systems and many lives of populations and transport crew members.

At present there is famine across a large part of Africa. Severe drought has caused crops to fail and cattle to die in Ethiopia, Somalia and Kenya, leading to severe food and water shortages, affecting more than

17 million people. In addition, conflict in South Sudan, Yemen and Nigeria has created millions of refugees and more than 1.4 million children are at risk. The United Nations needs $5.6 million to avert a crisis (Stephen O'Brien, UN Secretary General for Humanitarian Affairs, 11 March 2017). When such a crisis develops, the normal supply chain is disrupted in three ways. Firstly, primary produce is no longer available. Secondly, by having to flee in search of food and water, people are displaced from their normal systems of supply and must travel to feeding stations at relief centres. Finally, the centres themselves are confronted by unknown numbers of refugees and must attempt to establish an adequate supply chain that can be able to meet fluctuating demands. The refugees themselves are also deprived of any vestige of food choice in trying simply to survive. All these factors create food insecurity.

There is evidence that natural disasters are increasing in frequency, as has been discussed at length by Rosen (2016). That such weather events are a result of climate change is supported by Cornwall (2016) in his explanation of the development of one product of the emerging event attribution science. Society should be better at its preparations to ameliorate the effects of disasters, as has been seen in the successful development of earthquake-resistant buildings in Japan (McNutt, 2016), but not all societies are equal in their capacities to develop and implement appropriate strategies and even sound strategies may fail, as with hurricane Katrina in New Orleans. An investigation by the US Congress (19 February 2006) concluded that there had been "a failure of initiative, mismanagement and lack of leadership in relief efforts" and that "FEMA and the Red Cross did not have a logistical capacity sophisticated enough to fully support the massive number of Gulf Coast victims". The effects of hurricane Sandy, which hit New York in 2012, were better managed as a result of good cooperation between the mayor and the

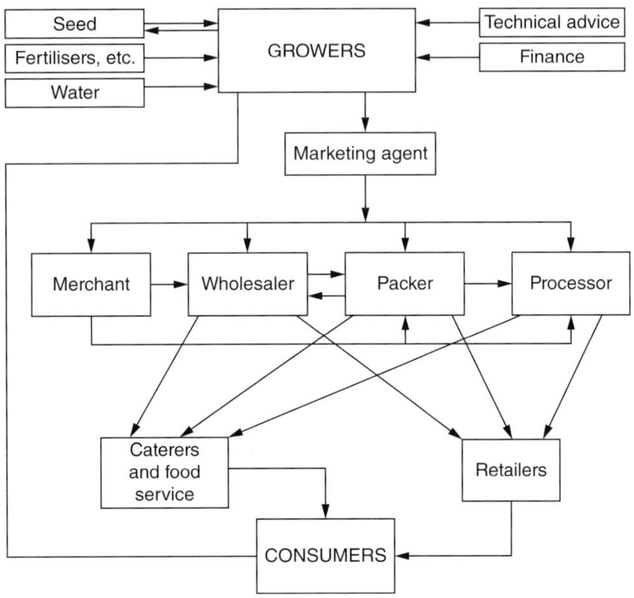

Figure 20.5 The UK potato supply chain.

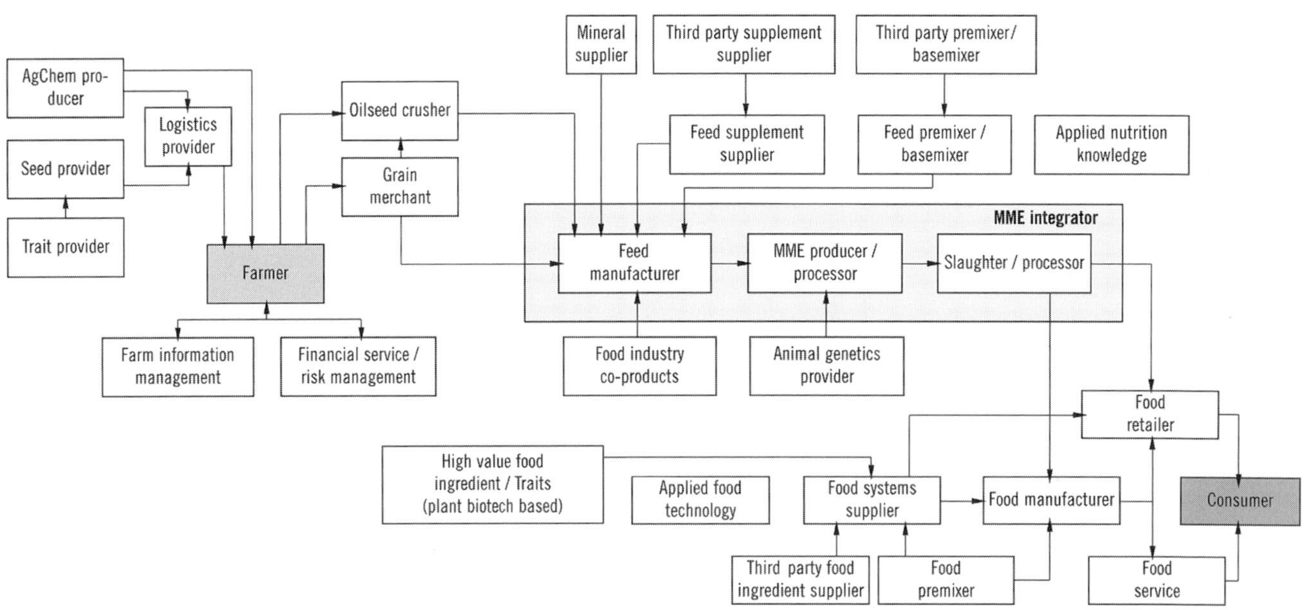

Figure 20.6 The complexity of food chains. Adapted from the publication 'The Vision for 2020 and beyond' of the European Technology Platform 'Food for Life'.

federal government through the direct intervention of President Obama. Although the damage caused was very great, pre-disaster planning and targeted evacuations helped to minimize loss of life. Much has been learned since Katrina.

When prevention strategies fail and disasters occur the first response is to prevent further loss of life, then to treat the injured, providing safe shelter and clothing. The continuing supply of food and water not only sustains life but provides some comfort and a sense of greater normality to those affected and displaced.

The first responders to disasters are usually neighbours and other members of the local community, who not only understand those affected, but can also supply support that is culturally acceptable to and will be welcomed by victims. The resources of first responders will be limited in the range of what is provided and the duration of their support. Charity organizations, supported by the local and regional Red Cross/Crescent, are usually next on the scene and can provide greater resources over a longer period. Subsequently state/regional emergency response will be engaged, and ultimately international support will become available from international charities and appropriate UN organizations. Although the majority of the aid responses will be contracted by both UN and international charities, significant support is often given by multi-national corporations in both materials and logistical support using their successful global supply chain networks (Thomas & Kopczak, 2005; Tatham & Pettit, 2010). Companies such as Coca-Cola, Cargill, Unilever, TNT, DHL and others see such support as part of their corporate policies. A recent article by Mermelstein (2016) discusses how the impact of disasters on the food supply is being addressed, especially in the United States. All responses that bring resources or information to bear on disasters will require the establishment and maintenance of effective and efficient supply chains.

20.5 Food Security

Not all aspects of climate change are seen as negative events for populations, the environment and hence, food security. Smith (2013) offers an expanded discussion of the co-benefits of good climate policies linking reduction in the use of fossil fuels with lower levels of greenhouse gases (GHGs) in the environment.

In examining the role of supply chains in the aims and activities of Future Earth, the focus has been mainly on the effective delivery of essential goods and services, including food and water. It is important to appreciate that as these goods and services flow to consumers, information flows in the opposite direction, providing strategic advice for those associated with each link in the chains across the networks. So consumer choice, another aspect of Food Security, not only affects their own health and well-being, but also directs investment in and planning for future activities. Such choice can be influenced by those who stand to profit from choices made but can also be used to influence a better approach to health and well-being. An approach to the promotion of healthier eating through a process of *Mind Genomics* has been described by King et al. (2016) and examines a positive collaboration on this approach between academia and the food industry.

To return to the selected components from the key focal challenges of Future Earth the critical role of supply chains has been demonstrated:

1. "The delivery of water, energy and food for all" cannot be accomplished without the development and maintenance of effective supply chains.
2. "Build healthy, resilient and productive cities" will not take place unless their inhabitants can be provided with the essential goods and services to support them and will require efficient supply chain systems.
3. "Promote sustainable rural futures to feed rising and more affluent populations" can occur only if and when the necessary support, information, technology and improved agricultural science are deployed. These require an effective supply chain network.
4. "Improve human health": the essential components required to enable these improvements are a complex mixture of material, educational and psychological support and supply chains to assist these improvements.

The key elements of food security for all populations are availability, access, affordability and preference/choice of the foods necessary for their continuing development of good health and well-being. Effective and efficient supply chain networks, even at their most simplistic level, are essential to the success of these and the broader aims of Future Earth.

20.6 Summary

Of the three broad research themes of Future Earth, global development was identified as most closely aligned with global food security and very dependent on supply chains. The four focal challenges most sensitive to the influence of supply chains have been identified and described.

The concept of supply chains has been explained from first principles and expanded to cover the global food supply. Potential effects of climate change on food supply chains have been examined and the effects of disasters resulting from extreme weather conditions considered. The importance of the role of supply chains in supporting the key aspects of global food security (availability, access, affordability and preference/choice) in providing for the continuing development of good health and well-being has been demonstrated. The importance of effective and efficient supply chain networks to the success of the broader aims of Future Earth has been established.

References

Beer, T, Lin, B L & McGill, A E J (2016). "Urbanisation, nutrition and food security: a climatological perspective" in Maheshswari, B, Singh, V P & Thoradeniya, B (2016). "Balanced Urban Development: Options and Strategies for Liveable Cities". Springer Verlag. ISBN: 978–3–319–28110.

Bhatt V & Farah L M (2009). "Designing Edible Landscapes". Open House International, **34**: 5–7

Bourlakis, M A & Weightman, P W H (2004). "Introduction to the UK Food Supply Chain" In Bourlakis, M A & Weightman, P W H (eds) "Food Supply Chain Management", Figure 1.1, p 6. Blackwell, ISBN: 1–4051-0168–7

Boyer, K K & Verma, R (2010). "Operations and Supply Chain Management for the 21st Century". South Western Cengage Learning. ISBN-13: 978–0–618–74933-1, p 19.

Cornwall, W (2016). "Efforts to link climate change to severe weather gain ground". Science, **351**, 1249–1250.

Dani, S & Deep, A (2010). "Fragile food supply chains: reacting to risks". International Journal of Logistics: Research and Applications, **13**(5), 395–410.

Dumke, D (2010). "Fragile Food Supply Chains: Reacting to Risks". http://scrimblog.com/review/fragile-food-supply-chains-reacting-to-risks, accessed 26/07/2016.

ETP (European Technology Platform on Food for Life) (2005). The Vision for 2020 and beyond. Brochure published by the European Union through the Confederation of the Food ad Drink Industries (CIAA). Brussels: European Technology Platform on Food for Life, p. 29.

FAO (Food and Agriculture Organization) (1996). World Food Summit, Rome, Italy. 13–17 November.

Future Earth (2013). Initial Design Report, ICSU. ISBN: 978-0-930357-92-4.

Future Earth (2014). Vision – Executive Summary, ICSU. ISBN:978-0-9330357-95-5.

Garfield, L & Jacobs, S (2017). "Kimbal Musk - Elon's brother - just opened a shipping container farm compound in New York". www.businessinsider.com/kimbal-musk-shipping-container-farms-new-york-city-2016-12/.

King, J, Wicker, L & Moskowitz, H (2016). "Influencing consumer choice for healthier products". *Food Technology*, **3**, 41–47.

Maslow, A H (1943). "A theory of human motivation". *Psychological Review*, **50** (4), 370–396.

Mermelstein, N H (2016). "When natural disasters affect food supply". *Food Technology*, **3**, 64–67.

McNutt, M (2016). "Hazards without disasters". *Science*, **353**, 201.

Mougeot, L A J (2005). "Introduction". In: Mougeot L A J (ed) Agropolis. Earthscan, London.

Premat, A (2005). "Moving between the plan and the ground: shifting perspectives on urban agriculture in Havana, Cuba". In: Mougeot, L A J (ed) Agropolis. Earthscan, London.

Rio+20, (2012). United Nations Conference on Sustainable Development. Rio de Janeiro, June.

Rosen, J (2016). "Thinking the unthinkable". Scienc*e*, **353**, 233–237.

Smit, J, Nasr, J & Ratta, A (1996). "Urban agriculture: food jobs and sustainable cities". United Nations Development Programme (UNDP), New York.

Smith, A (2013). "The climate bonus: co-benefits of climate policy". Routledge, London and New York. ISBN: 978-1-84971-341-2.

Tatham, P H & Pettit, S J (2010). "Transforming Humanitarian Logistics: The journey to Supply Network Management." International Journal of Physical Distribution and Logistics Management, **40** (8/9), 609–622.

Thomas, A S & Kopczak, L R (2005). "Understanding Humanitarian Supply Chains", pp. 1–15. In "From Logistics to Supply Chain Management: The Path Forward in the Humanitarian Sector" (2005), Fritz Institute, San Francisco.

Wilcockson, S (2004). "UK Crop Production", Figure 6.1, p. 90. In Bourkalis, M A &Weightman, P W H (eds) "Food Supply Chain Management", Blackwell. ISBN: 1-4051-0168-7.

21

The Impact of Global Climate Change on Nutrition Security: A Multidimensional Challenge

Ibrahim Elmadfa and Alexa Leonie Meyer

21.1 Introduction

The effects of global climate change due to human activities increasingly manifest themselves and pose a serious threat to global food and nutrition security. While their impact on agriculture, marine and aquatic ecosystems and food production is most obvious, the implications for human nutrition go much further because healthy and wholesome nutrition and adequate nutritional status are not just determined by the availability of food. Climate change, and particularly extreme weather events resulting from it, can also reduce food accessibility by impeding food distribution, destroying market and transportation infrastructure as well as leading to price increases. Moreover, climate change can have detrimental effects on human health directly through the exposure to extreme weather conditions as well as indirectly through its influence on food spoilage and pathogens. Poor health, in turn, compromises the ability of the body to absorb and use food and nutrients properly (Met Office & WFP, 2012; IPCC, 2014).

On the other hand, agriculture and food production are also contributors to climate change. The release of CO_2 from deforestation and land-use change, the use of fossil fuels for agricultural purposes and the production of CH_4 originating from ruminant livestock and rice cultures and of NO_2 from fertilizer use are all major sources of greenhouse gas emissions. Attempts to mitigate climate change therefore also have to target agricultural practices, with potential negative effects on productivity emerging as trade-offs (Tubiello et al., 2014; Elbehri et al., 2015).

This double effect of climate change – through its direct impact on food production as well as the need for a more sustainable agriculture – makes it difficult to predict the full effects of all aspects of climate change on global food and nutrition security (see Figure 21.1). However, considering the already-heavy burden of malnutrition, its consequences may prove devastating. Despite a growing prevalence of overweight, underweight and malnutrition currently still affect a large proportion of the global population with an estimated 795 million persons being unable to obtain sufficient food to meet their dietary energy requirements in 2014–2016. An overwhelming majority of them, 780 million, live in developing countries (FAO et al., 2015). Young children less than the age of five years are particularly endangered and threatened in their development with 159 million being stunted (UNICEF et al., 2015). Progress has been made over the last three decades to combat hunger and improve the global nutritional situation, albeit at a slower pace than desired. Since the early 1990s, the prevalence of undernourishment in developing countries has declined from 23% to 13%. The first Millenium Development Goal (MDG 1) of cutting by half the proportion of people suffering from hunger between 1990 and 2015 was achieved by seventy-two countries (FAO et al., 2015). However, great inequalities exist between regions and countries and Southern Asia and Sub-Saharan Africa remain the parts of the world most severely hit by hunger and food insecurity with a steady increase in absolute numbers in the latter (FAO, 2000; FAO et al., 2015).

Climate change and its effects on agriculture and food production are expected to interfere with the endeavours to eradicate hunger and therefore to worsen the nutritional situation. Particularly worrisome is the fact that some regions that are already severely affected will also be those that experience the most negative effects of climate change.

21.2 General Aspects of Food and Nutrition Security

The concept of food and nutrition security has evolved over time. While the production and availability of food

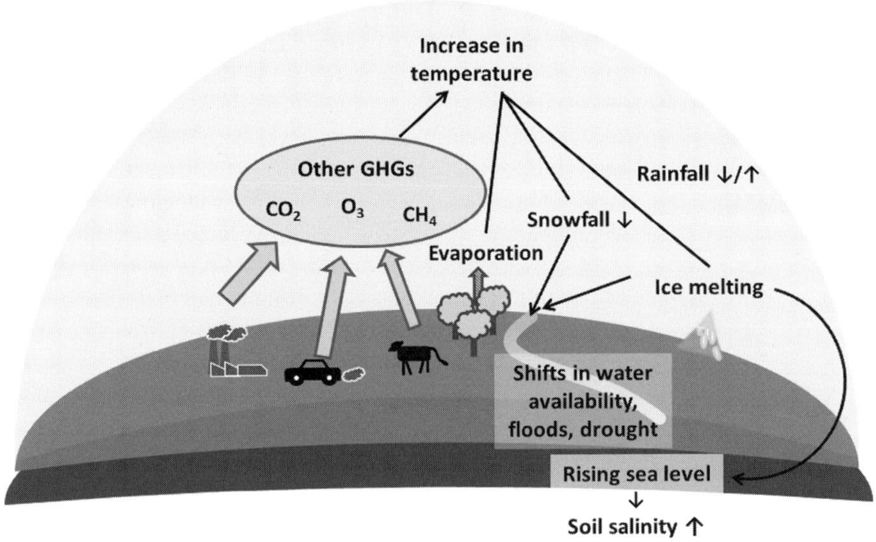

Figure 21.1 Environmental effects of climate change affecting food production.

were originally considered major determinants, the importance of adequate distribution and the economic, infrastructural, demographic and socio-cultural influences on it have been increasingly acknowledged (Pangaribowo et al., 2013).

This is reflected in the definition by the Food and Agriculture Organization of the United Nations (FAO) of food security as "a situation that exists when all people, at all times, have physical, social and economic access to sufficient, safe and nutritious food that meets their dietary needs and food preferences for an active and healthy life" (FAO, 2000). From this definition, four dimensions of food security have been identified: food availability, economic and physical access to food, food utilization and stability over time. In this context, utilization particularly applies to the physiological context and the absorption and metabolism of nutrients by the body (FAO et al., 2014). As depicted in Figure 21.2, the concept of food security is thus expanded to include health aspects as it has been recognized that the mere availability and accessibility of food are not sufficient to ensure an adequate nutritional status at the individual level. The health status of an individual can affect his or her nutrient requirements, food digestion and the absorption and metabolism of nutrients. This is taken into account when using the term "nutrition security". Other important factors are the right choice of food and its preparation, which are, among others, determined by nutritional knowledge. The quality of plant proteins can thus be increased by combining different sources, and appropriate processing techniques can improve the bioavailability of micronutrients. This is of particular

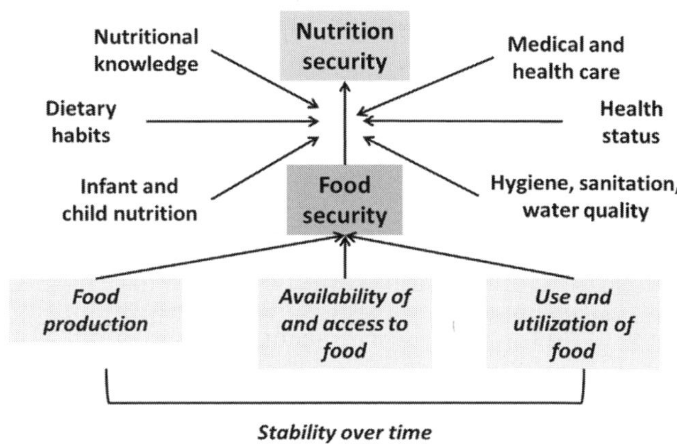

Figure 21.2 Determinants of and factors affecting food and nutrition security.

relevance when resources and the variety of available foods are limited as is the case in many poor communities. Nutrition security also relies on an environment enabling the adequate use of the food available – i.e., sanitary and safe living conditions and good health care (Committee on World Food Security, 2012).

21.3 The Impact of Climate and Atmospheric Changes on Food Production and Availability

21.3.1 Effects of Climate Change on Agricultural Crop Productivity

Among the most immediate threats of climate change to food security are its direct effects on agriculture and food production. Increases in temperature, as well as extreme weather events due to global warming, are already manifesting themselves and are expected to become more important in future decades though displaying large regional differences (Herring et al., 2015). Ambient temperature averages as well as extremes have shown a steady rise since the midst of the nineteenth century that is further accelerating (IPCC, 2007). With regard to agriculture, temperature variability over a year and during the growth cycle of individual crops specifically plays an important role and may exert stronger influences on yield than average temperature (Thornton et al., 2014). This is exemplified by the fact that the emergence of agriculture and our entire civilization was mainly enabled by the stable climate of the last ten thousand years following the highly variable climate of the Pleistocene glaciations (Richerson, 2001).

Between 1980 and 2011, increases of both maximum and minimum temperatures during the growing season were observed in most of the main global agricultural regions with an average magnitude of about 0.3°C and 0.2°C per decade, respectively (Lobell & Gourdji, 2012). As shown in Figure 21.3, drought and heat, and even more a combination of the two, reduce photosynthetic activity, nutrient assimilation, growth and development of plants (Prasad et al., 2008). Despite great differences between species, major crop plants such as wheat, maize, rice, barley, soybeans and peanuts have rather narrow individual optimum temperature ranges and low tolerance for excessive heat or cold during crucial development stages that impairs growth and ripening (Hatfield et al., 2011; Sánchez et al., 2014; García et al., 2015; Hatfield & Prueger, 2015). Growth and maturing, but also senescence are accelerated in various crops and yields are reduced because the plant has less time to develop and assimilate (Hatfield et al., 2011; Asseng et al., 2014). Earlier flowering together with more frequent extreme weather events may moreover increase the risk of frost exposure and damage, at least in certain regions, despite a general tendency towards reduced late frosts (Rochette et al., 2004; Trnka et al., 2011; Hoffmann & Rath, 2013; Legave et al., 2015). On the other hand, a decrease in frost days can have negative effects on agricultural production. Many plants originating in temperate climate zones such as major fruit tree crops require exposure to low temperatures for a certain period for adequate flowering and fruit development in the following spring, a phenomenon known as dormancy. By preventing dormancy, warmer winters may result in lower productivity (Luedeling et al., 2009). Studies from Canada show an association between decreases in cool season vegetable (potatoes, carrots and Brassicaceae varieties) yields and extreme temperatures during the growing season especially in the summer months. Yields of cabbage, rutabaga and cauliflower decreased by 10% for every ten days above 30°C during the growing season (Warland et al., 2006).

Because of the higher variability of precipitation, effects of altered rainfall patterns due to climate change on agricultural production are more difficult to predict. Changes in the occurrence of droughts vary between regions. Increases in intensity and duration have been reported for southern Europe, West Africa and southern Australia and decreases in central North America and northwestern Australia. There is also a probability for future increases in some regions of the world such as southern and central Europe, the Mediterranean region, central North America, Central America and southern Africa that are in part of anthropogenic origin (Seneviratne et al., 2012). The extreme heat and drought over wide parts of Europe in the summer of 2003 may serve

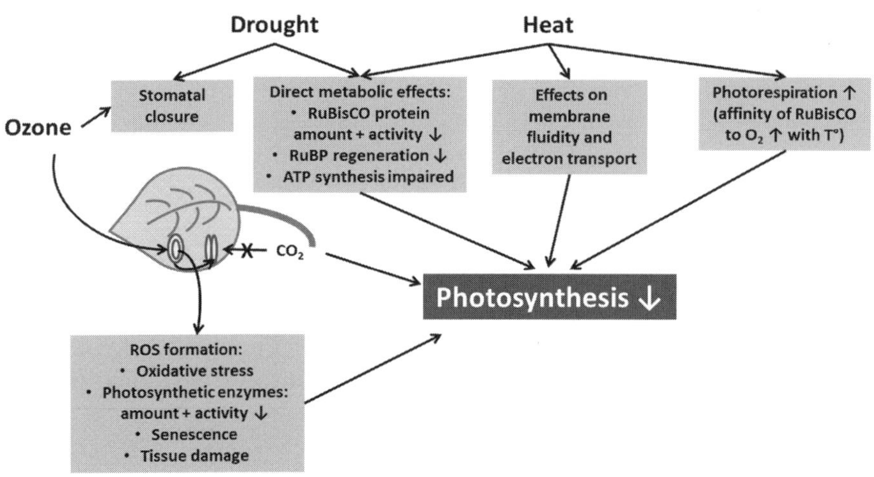

Figure 21.3 Effects of heat stress, drought and ozone on photosynthesis. RuBisCO: ribulose-1,5-bisphosphate carboxylase/oxygenase, RuBP: ribulose-1,5-bisphosphate, ROS: reactive oxygen species.

as an example. They resulted in significant reductions of net primary productivity of the general vegetation and agricultural crops (Ciais et al., 2005). It has been estimated that the global area of drylands has already increased since the 1950s and that most of Africa and large parts of the American continent, southern Europe, the Middle East, Australia and Southeast Asia could become increasingly arid (Dai, 2011). On the basis of a high greenhouse gas emission scenario (RCP8.5), it is projected to increase from an area of about 2.4 million km^2 over the 1991–2005 period to about 5.8 million km^2 by the end of the twenty-first century (Feng & Fu, 2013).

Nevertheless, in cooler regions of the world, warmer climate could have positive effects on agricultural production by increasing the duration of the growing season and enabling the cultivation of crops that are not adapted to low temperatures. This has, for example, been described and projected for regions in northern China with a north- and westward expansion of the area suitable for multiple cropping (Yang et al., 2015). Productivity could also increase in the North and some mountain areas of Europe. However, for the latter, the projected yield increases are thought to be small, as a result of other factors constraining cultivation in northern and mountain regions, so any realised augmentations are unlikely to compensate for losses in the highly productive regions of Europe more severely affected by climate change in the form of heat and drought (Trnka et al., 2011; Elsgaard et al., 2012). Indeed, observations from France suggest that heat exposure during summer has a stronger effect on wheat and barley yields than reduced cold exposure during winter, resulting in an overall trend to lower yields of these cereals (Moore & Lobell, 2015; Gammans et al., 2016). This is even more the case at the global level (Asseng et al., 2015; Moore & Lobell, 2015).

Crop yields are also reduced by the increasing atmospheric ozone (O_3) concentrations, a secondary effect of climate change. O_3 is a strong phytotoxin owing to its prooxidative effects, which cause the production of reactive oxygen species (ROS). ROS damage to proteins, lipids and DNA reduces photosynthesis and biomass production, promotes leaf senescence and impairs root growth (Booker et al., 2009) (see Figure 21.3).

The negative effects of temperature, drought and O_3 are mitigated by higher CO_2 concentrations. However, the extent to which productivity increases depends on additional factors such as the availability of water and nitrogen (Reich et al., 2014), ambient temperature and the type of crop considered. It is also limited by plant acclimation when the transport capacities of assimilates in a plant are surpassed (Long et al., 2004; Ainsworth & Rogers, 2007; Leakey et al., 2009). Moreover, higher CO_2 concentrations do not have only positive effects; they also reduce the nutritive quality of crops and their pest resilience, as will be discussed later. In addition, a stimulation of weed growth by CO_2 must be taken into account. Added to the variable effects of O_3 on different plant species, it could disturb the equilibrium between weeds and crops and shift the balance in favour of the former by making them more competitive (Peters et al., 2014).

At the global level, depending on the model systems used, predictions about future changes in crop yields as a measure of productivity give very variable results. When considering climate change in isolation, Müller and Robertson (2014) found that for the Representative Concentration Pathway (RCP) 8.5 scenario, with the highest projected greenhouse gas emission and radiative forcing, forecast reductions were up to 21% for global wheat production, up to 18% for rice, up to 24% for groundnuts, up to 30% for soybeans and up to as high as 38% for maize in a study based on combinations of two different climate models and two global agricultural growth models. These estimates overlap with the upper end of the range of outcomes according to Lobell and Gourdji (2012), who proposed global crop yields could decrease by 3% to 8% per 1°C of temperature increase, leading to a decline of 0.3% to 4% for a plausible range of 0.1°C to 0.5°C of temperature increase. It must, however, be kept in mind that these models consider average climate developments only. Potential effects of extreme weather events were not integrated, although they can have significant impacts on food and nutrition security, as pointed out later in more detail.

Nevertheless, integrating increases in CO_2 concentrations into projection models on crop production changes leads to a more positive outlook with increases in yield of 1% to 2.7% for a plausible range of CO_2 increase (Lobell & Gourdji, 2012). In cooler areas of the temperate zone, the joint effects of milder temperatures and CO_2 fertilization will likely cause larger increases in crop productivity. On the basis of four different climate change scenarios adapted to the European region, Iglesias et al. (2012) project increases in yield of wheat, maize and soybean of 34% to 54% for boreal Scandinavia, of 11% to 33% for the Continental South region and of 20% to 23% for the Alpine region but declines of up to 22% to 27% for the Mediterranean and South Atlantic regions (Iglesias et al., 2012). However, any positive effect of moderate climate change would probably be lost with temperatures increasing to or above 4°C when compensation through CO_2 fertilization and genetic crop improvement reach their limits (Rötter et al., 2011).

In its Fifth Assessment Report, the Intergovernmental Panel on Climate Change (IPCC) projects a mean rise in ambient temperature of 2°C or more for large parts of the African continent at a rate that will exceed the global average (Niang et al., 2014). Together with changing precipitation patterns, the temperature increases will have profound impacts on the production of major food and feed crops. Regardless of the variability among different climate scenarios, reductions in the production of important staple food crops have been projected, among others, for wheat in Sudan, sorghum in Nigeria and maize in Côte d`Ivoire and Ghana (Jalloh et al., 2013; Waithaka et al., 2013).

A major concern across this work is the uncertainty around whether future crop production will be able to cover the increasing demand for food due to population growth but also increasing life standards (Porter et al., 2014). Higher affluence in many emerging countries is associated with higher consumption of meat, for instance, thereby raising the demand for grain for feed (Myers & Kent, 2003). This will drive market price increases even further and worsen food access of the poorest people. It has been estimated that climate change could increase the global prevalence of hunger by 10% to 20% in 2050 and the number of malnourished children by 21% (Parry et al., 2009).

21.3.2 Effects on Animal Food Production

About a quarter (24% by mass) of major crops and about 36% of the cereal produced globally are used as animal feed (Alexandratos & Bruinsma, 2012; Cassidy et al., 2013). Effects of climate change on crop yields and composition will have impacts on animal food production as well by modifying feed availability and quality. Because of the variability of climate change–related changes in crop production projected by different models it is difficult to estimate the effects of climate change on animal food production. One of the most recent and most comprehensive attempts performed at global and regional levels concludes that while the overall effects of climate change on the livestock sector will be limited compared to other influencing factors, changes in grassland yields could have a stronger influence on livestock production, especially in the case of ruminants and in certain regions. This is of relevance as climate effects are projected to be more beneficial on grassland than on crops. Effects of increases in grass yields would be largest in arid areas of Sub-Saharan Africa and South Asia, where the greatest rise in ruminant meat demand is also projected. However, productivity in milk and meat would be rather low under these circumstances, making intensive animal husbandry impossible (Havlik et al., 2015).

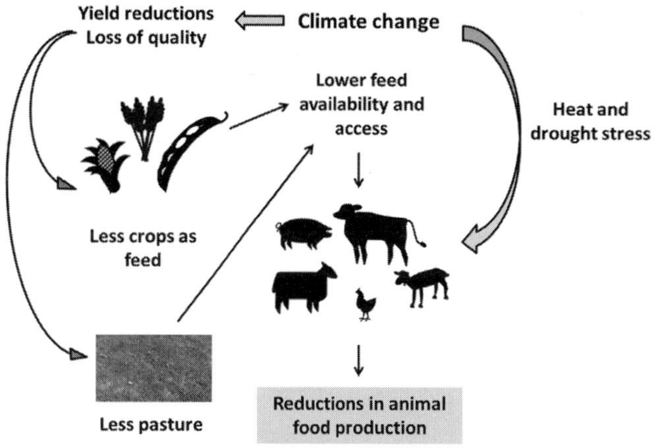

Figure 21.4 Impacts of climate change on animal food production.

For example, increases in grassland productivity due to climate change and CO_2 fertilization could balance if an expansion of pasture grounds predominantly occurs in lower-yielding areas though the opposite development under negative projections is very unlikely. However, this model projects a more positive climate effect on grassland than crop productivity that could promote production of milk and ruminant meat from grazing animals. The share of ruminants raised on grass could thus almost double (from 20% to 38%) under the most positive scenario (Havlik et al., 2015). An overview of the negative impacts of climate change on animal food production is given in Figure 21.4.

The projected temperature increases also have a direct impact on animal performance and health. While they are obviously a threat in low-income communities in arid regions they already occur in temperate regions. Increases in heat waves will pose pressure on farmers by exposing their animals to heat stress that particularly affects high-yielding animal breeds with a high muscle mass and rapid growth leading to greater metabolic heat production. High temperatures are also associated with reduced feed intake, higher infection rates and impaired disease resistance in animals (Nardone et al., 2010; Babinszky et al, 2011).

On the other hand, as exemplified in Figure 21.5, livestock production, especially of ruminants, itself accounts for more than 50% of anthropogenic greenhouse gas emissions from agriculture, forestry and land use change, with large parts resulting from energy and feed inputs (FAO, 2006). A reduction of the current high consumption level in most developed countries is therefore a major step towards climate change mitigation that would even be associated with beneficial health effects (Dietary Guidelines Advisory Committee, 2015). However, these foods are also a source of critical

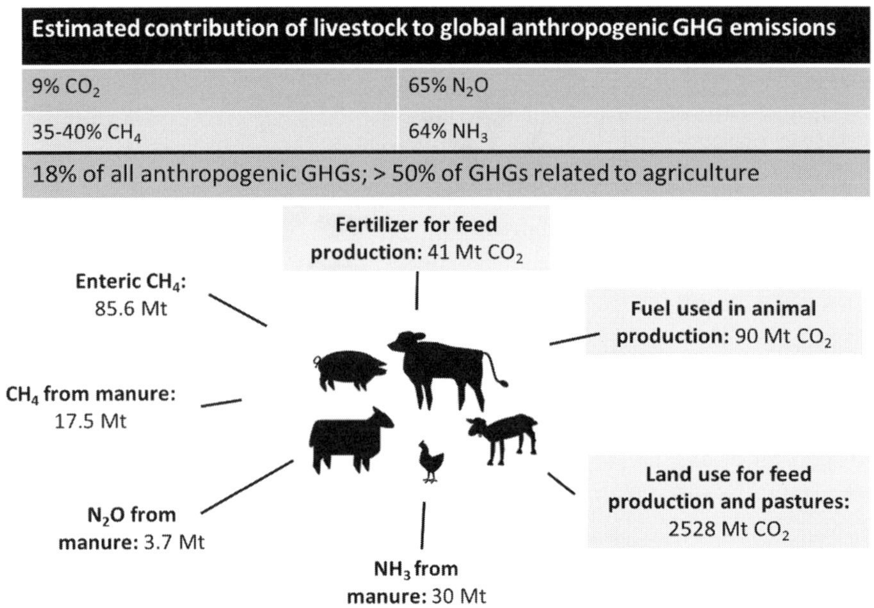

Figure 21.5 Livestock as a contributor to anthropogenic greenhouse gas emissions (based on data taken from: FAO: Livestock's long shadow, 2006).

nutrients such as iron, zinc and vitamins of the B group, of which intake is often not optimal (Elmadfa et al., 2009; Elmadfa & Meyer, 2017). This is even more the case in low-income countries where livestock is an essential basis of income and food security for certain communities (Herrero et al., 2013).

Moreover changes in agricultural and land-use practice also offer unique opportunities for climate change mitigation through sequestration of CO_2. This applies especially to grassland that can thus provide a basis for sustainable livestock husbandry (Herrero et al., 2016).

Potential threats to food security also arise from the effects of climate change on fish production. As ectothermic organisms fish and molluscs are particularly sensitive to temperature changes and many, especially those living in cold environments, have very narrow optimal temperature ranges (Pörtner, 2002). Besides direct effects on the physiology and behaviour of marine organisms, climate change acts indirectly by altering the ecosystem where a given species is living. Warming of the oceans is associated with changes in the distribution of fish and the composition of marine fauna, with a trend towards northward migration or into deeper waters leading to a subtropicalization of northern seas (Cheung et al., 2013; IPCC, 2014; Montero-Serra et al., 2015). Increased stratification leads to perturbations in nutrient supply with consequences for primary production (FAO, 2009; IPCC, 2014). While it is difficult to predict future trends of fish and seafood production, a number of projection models suggest major declines in many important fish production areas across the globe, namely, tropical and subtropical regions such as the Indo-Pacific, but also the Eastern Atlantic. In turn, productivity is thought to increase in cooler environments such as the Arctic mostly as a result of warming and a decline in ice coverage. These projections are supported by recent observations (FAO, 2009).

Besides increases of the water temperature, acidification caused by the uptake of atmospheric CO_2 by the oceans poses another threat to marine organisms. Despite a high spatial and temporal variability, the average pH of surface ocean water has declined from a preindustrial value of 8.21 to the current 8.1, corresponding to an increase in $[H^+]$ concentration by about 30% (The Royal Society, 2005). Ocean acidification was found to alter the biochemical composition of phytoplankton and copepod species and reduce the trophic transfer between these organisms (Cripps et al., 2016). Through its effects on calcium carbonate saturation of the water, ocean acidification for instance affects calcifying and shell-building marine organisms, among which are important human food sources and corals acting as habitat providers (The Royal Society, 2005; FAO, 2009).

Capture fishery is particularly vulnerable to climate change as it depends on the natural environment. However, aquaculture, which is playing an increasingly important role in the supply of fish and seafood for human direct consumption (FAO, 2016), is also affected by climate change. Notably, a large proportion of aquaculture production systems, especially in low-income countries, rely on natural environments such as lakes, coastal basins or rice fields and are therefore highly dependent on environmental stability. Drying of lakes and increased water use for irrigation pose a threat to both freshwater and aquaculture systems. The availability of fishmeal and fish oil as feed for aquaculture is also affected by climate change (FAO, 2009).

Even though ecosystems may be able to adapt, the effects of climate change on marine and aquatic food webs will have significant impacts on food security as fish and seafood are an important source of energy, protein, fatty acids and micronutrients for a large proportion of the global population, especially in many low-income countries (FAO, 2016). For more detail on this subject, the reader is referred to Chapter 23.

21.3.3 Food Losses and Food Safety Issues Due to Extreme Weather Events

In addition to the long-term effects of changes in mean ambient temperature, precipitation and atmospheric composition on crop yields, acute extreme weather events such as floods, storms and wild fires present a threat to food security and are also projected to increase as a result of climate change (Seneviratne et al., 2012). According to estimates from the FAO, damage and losses to agriculture made up about 25% of all damage and losses due to natural disasters related to climate change in developing countries between 2003 and 2013. As shown in Table 21.1, crops and livestock were by far the most affected (42% and 36% of all agricultural losses). The majority of crop losses (60% of total) was attributable to floods while drought caused the greatest losses among livestock (86% of total) (FAO, 2015).

In addition to crops in the fields, stores of already harvested food can be destroyed and losses of seeds for planting can endanger food security over the longer term. In this context, emergency help in the form of seed aid may even exacerbate insecurity and cause new problems when it is not appropriately timed, the distributed seed is not adapted to the new environment or it disrupts local seed procurement systems or introduces exogenous pests and weeds (Remington et al., 2002; Sperling & McGuire, 2010).

21.3.4 Climate Change and Its Impact on Pests and Diseases – Threats to Food Security

More losses of food are also projected to result from changing pest and parasite occurrence. Changing climatic conditions have effects on the distribution of insects including agricultural pests and animal parasites. Higher average temperatures will lead to larger populations of certain insects especially when winters become milder, enabling a better survival of insects and their larvae and an earlier appearance in spring. This has been described for some aphid species acting as major carriers of various crop viruses such as cereal yellow dwarf viruses or potato viruses, thereby causing yield losses (Gregory et al., 2009). Shifts in the insect population due to climate change are particularly relevant when pests gain an advantage over their natural enemies (Bale & Hayward, 2010; Thomson et al., 2010). Moreover, changing climate can create living conditions for species that were previously not able to subsist in a given habitat (Bebber et al., 2013; Olfert & Weiss, 2006). McMurray et al. (2013) warned that the cultivation of new crops as an adaptation to altered climate could also introduce new pests and diseases that might ultimately also affect the indigenous crops and ecosystem. Resident plants may not be able to cope with new herbivorous insects and plant resilience has been shown to be affected by heat stress (Zhu et al., 2010). This effect has also been observed for higher CO_2 concentrations under which the production of proteinase inhibitors was reduced in soybeans (Zavala et al., 2008). Although increased CO_2 also induces changes in crop physiology and morphology that make the plant less

Table 21.1 *Damage and Losses Caused by Natural Disasters to Agriculture in 2015*

		Storms	Floods	Droughts	Tsunamis	Earthquakes
Subsector	Share by subsector (% of total)	Average Contribution to Total Damage and Losses (%)				
Crops	42	22.8	59.6	15.1	1.7	0.7
Livestock	36	3.2	8.4	85.8	0.2	2.3
Fisheries	6	16.4	9.8	3.1	70.2	0.4
Forestry	2	89.1	5.0	0	3.2	2.7
Irrigation	4					
Other	10					

Source: FAO. The impact of natural hazards and disasters on agriculture and food security and nutrition, 2015.

nutritious to pests (lower nitrogen content, increased leaf toughness), many of these effects are diminished by higher temperature (Zvereva & Kozlov, 2006).

Exposure to drought also influences the contents of nutrients in plants, on the one hand, and that of phytochemicals serving as insect repellents, on the other hand. Thus, water deficiency is associated with increases in free amino acids and sugars in some crops that make them more nutritious for insects. In turn, concentrations of repellents are often reduced. Drought also has direct effects on the occurrence of insect pests such as locusts. For instance, it can negatively affect pathogens that would otherwise control the harmful species (Showler, 2013).

Increased occurrence of pests and fungi would entail higher uses of pesticides and fungicides and result in an increased exposure of consumers to these latter. Moreover, the climate – temperature, precipitation and wind – and other environmental factors such as soil properties, water and vegetation that are affected by climate change also influence the emission and distribution of contaminants such as polycyclic aromatic hydrocarbons or toxic heavy metals, influencing their occurrence in foods (Miraglia et al., 2009; Ziska et al., 2016).

Climate-related increases in pathogen occurrence and infectious diseases also threaten animal food production. Warming and altered rainfall directly influence the survival and development of pathogens and parasites and the insects transmitting them, thereby causing shifts in their geographical distribution. Rises in seafood poisoning due to warming of coastal regions have also been projected (van der Fels-Klerx et al., 2012). While extreme heat and drought can reduce the survival of some pathogens, floods are associated with a higher risk of disease outbreaks as they offer optimal conditions for a number of pathogens and vectors such as mosquitoes and spirochetes. Up to a certain level, higher temperatures can also hasten the reproduction of some viruses and microorganisms. Infections are further promoted by the negative effects of heat stress on the immune defence of animals. Moreover, the transmission of disease is also influenced by behavioural changes due to climate change, leading to increased migration and crowding of persons and their livestock in certain places. Adaptation strategies to climate change aimed at reducing the impact of animal keeping on greenhouse gas emission might also facilitate the spread of infections, e.g., by altered pasture schemes and exposure to pathogens (Skuce et al., 2013; Bett et al., 2017).

Increased contamination of feed and food by fungi is another potential threat arising from climate change. The climatic conditions in the field as well as during storage influence not only the growth of the fungi but also the production of mycotoxins. Higher temperature and reduced precipitation are projected to promote the expansion of species from warmer to cooler regions, changing the geographic distribution of different fungi and mycotoxins in foods (Paterson & Lima, 2011). Recently, contamination of commodities with aflatoxins from *Aspergillus* species has been reported in regions of Europe where it was previously uncommon (Miraglia et al., 2009; Battilani et al., 2016). This is of particular concern considering the genotoxicity and carcinogenicity of various aflatoxins produced by these fungi. Drought stress was found not only to make plants such as maize more susceptible to infections by mycotoxigenic fungi but also to stimulate toxin production in these latter (Medina et al., 2014). Increased precipitation could also favour fungal growth and pose particular problems for the storage of feed and food commodities (Paterson & Lima, 2011).

Higher ambient temperatures, changes in precipitation and water availability as well as extreme events such as floods drive the development and spreading of food-borne infectious diseases. The occurrence of infections caused by major microbial pathogens found in foods such as *Salmonella, Campylobacter* and *Escherichia coli* correlates with higher temperatures and season. It is also promoted by behavioural factors such as a higher consumption of raw foods, in particular fruits and salads, observed under warm weather conditions (Lake et al., 2009; Smith et al., 2014; Ziska et al., 2016). Flooding causes contamination of drinking water and food through spill-over of industrial and agricultural waste water (Ziska et al., 2016). Although recent improvements in food hygiene at the production as well as the household level may be able to defuse somewhat or at least control the threat from food-borne infections (Lake et al., 2009), the required technological, pharmacological, logistic and regulatory means are less available to low-income communities that are more affected by climate change (Kitinoja et al., 2011; Grace, 2015).

Infectious diseases, especially when associated with gastrointestinal symptoms such as diarrhoea and vomiting, have further negative effects on nutrition security by impairing the absorption of nutrients and water. Some also impair the immunological functions of the diseased, making them further susceptible to infections. Vulnerable groups such as children and elderly adults are at special risk (Katona & Katona-Apte, 2008).

21.4 Reduced Access to Food and Lower Stability Due to Climate Change

Access to food is another crucial pillar of food and nutrition security besides food production and availability. Climate change affects access to food in many ways. Decreases in crop yield and quality, on the one hand, and a higher demand for food, on the other, lead to price increases. The World Bank reported an increase of its food price index by 15% between October 2010 and January 2011 that was mostly attributable to increases for sugar (20%), fats and oils (22%), wheat (20%) and maize (12%) that were partly due to weather-induced yield reductions (World Bank, Poverty Reduction and Equity Group, 2011). More recently, prices have declined again, also in relation to gains in productivity (World Bank, 2015). Because of the many factors involved, it is difficult to predict price developments in the decades to come. Projected increases are generally obtained with more pessimistic scenarios not taking into account the effects of higher CO_2 levels on crop yields (Porter et al., 2014). Nevertheless, impacts of climate variability on global and local food prices are undeniable.

Projections of the extent of future increases for different crops have been attempted on the basis of different scenarios of future climate change as well as economic and socio-demographic developments. They give a very broad range of potential trends: Calculations by Nelson et al. (2010), for example, suggest price increases for the three major crops (wheat, maize and rice) until 2050 from 31.2% for rice up to 106.3% for maize, depending on the scenario. The authors note that prices are likely to rise in all models as opposed to the price declines seen in the twentieth century. The impact of climate change becomes maybe most obvious when price changes for some major food commodities between 2050 and 2010 are projected for different possible future climate scenarios: Based on different models taking climate change into account future crop price increases may range from 34.1% for rice under an optimistic scenario to 106.3% for maize under a pessimistic scenario. In turn, without the effect of climate change, price increases under different scenarios are projected to range from 18.4% to 34.1% only.

Higher prices for agricultural commodities mean higher income for the producing farmers and can thereby act as a driver for increases in crop yields by making farming more profitable, enabling higher gains for farmers and thereby greater investments in productivity. However, price increases worsen the situation of vulnerable population groups that spend a large part of their low income on food and are highly dependent on purchased food available at low prices. In this way, they particularly threaten food security in countries and communities that are not self-sufficient, typically hitting the poorest socioeconomic strata (who are the least able to cope) the hardest (Prakash, 2011). Accordingly, many developing countries have aspired to higher self-sufficiency and have restricted exports. However, this closure of markets inhibits inputs in productivity and results in increased vulnerability to climate change–related crop failures (Abbott, 2009). Indeed, global trade can help level out price increases due to local food crop failures and prevent food shortages, even though it exposes producers and consumers to global price fluctuations (Elbehri et al., 2015).

The increasing use of oilseed and cereal crops for biofuel production is another driver of food price increases leading to competition for arable lands. This situation poses an additional threat to food production, which is further enhanced by policies such as the European Union Biofuels Directive aimed at increasing its use (Banse et al., 2008; World Bank, 2011), necessitating that more agricultural land be used for fuel production (see Table 21.2).

Table 21.2 *Biofuel Production in 2015*

Total amount produced[1]	Proportion of different biofuels[1]	Leading producers, % contribution (major crops used)[1]	Proportion of globally produced crops used for biofuel production (2016)[2]
133 billion liters	Bioethanol: 62%	USA: 57% (maize) Brazil: 29% (sugarcane) EU: 4% (wheat) China: 3% (wheat, maize) Canada: 2% (wheat, maize)	Maize: 16% Wheat: 1% Sugar (cane and beet): 20% Plant oils: 13%
	Biodiesel: 24%	EU: 38% (esp. DE 9% and FR 8%) (rapeseed) USA: 16% (soybean) Brazil: 14% (soybean)	
	Others: 14%		

Sources: [1] World Bioenergy Association. Global Bioenergy Statistics 2016
[2] OECD-FAO Agricultural Outlook 2016–2025

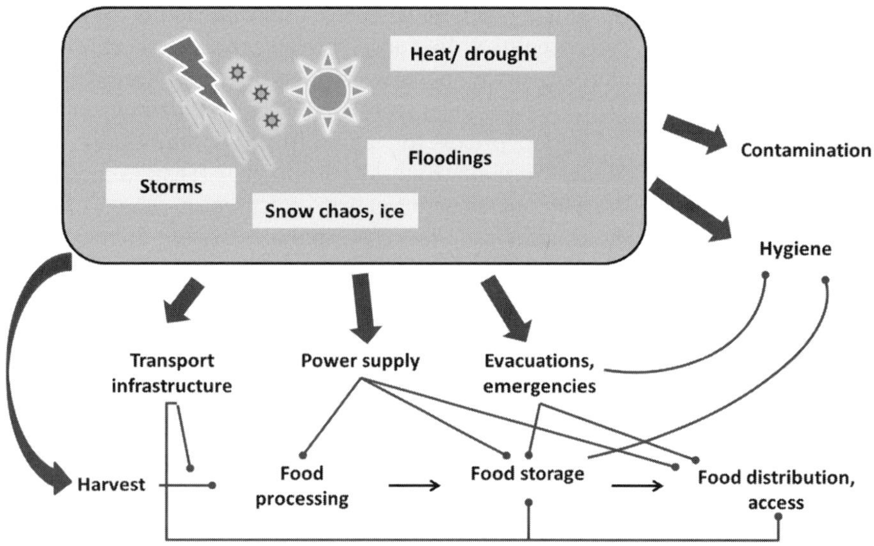

Figure 21.6 Impact of extreme weather events on food security; dotted lines symbolize negative effects on food supply processes.

While price increases resulting from lower and more unstable food production are unlikely to result in food shortages in developed countries, they would narrow the choice of foods available to many less affluent consumers. This could result in a lower consumption of healthy foods such as fruits and vegetables, which are generally more expensive than highly processed energy-dense foods with low micronutrient content (Drewnowski & Darmon, 2005; Lake et al 2012). Price increases for staple foods are even more problematic as they may result in Giffen behavior (Jensen & Miller, 2008) leading to an even greater substitution of these foods for higher-quality but more expensive commodities. Among the poorest people, lower consumption of cheap staple foods would eventually result in malnutrition (Jensen & Miller, 2008).

Increasing climate instability produces an associated rise in extreme weather events, some of which can already be observed, though future predictions of impacts are highly uncertain. As shown in Figure 21.6, extreme weather events restrict access to food in a more direct physical way, by destroying and disrupting roads and markets and interrupting food supply. This impact can have greater consequences for food security than losses of crops inasmuch as it also complicates delivery of food aid (Gregory et al., 2005; Ziska et al., 2016). While low-income countries typically face the greatest difficulties in maintaining food security under such circumstances, negative effects of extreme weather events on food supply and access present a threat also to developed countries, and the need for adaptation of transport and retail infrastructure has been recognised by governments (White et al., 2011; U.S. Department of Transportation, 2014; Ziska et al., 2016).

Food distribution and storage in retail stores are also affected by climate change, as higher ambient temperatures would see an increasing need for cooling even of food items that could previously be handled at ambient temperatures. While use of the cold chain is common in high-income countries, it is less available to low-income countries (where it is often actually more typically required to prevent extensive spoilage). Adoption of adequate technologies is recommended to reduce food waste but has to be planned thoroughly to be cost-effective, sustainable and acceptable (Kitinoja et al., 2011).

Access to food is even more limited to displaced persons. Migration and displacement due to climate change and natural disaster have increased by about 60% since the 1970s. In 2014 alone, it has been estimated that more than 19.3 million persons from one hundred countries lost their homes as a result of natural disasters. While the majority of cases are from developing countries, developed countries are also affected (Yonetani, 2015). Conflict arising from environmental deterioration acts as a further cause of displacement that could increase because of climate change (Adger et al., 2014). Low-lying countries also face displacement due to rising sea levels – which directly reduce available coastal lands and also cause rising salt tables, degrading soil productivity (Khanom, 2016; Teh & Koh, 2016).

Disasters such as floods along with rising sea level in coastal regions pose a threat to water hygiene and quality (Vineis et al., 2011). Rising sea level is associated with increased salinity of freshwater from underground resources and rivers in coastal areas, as evidenced by a study from Bangladesh. Salt content of the water was especially high during the dry season, resulting in daily intakes of up to 16 g of NaCl with 2 liters of drinking water compared to 1.2 g during the monsoon season. The estimated high intake from groundwater and river water was mirrored by high urinary Na excretion by pregnant women drinking shallow well water compared to rain and filtered water. There was also a seasonal increase in the occurrence of hypertension during the dry season that coincided with high water Na content (Khan et al., 2011).

21.5 Changes in Food Composition and Nutritional Quality Due to Climate Change

The environmental conditions under which plants grow influence their composition and, hence, the nutritional quality of food crops. Table 21.3 summarizes some of the climate change–related effects on nutrients and other components reported in the literature.

While exposure to sunlight is often associated with higher sugar content, lower acidity and hence, better taste in fruits as well as higher concentrations of certain vitamins, excessive irradiation and very high temperatures are less beneficial with regard to quality (Moretti et al., 2010). Abiotic stress such as heat and drought increases the generation of ROS in plants that, in excessive amounts, cause oxidative damage. Plants protect themselves by producing various antioxidative agents, among which are ascorbic acid (vitamin C), tocochromans (vitamin E), carotenoids and various secondary plant compounds including polyphenols (Das & Roychoudhury, 2014; Choudhury et al., 2016). Up to a certain level, heat, drought and excessive radiation increase the content of these substances (reviewed by Wang & Frei, 2011). However, direct comparisons between fruits and vegetables exposed to different temperatures during the growing season showed lower contents of antioxidants when the optimal temperature was exceeded. For example, vitamin C (ascorbic acid) and various carotenoids are reported to be lower in tomatoes at high temperatures (Rosales et al., 2006; Gautier et al., 2008). A study on the combined effects of light exposure and temperature on vitamin C synthesis and metabolism in tomatoes revealed that light induced the enzymes involved in the synthesis and recycling of ascorbic acid but that the expression and activity of some of the enzymes decreased with higher temperatures (27°C and 31°C) resulting in lower ascorbic acid contents in the fruits (Massot et al., 2013). The effects of temperature differ among plant varieties: Thus, concentrations of lutein and β-carotene in kale showed a significant linear increase when the air temperature was raised from 15°C to 30°C while an increase from 10°C to 25°C resulted in a significant decrease of carotenoids in spinach (Lefsrud & Kopsell, 2005).

The vitamin E profile of Emmer wheat seeds differed between plants grown in a hot and dry or a more temperate and mesic climate, showing higher total tocochroman contents and a higher proportion of tocotrienols in the seeds exposed to the stressful conditions (Watts et al. 2015). Higher α-tocopherol levels were measured following drought exposure in some potato cultivars (Andre et al., 2009). Changes in tocochroman pattern were also observed in barley, rice and soybeans grown under high temperature and/or limited water availability (Britz et al., 2007; Chennupati et al., 2011; Shen et al., 2013). In soybeans, higher growing temperatures were associated with higher contents of α-tocopherol and lower contents of γ-tocopherol while the opposite pattern was observed for lower temperatures (Carrera et al., 2011). While α-tocopherol is the vitamer with the highest vitamin E activity, the other forms have various biological effects as well, with potential implications for human health. The content of vitamin E and its composition are also relevant for the protection of fatty acids from oxidation and the stability of lipids (Yoshida et al., 2007).

Despite having no nutritive value, secondary plant compounds have shown a wide array of health effects in humans and animals that are only partly due to their antioxidant properties. Climate-related effects on these substances might therefore be relevant from a health perspective. In soybeans, total isoflavone contents decreased with rising ambient temperatures (Carrera et al., 2011). Heat stress (25°C at night and 33°C during the day) was related to lower contents of isoflavones in soybeans, especially following exposure during the reproductive stages (flowering, seed formation and maturation) (Chennupati et al., 2011). Exposure of the plants to 28°C during seed development resulted in a decrease of total isoflavone content by 87%, an effect that could be mitigated by higher CO_2 concentration and exposure to drought (Caldwell et al., 2005).

Exposure to elevated CO_2 concentrations was also associated with higher contents of other secondary plant compounds such as polyphenols, phenolic acids, flavonoids and anthocyanins in strawberries, lettuce and Chinese cabbage, grapes and tomatoes (Wang et al., 2003; Fu et al., 2015; Dannehl & Josuttis, 2014). Moreover, these studies also reported higher amounts of nutritive antioxidants, vitamin C and different carotenoids in the fruits and vegetables exposed to higher CO_2 levels (Wang et al., 2003; Fu et al., 2015; Dannehl et al., 2012). However, vitamin C contents of lettuce and Chinese cabbage declined again at very high levels of CO_2 (>2000 ppm) (Fu et al., 2015). Moreover, grains of rice

Table 21.3 *Effects of Environmental Consequences of Anthropogenic Climate Change on the Nutritional Composition and Quality of Some Crop Plants*

Component	Heat stress	Drought	Elevated [CO_2]
Protein	↑ (Concentration effect) Leaf senescence mobilizes N for protein synthesis[1–3]		↓ (Dilution effect)[1,3–5]
Lipids	↓ Total lipids, ↑ SFA, ↓ PUFA in soybeans[6] ↑ Total lipids in avocado and soybeans[7–9] ↓ Total lipids in soybeans (very high T°),[8] in rapeseed (esp. PUFA)[10] ↑ SFA ↑↓ MUFA ↓ PUFA[6,7,9]	↑ Lipid peroxidation[11]	
Carbohydrates	↓ Starch in cereals and pulses[1,12] ↓ Sugar contents in kiwi fruits[13]	↓ Starch in cereals[1] ↑ Sugars in tubers and some fruits and vegetables[1]	↑ Starch in potatoes,[14,15] slightly ↑ in wheat[4]
Minerals		↓ or ↑ Different minerals (altered absorption and transport)[1]	↓ Fe and Zn in cereals and pulses[16] ↓ Ca, Mg, Fe, Zn, Mn in wheat[4] ↓ Ca and K in potato[17]
Vitamin E	↑ Total content[1,18,19] Changed vitamer pattern[18–21] ↑ γ-Tocopherol ↓ α-Tocopherol in soybeans grown under cooler temperatures[9]	↑ Total content[1,18,19] Changed vitamer pattern[18–19] ↑ α-Tocopherol in some drought-stressed potato cultivars[22]	↓ In rice grains, ↑ in the bran and husk[23]
Vitamin C	↓ In tomatoes[25,25] and in kiwi fruits[13]	↑ In various crops[1]	↑ In strawberries[26] ↑ In Chinese cabbage, ↓ at very high levels (>2000 ppm)[27]
Carotenoids	↓ In tomatoes[11,24] and spinach[28] ↑ in kale[28]	↑ (↓) In some potato cultivars[22]	↑ In tomatoes at optimal temperature[29]
Bioactive secondary plant compounds	↑ Polyphenols in lettuce and spinach[1] ↓ Total isoflavones in soybeans[9]	↑ Phenolic acids in some potato cultivars[22]	↑ Phenolic acids, flavonols and anthocyanins in strawberries[26] ↑ Polyphenols and flavonoids in lettuce and Chinese cabbage but decrease at very high levels (>2000 ppm)[27] ↑ Polyphenols in various fruits[30]
Antinutritive substances		↑ Glycoalkaloids in some potato cultivars[22]	↑ Phytate in sorghum (trend)[16] ↑ Cyanogenic substances in cassava leaves[31]

Note: Superscript numbers indicate sources as listed in the following, ↑, increase of the component; ↓, decrease of the component; SFA, PUFA, .
[1]Wang & Frei, 2011; [2]Triboi & Blondel, 2002; [3]Pleijel et al., 1999; [4]Högy & Fangmeier, 2008; [5]Taub et al., 2007; [6]Dornbos & Mullen, 1992; [7]Woolf et al., 1999; [8]Thomas et al., 2003; [9]Carrera et al., 2011; [10]Namazkar et al., 2016; [11]Rosales et al., 2006; ; [12]DaMatta et al., 2010; [13]Richardson et al., 2004; [14]Donnelly 2001; [15]Vorne et al., 2002; [16]Myers et al., 2014; [17]Fangmeier et al., 2002; [18]Shen et al., 2013; [19]Watts et al., 2015; [20]Britz et al., 2007; [21]Chennupati et al., 2011; [22]Andre et al., 2009; [23]Goufo et al., 2014; [24]Gautier et al., 2008; [25]Massot et al., 2010; [26]Wang et al., 2003; [27]Fu et al., 2015; [28]Lefsrud & Kopsell, 2005; [29]Dannehl et al., 2012; [30]Dannehl & Josuttis 2014; [31]Gleadow et al., 2009

plants exposed to elevated CO_2 (550 ppm) contained lower levels of total polyphenols and flavonoids as well as of vitamin E (tocopherols and tocotrienols). Among the latter, reductions were most pronounced for α-tocotrienol (69%), γ-tocotrienol (46%) and α-tocopherol (38%). Changes were observed in all rice milling fractions (white and brown rice, bran and husk) with a lower total antioxidative capacity in the exposed samples. In the bran and husk, however, vitamin E levels increased upon higher CO_2 exposure (Goufo et al., 2014).

Effects of environmental stresses on macronutrient composition of important food crops have been reported for a number of crops, but are most notable for lipids and carbohydrates. Exposure to high temperatures and water scarcity were most often associated with lower concentrations of both nutrient classes, although increases have also been observed in some instances (reviewed by Wang & Frei, 2011 and DaMatta et al., 2010). Higher temperatures were associated with lower concentrations of some soluble sugars and starch in wheat and legumes. Heat treatment (up to 30°C) of kiwi plants during fruit development and starch accumulation resulted in greater vegetative growth and redirection of carbohydrates from the fruits to the vegetative parts. Sugar, starch and ascorbic acid levels were reduced in the fruits. Furthermore, there was a carryover effect to the next season with reduced flowering and higher vegetative growth probably due to altered carbohydrate storage and availability (Richardson et al., 2004). Notably, an increase in raffinose reported for red kidney beans might impair their digestibility (Thomas et al., 2009).

More pronounced than the effects on lipid quantity were those on quality and fatty acid composition, as shown in Figure 21.7 (Wang & Frei, 2011). Regardless of a high variability of results, there is an overall trend towards reductions in polyunsaturated fatty acids in favour of the monounsaturated and saturated forms by exposure to high temperatures and water scarcity. This has been reported for high-oleic sunflower varieties, soybeans, rapeseed and avocado (Flagella et al., 2002; Thomas et al., 2003; Dornbos & Mullen, 1992; Carrera et al., 2011; Woolf et al., 1999; Namazkar et al., 2016). This observation is in accordance with the greater susceptibility of polyunsaturated fatty acids to oxidative damage. Impacts of exposure to elevated CO_2 levels are more variable (Namazkar et al., 2016).

The composition of fatty acids in the diet has a number of health effects, namely, through its influence on the blood lipids, cardiovascular health, glucose metabolism and inflammatory events in the body. Polyunsaturated fatty acids are associated with beneficial effects and a lower risk for metabolic diseases contrary to certain saturated fatty acids (FAO, 2010).

In turn, the amount of protein in food plants has been reported to increase under stressful environmental conditions. However, this has been ascribed to the fact that heat, drought and elevated O_3 exposure, for instance, are also associated with lower biomass production,

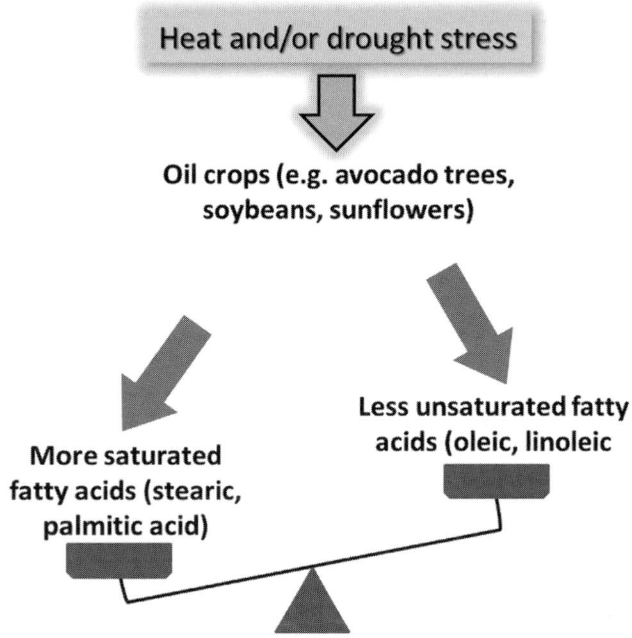

Figure 21.7 Effects of heat and drought on the fat quality of plant oils.

resulting in a concentration of protein (Pleijel et al., 1999; Wang & Frei, 2011). This is supported by the observation that protein synthesis is less affected than that of the other macronutrients and that enhanced leaf senescence caused by heat, drought and high O_3 exposure mobilizes nitrogen that can be reutilized for protein synthesis by the plant compensating for impaired nitrogen uptake by the root (Triboi & Triboi-Blondel, 2002).

The opposite effect, i.e., a dilution of nutrients, was observed under high CO_2 levels as a result of their stimulating effect on photosynthesis and biomass production (Pleijel et al., 1999). However, dilution does not fully account for lower nitrogen contents in plants under high CO_2, suggesting that nitrogen assimilation cannot keep pace with the accelerated carbon fixation. This is also supported by the fact that leguminous plants with their greater nitrogen fixation capacity are less affected by these effects (Rogers et al., 2009). A reduction of protein contents was seen in cereals (wheat, maize and rice) grown under higher- than-normal CO_2 concentrations but was absent or much lower in legumes (soybeans and field peas) (Högy & Fangmeier, 2008; Taub et al., 2008; Myers et al., 2014).

Rising CO_2 levels have also been related to lower concentrations of essential bulk and trace minerals in crops such as wheat and rice that are most likely related to the higher biomass growth (DaMatta et al., 2010).

The stimulating effect of elevated CO_2 concentrations has been related to the higher concentrations of starch that have been reported in potatoes and cereals grown under such conditions (Donnelly et al., 2001; Vorne et al., 2002; Högy & Fangmeier, 2008).

In line with the stress-induced increase in protective molecules, higher production of antinutritive compounds that protect the plant from oxidative stress and attacks by herbivores, bacteria, viruses or fungi was also described under unfavourable climate conditions. For example, late sowing of wheat that resulted in higher mean maximum temperature during grain filling was associated with higher grain concentrations of phytate, which interferes with the absorption of minerals and trace elements. The contents were further increased by drought (Singh et al., 2012). In sorghum and rice, exposure to higher CO_2 was also linked to markedly increased contents of phytate (Myers et al., 2014). Exposure to drought and to elevated O_3 levels was associated with increased concentrations of glycoalkaloids (α-solanine and/or α-chaconine) in potatoes (Andre et al., 2009; Donnelly et al., 2001).

A negative effect of elevated CO_2 concentrations was also reported in cassava, in which it was associated with higher contents of cyanogenic glycosides in the leaves but not in the tubers. Considering that in some parts of Africa and Southeast Asia, the leaves are consumed as a protein source to complement the starch-rich tuber (Montagnac et al., 2009), this finding has important health implications in regions where cassava is a staple food (Gleadow et al., 2009).

Effects of climate change on composition and nutritional quality were also observed in animal food. Lower meat quality was reported in chicken, pigs and cattle exposed to heat and drought. Protein content of muscle tissue was reduced and fat content tended to increase (Babinszky et al., 2011; Zhang et al., 2012). Poultry has shown a particular sensitivity to heat stress. Effects are not limited to meat quality but also to that of eggs. Heat stress increased lipid peroxidation in the egg yolk (Fouad et al., 2016). In dairy cattle, heat exposure was associated with lower protein and fat content of the milk. The fatty acid pattern was also altered, with lower proportions of short- and middle-chain fatty acids and higher relative amounts of long-chain fatty acids. A positive finding was the higher content of polyunsaturated fatty acids, which have beneficial health effects. However, milk from heat-stressed animals also had higher amounts of trans-monounsaturated fatty acids (Renna et al., 2010; Hammami et al., 2015) that have been associated with a higher risk of cardiovascular diseases despite some controversy whether this also applies to trans-fatty acids from dairy sources (Gebauer & Baer, 2013).

Changes of the temperature and pH level of seawater have also been associated with alterations in the fatty acid composition of plankton (Galloway & Winder, 2015; Garzke et al., 2016; Jayalakshmi et al., 2016). These changes were transferred to other marine organisms feeding on plankton such as clams and fish (Anacleto et al., 2014; Diaz-Gil et al., 2015). Even though these observations were not reproduced in other studies and are probably dependent on the composition of the plankton community (King et al., 2015; Bermúdez et al., 2016), they might still have relevant nutritional implications as fish and shellfish are the major source of essential n-3 long-chain polyunsaturated fatty acids (n-3 LC-PUFAs). The positive effects of n-3 LC-PUFAs such as eicosapentaenoic acid (EPA) and docosahexaenoic acid (DHA) on cardiovascular health and neurological functions as well as their immune-modulating properties are widely acknowledged, and a higher intake of these nutrients is generally recommended, not only in the form of fish, but also from supplements that may especially be needed to achieve the higher intake levels required for clinical effects (FAO, 2010). This underscores the importance of fatty acid composition of seafood also as a raw material of fish oil production as terrestrial sources of n-3 fatty acids are limited and the conversion of α-linolenic acid from plants to EPA and DHA appears insufficient to meet the requirements (Harris, 2014).

21.6 Conclusion

Climate change is likely to have far-reaching effects on food production and access to it and is consequently posing a threat to nutrition security, especially in developing countries. The effects are not limited to the amount of food produced but also affect the composition of crops, leading to lower contents of protein and some micronutrients. However, the composition of natural food is influenced by many environmental factors regardless of climate change. With a diverse balanced diet, any variability in food composition due to climate change is unlikely to exert a major negative health effect with regard to nutrient supply or antinutritive components. In turn, it may become an issue when the diet is based on a very limited number of staple crops possibly even grown under the same conditions. This situation may arise from reductions in diet diversity and poorer food choice under climate change and may particularly affect vulnerable groups such as young children or elderly adults.

Adaptation to climate change is imperative to prevent a further increase in global hunger and malnutrition. It has been estimated that between 2010 and 2050 the prevalence of moderate stunting as an indicator of chronic malnutrition might increase by up to 29% in South Asia and Sub-Saharan Africa compared to the projected development without climate change. The outlook was even worse for the prevalence of severe stunting, which might rise by as much as 61% in South Asia. The proportion of undernourished preschool children (based on per capita calorie availability) was predicted to increase by 52% to 142% as a result of climate change (Lloyd et al., 2011) (see Figure 21.8).

Nutrition security in the future will also strongly depend on economic development as exemplified by the International Food Policy Research Institute's (IFPRI's) model analysis on Food Security, Farming, and Climate Change to 2050 (Nelson et al. 2010). The optimistic scenario (with high GDP growth and low population growth) features an increase of the average per capita kilocalorie availability in developing countries to almost 85% of the level in developed countries by 2050, and a reduction of the number of malnourished children by as much as 45.9%. Even in low-income developing countries, a 36.6% reduction could be achieved. In turn, under a pessimistic scenario, the number of malnourished children would only decline by 1.8% in developing countries and for low-income developing countries an increase by 18.1% is projected. All these trends assume that there would be no further climate deterioration until 2050. Without this achievement, the numbers of malnourished children in 2050 are projected to be about 9% to 11.5% higher compared to the perfect mitigation scenarios (Nelson et al., 2010).

Moreover, economic development can further exacerbate climate change. Greenhouse gas emissions are highly correlated to GDP and consumption level (Fleurbaey et al., 2014).

A general problem arises from the difficulty of predicting the extent and effects of climate change and its regional variability. Indeed, the various models that have been developed so far vary largely in the projected change in temperature and rainfall as well as the increase in greenhouse gas emissions and other factors, complicating the prediction of the impact on food production and security. Technological progress and adaptation approaches with regard to selection of crop varieties and farming practices together with the fertilizing effect of higher atmospheric CO_2 levels are thought to compensate to a certain degree for yield losses in many regions not just in high-income countries. However, the success of adaptation depends on the future development of climate and will not be a given for the more pessimistic projections. In the light of the rising food demand due to population growth and economic development, sustainable solutions to increase food production are urgently needed.

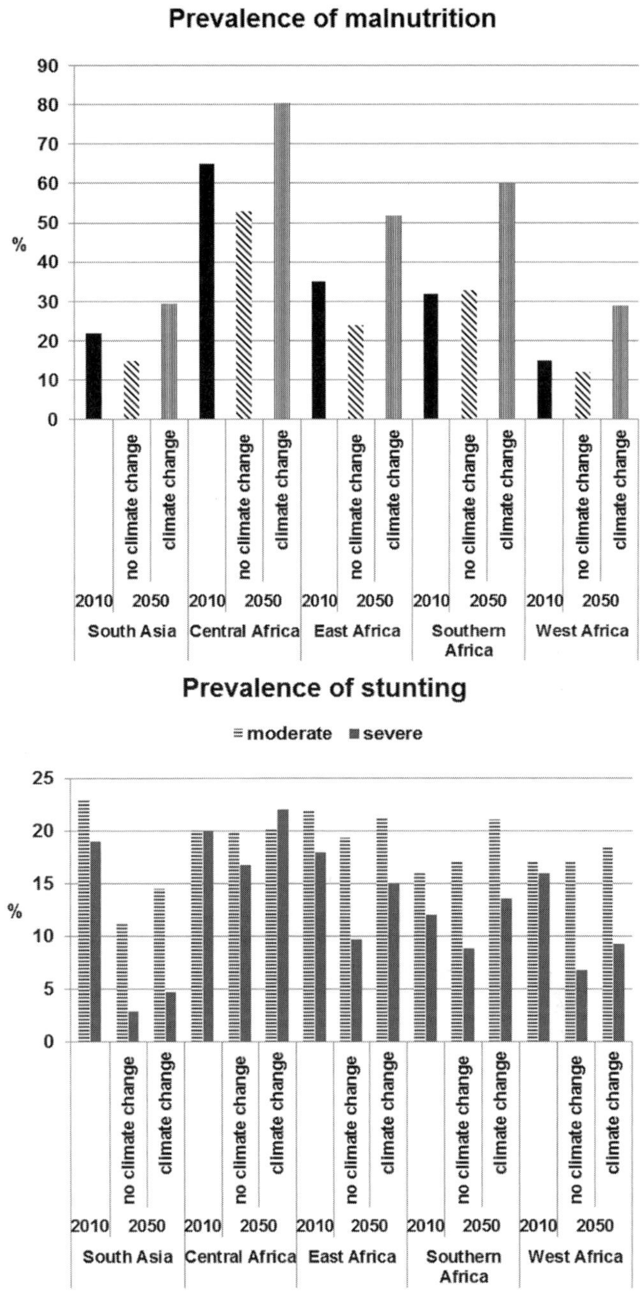

Figure 21.8 Impact of climate change on trends in the prevalence of malnutrition and stunting in children less than five years in South Asia and Sub-Saharan Africa (source: adapted from Lloyd et al., 2011).

References

Abbott, P. (2009). Development dimensions of high food prices. In *OECD Food, Agriculture and Fisheries Working Papers*, No. 18, Paris: OECD.

Adger, W. N., Pulhin, J. M., Barnett, J., Dabelko, G. D., Hovelsrud, G. K. et al. (2014). Human security. In Field,

C. B., Barros, V. R., Dokken, D. J. et al. (Eds.), Climate change 2014: Impacts, adaptation, and vulnerability. Part A: Global and sectoral aspects. Contribution of Working Group *II to the fifth assessment report of the Intergovernmental Panel on Climate Change*. Cambridge and New York: Cambridge University Press, pp. 755–791.

Ainsworth, E. A. and Rogers, A. (2007) The response of photosynthesis and stomatal conductance to rising [CO_2]: mechanisms and environmental interactions. *Plant Cell Environ.*, **30**, 258–270.

Alexandratos, N. and Bruinsma, J. (2012). World agriculture towards 2030/2050: the 2012 revision. *ESA Working paper* No. 12–03. Rome: FAO.

Anacleto, P., Maulvault, A. L., Bandarra, N. M. et al. (2014). Effect of warming on protein, glycogen and fatty acid content of native and invasive clams. *Food Res. Int.*, **64**, 439–445.

Andre, C. M., Schafleitner, R., Guignard, C. et al. (2009). Modification of the health-promoting value of potato tubers field grown under drought stress: Emphasis on dietary antioxidant and glycoalkaloid contents in five native Andean cultivars (Solanum tuberosum L.). *J. Agric. Food Chem.*, **57**, 599–609.

Asseng, S., Foster, I. and Turner, N. C. (2014). The impact of temperature variability on wheat yields. *Glob. Change Biol.*, **17**, 997–1012.

Asseng, S., Ewert, F., Martre, P. et al. (2015) Rising temperatures reduce global wheat production. *Nat. Clim. Change*, **5**, 143–147.

Babinszky, L., Halas, V. and Verstegen, M. W. A. (2011). Impacts of climate change on animal production and quality of animal food products. In Blanco, J. and Kheradmand, H. (Eds.), *Climate Change - Socioeconomic Effects*. Rijeka: InTech, pp. 165–190.

Bale, J. S. and Hayward, S. A. L. (2010). Insect overwintering in a changing climate. *J. Exp. Biol.*, **213**, 980–994.

Banse, M., van Meijl, H., Tabeau, A. and Woltjer, G. (2008). Will EU biofuel policies affect global agricultural markets? *Eur. Rev. Agric. Econ.*, **35**, 117–141.

Battilani, P., Toscano, P. and Van der Fels-Klerx, H. J. (2016) Aflatoxin B1 contamination in maize in Europe increases due to climate change. *Sci. Rep.*, **6**, 24328.

Bebber, D. P., Ramotowski, M. A. T. and Gurr, S. J. (2013). Crop pests and pathogens move polewards in a warming world. *Nat. Clim. Change*, **3**, 985–988.

Bermúdez, R., Winder, M., Stuhr, A. et al. (2016). Effect of ocean acidification on the structure and fatty acid composition of a natural plankton community in the Baltic Sea. *Biogeosciences*, **13**, 6625–6635.

Bett, B., Kiunga, P., Gachohi, J. et al. (2017) Effects of climate change on the occurrence and distribution oflivestock diseases. *Prev. Vet. Med.*, **137**(Pt B), 119–129.

Booker, F., Muntifering, R., McGrath, M. et al. (2009) The ozone component of global change: potential effects on agricultural and horticultural plant yield, product quality and interactions with invasive species. *J. Integr. Plant Biol.*, **51**, 337–351.

Britz, S. J., Prasad, P. V. V., Moreau, R. A. et al. (2007) Influence of growth temperature on the amounts of tocopherols, tocotrienols, and γ-oryzanol in brown rice. *J. Agric. Food Chem.*, **55**, 7559–7565.

Caldwell, C. R., Britz, S. J. and Mirecki, R. M. (2005). Effect of temperature, elevated carbon dioxide, and drought during seed development on the isoflavone content of dwarf soybean [*Glycine max* (L.) Merrill] grown in controlled environments. *J. Agric. Food Chem.*, **53**, 1125–1129.

Carrera, C., Martínez, M. J., Dardanelli, J. and Balzarini, M. (2011). Environmental variation and correlation of seed components in nontransgenic soybeans: protein, oil, unsaturated fatty acids, tocopherols, and isoflavones. *Crop Sci.*, **51**, 800–809.

Cassidy, E. S., West, P. C., Gerber, J. S. and Foley, J. A. (2013). Redefining agricultural yields: from tonnes to people nourished per hectare. *Environ. Res. Lett.*, **8**, 034015.

Chennupati, P., Seguin, P. and Liu, W. (2011). Effects of high temperature stress at different development stages on soybean isoflavone and tocopherol concentrations. *J. Agric. Food Chem.*, **59**, 13081–13088.

Cheung, W. W. L., Watson, R. and Pauly, D. (2013) Signature of ocean warming in global fisheries catch. *Nature*, **497**, 365–369.

Choudhury, F. K., Rivero, R. M., Blumwald, E. and Mittler, R. (2016) Reactive oxygen species, abiotic stress and stress combination. *The Plant Journal*, [e-pub ahead of print] doi: 10.1111/tpj.13299.

Ciais, P., Reichstein, M., Viovy, N. et al. (2005). Europe-wide reduction in primary productivity caused by the heat and drought in 2003. *Nature*, **473**, 529–533.

Committee on World Food Security (CFS). (2012). *Coming to terms with terminology: food security, nutrition security, food security and nutrition, food and nutrition security*, thirty-ninth session, Rome, 15–20 October 2012. Available at: http://www.fao.org/docrep/meeting/026/MD776E.pdf (Accessed 21 June 2016).

Cripps, G., Flynn, K. J., and Lindeque, P. K. (2016) Ocean acidification affects the phyto-zoo plankton trophic transfer efficiency. *PLoS ONE*, **11**, e0151739.

Dai, A. (2011) Drought under global warming: a review. *WIREs Clim. Change*, **2**, 45–65.

DaMatta, F. M., Grandis, A., Arenque, B. C. and Buckeridge, M. S. (2010). Impacts of climate change on crop physiology and food quality. *Food Res. Int.*, **43**, 1814–1823.

Dannehl, D., Huber, C., Rocksch, T., Huyskens-Keil, S. and Schmidt, U. (2012). Interactions between changing climate conditions in a semi-closed greenhouse and plant development, fruit yield, and health-promoting plant compounds of tomatoes. *Sci. Hortic.*, **138**, 235–243.

Dannehl, D. and Josuttis, M. (2014). Cultivar and production effects on bioactive polyphenols. In R. Ross Watson (Ed.), *Polyphenols in Plants. Isolation, Purification and Extract Preparation*. London: Academic Press, pp. 3–13.

Das, K. and Roychoudhury, A. (2014). Reactive oxygen species (ROS) and response of antioxidants as ROS-scavengers during environmental stress in plants. *Front. Environ. Sci.*, **2**, 53.

Díaz-Gil, C., Catalán, I. A., Palmer, M., Faulk, C. K. and Fuiman, L. A. (2015) Ocean acidification increases fatty acids levels of larval fish. *Biol. Lett.*, **11**, 20150331.

Dietary Guidelines Advisory Committee (DGAC). (2015). *Scientific report of the 2015 Dietary Guidelines Advisory Committee to the Secretaries of the U.S. Departments of Health and Human Services (HHS) and Agriculture (USDA)*. Washington, DC: U.S. Department of Agriculture, Agricultural Research Service.

Donnelly, A., Lawson, T., Craigon, J. et al. (2001). Effects of elevated CO_2 and O_3 on tuber quality in potato. *Agric. Ecosyst. Environ.*, **87**, 273–285.

Dornbos, D. L. and Mullen, R. E. (1992). Soybean seed protein and oil contents and fatty acid composition – adjustments by drought and temperature. *J. Am. Oil Chem. Soc.*, **69**, 228–231.

Drewnowski, A. and Darmon, N. (2005). The economics of obesity: dietary energy density and energy cost. *Am. J. Clin. Nutr.*, **82**, 265S–273S.

Elbehri, A., Elliott, J. and Wheeler, T. (2015). Climate change, food security and trade: An overview of global assessments and policy insights. In FAO (Ed.). *Climate Change and Food Systems: Global Assessments and Implications for Food Security and Trade*. Geneva: Food Agriculture Organization of the United Nations (FAO), pp. 1–27.

Elmadfa, I., Meyer, A., Nowak, V. et al. (2009). European nutrition and health report 2009. *Forum Nutr.*, **62**, 1–405.

Elmadfa, I. and Meyer, A. L. (2017). Animal proteins as important contributors to a healthy human diet. *Annu. Rev. Anim. Biosci.*, **5**, 1111–131.

Elsgaard, L., Børgesen, C. D., Olesen, J. E. et al. (2012) Shifts in comparative advantages for maize, oat and wheat cropping under climate change in Europe. *Food Addit. Contam. Part A – Chem.*, **29**, 1514–1526.

Fangmeier, A, De Temmerman, L., Black, C., Persson, K. and Vorne, V. (2002). Effects of elevated CO_2 and/or ozone on nutrient concentrations and nutrient uptake of potatoes. *Eur. J. Agron.*, **17**, 353–368.

FAO. (2000) *The State of Food Insecurity in the World 2000*. Rome: United Nations Food and Agriculture Organization.

FAO. (2006). *Livestock's Long Shadow: Environmental Issues and Options*. Rome: Food and Agriculture Organization.

FAO. (2009) Climate change implications for fisheries and aquaculture: Overview of current scientific knowledge. In: *FAO Fisheries and Aquaculture Technical Paper 530*. Rome: United Nations Food and Agriculture Organization.

FAO (2010) Fats and fatty acids in human nutrition. Report of an expert consultation. Geneva, 10–4 November, 2008. *FAO Food and Nutrition Paper* **91**. Rome: United Nations Food and Agriculture Organization.

FAO. (2015). *The Impact of Natural Hazards and Disasters on Agriculture and Food Security and Nutrition: A Call for Action to Build Resilient Livelihoods*. Rome: United Nations Food and Agriculture Organization (FAO).

FAO (2016) *The State of World Fisheries and Aquaculture 2016: Contributing to food security and nutrition for all*. Rome: United Nations Food and Agriculture Organization.

FAO, IFAD and WFP. (2014). *The State of Food Insecurity in the World 2014: Strengthening the Enabling Environment for Food Security and Nutrition*. Rome: United Nations Food and Agriculture Organization.

FAO, IFAD and WFP. (2015) The state of food insecurity in the world 2015. In: *Meeting the 2015 international hunger targets: taking stock of uneven progress*. Rome: United Nations Food and Agriculture Organization.

Feng, S. and Fu, Q. (2013) Expansion of global drylands under a warming climate. *Atmos. Chem. Phys.*, **13**, 10081–10094.

Flagella, Z., Rotunno, T., Tarantino, E., Di Caterina, R. and De Caro, A. (2002). Changes in seed yield and oil fatty acid composition of high oleic sunflower (*Helianthus annuus* L.) hybrids in relation to the sowing date and the water regime. *Eur. J. Agron.*, **17**, 221–230.

Fleurbaey M., Kartha, S., Bolwig, S. et al. (2014) Sustainable development and equity. In Edenhofer, O., Pichs-Madruga, R., Sokona, Y. et al. (Eds.), Climate Change 2014: Mitigation of Climate Change. Contribution of Working Group III to the Fifth *Assessment Report of the Intergovernmental Panel on Climate Change*. Cambridge: Cambridge University Press, pp. 283–350.

Fouad, A. M., Chen, W., Ruan, D. et al. (2016). Impact of heat stress on meat, egg quality, immunity and fertility in poultry and nutritional factors that overcome these effects: A review. *Int. J. Poultry Sci.*, **15**, 81–95.

Fu, Y., Shao, L., Liu, H. et al. (2015) Unexpected decrease in yield and antioxidants in vegetable at very high CO_2 levels. *Environ. Chem. Lett.*, **13**, 473–479.

Galloway, A. W. E. and Winder, M. (2015) Partitioning the relative importance of phylogeny and environmental conditions on phytoplankton fatty acids. *PLoS ONE*, **10**, e0130053.

Gammans, M., Mérel, P. and Ortiz-Bobea, A. (2016) The impact of climate change on cereal yields: Statistical evidence from France. Selected Paper prepared for presentation at the Agricultural and Applied Economics Association's 2016 AAEA Annual Meeting, Boston, MA, July 31–August 2, 2016. Available at: www.uni-goettingen.de/de/document/download/98d80dc02fbdf7cfe708b3a29eaf987d.pdf/Session%208b%20M%C3%A9rel.pdf (Accessed 10 August 2016).

García, G. A., Dreccer, M. F., Miralles, D. J. and Serrago, R. A. (2015). High night temperatures during grain number determination reduce wheat and barley grain yield: a field study. *Glob. Change Biol.*, **21**, 4153–4164.

Garzke, J., Hansen, T., Ismar, S. M. H. and Sommer, U. (2016). Combined effects of ocean warming and acidification on copepod abundance, body size and fatty acid content. *PLoS ONE*, **11**, e0155952.

Gautier, H. Diakou-Verdin, V., Bénard, C. et al. (2008) How does tomato quality (sugar, acid, and nutritional quality) vary with ripening stage, temperature, and irradiance? *J. Agric. Food Chem.*, **56**, 1241–1250.

Gebauer, S. K. and Baer, D. J. (2013) Trans-fatty acids: health effects, recommendations, and regulations. *Encyclopedia of Human Nutrition*, 3rd ed., Vol. **4**, 288–292.

Gleadow, R. M., Evans, J. R., McCaffery, S. and Cavagnaro, T. R. (2009). Growth and nutritive value of cassava (Manihot esculenta Cranz.) are reduced when grown in elevated CO_2. *BMC Plant Biol.*, **11** (Suppl. 1), 76–82.

Goufo, P., Pereira, J., Figueiredo, N. et al. (2014). Effect of elevated carbon dioxide (CO_2) on phenolic acids, flavonoids, tocopherols, tocotrienols, γ-oryzanol and antioxidant capacities of rice (*Oryza sativa* L.). *J. Cereal Sci.*, **59**, 15–24.

Grace, D. (2015). Food safety in low and middle income countries. *Int. J. Environ. Res. Publ. Health.*, **12**, 10490–10507.

Gregory, P. J., Ingram, J. S. I. and Brklacich, M. (2005). Climate change and food security. *Phil. Trans. R. Soc. B*, **360**, 2139–2148.

Gregory, P. J., Johnson, S. N., Newton, A. C. and Ingram, J. S. I. (2009). Integrating pests and pathogens into the climate change/food security debate. *J. Exp. Bot.*, **60**, 2827–2838

Hammami, H., Vandenplas, J., Vanrobays, M.-L. et al. (2015) Genetic analysis of heat stress effects on yield traits, udder health, and fatty acids of Walloon Holstein cows. *J. Dairy Sci.*, **98**, 4956–4968.

Harris, W. S. (2014). Achieving optimal n–3 fatty acid status: the vegetarian's challenge ... or not. *Am. J. Clin. Nutr.*, **100**, 449S–452S.

Hatfield, J. L. and Prueger, J. H. (2015). Temperature extremes: effect on plant growth and development. *Weather & Climate Extr.*, **10**, 4–10.

Hatfield, J. L., Boote, K. J., Kimball B. A. et al. (2011) Climate impacts on agriculture: Implications for crop production. *Agron. J.*, **103**, 351–370.

Havlík, P., Leclère, D., Valin, H. et al. (2015). Global climate change, food supply and livestock production systems: A bioeconomic analysis. In FAO (Ed.), *Climate Change and Food Systems: Global Assessments and Implications for Food Security and Trade*. Geneva: Food Agriculture Organization of the United Nations, pp. 176–201.

Herrero, M., Grace, D., Njuki, J. et al. (2013). The roles of livestock in developing countries. *Animal*, **7**, 3–18.

Herrero, M., Henderson, B., Havlík, P. et al. (2016). Greenhouse gas mitigation potentials in the livestock sector. *Nat. Clim. Change*, **6**, 452–461.

Herring, S. C., Hoerling, M. P., Kossin, J. P. et al. (Eds.) (2015). Explaining extreme events of 2014 from a climate perspective. *B. Am. Meteorol. Soc.*, **96**, S1–S172.

Hoffmann, H. and Rath, T. (2013). Future bloom and blossom frost risk for *Malus domestica* considering climate model and impact model uncertainties. *PLoS ONE*, **8**, e75033.

Högy, P. and Fangmeier, A. (2008). Effects of elevated atmospheric CO_2 on grain quality of wheat. *J. Cereal Sci.*, **48**, 580–591.

Iglesias, A., Garrote, L., Quiroga, S. and Moneo, M. (2012). A regional comparison of the effects of climate change on agricultural crops in Europe. *Clim. Change*, **112**, 29–46.

IPCC. [Solomon, S., Qin, D., Manning, M. et al. (Eds.)] (2007). *Climate Change 2007: The Physical Science Basis. Contribution of Working Group I to the Fourth Assessment Report of the Intergovernmental Panel on Climate Change*. Cambridge: Cambridge University Press.

IPCC (Intergovernmental Panel on Climate Change). [Field, C. B., Barros, V. R., Dokken, D. J. et al. (Eds.)]. (2014). *Climate Change 2014: Impacts, Adaptation, and Vulnerability. Part a: Global and Sectoral Aspects*. Contribution of Working Group II to the Fifth Assessment Report of the Intergovernmental Panel on Climate Change. Cambridge: Cambridge University Press.

Jalloh, A., Nelson, G. C., Thomas, T. S. et al. (Eds.) (2013). *West African Agriculture and Climate Change: A Comprehensive Analysis*. IFPRI Research Monograph. Washington, DC: International Food Policy Research Institute.

Jayalakshmi, T., Nandakumar, R., Balaji Prasath, B. and Santhanam, P. (2016). Effect of acidification on fatty acids profiling of marine benthic harpacticoid copepod Parastenhelia sp. *Ann. Agrar. Sci.*, **14**, 278–282.

Jensen, R. and Miller, N. (2008) Giffen behavior and subsistence consumption. *Am. Econ. Rev.*, **97**, 1553–1577.

Katona, P. and Katona-Apte, J. (2008). The interaction between nutrition and infection. *Clin. Infect. Dis.*, **46**, 1582–1588.

Khan, A. E., Ireson, A. and Kovats, S. (2011). Drinking water salinity and maternal health in coastal Bangladesh: implications of climate change. *Environ. Health Persp.*, **119**, 1328–1332.

Khanom, T. (2016) Effect of salinity on food security in the context of interior coast of Bangladesh. *Ocean & Coastal Management*, **130**, 205–212.

King, A. L., Jenkins, B. D., Wallace, J. R. et al. (2015). Effects of CO_2 on growth rate, C:N:P, and fatty acid composition of seven marine phytoplankton species. *Marine Ecology Progress Series*, **537**, 59–69.

Kitinoja, L., Saran, S., Roy, S. K. and Kader, A. A. (2011). Postharvest technology for developing countries: challenges and opportunities in research, outreach and advocacy. *J. Sci. Food Agr.*, **91**, 597–603.

Lake, I. R., Gillespie, I. A., Bentham, G. et al. (2009). A re-evaluation of the impact of temperature and climate change on foodborne illness. *Epidemiol. Infect.*, **137**, 1538–1547.

Lake, I. R., Hooper, L., Abdelhamid, A. et al. (2012). Climate change and food security: health impacts in developed countries. *Environ. Health Persp.*, **120**, 1520–1526.

Leakey, A. D. B., Ainsworth, E. A., Bernacchi, C. J. (2009) Elevated CO_2 effects on plant carbon, nitrogen, and water relations: six important lessons from FACE. *J. Exp. Bot.*, **60**, 2859–2876.

Lefsrud, M. G., and Kopsell, D. A. (2005). Air temperature affects biomass and carotenoid pigment accumulation in kale and spinach grown in a controlled environment. *HortSci.*, **40**, 2026–2030.

Legave, J.-M., Guédon, Y., Malagi, G., El Yaacoubi, A. and Bonhomme, M. (2015). Differentiated responses of apple tree floral phenology to global warming in contrasting climatic regions. *Front. Plant Sci.*, **6**, 1054.

Lloyd, S. J., Kovats, S. R. and Chalabi, Z. (2011). Climate change, crop yields, and undernutrition: development of a model to quantify the impact of climate scenarios on child undernutrition. *Environ. Health Perspect.*, **119**, 1817–1823.

Lobell, D. B., Gourdji, S.M. (2012) The influence of climate change on global crop productivity. *Plant Physiol.*, **160**, 1686–1697.

Long, S. P., Ainsworth, E. A., Rogers, A. and Ort, D. R. (2004). Rising atmospheric carbon dioxide: Plants FACE the future. *Annu. Rev. Plant Biol.*, **55**, 591–628.

Luedeling, E., Zhang, M. and Girvetz, E. H. (2009). Climatic changes lead to declining winter chill for fruit and nut trees in California during 1950–2099. *PLoS ONE* **4**, e6166.

Massot, C., Bancel, D., Lopez Lauri, F. et al. (2013). High temperature inhibits ascorbate recycling and light stimulation of the ascorbate pool in tomato despite increased expression of biosynthesis genes. *PLoS ONE*, **8** (12), e84474.

McMurray, G., Arruda, C., Britton, D. et al. (2013). *Food Security: a Systems Approach*. White paper presented at: EU Science: Global Challenges and Global Collaboration, Brussels, March 4 to 9, 2013.

Medina, A., Rodriguez, A. and Magan, N. (2014) Effect of climate change on *Aspergillus flavus* and aflatoxin B1 production. *Front. Microbiol.*, **5**, 348.

Met Office and World Food Programme (WFP). (2012). *Climate Impacts on Food Security and Nutrition*. Devon: Met Office and Rome: WFP.

Miraglia, M., Marvin, H. J. P., Kleter, G. A. et al. (2009). Climate change and food safety: an emerging issue with special focus on Europe. *Food Chem. Toxicol.*, **47**, 1009–1021.

Montagnac, J. A., Davis, C. R. and Tanumihardjo, S. A. (2009). Nutritional value of cassava for use as a staple food and recent advances for improvement. *Compr. Rev. Food Sci. Food Saf.*, **8**, 181–194.

Moore, F. C. and Lobell, D. B. (2015) The fingerprint of climate trends on European crop yields. *Proc. Natl. Acad. Sci. U.S.A.*, **112**, 2670–2675.

Moretti, C. L., Mattos, L. M., Calbo, A. G. and Sargent, S. A. (2010). Climate changes and potential impacts on postharvest quality of fruit and vegetable crops: a review. *Food Res. Int.*, **43**, 1824–1832.

Montero-Serra, I., Edwards, M. and Genner, M. J. (2015) Warming shelf seas drive the subtropicalization of European pelagic fish communities. *Global Change Biol.*, **21**, 144–153

Müller, C. and Robertson, R. D. (2014). Projecting future crop productivity for global economic modeling. *Agr. Econ.*, **45**, 37–50, DOI: 10.1111/agec.12088.

Myers, N. and Kent, J. (2003) New consumers: The influence of affluence on the environment. *Proc. Natl. Acad. Sci. U.S.A.*, **100**, 4963–4968.

Myers, S. S., Zanobetti, A., Kloog, I. et al. (2014) Increasing CO_2 threatens human nutrition. *Nature*, **510**, 139–142.

Namazkar, S., Stockmarr, A., Frenck, G. et al. (2016). Concurrent elevation of CO_2, O_3 and temperature severely affects oil quality and quantity in rapeseed. *J. Exp. Bot.*, **67**, 4117–4125.

Nardone, A., Ronchi, B., Lacetera, N., Ranieri, M. S. and Bernabucci, U. (2010). Effects of climate changes on animal production and sustainability of livestock systems. *Livest. Sci.*, **130**, 57–69.

Nelson, G. C., Rosegrant, M. W., Palazzo, A., Gray, I., Ingersoll, C. et al. (2010). *Food Security, Farming, and Climate Change to 2050: Scenarios, Results, Policy Options*. Washington, DC: International Food Policy Research Institute (IFPRI).

Niang, I., Ruppel, O. C., Abdrabo, M. A. et al. (2014). Africa. In Field, C. B., Barros, V. R., Dokken, D. J. et al. (Eds.), *Climate Change 2014: Impacts, Adaptation, and Vulnerability. Part B: Regional Aspects. Contribution of Working Group II to the Fifth Assessment Report of the Intergovernmental Panel on Climate Change*. Cambridge: Cambridge University Press, pp. 1199–1265.

Olfert, O. and Weiss, R. M. (2006). Impact of climate change on potential distributions and relative abundances of *Oulema melanopus*, *Meligethes viridescens* and *Ceutorhynchus obstrictus* in Canada. *Agric. Ecosyst. Environ.*, **113**, 295–301.

Pangaribowo, E. H., Gerber, N. and Torero, M. (2013). Food and nutrition security indicators: a review. *ZEF Working Paper Series*, **108**.

Parry, M., Evans, A., Rosegrant, M. W. and Wheeler, T. (2009). *Climate Change and Hunger: Responding to the Challenge*. Rome: World Food Programme.

Paterson, R. R. M. and Lima, N. (2011) Further mycotoxin effects from climate change. *Food Res. Int.*, **44**, 2555–2566.

Peters, K., Breitsameter, L. and Gerowitt, B. (2014). Impact of climate change on weeds in agriculture: a review. *Agron. Sustain. Dev.*, **34**, 707–721.

Pleijel, H., Mortensen, L., Fuhrer, J., Ojanperä, K., and Danielsson, H. (1999). Grain protein accumulation in relation to grain yield of spring wheat (Triticum aestivum L.) grown in open-top chambers with different concentrations of ozone, carbon dioxide and water availability. *Agric. Ecosyst. Environ.*, **72**, 265–270.

Pörtner, H. O. (2002) Climate variations and the physiological basis of temperature dependent biogeography: systemic to molecular hierarchy of thermal tolerance in animals. *Comp. Biochem. Physiol. A Mol. Integr. Physiol.*, **132**, 739–761.

Porter, J. R., Xie, L., Challinor, A. J. et al. (2014) Food security and food production systems. In Field, C. B., Barros, V. R., Dokken, D. J. et al. (Eds.), *Climate*

Change 2014: Impacts, Adaptation, and Vulnerability. Part A: Global and Sectoral Aspects. Contribution of Working Group II to the Fifth Assessment Report of the Intergovernmental Panel on Climate Change. Cambridge: Cambridge University Press, pp. 485–533.

Prakash, A. (Ed.) (2011) *Safeguarding Food Security in Volatile Global Markets*. Rome: Food and Agriculture Organization of the United Nations.

Prasad, P. V. V., Staggenborg, S. A. and Ristic, Z. (2008) Impacts of Drought and/or heat stress on physiological, developmental, growth, and yield processes of crop plants. In L. R. Ahuja, V. R. Reddy, S. A. Saseendran and Q. Yu (Eds.), *Response of Crops to Limited Water. Understanding and Modeling Water Stress Effects on Plant Growth Processes*. Advances in Agricultural Systems Modeling 1, Madison, WI: American Society of Agronomy, Crop Science Society of America, Soil Science Society of America, pp. 301–355.

Reich, P. B., Hobbie, S. E. and Lee, T. D. (2014). Plant growth enhancement by elevated CO_2 eliminated by joint water and nitrogen limitation. *Nat. Geosci.*, **7**, 920–924.

Remington, T., Maroko, J., Walsh, S., Omanga, P. and Charles, E. (2002). Getting off the seeds-and-tools treadmill with CRS seed vouchers and fairs. *Disasters*, **26**, 316–328.

Renna, M., Lussiana, C., Malfatto, V., Mimosi, A. and Battaglini, L. M. (2010). Effect of exposure to heat stress conditions on milk yield and quality of dairy cows grazing on Alpine pasture. *Proceedings of the 9th European IFSA Symposium*, 4–7 July 2010, Vienna, Austria, pp. 1338–1348.

Richardson, A. C., Marsh, K. B., Boldingh, H. L. et al. (2004). High growing temperatures reduce fruit carbohydrate and vitamin C in kiwifruit. *Plant Cell Environ.*, **27**, 423–435.

Richerson, P. J., Boyd, R. and Bettinger, R. L. (2001) Was agriculture impossible during the Pleistocene but mandatory during the Holocene? A climate change hypothesis. *Am. Antiq.*, **66**, 387–411.

Rochette, P., Bélanger, G., Castonguay, Y. et al. (2004). Climate change and winter damage to fruit trees in eastern Canada. *Can. J. Plant Sci.*, **84**, 1113–1125.

Rogers, A., Ainsworth, E. A. and Leakey, A. D. B. (2009). Will elevated carbon dioxide concentration amplify the benefits of nitrogen fixation in legumes? *Plant Physiol.*, **151**, 1009–1016.

Rosales, M. A., Ruiz, J. M., Hernández, J., Soriano, T., Castilla, N. and Romero, L. (2006). Antioxidant content and ascorbate metabolism in cherry tomato exocarp in relation to temperature and solar radiation. *J. Sci. Food Agr.*, **86**, 1545–1551.

Rötter, R. P., Palosuoa, T., Pirttioj, N. K. et al. (2011). What would happen to barley production in Finland if global warming exceeded 4 °C? A model-based assessment. *Eur. J. Agron.*, **35**, 205–214.

Sánchez, B. A., Rasmussen, A. and Porter, J. R. (2014). Temperatures and the growth and development of maize and rice: a review. *Global Change Biol.*, **20**, 408–417.

Seneviratne, S. I., Nicholls, N., Easterling, D. et al. (2012). Changes in climate extremes and their impacts on the natural physical environment. In Field, C. B., Barros, V., Stocker, T. F. et al. (Eds.), *Managing the Risks of Extreme Events* and Disasters to Advance Climate Change Adaptation, A Special Report of Working Groups I and II of the Intergovernmental Panel on Climate Change (IPCC). Cambridge: Cambridge University Press, pp. 109–230.

Shen, Y., Lansky, E., Traber, M. and Nevo, E. (2013). Increases in both acute and chronic temperature potentiate tocotrienol concentrations in wild barley at "Evolution Canyon". *Chem Biodivers.*, **10**, 1696–1705.

Showler, A. T. (2013) Drought and arthropod pests of crops. In Neves, D. F. and Sanz, J. D. (Eds.), *Droughts: New Research*. Hauppauge, NY: Nova Science, pp. 131–156.

Singh, S., Gupta, A. K. and Kaur, N. (2012). Influence of drought and sowing time on protein composition, antinutrients, and mineral contents of wheat. *Scientific World J.*, 2012, 485751.

Skuce, P. J., Morgan, E. R., van Dijk, J. and Mitchell, M. (2013). Animal health aspects of adaptation to climate change: beating the heat and parasites in a warming Europe. *Animal*, **7** (s2), 333–345.

Smith, K. R., Woodward, A., Campbell-Lendrum, D. et al. (2014). Human health: impacts, adaptation, and co-benefits. In Field, C. B., Barros, V. R., Dokken, D. J. et al. (Eds.), *Climate Change 2014: Impacts, Adaptation, and Vulnerability. Part A: Global and Sectoral Aspects*. Contribution of Working Group II to the fifth assessment report of the Intergovernmental Panel on Climate Change. Cambridge: Cambridge University Press, pp. 709–754.

Sperling, L. and McGuire, S. J. (2010). Persistent myths about emergency seed aid. *Food Policy*, **35**, 195–201.

Taub, D. R., Miller, B. and Allen, H. (2008). Effects of elevated CO_2 on the protein concentration of food crops: a meta-analysis. *Global Change Biol.*, **14**, 565–575.

Teh, S. Y. and Koh, H. L. (2016) Climate change and soil salinization: impact on agriculture, water and food security. *International Journal of Agriculture, Forestry and Plantation*, **2**, 1–9.

The Royal Society (2005) *Ocean Acidification Due to Increasing Atmospheric Carbon Dioxide*. London: The Royal Society.

Thomas, J. M. G., Prasad, P. V. V., Boote, K. J. and Allen, L. H. Jr. (2009). Seed composition, seedling emergence and early seedling vigour of red kidney bean seed produced at elevated temperature and carbon dioxide. *J. Agron. Crop Sci.*, **195**, 148–156.

Thomas, J. M. G., Boote, K. J., Allen, L. H. Jr., Gallo-Meagher, M. and Davis, J. M. (2003). Elevated temperature and carbon dioxide effects on soybean seed composition and transcript abundance. *Crop Sci.*, **43**, 1548–1557.

Thomson, L. J., Macfadyen, S. and Hoffmann, A. A. (2010). Predicting the effects of climate change on natural enemies of agricultural pests. *Biol. Control*, **52**, 296–306.

Thornton, P. K., Ericksen, P. J., Herrero, M. and Challinor, A. J. (2014) Climate variability and vulnerability to climate change: a review. *Global Change Biol.*, **20**, 3313–3328.

Triboi, E. and Triboi-Blondel, A. M. (2002). Productivity and grain or seed composition: a new approach to an old problem. *Eur. J. Agron.*, **16**, 163–186.

Trnka, M., Olesen, J. E., Kersebaum, K. C. et al. (2011) Agroclimatic conditions in Europe under climate change. *Global Change Biol.*, **17**, 2298–2318.

Tubiello, F. N., Salvatore, M., Cóndor Golec, R. D. et al. (2014). *Agriculture, forestry and other land use emissions by sources and removals by sinks. 1990 – 2011 analysis.* Rome: FAO Statistics Division Working Paper Series ESS/14–02.

UNICEF, WHO and The World Bank Group. (2015). *Joint Child Malnutrition Estimates – Levels and Trends (2015 Edition): Key Findings.* New York: UNICEF, WHO, World Bank. Available at: http://www.who.int/entity/nutgrowthdb/jme_brochure2015.pdf?ua=1 (Accessed 29 June 2016).

U.S. Department of Transportation. (2014). *Climate Adaptation Plan: Ensuring Transportation Infrastructure and System Resilience*. Washington, DC: DoT.

van der Fels-Klerx, H. J., Olesen, J. E., Naustvoll, L.-J., Friocourt, Y., Mengelers M. J. B. and Christensen, J. H. (2012).Climate change impacts on natural toxins in food production systems, exemplified by deoxynivalenol in wheat and diarrhetic shellfish toxins. *Food Addit. Contam. Part A – Chem.*, **29**, 1647–1659

Vineis, P., Chan, Q. and Khan, A. (2011) Climate change impacts on water salinity and health. *J. Epidemiol. Global Health.*, **1**, 5–10.

Vorne, V., Ojanperä, K., De Temmerman, L. et al. (2002). Effects of elevated carbon dioxide and ozone on potato tuber quality in the European multiple-site experiment 'CHIP-project'. *Eur. J. Agron.*, **17**, 369–381.

Waithaka, M., Nelson, G., Thomas T. and Kyotalimye, M. (Eds.) (2013). *East African agriculture and climate change: A comprehensive analysis. IFPRI Research Monograph.* Washington, DC: International Food Policy Research Institute.

Wang, Y. and Frei, M. (2011). Stressed food – the impact of abiotic environmental stresses on crop quality. *Agric. Ecosyst. Environ.*, **141**, 271– 286.

Wang, S. Y., Bunce, J. A. and Maas, J. L. (2003). Elevated carbon dioxide increases contents of antioxidant compounds in field-grown strawberries. *J. Agric. Food Chem.*, **51**, 4315–4320.

Warland, J., McKeown, A. W. and McDonald M. R. (2006). Impact of high air temperatures on Brassicaceae crops in southern Ontario. *Can. J. Plant Sci.*, **86**, 1209–1215.

Watts, E. J., Shen, Y., Lansky, E. P. et al. (2015) High environmental stress yields greater tocotrienol content while changing vitamin E profiles of wild emmer wheat seeds. *J. Med. Food*, **18**, 216–223.

White, R., Stewart, B. and O'Neill, P. (2011). *Access to Food in a Changing Climate: A Report to DEFRA*. University of Oxford, Environmental Change Institute. Available at: http://www.eci.ox.ac.uk/publications/downloads/2011-OIA-ECI-report-access-to-food.pdf. (Accessed 28 September 2016)

Woolf, A. B., Ferguson, I. B., Requejo-Tapia, L. C., Boyd, L., Laing, W. A. and White, A. (1999). Impact of sun exposure on harvest quality of 'Hass' avocado fruit. *Rev. Chapingo Ser. Hortic.*, **5**, 353–358.

World Bank, Poverty Reduction and Equity Group. (2011). *Food Price Watch*, **5**, February 2011.

World Bank, Poverty Reduction and Equity Group. (2015). *Food Price Watch*, **19**, June 2015.

Yang, X., Chen, F., Lin, X. et al. (2015) Potential benefits of climate change for crop productivity in China. *Agricultural and Forest Meteorology*, **208**, 76–84.

Yonetani, M., Lavell, C., Bower, E. et al. (2015). *Global Estimates 2015: People Displaced by Disasters.* Geneva: Internal Displacement Monitoring Centre (IDMC).

Yoshida, Y., Saito, Y., Jones, L. S. and Shigeri, Y. (2007). Chemical reactivities and physical effects in comparison between tocopherols and tocotrienols: physiological significance and prospects as antioxidants. *J. Biosci. Bioeng.*, **104**, 439–45, doi: 10.1263/jbb.104.439.

Zavala, J. A., Casteel, C. L., DeLucia, E. H. and Berenbaum, M. R. (2008). Anthropogenic increase in carbon dioxide compromises plant defense against invasive insects. *Proc. Natl. Acad. Sci. U.S.A.*, **105**, 5129–5133.

Zhang, Z. Y., Jian, G. Q., Zuo, J. J. (2012). Effects of constant and cyclic heat stress on muscle metabolism and meat quality of broiler breast fillet and thigh meat. *Poultry Sci.*, **91**, 2931–2937.

Zhu, Y., Qian, W. and Hua, J. (2010). Temperature modulates plant defense responses through NB-LRR proteins. *PLoS Pathogens*, **6**, e1000844.

Ziska, L., Crimmins, A., Auclair, A. et al. (2016). Food Safety, Nutrition, and Distribution. In: The Impacts of Climate Change on Human Health in the United States: A Scientific Assessment. Washington, DC: U.S. Global Change Research Program, pp. 189–216. Available at: http://dx.doi.org/10.7930/J0ZP4417 (Accessed on 29 September 2016)

Zvereva, E. L. and Kozlov, M. V. (2006). Consequences of simultaneous elevation of carbon dioxide and temperature for plant–herbivore interactions: a metaanalysis. *Glob. Change Biol.*, **12**, 27–41.

22

Marine Systems, Food Security, and Future Earth

Elizabeth A. Fulton, Éva Plagányi, William Cheung, Julia Blanchard, and Reg Watson

22.1 Current Dependence

Seafood,[1] wild caught and farmed, has been a staple of human diets for millennia. Dietary (isotope) and archaeological evidence indicates fish were a significant part of the diet of people in Africa and Eurasia in the Mid-Palaeolithic (Hu et al., 2009), with the earliest South African middens more than 150,000 years old. Tunas and other such pelagic fish have been exploited for at least forty-two thousand years (O'Connor et al., 2011) and fish hooks carved from *Trochus* shell have been found from Timor to Okinawa, indicating the widespread use of such fishing gear by 23000 BP (Fujita et al., 2016). Written evidence of the importance of seafood to humanity dates to 4000 BC in Egypt (Chimits, 1957), with examples of texts on fishing methods and management found in all major ancient human cultures (Radcliffe, 1921; Costa-Pierce 2002). This love of seafood has not dwindled through time and today the total annual global fish production stands at more than 167 million tonnes of fish, with 146.3 million tonnes for human consumption and more than half of that from aquaculture (FAO, 2016). It is a major source of dietary protein, with more than 3 billion people dependent upon it. Fish has been the major source of animal protein for human consumption since 1960 and this per capita consumption is growing globally, trebling in the last fifty years (Béné et al., 2015). The affluent industrialised nations now consume at least 26.8 kg per capita annually; consumption is slightly lower in developing nations (at 20 kg per capita annually) and even the citizens of the poorest and food-deficit nations have doubled their consumption of seafood over the last thirty years, now consuming 7.6 kg per capita annually (FAO, 2016). This growth in consumption has outpaced that of all other livestock sectors (including chicken, pork and beef) (Béné et al., 2015). Challenges to land-based production such as shortages of water for irrigation (Elliot et al., 2014), global warming and loss of fertile soils (Rosenzweig et al., 2014) will all potentially increase the role of seafood in food security.

The dependence of humanity on seafood goes far beyond simple protein consumption. Fish have a reputation as a source of long-chain n-3 (or omega-3) polyunsaturated fatty acids, but they are also a crucial source of readily accessible (bioavailable) micronutrients – such as selenium, zinc, iron, calcium, iodine and vitamins A, B and D (FAO, 2016; Golden et al., 2016). A reduction in the availability of seafood could see 19% of the global population at risk of increased perinatal, child and maternal mortality; growth and cognitive retardation; reduced immune function and other conditions associated with malnutrition (Golden et al., 2016). Many of these countries are situated in equatorial South East Asia, Africa and Latin America. There is little capacity to look to other food sources in these countries, as poverty means they have a limited capacity to switch to alternative dietary sources. A livelihood feedback effect entrenches this dependence as the same countries also have a heavy reliance on fisheries as sources of employment and are economically sensitive to shifts in seafood production (Allison et al., 2009).

22.2 Pressures on Seafood Production

The current state of the world's fisheries is not completely transparent, as many species are not directly assessed. The FAO states that approximately 68% of

[1] Here we use the word seafood to refer to materials from both freshwater and marine flora and fauna. While inland, freshwater, sources of production do account for 65% of aquaculture's annual tonnage, it still only makes up 35% of total production. Thus we have decided to use the term seafood in the general sense unless a clear distinction is needed for clarity.

fish stocks are within biologically sustainable levels, though the vast majority of these (58.1% of all stocks) have no capacity for increased production as they are already at (or close to) their maximum sustainable production (FAO 2016). Within the broader scientific literature, multiple lines of evidence suggest that approximately two-thirds of wild stocks are still currently overfished or recovering from past over-exploitation and depletion below sustainable levels (Worm et al., 2009, Costello et al., 2012, Costello et al., 2016). However, it is not all bad news. The livestock and aquaculture feed industries, which are significant consumers of global fishmeal and fish oil production (Shepherd and Jackson, 2013), are beginning to shift to fishmeal-free feeds (Skretting, 2015), helping to reduce absolute pressure on marine fish stocks. In terms of fisheries activities, the recent introduction of import controls in the EU and the United States, which require the source nations to be following sustainable fisheries and conservation practices, has seen rapid responses within fisheries and management agencies in many of the most problematic fishing nations (e.g., in Asia; EJF et al., 2016). Moreover, there are new audit processes relating to local and distant water fleets (e.g. for European fleets) that are also acting to move global and high seas fishing practices to a more sustainable footing. This occurs hand in hand with the increasing use of satellite technology to identify illegal, unreported and unregulated (IUU) fishing. Management agencies can now make informed decisions in light of these IUU activities (and the pressures they add to stocks); they also help in the eradication of such activities. All of this reflects growing environmental awareness that is pushing for more equitable and sustainable ocean solutions.

Unfortunately, direct fishing pressure is but one of a litany of stressors currently affecting aquatic ecosystems and their production. Human activity now touches all areas of the globe, even the most remote oceanic locations (Halpern et al., 2008; Halpern et al., 2015). Pollution and eutrophication may have been the most evident non-fishing effects of human activities on aquatic systems prior to the twenty-first century, but large-scale environmental change has since added shifting currents and (typically) rising temperatures, sea level rise due to the thermal expansion of the oceans and ice melt, thermal stratification, deoxygenation, ocean acidification, and changes in the periodicity and amplitude of extreme weather events and multi-year and decadal climate cycles, such as El Niño (IPCC, 2013). Of these, ocean warming has garnered the greatest attention. The physical extent of the oceans means they have a large role in climate buffering, accounting for about 93% of the additional energy that has seen the surface layers warm on aggregate by more than 0.3 since 1971 (IPCC, 2013) and hotspot locations by as much as 1.5 °C–2.5 °C over the past hundred years (Hobday and Pecl, 2014). Extreme weather events (e.g., droughts, floods, storms and heatwaves on land and in the sea), which are becoming more frequent (Perkins et al., 2012), can exacerbate thermal stresses and cause step-wise changes in systems (Hobday et al., 2016). Temperature shifts and ranges are considered to be among the major drivers of shifts in phenology and species distributions that have already been observed globally across all taxa (Poloczanska et al., 2013). These shifts are not restricted to wild species; aquaculture is seeking to shift to deeper waters (e.g., salmon farms in Tasmania, Australia) or into enclosed land-based facilities as near-shore waters warm past thermal optima. Again, the picture is not uniformly bleak. Acclimatization and evolution will likely see some species cope much better than currently predicted. Furthermore, some species and locations benefit from the coming changes. For example, catches in the high latitudes could increase, by as much as 10%–40%, potentially even doubling in the high Arctic (Cheung et al., 2016). Uncertainty remains high around the true magnitude of such change, however. While warming is predicted to lead to poleward shifts in productive areas (Yool et al., 2015), the ongoing loss of sea ice – which has contracted, on average, by 35,000 km^2 annually since 1979 (Parkinson, 2014) – may lead to increased stratification, the suppression of the spring plankton blooms (Yool et al., 2015) and potential shifts in the plankton food webs to more complicated networks with less available energy for fish and other top predators (Murphy et al., 2016). This means that the net outcome for polar fisheries production remains unclear (Pecl et al., 2017).

Ocean acidification appears to be the other important driver of major faunal shifts to date. The oceans have absorbed 50% of the emitted atmospheric carbon (Sabine et al., 2004), which has acidified these water bodies. In addition, the rate of anthropogenic CO_2 release is leading to not only a relatively rapid decrease in ocean pH, but also a decreased calcium carbonate saturation (which is critical for shell and skeletal formation by marine organisms). Together this is creating conditions potentially unmatched for at least 300 million years (Hönisch et al., 2012) and presenting a range of physiological challenges to marine flora and fauna, particularly calcifying organisms (Doney et al., 2009). The full extent of potential effects of ocean acidification on marine habitats, ecosystems and their productive capacity is quite broad – as some species seem relatively unaffected while others are quite sensitive (Branch et al., 2013). Impact mechanisms include effects on larval and juvenile survival, shifts in sensory capacity and resulting behavioural responses, modified growth rates and habitat losses (Branch et al., 2013).

The combined effects of temperature and acidification on fish production have been explored in a number of modelling studies, which highlight the potential for complex interactions between the drivers (Griffith et al., 2012) and spatial heterogeneity in outcomes – with an expansion of biodiversity and catches in temperate zones, of as much as 80% or more, and a loss in the tropics of up to 50% (Cheung et al., 2010; Jones and Cheung, 2015). Ocean acidification has as much potential to influence aquaculture practices, in particular those that have been traditionally reliant on water sourced from oceanic or near-shore waters. This has been the case in the US Pacific Northwest, where coastal oyster growers and local shellfish fisheries have already lost nearly US$110 million (jeopardizing approximately thirty-two hundred jobs) as a result of acidified water (caused by upwelling of CO_2-enriched waters and local factors such as eutrophication and river discharge; Ekstrom et al., 2015).

The implications of changing precipitation (and thus near-shore and estuarine salinity patterns), winds, ocean thermal stratification and hypoxia have received less attention in the literature, with regard to their potential ecosystem and food security effects. What is known indicates that their influence will be complicated and situation-specific. For example, increased stratification in the Peruvian upwelling ecosystems, due to weakening winds, impairs the delivery of nutrients to surface waters, leading to a substantial drop in primary production (by a factor of between 5 and 20), in turn leading to a drop in anchovy biomass and catch (Barber and Chavez, 1983) and widespread mortality of seabirds and other top predators (Apaza and Figari, 1999). In contrast, long-term warming and reduced ventilation (due to stratification) in the North Pacific are projected to reduce oxygen concentrations in the California current system; this (along with enhanced local production due to increased delivery of upwelled nitrate) is expected to lead to an expansion of underlying oxygen minimum zones, where oxygen levels are low enough to constrain the presence of fish and other species (Capone and Hutchins, 2013). Whilst the additional surface production initially sounds positive, there is a good chance it will be of a kind that is directed away from food webs that ultimately supply species targeted by fisheries and will instead be trapped in the microbial food web (Cotner and Biddanda 2002). The subsurface deoxygenation is also problematic, as it has been shown that the lower oceanic oxygen content (in combination with increasing ocean temperatures) could see assemblage-averaged maximum body weight shrink by as much as 14%–24% (Cheung et al., 2013), with implications for fecundity (which is strongly size-related in many fish; Beldade et al., 2012), size-based feeding interactions and ultimately potential production. Where deoxygenation continues, extreme hypoxia can result and sulphide production can occur, killing fish and shellfish and damaging fisheries capacity (Capone and Hutchins, 2013), as observed after particularly intense upwelling events off the coasts of Peru (Dugdale et al., 1977) and Namibia (Brüchert et al., 2006). Furthermore, the shoaling of deoxygenated water is squeezing marine ecosystems into shallower depth bands (Koslow and Davison, 2016), changing the nature and extent of food web interactions, with as-yet-unknown consequences.

The nature and extent of some of the historical pressures on marine ecosystems (and thus the species that support marine food security) have grown in pervasiveness. Pollution of oceanic and near-shore waters is now widespread (Halpern et al., 2008), whether in the form of fertilizers, herbicides, pesticides, pharmaceuticals, heavy metals, textile fibres, plastics or other debris. These pollutants not only directly affect fish (e.g., via immunosuppression and malformation; Nicklisch et al., 2016), the broader ecosystem and their productivity, but can accumulate and create health risks for any human consumers (Whitmee et al., 2015). Furthermore, much of this pollution originates (or is most evident) in countries most dependent on seafood – e.g., China, Indonesia and many other low-latitude countries in Southeast Asia, Africa and Latin America. Take plastic pollution as a case in point. With more than 300 million tonnes of plastic produced per annum (Lönnstedt and Eklöv, 2016), it is unsurprising that a substantial tonnage, of the order of 4.8–12.7 million tonnes, enters the ocean each year (Jambeck et al., 2015). Plastic contamination directly kills a number of marine species (e.g., turtles, sunfish, seabirds and seals; Derraik, 2002), through either direct entanglement or consumption and consequent physical damage (Butterworth, 2016) or toxic leaching (Lönnstedt and Eklöv, 2016). Microplastics (tiny plastics fragments and beads less than a millimetre in size) are even more insidious, as these microscopic products of the slow breakdown of plastic fibres are now accumulating throughout the world's oceans (Wilcox et al., 2015) and have been shown to produce similar effects to acidification – i.e., reduced larval survival and shifts in sensory capacity and resulting feeding and predator avoidance behaviours (Lönnstedt and Eklöv, 2016).

Human activities are also changing ocean habitats in other ways. Near-shore industries and terrestrial land use have increased turbidity (cloudiness) in many waterways where coastlines are developed – for example, in the coastal waters off the Great Barrier Reef (Brodie et al., 2012). The impacts of increased turbidity, sedimentation and concomitant reductions in light for shallow water habitats, such as corals and seagrass, are reasonably well appreciated (Sofonia and Anthony, 2008), and have been among the motivating forces for improved waste water treatment in coastal areas (Evans et al., 2017). The influence of noise and light pollution (e.g., due to increasing levels of shipping, where container port traffic has more than doubled every decade for the last forty years) is less well understood. Awareness around how light pollution can increase the mortality of turtle hatchlings has led to conservation advances in that arena (Thums et al., 2016), but the ultimate effects of the use of sub-surface lights from deep sea mining equipment remain unclear (Van Dover, 2014) because of the nascent nature of that industry. The individual species level implications of noise pollution are beginning to be synthesised, although ecosystem-level effects have not yet received the same attention. Nevertheless, the World Health Organization has classified omnipresent anthropogenic noise as one of the most hazardous forms of pollution (Kunc et al., 2016). Noise is intentionally produced for seismic exploration and sonar, but is also an unintentional by-product of marine industries – such as mining, shipping, fishing – and leisure activities such as recreational boating (Hildebrand, 2009). Noise is known to impact upon behaviour (e.g. foraging or communication) and can influence social bonds and population health, even without the additional direct effects on physiology and anatomy and consequent mortality and growth effects (Popper and Hastings, 2009; Kunc et al., 2016).

These many pressures are putting cumulative stress on the world's oceans and their capacity to supply a sustainable food supply. Whereas evolution, acclimatisation and behavioural changes are possible adaptation responses (at least up to a point depending on physiological optima and plasticity; Hoegh-Guldberg and Bruno, 2010), the rate of change and the level of stress induced are important constraints. Adaptation has already been observed, for instance, changes in vocal profiles in response to noise (Francis and Barber, 2013), with the most commonly observed responses being shifted phenology and spatial distributions driven by temperature, habitat loss (e.g., sea ice), acidification and their interaction with the other stressors (Poloczanska et al., 2013). These responses are not uniform, however. For instance, species that begin with large latitudinal range sizes, are omnivorous or have high adult mobility have shown faster poleward range extensions (Sunday et al., 2015). This is leading to the formation of novel ecosystems, with species compositions not previously observed (Hoegh-Guldberg and Bruno, 2010). The nature of the interactions in these ecosystems and the connectivity between them (given changing ocean currents) have as-yet-uncertain implications for global seafood production. It is likely that sustainable fisheries with production on a par with today, or even greater, will be possible in some locations (though the composition of the catch may need to change), while other locations will lose productive potential. Whilst there have been projections of potential production based on additive species distribution models (e.g., Cheung et al., 2010), few such projections have taken into account the implications of the loss of habitats (or their complexity) that appears likely (Hoegh-Guldberg and Bruno, 2010). This is despite the critical dependence of many exploited species on suitable habitat. Similarly, the implications of the effects of exotic species on ecosystems (e.g., witness the effect of the range-extending *Centrostephanus rodgersii* sea urchin on Tasmanian kelp forests, which has led to a reduction in biodiversity and urchin barren formation, a phenomenon not previously seen in the region; Ling et al., 2009) have yet to be considered in depth in projections, despite being some of the first directly attributable effects of global change on the oceans.

Another little-considered influence of climate on the oceans is the increased potential for severe marine disease outbreaks (Harvell et al., 2002). Marine pathogens affect all of the species targeted by fisheries or used in aquaculture (Burge et al., 2014). Aquaculture already faces significant sanitary costs with outbreaks of viruses and parasites having devastating effects on production. In 2007 an outbreak of viral Infectious salmon anaemia (ISA) began in Chile; over the next couple of years it saw a 85% reduction in the production of cultured salmon in Chile, resulting in the loss of 570,000 t of fish (Dhar et al., 2014), valued at USD$2 billion, and the loss of approximately twenty-five thousand jobs (World Bank, 2014). Intensive production practices exacerbate pathogen risks, but climate drivers inflate the risk still further – as the pathogens are found in new locations or with new virulence, while stressed hosts are more vulnerable to infection (Burge et al., 2014). Wild fish kills resulting from epizootic infection also occur, but how these will interact with climate and other stressors is not well understood (Burge et al., 2014). What is known is that terrestrial disease management procedures (e.g., quarantining, culling and vaccinating) are not particularly successful in the ocean (Burge et al., 2014). Consequently, it will be important going forward for fisheries

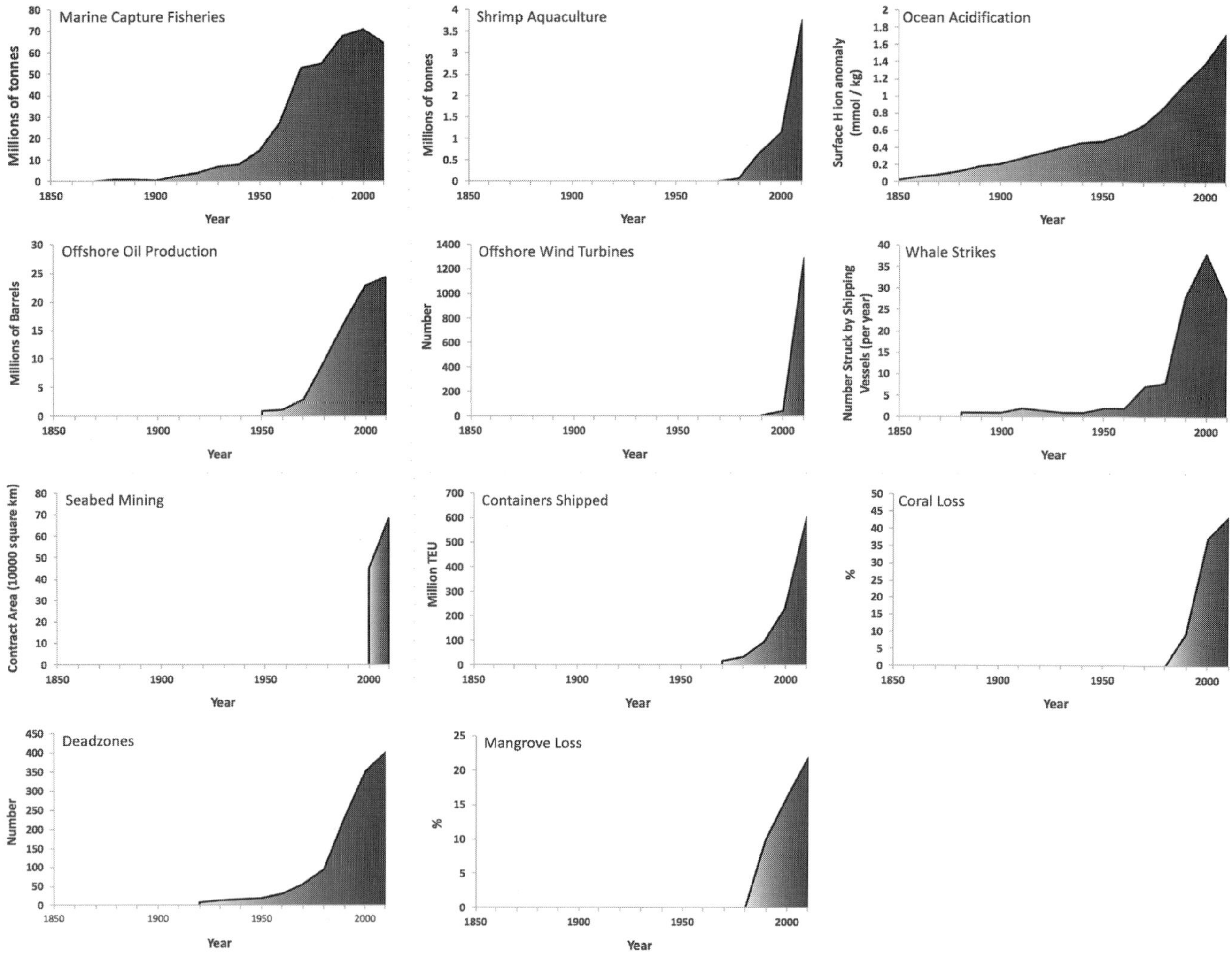

Figure 22.1 Trends in ocean variables – stressors and habitat loss. Data sourced from World Bank (http://data.worldbank.org), EIA (2016), FAO (2016), ISA (www.isa.org.jm), NRC (2014), REN21 (2015), McCauley et al. (2015).

and aquaculture to integrate disease monitoring and control into management practices – beginning with efforts to improve understanding of disease and susceptibility, followed by the development of new control measures and improved capacity to forecast outbreak conditions (Burge et al., 2014). This will help protect not only production, but also human health, as there can be devastating effects (acute and chronic) of the accidental ingestion of marine toxins (Roberts et al., 2016).

A final pressure being put on marine systems that may influence seafood production is the growing number of artificial structures in the coastal and marine space (e.g., artificial reefs, oil platforms or wind turbines; see Figure 22.1). The number of such structures is set to continue growing as humans move more industrial activity into the oceans, but also as they increase man-made coastal defences to protect against rising sea levels and increasingly severe storms (Asif and Muneer, 2007). A growing appreciation of the dual roles of such structures is seeing their designs incorporate ecological engineering solutions that can provide additional habitat and support multiple ecosystem functions (Dafforn et al., 2015).

22.3 Human Pressures

This inventory of impact and how it may change the productivity of the oceans and the security of those dependent upon them stands against a backdrop of accelerating human use of marine and coastal zones (Figure 22.1). Not only does the global human population continue to grow to historically unprecedented levels (projected to reach up to 12.3 billion by 2100; Gerland et al., 2014), but it will be increasingly concentrated in urban areas along the coastal zones (Crossland et al., 2005) and in the very countries identified as being the most dependent on fisheries and vulnerable to marine production losses (Bradshaw and Brook, 2014). In parallel with this increasing population is an accelerating expansion of marine industries (McCauley et al., 2015). There is growth in all marine sectors – across industries long established as major marine players (e.g., shipping), smaller marine users that have recently grown to dominate economically (e.g., oil and gas) and uses that are only just beginning in earnest (e.g., seabed mining and energy generation). For example, ship-based trade has always been a major marine user and now tops 600 million containers per year and continues to grow exponentially (World Bank data available from http://data.worldbank.org/indicator/IS.SHP.GOOD.TU). Oil and gas extraction has grown to be the largest marine sector by value, at $396 billion per year globally. While extraction appears to have stabilised at about 25 million barrels per day (because of production costs; EIA 2016) it is pushing into new spatial areas – such as the Arctic and deep ocean – where effective spill response strategies will be more challenging, especially as it is currently unclear whether clean-up technologies developed in temperate or tropical waters will be effective in polar latitudes, where oil weathering behaves differently (NRC 2014). Seabed mining is an example of an industry just beginning to grow, with 1.2 million km^2 under contract (International Seabed Authority, www.isa.org.jm). Similarly, ocean renewable energy is still in its earliest stages, but already produces in excess of 530 MW per year with investment currently doubling year on year (REN21, 2015).

The large-scale expansions of use and footprint witnessed terrestrially following the Industrial Revolution, and again in the great acceleration of the 1950s, appear to have finally reached the oceans. Terrestrial experience indicates that large-scale changes in the form and productivity of those ecosystems will follow. Indeed, the tension between hunter-gathering approaches to marine harvest and farming-style practices is already manifest, with marine wild capture harvest stagnating (even declining slowly) and the growth of aquaculture topping 5% annually (FAO, 2016). This rapid growth rate has been necessary as human consumption rates have been growing by 10%–25% per decade, driven by rising urbanisation (city dwellers spend more of their income on food stuffs, particularly proteins), affluence and the emergence of a middle class in developing nations, and globalisation of food systems with a concomitant tendency to increased protein consumption worldwide (FAO, 2016). This globalisation and homogenisation of diets have been made possible by trade liberalisation and technological innovations and improvements along the entire supply chain (processing and preparation, transportation, distribution and marketing; FAO, 2016). Further changes have been driven by shifting consumer behaviour and habits, with convenience and price drivers now competing with sustainability and safety as determinants of choice, supported by the rise in eco-labelling programs such as the Marine Stewardship Council and Oceanwatch (Leadbitter and Benguerel, 2014). All of this has boosted the consumption of seafood, as has the growing dominance of transnational retail chains and supermarkets – which now represent the site of more than 70%–80% of retail purchases of seafood, even in developing countries, as they offer wider and (often) safer choices and reduced seasonal fluctuation in product availability (FAO, 2016). This has seen the gap on fish consumption per capita begin to converge on the rate in industrialised nations, and while these trends in consumption and drivers will ultimately plateau (as they have in saturated markets in the developed world) they are unlikely to reverse and so the demand for seafood will remain high into the future.

Seafood will also retain an important position in general food security. Despite economic statements regarding substitutability of seafood (Chern et al., 2002), it is an important staple for many island nations (who have no capacity to switch to the production of other food stuffs) and for the poorest people on the planet. Seafood will remain an important means of avoiding the worst effects of malnutrition, with aquaculture becoming increasingly important for food security as a means of providing significant production of inexpensive protein for domestic, rural consumption (FAO et al., 2014; FAO, 2016). These low-value species are typically not as nutritious as the species preferred by wealthier consumers; nor are they as nutritionally valuable as the oil-rich forage fish currently accessible to impoverished communities at local markets (Golden et al., 2016). Nevertheless, integrated aquaculture remains a vital means of contributing to local diets and economies via marrying aquaculture and small-scale agriculture (e.g., fish raised in rice paddies; Golden et al., 2016).

At present, nearly 11% of the world's population is undernourished, with 98% of those people concentrated in Africa and Southern and Eastern Asia. Population projections indicate that these countries will continue to face poverty (World Bank 2016) and food security issues into the future (Allison et al., 2009). These nations seem to be caught in the perfect storm of greatest sensitivity to lost production (Allison et al., 2009) and exposure to some of the most negative effects of climate change on marine and coastal ecosystems – both with respect to potential biodiversity (Molinos et al., 2015) and fish production (Cheung et al., 2010; Jones and Cheung, 2015; Cheung et al., 2016), and in terms of lost agricultural production (Battisti and Naylor 2009; Gornall et al., 2010) with potential cereal losses of up to 50% (Rosenzweig et al., 2014). As climate stress on crops and livestock becomes global in character, as projected for the end of the century, these hard-hit areas will find it increasingly difficult to rely on trade to make up their deficits as there will be fewer food surpluses to draw upon, unless there is significant adaptation of agroecosystems (Battisti and Naylor, 2009). These trade pressures will also be felt in wealthier nations, which currently subsidize their own sustainability by importing large proportions of their seafood (CBI, 2015). The situation is likely to worsen unless fisheries management reforms are enacted, as continuation of business as usual in fisheries management at a global scale would see further declines in the number of sustainable stocks such that an estimated 88% of all stocks would be overfished (Costello et al., 2016). This will be a significant challenge in those same nations repeatedly identified as the most vulnerable as they also represent the locations of the highest numbers of fishing vessels – Asia alone is estimated to have 3.5 million vessels, 75% of the global fleet (FAO, 2016).

22.4 Future Options

The many pressures on the global oceans potentially paint a pretty bleak picture of future fish production potential and its role in global food security. However, it need not be that way, necessarily. Regional projections (e.g., for south eastern Australia) indicate that sustainable fisheries at or above those levels seen today are possible so long as carbon emissions can be kept below the levels seen in the highest IPCC scenarios, although catch composition would need to shift to match ecosystem shifts (Fulton and Gorton, 2014). Work in the Pacific (Bell et al., 2013) has shown that even with large-scale shifts in the distribution of tropical tunas and the decline of reef fish, Pacific Island nations should be able to maintain food security by building on existing transnational cooperative partnerships, progressively shifting from coral reef fish to pelagic fish (particularly tuna) and making the most of increased rainfall to support more freshwater aquaculture. At a global scale, Costello et al., (2016) showed that if institutional reforms could be implemented, 98% of all stocks would be at sustainable levels by 2050, increasing supply by 16 million tonnes (and improving economic performance by $53 billion). Further improvements are possible if the current losses along the supply chain are reduced by improved efficiencies, technologies and behavioural practices (Plagányi et al., 2014).

The potential total capacity of wild ocean production remains a topic of considerable interest, with estimates ranging from 50 million (Graham and Edwards, 1962) to 410 million tonnes (Wilson et al., 2009), with many converging on estimates in the range of 100–250 million tonnes (see Table 22.1). The realised answer will no doubt depend on what mix of species is ultimately harvested (Fogarty et al., 2016). For example, whether uses are found for currently unfished species – such as the enormous biomass of mesopelagic fish, which are ubiquitous at depths of 200–1000 m around the globe and may total 115 billion tonnes (Irigoien et al., 2014). Total harvest will also depend on whether stocks are fished sustainably, as depleted stocks do not produce as much catch as stocks maintained at sustainable target levels (Costello et al., 2012). As much as 50% of the catch could be lost if stocks remain depleted into the future (Costello et al., 2012). There is a strong contrast in the success of fisheries management in different locations around the world, with many developing nations having several stocks still in a poor state and overfishing still widespread (FAO, 2016). However, in other locations (primarily in the developed world) significant advances have been made in the last decade. For example, adoption of harvest strategies in Australia has led to significantly improved biological and economic performance of fisheries (with >80% of stocks now sustainable). This has involved significant upfront costs ($220 million AUD to restructure the fleets and remove excess capacity and then significant further investment in research to design, test and implement the strategies) and is likely not easily replicated in all locations (Smith et al., 2014). Nevertheless, lessons learnt regarding the value of transparency and clear policy are transferable, as are the growing number of assessment methods for use in data-poor situations, and so there can be optimism that stock recoveries in the majority of locations are not implausible. Indeed, the rapid response to the change in EU laws on fisheries imports indicates how responsive fisheries can be when clear incentives are laid out.

The pattern of fishing may also prove important (Garcia et al., 2012). There has long been a cultural difference in the fishing approaches in different countries. Some countries have chosen to "fish down the food web" (Pauly et al., 1998), targeting large fish first and moving on to smaller species when the larger fish

Table 22.1 *Example Estimates of Total Marine Exploitable Production (Assuming 20% of Total Biomass Can Be Sustainably Exploited at An Ecosystem Scale)*

Estimate of potential total exploitable production (millions of tonnes)	Comments on composition or assumptions	Reference
50–115	Transfer efficiency (from trophic level to trophic level) method	Graham and Edwards 1962
200	Transfer efficiency (from trophic level to trophic level) method	Schaefer 1965
150	Transfer efficiency (from trophic level to trophic level) method	Ricker 1969
100	Transfer efficiency (from trophic level to trophic level) method, partitioning spatially and by ecosystem and food web type	Ryther 1969
100, 260–350	Catch extrapolation (lower rate is for existing target species, higher if targeting extended to lightly exploited species: krill, mesopelagics etc)	Gulland 1970
180–205	Size-based macroecological approach	Jennings et al. 2008
220	Based on large marine ecosystem scale trophic models	Christensen et al. 2008
180–410	Used two independent methods (size-based macroecological approach and a combination of 66 LME scale trophic models).	Wilson et al. 2009
146	Empirical relationship between primary production and fish catch	Chassot et al. 2010
180–230	Transfer efficiency (from trophic level to trophic level) method, with extended food web, spatial parameterisation and conditioned on satellite primary production	Rosenberg et al. 2014
14–100	Macroecological, size- and trait-based models	Jennings and Collingridge 2015

Source: Rosenberg et al 2014

begin to be depleted; other countries have "fished through the food web" (Essington, et al., 2006), adding ever more species. While it is clear that fishing for what is accessible and has a market value has been a universal characteristic of fisheries (Sethi et al., 2010), it is equally clear that discarding and targeting patterns are not the same in different locations. People visiting a fish market in Southeast Asia will find many more species, fish and invertebrates, than if they wander through a similar market in Australia or North America. Consideration of alternative fishing patterns across food webs has shown that broader (but still sustainable) targeting of species, so that ecosystem structure is maintained, has the potential to support higher catches than more selective fishing approaches (Garcia et al., 2012). This does not mean that this approach is free of impact, or of ecosystem considerations; nor would it necessarily be easy to implement, but it does highlight that the topic of sustainable multispecies harvesting will likely receive significant future attention.

Model-based projections of future fisheries production are becoming more common and are beginning to accrue across a broader diversity of approaches – from trophic-based modelling platforms commonly used at regional scales (e.g., EcoOcean; Alder et al., 2007) through to global models of species distribution (e.g., Cheung et al., 2010), macroecological or size-based formulations (Barange et al., 2014). Very few of these projections suggest there will be a significant drop in wild capture production in the next forty years at a global level – Alder et al., (2007) and Costello et al., (2016) forecast a <7% drop under BAU or increasingly intense fishing pressure in combination with a high-emissions climate scenario. The majority of global projections indicate global catches will remain steady or marginally increase (by <3%–6%; Cheung et al., 2010; World Bank, 2013; Merino et al., 2012; Barange et al., 2014, Cheung et al., 2016). However, uncertainty in the estimates was not considered in any great depth until recently when Cheung et al., (2016) found results were sensitive to the species distribution model used (indicating global catch decreases up to 13% for some model types). Moreover, the relatively robust global catch estimates have high spatial variability, with all models indicating quite large changes regionally – with 30%–50% drops seen in many tropical locations and increases in the temperate and sub-polar regions of similar magnitudes (though these increases may not be fully realised as a result of ocean acidification; Cheung et al., 2011). Work to date indicates that at large scales (e.g., global) total biomasses may be relatively predictable, because of primary production constraints and macroecological principles, but composition will be uncertain – meaning that flexibility to redirect future fishing capacity as needed given realised species shifts will be crucial (Jennings and Brander 2010). This may necessitate (i) encouraging engineers to design vessels capable of easily switching among fisheries and target species, (ii) developing markets that can cope with changing and diverse fish products and (iii) supporting local communities to shift behaviours to increase their adaptive capacity (Allison et al., 2009; Fulton and Gorton 2014).

Given growing recognition of the role of social factors in exacerbating or ameliorating environmental problems, additional solutions may result from a greater focus on political ecology or environmental justice (e.g., Dodds, 1997) and the role that policy interventions can play in addressing ecology and equity concerns in global fisheries (Rammelt and Schie, 2016). As an example, Cinner et al. (2016) identify characteristics necessary to sustain healthy ecosystems in the face of changing environmental conditions and socioeconomic drivers, with so-called bright spots identified as places where ecosystems are substantially better than expected. In addition to environmental conditions, they highlight several socioeconomic drivers that contribute to the success of bright spots, including strong sociocultural institutions, customary taboos and marine tenure and high levels of local engagement in management. Recently Doyen et al., (2016) revisited the notion of Tragedy of Open Access and over-exploitation and used a modelling approach to demonstrate the beneficial role of cooperation in reducing harvests and promoting sustainable harvesting. The role of gender in natural resource management is also gaining increasing attention, and some recent work suggests that increasing participation by women in fisheries and decision making may lead to more sustainable use of natural resources, as they appear to change their behavior more readily to reduce resource extractions (Revollo-Fernández et al., 2016).

For all the attention that fisheries projections have received, trends in production over the last half-century clearly indicate that aquaculture will be the dominant producer of seafood. Currently, fifteen countries supply 93% by weight of cultured fish products (FAO, 2016). Sustainable expansion into other geographic areas, as well as potential expansion into other species, will likely be a key feature of local and regional food security. A number of largely economic projections for the growth of aquaculture have been undertaken in the last decade – indicating an increase of somewhere between 1% and 35% in the next fifteen years (World Bank, 2013; FAO, 2016), although the lower end of the estimates is no longer plausible as it is already close to being exceeded. The true capacity for (and modes of) sustainable aquaculture remains unclear, however. The industry as a whole is currently in a rapid growth phase of 5% per annum, though this is down from the historic highs of >10% per annum in the 1980s (World Bank 2013). Growth in the aquaculture production of seaweeds is also opening up more food and livelihood options (Buschmann et al., 2008). New more efficient technologies and integrated multi-trophic aquaculture methods are also being introduced (Troell et al., 2009), where the nutrients and waste particles from finfish cages are captured by nearby algal and bivalve cultures, transforming the waste into additional commercially valuable crops, reducing the overall ecological footprint (Neori et al., 2004) and in some instances reducing costs (Troell et al., 2009) – e.g., where biofuel is generated from algae grown on nutrients from fish pens. The growth in interest in engineered offshore platforms could also see production rise (Troell et al., 2009). A further complicating factor is the capacity of the aquaculture industry to manage sanitary threats and to shift to a more sustainable footing – at present expansion in some locations and for some species has come at the expense of severe environmental or social problems (Goldburg et al., 2001; Baines et al., 2015). There are examples of low-footprint aquaculture production systems (e.g., for bivalves), and the proportion of production made up of these species and species with high production efficiency is growing (FAO, 2016) so the high-end estimates do not appear unreasonable. Importantly, given the projected potential climate-induced reduction in agricultural (Rosenzweig et al., 2014) and wild capture fisheries production (Cheung et al., 2016) in Asia and Africa there is significant potential for expansion of aquaculture, for domestic and international markets, if they are prepared to invest in sustainable aquaculture (World Bank, 2013).

By and large these projections of fisheries and aquaculture production have been done without consideration of the potential influences of the cumulative effects of global change, focusing primarily on climate effects and direct potential drivers of the expanding human population on fisheries demand. Similarly, truly novel production platforms have so far been given little consideration. One area of marine production that has received little attention to date is marine horticulture. Nonetheless, trial commercial enterprises suggest that this technology may be a feature of future food security. This technology has a few different guises. The most conventional is the use of aquaponics, or variants on that theme, where the waste from aquaculture ponds fertilises hydroponic gardens or halophyte (salt-tolerant) crops, which either go direct to market for human consumption or are fed to livestock (Rakocy, 2012). A somewhat related approach is to build hothouselike structures on floating platforms, a simple extension of land-based methods to oceangoing platforms that act as reclaimed land (Maeda, 2015). However, a slightly different approach is to have fully submerged structures that are attached to the seabed in relatively shallow waters (less than ten metres). Inside are hydroponic beds that are watered using desalination and are kept pest free and in relatively constant temperatures due to the stability of the surrounding seawater. Such novel structures may feature more widely in the production systems of the future; only time and human ingenuity will tell.

22.5 Conclusions

Seafood appears to have an important on-going role in world food security, though it may be from new sources – existing estimates suggest as much as two-thirds of future seafood for human consumption will be farm-raised (World Bank, 2013). Traditional seafood production schemes are coming under increasing pressure as the Anthropocene reaches the oceans in full force. Integrated system-level understanding and novel solutions are required to find and support the new options needed to deal with an ocean as we have never experienced before.

Acknowledgements

The authors would like to acknowledge support for work covered here from the CSIRO, the Fisheries Research and Development Corporation and Australian Research Council Discovery project support (DP140101377).

References

Alder, J., Guénette, S., Beblow, J., Cheung, W., and Christensen, V. (2007) Ecosystem-based Global Fishing Policy Scenarios. *Fisheries Centre Research Reports* 15(7).

Allison, E.H., Perry, A.L., Badjeck, M.-C., Adger, W.N., Brown, K., Conway, D., Halls, A.S., Pilling, G.M., Reynolds, J.D., Andrew, N.L., and Dulvy, N.K. (2009) Vulnerability of national economies to the impacts of climate change on fisheries. *Fish and Fisheries*, 10, 173–196.

Apaza, M., and Figari, A. (1999) Mortandad de aves marinas durante El Niño 1997–98 en el litoral sur de San Juan de Marcona, Ica, Perú. Revista Peruana de Biología, Vol. Extraordinario, pp. 110–117.

Asif, M., and Muneer, T. (2007) Energy supply, its demand and security issues for developed and emerging economies. *Renewable and Sustainable Energy Reviews*, 11, 1388–1413.

Barange, M., Merino, G., Blanchard, J.L., Scholtens, J., Harle, J., Allison, E.H., Allen, J.I., Holt, J., and Jennings, S. (2014) Impacts of climate change on marine ecosystem production in societies dependent on fisheries. *Nature Climate Change*, 4, 211–216.

Barber, R.T., and Chavez, F.P. (1983) Biological consequences of El Niño. *Science* 222, 1203–1210.

Battisti, D.S., and Naylor, R.L. (2009) Historical warnings of future food insecurity with unprecedented seasonal heat. *Science*, 323, 240–244.

Baines, J., Baines, T., and Quigley, R. (2015) *The Social and Community Effects of Aquaculture: A Case Study of Southland Aquaculture*. New Zealand Ministry for Primary Industries report.

Beldade, R., Holbrook, S.J., Schmitt, R.J., Planes, S., Malone, D., and Bernardi, G. (2012) Larger female fish contribute disproportionately more to self-replenishment. *Proceedings of the Royal Society B*, 279, 2116–2121.

Bell, J.D., Ganachaud, A., Gehrke, P.C., Griffiths, S.P., Hobday, A.J., Hoegh-Guldberg, O., Johnson, J.E., Le Borgne, R., Lehodey, P., Lough, J.M., Matear, R.J., Pickering, T.D., Pratchett, M.S., Gupta, A.S., Senina, I., and Waycott, M. (2013) Mixed responses of tropical Pacific fisheries and aquaculture to climate change. *Nature Climate Change*, 3, 591–599.

Béné, C., Barange, M., Subasinghe, R., Pinstrup-Andersen, P., Merino, G., Hemre, G.-I., and Williams, M. (2015). Feeding 9 billion by 2050—putting fish back on the menu. *Food Security*, 7, 261–274.

Bradshaw, C.J.A., and Brook, B.W. (2014) Human population reduction is not a quick fix for environmental problems. *Proceedings of the National Academy of Sciences*, 111, 16610–16615.

Branch, T.A., DeJoseph, B.M., Ray, L.J., and Wagner, C.A. (2013) Impacts of ocean acidification on marine seafood. *Trends in Ecology & Evolution*, 28, 178–186.

Brodie, J.E., Kroon, F.J., Schaffelke, B., Wolanski, E.C., Lewis, S.E., Devlin, M.J., Bohnet, I.C., Bainbridge, Z.T., Waterhouse, J., and Davis, A.M. (2012) Terrestrial pollutant runoff to the Great Barrier Reef: An update of issues, priorities and management responses. *Marine Pollution Bulletin*, 65, 81–100.

Brüchert, V., Currie, B., Peard, K.R., Lass, U., Endler, R., Dübecke, A., Julies, E., Leipe, T., and Zitzmann, S. 2006. Biogeochemical and physical control on shelf anoxia and water column hydrogen sulphide in the Benguel: A coastal upwelling off Namibia. In: *Past and Present Water Column Anoxia*, L.N. Neretin (Ed.). Springer. pg. 161–193.

Burge, C.A., Eakin, C.M., Friedman, C.S., Froelich, B., Hershberger, P.K., Hofmann, E.E., Petes, L.E., Prager, K.C., Weil, E., Willis, B.L., Ford, S.E., and Harvell, C.D. (2014) Climate change influences on marine infectious diseases: Implications for management and society. *Annual Reviews in Marine Science*, 6, 249–77.

Buschmann, A.H., Varela, D.A., Hernández-González, M.C., Huovinen, P., (2008)Opportunities and challenges for the development of an integrated seaweed based aquaculture activity in Chile: Determining the physiological capabilities of *Macrocystis* and *Gracilaria* as biofilters. *Journal of Applied Phycology*, 20, 571–577.

Butterworth, A. (2016) A Review of the welfare impact on pinnipeds of plastic marine debris. *Frontiers in Marine Science*, 3,149. doi: 10.3389/fmars.2016.00149.

Capone, D.G., and Hutchins, D.A. (2013) Microbial biogeochemistry of coastal upwelling regimes in a changing ocean. *Nature Geoscience*, 6, 711–717.

CBI (2015) *CBI Trade Statistics: Fish and Seafood*. CBI.

Chassot, E., Bonhommeau, S., Dulvy, N.K., Mélin, F., Watson, R., Gascuel, D. and Le Pape, O. (2010) Global marine primary production constrains fisheries catches. *Ecology Letters*, 13, 495–505.

Chern, W.S., Ishibashi, K., Taniguchi, K., and Tokoyama, Y. (2002) Analysis of food consumption behavior by Japanese households. ESA Working Paper No. 02–06.

Cheung, W.W.L., Lam, V.W.Y., Sarmiento, J.L., Kearney, K., Watson, R., Zeller, D., and Pauly, D. (2010). Large-scale redistribution of maximum fisheries catch potential in the global ocean under climate change. *Global Change Biology*, 16, 24–35.

Cheung, W.W.L., Dunne, J., Sarmiento, J.L., and Pauly, D. (2011) Integrating ecophysiology and plankton dynamics into projected maximum fisheries catch potential under climate change in the Northeast Atlantic. *ICES Journal of Marine Science*, 68, 1008–1018.

Cheung, W.W.L., Sarmiento, J.L., Dunne, J., Frölicher, T.L., Lam, V.W.Y., Palomares, M.L.D., Watson, R., and Pauly, D. (2013) Shrinking of fishes exacerbates impacts of global ocean changes on marine ecosystems. *Nature Climate Change*, 3, 254–258.

Cheung, W.W.L., Jones, M.C., Reygondeau, G., Stock, C.A., Lam, V.W.Y., and Frölicher, T.L. (2016) Structural uncertainty in projecting global fisheries catches under climate change. *Ecological Modelling*, 325, 57–66.

Cheung, W.W.L., Reygondeau, G., and Frölicher, T.L. (2016) Large benefits to marine fisheries of meeting the 1.5 °C global warming target. *Science*, 354, 1591–1594.

Chimits, P. (1957) Tilapia in ancient Egypt. *FAO Fisheries Bulletin*, 10, 211–215.

Christensen, V. Walters, C.J., Ahrens, R., Alder, J., Buszowski, J., Christensen, L.B., Cheung, W.W.L., Dunne, J., Froese, R., Karpouzi, V., Kastner, K., Kearney, K., Lai, S., Lam, V., Palomares, M.L.D., Peters-Mason, A., Piroddi, C., Sarmiento, J.L., Steenbeek, J., Sumaila, R., Watson, R., Zeller, D. and Pauly, D. (2008) Models of the world's large marine ecosystems. GEF/LME global project Promoting Ecosystem-based Approaches to Fisheries Conservation and Large Marine Ecosystems. *IOC Technical Series* No. 80. Paris, UNESCO.

Cinner, J.E., Huchery, C., MacNeill, M.A., Graham, N.A.J., McClanahan, T.R., Maina, J.,Maire, E., Kittinger, J.N., Hicks, C.C., Mora, C., Allison, E.H., D'Agata, S., Hoey, A., Feary, D.A., Crowder, L., Williams, I.D., Kulbicki, M., Vigliola, L., Wantiez, L., Edgar, G., Stuart-Smith, R.D., Sandin, S.A., Green, A.L., Hardt, M.J., Beger, M., Friedlander, A., Campbell, S.J., Holmes, K.E., Wilson, S.K., Brokovich, E., Brooks, A.J., Cruz-Motta, J.J., Booth, D.J., Chabanet, P., Gough, C., Tupper, M., Ferse, S.C.A., Sumaila, U.R., and Mouillot, D. (2016) Bright spots among the world's coral reefs. *Nature*, 535, 416–419.

Costa-Pierce, B.A. (2002). *Ecological Aquaculture: The evolution of the blue revolution*. Oxford: Wiley-Blackwell. pp 1–29.

Costello, C., Ovando, D., Hilborn, R., Gaines, S.D., Deschenes, O., Lester, S.E. (2012) Status and solutions for the world's unassessed fisheries. *Science*, 338, 517–520.

Costello, C., Ovando, D., Clavelle, T., Strauss, C.K., Hilborn, R., Melnychuk, M.C., Branch, T.A., Gaines, S.D., Szuwalski, C.S., Cabral, R.B., Rader, D.N., and Leland, A. (2016) Global fishery prospects under contrasting management regimes. *Proceedings of the National Academy of Sciences*, 113, 5125–5129.

Cotner, J.B., and Biddanda, B.A. (2002) Small players, large role: Microbial influence on biogeochemical processes in pelagic aquatic ecosystems. *Ecosystems*, 5, 105–121.

Crossland, C.J., Kremer, H.H., Lindeboom, H.J., Marshall Crossland, J.I., and Le Tissier, M.D.A. (2005) *Coastal Fluxes in the Anthropocene: The Land-Ocean Interactions in the Coastal Zone Project of the International Geosphere-Biosphere Programme*. Springer.

Dafforn, K.A., Glasby, T.M., Airoldi, L., Rivero, N.K., Mayer-Pinto, M., and Johnston, E.L. (2015). Marine urbanization: An ecological framework for designing multifunctional artificial structures. *Frontiers in Ecology and the Environment*, 13, 82–90.

Derraik, J.G.B. (2002) The pollution of the marine environment by plastic debris: A review. *Marine Pollution Bulletin*, 44, 842–852.

Dhar, A.K., Manna, S.K., and Allnutt, F.C.T. (2014) Viral vaccines for farmed finfish. *Virus Disease*, 25, 1–17.

Dodds, S. (1997) Towards a 'science of sustainability': Improving the way ecological economics understands human well-being. *Ecological Economics*, 23, 95–111.

Doney, S.C., Fabry, V.J., Feely, R.A., and Kleypas, J.A. (2009) Ocean acidification: The other CO_2 problem. *Annual Reviews in Marine Science*, 1, 169–92.

Doyen, L., Pereau, J.-C., and Cissé, A. (2016) The tragedy of open ecosystems. *Dynamic Games and Applications*. doi: 10.1007/s13235-016-0205-3.

Dugdale, R., Goering, J., Barber, R., Smith, R., and Packard, T. (1977) Denitrification and hydrogen sulfide in the Peru upwelling region during 1976. *Deep Sea Research*, 24, 601–608.

EIA (U.S. Energy Information Administration) (2016) *Monthly Energy Review October 2016*. Washington, DC: EIA.

Elliott, J., Deryng, D., Müller, C., Frieler, K., Konzmann, M., Gerten D., Glotter, M., Flörke, M., Wada, Y., Best, N., Eisner, S., Feket, B.M., Folberth, C., Foster, I., Gosling, S.N., Haddeland, I., Khabarov, N., Ludwig, F., Masaki, Y., Olin, S., Rosenzweig, C., Ruane, A.C., Satoh, Y., Schmid, E., Stack, T., Tang, W., and Wisser, D. (2014). Constraints and potentials of future irrigation water availability on agricultural production under climate change. *Proceedings of the National Academy of Sciences*, 111, 3239–3244.

EJF (2016) *The EU IUU Regulation: Building on success, EU progress in the global fight against illegal fishing.* Report by The Environmental Justice Foundation (EJF), Oceana, The Pew Charitable Trusts and WWF.

Ekstrom, J.A., Suatoni, L., Cooley, S.R., Pendleton, L.H., Waldbusser, G.G., Cinner, J.E., Ritter, J., Langdon, C., van Hooidonk, R., Gledhill, D., Wellman, K., Beck, M.W., Brander, L.M., Rittschof, D., Doherty, C., Edwards, P.E.T., and Portela, R. (2015) Vulnerability and adaptation of US shellfisheries to ocean acidification. *Nature Climate Change*, 5, 207–214.

Essington, T.E., Beaudreau, A.H., and Wiedenmann, J. (2006) Fishing through marine food webs. *Proceedings of the National Academy of Sciences*, 103, 3171–3175.

Elliot, J., Deryng, D., Müller, C., Frieler, K., Konzmann, M., Gerten, D., Glotter, M., Flörke, M., Wada, Y., Best, N., Eisner, S., Fekete, B.M., Folberth, C., Foster, I., Gosling, S.N., Haddeland, I., Khabarov, N., Ludwig, F., Masaki, Y., Olin, S., Rosenzweig, C., Ruane, A.C., Satoh, Y., Schmid, E., Stacke, T., Tang, Q., and Wisser, D. (2014) Constraints and potentials of future irrigation water availability on agricultural production under climate change. *Proceedings of the National Academy of Sciences*, 111, 3239–3244.

Evans, K., Bax, N., and Smith, D.C. (2017) *Australia State of the Environment 2016: Marine Environment, Independent Report to the Australian Government Minister for the Environment and Energy*. Canberra: Australian Government Department of the Environment and Energy.

FAO (2016) The State of World Fisheries and Aquaculture *2016. Contributing to Food Security and Nutrition for All.* Rome: FAO.

FAO, IFAD and WFP (2014) The State of Food Insecurity in the World *2014. Strengthening the Enabling Environment for Food Security and Nutrition.* Rome: FAO.

Fogarty, M.J., Rosenberg, A.A, Cooper, A.B., Dickey-Collas, M., Fulton, E.A., Gutiérrez, N.L., Hyde, K.J.W., Kleisner, K.M., Kristiansen, T., Longo, C., Minte-Vera, C.V., Minto, C., Mosqueira, I., Osio, G.C., Ovando, D., Selig, E.R., Thorson, J.T., and Yimin, Y (2016) Fishery production potential of large marine ecosystems: A prototype analysis. *Environmental Development*, 17, 211–219.

Francis, C.D., and Barber, J.R. (2013) A framework for understanding noise impacts on wildlife: An urgent conservation priority. *Frontiers in Ecology and the Environment*, 11, 305–313.

Fujita, M., Yamasaki, S., Katagiri, C., Oshiro, I., Sano, K., Kurozumi, T., Sugawara, H., Kunikita, D., Matsuzaki, H., Kano, A., Okumura, T., Sone, T., Fujita, H., Kobayashi, S., Naruse, T., Kondo, M., Matsu'ura, S., Suwa, G., and Kaifu, Y. (2016) Advanced maritime adaptation in the western Pacific coastal region extends back to 35,000–30,000 years before present. *Proceedings of the National Academy of Sciences*, 113, 11184–11189.

Fulton, E.A., and Gorton, R. (2014) *Adaptive Futures for SE Australian Fisheries & Aquaculture: Climate Adaptation Simulations*. Hobart: CSIRO.

Garcia, S.M., Kolding, J., Rice, J., Rochet, M.J., Zhou, S., Arimoto, T., Beyer, J.E., Borges, L., Bundy, A., Dunn, D., Fulton, E.A., Hall, M., Heino, M., Law, R., Makino, M., Rijnsdorp, A.D., Simard, F., and Smith, A.D.M. (2012) Reconsidering the consequences of selective fisheries. *Science*, 335, 1045–1047.

Gerland, P., Raftery, A.E., Ševčíková, H., Li, N., Gu,D., Spoorenberg,T., Alkema, L., Fosdick, B.K., Chunn, J., Lalic, N., Bay, G., Buettner, T., Heilig, G.K., and Wilmoth, J. (2014) World population stabilization unlikely this century. *Science*, 346, 234–237.

Goldburg, R.J., Elliott, M.S., and Naylor, R.L. (2001) *Marine Aquaculture in the United States: Environmental Impacts and Policy Options*. Wasington, DC: Pew.

Golden, C.D., Allison, E.H., Cheung, W.W.L., Dey, M.M., Halpern, B.S., McCauley, D.J., Smith, M., Vaitla, B., Zeller, D., and Myers, S.S. (2016) Fall in fish catch threatens human health. *Nature*, 534, 317–320.

Gornall, J., Betts, R., Burke, E., Clark, R., Camp, J., Willett K., and Wiltshire, A. (2010) Implications of climate change for agricultural productivity in the early twenty-first century. *Philosophical Transactions of the Royal Society B*, 365, 2973–2989.

Graham, H.W., and Edwards, R.L. (1962) The world biomass of marine fishes. In: *Fish in Nutrition*, E. Heen and R. Kreuzer, (eds). London: Fishing News (books), pp. 3–8.

Griffith, G.P., Richardson, A.J., Fulton, E.A., and Gorton, R. (2012) Evaluating the interaction effects of ocean warming, ocean acidification and fisheries. *Conservation Biology*, 6, 1145–1152.

Griffith, G.P., and Fulton, E.A. (2014) New approaches to simulating the complex interaction effects of multiple human impacts on the marine environment. *ICES Journal of Marine Science*, 71, 764–774.

Gulland, J.A. (1970) Summary In: *The Fish Resources of the Ocean*, J.A. Gulland (ed). FAO Fisheries Technical Paper 97.

Halpern, B.S., Walbridge, S., Selkoe, K.A., Kappel, C.V., Micheli, F., D'Agrosa, C., Bruno, J.F., Casey, K.S., Ebert,C., Fox, H.E., Fujita, R., Heinemann, D., Lenihan, H.S., Madin, E.M.P., Perry, M.T., Selig, E.R., Spalding, M., Steneck, R., and Watson, R., (2008) A global map of human impact on marine ecosystems. *Science*, 319, 948–952.

Halpern, B.S., Frazier, M., Potapenko, J., Casey, K.S., Koenig, K., Longo, C., Stewart Lowndes, J., Rockwood,

R.C., Selig, E.R., Selkoe, K.A., and Walbridge, S. (2015) Spatial and temporal changes in cumulative human impacts on the world's ocean. *Nature*, 6, 1–7.

Harvell, C.D., Mitchell, C.E., Ward, J.R., Altizer, S., Dobson, A.P., Ostfeld, R.S., and Samuel, M.D. (2002) Climate warming and disease risks for terrestrial and marine biota. *Science*, 296, 2158–2162.

Hildebrand, J.A. (2009) Anthropogenic and natural sources of ambient noise in the ocean. *Marine Ecology Progress Series*, 395, 5–20.

Hobday, A.J., and Pecl, G.T. (2014) Identification of global marine hotspots: Sentinels for change and vanguards for adaptation action. *Reviews in Fish Biology and Fisheries*, 24, 415–425.

Hobday, A.J., Alexander, L.V., Perkins, S.E., Smale, D.E., Straub, S.C., Oliver, E.C.J., Benthuysen, J.A., Burrows, M.T., Donat, M.G., Feng, M., Holbrook, N.J., Moore, P.J., Scannell, H.A., Gupta, A.S., and Wernberg, T. 2016. A hierarchical approach to defining marine heatwaves. *Progress in Oceanography*, 141, 227–238.

Hoegh-Guldberg, O., and Bruno, J.F. (2010) The impact of climate change on the world's marine ecosystems. *Science*, 328, 1523–1528.

Hönisch, B., Ridgwell, A., Schmidt, D.N., Thomas, E., Gibbs, S.J., Sluijs, A., Zeebe, R., Kump, L., Martindale, R.C., Greene, S.E., Kiessling, W., Ries, J., Zachos, J.C., Royer, D.L., Barker, S., Marchitto Jr., T.M., Moyer, R., Pelejero, C., Ziveri, P., Foster, G.L., Williams, B., 2012. The geological record of ocean acidification. *Science*, 335, 1058–1063.

Hu, Y., Shang, H., Tong, H., Nehlich, O., Liu, W., Zhao, C., Yu, J., Wang, C., Trinkaus, E., and Richards, M.P. (2009) Stable isotope dietary analysis of the Tianyuan 1 early modern human. *Proceedings of the National Academy of Sciences*, 107, 10971–10974.

IPCC. (2013) Summary for Policymakers. In: Climate Change 2013: The Physical Science Basis. Contribution of Working Group I to the Fifth Assessment Report of the Intergovernmental Panel on Climate Change, T.F. Stocker, D. Qin, G.-K. Plattner, M. Tignor, S.K. Allen, J. Boschung, A. Nauels, Y. Xia, V. Bex and P.M. Midgley (eds). Cambridge: Cambridge University Press.

Irigoien, X., Klevjer, T.A., Røstad, A., Martinez, U., Boyra, G., Acuña, J.L., Bode, A., Echevarria, F., Gonzalez-Gordillo, J.I., Hernandez-Leon, S., Agusti, S., Aksnes, D.L., Duarte, C.M., and Kaartvedt, S. (2014) Large mesopelagic fishes biomass and trophic efficiency in the open ocean. *Nature Communications*, 5, 1–10.

Jambeck, J.R., Geyer, R., Wilcox, C., Siegler, T.R., Perryman, M., Andrady, A., Narayan, R., and Law, K.L. (2015) Plastic waste inputs from land into the ocean. *Science*, 347, 768–771.

Jennings, S., Mélin, F., Blanchard, J.L., Forster, R.M., Dulvy, N.K., and Wilson, R.W. (2008) Global-scale predictions of community and ecosystem properties from simple ecological theory. *Proceedings of the Royal Society B Biological Sciences*, 275, 1375–1383.

Jennings, S., and Brander, K. (2010) Predicting the effects of climate change on marine communities and the consequences for fisheries. *Journal of Marine Systems*, 79, 418–426.

Jennings, S., and Collingridge, K. (2015) Predicting consumer biomass, size-structure, production, catch potential, responses to fishing and associated uncertainties in the world's marine ecosystems. *PLoS One*, 10, e0133794.

Jones, M.C., and Cheung, W.W.L. (2015) Multi-model ensemble projections of climate change effects on global marine biodiversity. *ICES Journal of Marine Science*, 72, 741–752.

Koslow, J.A., and Davison, P.C. (2016) Productivity and biomass of fishes in the California Current Large Marine Ecosystem: Comparison of fishery-dependent and -independent time series. *Environmental Development*, 17, 23–32.

Kunc, H.P., McLaughlin, K.E., and Schmidt, R. (2016) Aquatic noise pollution: Implications for individuals, populations, and ecosystems. *Proceedings of the Royal Society B*, 283, 20160839.

Leadbitter, D., Benguerel, R. (2014). Sustainable tuna–can the marketplace improve fishery management? *Business Strategy and the Environment*, **23**, 417–432.

Ling, S.D., Johnson, C.R., Ridgway, K., Hobday, A.J., and Haddon, M. (2009). Climate-driven range extension of a sea urchin: Inferring future trends by analysis of recent population dynamics. *Global Change Biology*, 15, 719–731.

Lönnstedt, O.M., and Eklöv, P. (2016) Environmentally relevant concentrations of microplastic particles influence larval fish ecology. *Science*, 352, 1213–1216.

Maeda, H. (2015) *The Past, Present and Future of the Ocean Engineering Activities: Maritime Technology and Engineering* – Guedes Soares & Santos (eds). Taylor & Francis Group, London. pgs 3–9.

McCauley, D.J., Pinsky, M.L., Palumbi, S.R., Estes, J.A., Joyce, F.H., and Warner, R.R. (2015) Marine defaunation: Animal loss in the global ocean. *Science*, 347, 1255641.

Merino, G., Barange, M., Blanchard, J.L., Harle, J., Holmes, R., Allen, I., Allison, E.H., Badjeck, M.C., Dulvy, N.K., Holt, J., Jennings, S., Mullon, C., Rodwell, L.D. (2012) Can marine fisheries and aquaculture meet fish demand from a growing human population in a changing climate? *Global Environmental Change*, 22, 795–806.

Molinos, J.G., Halpern, B.S., Schoeman, D.S., Brown, C.J., Kiessling, W., Moore, P.J., Pandolfi, J.M., Poloczanska, E.S., Richardson, A.J., and Burrows, M.T. (2015) Climate velocity and the future global redistribution of marine biodiversity. *Nature Climate Change*. doi: 10.1038/nclimate2769.

Murphy, E.J., Cavanagh, R.D., Drinkwater, K.F., Grant, S.M., Heymans, J.J., Hofmann, E.E., Hunt Jr. G.L., and Johnston, N.M. (2016) Understanding the structure and functioning of polar pelagic ecosystems to predict the impacts of change. *Proceedings of the Royal Society B*, 283, 20161646.

Neori, A., Chopin, T., Troell, M., Buschmann, A., Kraemer, G.P., Halling, C., Shpigel, M., and Yarish, C. (2004). Integrated aquaculture: Rationale, evolution and state of the art emphasizing seaweed biofiltration in modern mariculture. *Aquaculture*, 231, 361–391.

Nicklisch, S.C.T., Rees, S.D., McGrath, A.P., Gökirmak, T., Bonito, L.T., Vermeer, L.M., Cregger, C., Loewen, G., Sandin, S., Chang, G., and Hamdoun, A. (2016) Global marine pollutants inhibit P-glycoprotein: Environmental levels, inhibitory effects, and cocrystal structure. *Science Advances*, 2, e1600001.

NRC (National Research Council)(2014) *Responding to Oil Spills in the U.S. Arctic Marine Environment*. Washington, DC: The National Academies Press.

O'Connor, S., Ono, R., and Clarkson, C. (2011) Pelagic fishing at 42,000 years before the present and the maritime skills of modern humans. *Science*, 334, 1117–1121.

Parkinson, C. (2014) Global sea ice coverage from satellite data: Annual cycle and 35-yr trends. *Journal of Climate*, 27, 9377–9382.

Pauly, D., Christensen, V., Dalsgaard, J., Froese, R., and Torres Jr., F. (1998) Fishing down marine food webs. *Science*, 279, 860–863.

Pecl, G.T., Araújo, M.B., Bell, J.D., Blanchard, J., Bonebrake, T.C., Chen, I-C., Clark, T.D., Colwell, R.K., Danielsen, F., Evengård, B., Falconi, L., Ferrier, S., Frusher, S., Garcia, R.A., Griffis, R.B., Hobday, A.J., Janion-Scheepers, C., Jarzyna, M.A., Jennings, S., Lenoir, J., Linnetved, H.I., Martin, V.Y., McCormack, P.C., McDonald, J., Mitchell, N.J., Mustonen, T., Pandolfi, J.M., Pettorelli, N., Popova, E., Robinson, S.A., Scheffers, B.R., Shaw, J.D., Sorte, C.J.B., Strugnell, J.M., Sunday, J.M., Tuanmu, M.-N., Vergés, A., Villanueva, C., Wernberg, T., Wapstra, E., and Williams, S.E. (2017) Biodiversity redistribution under climate change: Impacts on ecosystems and human well-being. *Science*, 355, eaai9214.

Perkins, S.E., Alexander, L.V., and Nairn, J., (2012) Increasing frequency, intensity and duration of observed global heat waves and warm spells. *Geophysical Research Letters*, 39, 20. http://dx.doi.org/10.1029/2012GL053361.

Plagányi, É.E., van Putten, I., Thébaud, O., Hobday, A.J., Innes, J., Lim-Camacho, J., Norman-Ló pez, A., Bustamante, R.H., Farmery, A., Fleming, A., Frusher, S., Green, B., Hoshino, E., Jennings, S., Pecl, G., Pascoe, S., Schrobback, P., and Thomas, L. (2014) A quantitative metric to identify critical elements within seafood supply networks. *PLoS ONE*, 9: e91833. doi: 10.1371/journal.pone.0091833.

Poloczanska, E.S., Brown, C.J., Sydeman, W.J., Kiessling, W., Schoeman, D.S., Moore, P.J., Brander, K., Bruno, J.F., Buckley, L.B., Burrows, M.T., Duarte, C.M., Halpern, B.S., Holding, J., Kappel, C.V., O'Connor, M.I., Pandolfi, J.M., Parmesan, C., Schwing, F., Thompson, S.A., and Richardson, A.J. (2013) Global imprint of climate change on marine life. *Nature Climate Change*, 3, 919–925.

Popper, A.N., and Hastings, M.C. (2009) The effects of human-generated sound on fish. *Integrative Zoology*, 4, 43–52.

Radcliffe, W. (1921) *Fishing from the Earliest Times*. London: Murray.

Rakocy, J.E. (2012) Aquaponics – Integrating fish and plant culture. In: *Aquaculture Production Systems*, J. Tidwell (Ed.). John Wiley & Sons.

Rammelt, C.F., and M. van Schie. (2016) Ecology and equity in global fisheries: Modelling policy options using theoretical distributions. *Ecological Modelling*, 337, 107–122.

Revollo-Fernández, D., Aguilar-Ibarra, A., Micheli, F., and Sáenz-Arroyo, A. (2016) Exploring the role of gender in common-pool resource extraction: Evidence from laboratory and field experiments in fisheries. *Applied Economics Letters*, 23, 912–920.

REN21 (Renewable Energy Policy Network for the 21st Century) (2015) *Renewables 2015 Global Status Report*. Paris: REN21 Secretariat.

Ricker, W.E. (1969) Food from the sea. In: *Resources and Man, the report of the Committee on Resources and Man to the U.S. National Academy of Sciences*. San Francisco: W.H. Freeman,

Roberts, S.M., Grattan, L.M., Toben, A.C., Ausherman, C., Trainer, V.L., Tracy, K., Morris, J.G. Jr. (2016) Perception of risk for domoic acid related health problems: A cross-cultural study. *Harmful Algae*, 57, 39–44.

Rosenzweig, C., Elliott, J., Deryng, D., Ruane, A.C., Müller, C., Arneth, A., Boote, K.J., Folberth, C., Glotter, M., Khabarov, N., Neumann, K., Piontek, F., Pugh, T.A.M., Schmid, E., Stehfest, E., Yang, H., and Jones, J.W. (2014) Assessing agricultural risks of climate change in the 21st century in a global gridded crop model intercomparison. *Proceedings of the National Academy of Sciences*, 111, 3268–3273.

Ryther, J. (1969) Photosynthesis and fish production in the sea. *Science*, 166, 72–76.

Sabine, C.L., Feely, R.A., Gruber, N., Key, R.M., Lee, K., Bullister, J.L., Wanninkhof, R., Wong, C.S., Wallace, D.W.R., Tilbrook, B., Millero, F.J., Peng, T.-H., Kozyr, A., Ono, T., and Rios, A.F. (2004) The oceanic sink for anthropogenic CO_2. *Science*, 305, 367–71.

Schaefer, M.B. (1965) The potential harvest of the sea. *Transactions of the American Fisheries Society*, 94, 123–128.

Sethi, S.A., Branch, T.A., and Watson. R. (2010) Global fishery development patterns are driven by profit but not trophic level. *Proceedings of the National Academy of Sciences*, 107, 12163–12167.

Shepherd, C.J., and Jackson, A.J. (2013). Global fishmeal and fish-oil supply: Inputs, outputs and marketsa. *Journal of Fish Biology*, 83, 1046–1066.

Skretting. (2015). *Annual Sustainability Report 2015.* Skretting *Australia.* Cambridge Tasmania, Skretting. 24 pages.

Smith, A.D.M., Smith D.C., Haddon, M., Knuckey, I., Sainsbury, K.J., and Sloan, S. (2014) Implementing harvest strategies in Australia: 5 years on. *ICES Journal of Marine Science*, 71, 195–203.

Sofonia, J.J., and Anthony K.R.N. (2008) High-sediment tolerance in the reef coral *Turbinaria mesenterina* from the inner Great Barrier Reef lagoon (Australia). *Estuarine, Coastal and Shelf Science*, 78, 748–752.

Sunday, J.M., Pecl, G.T., Frusher, S., Hobday, A.J., Hill, N., Holbrook, N.J., Edgar, G.J., Stuart-Smith, R., Barrett, N., Wernberg, T., Watson, R.A., Smale, D.A., Fulton, E.A., Slawinski, D., Feng, M., Radford, B.T., Thompson, P.A., and Bates, A.E. (2015). Species traits and climate velocity explain geographic range shifts in an ocean-warming hotspot. *Ecology Letters*, 18, 944–953.

Thums, M., Whiting, S.D., Reisser, J., Pendoley, K.L., Pattiaratchi, C.B., Proietti, M., Hetzel, Y., Fisher, R., and Meekan, M.G. (2016) Artificial light on water attracts turtle hatchlings during their near shore transit. *Royal Soceity Open Science*, 3, 160142.

Troell, M., Joyce, A., Chopin, T., Neori, A., Buschmann, A.H., and Fang, J.G. (2009). Ecological engineering in aquaculture—potential for integrated multi-trophic aquaculture (IMTA) in marine offshore systems. *Aquaculture*, 297, 1–9.

Van Dover, C.L. (2014) Impacts of anthropogenic disturbances at deep-sea hydrothermal vent ecosystems: A review. *Marine Environmental Research*, 102, 59–72.

Whitmee, S., Haines, A., Beyrer, C., Boltz, F., Capon, A.G., de Souza Dias, B.F., Ezeh, A., Frumkin, H., Gong, P., Head, P., Horton, R., Mace, G.M., Marten, R., Myers, S.S., Nishtar, S., Osofsky, S.A., Pattanayak, S.K., Pongsiri, M.J., Romanelli, C., Soucat, A., Vega, J., and Yach, D. (2015) Safeguarding human health in the Anthropocene epoch: Report of The Rockefeller Foundation–Lancet Commission on planetary health. *Lancet*, 386, 1973–2028.

Wilcox, C., Van Sebille, E., and Hardesty, B.D. (2015) Threat of plastic pollution to seabirds is global, pervasive, and increasing. *Proceedings of the National Academy of Sciences*, 112, 11899–11904.

Wilson, R.W., Millero, F.J., Taylor, J.R., Walsh, P.J., Christensen, V., Jennings, S., and Grosell, M. (2009) Contribution of fish to the marine inorganic carbon cycle. *Science*, 323, 359–362.

World Bank (2013) *Fish to 2030: Prospects for Fisheries and Aquaculture*. Washington, DC: World Bank.

World Bank (2014) *Reducing Disease Risk in Aquaculture*. Washington, DC: World Bank.

World Bank (2016) *Global Monitoring Report 2015/2016: Development Goals in an Era of Demographic Change*. Washington, DC: World Bank.

Worm, B., Hilborn, R., Baum, J., Branch, T., Collie, J., Costello, C., Fogarty, M., Fulton, E.A., Hutchings, J., Jennings, S., Jensen, O., Lotze, H., Mace, P., McClanahan, T., Minto, C., Palumbi, S., Parma, A., Ricard, D., Rosenberg, A., Watson, R., and Zeller, D. (2009) Rebuilding global fisheries. *Science*, 325, 578–585.

Yool, A., Popova, E.E., and Coward, A.C. (2015) Future change in ocean productivity: Is the Arctic the new Atlantic? *Journal of Geophysical Research*, 120, 7771–7790.

PART VII

Future Earth and Risk, Safety and Security

23

Geophysical Studies, Natural Hazards, and Climate Change

Jaime Urrutia-Fucugauchi and Ligia Pérez-Cruz

23.1 Introduction

The impacts of natural hazards and climate change on agriculture and food security are being intensely studied. In general, there are significant advances in understanding the phenomena and in mitigation, prevention and forecasting capacities. However, disasters continue to affect large sectors of the society, emphasizing the need for improvement in accuracy of disaster risk assessments and prevention and mitigation programs. Disaster risk depends on a convolution of natural hazards and vulnerability and society resilience, involving a complex array of variables (Beer and Ismail-Zadeh, 2002; Ismail-Zadeh et al., 2014). Risk analysis involves a complex set of variables, in addition to the geophysical phenomena (Figure 23. 1). Risks depend on interplay of factors, including location, geology, vulnerability and economic and social factors. Effects are geographically unevenly distributed, with severity of impacts correlating with population size, density, demographics and economic development affecting particularly regions in the developing and less developed nations (Beer, 2010).

The 2015 report of the Food and Agriculture Organization (FAO) in the monitoring period for the Millennium Development Goal (MDG) targets estimates that around 795 million people remain undernourished. Although the percentage of the global population facing undernourishment has decreased, with more than 50 percent of the countries achieving the MDG targets, progress has been slow and uneven with regions mainly in the developing world still facing severe problems. Natural disasters in the 2003–2013 decade affected more than 1.9 billion people with estimated losses exceeding US$490 billion. The agriculture sector, which includes crops, livestock, fisheries and forestry, represents ~22 percent of the economic impact in developing countries. Food production, and in particular agriculture, is affected by floods, droughts and tropical storms, with effects likely to be more severe in the years to come as a result of climate change. Of those, agricultural effects due to droughts are the largest, accounting for about 80 percent of the losses, particularly in crops and livestocks (FAO, IFAD and WFP, 2015).

The coverage and spatial-temporal resolution of geophysical methods and multi-sensor satellite networks have increased over the past years, providing coverage at local to global scales. The geophysical data available have a wide range of applications, opening new possibilities for addressing long-standing issues related to natural hazards, land use, agriculture, pollution, soil erosion and climate change impacts, as well as for exploration and exploitation of mineral and energy resources and groundwater. In the developing and less developed nations, use of and access to geophysical data and survey capabilities, theoretical developments, laboratory studies and data have not yet reached their full potential, with partial access to the new technologies, inadequate infrastructure and lack of expertise in data processing, analysis and interpretation, despite their potential to address the pressing challenges.

Here we discuss geophysical methods and applications as related to natural hazards, climate change and food security. The basic premise is that expanded use of geophysical data and capabilities can significantly contribute to address the challenges from natural hazards, from increasing demand for natural resources and from climate change impacts.

23.2 Climate Change and Natural Hazards

Climate change has emerged as a major factor at a global scale, negatively impacting the food security on availability, stability, utilization and access (Rosenzweig and Parry, 1994; Parry and Livermoore, 2002; Allen and Ingram, 2002; Schmidhuber and Tubiello,

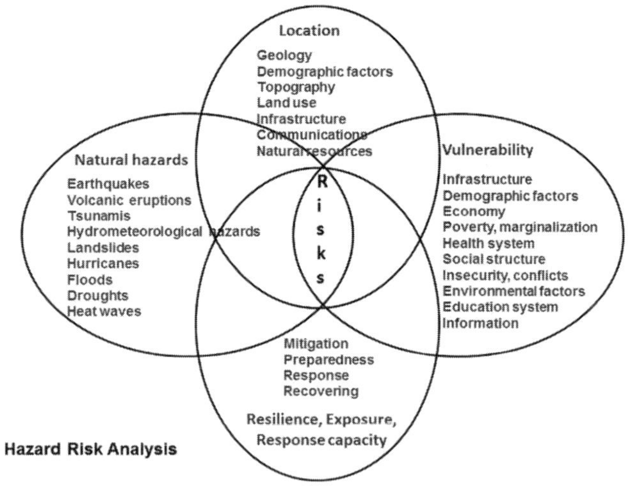

Figure 23.1 Schematic representation of variables involved in risk analysis (see also Beer, 2010; Ismail-Zadeh et al., 2014).

2007). Climate models have increased their spatial-temporal resolution, investigating different scenarios for the increase of CO_2 and other greenhouse gas emissions (Manning et al., 2004; Moss et al., 2010). Climate change affects the hydrological cycle over a wide range of spatial-temporal scales in an uneven way (Gleick, 1989; Ramanathan et al., 2001; Emori et al., 2005; Milly et al., 2005; Trenberth, 2011; Cook et al., 2015, 2016; Diffenbaugh et al., 2015). Effects on agriculture amplify, increasing vulnerability in critical regions already under severe stress.

Impacts of climate change are being intensely investigated and discussed from a wide range of viewpoints, with special attention on modeling and predictions for the magnitude and spatial-temporal patterns across the world in the coming decades (IPCC, 2014). The changing hydrologic cycle is impacting agriculture, with modification of precipitation patterns, aridity, soil erosion, aquifers, sea level, coastal processes, deforestation and ecosystems. These changes pose short- and long-term risks to agriculture, ecosystems and land use that translate into risks for food security, nutrition and the well-being of human society.

The complexity of the climate system, including abrupt events, teleconnections and non-linear behavior, is investigated using numerical simulations. Global Climate Models (GCMs) have increased resolution while incorporating a larger number of climate variables (Lambert and Boer, 2001; Griffies, 2004; Doutriaux-Boucher and Quass, 2004; Emori et al., 2005; Flato, 2005; Power and Colman, 2006). GCMs are being used to model distinct scenarios and investigate impacts of discrete events, state changes and teleconnections (Davey et al., 2002; Alexeev, 2003; Stephens, 2005; AchutaRao and Sperber, 2006; Latif and Keenlyside, 2009). Climate variables encompass the different processes represented in the surface temperature, humidity, winds, precipitation, ocean currents, cloud cover and solar radiation. Major efforts are directed to investigate forcing mechanisms, natural variability, energy balance and spatial-temporal scales, as well as the role of component interactions, feedback processes, cyclicity, non-linear effects, extreme events, tipping points and anthropogenic-induced processes. Internal climate variations form a major component in the natural system variability, resulting from distinct response processes, feedback mechanisms and non-linear interactions to external forcing. They include the ENSO events from the atmosphere–ocean interactions in the tropical Pacific Ocean, latitudinal migration of the Inter-Tropical Convergence Zone, decadal oscillations and monsoon variability (Haug et al., 2001; Soden, 2000; AchutaRao and Sperber, 2006; Pérez-Cruz, 2006, 2013; McPhadden et al., 2006; Latif and Keenlyside, 2009).

Potential impacts of events such as floods, droughts, tropical storms and hurricanes are a major concern. This is well exemplified by multi-year droughts, fires and floods in several regions and the impacts of volcanic eruptions, earthquakes, tsunamis, landslides and hurricanes. They affect regions in developed and developing nations as in the 1991 Pinatubo eruption, 2004 Sumatra earthquake and Indian Ocean tsunami, 2004 Hurricane Katrina and Hurricane Wilma, 2010 Haiti earthquake, 2011 Tohoku earthquake and tsunami, 2012 Hurricane Sandy and many others. The impact of natural disasters has increased with population growth and demographic changes, emphasizing the need for disaster risk analysis and mitigation and prevention programs (Beer, 2010). The United Nations International Strategy for Disaster Risk Reduction (UNISDR) reported that in the period 2000–2012, disasters resulted in losses up to US \$1.7 trillion (Southgate et al., 2013). Disasters impact both developed and developing nations, with the number of recorded events increasing over the years (Wirtz et al., 2014; Munich Re, 2015; Figure 23.2).

The FAO, IFAD and WFP (2015) report on the impact of natural disasters on agriculture reviewed impacts from seventy-eight events that occurred in 2003–2013 in developing nations in Africa, Asia and Latin America. The estimated losses reached US\$140 billion, with about 22 percent in agriculture. Most of the losses concentrated in crops and livestock, amounting to ~78 percent (Figure 23.3a). Largest impacts on crops were those from flooding, followed by storms and droughts, whereas, livestock was mostly affected by

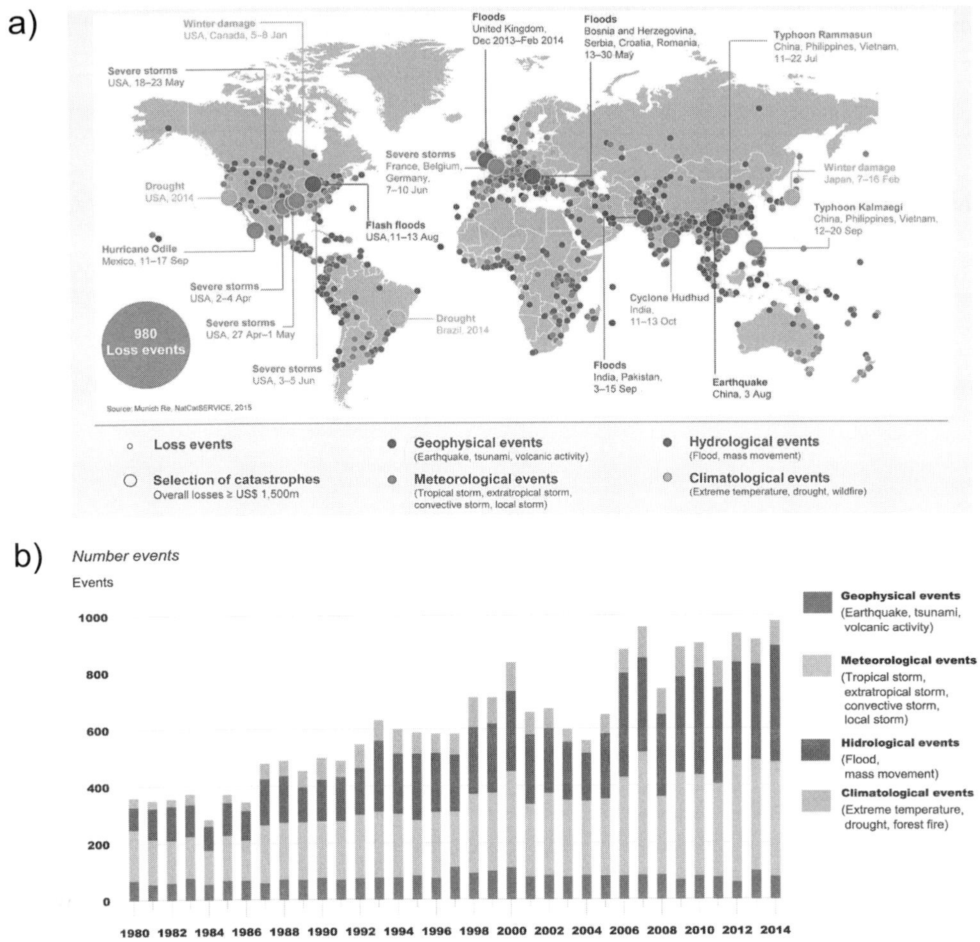

Figure 23.2 Natural disasters recorded in the NatCatSERVICE of Munich Re. (a) Worldwide major loss events in 2014 (from Munich Re, 2015). (b) Histogram of number of disasters worldwide between 1980 and 2014, divided into geophysical (earthquakes, tsunamis, volcanic eruptions), meteorological (storms), hydrological (floods, landslides) and climatological (drought, hot waves, forest wildfires) events (from Wirtz et al., 2014). For the color version, please refer to the plate section.

droughts. In the decade reviewed, the losses on crops and livestock reached about US$70 billion (Figure 23.3b), with 83 percent due to droughts and floods. The production losses in crops and livestock mainly affected countries in Asia (US$28 billion) and Africa (US$26 billion) (Figure 23.3c). The impacts on given countries varied widely, with the region of Sub-Saharan Africa severely affected by droughts.

23.3 Geophysical Data and Methods

The basic premise developed in this chapter is that increased effective use of geophysical data and capabilities will provide tools to address climate change and natural hazards impacts. Geophysical methods have been developed to investigate Earth's processes, land features, underground structures and geophysical phenomena using measurements on natural and induced fields and physical properties, which can be applied to a wide range of problems at local, regional and global scales (Reynolds, 2011; Keary et al., 2013). Advances in instrumentation, data acquisition, processing and modeling allow applications for exploration/exploitation of mineral and energy resources, groundwater, soils and pollution as well as for monitoring environmental processes and phenomena.

National and international programs that incorporate geophysical studies have been developed as part of mitigation, prevention and technical aid projects. Projects focus on groundwater, contamination and soil erosion, as well as on monitoring of active volcanoes and seismic zones and hurricane-prone areas. These projects could be expanded by taking advantage of new capabilities. Improved understanding of the

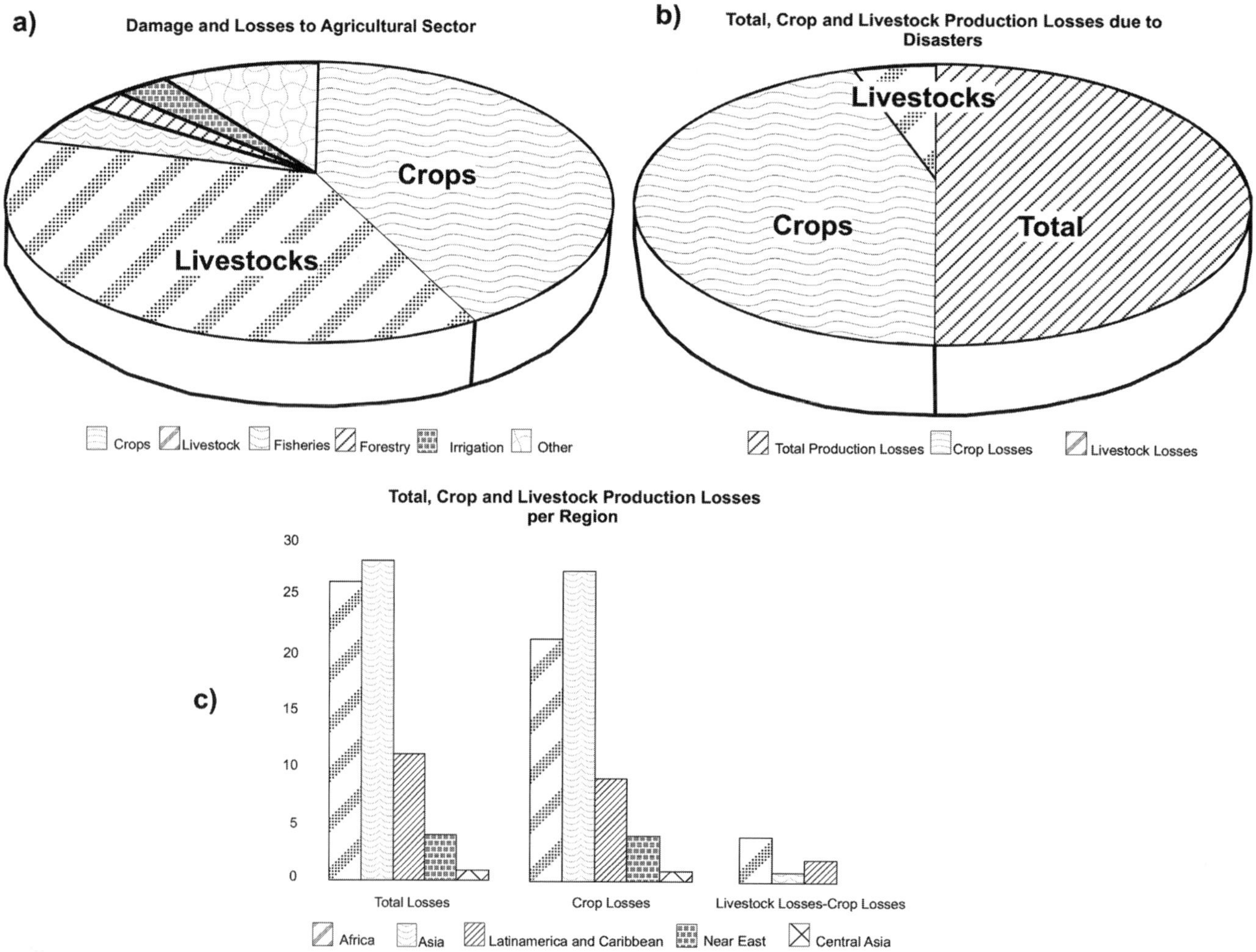

Figure 23.3 Impact of natural disasters on the agriculture sector (taken from FAO, 2015). (a) Estimated post-disaster damage and losses in the period 2003–2013 due to 78 disaster events, divided by subsector (crops, livestock, fisheries, forestry, irrigation and others). (b) Estimated production losses in crops and livestock. (c) Estimated production losses in crops and livestock divided by region: Africa, Asia, Latin America and Caribbean, Near East and Central Asia.

climate system has required integrative analyses of the interactions among the Earth´s components of the atmosphere, hydrosphere, cryosphere, lithosphere, ionosphere and biosphere. Interactions involve component flow and feedback mechanisms acting with distinct spatial-temporal scales in the geochemical cycles and surface processes. In this context, geophysical methods provide tools and data such as satellite networks, global positioning system (GPS), global instrumental monitoring networks, geophysical data centers, remote sensing, tomography, big data analysis and high-performance computing (Hwang et al., 2014; Urrutia-Fucugauchi and Pérez-Cruz, 2016). Developments in instrumentation, modeling and observation capabilities provide new tools for local to global surveys and process monitoring.

Geophysical methods developed for exploration of mineral and energy resources have direct application to environmental studies of soils, erosion, contamination, groundwater etc (Reynolds, 2011). Introduction of airborne surveys increases the capacity for surveying larger areas, reducing the costs and time. Remote sensing techniques that can be used from aircraft and satellites or from remote ground stations also contribute to enhance the capabilities for rapid surveying of large areas with increased precision and resolution (e.g., Francis and Rothery, 2000; WCDRR, 2015). These geophysical techniques and capabilities are still underutilized in many parts of the world, including those regions where they are most needed.

The NASA Earth Observing System (EOS), formed by a series of satellites launched over several years, monitors the Earth's climate system with observations on the atmosphere, cloud cover, oceans, ocean currents, surface temperatures, ice caps, aerosols and vegetation. The coming generation of Earth's observing system includes the NPOESS satellite, directed to land cover, vegetation productivity, atmospheric ozone and aerosols, sea and land surface temperatures and ice cover (NASA, 2011). Vegetation coverage, growth, crop types, crop yields and related meteorological conditions and soil moisture are now systematically surveyed. The MODIS (Moderate Resolution Imaging Spectroradiometer) and the ASTER (Advanced Spaceborne Thermal Emission and Reflection Radiometer) projects on the EOS Terra platform provide high-resolution image data on the visible-shortwave infrared and thermal-infrared channels. The Earth Observation Satellite System has increased its capabilities and spatial-temporal coverage, which integrates with aerial and ground-based surveys (e.g., Hook et al., 2001; Paul et al., 2004). Multi-sensor satellites are capable of recording data at different altitudes with increased resolution and coverage, which permit, for instance, accurate forecasts of weather conditions, monsoon patterns, tropical storms and hurricanes. The LANDSAT system has been highly successful in providing multi-spectral data useful for a wide range of applications for geologic mapping, mineralogical and lithological characterization, exploration of mineral and energy resources, groundwater, contamination, environmental degradation and land use.

Satellite data have a wide range of applications, in addition to the monitoring programs, providing information on the Earth's system components. The satellite network provides global data on the climate system, oceans, atmosphere, continents, ionosphere and magnetosphere (Tanré et al., 2001; Herman et al., 1997; Ramanathan et al., 2001; Wentz and Schabel, 2000; Hook et al., 2001; Paul et al., 2004). The multi-spectral systems, GPS, gravity, magnetics and radar satellites provide geophysical data with increasing resolution, permitting studies at regional to local scales, which can be integrated to aerial and ground-based surveys.

23.4 Hazards and Climate Change in Mexico

The territory of Mexico is characterized by a relatively complex geology, which translates into a mosaic of regions with distinct characteristics in stratigraphy, tectonics, topography, geomorphology and surface geology. Mexico forms the southern end of the North American plate and is bounded on the Pacific Ocean side by the Middle America subduction zone and the ridge/transform zone of the Gulf of California that connects with the San Andreas Fault. Central and northern Mexico have an elevated topography with the volcanic sierras of the Sierra Madre Occidental and the Trans Mexican volcanic belt and the thrust belt of the Sierra Madre Oriental. Southern Mexico is crossed by the continuation of the Sierra Madre del Sur thrust belt and the Chiapas volcanic arc (Figure 23.4). The territory extends from the subtropical regions to the temperate zone across the Tropic of Cancer. Northern Mexico is a semi-arid to arid region, part of the Chihuahua desert. In contrast southern Mexico is characterized by high humidity with high mean annual precipitation. The area includes the country's largest river, a different physiographic province, and thus presents distinct characteristics that respond differently to climate change. This in turn translates to distinct problems and distinct responses by the vegetation, agriculture, water supply, hydrologic cycle and surface temperatures.

Climate change is affecting the precipitation patterns across the country in different forms, with marked differences. In the past years, severe drought affected central and northern Mexico, while tropical storms and flooding affected areas in southern Mexico. Those events were devastating, affecting agriculture and food production, with major repercussions in the economy and to the local population. Population density, demographics and diversity vary widely across the different regions, particularly between northern, central and southern states. Correlation of the different social and economic indicators with the physiographic and geological maps suggests that geographic, environmental and climatic factors have implications for the agriculture, land use, type and availability of natural resources, and the vegetation, fauna, communications, transportation of commodities and persons, etc. Geographic barriers have been important in limiting contacts and isolating populations. A large percentage of the population lives in the central area, in the large cities of Mexico, Guadalajara, Puebla, Queretaro, Toluca and Morelia. The population of the metropolitan area of Mexico City has been growing for several decades, reaching about 22 million people. The demand for food products, energy, water and other commodities has also significantly increased. Water supply is from the valley aquifers and from the Lerma basin. Exploitation of groundwater resources has resulted in differential sinking of sectors, with compaction of the lake sediments and pyroclastic deposits. Mexico City is built on a high-altitude 2,200–2,300-m asl closed volcanic basin within the Trans Mexican volcanic belt. The basin was artificially drained and the downtown area was

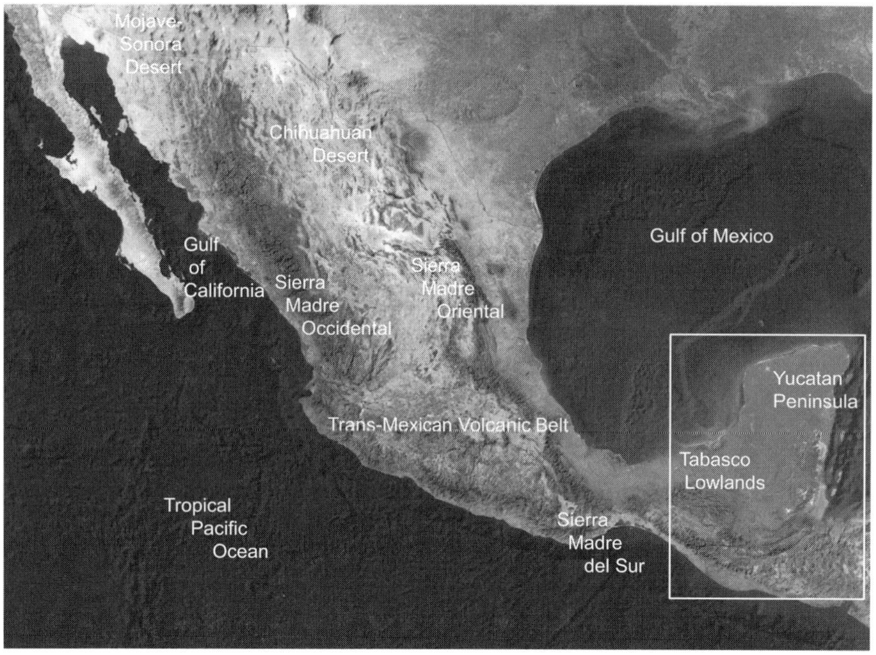

Figure 23.4 LANDSAT mosaic image of Mexico (NASA satellite image). Territory is characterized by elevated topography, with the volcanic sierras of the Sierra Madre Occidental and Trans Mexican volcanic belt and the thrust belt of the Sierra Madre Oriental and Sierra Madre del Sur. Area of southern Mexico marked in the rectangle is shown in Figure 23.8. For the color version, please refer to the plate section.

built on lake sediments and volcanic terrains. The Mexico City infrastructure is susceptible to damage from earthquakes occurring in the volcanic belt and the plate subduction zone to the south. The 1985 M 8.5 earthquake with epicenter in the Michoacan sector some four hundred kilometres away resulted in severe damage, with edifices collapsed and loss of lives. The damage was more severe in the lake sediment area and the transition zones to the volcanic terrain.

With the increase of greenhouse gases and warmer global temperatures, changes in precipitation patterns worldwide are being monitored. Climate models are used to forecast changes at regional scales. Warmer temperatures will affect the evapotranspiration rates, vegetation and hydrologic cycles, which for southwestern North America will increase the frequency, duration and severity of droughts (Seager et al., 2007, 2009; Woodhouse et al., 2010: Ault et al., 2014; Swain et al., 2014; Cook et al., 2015). Global warming is increasing the sea surface temperature, changing the ocean–atmosphere interactions and energy state with effects on frequency and magnitude of ENSO, hurricanes and tropical storms. ENSO is the strongest effect on precipitation patterns in the Pacific margin of the Americas and worldwide (McPhadden et al., 2006; Latif and Keenlyside, 2009). ENSO triggers major changes, with droughts in eastern Asia, Indonesia and Australia and increased rainfall in western North America. In the tropical Pacific and western North America ENSO variability exerts a strong effect on the hydrologic cycle, in addition to the monsoon variability, ITCZ migration, decadal oscillations and teleconnections.

In North America, the Drought Monitor program, a Canada–United States–Mexico collaboration, has been recording precipitation changes through the region for the past decade (Lawrimore et al., 2002). The monitoring program is a useful tool for planning and designing mitigation and prevention strategies (Figure 23.5). Northern Mexico and the Southwest United States are characterized by arid-semiarid climate, with relatively high interannual to decadal variability. Studies have long been investigating the precipitation trends in the region at the different spatial-temporal scales using meteorological data and paleoclimatic proxies (e.g., Seager et al., 2009; Woodhouse et al., 2010; Arriaga-Ramirez and Cavazos, 2010). Occurrence of arid trends and severe droughts has been documented at various periods in the past thousand years and recent historical times (Cook et al., 2016). Paleoclimate reconstructions show that severe multi-year droughts have affected northern Mexico and the southwestern United States in the recent past. Woodhouse et al. (2010) reviewed the paleoclimate record for the last twelve hundred years, finding evidence for intense droughts in the mid-twelfth century, which were associated with warm temperatures. In their analysis, they suggest that climate scenarios for current warmer trends should consider possible associated future drought. Seager et al. (2014)

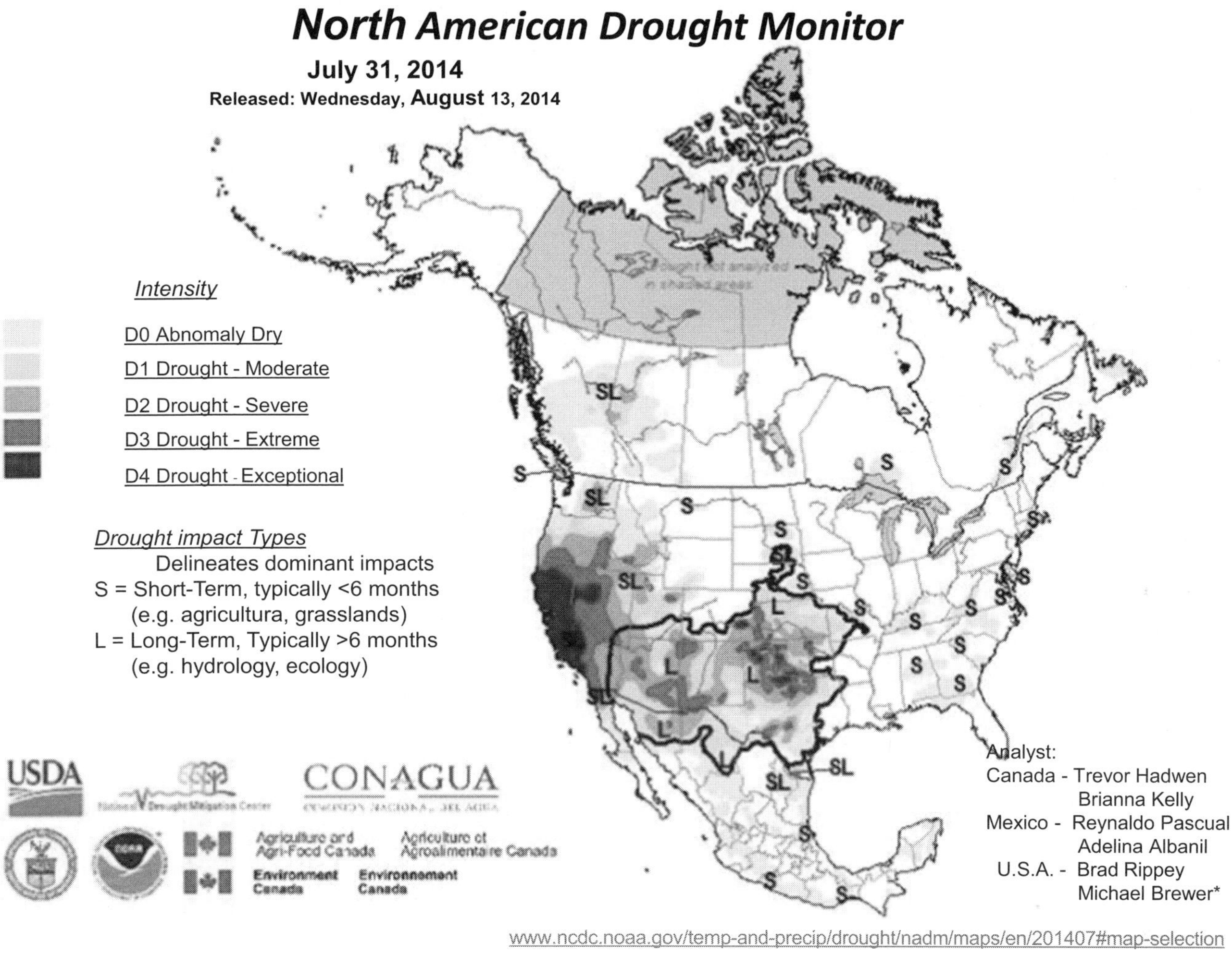

Figure 23.5 The North America Drought Monitor drought conditions and impact in North America for July 2014 (from North America Drought Monitor, August 2014). For the color version, please refer to the plate section.

examined the causes of the 2010/2011 drought in northern Mexico and the southern United States, showing a connection with development of La Niña events and the interplay of ocean forcing and internal atmospheric variability. The simulations and data observations showed the complexity and temporal changes during the two-year drought, highlighting the forecast challenge. Analyses by Ault et al. (2014), Cook et al. (2015) and Diffenbaugh et al. (2015) show that the risk of severe multi-year droughts will significantly increase as temperatures continue to rise. Instrumental data with higher resolution coverage are needed to improve the modeling accuracy and reduce uncertainties for developing climate scenarios.

Planning strategies require incorporating local and regional geophysical data. Results from meteorological instrumental networks and GCMs can be integrated with satellite observations and ground surveys, prevention and mitigation programs. Use of monitoring data, climate modeling and geophysical survey data in the mitigation and prevention programs permits development of scenarios (McBean, 2002). Aid programs that do not incorporate risk analysis and prediction and are based on limited long-term planning are losing critical information (McBean, 2002; Beer and Ismail-Zadeh, 2003). Use of early warning systems that integrate monitoring and geophysical data, with statistical analyses and simulations, permits optimizing available capacities (e.g., Medina-Cetina and Nadim, 2008). Adequate planning, incorporating available data, prediction and risk analysis, permits optimizing of strategies and implementing of actions to reduce vulnerability.

The flooding events in southern Mexico exemplify the need to incorporate geophysical data, geographical

Figure 23.6 View of area affected by flooding in November 2007 in the Villahermosa region, Tabasco, southern Mexico.

information systems and satellite and ground-based monitoring networks. Increased precipitation within the Grijalva River basin in late 2007 triggered runoff and flooding in the Tabasco lowlands. Flooding affected an extensive zone including the city of Villahermosa (Figure 23.6). Geophysical analyses assist the design and provide information to build infrastructure and identify and characterize the risk zones, urban planning and regional actions in the river basins and coastal area. Poorly planned emergency measures to aid the affected areas do not assist in a longer-term program. The emergency measures were implemented each time, did not last, provided limited aid and did not contribute to building a protection infrastructure for future events. Such flooding events will happen again and the civil protection agencies and health and urban planning systems can be better prepared if they do not address them as unforeseen catastrophes but incorporate risk analysis and prediction modeling in the planning and response strategies (e.g., McBean, 2002).

Expanded and higher-density meteorological data sets are also needed for improved weather forecast modeling and for assessing short- and long-terms trends. In the region, precipitation spatial-temporal patterns are affected by several factors, including the ITCZ latitudinal migration, ENSO events, PDO and the tropical Atlantic and Pacific sea surface temperatures. Analyses of meteorological data for Mexico and Central America show a trend to warmer temperatures. Precipitation shows spatial variability and apparent occurrence of extreme events (Aguilar et al., 2005; Peterson et al., 2008). Mexico is characterized by a marked latitudinal variability, with an arid northern region and a high-precipitation southern region, which are associated with the influence of sea surface temperature anomalies in the Atlantic and Pacific Oceans and teleconnections. Droughts in western North America occur at times of increased precipitation in southern Mexico and Central America (Mendez and Magaña, 2010).

Satellite data and numerical simulation modeling document the areas affected and susceptible to be affected (Rivera-Trejo et al., 2010; Santos-Reyes and Beard, 2011; Silva et al., 2011). Aparicio et al. (2008) identified several factors triggering the flooding: heavy precipitation, high water levels in upstream dams, river runoffs, high soil moisture levels and high tides in the coast. From the analysis, they emphasized a need for integral planning, hydro-meteorological monitoring, dam control and precipitation-runoff modeling and recommended revision of the territorial ordinance regulations. The satellite multi-spectral and radar data permit identification of risk zones linked to flooding levels. Figure 23.7 shows flooding maps from the TerraSAR-X high-resolution radar satellite of the affected areas in Tabasco State during the 2007 disaster. Integration of satellite data into the monitoring network provides an expanded coverage with high spatial-temporal resolution. The satellite data can also be integrated with other geophysical methods, including LIDAR surveys and ground-based studies (Aparicio et al., 2008; Ramos et al., 2009; Rivera-Trejo et al., 2010; Santos-Reyes and Beard, 2011; Silva et al., 2011).

Integration of satellite systems and geophysical methods can be applied to investigate the aquifers and soil erosion and evaluate the areas affected by drought and wildfires including the state of the vegetation cover. Figure 23.8a shows a satellite TOMS of a smoke plume of aerosols as they move through North America. The plume was produced by wildfires in southern Mexico on 15 May 1998 (Kaufman et al., 2002). Figure 23.8b shows a MODIS satellite image of southern Mexico and northern Central America, including the Grijalva and Usumacinta River basins. Image analyses, with integrated geological data, permit one to characterize the basins, sensitive areas for flooding control and land changes. The image also shows the wildfire locations spread in the area. Groundwater characterization provides data for water management, exploration of aquifers, water quality and recharge zones, which are important for short- and long-term mitigation programs. Urban planning can link to response system, health, educational center, nursery and other facilities that constitute especially sensitive infrastructure. Overall costs of droughts and heat waves in the past decades amount to hundreds of billions of dollars, plus the damage to agriculture and population. Adequate planning, mitigation and preparedness strategies can significantly reduce the impacts.

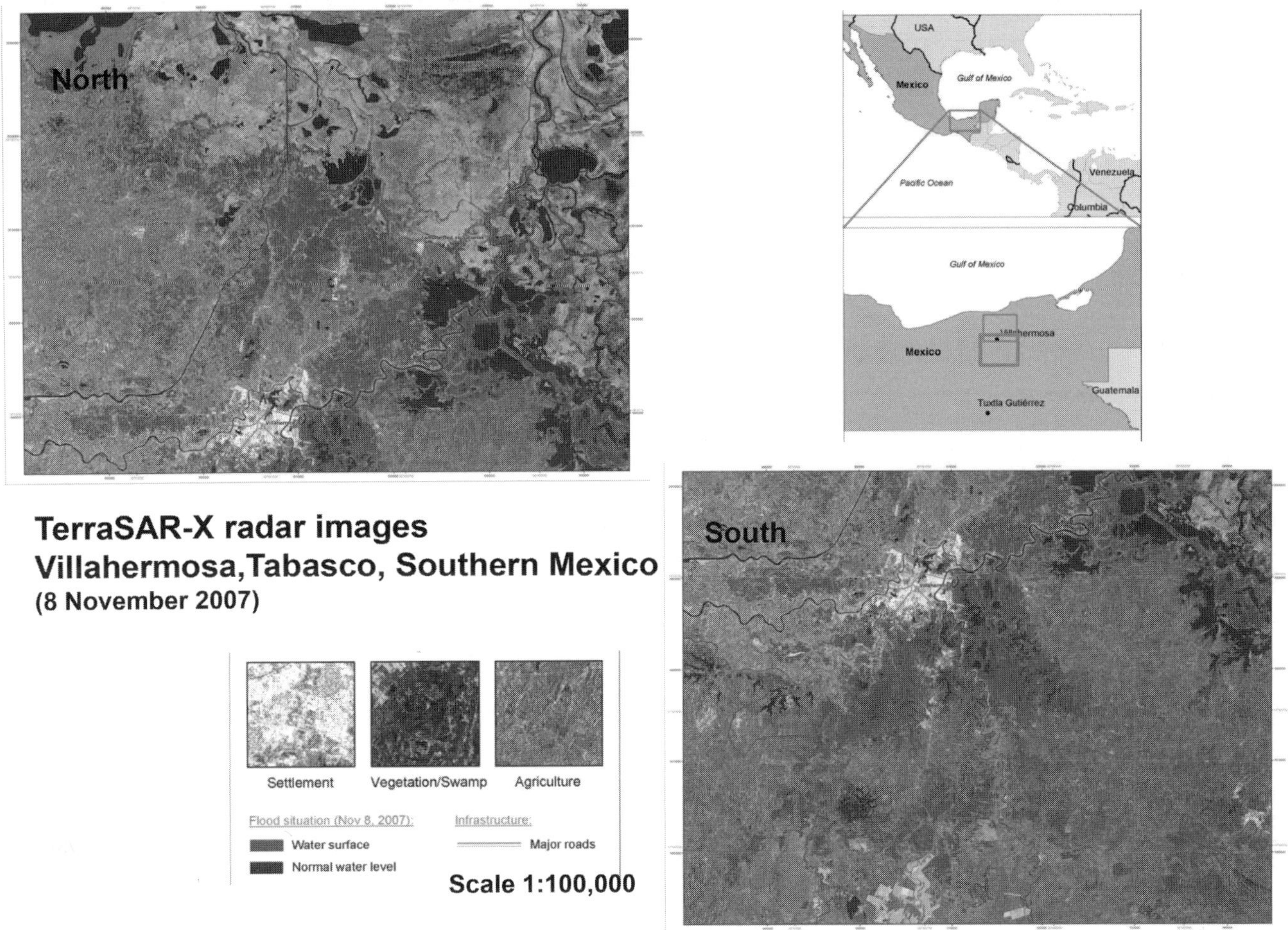

Figure 23.7 TerraSAR-X high-resolution radar satellite images of the Villahermosa region, Tabasco State, southern Mexico. Images of northern and southern Villahermosa area in scan mode taken on November 8, 2007 (SAR images from German Aerospace Center DLR, Center for Satellite Based Disaster Information ZKI flood maps). For the color version, please refer to the plate section.

23.5 Discussion

The impact of natural hazards and climate change on agriculture and food security is a major concern to developed and developing nations. The population growth, demographic changes, increased global demand for minerals, water and energy resources, pollution, land use, deforestation, environmental degradation, biodiversity losses, changes in atmospheric gas composition and global warming highlight the importance of understanding Earth's system interconnections (ICSU, 2010). Expanded use of geophysical tools, including satellite, aerial, marine and ground-based methods, needs to occur at local, regional and global scales and to be integrated into mitigation and adaptation strategies. Study of spatial-temporal scales, interconnections and feedback mechanisms requires integrative multi-disciplinary approaches and big data analysis. Developments in high-performance computing, telecommunications, new electronics and the global web permit analysis of massive data sets and implementation of numerical simulations to tackle the complex problems of the climate system and natural hazards. Study of the interactions, feedback mechanisms, non-linear behavior, extreme events and state changes requires expanded observational capabilities and high-resolution models.

Global warming is changing weather patterns, resulting in loss of natural resources and biodiversity, at different interconnected spatial-temporal scales (Alley et al., 2003; Milly et al., 2005; Trenberth, 2011). Effects combine in different ways, affecting agriculture and food production. As global temperature increases, effects become more marked, affecting larger areas, including the heavily populated regions in the

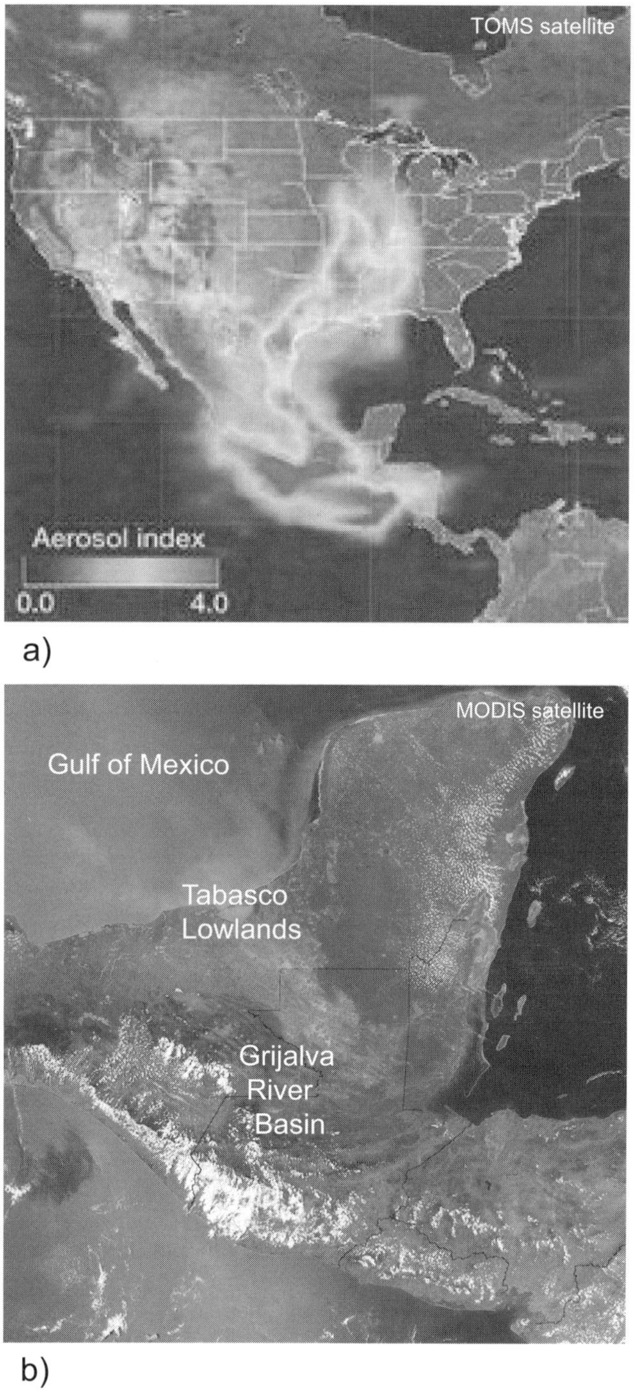

Figure 23.8 (a) Satellite TOMS image showing a smoke plume as it moved through North America (Kaufman et al., 2002). Plume raised from wildfires in southern Mexico (image on 15 May 1998; by J. Herman, NASA/GSFC; Scores of fires were burning in Mexico and Central America on 16 April 2003. (b) Satellite MODIS image (moderate resolution imaging spectroradiometer from Aqua satellite) of southern Mexico and northern Central America. Image shows location of wildfires, marked in red (image on 16 April 2003; by Jeff Schmaltz, MODIS Rapid Response Team, NASA/GSFC). For the color version, please refer to the plate section.

developing nations where there is less capacity, infrastructure and resilience (FAO, IFAD and WFP, 2015). Poverty and hunger affect a significant percentage of the world population with huge inequalities including economies with high GDP, such as the emerging economies of India, Brazil and Mexico. In Mexico, large sectors of the population live below the poverty line, with >7.4 million in extreme poverty as defined by the SEDESOL Social Development Ministry. Poverty is defined in terms of the daily income, as well as factors such as access to education, health care, clean water, shelter, social security and nutrition.

The paleoclimate record shows that abrupt climate changes occur repeatedly, related to thresholds and state shifts. Alley et al. (2003) analyzed the impact of abrupt events due to natural and human-induced forcing. Records and numerical modeling for the twentieth and twenty-first centuries illustrate the uncertainties, showing juxtaposition of inter-decadal variability and anthropogenic forcing. Severe droughts in the Sahel and the western and southern United States (Griffin et al., 2014; Cook et al., 2015, 2016; Diffenbaugh et al., 2015) show such spatial-temporal variability. Abrupt changes in precipitation are observed in different regions, likely becoming more frequent as global surface temperatures increase (e.g., Milly et al., 2002, 2005; Trenberth, 2011; Swain et al., 2014; Mann and Gleick, 2015). Studies examine different internal and external forcing mechanisms, including ENSO events, decadal oscillations, teleconnections, solar radiation and anthropogenic induced effects. The potential impact of abrupt changes and destabilizing capacity needs to be accounted for in climate change scenarios, which emphasize the importance of satellite and instrumental monitoring networks, climate modeling and risk analysis (Wentz and Schabel, 2000; Alley et al., 2003).

Analyses of agricultural productivity derived from climate modeling show that productivity and changes vary from region to region. The projected changes incorporating effects of carbon fertilization show that large regions will experience significant reductions. There are many variables involved, resulting in uncertainties for long-term prediction. This emphasizes the importance of science in disaster risk reduction and the challenges in having science-based policy decision processes in a widest sense (Southgate et al., 2013). The UNISDR program aims for an effective use of science knowledge and capabilities, with analyses showing that effective disaster risk reduction programs require inclusion of science advances, increased investment and capacity building within improved policy development and long-term strategies. The program has been highlighting case studies that illustrate the potential of using scientific approaches for disaster risk reduction.

Additional efforts need to be incorporated such as in known disaster-prone areas in the developing nations, which continue to have major impacts in economic and life losses. Lack of adequate mitigation and prevention programs is exemplified by the repeat events; Southgate et al. (2013) remarked that the high costs of rescue and rebuild actions did not contribute to reduce the vulnerability in the likelihood of future events.

The aid programs built in response to the drought and flooding disasters in northern and southern Mexico did not reduce future vulnerability for repeat events. Risk reduction efforts should aim for effective long-term mitigation programs, incorporating the scientific information. In Mexico, the National Disaster Fund FONDEN Program provides post-disaster funds, after the assessment and declaration of natural disaster in given regions. In the period 1999–2011 (FONDEN, 2012), there were on average about thirty disaster declarations per year (Figure 23. 9a). In the year 2010, reconstruction costs reached about US$55 billion. Major disasters were triggered by flooding and storms, and reconstruction costs focused on hydraulic and road infrastructure (Figure 23.9b). Disasters occurred in known disaster-prone regions and reconstruction costs did not contribute to reducing vulnerability for future repeat events.

Geophysical tools available for aerial and ground studies provide surface and underground information at various spatial scales. Improved processing capabilities are developed and integrative geophysical surveys can be used as part of management systems for natural resources and risk analyses. There are several areas where integrated geophysical observations using satellite, aerial and ground data can be especially useful: in particular, for regional mapping and geological characterization needed for exploration of mineral resources, groundwater, coal and hydrocarbons. Related to agriculture activities and crop production, integrated

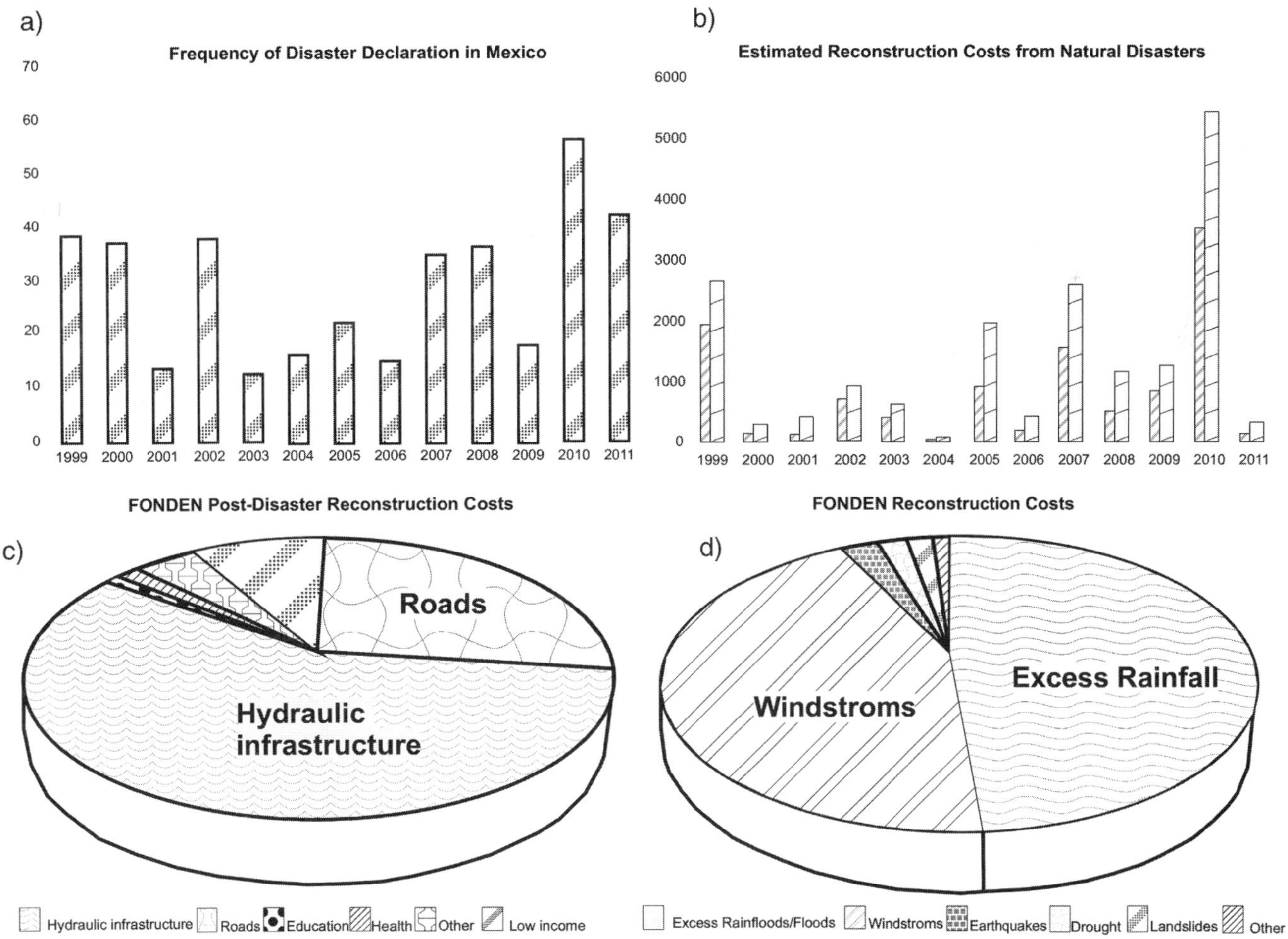

Figure 23.9 FONDEN data on natural disasters in Mexico during the period 1999–2011. (a) Number of declarations of natural disasters in Mexico, (b) estimated costs of natural disasters in Mexico, (c) causes of natural disasters, (d) allocation of funds for the infrastructure and aid programs (from FONDEN, 2012).

geophysical studies provide information on the soils, extent, type, distribution, thickness and bedrock, as well as on erosion, deforestation and land use. Satellite data provide high-resolution monitoring tools at regional and local scales. The Earth Observation Satellite Network provides detailed information useful for climate modeling and hazard risk assessment (Wentz and Schabel, 2000; Hook et al., 2001; Doutriaux-Boucher and Quass, 2004; Paul et al., 2004; Emori et al., 2005). The data and processing capabilities need to be transferred to developing nations, where they can contribute to capacity building. Resolution of climate models has been increasing over the past years. GCMs permit evaluation of climate impacts under distinct scenarios (Emori et al., 2005). Coupled models using high temporal-spatial resolution provide critical data for developing mitigation and adaptation programs, which are especially needed for developing nations where economic resources and infrastructure are limited.

The multi-year drought in northern Mexico and flooding disasters in southern Mexico show the regional variability and the potential impact of hydro-meteorological phenomena. Analyses highlight the interrelations among the different factors involved, including the monitoring, mitigation, prevention; economic, social and response capacity and resilience. In this context, studies emphasize the importance of long-term planning with science-based strategies. In the case of flooding risks in southern Mexico, Aparicio et al. (2008) and Rivera-Trejo et al (2010) emphasized needs for non-structural mitigation measurements, urban growth regulations, planning, early-warning systems and monitoring networks. Integral analysis should be a major component in strategic planning and decision making. Needs for long-term planning strategies are clearly emphasized by the case studies documented all over the world, especially in disaster-prone areas (UNISDR, 2009; ICSU, 2010; Southgate et al., 2013; WCDRR, 2015; FAO, IFAD, WFP, 2015).

A global view of assessment on food security, poverty and susceptibility and resilience to natural hazards shows strong correlation to social and economic variables such as GDP per capita, population size, education, demography and health services. Most of the fast-growing urban centers are in developing countries and fast-growing megacities are located in countries with inadequate infrastructures. High indices of food insecurity are accompanied by inadequate health systems, social unrest and conflicts; international collaboration efforts need to focus on building local capacities within structured long-term programs. The strong economies have the largest CO_2 emissions and waste and water consumption per capita. Climate change affects the different regions of the world unevenly; as a result some regions are affected more (Allen and Ingram, 2002). Climate change poses threats to developed and developing countries, although the response capacity and resilience vary widely.

23.6 Conclusions

The importance of long-term effective prevention programs is emphasized by analyses of disaster-prone areas, where actions are mainly directed to rescue and rebuilding, with limited actions towards capacity building and reduction of vulnerability in case of future repeat events (Southgate et al., 2013). The destructive 2011 floods in Thailand, the multi-year drought and wildfires affecting large areas in northern Mexico and the southern United States and the 2007 flooding in Tabasco State in southern Mexico illustrate the impacts and challenges of repeat events in known disaster-prone areas. The multi-year drought in northern Mexico spanning 2010–2011 was the most severe in the past seventy years, affecting about 40 percent of the territory. Droughts of varying intensity and duration characterize the region, affecting the agriculture, soils and aquifers and producing severe economic losses over the years. These case examples highlight the need to incorporate science-based information in the policy making for natural hazards and climate change.

Earth observation systems, geophysical instrumental networks, satellite and ground-based observatories, global and regional databases, geographic information systems and high-resolution geophysical methods have multiple potential applications. These tools, coupled with improved understanding of phenomena, higher capacity for computer modeling and numerical simulations, provide a scientific-technical framework for developing strategies. Hazard prevention and mitigation programs result in high costs globally, with major effects concentrated on the less developed nations already affected by poverty, famines, health problems, social inequalities, poor infrastructure, low life expectancy, high population growth, inadequate education systems, immigration, economic crises and social and political conflicts.

Integration of scientific and technological advances with efficient transfer mechanisms and capacity building to developing nations is needed. Expanded use of geophysical data, studies and integrated risk analyses can significantly contribute, in addition to natural resources exploration/exploitation programs, to developing and implementing more effective forecast models, mitigation strategies and prevention programs.

Acknowledgments

We acknowledge the critical comments by two reviewers and editor Tom Beer, which have been useful in preparing the chapter. This study forms part of the PALEOMAR network program and National University of Mexico UNAM Continental and Ocean Drilling Program. Preliminary reports on geophysical studies, natural hazards and climate change were presented at the 2013 AGU Joint Assembly Meeting of the Americas, 2014 Disaster Risk SMF-ICSU workshop and the 2016 EGU GIFT Workshop on Natural Hazards. Partial support for the studies is provided by the UNAM-PAPIIT grant IG-101115 and UNAM Oceanographic Platforms CABO projects.

References

AchutaRao, K. and Sperber, K.R. (2006). ENSO simulation in coupled ocean-atmosphere models: Are the models better? *Climate Dynamics*, 27, 1–15.

Aguilar, E., Peterson, T.C., Obando, P.R., Frutos, R., Retana, J.A., Solera, M., Soley, J., Garcia, I.G., Araujo, R.M., Santos, A.R. and Valle, V.E. (2005). Changes in precipitation and temperature extremes in Central America and northern South America, 1961–2003, *Journal of Geophysical Research – Atmospheres*, 110, D23.

Alexeev, V.A. (2003). Sensitivity to CO_2 doubling of an atmospheric GCM coupled to an oceanic layer: A linear analysis, *Climate Dynamics*, 20, 775–787.

Allen, M.R. and Ingram, W.J. (2002). Constraints on future changes in climate and the hydrologic cycle, *Nature*, 419, 224–231.

Alley, R.B., Marotzke, J., Nordhaus, W.D., Overpeck, J.T., Peteet, D.M., Pielke Jr., R.A., Pierrehumbert, R.T., Rhines, P.B., Stocker, T.F., Talley, L.D., Wallace, J.M. (2003). Abrupt climate change, *Science*, 299, 2005–2010.

Aparicio, J., Martinez-Austria, P.F., Guitron, A. and Ramirez, A.I. (2008). The October-November 2007 floods in Tabasco, Mexico: An interim diagnosis and courses of action. 4th International Symposium on Flood Defense, Toronto, Ontario, Canada, 58–1–7.

Arredondo-Moreno, T. and Huber-Sannwald, E. (2011). Impact of drought in agriculture in northern Mexico. *Coping of Global Environmental Change, Disasters and Security*, 5, 875–891.

Arriaga-Ramirez, S. and Cavazos, T. (2010). Regional trends of daily precipitation indices in northern Mexico and southwest United States. *Journal of Geophysical Research: Atmospheres*, 115, D14. doi:10.1029/2009/DOI3248.

Ault, T.R., Cole, J.E., Overpeck, J.T., Pederson, G.T. and Meko, D.M. (2014). Assessing the risk of persistent drought using climate model simulations and paleoclimate data. *Journal of Climate*, 27, 7529–7549.

Beer, T. (Ed) (2010). *Geophysical Hazards Minimizing Risk Maximizing Awareness*. International Year of Planet Earth Series. Dordrecht: Springer Science and Business Media.

Beer, T. and Ismail-Zadeh, A. (Eds) (2003). *Risk Science and Sustainability: Science for Reduction of Risk and Sustainable Development of Society*. Dordrecht: Kluwer Academic.

Bronstert, A. (2003). Floods and climate change: Interactions and impacts. *Risk Analysis*, 23, 545–557.

Chavez, M., Ghil, M. and Urrutia-Fucugauchi, J. (Eds) (2016). Extreme Events: Observations, Modeling and Economics. *Wiley, American Geophysical Union Monograph* 214.

Cook, B.I., Ault, T.R. and Smerdon, J.E. (2015). Unprecedented 21st century drought risk in the American Southwest and Central Plains. *Science Advances*, 1(1), e1400082.

Cook, B.I., Cook, E.R., Smerdon, J.E., Seager, R., Williams, A.P., Coats, S., Stahle, D.W. and Diaz-Villanueva, J. (2016). North American megadroughts in the Common Era: Reconstructions and simulations. *Wiley Interdisciplinary Reviews: Climate Change*, 7(3), 411–432.

Davey, M. et al. (2002). STOIC: A study of coupled GCM climatology and variability in tropical ocean regions. *Climate Dynamics*, 18, 403–420.

Diffenbaugh, N.S., Swain, D.L. and Touma, D. (2015). Anthropogenic warming has increased drought risk in California. *Proceedings of the National Academy of Sciences*, 112(13), 3931–3936.

Doutriaux-Boucher, M. and Quass, J. (2004). Evaluation of cloud thermodynamic phase parametrization in the LMDZ GCM by using POLDER satellite data. *Geophysical Research Letters*, 31, L06126.

Emori, S., Hasegawa, A., Suzuki, T. and Dairaku, K. (2005). Validation, parametrization dependence and future projection of daily precipitation simulated with an atmospheric GCM. *Geophysical Research Letters*, 32, L06708.

FAO, IFAD and WFP (2015). *The State of Food Insecurity in the World*. Meeting the 2015 international hunger targets: Tacking stock of uneven progress. Rome: Publ. Food and Agriculture Organization, United Nations, FAO.

Flato, G.M. (2005). *The Third Generation Coupled Global Climate Model (CGCM3)*. www.cccma.bc.ec.gc.ca/models/cgcm3.shtml.

FONDEN. (2012). *FONDEN El Fondo de Desastres Naturales de México – Una Reseña. Libro Fondo Nacional de Desastres Naturales*. Banco Mundial, México: Sistema de Gestión de Riesgo de Desastres.

Francis, P. and Rothery, D. (2000). Remote sensing of active volcanoes. *Annual Review of Earth and Planetary Sciences*, 28(1), 81–106.

Gleick, P.H. (1989). Climate change, hydrology, and water resources. *Reviews of Geophysics*, 27(3), 329–344. doi:10.1029/RG027i003p00329.

Griffies, S.M. (2004). *Fundamentals of Ocean Climate Models*. Princeton, NJ: Princeton University Press.

Griffin, D. and Anchukaities, K.J. (2014). How unusual is the 2012–2014 California drought? *Geophysical Research Letters*, 41(24), 9017–9023.

Hagemann, S. (2002). An Improved Land Surface Parameter Dataset for Global and Regional Climate Models. Max Planck Institute Meteorology Report 162, Hamburg, Germany 21 pp.

Hallack-Alegria, M., Ramirez-Hernandez, J. and Watkins, D.W. (2012). ENSO-conditions rainfall drought frequency analysis in northwest Baja California, Mexico. *International Journal of Climatology*, 32(6), 831–842.

Haug, G. et al. (2001). Southward migration of the Intertropical Convergence Zone through the Holocene. *Science*, 293, 1304–1308.

Herman, J. R., Bhartia, P. K., Torres, O., Hsu, C., Seftor, C. and Celarier, E. (1997). Global distribution of UV-absorbing aerosols from Nimbus 7/TOMS data. *Journal of Geophysical Research: Atmospheres*, 102(D14), 16911–16922.

Hook, S.J., Myers, J.J., Thorne, K.J., Fitzgerald, M. and Kahle, A.B. (2001). The MODIS/ASTER airborne simulator (MASTER) – a new instrument for earth science studies. *Remote Sensing of Environment*, 76, 95–102.

Hwang, L., Jordan, T., Kellog, L., Tromp, J. and Wiellemann, R. (2014). *Advancing solid Earth system science through high-performance computing*. Computational Infrastructure for Geodynamics Publ., University of California, Davis.

ICSU. (2010). *Earth System Science for Global Sustainability: The Grand Challenges*. Paris: International Council for Science.

IPCC. (2014). *Climate Change 2014: Synthesis Report*. Contribution of Working Groups I, II and III to the Fifth Assessment Report of the IPCC (Core Writing Team, R.K. Pachuri and L.A. Meyer Eds, Geneva, Switzerland.

Ismail-Zadeh, A., Urrutia-Fucugauchi, J., Kijko, A., Takeuchi, K. and Zialapin, I. (Eds) (2014). *Extreme Natural Hazards, Disaster Risks and Societal Implications*. Cambridge: Cambridge University Press.

Kaufman, Y.J., Tanré, D. and Boucher, O. (2002). A satellite view of aerosols in the climate system. *Nature*, 419, 215–223.

Kearey, P., Brooks, M. and Hill, I. (2013). *An Introduction to Geophysical Exploration*. John Wiley & Sons.

Kleidon, A. (2004). Global datasets of rooting zone depth inferred from inverse methods. *Journal of Climate*, 17, 2714–2722.

Lal, R. et al. (2004). Soil carbon sequestration impacts on global climate change and food security. *Science*, 304, 1623–1627.

Lambert, S.J. and Boer, G.J. (2001). CMIP1 evaluation and intercomparison of coupled climate models. *Climate Dynamics*, 17, 83–106.

Latif, M. and Keenlyside, N.S. (2009). El Niño/Southern Oscillation response to global warming. *Proceedings National Academy of Sciences*, 106, 20578–20583.

Lawrimore, J., Heim Jr, R. R., Svoboda, M., Swail, V. and Englehart, P. J. (2002). Beginning a new era of drought monitoring across North America. *Bulletin of the American Meteorological Society*, 83(8), 1191–1192.

Mann, M.E. and Gleick, P.H. (2015). Climate change and California drought in the 21st century. *Proceedings of the National Academy Sciences*, 112(13), 3858–3859.

Manning, M.R., Petit, M., Easterling, D., Murphy, J., Patwardhan, A., Rogner, H., Swart, R. and Yohe, G. (Eds) (2004). *IPCC workshop on describing scientific uncertainties in climate change to support analysis of risks of options*. Geneva: IPCC Workshop Report.

McBean, G.A. (2002). Prediction as the basis for planning and response. *Water International*, 7, 70–76.

McPhadden, M., Zhang, X., Hendon, H.H. and Wheeler, M.C. (2006). Large scale dynamics and MJO forcing of ENSO variability. *Geophysical Research Letters*, 33, L16702.

Mendez, M. and Magaña, V. (2010). Regional aspects of prolonged meteorological droughts over Mexico and Central America. *Journal of Climate*, 23, 1175–1188.

Medina-Cetina, Z. and Nadim, F. (2008). Stochastic design of an early warning system. *Georisk*, 2, 223–236.

Milly, P.C.D., Wetherald, R.T., Dunne, K.A. and Delworth, T.L. (2002). Increasing risk of great floods in a changing climate. *Nature*, 415, 514–517.

Milly, P.C.D., Dunne, K.A. and Vecchia, A.V. (2005). Global pattern of trends in streamflow and water availability in a changing climate. *Nature*, 438, 347–350.

Moss, R., Edmonds, J.A., Hibbard, K.A., Manning, M.R., Rose, S.K., van Vuueren, D.P., Carter, T.R., Emori, S., Kainuma, M., Kram, T., Meehl, G.A., Mitchel, J., Nakicenovic, N., Rihai, K., Smith, S.J., Stouffer, R.J., Thompson, A.M., Weyyant,.P. and Wilbanks, T.J. (2010). The next generation of scenarios for climate change research and assessment. *Nature*, 463, 747–756.

Mourtzinis, S., Ortiz, B.V. and Damianidis, D. (2016). Climate change and ENSO effects on southwestern US climate patterns and maize yield. *Scientific Reports*, 6(29777), 1–7.

MunichRe. (2015). NatCatSERVICE Download Center. Natural catastrophes 2014 Report. www.businesswire.com/news/home/20150107005586/en/Review-natural-catastrophes-2014-losses-weather-extremes.

NASA. (2011). National Aeronautics and Space Administration, NPP NPOESS Preparatory Project. www.nasa.gov/NPP.

Parry, M. and Livermoore, M. (2002). Climate change, global food supply and risk of hunger. *Issues in Environmental Science and Technology*, 17, 109–136.

Paul, F., Huggel, C. and Kääb, A. (2004). Combining satellite multispectral image data and digital elevation model for mapping debris-covered glaciers. *Remote Sensing of Environment*, 89, 510–518.

Pérez-Cruz, L. (2006). Climate and ocean variability during mid-late Holocene recorded in laminated sediments from Alfonso basin, Gulf of California, Mexico. *Quaternary Research*, 65, 401–410.

Pérez-Cruz, L. (2013). Hydrological changes and paleoproductivity in the Gulf of California during middle and late

Holocene and their relationship with ITCZ and North American Monsoon variability. *Quaternary Research*, 79, 138–151.

Peterson, T.C., Zhang, A., Brunet-India, M. and Vazquez-Aguirre, J.L. (2008). Changes in North American extremes derived from daily weather data. *Journal of Geophysical Research: Atmospheres*, 113(D7).

Power, S.B. and Colman, R. (2006). Multi-decadal predictability in a coupled GCM. *Climate Dynamics*, 26, 247–272.

Ramanathan, V.C.P.J., Crutzen, P.J., Kiehl, J.T. and Rosenfeld, D. (2001). Aerosols, climate, and the hydrological cycle. *Science*, 294(5549), 2119–2124.

Ramos J., Marrufo, L. and Gonzalez, F.J. (2009). Use of Lidar data in floodplain risk management planning: The experience of Tabasco 2007 flood. In: *Advances in Geoscience and Remote Sensing*, Gary Jedlovec (Ed). Available from: www.intechopen.com/books/advances-in-geoscience-and-remote-sensing/use-of-lidar-data-in-floodplain-risk-management-planning-the-experience-of-tabasco-2007-flood.

Reynolds, J.M. (2011). *An Introduction to Applied and Environmental Geophysics*. John Wiley & Sons.

Rivera-Trejo, F., Soto-Cortés, G. and Méndez-Antonio, B. (2010). The 2007 flood in Tabasco, Mexico: An integral analysis of a devastating phenomenon. *International Journal of River Basin Management*, 8(3–4), 255–267.

Rosenzweig, C. and Parry, M.L. (1994). Potential impacts of climate change in world food supply. *Nature*, 367, 133–138.

Santos-Reyes, J. and Beard, A.N. (2011). Applying the SDMS model to the analysis of the Tabasco flood disaster in Mexico. *Human Ecol. Risk Assessment: An International Journal*, 17, 643–677.

Schmidhuber, J. and Tubiello, F.R. (2007). Global food security under climate change. *Proceedings of the National Academy of Sciences*, 104(50), 19703–19708.

Seager, R. Goddard, L., Nakamura, J., Henderson, N. and Eun Lee, D. (2014). Dynamical causes for the 2010/11 drought in Texas–northern Mexico. *Journal of Hydrometereolgy*, 15, 39–68.

Seager, R., Ting, M., Held, I., Kushnir, Y., Lu, J., Vecchi, G., and Li, C. (2007). Model projections of an imminent transition to a more arid climate in southwestern North America. *Science*, 316(5828), 1181–1184.

Seager, R., Ting, M., Davis, M., Cane, M., Naik, N., Nakamura, J., Li, C., Cook, E. and Stahle, D.W. (2009). Mexican drought: An observational modeling and tree ring variability and climate change. *Atmosfera*, 22, 1–31.

Silva, I.H., Miranda, F, Beisl, C.H. and Landau, L. (2011). System for flooding alert in tropical coastal zones using GIS and remote sensing: A case study Villahermosa, Mexico. *Journal of Coastal Research*, 64, 1734–1736.

Soden, B.J. (2000). The sensitivity of the tropical hydrological cycle to ENSO. *Journal of Climate*, 13, 538–549.

Southgate, R.J., Roth, C., Schneider, J., Shi, P., Onishi, T., Wenger, D., Amman, W., Ogallo, L., Beddington, J. and Murray, V. (2013). Using Science for Disaster Risk Reduction. UNISDR Report, www.preventionweb.net/go/scitech.

Stephens, G.L. (2005). Cloud feedbacks in the climate system: A critical review. *Journal of Climate*, 18, 237–273.

Svoboda, M. et al., (2002). The Drought Monitor. *Bulletin American Metereological Society*, 83, 1181–1190.

Swain, D. L., Tsiang, M., Haugen, M., Singh, D., Charland, A., Rajaratnam, B. and Diffenbaugh, N. S. (2014). The extraordinary California drought of 2013/2014: Character, context, and the role of climate change. *Bulletin of the American Meteorological Society*, 95(9), S3.

Tanré, D., Kaufman, Y. J., Holben, B. E. A., Chatenet, B., Karnieli, A., Lavenu, F., and Smirnov, A. (2001). Climatology of dust aerosol size distribution and optical properties derived from remotely sensed data in the solar spectrum. *Journal of Geophysical Research: Atmospheres*, 106(D16), 18205–18217.

Trenberth, K.E. (2011). Changes in precipitation with climate change. *Climate Research*, 47(1–2), 123–138.

UNISDR. (2009). Reducing Disaster Risks Through Science: Issues and Actions. ISDR Scientific and Technical Committee Full Report, Geneva www.unisdr.orgfiles/11543 STCReportlibrary.pdf.

Urrutia-Fucugauchi, J. (2014). Magnetic studies of active volcanoes in Mexico: Implications for volcanic hazards and volcano monitoring. In: Ismail-Zadeh, A. et al. (Eds), *Extreme Natural Hazards, Disaster Risks and Societal Implications*. Cambridge: Cambridge University Press, 152–166.

Urrutia-Fucugauchi, J. and Pérez-Cruz, L. (2016). Planetary sciences, geodynamics, impacts, mass extinctions and evolution: Developments and interconnections. *International Journal Geophysics*, 2016, ID 4703168. doi:10.1155/2016/4703168.

WCDRR. (2015). SATELLITE Earth Observations in Support of Disaster Risk Reduction. Special 2015 *WCDRR edition*. European Space Agency, CEOS Earth Observation Handbook for WCDRR.

Wentz, F.J. and Schabel, M. (2000). Precise climate monitoring using complementary data sets. *Nature*, 403, 414–416.

Wirtz, A., Löw, P., Mahl, T. and Yildrim, S. (2014). Hitting the poor: Public-private partnership as an option. In: Ismail-Zadeh, A. et al. (Eds), *Extreme Natural Hazards, Disaster Risks and Societal Implications*. Cambridge: Cambridge University Pres, 386–398.

Woodhouse, C.A., Meko, D.M., MacDonald, G.M., Stahle, D.W. and Cook, E.R. (2010). A 1,200-year perspective of 21st century drought in southwestern North America. *Proceedings National Academy Science*, 107, 21283–21288.

24

Climatic Consequences and Agricultural Impacts of Nuclear Conflicts

Owen B. Toon, Alan Robock, Michael Mills, Lili Xia, and Charles Bardeen

Public debate about nuclear weapons is no longer common or sustained. It was barely a news story when Donald Trump, now president of the United States, argued during the presidential election that proliferation is bound to continue, and therefore the United States could save money by allowing Japan and South Korea to become nuclear states to counter North Korea (Sanger and Haberman, 2016). It was merely another daily story for President Trump to argue during the campaign that nuclear weapons should be considered for use against terrorist organizations such as ISIS, or in Europe because it has plenty of space (Sanger and Haberman, 2016). The U.S. population did not react when President Obama, who advocated the elimination of nuclear weapons, set in motion projects to upgrade weapons costing hundreds of billions of dollars in a time of financial constraint. Unfortunately, this lack of public interest or concern is not justified by the facts. We are only an accident, a mistake, or a deranged politician away from killing most of the world's population now. As arsenals and the number of nuclear weapons states continue to grow, the threat of a nuclear exchange is bound to grow in the future.

Nuclear-armed countries are a threat to people everywhere partly because of the destructive power of single weapons – one weapon is enough to destroy a small city – and partly because of the growing ability of nations to launch missiles across the globe. However, it is not just the brute force attack killing people in the geographically limited target zone that threatens people everywhere. Most people have forgotten nuclear winter. For example, Lepore (2017) states that "the nuclear-winter debate has long since been forgotten." However, she fails to cite any of the extensive scientific literature on this topic after about 1986. Of course, the debate is forgotten once the press decides not to report on the modern status of the debate. Many think that the theory was disproven, or that the end of the nuclear arms race and subsequent reduction of Russian and American nuclear arsenals eliminated the global dangers of nuclear war. But they are wrong.

Nuclear winter is an assault on the global climate system caused by smoke from fires ignited by the bombs. As the smoke rapidly spreads globally in the stratosphere it will reduce temperatures and rainfall, and destroy the global ozone layer, which shields us from harmful ultraviolet radiation from the Sun. Recently it has been shown that even the smoke created by the use of one hundred weapons of the size used on Hiroshima in the Second World War, comparable to the arsenals of India or Pakistan, could cause environmental damage that would extend globally, threatening the world food supply and creating mass starvation worldwide (Robock and Toon, 2012; Xia and Robock, 2013; Özdoğan et al., 2013; Mills et al., 2014; Xia et al., 2015). The effects of food loss would also be felt in the aggressor nation. Hence, being a nuclear aggressor is suicidal. The deaths from these environmental changes would likely be a factor of 10 or more larger than the direct casualties from the explosions – potentially threatening the bulk of the human population – and would not be limited to the combatants.

The Cold War nuclear standoff between the West and the East, which was based on mutual assured destruction (MAD), may be slowly evolving into a nuclear free-for-all (e.g., Evans, 2014). Former U.S. Secretaries of State George Shultz and Henry Kissinger, former U.S. Secretary of Defense William Perry and former U.S. Senator Sam Nunn founded the Nuclear Security Project, "an effort to galvanize global action to reduce urgent nuclear dangers and build support for reducing reliance on nuclear weapons, ultimately ending them as a threat to the world."[1] These warriors of the Cold War argue in a series of editorials in the *Wall Street Journal* (Shultz et al., 2007, 2008, 2010, 2011) that in a world with many nuclear powers, each capable of destroying any country on Earth, and terrorist groups with increasing potential to gain control of nuclear weapons, MAD no longer works and one can no longer assume that peace can be maintained through rational analysis, careful control of weapons or successful negotiation between well-defined states with clear national interests. The only solution, they argue, is to create a world without nuclear weapons.

The newer nuclear states such as India and Pakistan have nuclear arsenals at the levels of the Soviet Union and the Western powers in the early 1950s (Kristensen and Norris, 2013, 2014). If the weapons were used, they could destroy a large fraction of the infrastructure of any country on Earth. India, North Korea and Iran are developing ballistic weapons–delivery systems that can send nuclear warheads over intercontinental scales. In coming decades it is possible that there will be a global nuclear gridlock caused by multiple nuclear-armed Asian, American and European states each with differing goals and aspirations, and each with the capability to destroy any adversary purposefully and to destroy much of the rest of the world inadvertently.

Against this backdrop of expanding nuclear states, the concept of SAD, self-assured destruction, has been introduced (Robock and Toon, 2012). According to this concept it is suicidal to use nuclear weapons, even the not-so-modest arsenals of newer nuclear states such as Pakistan and India. Everyone recognizes that immense damage and loss of life will occur in the combatant countries as a result of the explosions of the weapons. However, it is not as well understood that worldwide environmental damage from even a regional conflict could be much worse than the direct effects of the explosions. The fires started from weapons exploding in cities will flood the upper atmosphere with smoke. Smoke absorbs sunlight, heating the upper atmosphere and destroying the protective ozone layer. The light-absorbing smoke also prevents sunlight from reaching the Earth's surface, driving global temperatures down enough to damage agriculture at mid-latitudes. The loss of agricultural productivity will create mass starvation globally, including in those countries that used the nuclear weapons. To reiterate, the deaths from these environmental changes would likely be a factor of 10 or more larger than the direct casualties from the explosions – potentially threatening the majority of the human population – and would not be limited to the combatants.

24.1 World Nuclear Arsenals

Figure 24.1 illustrates the history of the number of warheads on the planet, and those in Russia and the United States, which currently control more than 90% of the weapons (Kristensen and Norris, 2014). The world total peaked at around seventy thousand warheads in 1986, five years before the disintegration of the Soviet Union. The downward trend in nuclear weapons, begun in 1986, has continued to this day. However, most of the downward trend occurred in the 1990s, and the rate of decline has slowed. The 2016 world arsenal was near fifteen thousand weapons, about 22% of the peak in 1986. The Strategic Offensive Reductions Treaty (SORT) reached by Presidents George W. Bush and Vladimir Putin, and in place from 2003 to 2011, regulated the number of strategic warheads not to exceed seventeen hundred to twenty-two hundred for each country by 31 December 2012. The New Strategic Arms Reduction Treaty (New-START), developed under Presidents Barack Obama and Dmitry Medvedev and in place after 2011, results in similar numbers of deployed strategic warheads to those in the SORT treaty. Certain types of warheads were not counted in either the SORT or New START. Of the fifteen thousand total warheads believed to exist among all nuclear powers in 2016, only about thirty-seven

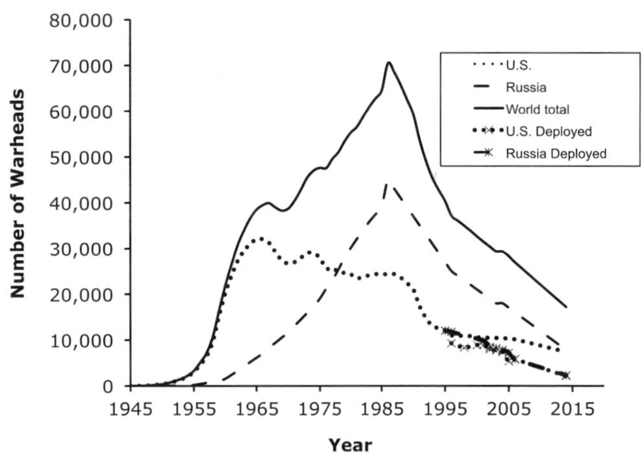

Figure 24.1 The number of nuclear warheads since World War II on the basis of data from Kristensen and Norris (2013) is shown. The arsenals of the United States, Russia and the world are depicted. After about 2000, Russia and the United States began to base treaties on the number of deployed strategic warheads, rather than the total number of warheads. Deployed strategic warheads are shown with symbols. In 2013, there were about 16,300 warheads in the world. About 6,200 of these were retired and ostensibly waiting to be destroyed. The United States and Russia had about 3,750 deployed strategic warheads, which are covered by treaties negotiated under the Bush and Obama administrations. Another 5,380 warheads were in storage, or considered not to be strategic, and therefore not covered in recent treaties (updated from box 2 in Toon et al., 2008).

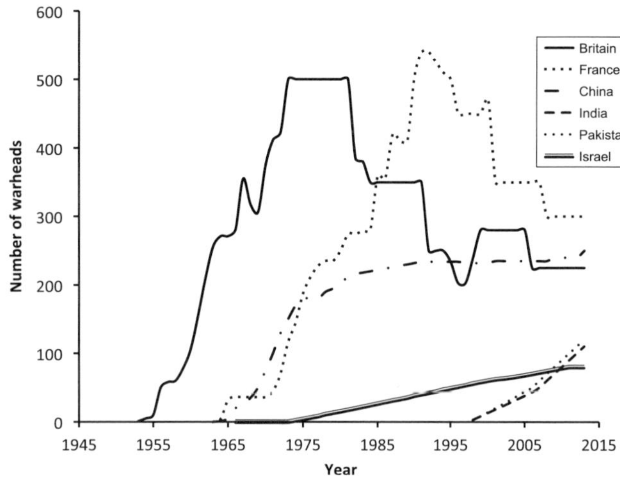

Figure 24.2 The number of warheads thought to be in the arsenals of Britain, France, China, India, Pakistan and Israel on the basis of data from Kristensen and Norris (2013) is illustrated. North Korean weapons are not shown because it is uncertain that they have an arsenal of usable weapons (from Toon et al., 2017).

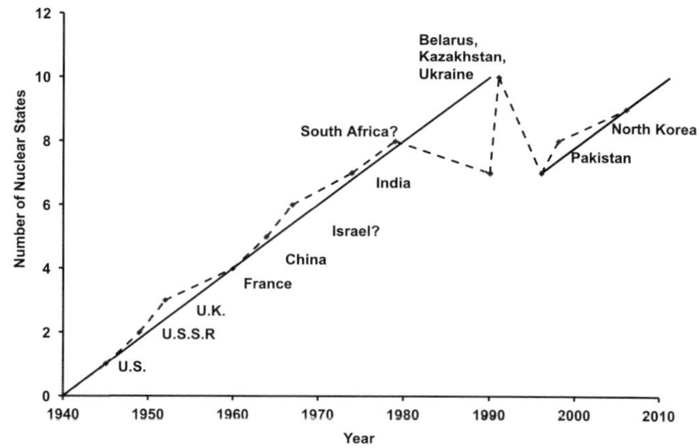

Figure 24.3 The dates when various nations obtained a nuclear warhead, mainly based on when they first tested a weapon, are shown. For Israel and South Africa the evidence for tests is controversial, so an estimate for when they had a usable weapon is given. Belarus, Kazakhstan and Ukraine inherited weapons from the Soviet Union and transferred them to Russia in the 1990s. South Africa gave up its weapons in the 1990s. The solid lines represent one new nuclear state every five years (updated from Box 2 in Toon et al., 2008).

hundred deployed strategic weapons are regulated under New START. About one thousand unregulated weapons are owned by France, China, the United Kingdom, Israel, Pakistan, India and North Korea together. Most of these weapons are not deployed, but are in storage. The remainder includes about seven thousand weapons in the United States and Russia that are considered either not deployed or not strategic (many are strategic warheads in storage; others are tactical weapons). There may be another five thousand warheads waiting to be dismantled in Russia and the United States. The United States also has twenty thousand plutonium pits in storage. Both Russia and the United States are embarked on expensive upgrades to their nuclear capabilities.

Figure 24.2 illustrates the arsenals of the countries with nuclear weapons other than the United States and Russia (Kristensen and Norris, 2013). It is very difficult to determine the numbers of weapons in most of these countries. While Britain and France have been slowly reducing their arsenals, and China seems to have maintained a constant level, Pakistan, India and possibly Israel seem to be increasing their arsenals. The yields of the weapons in most of the programs are not known. Britain, France and China have weapons with yields above one hundred kilotons (kt);[2] however, it is likely, on the basis of their nuclear tests, that India and Pakistan have weapons with yields similar to those of the U.S. weapons used in World War II, around ten to twenty kilotons, or even less.

Figure 24.3 outlines the history of nuclear proliferation. Up until the mid-1980s about one new nuclear state appeared every five years. Following the start of the build-down of nuclear weapons by the United States and Russia in 1986, a number of states abandoned their arsenals, or stopped nuclear weapons programs that were under consideration or development. South Africa developed a small nuclear arsenal and then abandoned it in the 1990s, the only country ever to abandon a self-developed nuclear arsenal. A number of countries inherited nuclear weapons from the former Soviet Union, but gave them up, including Kazakhstan, Belarus and Ukraine. Unfortunately, proliferation was renewed in 1998 when India and then Pakistan tested nuclear weapons. As Figure 24.3 indicates, the world now may be back on the trend of one new nuclear state about every five years.

Parts of the nuclear story are encouraging. Much of the world is ridding itself of nuclear weapons. Figure 24.4 shows the history of the development of nuclear-weapon-free zones. Almost one-third of the human population now lives in regions in which the United Nations has recognized treaties banning nuclear weapons. The Treaty for the Prohibition of Nuclear Weapons in Latin America and the Caribbean, also known as the Treaty of Tlatelolco, went into effect in 1968 and was eventually signed by all thirty-three independent nations of Latin America and the Caribbean, including Cuba. However, Brazil and Argentina reserved the right to conduct "peaceful nuclear weapons

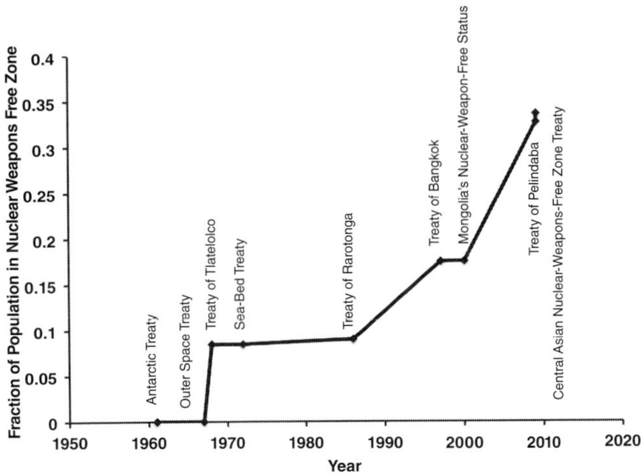

Figure 24.4 The fraction of the world's current population that lives in nuclear-weapon-free zones, as recognized through UN treaties, is shown. The names of the various treaties, as described in the text, are listed. Currently about one-third of Earth's population lives in a nuclear-free zone (from Toon et al., 2017).

explosions," and islands such as Puerto Rico, British Virgin Islands and Guadeloupe, associated with nuclear weapons states, are excluded. Africa established a nuclear-weapon-free zone under the Treaty of Pelindaba, which took effect in 2009. As a result of the additional treaties for Antarctica and the Rarotonga Treaty involving Australia, New Zealand and a number of island nations in the Pacific the entire Southern Hemisphere (with the exception of islands associated with states with nuclear weapons and international waters) is a nuclear-weapons-free zone. The Association of Southeast Asian Nations, including Brunei, Darussalam, Cambodia, Indonesia, Laos, Malaysia, Myanmar, Philippines, Singapore, Thailand and Vietnam, became a nuclear-weapon free-zone in 1997 under the Bangkok Treaty. There is also a Central Asian nuclear free treaty signed by Kazakhstan, Kyrgyzstan, Tajikistan, Turkmenistan and Uzbekistan, which entered into force in 2009. Mongolia declared itself a nuclear-weapon-free zone in 1992, and it was formally recognized as such in 2012 by the five original nuclear weapons states: the United States, China, France, the United Kingdom and Russia. In total there are at least sixteen countries in Asia that are free of nuclear weapons, which is almost three times as many as Asian nuclear weapons states.

24.2 How Many Weapons Are Enough?

Comparison of Figure 24.1 and Figure 24.2 raises the question, How do you know how many weapons are enough? None of the nuclear weapons states has stated its criteria for answering this question.

In the case of the United States, analyses in the late 1940s reported by Eden (2004) identified militarily important industrial sites in the Soviet Union and concluded these could be destroyed by the delivery of one hundred atomic weapons, which were similar in yield to the weapons in the arsenals of present-day Pakistan, India and North Korea. Of course, weapons may not explode, or may never reach their target, so it was estimated that two hundred weapons would be needed for the United States to destroy the Soviet Union. Despite this estimate, the U.S. arsenal rose to more than 150 times this many weapons, and with typically more than ten times the average yield per weapon. One reason for the bloated number of weapons may be the competition between Russia and the United States and attempts by each to dominate the other so that a first strike might overwhelm the adversary (e.g., Lieber and Press, 2006). In the case of a first strike, each missile of the adversary must be targeted with multiple warheads. The opponent must then obtain even more warheads to be able to attack each of the enemy's missiles with more than one warhead. This competition to outnumber the other side leads to exponential growth in warheads. Alternative suggestions for the large numbers of weapons include competition for funding to support the nuclear infrastructure, political posturing, lack of planning or lack of thought about the size of the arsenal needed or concern that there would be a high failure rate of the weapons.

Toon et al. (2008) analyzed a modern attack on the urban areas in a number of countries. They found that in an attack on U.S. urban areas based on population density with one thousand weapons of one-hundred-kiloton yield, 48% of the U.S population would be within five kilometers of ground zero, 20% of the population (about 60 million people) would be killed outright, and another 16% injured (about 40 million people). They also found that a war between India and Pakistan using one hundred weapons (about half their current arsenals) with fifteen-kiloton yield, exploded on cities on the basis of population, would kill or injure about 45 million people.

The eventual build-up to more than thirty thousand American warheads and forty thousand Soviet warheads, most with more than ten times the yield of the Hiroshima weapon, was gross overkill. There are simply not that many targets. U.S. targeting strategy is highly classified. However, recently revealed nuclear targeting documents show that in 1959 population was treated as a distinct target for U.S. nuclear weapons (Shane, 2015). Yet, even in a counterforce war, in which population is not directly attacked, substantial civilian casualties will occur because military and industrial targets are both located in urban areas. For instance, Bush et al. (1991) and Small (1989) considered a counterforce[3] attack by Russia on 3030 leading U.S. targets such as U.S. Army, Navy, and Air Force bases; fuel storage locations; refineries and harbors (but not missile silos or launch-control facilities) with five-hundred-kiloton warheads. In fact, because of the significant area destroyed by a nuclear bomb explosion, there were only 348 unique targets in these studies that were on average each attacked by 8.7 weapons when 3030 weapons were used. While cities were not targeted directly, 50% of the U.S. urban areas were destroyed.

There are only about 300 cities in the United States, and 180 in Russia with a population greater than 100,000. There are about 60 cities in Pakistan, 365 in India and 445 in China with a population greater than 100,000. It might require more than one weapon to destroy some large cities, and there are some important targets in rural areas. Nevertheless, the two thousand strategic deployed weapons in both the American and the Russia arsenals under current treaty still allow them to destroy the bulk of any adversary's infrastructure and population given the likely assumption that the majority of the weapons detonate. The current arsenal of India is great enough to attack almost every moderate-sized city in Pakistan two times.

This sort of gruesome calculation of the numbers of weapons needed to destroy an adversary has likely been repeated in many nations. In terms of the numbers of weapons in Figures 24.1 and 24.2, and the numbers of cities in the world that might be targets, the number of weapons may be chosen by most countries to ensure they can bomb every moderate-sized city of any country in the world that they choose. The emphasis in sizing nuclear arsenals is on ensuring a level of destruction, not a level of survival.

Unfortunately, the collateral damage from using these weapons may be much greater than the direct damage. Therefore, ensuring the destruction of any adversary, and giving no consideration to how many might survive, may lead to the accidental deaths of most of the world's population.

24.3 Self-Assured Destruction

Nuclear weapons cause damage through prompt and delayed nuclear radiation, thermal radiation and shock waves. Prompt radiation arises from the nuclear fireball and the nuclear reactions producing the detonation. Delayed radiation is from the fallout containing the radioactive daughter products of the nuclear explosion. Thermal radiation is a bright pulse of light emitted by the explosion. Shock waves are high winds and associated pressure fluctuations that occur over very short periods. It is difficult to untangle the direct damage from prompt radiation, thermal radiation and shock waves because their zones of influence overlap near ground zero. Delayed radiation could kill large numbers of people in areas downwind of ground bursts, which might be used to attack missile silos. Ground bursts near the surface (as opposed to being deeply buried) lift relatively large soil particles into the air. Radioactive particles attach themselves to these dust particles, and when they fall out of the atmosphere onto the ground (hence the name "fallout") within hours or days, they can expose the population at the surface to dangerous levels of radiation. However, even with ground bursts, the casualties from delayed exposure to radiation are likely to be less than those from the other direct effects in an attack on a city. An attack on a missile silo, or other unpopulated area, might have the majority of casualties from delayed radiation. Airbursts are generally more destructive than ground bursts because the area impacted by prompt radiation, thermal radiation and shock waves is larger. Therefore, combatants are likely to use airbursts unless they are attacking buried targets. The debris from airbursts generally contains little material of large enough size to fall out of the atmosphere quickly. Therefore, the nuclear radiation largely decays before it reaches the surface and is less of a hazard to people than material that falls out rapidly. If your goal is killing people, an airburst over a city is more effective than a ground burst that kills fewer with prompt radiation, thermal radiation and shock waves and more with delayed radiation.

Following a nuclear explosion the shock waves knock down structures. These ruptured buildings will experience myriad small fires, which can coalesce into a firestorm. The thermal radiation from the fireball can also set fires over a wide area. One does not need a nuclear weapon to initiate a firestorm. For example, the United States and its allies purposefully set many firestorms in urban areas in World War II, including at Dresden, Hamburg, Tokyo and several other Japanese cities, using hundreds of aircraft carrying incendiary bombs. There have also been damaging firestorms after earthquakes, such as the 1906 San Francisco earthquake. It is believed that fires rather than the earthquake caused the vast majority of the damage to that city. The energy released in firestorms is immense. For instance, Toon et al. (2007) estimated that the energy released in the Hiroshima firestorm, which reached full strength several hours after the nuclear explosion, was about one thousand times greater than the energy released in the nuclear explosion.

Firestorms are self-feeding fires that suck air into themselves, and generate immense columns of rising smoke. There are many observations of stratospheric injections of smoke from large intense forest fires, which are similar to urban firestorms (e.g., Fromm and Servranckx, 2003). Once the smoke is about ten kilometers (approximately six miles) above the ground, sunlight will heat the smoky air and it will rise; this is called self-lofting. Observations may have recorded smoke from forest fires rising by self-lofting to twenty kilometers above the surface (de Laat et al., 2012), and models suggest large smoke injections in a nuclear conflict could reach fifty kilometers above the ground by self-lofting. At altitudes above about twenty kilometers it never rains, so the smoke would remain in the air for years. The clouds from very large volcanic eruptions are observed to remain in the stratosphere for a few years, but they have only a small amount of self-lofting because their semi-transparent particles do not absorb much sunlight.

Toon et al. (2007, 2008) discuss the amount of smoke produced by varying numbers of nuclear weapons exploding in different countries around the world. The amount of black carbon (or soot) is the key factor in the climate calculations. Toon et al. (2007) suggest that in a war between India and Pakistan involving one hundred Hiroshima-sized weapons about 7 million tons of black carbon could be produced. Toon et al. (2008) suggested that a war between the United States and Russia involving forty-four hundred strategic weapons, close to what is allowed by current treaties, with 100,000-ton yield each would produce about 180 million tons of black carbon. Robock et al. (2007) investigated the climate changes after the injections of 50 or 150 million tons of black carbon. The 150-million-ton case produced global average temperatures more than 8 °C cooler than normal, which is well below the average temperature found in the last ice age. Mid-latitude grain growing regions had simulated temperatures below freezing every day for one or two years. Clearly, injecting 150 million tons of black carbon into the stratosphere creates a climate catastrophe.

A recent study of the aftermath of a nuclear conflict (Mills et al., 2014) used an Earth system climate model including atmospheric chemistry, ocean dynamics as well as interactive sea ice and land components to investigate the environmental damage from a limited regional nuclear war between India and Pakistan. Stenke et al. (2013) performed a similar study. Pausata et al. (2015) considered the importance of the length of the conflict, and of adding organic material to the 5 million tons of black carbon. Weakly light-absorbing organic material dominates the black carbon in smoke by a large factor. For the same assumptions about the length of conflict and materials injected these three studies produced similar climate changes. Mills et al. (2014) assumed each side detonated fifty weapons with fifteen kilotons of yield for each weapon over urban areas. One hundred total weapons is about half of the total current arsenals of India and Pakistan. These urban explosions are assumed to start one hundred firestorms producing a total of 5 million tons of smoke. In the computer simulation this smoke self-lofted from the upper troposphere to the stratosphere, where it spread globally. The smoke produced a sudden drop in surface temperatures of about 1.5 °C, because the smoke blocked sunlight from reaching the surface, and intense heating of the stratosphere, because the smoke absorbed sunlight. The results showed that about one-third of the smoke was removed after about nine years. In the hot stratosphere, ozone was destroyed by chemical reactions. Global ozone losses of 20%–50% over populated areas, levels unprecedented in human history, would accompany the coldest average surface temperatures since the waning phases of the last ice age, thousands of years ago. Ozone in the stratosphere protects us from the harmful effects of solar ultraviolet light. Mills et al. (2014) calculated summer enhancements in ultraviolet solar exposure indices of 30%–80% over mid-latitudes, suggesting widespread damage to human health and agriculture, as well as terrestrial and aquatic ecosystems, though the effects of enhanced ultraviolet light are poorly understood since such enhancements have never been observed. Killing frosts would reduce growing seasons by ten to forty days per year for five years at mid-latitudes. Surface temperatures would be reduced for more than twenty-five years. The long period of cold temperatures is due to thermal inertia from the cooled ocean waters and to extra reflection of sunlight back to space by expanded sea ice.

Large global decreases in ozone have not been observed in human experience, though there is a significant latitudinal gradient in ultraviolet intensity. It is known, for instance, that the incidence of skin cancer in people with light-colored skin increases with decreasing latitude as a result of increasing ultraviolet light (e.g., Fears et al., 1976). The ultraviolet light is greatest in the tropics because the sun is more intense there and because there is less ozone in the tropics than elsewhere. While we know that enhanced ultraviolet light is hazardous, its effects are not yet included in agricultural models, so it is not possible to gauge its impact on food supplies or the environment. On the other hand, we have a lot of experience with the effects of changes in

Table 24.1 *Loss of Agricultural Productivity Following a Regional Nuclear Conflict with 100 Warheads*

	First 5 years	Second 5 years
U.S. maize	–20%	–10%
U.S. soybeans	–15%	–10%
China maize	–15%	–10%
China middle season rice	–25%	–20%
China spring wheat	–25%	–20%
China winter wheat	–40%	–25%

Özdoğan et al., 2013: Xia and Robock, 2013; Xia et al., 2015

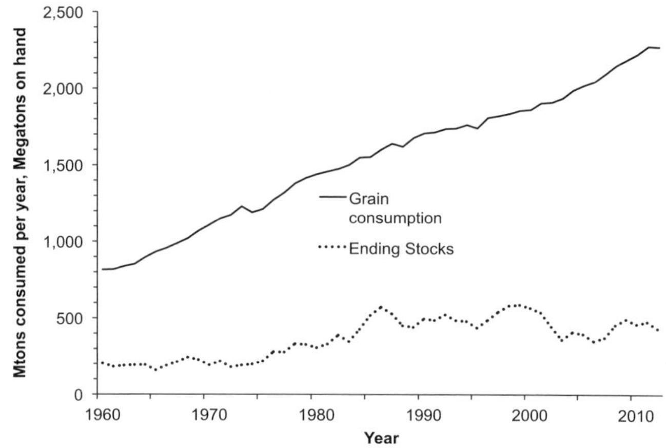

Figure 24.5 The worldwide consumption of grains and the ending stocks from 1960 until 2012 are shown. Data from Earth Policy Institute (2012) (from Toon et al., 2017).

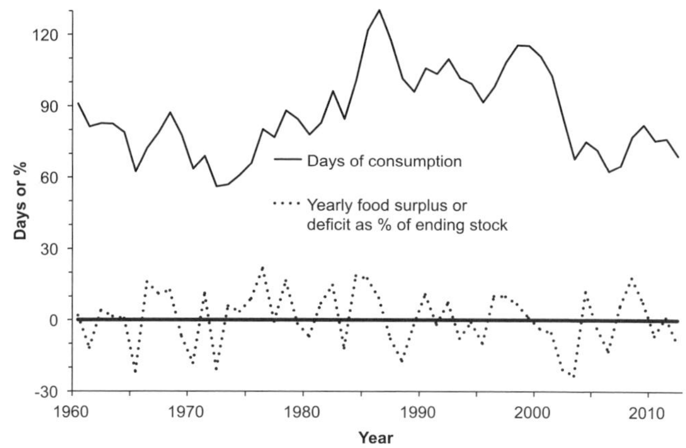

Figure 24.6 The number of days that the ending stocks could supply world food consumption, and the surplus or deficit of food as a percentage of the ending stocks are shown. Data from Earth Policy Institute (2012) (from Toon et al., 2017).

temperature and precipitation on agriculture. Table 24.1 provides some calculated crop losses due to temperature and precipitation changes following a regional war with one hundred weapons. The crop declines are mainly due to reduced temperatures and shortened growing seasons. These crop declines do not consider social feedback; for instance, farmers may abandon their land if they think it is poisoned by radiation or may concentrate on feeding their local communities rather than exporting food. Harwell et al. (1986) provide a detailed discussion of the effects of breakdown in transportation, difficulty obtaining fuel for farm machinery, disruption of grain markets and economic aspects of farming such as obtaining loans for planting food as well as many other issues. These social responses can be extremely important to the food supply, possibly even more important that the direct effects of climate on agriculture.

There are no studies on the effect of crop losses such as those shown in Table 24.1 on human society. However, there are several reasons to think they would be devastating. The world does not recognize the biblical warning by Joseph to the pharaoh of Egypt to store grain for seven years of famine. Instead, the total world grain storage, as shown in Figure 24.5, would be consumed in around seventy days at the current rate of consumption (Figure 24.6). As shown in Figure 24.5, consumption has steadily risen since 1960 as world population increased, but ending food stocks, which represent food storage, have leveled out since about 1985. Food production and consumption are closely balanced, and the food surplus or deficit is a small fraction of ending stocks (<23%) in the past fifty years, as shown in Figure 24.6. The small ups and downs of the ending stocks curve in Figure 24.5 represent the accumulation of the year-to-year variations in the production of grains relative to their consumption. Recently food consumption has tended to exceed production slightly, so ending stocks are slowly trending downward in Figure 24.5. Given the rising population, the days of food on hand figure has declined about a factor of 2 in the past decade to less than seventy days at present as shown in Figure 24.6.

Table 24.1 suggests, from the limited studies done so far, that a regional nuclear conflict between India and Pakistan might reduce global grain production 20% for five years and 10%–15% for another five years. A 20% reduction in grain production today would represent about 450 Mton of grain each year, which is comparable to the ending stocks on hand in 2012. This would be an unprecedented loss of food, exhausting the global food storage in one year, and could not be made up, since the loss would continue for a decade.

Figure 24.7 U.N. Food and Agriculture Food Price Index, based on data from Earth Policy Institute, Monthly Food Price Indexes, January 1990–January 2014, updated 6 February 2014 is shown (from Toon et al., 2017).

How would the world respond to a sudden global decline in food? Some clues are given in Figure 24.7, which illustrates an inflation-corrected food price. The Food Price Index is an estimate of the relative costs of meat, dairy, grains, oils and sugar. Historically the highest price was in the early 1970s, which corresponds to a time when Figure 24.6 indicates there was a minimum of days of food consumption near fifty-six days. There is a recent period of a relatively high Food Price Index with peaks in 2008 and 2011 and a new minimum of days of food consumption.

Analysis of the 1970s peak in the price of food indicates that a major cause was a drop in agricultural production, particularly in Russia, that was partly compensated by the United States' selling Russia the equivalent of 30% of the U.S. wheat production in the previous few years. The food price jump was exacerbated by decisions made by various political bodies (e.g., Schnittker, 1973). Although apparently unrelated in cause, the Arab oil embargo coincided in time, and drove energy prices to record levels, causing fuel rationing and other disruptions across the globe.

The 2008 food price peak followed another low point in the days of food consumption near sixty-three, and another runup in energy prices. These changes were coupled with the increased use of biofuels, which diverted about 100 million tons of maize away from food; wheat growing failures in Australia; an increased consumption of meat in China and various types of speculation and political manipulation. In the case of rice, India and Vietnam stopped exporting rice, a policy that may have been due to internal political manipulations in India and speculation in Vietnam. The loss of two of the world's largest rice exporters caused a cascade of panic in rice-importing nations (Slayton, 2009). Whatever the cause, the rise in food prices coincided with food riots in a number of countries and significant numbers of deaths in those riots (Lagi et al., 2011). The government of Haiti was overthrown in 2008 during food-related riots, and there were bread riots in Egypt in 2008 (Sternberg, 2013).

The 2011 spike in the Food Price Index may have had even more dramatic impacts on world affairs. In 2010 Russian wheat production fell 32.7%, Ukrainian wheat production fell 19.3%, Canadian wheat production fell 13.7% and Australian wheat production fell 8.7% (Sternberg, 2013). In the fall of 2010, China experienced drought and began to purchase wheat, driving up prices. People in Middle Eastern and North African countries spend large fractions of their income on food. Libya, Jordan, Algeria, Tunisia, Yemen and Egypt are each in the range of 35% to 44% of income spent on food. A large fraction of the food is bread. A significant causal factor in the Arab Spring uprisings, beginning in December 2010 and still ongoing today, was food insecurity, which was due to the wheat failures in 2010 (Sternberg, 2013). Since 2010 rulers have been forced out of power in Tunisia, Egypt, Libya and Yemen. There have been uprisings in Bahrain and Syria and major protests in Algeria, Iraq, Jordan, Kuwait, Morocco and Sudan. The ongoing conflict in Syria may have had its roots in agricultural losses within Syria.

The production failures in recent decades pale in comparison to those that might follow a nuclear conflict between India and Pakistan. In that case, the world grain storage might be eliminated in the first year after a war, and a deficit of 10%–20% might then occur for a decade. Helfand (2013) estimated that 2 billion people who are now only marginally fed might die from starvation and disease in the aftermath of a nuclear conflict between India and Pakistan. A conflict in the future with more weapons or between other powers with more weapons could be much worse.

Scientists working with nuclear weapons have understood since the dawn of the nuclear age that nuclear explosions would create fires if there were flammable materials at the site of the explosion (e.g., Eden, 2004). For this reason the more than five hundred aboveground nuclear tests were conducted in deserts, or on islands, so that there was little fuel to ignite a fire. A firestorm did occur at Hiroshima after the nuclear explosion, and there were numerous firestorms set by conventional weapons during World War II. However, during World War II the number of firestorms was much smaller than the one hundred considered by Mills et al. (2014) as being important. Also the firestorms in World War II were spread out over a year or more, so their impacts were not cumulative. Finally, during World War II there was little priority put on stratospheric observations, and instruments of that day were not sensitive enough to observe any injections of smoke into the stratosphere. The 1.5 °C cooling found by Mills et al. (2014) for a war between India and Pakistan would be much larger than the natural year-to-year variations in global average temperature and easily observed. However, the injection during World War II of only a few percent of the smoke assumed by Mills et al (2014) would produce a temperature change that would be difficult to detect even now with a functioning global temperature measurement system. For these reasons we do not have any experimental confirmation of climate effects from firestorms in World War II.

24.4 Scenarios for War

The United States and Russia remain in "launch on warning status" which means that the leaders of these countries are constantly prepared to launch some of their nuclear missiles within a few moments of learning that missiles from the other side are headed their way. This is a very dangerous situation. There are numerous examples of each side's mistakenly thinking the other has initiated an attack, and war has been narrowly averted by identifying or discounting the mistake (Union of Concerned Scientists, 2012). For example, on 25 January 1995 the Russians detected a missile launch off the coast of Norway, which they interpreted to be the start of a nuclear attack. Fortunately they did not observe other missile launches and declared it a false alarm. It was a routine scientific rocket launch, which had not been reported properly in advance through the Russian system. On 26 September 1983 Soviet early warning satellites detected five land-based missile launches. Fortunately, the officer on duty decided this must be an error since it seemed improbable that an attack would involve only five launches. It was later found that the false signals were due to reflections from clouds. It is fortunate that only five reflections were detected. Launch on warning is partly based on the need to launch land-based missiles before they are attacked by a nuclear adversary. Such missiles, whose locations are well known, are destabilizing because of this need to use them on short warning.

In Asia, a number of analyses have been conducted of how a war might start. For instance, Lavoy and Smith (2003) discuss three plausible scenarios for a nuclear war between India and Pakistan. India has conventional military superiority. India is also geographically much larger than Pakistan so parts of its nuclear arsenal are currently out of range of Pakistan's forces, while all of Pakistan is easily reached from India. One possible route to nuclear war involves a conventional conflict between India and Pakistan. If Pakistan perceived that India were about to invade, that could put pressure on Pakistan to launch its nuclear weapons before they were overrun by the superior Indian forces. Another possibility for starting a nuclear conflict is that India or Pakistan could lose control of its command and control structures as a result of an attack on them by the other side, or possibly an attack by terrorists. In such a scenario it is not clear who might be in control of the nuclear forces and what steps they might take. A third possibility for starting a nuclear conflict is that India or Pakistan might mistake an attack by conventional forces, or even military exercises, for an attack by nuclear forces. Both countries have ballistic missiles and aircraft that are potentially dual-use between conventional and nuclear weapons, making any attack ambiguous.

Another point of future conflict could involve North Korea. It is not clear that North Korea currently has operational nuclear weapons, though it has tested nuclear explosives. It is also not certain that North Korea has nuclear-capable missiles (Kristensen and Norris, 2014), though it continues to test long-range missiles and submarine-launched missiles and does have nuclear-capable aircraft. Fortunately, no other countries with nuclear weapons are near North Korea, except China, with which North Korea is allied. Factors that might trigger a nuclear conflict involving North Korea have some parallels with the Pakistan/India situation. North Korea has a much larger military than South Korea, including the U.S. troops stationed in South Korea, and Seoul is only fifty-five kilometers from the North Korean border. It would, therefore, be possible, even easy, for North Korea to overrun South Korea in a sudden conventional attack, possibly triggering a nuclear attack by the United States. Hayes and Cavazos (2015) have analyzed some of the many possible scenarios for a nuclear conflict involving North Korea. Generally, they conclude that North Korea does not have the capability to launch a nuclear attack at present, but may have in the near-future. While North Korea is expanding its capabilities it is difficult to find a scenario in which they could launch a first-strike conventional or nuclear conflict and not expose themselves to devastating nuclear retaliation, even in the future. However, the North Koreans may not analyze the situation in the same ways that are used in the West; their leadership might not be rational; there is the potential for the regime to be overthrown, leading to nuclear weapons use in North Korea itself and it is possible that North Korea's continuing provocations of their neighbors might lead to an attack on North Korea. For example, North Korea torpedoed and sank the South Korean warship *Cheonan* on 25 March 2010, which might have been considered an act of war (BBC News, 2010). It is also possible in the near-future that North Korea may calculate that they can launch a conventional attack on South Korea, while preventing a retaliation from the United States by threatening the United States with a nuclear attack on its territory. Any of these possibilities, and several others, might lead to the use of nuclear weapons.

There are numerous other scenarios that could lead either to a nuclear conflict or to further destabilization of Asia. For example, Japan or South Korea could develop nuclear weapons to defend against North Korea, likely because they lose confidence that the United States will continue to protect them. In the Middle East, Iran could obtain nuclear weapons, which would likely trigger Saudi Arabia and other countries in the Middle East to obtain them. It is hoped that the 2015 agreement with Iran will prevent the Middle East from becoming a warren of nuclear states.

Unless the world does something to build down the existing arsenals further, and stops proliferation, the remainder of the twenty-first century is likely to involve increasing risk of nuclear confrontations.

24.5 What Can Be Done to Prevent Nuclear Conflict?

No government, anywhere, is openly conducting studies to determine the damage that might be done to them in a nuclear conflict in which they are a combatant, or a bystander. There is no evidence that any current leader of any nuclear power is aware of the potential harm that might occur as a result of the environmental damage from a nuclear conflict. For example, the U.S. Nuclear Posture Review of 2010 does not mention the words "nuclear winter" nor any environmental damage that might be caused by a nuclear conflict (U.S. Government, 2010). However, the cold temperatures and the destruction of the ozone layer caused by the smoke from burning cities following a nuclear conflict, even between powers such as India and Pakistan with modest numbers of weapons compared with other nuclear weapons states, could devastate agricultural productivity, leading to mass starvation across the globe. In addition, changes in governmental stability, transportation of food, availability of fuel, functioning of economic organizations such as banks, stability of police and military authority and many other changes induced by war could be devastating to civilization. It is dangerous not to discuss these issues and risks openly in society. Such discussions might lead to lessening nuclear tensions and reductions in nuclear stockpiles.

Calculating the direct casualties from a nuclear conflict is very straightforward. The major uncertainties are the number and yields of the weapons used and their targets. There is also uncertainty about extrapolating the casualties from Hiroshima and Nagasaki to those that would occur in modern cities at various distances from ground zero. It is more difficult to compute the environmental impacts from nuclear conflicts and their consequences. However, the major uncertainties are the same as for the direct effects, the number and yields of the weapons used and their targets. The size and duration of resulting firestorms are uncertain since we have no modern examples. In addition, the amount of fuel burned, the amount of smoke emitted and the amount of smoke removed in local precipitation are uncertain. Uncertainty does not mean that the impacts have been overestimated. It is just as likely that the effects have been underestimated. As in many other areas of human endeavor, uncertainty is unavoidable when predicting the outcome of possible nuclear wars. However, when gambling with the future of human civilization it is not wise to ignore the possibility of nuclear winter by hoping that no country will ever use its arsenal.

Goldstein (2017) argues that politicians do not care about the predictions of models, and use science only to manipulate events. However, the Montreal Protocol, which is based on modeling showing that the release of certain chemicals is destroying the ozone layer, has eliminated industrial emissions of ozone-destroying chemicals. The Intergovernmental Panel on Climate Change, which is largely based on models for future climate, is currently pushing a revision of the Earth's energy production system. These are very large changes caused by scientific recognition of a problem, and wrought by statesmen who have taken science very seriously. Of course, in both these examples there was also evidence that the Earth was being impacted, such as the Antarctic ozone hole, melting glaciers, rising sea level and rising temperatures. Nuclear war is not approachable by experiment or observation. Nevertheless, there is no reason to be pessimistic and assume that politicians cannot understand the advice about the dangers of nuclear wars from their scientific establishments, or will not act on that advice. History shows, even in the case of nuclear weapons, that statesman and stateswomen will understand these dangers if they are told of them, and act appropriately.

In the early 1980s studies of the environmental effects of a nuclear conflict played a role in causing the leaders of the United States and of the Soviet Union to reduce their arsenals. Soviet leader Mikhail Gorbachev observed "Models made by Russian and American scientists showed that a nuclear war would result in nuclear winter that would be extremely destructive to all life on Earth: the knowledge of that was a great stimulus to us, to people of honor and morality, to act" (Hertsgaard, 2000). U.S. President Ronald Reagan (New York Times, 1985) noted that

> a great many reputable scientists are telling us, that such a war could just end up in no victory for anyone because we would wipe out the earth as we know it. And if you think back to . . . natural calamities - back in the last century, in the 1800's, . . . volcanoes – we saw the weather so changed that there was snow in July in many temperate countries. And they called it the year in which there was no summer. Now if one volcano can do that, what are we talking about with the whole nuclear exchange, the nuclear winter that scientists have been talking about? It's possible.

As an inital step forward from our present status toward a world that is free of the threat of a global nuclear catastrophe, we suggest that the nuclear-armed states collectively engage in a dialog on the impacts of nuclear conflicts. There is precedence for such a dialog in two formats. Just prior to the build-down in nuclear weapons illustrated in Figure 24.1 and the pause in the creation of new nuclear states illustrated in Figure 24.3 the scientific academies of the world conducted a study of nuclear conflicts under the umbrella of the U.N. Scientific Committee on Problems of the Environment, as reported by Pittock et al. (1986) and Harwell and Hutchinson (1986). Of course, it is not just scientists who need to participate in this discussion. During the early 1980s a vibrant debate about nuclear weapons took place among a wide variety of interest groups including physicians, politicians, philosophers and poets. These scientific and public debates led to the build-down in nuclear weapons beginning in the mid-1980s and illustrated in Figure 24.1. We suggest that there be a new study of the consequences of nuclear conflict, as also recommended by Jeanloz (2015).

The study should not only consider the consequences of conflicts, but also address whether the reasons cited to develop nuclear weapons are valid. For example, Japan and South Korea do not have nuclear arsenals, while North Korea does. There is no evidence that North Korea is technologically superior in any way, has a greater place in any facet of world decision making or has a greater influence on any aspect of world affairs. It could also be argued that North Korea is no safer because of nuclear technology. The same opposing army is still in South Korea that was there prior to North Korea testing nuclear weapons. Indeed, sanctions placed on North Korea because of its nuclear weapons program, which may not have impacted its nuclear weapons program, have damaged its economy (e.g., Kim, 2014). Likewise, India's possession of a nuclear arsenal has not helped it gain entry as a permanent member to the U.N. Security Council. While India is recognized for its many prominent scientists and thinkers in a variety of peaceful fields of study, its nuclear program has done nothing to gain it respect as a technological power in the rest of the world. Pakistan's nuclear weapons have not helped it solve its dispute with India over Kashmir; nor have they prevented other nations from bombing its citizens and making armed military raids to seize militants located in its territories. Each of these examples, and many others, of whether the possession of nuclear weapons has achieved the goals of the country with the weapons is worthy of debate.

India and Pakistan are engaged in a rapid build-up of nuclear weapons, which is a danger to them both, and to the rest of the world. While dialogs between these nations on a variety of issues, such as Kashmir, have taken place for decades, little progress has been made. Possibly a detailed analysis by their own scientists and scientists of other nations of the consequences of a war will aid them in constructive discussions on reducing nuclear weapons instead of expanding their arsenals. Of course, these studies should include an analysis of conflicts involving all the nuclear-armed nations and consider all of their arsenals.

History shows that nuclear weapons do not achieve the goals envisioned by the states that have them. Not even a single weapon can be used given the potential for escalation to the use of many more weapons that could lead to a nuclear winter. Such escalation is particularly likely because of the eighteen hundred weapons on high alert in the United States and Russia that could be used on short notice. A war cannot be fought given the adverse impacts on the food supply of all nations of the world, including the ones who used them. The case of Pakistan's invasion of Kashmir after its test of nuclear weapons, and the entire Cold War, show that nuclear arsenals do not provide cover for unopposed conventional warfare. The Argentinean invasion of the Falklands and the Yom Kippur War show that possession of nuclear weapons does not stop invasions of nuclear nations by non-nuclear nations. The Vietnam War and the Soviet–Afghan War show that nuclear weapons do not help win wars, since in both wars the nuclear power lost. The weapons do not confer immunity from meddling by others in internal affairs, nor give countries with them enhanced access to world government. On the contrary, the weapons are very expensive to obtain and maintain. The weapons expose their possessors, and the rest of the world, to the potential for terrorists to obtain and use them. There is a significant risk, particularly between India and Pakistan, of nuclear conflict starting from misunderstanding, or misinterpretation of the other country's actions. A large fraction of the world population, inside and outside the combat zone, could die from such errors. It would behoove each nuclear weapon state to get rid of its nuclear weapons before it is too late.

Acknowledgments

Owen B. Toon, Alan Robock, Lili Xia, and Charles Bardeen are supported by a grant from the Open Philanthropy Project. Parts of this paper are excerpted from Toon et al. (2017).

Internet Resources and Notes

[1] http://www.nuclearsecurityproject.org. Quote obtained May 28, 2014.

[2] The explosive energy released in a nuclear detonation is measured in relation to an explosion of the equivalent weight of the conventional explosive trinitrotoluene (TNT). The explosive power in all bombs used in WWII combined, including the atomic bombs dropped on Hiroshima and Nagasaki, was several megatons, which could now be released by just one high-yield nuclear weapon.

[3] In the lexicon of military targeting there are countervalue, counterforce, and rational wars. Rational wars use a small number of weapons to attack targets of symbolic value. Countervalue wars use massive attacks on urban areas to destroy economic and social infrastructure. Counterforce wars involve massive attacks on military, economic and political targets. In reality, countervalue and counterforce wars end up attacking much the same targets, especially for large numbers of high yield weapons.

References

BBC News (2010). 'North Korean torpedo' sank South's navy ship-report, www.bbc.com/news/10129703, Downloaded February 8, 2017.

Bush, B. W. Dore, M. A. and Small, R. D. (1991). Nuclear Winter Source-Term Studies: Smoke Produced by a Nuclear Attack on the United States. In rep. no. DNA- TR-86–220-V6, 6. Alexandria, VA: Defense Nuclear Agency.

De Laat, J., Stein-Zweers, D. C., Boers, R. and Tuinder, O. N. E. (2012). A solar escalator: Observational evidence of the self-lofting of smoke and aerosols by adsorption of solar radiation in the February 2009 Australian Black Saturday plume. *J. Geophys. Res.*, **117**(D04204), doi:10.1029/2011JD017016.

Eden, L. (2004). *Whole World on Fire*. Ithaca, NY: Cornell University Press.

Earth Policy Institute Data Center. (September 14, 2012). World grain consumption and stocks as days of consumption, 1960–2012, Available at www.earth-policy.org/data_center/C24. Downloaded May 24, 2014.

Evans, G. (2014). Nuclear deterrence in Asia and the Pacific. *Asia and the Pacific Policy Studies*, **1**, 91–111, doi:10.1002/app5.00011.

Fears, T. R., Scotto, J., and Schneiderman, M. A. (1976). Skin cancer, melanoma and sunlight. *Am. J. Public Health*, **66**, 461–464.

Fromm, M. D. and Servranckx, R. (2003). Transport of forest fire smoke above the tropopause by supercell convection. *Geophys. Res. Lett.* **30**(10), doi: 10.1029/2002GL016820.

Goldstein, D. (2017). Commentary: Science, politics, and risk: Catastrophic Asia from the perspective of a Brazilianist anthropologist. *J. Asian Studies*, doi:10.1017/S0021911817000109.

Harwell, M. A. and Hutchinson, T. C. (1986). *Environmental Consequences of Nuclear War*: Volume II: *Ecological and Agricultural Effects* (SCOPE SERIES VOL. 2). New York: Wiley.

Hayes, P. and Cavazos, R. (2015). North Korea's nuclear force roadmap: Hard choices. *NAPSNet Special Reports, March* 2, 2015, http://nautilus.org/napsnet/napsnet-special-reports/north-koreas-nuclear-force-roadmap-hard-choices.

Helfand, I. (2013). Nuclear famine: Two billion people at risk? *International Physicians for the Prevention of Nuclear War*. Available at http://www.ippnw.org/pdf/nuclear-famine-two-billion-at-risk-2013.pdf. Downloaded May 24, 2014.

Hertsgaard, M. (2000). Mikhail Gorbachev explains what's rotten in Russia. *Salon*, September 7. Available at www.salon.com/2000/09/07/gorbachev/, downloaded May 28, 2014.

Jeanloz, R. (2015). Environmental effects of nuclear war, in *Andrei Sakharov, The Conscience of Humanity*, ed Sidney D. Drell and George P. Shultz. Stanford, CA: Hoover Institution Press.

Kim, H. (2014). Stifled growth and added suffering: Tensions inherent in sanctions policies against North Korea, *Critical Asian Studies*, **46**, 91–112, doi:10.1080/14672715.2014.863579.

Kristensen, H. M. and Norris, R. S. (2013). Global nuclear weapons inventories, 1945–2013. *Bull. Atomic Scientists* **60**(5): 75–81, doi:10.1177/0096340213501363.

Kristensen, H. M. and Norris, R. S. (2014). Worldwide deployments of nuclear weapons, 2014, *Bull. Atomic Scientists*, **70**(5), 96–108, doi: 10.1177/0096340214547619. http://bos.sagepub.com/content/70/5/96.

Lavoy, P. R. and Smith, S. A. (2003). *The Risk of Inadvertent Nuclear Use Between India and Pakistan*. Strategic Insight II. ADA525408, available at http://www.dtic.mil/dtic/tr/fulltext/u2/a525408.pdf, downloaded May 18, 2015.

Lagi, M., Bertrand, K. Z. and Bar-Yam, Y. (2011). The food crises and political instability in North Africa and the Middle East. Available at arXiv:1108.2455 [physics.soc-ph], downloaded on May 28, 2014.

Lepore, J. (2017). The atomic origins of climate science. *The New Yorker*, http://www.newyorker.com/magazine/2017/01/30/the-atomic-origins-of-climate-science.

Lieber, K. A. and Press, D. G. (2006). The rise of U.S. nuclear primacy. *Foreign Affairs,* Available at http://www.foreignaffairs.com/articles/61508/keir-a-lieber-and-daryl-g-press/the-rise-of-us-nuclear-primacy, downloaded May 28, 2014.

Mills, M. J., Toon, O. B., Lee-Taylor, J. and Robock, A. (2014). Multidecadal global cooling and unprecedented ozone loss following a regional nuclear conflict. *Earth's Future*, **2**, 161–176, doi:10.1002/2013EF000205.

New York Times. (1985). Transcript of interview with President on a range of issues, *New York Times*, February 12. Available at: http://www.nytimes.com/1985/02/12/world/transcript-of-interview-with-president-on-a-range-of-issues.html?pagewanted=all, downloaded May 28, 2014.

Özdoğan, M., Robock A. and Kucharik, C. (2013). Impacts of a nuclear war in South Asia on soybean and maize production in the Midwest United States. *Climatic Change*, **116**, 373–387, doi:10.1007/s10584-012-0518-1.

Pausata, F. S. R., Lindvall, J., Eckman, A. M. L., and Svensson, G. (2016). Climate effects of a hypothetical regional nuclear war: Sensitivity to emission duration and particle composition. *Earth's Future*, **4**, 498–511, doi:10.1002/2016EF000415.

Pittock, A. B., et al. (1986). *Environmental Consequences of Nuclear War:* Volume I: *Physical and Atmospheric Effects* (SCOPE SERIES VOL. 1). New York: Wiley.

Robock, A., Oman, L., and Stenchikov, G. L., (2007). Nuclear winter revisited with a modern climate model and current nuclear arsenals: Still catastrophic consequences. *J. Geophys. Res.* **112**, D13107, doi:10.1029/2006JD008235.

Robock, A. and Toon O. B. (2012). Self-assured destruction: The climate impacts of nuclear war, *Bull. Atomic Scientists*, **68**, 66–74, doi:10.1177/0096340212459127.

Sanger, D. E. and Haberman, M. (2016). In Donald Trump's worldview, America comes first and everybody else pays. *New York Times*, March 26, 2016. Transcript www.nytimes.com/2016/03/27/us/politics/donald-trump-transcript.html?_r=0.

Schnittker, J. (1973). The 1972–73 food price spiral. *Brookings papers on economic activity*, Available at Brookings Institute, http://www.brookings.edu/~/media/Projects/BPEA/1973%202/1973b_bpea_schnittker.PDF, downloaded May 28, 2014.

Shane, S. (2015). 1950s U.S. Nuclear Target list offers chilling insight. *New York Times*, Dec. 23, A10, Available at www.nytimes.com/2015/12/23/us/politics/1950s-us-nuclear-target-list-offers-chilling-insight.html, downloaded February, 1, 2016.

Shultz, G. P., Perry, W. J., Kissinger, H. A. and Nunn, S. (2007). A world free of nuclear weapons. *Wall Street Journal*, January 4. Available at: http://online.wsj.com/article/SB116787515251566636.html, downloaded May 28, 2014.

Shultz, G. P., Perry, W. J., Kissinger, H. A. and Nunn, S. (2008). Toward a nuclear-free world. *Wall Street Journal*, January 15. Available at: http://online.wsj.com/article/SB120036422673589947.html, downloaded May 28, 2014.

Shultz, G. P., Perry, W. J., Kissinger, H. A. and Nunn, S. (2010). How to protect our nuclear deterrent. *Wall Street Journal*, January 19. Available at: http://online.wsj.com/article/SB10001424052748704152804574628344282735008.html, downloaded May 28, 2014.

Shultz, G. P., Perry, W. J., Kissinger, H. A. and Nunn, S. (2011). Deterrence in the age of nuclear proliferation. *Wall Street Journal*, March 7. Available at: http://online.wsj.com/article/SB10001424052748703300904576178760530169414.html, downloaded June 23, 2018.

Slayton, T. (2009). Rice crisis forensics: How Asian governments carelessly set the world rice market on fire. *CGD Working Paper 163*. Washington, D.C.: Center for Global Development. www.cgdev.org/content/publications/detail/1421260/.

Small, R. D. (1989). Atmospheric smoke loading from a nuclear attack on the United States. *Ambio*, **18**, 377–383.

Stenke, A., Hoyle, C. R., Luo, B., Rozanov, E., Gröbner, J., Maag, L., Brönnimann, S., and Peter, T. (2013). Climate and chemistry effects of a regional scale nuclear conflict, *Atmos. Chem. Phys.*, **13**(19), 9713–9729, doi:10.5194/acp-13-9713-2013.

Sternberg, T. (2013). Chinese drought, wheat, and the Egyptian uprising: how a localized hazard became globalized. In *The Arab Spring and Climate Change: A Climate and Security Correlations Series*, ed. Caitlin E. Werrell and Francesco Femia, Center for American Progress, 7–14. Available at http://americanprogress.org/issues/security/report/2013/02/28/54579/the-arab-spring-and-climate-change/, downloaded June 23, 2018.

Toon, O. B., Turco, R. P., Robock, A., Bardeen, C., Oman, L. and Stenchikov, G. L. (2007). Atmospheric effects and societal consequences of regional scale nuclear conflicts and acts of individual nuclear terrorism. *Atm. Chem. Phys.*, **7**, 1973–2002.

Toon, O. B., Robock, A. and Turco, R. P. (2008). Environmental consequences of nuclear war. *Physics Today*, **61**(12), 37–42 Available at http://dx.doi.org/10.1063/1.3047679.

Toon, O. B., Robock, A., Mills, M. and Xia, L. (2017). Asia treads the nuclear path, unaware that self-assured destruction would result from nuclear war. *J. Asian Studies*, **76**, 437–456, doi:10.1017/S0021911817000080.

Union of Concerned Scientists (2015). Close calls with nuclear weapons. www.ucsusa.org/weaponincidents, Downloaded June 23, 2018.

United States Government (2010). Nuclear posture review, www.defense.gov/Portals/1/features/defenseReviews/NPR/2010_Nuclear_Posture_Review_Report.pdf, downloaded June 23, 2018.

Xia, L. and Robock, A. (2013). Impacts of a nuclear war in South Asia on rice production in mainland China. *Climatic Change*, **116**, 357–372, doi:10.1007/s10584-012-0475-8.

Xia, L., Robock, A., Mills, M. J., Stenke, A. and Helfand, I. (2015). Decadal reduction of Chinese agriculture after a regional nuclear war. *Earth's Future*, **3**, 37–48, doi:10.1002/2014EF000283.

25

Advancing Spring Flood Risk Reduction in the Arctic through Interdisciplinary Research and Stakeholder Collaborations

Yekaterina Y. Kontar, Sarah F. Trainor, Tuyara N. Gavrilyeva, John C. Eichelberger, and Nikita I. Tananaev

25.1 Introduction: Spring River Ice Breakup and Floods

River ice thawing and breakup are an annual springtime phenomenon in the North.[1] Depending on regional weather patterns and river morphology, breakups can result in floods (Prowse et al., 2011). Breakup floods often cause catastrophic ice and water damage to exposed and vulnerable communities, and lead to socioeconomic and ecological crisis (Buzin et al., 2014; Burrell et al., 2015). Prowse et al. (2011) approximated the cost of annual spring floods in North America to be $280 million. Adverse ecological impacts of spring floods are indirect and commonly caused by spilled toxic and human waste (Burrell et al., 2015).

Spring floods commonly originate upstream from ice jams – accumulations of chunks and sheets of ice in the river channel that block or restrict stream flow during spring melting (Beltaos, 2007). Ice jams form anywhere in a river channel where its transport capacity is exceeded by obstructions, such as river bends or bridges (Beltaos et al., 2006). Once formed, ice jams produce rise of water levels of meters per day and can hold for days, forcing melt water and ice floes to back up for miles and cause flooding upstream (Beltaos, 2007). Sudden release of ice jams can also result in flash flooding downstream and cause additional damage by rapid withdrawal of water from flooded areas.

As with all natural disasters, spring floods result from a complex interaction between a series of natural and human processes and events that generate the three defining components of disaster risk – hazard, exposure, and vulnerability (United Nations International Strategy for Disaster Reduction [UNISDR], 2015). In the context of this research, *hazard* is regarded as a potentially harmful natural event (i.e., ice-jam flood) that may result in loss of life, injuries and other health impacts, damage and destruction of infrastructure and private property, social and economic disruption, and environmental degradation (Global Facility for Disaster Reduction and Recovery [GFDRR], 2014).

Exposure refers to the presence and number of people along with their environmental and economic resources and social and cultural assets in at-risk locations (GFDRR, 2014). If no individuals or assets are exposed to an ice-jam flood during its occurrence, then there is no risk, and subsequently no disaster. *Vulnerability* represents the characteristics and circumstances (e.g., lack of critical infrastructure, marginalization of populations, poor environmental management, and inadequate floodplain use) of a population that make it susceptible to the adverse impacts of a flood (Cutter et al., 2015). Not all people in flood-exposed communities are equally vulnerable.

Spring floods are complex natural and social phenomena. Understanding spring flood risk requires understanding of a wide spectrum of physical, climatological, political, economic, cultural, social, and psychological processes. Thus, accurately identifying and assessing flood risk require interdisciplinary research. Additionally, designing and implementing effective flood management strategies require stakeholder collaboration.

25.2 Spring Flood Management: Risk Reduction vs. Crisis Management

Spring flood risk is especially high in rural and remote northern communities where flood relief and recovery are challenged by the regions' unique climatological, geographical, and logistical features (Kravitz & Gastaldo, 2013). The most common challenges include the early onset of winter, seasonal lack of daylight, vast

[1] In the context of this paper, North is referred to the area traditionally covered by the terms Arctic and Subarctic.

distances, and limited physical and communication infrastructure (Benoit, 2014; Kontar et al., 2015).

Inadequate risk assessment and emergency training further complicate disaster response in northern regions (Benoit, 2014; Institute for Risk and Disaster Reduction [IRDR], 2014). For instance, disaster practitioners' reports from Alaska (United States) and the Northwest Territories (Canada) have repeatedly indicated many complications and delays during disaster relief operations. Major emergency responses (i.e., national disaster responses) are launched from the southern hubs in lower latitudes, which are relatively long distances away from the impacted communities. Responders from the South are often unfamiliar with the geographic area, as well as the unique logistical and cultural features of the North. Moreover, processes used to trigger federal assistance vary between jurisdictions, creating additional complications and delays in disaster relief (Benoit, 2014; IRDR, 2014).

The aforementioned challenges in relief and recovery operations call for the advancement of proactive spring flood management in the North. Proactive disaster management incorporates a series of disaster prevention, mitigation, and preparedness measures that are aimed at reducing or preventing the adverse impacts of likely, imminent, or current hazard events (UNISDR, 2015). Although often used interchangeably, *disaster mitigation* and *disaster prevention* are not synonyms. *Mitigation* refers to lessening the adverse impacts of potential disasters through a range of structural (e.g., flood dikes) and nonstructural (e.g., land use zoning) measures, while *prevention* refers to the complete avoidance of negative disaster impacts (e.g., full relocation of at-risk populations) (IPCC, 2012; UNISDR, 2015). *Preparedness* measures comprise activities (e.g., risk assessment, disaster forecast, and establishment of evacuation routs) that increase people's ability to predict, prepare for, as well as respond to and recover from disasters (UNISDR, 2015).

Unlike most natural disasters, opportunities for reducing spring flood risk lie not only in reducing vulnerability and exposure of the at-risk population, but also in reducing the severity of the hazard itself. Throughout history, an array of ice jam prevention and mitigation measures have been implemented in the U.S. and Russian North with varying degrees of success (e.g., Belore et al., 1990; Kusatov et al., 2012; Buzin et al., 2014; Burrell et al., 2015).

Every mitigation and prevention technique has potential benefits and drawbacks (Table 25.1). The selection of a particular technique or a combination of techniques should depend upon not only the availability of resources and time, but also the careful analysis of the causative factors of ice jams and unique geographic, geological, and morphological features of the river channel and the floodplain (Kusatov et al., 2012; Burrell et al., 2015). The history of floodplain management and progression of a population's vulnerability to floods, along with regional environmental constraints, should also be included in the analysis (Burrell et al., 2015).

25.3 Case Study: Comparative Analysis of Spring Flood Management in Alaska, United States, and the Sakha Republic, Russia

With the goal to identify best practices in spring flood risk reduction, we conducted a comparative case study analysis between the two flood-prone communities of Galena in Alaska, United States, and Edeytsy in the Sakha Republic (Northeast Siberia), Russia. Case study research has a long tradition in disaster research as it facilitates the discovery of cultural, political, and social factors relevant to the phenomenon of interest (Bhattacherjee, 2012). It aids disaster scholars in identifying driving forces of disaster risk in a specific community. Via two case studies, we assessed each component of the flood risk, and analyzed existing practices in spring flood risk reduction. Specific points of comparison are outlined in Table 25.2.

25.3.1 Methodology

A bilateral and multidisciplinary team of spring flood management experts and stakeholders was assembled to identify and analyze the complex social and natural processes that generate the three defining components of spring flood risk – hazard, exposure, and vulnerability – and their interactions. The team consisted of U.S. and Russian geoscientists, social scientists, and students, as well as representatives from local and tribal administration, and federal and state agencies responsible for flood management.

The data were acquired and cross-referenced through the following methods:

- *focus groups* with the representatives from regional and national agencies responsible for flood risk reduction and emergency management in Alaska and the Sakha Republic;
- *surveys* administered to the populations that were impacted by the recent floods in both research sites;
- *reviews of documents and archives* (secondary data review) that included governmental and institutional reports, news articles, and feature stories that depicted natural and human factors during the recent and historical catastrophic floods in Galena and Edeytsy.

Table 25.1 *Spring Flood Mitigation Strategies: Existing Ice Jam Flood Mitigation and Prevention Techniques in Alaska and the Sakha Republic, and Their Potential Positive and Negative Impacts*

Ice Jam Prevention

Technique	**Description and Function**	**Region**	**Drawbacks**
Ice dusting/ sanding	Spreading a thin layer of a dark solid substance (e.g., coal) to promote the deterioration of an ice cover by decreasing the albedo of the ice, and thus increasing its melting rate	Sakha	• Often ineffective because of fresh snow covering dusted areas, or new periods of freezing • Not effective in persistently cloudy regions • Possible water and soil contamination • No statistical evidence of effectiveness
Blasting (pre-breakup)	Fracturing ice cover to promote early release of small ice floes prior to high spring flow	Sakha	• Expensive • Dangerous for practitioners • Negative environmental impacts • No statistical evidence of effectiveness
Ice cutting (saws)	Creating lines of weakness to promote deterioration of ice cover and release of stationary ice prior to high spring flow Decreasing size of ice floes to promote release and flow of ice	Sakha	• No statistical evidence of effectiveness

Flood Impact Reduction

Technique	**Description and Function**	**Region**	**Drawbacks**
Floodplain management	Prevention/discouragement of communities in flood-prone areas	Alaska Sakha	Can lead to interagency and intergovernmental conflicts
Flood-proofing • Dikes • Elevation of structures	Construction of dikes to confine stream flow to the river channel, thus to protect infrastructure and homes from inundation and ice debris damage Corrective elevation of existing structures or preventative elevation of new structures to a height above the largest flood to date.	Alaska Sakha	These structural measures require large initial investments and, in some cases, continuous future investments for after-flood repairs and maintenance
Relocation	Partial or full relocation of the at-risk communities away from the floodplain.	Alaska Sakha	• If relocation is imposed on communities, it is often unsuccessful because residents tend to return to their original settlements • Disaster assistance funds often exclude money for preventive measures
Flood forecasting and early warning system (EWS)	Predicting timing, duration, and magnitude of water level rise Encouraging timely evacuation of people, their assets, and cattle to prevent life loss and minimize flood damage.	Alaska Sakha	Inaccurate forecasts could lead to unnecessary evacuations, and hence discourage further future evacuation.
Blasting (ice jam removal)	Release of already formed ice jams by fracturing ice cover	Sakha	• Expensive • Dangerous for the practitioners • Negative environmental impacts • No statistical evidence of effectiveness

Focus groups and surveys were conducted during the site visits in the Sakha Republic in November 2015 and in Alaska in March 2016. The key goal of the focus group discussions was to identify the existing practices in flood risk reduction, and the key challenges (e.g., logistical and financial) various stakeholders face during flood management in both regions. The participants shared their expertise on the following topics: ice jam and flood mitigation, breakup monitoring and flood forecasting, flood preparedness, response and recovery, benefits and disadvantages of flood prevention versus flood response measures, and allocation of resources.

The goals of the surveys were to identify prevalent opinions and attitudes among the populations at risk regarding the socioeconomic impacts of the 2013 floods,

Table 25.2 *Points of Comparison.*

	Galena, Alaska	Edeytsy, Sakha Republic
Population size and demographics	467 (2015) vs. 477 (2013) • Native Alaskans (63.62%) • White (29.36%) • Other races (7.02%)	1,321 (2015) vs. 1,505 (2013) • Native Sakha (99.99%)
Livelihood	Subsistence and cash economy	Cattle ranching and farming
Mean annual income (household)	$64,067	$22,036
Average damage caused by flood in May 2013	Good	Good
Flood awareness, monitoring, and warning	Good	Good
Flood preparedness	Poor	Good
Flood management	Reactive	Proactive
Flood governance	Decentralized	Centralized
Satisfaction ratings with flood management	Poor	Good

Table 25.3 *Survey Design and Content.*

GRP 1	Qs 1–4	History of floods in Galena and Edeytsy
GRP 2	Qs 5–13	Socioeconomic impacts of the 2013 floods
GRP 3	Qs 14–18	Local ice jam and flood mitigation and preparation efforts
GRP 4	Qs 19–20	Flood information and warning channels
GRP 5	Qs 21–25	Assistance efforts from governmental and nongovernmental agencies involved in flood response and recovery
GRP 6	Qs 26–28	Correlation between floods and population relocation
GRP 7	Qs 29–31	Quality of drinking water during and after floods
GRP 8	Qs 32–34	Sanitary-epidemiological and health impacts of floods
GRP 9	Qs 35–38	Socioeconomic demographic characteristics of the respondents

and the effectiveness of the flood management strategies in both regions. The surveys comprised thirty-eight questions split into nine groups (Table 25.3). The samples for this study include 32 families (~17%) out of the total 190 families in Galena, and 60 families (~22%) out of the total 276 families in Edeytsy. Non-probability voluntary samples were used to collect data in both communities. All of the survey participants were literate and more than eighteen years old.

Review of historical hydrological reports assembled by the U.S. National Weather Service (NWS) and the Russian Federal Service for Hydrometeorology and Environmental Monitoring (Roshydromet) provided an in-depth understanding of the frequency and severity of the breakup floods in Galena and interior Alaska, and Edeytsy and the central Sakha Republic. Review of archived news articles, feature stories, memoirs, and governmental reports revealed the evolution of spring flood management efforts, and stakeholders' perspectives of flood events throughout the last seven decades in Alaska and the Sakha Republic.

25.3.2 Background: Progression of Flood Risk in Galena and Edeytsy

In May 2013, a series of ice jam floods inundated multiple communities along the Lena and Yukon Rivers in the Sakha Republic and Alaska, respectively. Within a week of each other, the communities of Galena and Edeytsy sustained major ice jam floods. In both communities, floodwaters and ice floes destroyed or severely damaged nearly all homes and key infrastructure and displaced hundreds of people (Kontar et al., 2015).

Galena is a remote community of 467 people, located in traditional Koyukon Athabascan Indian territory in central Alaska (U.S. Census Bureau, 2015). It is situated on the northern bank of the Yukon River, more than 430 kilometers away from the nearest urban center in Fairbanks, and 1600 km from the state's capital, Juneau. The community can be reliably accessed only via aircraft year-round. Galena's remoteness significantly delays flood relief and recovery efforts.

Propelled by the need to earn cash wages, Galena's first permanent population settled in town in 1920 to work at the ore mine (Sprott, 2000). Flood risk was not factored into the initial community settlement in its current floodplain location. The few original Galena families lived in log houses, which were easy to dismantle and dry out after the floods (Sprott, 2000). Establishment of the U.S. Air Force (USAF) base in Galena provided economic and political incentives for the community's expansion (Sprott, 2000). As the community grew, so did its dependence on the life-sustaining infrastructure and housing. Continuing not to invest in flood risk reduction measures left Galena's residents and their assets exposed to ice jam floods.

Edeytsy is a rural community of 1,261 people, located in central Sakha Republic in Northeast Siberia (Old Sakha [no date]). It is situated on the eastern bank of the Lena River, thirty-five kilometers from the regional center Namtsy and sixty kilometers from the republic's capital of Yakutsk. Unlike Galena, Edeytsy can be reliably accessed year-round via a dirt road, thus facilitating flood relief and recovery efforts.

Similarly to Galena, flood risk was not factored into the investment decisions during the initial settlement of Edeytsy in the current floodplain location. Edeytsy, as a municipality, was formed in 1930 by the Soviet state government (Old Sakha [no date]). Establishment of the Soviet regime and collectivization (a policy of forced consolidation of individual peasant households into collective farms) throughout the Sakha lands coocurred with forced settlements of populations into permanent locations (Osofsky, 1974). Several Native groups were integrated into Edeytsy to advance local kolkhozy (collective farms) (Old Sakha [no date]). Edeytsy was settled on the river to facilitate irrigation and water supplies for large cattle herds. Continuous population increase and kolkhoz expansion have amplified the initial flood risk in Edeytsy.

25.3.3 Spring Flood Risk Reduction in Galena and Edeytsy

The original absence of flood risk governance in both regions resulted in a lack of building codes, and flood risk reduction measures. For decades, flood-ravaged houses in Galena and Edeytsy were rebuilt on the same sites with no intent to reduce populations' exposure to spring floods (Morgan, 1972; Lindenau, 1983; Sprott, 2000). As Galena and Edeytsy became strategically important by the mid-1940s, state governments began to implement flood risk management strategies to protect valuable assets from the adverse impacts of spring floods. For example, a dike was constructed in Galena in 1944 to protect the airport infrastructure (Mongin et al., 1972). Old Town, the original native settlement, is located between the dike and the river, and thus is completely exposed to ice jam floods (Kontar et al., 2015).

A large portion of Galena residents relocated to a new site after a major flood in May 1971 (Morgan, 1972). As revealed by archived news articles and feature stories, opinions of the local population regarding relocation were not accounted for when choosing the new site's location (e.g., News Miner, 1971a; News Miner, 1971b; Morgan, 1972; Arundale, 1985; Sprott, 2000). For the construction of Galena's New Town, the Alaska State Housing Authority (ASHA) allocated a site near Alexander Lake approximately one mile away from Old Town. ASHA surveyed the proposed relocation site and determined that it was appropriate for relocation as "it has received very little water during this record high flood" (News Miner, 1971b). Galena residents thought the proposed site was too low, and wanted to move the village onto high ground on the USAF dumpsite, Campion (News Miner, 1971b). Unable to receive federal approval, New Town was settled near Alexander Lake. More than 90 percent of New Town's buildings were destroyed or severely damaged by the 2013 flood (Kontar et al., 2015). The Campion site remained dry during all flood years, including 2013.

To protect the farm and pasture lands in Edeytsy and other riverine communities from spring flooding, the Soviet state agencies responsible for disaster management implemented a series of mechanical ice jam and flood mitigation and prevention strategies, such as river ice dusting, cutting, and blasting (Table 25.1). Mechanical mitigation efforts are proactive, but not necessarily effective. They are designed according to hydrological models developed by academic and state scientists; but no criteria for the effectiveness of these models have yet been established (Buzin et al., 2014). There is no published evidence of these methods' effectiveness, only anecdotal.

To reduce Edeytsy's exposure to spring floods, construction of a dike had been initiated after the flood in May 2010. According to the survey responses, Edeytsy's administration and residents believe that the completed dike will provide the necessary protection for their homes, infrastructure, and pasture and farmlands. However, the construction continued six years later as a result of incremental funding from the government.

After the 2013 floods, most houses in Galena were elevated on steel pilings at least one foot above the highest flood level. Despite their unaesthetic appearance and inconvenience for the elders, Galena residents believe that their houses and their property inside houses are protected from the spring floods. However, their movable assets, such as cars and snowmobiles, remain exposed to floods. In Edeytsy, six houses were

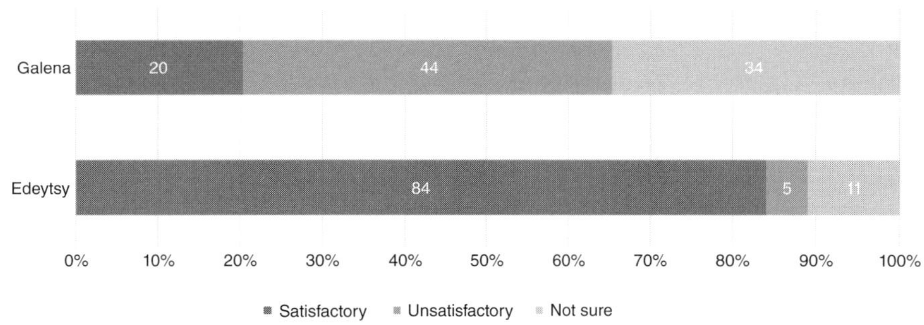

Figure 25.1 Comparison of flood management efforts: evaluation of the effectiveness of the key flood response and recovery efforts in Galena and Edeytsy.

elevated on wood pilings. An effective mitigation technique in regular floods, wood pilings, have proven to be unreliable at withstanding ice damage (Belore et al., 1990; Buzin et al., 2014). Their effectiveness in Edeytsy will be determined during the next flood.

As mentioned, flood preparedness is another integral component of flood risk reduction. During the 2013 floods, preparedness measures in Galena and Edeytsy included early warnings, evacuation of at-risk population and livestock to higher ground (only in Edeytsy), and facilitation of rapid response. Since a large portion of Galena's population (i.e., New Town residents) had not experienced a major ice jam flood in their lifetime, they developed a false sense of security. Caught off guard by the severity of the 2013 flood, most Galena residents did not take precautionary measures to protect their houses, movable assets, and animals. Despite the time constraints, the evacuation measures were successful and resulted in no fatalities (Kontar et al., 2015).

Under the leadership of the local administration, Edeytsy residents were more proactive in relocating their cattle, farm equipment, and personal vehicles to higher ground a day prior to the flood onset. As revealed through the focus group discussions, annual flood preparedness measures begin in Edeytsy and other rural Sakha communities with the evacuation of special risk groups approximately two to three days prior to the potential flood.

The overall coordination of spring flood risk reduction efforts in the Sakha Republic is organized and initiated at the federal level. This unified and centralized system encourages ongoing interagency communication, and facilitates the availability of the necessary financial and human resources to execute flood mitigation and preparedness measures in a timely and effective manner. However, the communities at risk are not included in the flood management decision making. Local administrations relay flood risk reduction measures from the federal and state governments.

Focus group discussions revealed that no centralized flood risk reduction efforts are conducted prior to the breakup onset in Alaska. Spring flood management is predominantly reactive (Taylor et al., 2016). The survey responses indicate the frustration of Galena residents with the lack of the preventative and preparedness measures (Figure 25.1). Galena responders were disappointed by the absence of short-term ice jam mitigation efforts. Watching their communities ravaged by ice floes, people hoped for prevention efforts.

Both case studies highlighted drawbacks and advantages in the existing flood risk reduction strategies. No specific strategy alone will effectively reduce the flood risk of every exposed and vulnerable region or community. To identify effective flood mitigation and preparedness strategies, it is crucial to evaluate the causative factors of spring floods, as well as the progression of flood risk and the history of floodplain management. To ensure that flood management strategies fit the unique features and needs of the at-risk communities, local stakeholders and their opinions and knowledge should be incorporated in the decision-making process.

25.3.4 Key Findings and Recommendations

The historical analysis of the risk progression in Edeytsy and Galena has demonstrated that the vulnerability of both communities to spring floods is deeply rooted in socioeconomic and political processes. It points to common root causes between the two research sites, including colonial heritage, top-down governance, and limited inclusion of local communities in the decision-making process.

The communities of Galena and Edeytsy were settled in their permanent floodplain locations with no consideration of flood risk. Rapid expansion of both communities, along with miscommunication or lack of communication between stakeholders responsible for managing flood risks and those facing flood risks, resulted in inadequate flood risk governance and management measures. Reducing the overall risk in flood-prone communities is only possible through integration of local knowledge and stakeholders in the decision-making processes regarding spring flood management.

Unlike in most natural disasters, opportunities for reducing spring flood risk lie not only in reducing vulnerability and exposure drivers, but also in reducing the severity of the hazard itself. Throughout history, an array of ice jam prevention and mitigation measures have been implemented in the U.S. and Russian North with varying degrees of success (Table 25.1). A series of mechanical ice jam mitigation and prevention techniques (e.g., ice dusting, blasting, and cutting) are implemented in the Russian North during every spring breakup. Although these measures are more proactive, there is no published evidence of their effectiveness. Advances in ice jam and flood risk mitigation would improve the overall flood risk reduction in both regions. Further in-depth comparative analysis is needed to determine the costs, benefits, and effectiveness of spring flood mitigation and prevention measures.

25.4 Conclusion

Spring floods are complex natural and social phenomena. Two main causes of spring floods are ice jams and floodplain use. In Alaska and the Sakha Republic, ice jam floods turned into disasters at the beginning of the twentieth century, when government programs forced the originally nomadic Indigenous Peoples to settle permanently on floodplains with life-sustaining infrastructure and property. The original absence of flood risk governance in Alaska and the Sakha Republic resulted in a lack of building codes and flood risk reduction measures. The economic and political incentives for the communities' expansion further outweighed considerations for flooding.

To minimize flood risk, a range of flood risk reduction measures have been implemented in Alaska and the Sakha Republic. The analysis of focus group discussions with U.S. and Russian representatives from the agencies responsible for flood management revealed two key categories of ice jam flood prevention and mitigation strategies in Alaska and the Sakha Republic:

1. *Ice jam prevention*: preventive measures to eliminate or lessen the likelihood of a damaging ice jam event (e.g., river ice dusting, cutting, and detonation). These short-term, mechanical mitigation measures are implemented to weaken or breach ice jams that have already formed.
2. *Reduction of ice jam flood impact*: preventive measures undertaken to reduce the potential damage from floodwater and ice debris. These measures are further divided into structural (e.g., construction of dikes and elevation of houses on steel pilings) and nonstructural (e.g., floodplain management and population relocation). These long-term flood risk mitigation efforts are usually initiated right after the floods. The implementation of these measures usually requires large initial investments and might take several years.

Ice jam prevention measures are more proactive and provide a placebo effect to the adverse population in the midst of crisis. Although they are implemented in the Sakha Republic on a regular basis, no correlation between these measures and the frequency and severity of spring floods has been established. Further comparative analysis is needed to determine the effectiveness of the existing spring flood mitigation measures.

Overall, the scale, frequency, and impact of spring floods can be effectively addressed only through interdisciplinary research and holistic policy solutions. Integrated flood risk reduction requires interagency collaborations with a diverse group of stakeholders. To facilitate integrated spring flood risk reduction in the rural North, it is crucial to provide a space for a dialogue among key stakeholders, including communities facing flood risk.

Acknowledgments

The authors thank two anonymous reviewers for their helpful comments on earlier drafts of the manuscript.

This manuscript was funded in part by a grant from the United States Department of State. The opinions, findings, and conclusions stated herein are those of the authors and do not necessarily reflect those of the United States Department of State.

This manuscript was also funded in part by a grant from the United States Department of State and by the Alaska Center for Climate Assessment and Policy, a Regional Integrated Sciences and Assessments Team (RISA), of the National Oceanic and Atmospheric Administration, Award NA16OAR4310162.

References

Arundale, W. (1985). *Edgar Nollner's Biography.* Available from: http://jukebox.uaf.edu/RavenStory/gif/Edgar%20Nollner's%20Biography.pdf [Accessed 20th March 2017].

Belore, H. S., Burrell, B. C. and Beltaos, S. (1990). Ice jam mitigation. *Canadian Journal of Civil Engineering*, **17**(5), 675–685. Available from: doi:10.1139/l90-081 [Accessed 20th March 2017].

Beltaos, S. (2007). River ice breakup processes: Recent advances and future directions. *Canadian Journal of Civil Engineering*, **34**(6), 703–716. Available from: doi:10.1139/l06-021[Accessed 20th March 2017].

Beltaos, S., Miller, L., Burrell, B. C. and Sullivan, D. (2006). Formation of breakup ice jams at bridges. *Journal of Hydraulic Engineering*, **123**(11), 12–29. Available from:

doi:10.1061/(ASCE)0733-9429(2006)132:11(1229) [Accessed 18th March 2017].

Benoit, L. (2014). *Perspectives on Emergency Response in the Canadian Arctic: Sinking of the MS Arctic Sun in Cumberland Sound, Nunavut*. Available from: http://gordonfoundation.ca/publication/732 [Accessed 16th March 2017].

Bhattacherjee, A. (2012). *Social Science Research: Principles, Methods, and Practices*. Available from: http://scholarcommons.usf.edu/oa_textbooks/3 [Accessed 20th February 2017].

Burrell, B. C., Huokuna, M., Beltaos, S., Kovachis, N., Turcotte, B. and Jasek, M. (2015). *Flood Hazard and Risk Delineation of Ice-Related Floods: Present Status and Outlook*. Paper presented at the 18th Workshop on the Hydraulics of Ice Covered Rivers, Quebec, Canada. Available from: www.researchgate.net/publication/308688387_Flood_Hazard_and_Risk_Delineation_of_Ice-Related_Floods_Present_Status_and_Outlook [Accessed 20th March 2017].

Buzin, V. A., Goroshkova, N. I. and Strizhenok, A. V. (2014). Maximum ice-jam water levels on the northern rivers of Russia under conditions of climate change and anthropogenic impact on the ice jamming process. *Russian Meteorology and Hydrology*, **39**(12), 823–827. Available from: doi:10.3103/s1068373914120061[Accessed 20th March 2017].

Cutter, S.L., Ismail-Zadeh, A., Alcántara-Ayala, I., Altan, O., Baker, D. N., Briceño, S. and Wu, G. (2015). Global risks: Pool knowledge to stem losses from disasters. *Nature* **522**, 277–279. Available from: doi:10.1038/522277a [Accessed 20th March 2017].

Global Facility for Disaster Reduction and Recovery (GFDRR). (2014). *Understanding Risk Review of Open Source and Open Access Software Packages Available to Quantify Risk from Natural Hazards*. Available from: www.gfdrr.org/sites/gfdrr/files/publication/UR-Software_Review-Web_Version-rev-1.1.pdf [Accessed 20th March 2017].

Institute for Risk and Disaster Reduction (IRDR). (2014). *Arctic Risk: A Discussion of the Possible Outcomes of Two Disaster Scenarios* (IRDR Report 2014–02). Available from: www.ucl.ac.uk/rdr/publications/irdr-special-reports/irdr-special-report-2014-01 [Accessed 18th March 2017].

Intergovernmental Panel on Climate Change (IPCC). (2012). *Managing the risks of extreme events and disasters to advance climate change adaptation*. Available from www.ipcc.ch/pdf/special-reports/srex/SREX_Full_Report.pdf.

Kontar, Y. Y., Bhatt, U. S., Lindsey, S. D., Plumb, E. W. and Thoman, R. L. (2015). Interdisciplinary approach to hydrological hazard mitigation and disaster response and effects of climate change on the occurrence of flood severity in central Alaska. *Proceedings of IAHS*, **369**, 13–17. Available from: doi:10.5194/piahs-369-13-2015 [Accessed 20th March 2017].

Kravitz, M. and Gastaldo, V. (2013). *Emergency Management in the Arctic: The context explained*. Available from: www.gordonfoundation.ca/publication/686 [Accessed 18th March 2017].

Kusatov, K. I., Ammosov, A. P., Kornilova, Z. G. and Shpakova, R. N. (2012). Anthropogenic factor of ice jamming and spring breakup flooding on the Lena River. *Russian Meteorology and Hydrology*, **37**(6), 392–396. Available from: doi:10.3103/S1068373912060064 [Accessed 18th March 2017].

Lindenau, Y. I. (1983). *Историко-этнографические материалы о народах Сибири и Северо-Востока: первая половина XVIII века* [Historical and ethnographic materials about peoples of Siberia and the Russian far east: First half of the 18th century]. Magadan, Russia: Magadan.

Mongin, A., Mesloh, D. E. and Beck, P. L. (1972). History of Alaskan Air Command, 1 July 1970–30 June 1972.

Morgan, L. (1972). *Galena – How to Win a Flood*. Anchorage, AK: The Alicia Patterson Fund.

News Miner. (1971a). Galena residents flee high waters of the Yukon River. *News Miner*, May 24.

News Miner. (1971b). Galena flood damage may reach $2 million. News Miner, May 29.

Old Sakha. (no date). *Географическая и историческая справка о наслеге [Geographical and historical information about nasleg]*. Available from: http://old.sakha.gov.ru/node/7353 [Accessed 18th March 2017].

Osofsky, S. (1974). *Soviet agricultural policy: toward the abolition of collective farms*. Westport, CT: Greenwood.

Prowse, T. D., Alfredsen, K., Beltaos, S., Bonsal, B., Duguay, C., Korhola, A., and Wrona, F. (2011). *Changing Lake and River Ice Regimes: Trends, Effects and Implications*. Available from: www.amap.no/documents/doc/snow-water-ice-and-permafrost-in-the-arctic-swipa-climate-change-and-the-cryosphere/743 [Accessed 18th March 2017].

Sprott, J. E. (2000). *Neelghu neets'edeneyh – We work together, we help each other: the story of Louden Tribal Council's Self-Governance process, 1993–2000*. Anchorage, AK: Environment and Natural Resources Institute.

Taylor, K. M., Hum, R. and Kontar, Y. Y. (2016). Comparative analysis of virtual relief networks and communication channels during disaster recovery after a major flood in Galena, Alaska, spring 2013. In *Communicating Climate-Change and Natural Hazard Risk and Cultivating Resilience: Case Studies for a Multi-disciplinary Approach*, ed. J. L. Drake, Y. Y. Kontar, J. C. Eichelberger, T. S. Rupp and K. M. Taylor. Cham, Switzerland: Springer, pp. 151–171.

United Nations International Strategy for Disaster Reduction (UNISDR). (2015). *Making development sustainable: The future of disaster risk management. Global assessment report on disaster risk reduction*. Available from: www.preventionweb.net/english/hyogo/gar/2015/en/gar-pdf/GAR2015_EN.pdf [Accessed 18th March 2017].

U.S. Census Bureau. (2015). *Annual Estimates of the Resident Population in Galena, Alaska: April 1, 2010-July 1, 2015*. Available from: https://factfinder.census.gov/faces/tableservices/jsf/pages/productview.xhtml?src=bkmk.

26

Geohazard Analysis for Disaster Risk Reduction and Sustainability

Alik Ismail-Zadeh

26.1 Introduction

Geohazards are defined as potentially damaging geophysical events (e.g. earthquakes, volcanic eruptions and emissions, landslides, and associated tsunamis), which "may cause loss of life, injury or other health impacts, property damage, social and economic disruption or environmental degradation" (UN, 2017). A geohazard event is characterized by its location, magnitude/intensity, time, and the probability of the occurrence. Triggered by geohazard event(s) a disaster (that is, a serious disruption of the normal functioning of a community/society) causes widespread human, economic, and/or environmental losses, which exceed the ability of the affected community/ society to cope using its own resources (UN, 2017). A disaster is a function of the risk process and results from the convolution of hazards with conditions of exposure, vulnerability, and insufficient capacity or measures or even interest to reduce the potential negative consequences of risk.

Disasters continue to grow in number and impact, although geohazard events are rather stable in the number of occurrences and follow a linear distribution of the logarithm of the number of the events with respect to their magnitude (e.g. the Gutenberg-Richter law in the case of earthquakes; Gutenberg and Richter, 1944; Malamud and Turcotte, 1999). Geohazards become direct threats to national security in many regions, because their impacts are amplified by rapid growth of population and unsustainable development practices, both of which increase exposure and vulnerabilities of communities, capital, and environmental assets. Reducing disaster risk using scientific knowledge is a foundation for sustainable development (Cutter et al., 2015). Our knowledge on geohazards and their interaction with human systems is lacking in some important areas and is being challenged by the unforeseen or unknown repercussions of a rapidly changing and increasingly interdependent world. In such a tightly coupled world a disaster not only affects the immediate area where it occurs, but also may have cascading impacts that can affect other nations near and far.

Scientific understanding of lithosphere dynamics, earthquake occurrence, volcanic eruptions and lava flows, land sliding, tsunami generation, and other geophysical events as well as hazard assessments have greatly advanced for the last several decades. Meanwhile despite these major advancements, we are not seeing significant disaster risk reduction and a concomitant decline in disaster impacts and losses. For example, the 2005 Hyogo Framework for Action (HFA) pledged to reduce disaster losses by 2015. Unfortunately, this did not occur at the global scale, although some success was achieved at local or national scales: namely, of the ten costliest disasters between 1980 and 2014, those that occurred for the HFA decade (2005–2014) were more than four times more expensive than those before 2005 (Cutter et al., 2015; Munich Re, 2015). Increasing vulnerability and exposure account for some of the increase, but not for all. There are several issues that should be improved before there can be an essential reduction in disaster losses, and one of the issues is enhancing geohazard research and integrating it into disaster risk assessment. How can such integration take place within the broad disaster science community? Ismail-Zadeh et al. (2017) provide an answer to this question suggesting that integration should be formed and realized within a specific mission, specific target, and efficient management, and it should proceed together with a practicing sector to make science needed and useful.

In this chapter, some advances in analysis of seismic and lava flow hazards, contribution of geohazard analysis to disaster risk assessments, and implication of risk reduction for sustainability are presented. Geodynamic processes leading to earthquake, volcano, and landslide occurrences are discussed in Section 26.2.

Section 26.3 describes basic seismic hazard assessment methods highlighting their advantages and challenges, and shows how quantitative modelling of extreme seismicity can assist in hazard analysis. Advanced modelling of lava flow is reviewed in Section 26.4. An interaction of the main risk components (hazards, vulnerability, and exposure) is analysed in Section 26.5. Some past major geohazard events and their impact on society and sustainability are reviewed in Section 26.6. The progress in disaster risk reduction is discussed in Section 26.7 in the framework of sustainable development goals, and concluding remarks are presented in Section 26.8.

26.2 Geodynamics and Geohazards

Geodynamics deals with the dynamic, physical, and chemical processes in the Earth's interior. Phenomena subject to these fundamental geological processes include plate tectonics, mantle convection, and their interaction. According to plate tectonic theory (e.g. Turcotte and Schubert, 2014), the plates are continually created and consumed. At ocean ridges adjacent plates diverge in a process that is known as seafloor spreading. As the lithosphere moves away from ocean ridges, it cools, becomes more dense, and thickens. Once the lithosphere becomes sufficiently dense compared to the underlying mantle rocks, it bends, founders, and begins sinking into the hot mantle as a result of gravitational instability. Because the lithosphere behaves elastically at short time scales (seconds to years), it can transmit stresses. The downward buoyancy forces (generated by the excess density of the rocks of the descending lithosphere) promote the sinking of the lithosphere, but elastic, viscous, and frictional forces resist the descent. The combination of these forces produces shear stresses high enough to cause earthquakes. Other processes contributing to stress generation in the descending lithosphere and its release in earthquakes can be plastic instability at high temperature (e.g. Griggs and Baker, 1969), faulting due to metamorphic phase transitions (e.g. Green and Burnley, 1989), dehydration-induced embrittlement (e.g. Ismail-Zadeh et al., 2000), and some others. Earthquakes occur as a sudden release of stresses. When an earthquake occurs, part of the released energy generates elastic waves propagating through the Earth. These waves generate ground motions and shaking, which may result in building damage or collapse, landslides, tsunami wave generation, etc.

Ocean trenches are the sites of the world's largest earthquakes and volcanoes. The earthquakes at oceanic trench zones can occur along the descending lithosphere to the depths of about seven hundred kilometres depending on the thermal state in the mantle. Meanwhile large earthquakes occur not only in subduction zones, but also inside continents (so-called intraplate earthquakes), especially in the regions of continental collisions, rifts, and grabens, such as the Kutch Rift in India, where the 2001 Bhuj M7.7 earthquake occurred far away from plate boundaries, causing widespread damage and killing more than twenty thousand people (Gupta et al., 2001); the Vrancea region in Romania, producing several strong damaging earthquakes a century (Ismail-Zadeh et al., 2008; 2012b); and the New Madrid Rift complex (United States), where large earthquakes took place in 1811 and 1812 (Braile et al., 1986). According to the global risk analysis (Dilley et al., 2005), an area of about 10 million square kilometres (7.5% of the total area of the planet) is estimated to undergo significant shaking by earthquakes (10% probability of peak ground acceleration of at least two metres per square second in a fifty-year period). This area is inhabited by approximately 1.2 billion people.

Volcanoes are associated with lithospheric plates' both converging (near subduction zones) and diverging (e.g. at the Mid-Atlantic Ridge). Active volcanoes occur above the descending lithosphere and are situated almost parallel to ocean trenches. If volcanoes stand on the seafloor, they form an island arc, and if the ocean trench lies close to a continent, the volcanoes form on the land. Meanwhile volcanoes can also form in the areas of crustal stretching (rifting), such as in the East African Rift, and are considered to be associated with mantle plumes or hot spots. According to Simkin and Siebert (1994), there are 1585 active volcanoes in the world (although most submarine volcano eruptions are not accounted for), and 315 of these volcanoes are located within one hundred kilometres of urban areas with a population of about half a billion people, and many of them live in densely populated regions in less-affluent countries. About sixty volcanoes on the land erupt annually including continuously active volcanoes, e.g. Stromboli volcano in Sicily, which explodes roughly every eight minutes (Donovan and Oppenheimer, 2014). Erupting volcanoes may pose hazards because of lahars, pyroclastic currents, lava flows, ash fallouts, and other volcanic threats. Between 1900 and 2008, some ninety thousand people lost their lives, fourteen thousand were injured, and 4.7 million were affected by volcanic events (Doocy et al., 2013). Predicting a volcanic eruption is an interdisciplinary science involving continuous observation of volcanic earthquakes, changes in ground conditions using geodesy and magnetic studies, groundwater, etc., and providing a clue of the forthcoming eruption.

Volcanic ash propagation is capable of causing damage over a relatively large area (e.g. ash fallout smothers forests and agricultural crops and contaminates water supplies), whereas other volcanic processes (e.g. pyroclastic or lava flows) usually only threaten areas that are close to the volcano. Volcanic ash can be a threat to people, causing respiratory distress, and to aircraft jet engines, resulting in chaos for aviation (Adey et al., 2011; Donovan and Oppenheimer, 2011; 2014). Large volcanic eruptions can even affect the Earth's temperature and climate via the generation of sulphate aerosol from sulphur dioxide emissions and as a result of changes in volcanic outgassing of carbon dioxide (e.g. Oppenheimer, 2003; Huybers and Langmuir, 2009). For example, an eruption of similar magnitude to that of Tambora (M = 6.9), depending on the location of the volcano responsible and magnitude of sulphur release, would likely have impacts around the world (Donovan and Oppenheimer, 2014).

Landslides are ground movements that can be triggered by earthquakes, volcanic eruptions, rainfalls, and associated soil liquefaction. Landslides take place in different parts of the world. Approximately 3.7 million km^2 of land are susceptible to sliding, and the areas of high risk of landslides are inhabited by 66 million people. Even if the attention given to landslides is not as focused as that given to other geohazards, such as earthquakes or volcanoes, the social impact of landslide disasters is nonetheless significant. It may continue to increase in the future because of the expansion of human settlements in areas of risk. Landscape modifications, such as changes in slope geometry due to construction of infrastructure, terracing, slope excavations, loading, and mining, in addition to disturbances to the equilibrium of slopes produced by vibrations, water linkage, land-use changes, deforestation, are significant factors determining slope instability and landslide dynamics (Alcántara-Ayala, 2014). Moreover, climate change may intensify the risk of landslides through an increase in the frequency and magnitude of rainfall (IPCC, 2012).

Tsunami waves are generated as a result of earthquakes beneath the ocean floor, submarine landslides, violent submarine volcanic eruptions, caldera collapses, flank failure into a water source, pyroclastic flow discharge into the sea (e.g. Latter 1981), and other Earth disturbances below and above water (e.g. due to meteorite impacts). In the case of thrust faulting earthquakes, the waves are the result of ground uplift and subsequent push of the water column above the average sea level. While the earthquakes, volcanoes, or landslides take place on specific areas, the effects of tsunamis are widely distributed in space and time, and consequences can be global. For example, the 2004 Aceh-Sumatra M9.3 earthquake-induced tsunami affected a number of countries around the Indian Ocean. At global scale, 14 million people are exposed to tsunamis, and the major potentially affected areas are located along the coasts of countries facing the oceans and seas (UNISDR, 2009).

How can geosciences and the geohazard analysis contribute to the reduction of disaster risk? Sections 26.3 and 26.4 describe some advances in earthquake and lava flow hazard modelling.

26.3 Earthquake Hazard Analysis

Seismic hazard assessment (SHA) in terms of engineering parameters of strong ground motion (namely, peak ground acceleration and seismic intensity) is based on the information about the features of excitation of seismic waves at the source, seismic wave propagation (attenuation), and site effect in the region under consideration and combines the results of seismological, geomorphological, geological, and tectonic investigations and modelling (e.g. Ismail-Zadeh, 2014). Two principal methods are intensively used in seismic hazard assessment: probabilistic and deterministic SHA.

The probabilistic SHA determines the probability rate of exceeding various levels of ground motion estimated over a specified period (Cornell, 1968). The probabilistic assessment considers uncertainties in earthquake source, path, and site conditions. The uncertainties are classified as epistemic and aleatory. Epistemic uncertainties reflect the incomplete knowledge about input model parameters to the assessment and variability of interpretations of available data. Aleatory uncertainties consider the inevitable unpredictability of the parameters: the uncertainties are mainly quantified through the use of the standard deviation of the scatter around the mean values. Although the probabilistic SHA method is dominant in hazard assessment, there are several points of criticism related to this assessment: e.g., validity of a point source model, ground motion uncertainties in the mathematical formulation of the method, and lack of capability to model the dependencies between large numbers of uncertain random parameters correctly (Panza et al., 2010).

The deterministic SHA is an alternative method for hazard analysis and is based on a specified earthquake scenario. For a given earthquake, the deterministic assessment model analyses the attenuation of seismic energy with distance to determine the level of ground motion at a particular site. Ground motion calculations capture the effects of local site conditions and use the available knowledge on earthquake sources and wave propagation processes. Namely, attenuation relationships are used for a given earthquake magnitude to calculate ground shaking at bedrocks, which is then amplified by factors based on local soil conditions. Although the occurrence frequency of the ground motion is usually not addressed in the deterministic SHA, the method is robust for an assessment of seismic hazard due to specified events and remains a useful approach for decision making (e.g. Babayev et al., 2010).

Since the beginning of the twenty-first century several extreme seismic events (e.g. the 2004 Sumatra-Andaman, 2005 Kashmir, 2008 Wenchuan, 2010 Chile, 2010 Haiti, 2011 Tohoku, and 2015 Nepal earthquakes) demonstrated that the events were a surprise, although scientific knowledge of historical earthquakes was available in most of these areas. For example, scientists knew about historical devastating earthquakes and tsunamis that had occurred in the Indian Ocean region (e.g. a magnitude 8.7 earthquake in 1833 followed by a powerful tsunami killed thirty-six thousand people), yet did not inform regional planners or policy makers of these historical precedents. Many existing seismic hazard maps do not mark the regions as those of significant hazards, but instead present a level of ground shaking that can be exceeded with a certain probability within a certain period. These maps provide an estimate of seismic hazard useful for engineering purposes, but not very useful for people living in the earthquake-prone regions. Ground shaking due to recent large earthquakes was much higher than expected (e.g. Stein et al., 2011), and hence, the numbers of fatalities in recent earthquake disasters were underestimated by approximately two to three orders of magnitude (Wyss et al., 2012). This reveals a weakness in current seismic hazard assessment and mapping, which is mostly engineering-oriented and not human-oriented.

In many cases, large earthquakes are not accounted for in the probabilistic SHA because of the lack of information about them. Our present knowledge about characteristics of seismicity is based on observed (recorded) and sometimes on historical data (from paleoseismological and archaeological studies, written accounts of intensities of large earthquakes, and some other sources). The information about extreme events in a particular region is incomplete as the large events are rare. Modelling of extreme seismic events using earthquake simulators is a way to overcome the difficulties in seismic hazard assessment by combination of observations, historic data, and modelled results.

Studying seismicity employing the statistical and phenomenological analysis of earthquake catalogues has the disadvantage that instrumental observations cover a short time interval compared to the duration of the tectonic processes responsible for earthquakes. The patterns of earthquake occurrence identifiable in a catalogue may be apparent and yet may not be repeated in the future. Moreover, the historical data on seismicity are usually incomplete. Numerical modelling of seismogenic processes allows generation of synthetic earthquake catalogues covering very long time intervals and, therefore, providing a basis for reliable estimates of the parameters of the earthquake occurrences (e.g. Soloviev and Ismail-Zadeh, 2003; Ismail-Zadeh et al., 2012a). Earthquake simulators help to generate synthetic seismic events and to study the events, which are consistent with geophysical, geological, and geodetic observations. For example, Gabrielov et al. (1990) developed a block-and-fault dynamics (BAFD) model, which was designed to answer the following important questions: how upper crustal/lithosphere blocks react to the plate motions and to a flow of the lower ductile crust (or the viscous asthenosphere in cases when the lithosphere blocks are involved in the modelling); how earthquakes cluster in the system of major regional faults; at which part of a fault system large (extreme) events can occur, and what is the occurrence time of the extreme events; how the properties of the frequency–magnitude relationship change prior to extreme events; how fault zone properties influence the earthquake clustering, its magnitude, and fault slip rates.

Using the BAFD model, Ismail-Zadeh et al. (2007b) developed numerical experiments to analyse the earthquake clustering, frequency-to-magnitude relationships, earthquake focal mechanisms, fault slip rates, and frequency of large events in the Tibet-Himalayan (TH) region. Synthetic seismicity spanned more than four thpousand years, a considerably longer interval of time compared to the duration of the recorded regional seismicity. The highest magnitude of the synthetic events was found to be M = 8.9, occurring several times at the same normal fault. It was shown that the recurrence time of the largest events varies considerably. It is worth remarking that the BAFD model for the TH region identified a cluster of large synthetic events (magnitude about 8; see Figure 26.1a) along the Longmen Shan fault before the 2008 Wenchuan (China) M = 7.9 earthquake occurred at this fault (Ismail-Zadeh et al., 2007b).

Earthquake simulators not only are tools for studying seismic preparation processes, but also can be useful in hazard assessment by incorporating observed and modelled extreme events in the ground motion estimation and earthquake forecasting. Sokolov and Ismail-Zadeh (2015) developed a new approach to a Monte

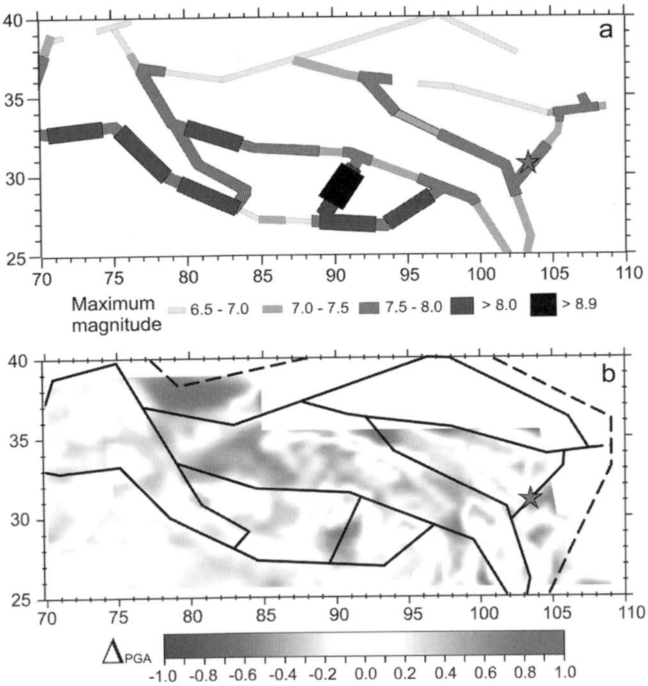

Figure 26.1 Seismic hazard assessment for the Tibet-Himalayan (TB) region. (a) Distribution of maximum magnitudes of the earthquakes predicted by the preferred block-and-fault dynamics (BAFD) models along the major faults of the TH region (Ismail-Zadeh et al., 2007b). (b) The difference Δ_{PGA} between two ground motion assessments for the TH region (see the text). Black lines are the regional fault system used in the modelling (dashed lines are the BAFD model boundary lines). The assessments were performed for rock-site condition (Vs30 = 800 m s^{-1}) related to the 475-yr return period. Because of the lack of information on earthquake source zones for the area located to the north from 35°N and to the east from 85°E, the ground shaking is not assessed in this area. The star shows the location of the 2008 Wenchuan M7.9 earthquake (modified after Sokolov and Ismail-Zadeh, 2015).

Carlo probabilistic SHA combining the observed regional seismicity with large magnitude synthetic events obtained by BAFD simulations. Earthquake scenarios for hazard assessment are generated stochastically to sample the magnitude and spatial distribution of seismicity, as well as the distribution of ground motion for each seismic event. This approach was employed for seismic hazard analysis in the TH region. Figure 26.1b presents a comparison of the results of the hazard assessment performed by the standard probabilistic SHA (the Global Seismic Hazard Assessment Program (GSHAP) model; Giardini et al., 1999) and the Data-Enhanced Assessment approach accounting for large synthetic events (DEA model; Sokolov and Ismail-Zadeh, 2015). It is evident that the difference in ground shaking predicted by the two models, or, strictly speaking, the difference $\Delta_{PGA} = \log_{10}(\text{PGA}_{DEA}/\text{PGA}_{GSHAP})$ between the relevant peak ground acceleration (PGA) estimates obtained by the DEA model (PGA_{DEA}) and by the GSHAP model (PGA_{GSHAP}), is significant ($\text{PGA}_{DEA} \geq 1.5\text{PGA}_{GSHAP}$, i.e. $\Delta_{PGA} \geq 0.176$) for several areas including the area of the 2008 Wenchuan earthquake. Therefore, the data-enhanced seismic hazard assessment by Sokolov and Ismail-Zadeh (2015) allows for better understanding of ground shaking and could be useful for earthquake risk assessment, engineering purposes, and emergency planning.

Current probabilistic SHA methods are based on point-wise (site by site) assessments of ground shaking and do not consider multiple-site effects. Sokolov and Ismail-Zadeh (2015, 2016) analysed some features of multiple-site (MS) probabilistic SHA, i.e. the annual rate of ground-motion level exceedance in at least one site of several sites of interest located within an area or along a linear extended object, and showed that the expected ground motion level in the selected area (multiple sites) is higher than that at individual sites. For example, the strong-motion records obtained in the epicentral zone of the 2008 Wenchuan earthquake were employed as an example of the ground motion exceeding the design level estimated using the conventional point-wise (PW) probabilistic SHA. To assess the difference between the point-wise and the multiple-site estimations of seismic hazards, Sokolov and Ismail-Zadeh (2015) considered a square area located near the epicentre of the Wenchuan earthquake or a rectangular area stretched along the causative fault. It was assumed that the entire area is characterized by the same design PGA value and that there are several sites where the multiple-site hazard should be evaluated. A low level of the ground motion correlation (e.g. the correlation distance of five kilometres) is assumed and allows for a high difference between the expected ground-motion parameters even for neighbouring sites.

Figure 26.2 illustrates that the greater is the area (and the larger is the number of considered sites), the greater is the difference between the hazard estimates for individual sites and the multiple sites. In the case considered, the level of multiple-site hazard estimated for the 2475-yr return period is larger than the highest recorded level of ground motion even for the relatively small area and for a small number of sites. Sokolov and Ismail-Zadeh (2016) proposed a multi-level approach to probabilistic SHA considering fixed reference probability of exceedance (e.g. 10% in fifty years): (i) a standard point-wise hazard assessment to be performed in a

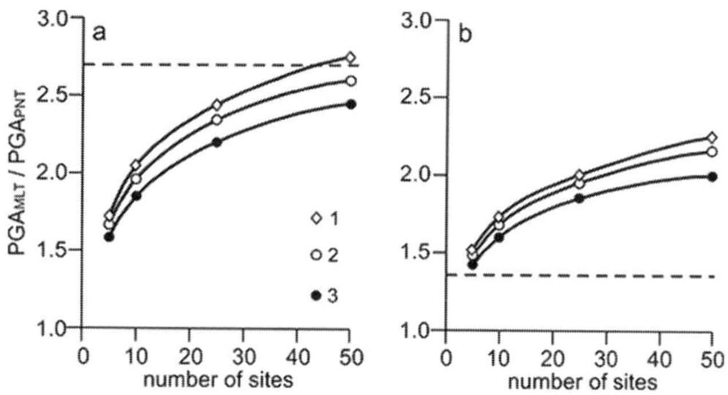

Figure 26.2 Relation between the peak ground acceleration obtained from multiple-site (PGA_{MS}) and point-wise (PGA_{PW}) hazard assessments for the area of the 2008 Wenchuan earthquake for the 475-yr (a) and 2475-yr (b) return periods. The dashed line shows the ratio between the maximum PGA recorded during the earthquake (about 810 cm s^{-2}) and the estimated design ground motion $PGA_{475} \sim 300$ cm s^{-2} (a) and $PGA_{2475} \sim 600$ cm s^{-2} (b). 1: area 600 km^2; 2: 400 km^2; 3: 100 km^2 (after Sokolov and Ismail-Zadeh, 2015).

seismic-prone region, and (ii) this analysis should be supplemented by a multiple-site hazard assessment for urban and industrial areas, or zones of a particular economic and social importance.

26.4 Lava Flow Analysis

During volcanic non-explosive (effusive) eruptions, partially molten rocks are erupted onto the Earth's surface, spread slowly on the surface from the volcanic edifice, and generate lava flows. The eruptions produce a variety of currents depending on the chemical composition and temperature of the magmatic rocks, and the topography of the surface over which the lava flows (Griffiths, 2000). Under relatively steady eruption conditions, lava flow rapidly forms a solid crust. Cooling and crystallization of the uppermost layer of the moving melt lead to a gravity current of lava under a solid crust, which insulates the lava flow interior. The crust preserves the lava against rapid cooling and permits the lava flow to extend to substantial distances. Eventual ruptures of the crust and generation of the crustal pieces result in a lava/debris flow-enhancing hazard.

Computer simulations play an important role in understanding the dynamics, thermal structures of lava flows, and hazard assessment (e.g. Costa and Macedonio, 2005, and references herein). Tsepelev et al. (2016) developed three-dimensional numerical models of a lava flow with rigid crustal fragments and performed a series of deterministic numerical simulations to analyse

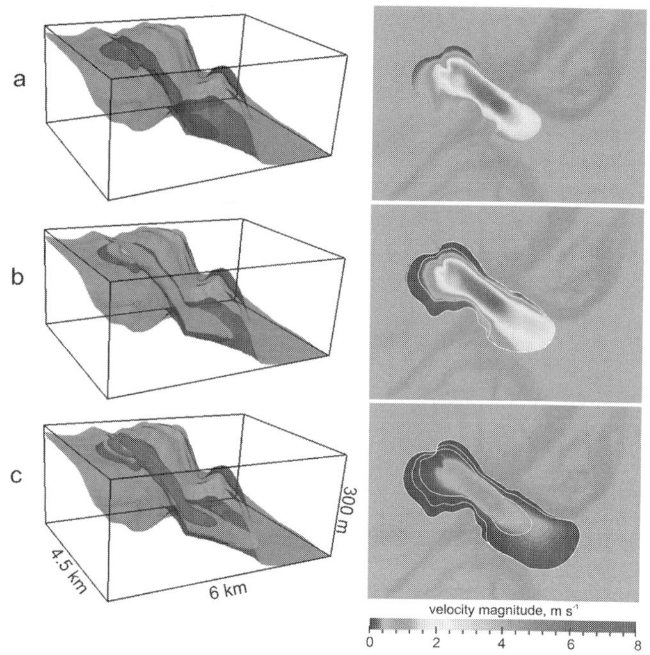

Figure 26.3 Model of the lava flow with compositional changes; snapshots at times 9000 s (a), 33000 s (b), and 63000 s (c). Blue-, green-, and red-coloured lavas are modelled fluids 1, 2, and 3, respectively (see Table 26.1 for their physical properties). Each time snapshot presents a 3-D view of the flow (left panels) and the velocity magnitude (right panels). The white curves in the right panels show the extent of lava flow at the end of the second and third episodes of lava effusion (after Tsepelev et al., 2016). For the color version, please refer to the plate section.

the spatial-temporal distribution of the crustal fragments depending on their size, a topographic slope, multiple sources of effusion, and the variation of physical properties of a lava. Particularly, a model describing lava flow was considered in the case when an eruption gives lava with changing composition with time. A solution to the model can provide useful insight into the evolution of the eruption and improve lava flow hazard assessment.

The evolution of lava flows during three episodes of the eruption is presented in Figure 26.3; the physical properties of the modelled lavas are presented in Table 26.1. Initially, lava (blue) flows down the slope for 2.5 hours. During the next episode of the eruption, a new portion of lava (green) with other physical properties starts to flow on the top of the blue lava for about 6.7 hours. As the viscosity of earlier lava becomes higher, it slowly moves down the slope together with a new portion of lava as a result of gravity and isostatic squeezing of older lava by hotter younger lava. The maximum rate of lava flow is estimated to be about

Table 26.1 *Physical Parameters of Modelled Lavas*

	Episode 1; 9000-s flow time	Episode 2; 24000-s flow time		Episode 3; 30000-s flow time		
Model parameters	Fluid 1	Fluid 1	Fluid 2	Fluid 1	Fluid 2	Fluid 3
Density, *kg* m^{-3}	2500	2500	2300	2500	2300	2300
Viscosity, *Pa s*	10^7	10^8	10^7	10^9	10^8	10^8
Rate, *m* s^{-1}	0.5	0	0.5	0	0	0.1

8 m s^{-1}, which is attained in the area of the model's steepest topography (the right panels in Figure 26.3 illustrates the distribution of velocity magnitudes). The third episode of the eruption starts from the same vent, but with an effusion rate that is lower than the initial rate by a factor of five. Therefore, the maximum fluid velocity drops to about 1.2 m s^{-1}. The newly injected lava flows on the top of two older and colder lavas pushing them down and further squeezing the underlying fluids. The computations are terminated after about 8.3 hours of lava flow. The scenario of lava flows described generates compositionally stratified lava flow containing several units. Because of lower viscosity green lava propagates faster and overcomes the initial blue lava. Thus, at any particular location the stratigraphy of the lava flow can depend on the eruption history in a complicated manner. For example, the front of the lava flow contains only the lava that was effused during the intermediate stage of eruption. Cooling of the flow can preserve such stratigraphy, if the upper unit has sufficient time for solidification due to low effusion rate.

An absolute temperature of the lava surface can be measured from space (e.g. using satellites, airplanes, or drones). Meanwhile the temperature and flow rate inside the lava flow remain unknown, and are difficult to determine, especially after solidification of lava's uppermost part and crust development. Korotkii et al. (2016) proposed a new quantitative approach to determine thermal and dynamic characteristics of a lava flow from thermal measurements at its surface, and showed that in the case of smoothed measured data the lava temperature and its flow velocity can be reconstructed with high accuracy, but noisy data degrade the accuracy of the solution. The proposed approach becomes important in studies of thick lava flows, and in assessments of the thermal budget of lava flows. Synthetic aperture radar satellite observations on lava thickness, volume, and flow extent, together with thermal measurement at the lava surface, facilitate research and data-driven modelling of lava flow.

Although lava flows are slow, allowing for evacuation of people from the areas of lava spreading, lava hazard is not negligible as hot lava kills vegetation; destroys houses, bridges, roads and other objects; and may melt snow/ice resulting in occasional flooding. The lava hazard can be mitigated if the flow patterns are known, and the complexity of the flow with debris is investigated to assist in diverting the flow (e.g. by placing barriers thath cut channels for the lava to follow) or in 'freezing' the lava flow with water (e.g. Fujita et al., 2009; Dietterich et al., 2015). Knowledge of lava flow paths and comprehensive lava hazard assessments can assist in disaster risk reduction efforts (Cutter et al., 2015).

26.5 Risk Analysis

Risk can be determined as the probability of harmful consequences or expected losses and damage due to a geophysical event resulting from interactions of hazards (*H*), vulnerability (*V*), and exposed values (*E*). Conventionally, risk (*R*) is expressed quantitatively by the convolution of these three parameters: $R = H \otimes V \otimes E$ (e.g. Kantorovich et al., 1973). Assuming that a natural hazard event happened, exposure and vulnerability are the key determinants of disaster risk and the main drivers of disaster losses. Changes in any of these alter the risk calculus by increasing or reducing the impacts of disaster risk on affected communities, regions, or countries. Exposure is the location of people, assets, and infrastructure in hazard-prone areas that could be affected, while vulnerability is the degree of susceptibility or sensitivity of people, assets, and infrastructure to suffer damages (UNISDR, 2013). There is temporal and spatial variability in exposure and vulnerability patterns from local to global scales. These geographic patterns are unevenly distributed across the globe and lead to the disproportionate impacts of disasters, especially in disadvantaged communities, regions, or countries. Also vulnerability depends on an individual's susceptibility: namely, populations may be less vulnerable or more resilient in the face of disasters because of former exposure to disasters that helped them to gain strength, or cultural traditions that improve collective action (Ismail-Zadeh and Cutter, 2015).

As geohazards cannot be significantly reduced and exposure increases with economic development, a major element in disaster risk management is the reduction of vulnerability. High vulnerability is associated with environmental degradation, rapid and unplanned urbanization in hazardous areas, failures of governance to reduce vulnerability, and the scarcity of livelihood options for the poor. Countries more effectively manage disaster risk if they include considerations of disaster risk in national development and sector plans, translating these plans into actions targeting vulnerable areas and groups (IPCC, 2012).

A rapid growth of population, intensive civil and industrial building, land and water instabilities, and lack of public awareness regarding hazards and risks contribute to the increase of vulnerability of big cities. For example, Babayev et al. (2010) assessed earthquake risk in Baku (Azerbaijan) on the basis of convolution of deterministic (scenario-based) seismic hazards, vulnerability (due to building construction fragility, population patterns, the gross domestic product per capita, and landslide occurrence), and exposure of infrastructure and critical facilities. Figure 26.4 presents the integrated earthquake risk assessment for three earthquake scenarios. One of the remarkable results of this assessment is the fact that the western-central part of the city is exposed to the highest risk independent on the earthquake scenarios: i.e. seismic risk does not depend much on the magnitude and the epicentral distance of the earthquakes employed for the assessment. This is because of physical vulnerabilities characterized by low quality of building construction, high density of population, and significant exposed values in this part of the city.

Many parts of the world are still vulnerable to earthquakes despite the progress made in earthquake research and engineering for the past several decades. For example, a deterministic earthquake risk assessment performed for Portugal illustrates that the highest risk is associated with building damage rather than with ground shaking or socio-economic vulnerability to disaster (Baker, 2013).

For the past several decades Iran has experienced devastating earthquakes, which caused high losses stemming from the vulnerability of the built environment as well as rapid population growth and seismically incompatible urban development. The 1990 Manjil M7.2 earthquake in Iran, which resulted in the loss of about fourteen thousand people and heavy damage of infrastructure, significantly impacted the socio-economic life in the country (Ghafory-Ashtiany, 2014). Fourteen years later, the 2003 Bam M6.6 earthquake in Iran killed at least twenty-six thousand and injured thirty thousand people mainly because of destruction and significant damage of poor-quality building constructions (Ghafory-Ashtiany, and Hosseini, 2008). Despite implementation of the governmental strategy for earthquake risk reduction (Ghafory-Ashtiany, 2014), existing physical vulnerability in Iran is a source for potential losses due to an earthquake occurrence.

Compared to Iran, Japan is a bright example of a successful and effective implementation of the national earthquake risk reduction policy. The 1995 Kobe M6.9 earthquake disaster caused at least six thousand deaths, many thousands of injuries, and USD 100 billion overall losses (Munich Re, 2015). Twenty-one years later, several people were injured and insignificant damage was reported in the aftermath of the 2016 Tottori M6.2 earthquake (Reuters, 2016). It provides evidence that substantial investment in risk reduction, particularly in building construction and reinforcement, pays off in significant decline in earthquake impact and losses.

With economic and technological development more and more structural elements (and sometimes the entire infrastructure) become exposed to risks due to geohazards. Risk assessment allows elaborating strategic countermeasure plans for the disaster risk mitigation. An estimation of risks may facilitate a proper choice in a wide variety of safety measures, ranging, in the case of earthquakes, from building codes and insurance to establishment of rescue-and-relief resources. Most of the practical problems require estimating risk for a territory as a whole, and within this territory separately for the objects of each type: areas, lifelines, sites of vulnerable constructions, etc. The choice of the territory and the objects is determined by the jurisdiction and responsibility of a decision maker. Each specific representation of risk is derived from the models of geohazards and the territorial distribution of population, property, vulnerable objects, and potential damage. Difficulties in decision making in the field of risk assessment are related to uncertainties in data (especially those related to social and physical vulnerabilities and exposure), imperfect methods for hazard assessment, and limitations in using mathematical tools for carrying out the historical analysis and forecasting.

Meanwhile risk assessment efforts assist in optimizing preventive mitigation measures to reduce losses from catastrophic disasters. "If about 5 to 10% of the funds, necessary for recovery and rehabilitation after a disaster, would be spent to mitigate an anticipated earthquake, it could in effect save lives, constructions,

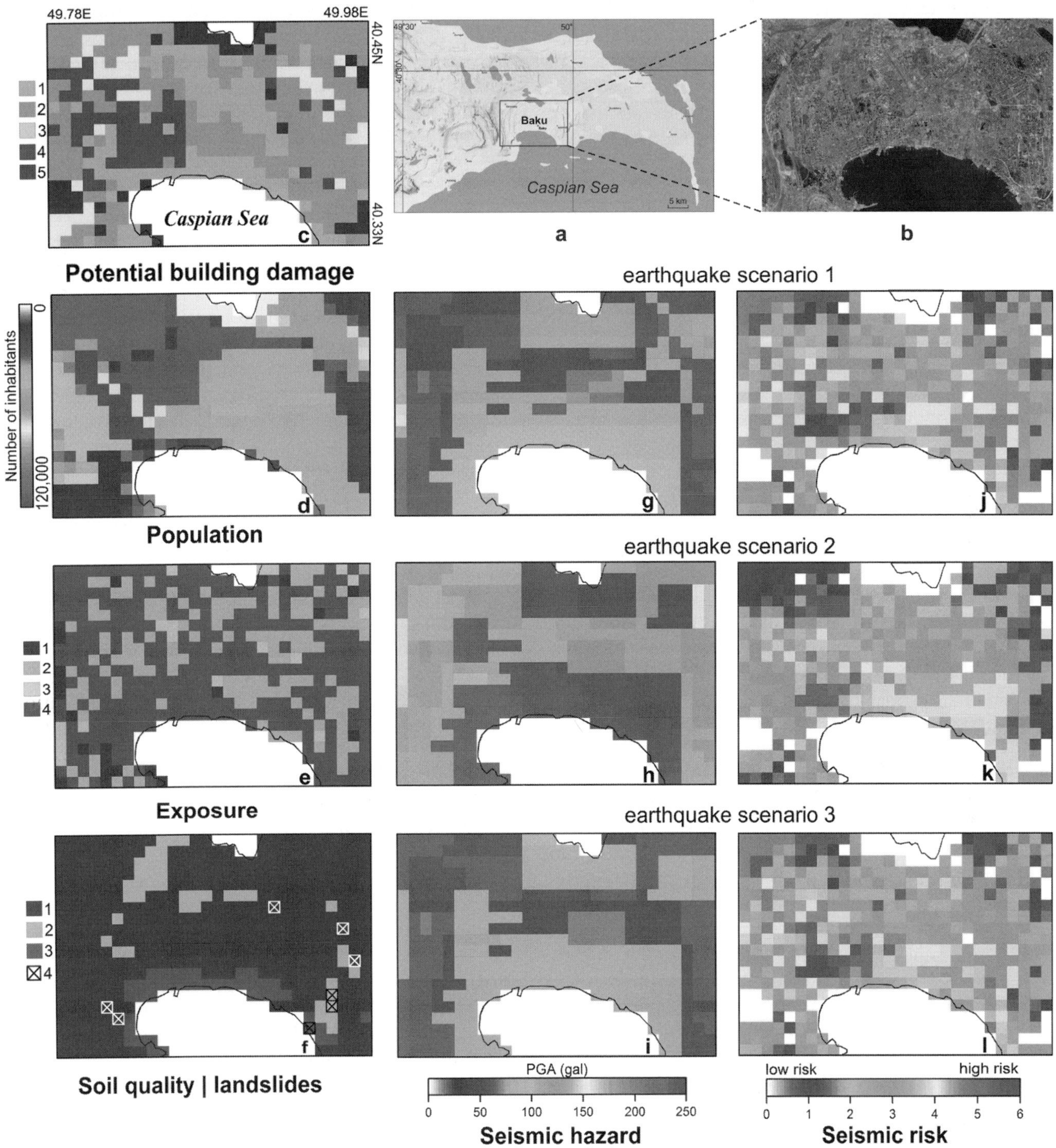

Figure 26.4 Earthquake risk for Baku, the capital city of Azerbaijan. (a) Topography of the Absheron peninsula and the location of Baku (the studied region is marked by a solid rectangle). (b) Satellite image of the studied region. (c) Distribution of building constructions: high (type 1), good (type 2), good to poor (type 3), and poor (type 4) quality. Construction-free areas are marked by 5. (d) Distribution of urban population. (e) Distribution of exposure. 1: low-value objects (park, lake, and other natural objects); 2: moderate-value objects (main road, square, and their intersections); 3: high-value objects (culture-valued objects); and 4: very-high-value objects (residential objects, hospital, school, commercial, governmental and state buildings, nation-valued object, important communication system). (f) Soil types. 1: Quaternary sand, clay, intrusions of limestone, and sandy clay; Upper Tertiary limestone and sandstone (minor amplification of shaking). 2: Quaternary sand, gravel-pebble, and limestone with clay intrusions (moderate amplification of shaking). 3: water-saturated sand, clay, and rubble (significant amplification of shaking). 4: landslide area. Computed surface peak ground acceleration (seismic hazard, g-i) and seismic risk (j-l) for three earthquake scenarios. Coastlines are marked by black curves (modified after Babayev et al., 2010). For the color version, please refer to the plate section.

and other resources" (Ismail-Zadeh and Takeuchi, 2007a). A new approach should be developed to mitigate disasters by investing in preventive measures. We should be prepared for extreme events and subsequent environmental, economic, and financial events associated with them; and recover easily from a disaster. We need stronger efforts to develop a society resilient to geohazards and related disasters.

26.6 Geohazard Events That Turned into Disasters

Not every geohazard event (an earthquake or landslide or volcano) results in loss of lives, properties, and infrastructure, only those occurring in close vicinity to vulnerable places. For example, a strong earthquake in the plate interior may lead to disaster if it strikes near a town, and it will become a disaster in the case of high physical and social vulnerability of the town to earthquakes. As mentioned in Section 26.5, vulnerability is a major driver of disasters, and the magnitude of a geohazard event is not very informative with respect to the disaster's severity. The following example is illustrative. The energy released by the 2010 Chile M8.8 earthquake was by a factor of about five hundred higher than that released by the 2010 Haiti M7.0 earthquake. However, the death toll showed the inverse proportionality: several hundred people lost their lives (primarily because of the triggered tsunami) in the case of the Chile earthquake versus several hundred thousand in the case of the Haiti earthquake resulting in a humanitarian catastrophe. Disasters happen mainly because of the "unwillingness of some local authorities to invest in resistant construction due to various reasons including irresponsibility, ignorance, corruption, the perceived requirement to balance the need for costs versus the increased costs of implementation, local politics, funding availability and other urgent and more politically competitive needs" (Ismail-Zadeh et al., 2017).

In May 2008, an earthquake of magnitude 7.9 occurred in Wenchuan, ichuan province, Southwest China. It became the most devastating earthquake in China of the twenty-first century with about 70,000 dead, more than 370,000 injured, and 18,000 missing (Wu and Ma, 2014). Although a comprehensive forensic investigation of the Wenchuan earthquake is still required to understand why the earthquake resulted in the disaster, an important lesson learned was "the necessity to strengthen the efforts to cope with earthquake disaster and to be prepared for the next big earthquake" (Wu and Ma, 2014).

The 2011 Tohoku M9.0 earthquake was the largest earthquake in the history of Japan, although the 1923 Kanto earthquake is considered to be the worst disaster, because it caused 105,000 casualties and destroyed the Tokyo metropolitan area (Moroi and Takemura, 2004). According to the Japanese National Policy Agency, the Tohoku earthquake and tsunami led to 15,894 deaths, 6,152 injuries, and 2,562 missing people. The earthquake and tsunami caused damage to Fukushima Dai-ichi nuclear power station, resulting in meltdown of the reactor, hydrogen explosion, and release of radioactive materials. Long-term forecasts of large earthquakes failed to foresee this giant earthquake, because it was based on a characteristic earthquake model and past historical data of a few centuries only. Similar tsunami disasters occurred in the past (e.g. tsunamis generated by the 1896 Sanriku earthquake caused 22,000 casualties; Satake, 2014), but the experience was not fully utilized to reduce the 2011 Great East Japan disaster. The lessons learned from the disaster include the importance of geophysical observations, improvement of long-term forecasting of earthquakes, effective tsunami warning, assessment for future tsunami hazard, and preparedness and education for infrequent hazards (Satake, 2014).

Societal impacts of volcanic eruptions (e.g. damage, disruptions, severe health problems) are associated with ash fallouts, gases, hot ash clouds and related hazard to aviation, lava flows, pyroclastic currents, and lahars. Disaster risks in areas of high lava flow and lahars hazard can be significantly reduced by reducing the existing and potential vulnerability; e.g. new infrastructure should not be developed in the areas, and people should move to safe places in advance. Governments can help. For example, in November 2014 the U.S. president declared a slowly moving lava flow from the Pu'u O'o vent of the Kilauea volcano in Hawaii to be a major disaster. Although the event was not a disaster in terms of loss of lives or property (no homes were destroyed and no injuries reported, unlike the 2018 eruption that caused major property damage), the declaration helped to release federal money to help in protection of local communities from the lava flow (Reuters, 2014). Lava flow or pyroclastic current can lead to evacuation of people or even a town under risk. The eruption of Soufrière Hills, Montserrat, in July 1995 threatened the capital city, Plymouth, and much of island's infrastructure. Plymouth was evacuated in April 1996 and was destroyed in the summer of 1997 by a pyroclastic current. About ten thousand people left the island during the eruption (Donovan and Oppenheinmer, 2014).

A volcanic super-eruption would have significant impacts on air quality across a large region. Agriculture and farming would be severely affected within the zone receiving ash fallout, compromising food security; for example, in the case of past eruptions of the

Yellowstone volcano, much of North America received several centimetres of ash fallout. "Ash would disrupt all forms of infrastructure, leaving hospitals without power and large areas without potable water due to contamination of water supplies by sulphate and halogens. This could lead to famine and disease at the continental scale, which would be difficult to mitigate because of the breakdown of transport links. Aviation could be grounded over much of the globe" (Donovan and Oppenheimer, 2014).

Development of human settlements in mountainous areas, urbanization processes, and particularly lack of adequate territorial planning and management, when combined with earthquakes and/or significant rainfall events, generate scenarios for landslide disasters to occur (Alcántara-Ayala, 2014). Historical cases of landslide disasters have shown their future potential impact given that patterns of population growth continue to rise dramatically in highly (land) unstable areas, vulnerability of population is increasing, and there is still a lack of landslide disaster risk management.

The most destructive earthquake-triggered landslide event took place in Tajikistan in July 1949. Hundreds of landslides constituting rock-slope failures and flow slides in loess were triggered by the Khait M7.4 earthquake near the Tien Shan mountains. A tragic movement was a rockslide transformed into a very rapid flow that travelled about 7.5 km with an estimated average velocity of about thirty metres per second over a vertical distance of 1421 m, in which some seven thousand to twelve thousand people died, according to different sources (Evans et al., 2009; Alcántara-Ayala, 2014). A crucial landslide disaster linked to human activities took place in Italy in 1963. After a series of instability episodes derived from the filling of the Vaiont reservoir, a mass of about 270 million m^3 was detached into the water and produced a tsunami-type wave several hundred metres high that overtopped the reservoir and swept away the towns of Longarone, Pirago, Rivalta, Villanova, and Faè, killing about two thousand people (e.g. Semenza and Ghiritti, 2000; Kilburn and Petley, 2003). Today, one of the serious concerns related to lake-outburst floods is that Lake Sarez in Pamir mountain, Tajikistan, formed as a natural dam as a result of the 1911 earthquake-triggered rockslide, which blocked the Murgab River. The maximum depth of the lake is 550 m, and its volume is about 17 billion m^3 (Schuster and Alford, 2004). In the case of dam failure (e.g. due to seismic shaking or overtopping caused by a landslide or other natural processes), more than a million people downstream along the Amu Daria river would be exposed to severe flooding (Stone, 2009).

Review of several geohazard events that triggered (or may trigger) disasters shows that a reduction of disaster risks caused by natural events requires an integrated approach coupling natural science, engineering, and social science. The development of a transdisciplinary, team-based approach is one key way of ensuring that the respective contributions of all disciplines and stakeholders are incorporated in a comprehensive risk management strategy (Ismail-Zadeh et al., 2017).

26.7 Disaster Risk Reduction and Implications for Sustainability

A sustainable society is one that progresses in its development while equitably meeting its present needs and not compromising the ability of future generations to develop and meet their own needs (UNGA, 1987). The challenges posed by disasters due to natural and/or human-induced environmental hazard events can result in negative impacts for development. Disasters can be complexly exacerbated by global poverty and can have very detrimental impacts on the sustainable development and on efforts to eradicate poverty. Effective and comprehensive knowledge of disaster risks can enable greater resilience to such stresses and enable development opportunities (McBean, 2014). There is need for focus on the "intrinsic relationship between disaster reduction, sustainable development and poverty eradication" (UNISDR, 2005), and this is a grand challenge of integrated research on disaster risk.

Since the 1980s the impacts of disasters have risen rapidly, affecting developed and developing countries and almost all sectors of the economy at local, national, and regional levels. Several hundred million people are affected annually, and losses reached more than USD 400 billion in 2011 mainly because of the 2011 Great East Japan earthquake and tsunami (MunichRe, 2014). At the World Conference on Disaster Risk Reduction in Sendai (Japan) in 2015, state representatives adopted the major agreement of the post-2015 development agenda, the Sendai Framework for Disaster Risk Reduction 2015–2030 – a new international framework for disaster risk reduction (Sendai Framework, 2015). The Sendai Framework aims to use targets and goals to bring about action. It calls on governments (i) to understand disaster risk; (ii) to strengthen risk governance to manage risks across all sectors; (iii) to invest in risk-reduction measures that promote resilience; and (iv) to enhance disaster preparedness and responses so that nations "build back better" in recovery.

At the gathering of high representatives of states in the UN Headquarters in New York, a set of global

Sustainable Development Goals (SDGs) to end poverty, protect the Earth's future, and ensure prosperity for all as a new sustainable development agenda was adopted. Several of the goals are directly related to disaster risk reduction: namely, by 2030, to (i) "build the resilience of ... those in vulnerable situations and reduce their exposure and vulnerability to climate-related extreme events and other economic, social and environmental shocks and disasters"; (ii) "strengthen capacity for adaptation to climate change, extreme weather, drought, flooding and other disasters"; (iii) "significantly reduce the number of deaths and the number of people affected and substantially decrease the direct economic losses relative to global gross domestic product caused by disasters, including water-related disasters, with a focus on protecting the poor and people in vulnerable situations"; (iv) "develop and implement, in line with the Sendai Framework for Disaster Risk Reduction 2015–2030, holistic disaster risk management at all levels", and (v) "strengthen resilience and adaptive capacity to climate-related hazards and natural disasters in all countries" (UN, 2015).

How can disaster risk research, including understanding of geohazards, exposure, and vulnerability as well as disasters as such, be leveraged to support the Sendai Framework and to reach the global SDGs? Although science is well recognized in the priorities of the Sendai Framework and the other political documents, there is a need to enhance the scientific basis for the policy agenda (Ismail-Zadeh and Cutter, 2015) and to incorporate scientific evidence in policy making, although it is "neither straightforward nor guaranteed" (Gluckman, 2016). To undertake this, a science-based approach to understanding and assessment of disaster risks at all levels should be developed. Risk assessments would provide a clear and unambiguous scientific view on the current state of knowledge in disaster risk, the potential socio-economic impacts of natural hazards, and the ways to reduce significant human and economic losses. Such a scientific view should present various knowledge inputs including local knowledge and business and various other group inputs (Ismail-Zadeh et al., 2017). Periodic disaster risk assessments would cover understanding natural hazards, exposure, and the vulnerability associated with disasters; capability of integrated monitoring and predictive systems to disseminate timely and accurate information needed for policy and decision making; methodologies and approaches for reducing vulnerability and increasing resilience of societies; and working with actors at various levels to imjprove understanding of the societal risk reduction (through the prevention, mitigation, and preparations for the increasing impact of natural events). Such scientific assessments would contribute to the significant enhancement of our knowledge of disaster risk at global, regional, and local levels and the awareness of those living with risk. The assessment should be undertaken by a high-level, transdisciplinary body of experts appointed by national governments together with international and intergovernmental scientific organizations dealing with disaster risks (Cutter et al., 2015).

According to the Budapest Manifesto (Beer and Ismail-Zadeh, 2003), scientists can contribute to decision making through a risk management framework with which to examine technical and social issues related to sustainability, namely: (i) anticipating disaster risks caused by natural events through wide-spread consultation; (ii) determining concerns by using risk assessment techniques for various scenarios; (iii) identifying the consequences by systematically cataloguing hazards; (iv) undertaking calculations with appropriate models; (v) evaluating the certainties, uncertainties, and probabilities involved in the calculations of the vulnerability and of the exposure; (vi) comparing with criteria to assess the need for further action; (vii) determining and acting on options to control, mitigate, and adapt to the risk; (viii) communicating the results to those who need to know; (ix) promoting and guiding monitoring systems to collect, assimilate and archive data relevant to the determination of sustainability and risk, now and in the future; (x) integrating the knowledge and understanding from all relevant disciplines to provide society with the tools to review the sustainability and the risks of proposed policies and plans. Though rational scientific methods hold the promise of an improved science of risk and sustainability, it must be remembered that the priorities for analyses are likely to be heavily influenced by the public and political agenda of the day. This means that implementation of risk management to achieve sustainability can be achieved only through an interaction of theory and praxis (Beer and Ismail-Zadeh, 2003).

Disaster risk reduction is essential in achieving sustainable development goals and in building resilience to natural hazards. This is because sustainability and disaster risk are complementary to the extent that policies and plans are sought to increase sustainability and reduce risk. Disasters undermine poverty eradication, link to unsustainable growth, magnify impacts for small developing countries, impact significantly sustainability of cities, become quite often concatenated, and affect (on average) disproportionately women, children, old people, and disabled people (IRDR, 2013). Disasters present a range of indirect and secondary impacts in addition to the direct losses. Individuals may suffer long-term disability, psychological harm, degraded living circumstances, interrupted education, increased disease occurrence, loss of employment and relocation, reduced nutrition, and stunting. Businesses and investment may fail and sectors may not reach their production and development targets. Government finances are often severely disrupted (Ismail-Zadeh and Cutter, 2015). Large monetary sums are expended on international emergency assistance after disasters. This is in effect a risk transfer mechanism, as it helps in smoothing the economic impacts on the affected communities, albeit at a very basic level. However, more timely interventions and sustained multi-year support to geohazards and risk research can pay handsomely (e.g. Ismail-Zadeh and Takeuchi, 2007a).

26.8 Conclusion

Geohazards have been recognized as a challenge a long time, and the scientific community has concentrated its efforts on solving the challenging problem with significant progress achieved. Progress in geohazard research and assessments is based on in-depth analysis of individual hazards. In seismic hazard analysis, recent advances are associated with a neo-deterministic approach to hazard assessment (Panza et al., 2010) and data-enhanced probabilistic SHA (Sokolov and Ismail-Zadeh, 2015). Particularly, tectonically realistic earthquake simulators help to generate seismicity for a significant duration of time and to employ large synthetic seismic events for hazard assessment. Similarly, quantifying lava flow hazards integrating field observations, satellite data, and advanced modelling of lava flows (e.g. Tsepelev et al., 2016; Korotkii et al., 2016) allows for tracing different processes of lava extent and forecasting the areas that could be threatened by lava. New scientific methods and approaches can enhance understanding of natural extreme events and vulnerability assessments at all levels including the gathering of a wide range of measures for increasing the resilience of society to geohazards.

With time, meanwhile, it was recognized that there are much more challenging problems requiring co-produced research on disaster risks that can enable understanding of the roots of potential disasters. A means of integration and co-production in disaster risk research is through the maturation of hazard and disaster science and through transdisciplinary studies aiming at in-depth investigations using a system analysis and at recommendations for actions to reduce risks and to improve resilience of society (e.g., Ismail-Zadeh et al., 2017). Such transdisciplinary research offers practice- and policy-oriented knowledge to reduce disaster risk and disasters (Hirsch Hadorn, 2008). Disaster losses due to geohazards will continue to impose a great challenge to sustainability unless challenging problems in geohazards research and in disaster science are resolved and implemented in social and political actions.

Acknowledgements

The author acknowledges support from the German Science Foundation (DFG grant IS-203/5) on earthquake hazard modelling and from the Russian Science Foundation (RSF grant 14–17-00520) on lava flow hazard modelling. Also the author thanks Oleg Melnik and Vladimir Sokolov for fruitful discussions, and two reviewers for their comments on the initial manuscript.

References

Adey, P., Anderson, B. and Guerrero, L. L. (2011). An ash cloud, airspace and environmental threat. *Transactions of the Institute of British Geographers*, **36**, 338–343.

Alcántara-Ayala, I. (2014). The spatial-temporal dimensions of landslide disasters. In: *Extreme Natural Hazards, Disaster Risks and Societal Implications*, eds. A. Ismail-Zadeh, J. U. Fucugauchi, A. Kijko, K. Takeuchi, and I. Zaliapin. Cambridge University Press, Cambridge, pp. 61–76.

Babayev, G., Ismail-Zadeh, A. and Le Mouël, J.-L. (2010). Scenario-based earthquake hazard and risk assessment for Baku (Azerbaijan). *Natural Hazard and Earth System Sciences*, **10**, 2697–2712.

Baker, J. (2013) Seismology: Quake catcher. *Nature*, **498**, 290–292.

Beer, T. and Ismail-Zadeh, A., eds. (2003). *Risk Science and Sustainability*. Dordrecht: Kluwer Academic.

Braile, L. W., Hinze, W. J., Keller, G. R., Lidiak, E. G. and Sexton, J. L. (1986). Tectonic development of the New Madrid Rift Complex, Mississippi Embayment, North America. *Tectonophysics*, **131**, 1–21.

Cornell, C. A. (1968). Engineering seismic risk analysis. *Bulletin of the Seismological Society of America*, **58**, 1583–1606.

Costa, A. and Macedonio, G. (2005). Numerical simulation of lava flows based on depth-averaged equations. *Geophysical Research Letters*, **32**, L05304, DOI: 10.1029/2004GL021817.

Cutter, S, Ismail-Zadeh, A., Alcántara-Ayala, I. et al. (2015). Pool knowledge to stem losses from disasters. *Nature*, **522**, 277–279.

Dietterich, H. R., Cashman, K. V., Rust, A. C. and Lev, E. (2015). Diverting lava flows in the lab. *Nature Geoscience*, **8**, 494–496.

Dilley, M., Chen, R. S., Deichmann, W., Lerner-Lam, A. L. and Arnold, M. (2005). *Natural Disaster Hotspots: A Global Risk Analysis*. Washington DC: The World Bank.

Donovan, A. R. and Oppenheimer, C. (2011). The 2010 Eyjafjallajökull eruption and the reconstruction of geography, *The Geographical Journal*, **177**, 4–11.

Donovan, A. R. and Oppenheimer, C. (2014). Extreme volcanism: disaster risks and societal implications. In: *Extreme Natural Hazards, Disaster Risks and Societal Implications*, eds. A. Ismail-Zadeh, J. U. Fucugauchi, A. Kijko, K. Takeuchi, and I. Zaliapin. Cambridge University Press, Cambridge, pp. 29–46.

Doocy, S., Daniels, A., Dooling, S., and Gorokhovich, Y. (2013). The human impact of volcanoes: a historical review of events 1900–2009 and systematic literature review. *PLOS Currents Disasters*, **1**, DOI: 10.1371/currents.dis.841859091a706efebf8a30f4ed7a1901.

Evans, S., Roberts, N., Ischuk, A., Delaney, K., Morozova, G. and Tutubalina, O. (2009). Landslides triggered by the 1949 Khait Earthquake, Tajikistan, and associated loss of life. *Engineering Geology*, **109**, 195–212.

Fujita, E., Hidaka, M., Goto, A. and Umino, S. (2009). Simulations of measures to control lava flows. *Bulletin of Volcanology*, **71**, 401–408.

Gabrielov, A. M., Levshina, T. A. and Rotwain, I. M. (1990). Block model of earthquake sequence. *Physics of the Earth and Planetary Interiors*, **61**, 18–28.

Giardini, D., Grünthal, G., Shedlock, K. M. and Zhang, P. (1999). The GSHAP Global Seismic Hazard Map. *Annali di Geofisica*, **42**, 1225–1228.

Ghafory-Ashtiany, M. (2014). Earthquake risk and risk reduction capacity building in Iran. In: *Extreme Natural Hazards, Disaster Risks and Societal Implications*, eds. A. Ismail-Zadeh, J. U. Fucugauchi, A. Kijko, K. Takeuchi and I. Zaliapin. Cambridge University Press, Cambridge, pp. 267–278.

Ghafory-Ashtiany, M. and Hosseini, M. (2008). Post Bam recovery and reconstruction. *Natural Hazard*, **44**, 229–241.

Gluckman, P. (2016). The science–policy interface. *Science*, **353**, 969, DOI: 10.1126/science.aai8837.

Green, H. W. II and Burnley, P. C. (1989). A new self-organizing mechanism for deep-focus earthquakes. *Nature*, **341**, 733–737.

Griffiths, R. W. (2000). The dynamics of lava flows. *Annual Review of Fluid Mechanics*, **32**, 477–518.

Griggs, D. T. and Baker, D. W. (1969). The origin of deep-focus earthquakes. In *Properties of Matter Under Unusual Conditions*, eds. H. Mark and S. Fernbach. New York: Wiley, pp. 23–42.

Gupta, H. K., Purnachandra Rao, N., Rastogi, B. K., and Sarkar, D. (2001). The deadliest intraplate earthquake. *Science*, **291**, 2101–2102.

Gutenberg, B. and Richter, C. F. (1944). Frequency of earthquakes in California. *Bulletin of the Seismological Society of America*, **34**, 185–188.

Hirsch Hadorn, G., Hoffmann-Riem, H., Biber-Klemm, S. et al., eds. (2008). *Handbook of Transdisciplinary Research*. Dordrecht: Springer.

Huybers, P. and Langmuir, C. (2009). Feedback between deglaciation, volcanism, and atmospheric CO_2. *Earth and Planetary Science Letters*, **286**, 479–491.

IPCC (2012). *Managing the Risks of Extreme Events and Disasters to Advance Climate Change Adaptation*. A Special Report of Working Groups I and II of the Intergovernmental Panel on Climate Change [Field, C. B., Barros, V., Stocker, T. F. et al., eds.]. Cambridge, UK, and New York, NY, USA: Cambridge University Press.

IRDR (2013). *Issue Brief: Disaster Risk Reduction and Sustainable Development*. Beijing: Integrated Research on Disaster Risk. Available at: www.irdrinternational.org/wp-content/uploads/2013/12/IRDR-ICSU-Brief-DRR-SD.pdf (retrieved on 05.08.2016).

Ismail-Zadeh, A. T., Panza, G. F. and Naimark, B. M. (2000). Stress in the descending relic slab beneath Vrancea, Romania. *Pure and Applied Geophysics*, **157**, 111–130.

Ismail-Zadeh, A. and Takeuchi, K. (2007a). Preventive disaster management of extreme natural events. *Natural Hazards*, **42**, 459–467.

Ismail-Zadeh, A. T., Le Mouël, J. L., Soloviev, A., Tapponnier, P. and Vorobieva, I. (2007b). Numerical modeling of crustal block-and-fault dynamics, earthquakes and slip rates in the Tibet-Himalayan region. *Earth and Planetary Science Letters*, **258**, 465–485.

Ismail-Zadeh, A., Schubert, G., Tsepelev, I., and Korotkii, A. (2008). Thermal evolution and geometry of the descending lithosphere beneath the SE-Carpathians: An insight from the past. *Earth and Planetary Science Letters*, **273**, 68–79.

Ismail-Zadeh, A., Le Mouël, J.-L. and Soloviev, A. (2012a). Modeling of extreme seismic events. In *Extreme Events and Natural Hazards: The Complexity Perspective*, eds. S. A. Sharma, A. Bunde, V. P. Dimri and D. N. Baker, Geophysical Monograph 196. Washington, DC: American Geophysical Union, pp. 75–97.

Ismail-Zadeh, A., Matenco, L., Radulian, M., Cloetingh, S. and Panza, G. (2012b). Geodynamic and intermediate-depth seismicity in Vrancea (the south-eastern Carpathians): Current state-of-the-art. *Tectonophysics*, **530–531**, 50–79.

Ismail-Zadeh, A. (2014). Extreme seismic events: from basic science to disaster risk mitigation. In *Extreme Natural Events, Disaster Risks and Societal Implications*, eds. A. Ismail-Zadeh J. Fucugauchi, A. Kijko, et al. Cambridge: Cambridge University Press, pp. 47–60.

Ismail-Zadeh, A., and Cutter, S. eds. (2015). *Disaster Risks Research and Assessment to Promote Risk Reduction and Management*. Paris: ICSU-ISSC.

Ismail-Zadeh, A., Cutter, S. L., Takeuchi, K., and Paton, D. (2017). Forging a paradigm shift in disaster science, *Natural Hazards*, **86**(2), 969–988.

Kantorovich, L., Keilis-Borok, V. I. and Molchan, G. (1973). Seismic risk and principles of seismic zoning. In *Computational and Statistical Methods for Interpretation of Seismic Data*, ed. V. I. Keilis-Borok. Moscow: Nauka, pp. 3–20 (in Russian).

Kilburn, C. R. J. and Petley, D. N. (2003). Forecasting giant, catastrophic slope collapse: lessons from Vajont, Northern Italy. *Geomorphology*, **54**, 21–32.

Korotkii, A., Kovtunov, D., Ismail-Zadeh, A., Tsepelev, I. and Melnik, O. (2016). Quantitative reconstruction of thermal and dynamic characteristics of lava from surface thermal measurements. *Geophysical Journal International*, **205**, 1767–1779.

Latter, J. H. (1981). Tsunamis of volcanic origin: summary of causes, with particular reference to Krakatoa, 1883. *Bulletin of Volcanology*, **44**, 467–490.

Malamud, B. D. and Turcotte, D. L. (1999). Self-organized criticality applied to natural hazards. *Natural Hazards*, **20**, 93–116.

McBean, G. (2014). The grand challenges of integrated research on disaster risk. In *Extreme Natural Hazards, Disaster Risks and Societal Implications*, eds. A. Ismail-Zadeh, J. U. Fucugauchi, A. Kijko et al. Cambridge: Cambridge University Press, pp. 15–25.

Moroi, T. and Takemura, M. (2004). Mortality estimation by causes of death due to the 1923 Kanto earthquake (in Japanese with English abstract). *Journal of the Japan Association of Earthquake Engineers*, **4**(4), 21–45.

Munich Re (2014). Topics Geo - Annual review of natural catastrophes 2013. Analyses, assessments, positions. Munich: MunichRe (retrieved on 05.08.2016).

Munich Re (2015). Loss events worldwide 1980 – 2014: 10 costliest events ordered by overall losses. MunichRe, NatCatSERVICE. Avaialble at: www.munichre.com/site/corporate/get/documents_E-567437233/mr/assetpool.shared/Documents/5_Touch/_NatCatService/Significant-Natural-Catastrophes/2014/10-costliest-events-ordered-by-overall-losses.pdf (retrieved on 12.02.2017).

Oppenheimer, C. (2003). Climatic, environmental and human consequences of the largest known historic eruption: Tambora volcano (Indonesia) 1815. *Progress in Physical Geography*, **27**, 230–259.

Panza, G. F., Irikura, K., Kouteva, M., Peresan, A., Wang, Z. and Saragoni, R. (2010). Advanced seismic hazard assessment. *Pure and Applied Geophysics*, **168**, DOI: 10.1007/s00024–010-0179–9.

Reuters (2014). Obama declares Hawaiian lava flow to be major disaster. Reuters, November 4, 2014. Available at: www.reuters.com/article/us-usa-lava-hawaii-idUSKBN0IO07E20141104 (retrieved on 12.02.2017).

Reuters (2016). Magnitude 6.6 quake jolts western Japan, no tsunami warning or major damage. Reuters, October 21, 2016. Available at: www.reuters.com/article/us-earthquake-japan-idUSKCN12L0F3 (retrieved on 12.02.2017).

Satake, K. (2014). The 2011 Tohoku, Japan, earthquake and tsunami. In: *Extreme Natural Hazards, Disaster Risks and Societal Implications*, eds. A. Ismail-Zadeh, J. U. Fucugauchi, A. Kijko, K. Takeuchi, and I. Zaliapin. Cambridge University Press, Cambridge, pp. 310–321.

Schuster, R. L., and Alford, D. (2004). Usoi landslide dam and lake Sarez, Pamir Mountains, Tajikistan. *Environmental & Engineering Geoscience* **10**, 151–168.

Semenza, E. and Ghirotti, M. (2000). History of the 1963 Vaiont slide: the importance of geological factors. *Bulletin of Engineering Geology and the Environment*, **59**, 87–97.

Sendai Framework (2015). Sendai Framework for Disaster Risk Reduction 2015–2030. Available at: www.unisdr.org/we/inform/publications/43291 (retrieved on 05.08.2016)

Simkin, T. and Siebert, L. (1994). *Volcanoes of the World: A Regional Directory, Gazetteer, and Chronology of Volcanism during the Last 10,000 Years*. Tucson, AZ: Geoscience Press.

Sokolov, V. and Ismail-Zadeh, A. (2015). Seismic hazard from instrumentally recorded, historical and simulated earthquakes: Application to the Tibet-Himalayan region. *Tectonophysics*, **657**, 187–204.

Sokolov, V. and Ismail-Zadeh, A. (2016). On the use of multiple-site estimations in probabilistic seismic hazard assessment. *Bulletin of the Seismological Society of America*, **106**(5), DOI: 10.1785/01201503062016.

Soloviev, A. A. and Ismail-Zadeh, A. T. (2003). Models of dynamics of block-and-fault systems. In *Nonlinear Dynamics of the Lithosphere and Earthquake Prediction*, eds. V. I. Keilis-Borok and A. A Soloviev. Heidelberg: Springer, pp. 69–138.

Stein, S., Geller, R. and Liu, M. (2011). Bad assumptions or bad luck: why earthquake hazard maps need objective testing. *Seismological Research Letters*, **82**, 623–626.

Stone, R. (2009). Peril in the Pamirs. *Science*, **326**, 1614–1617.

Tsepelev, I., Ismail-Zadeh, A., Melnik, O. and Korotkii, A. (2016). Numerical modelling of fluid flow with rafts: An application to lava flows. *Journal of Geodynamics*, **97**, 31–41.

Turcotte, D. L. and Schubert, G. (2014). *Geodynamics*, 3rd edn. Cambridge: Cambridge University Press.

UN (2015). *Transforming our World: The 2030 Agenda for Sustainable Development*. A/RES/70/1. United Nations, New York.

UN (2017). Report of the open-ended intergovernmental expert working group on indicators and terminology relating to disaster risk reduction. Available at: www.preventionweb.net/files/50683_oiewgreportenglish.pdf (retrieved on 12.02.2017)

UNGA (1987). United Nations General Assembly Report of the World Commission on Environment and Development: Our Common Future. New York: UN Headquarters. Available at: www.un-documents.net/wced-ocf.htm (retrieved on 05.08.2016).

UNISDR (2005). Hyogo Framework for Action 2005–2015: Building the Resilience of Nations and Communities to Disasters. World Conference on Disaster Reduction. Available at: www.unisdr.org/2005/wcdr/intergover/official-doc/L-docs/Hyogo-declaration-english.pdf (retrieved on 05.08.2016).

UNISDR (2009). *The Global Assessment Report on Disaster Risk Reduction (GAR2009)*. Geneva: UN Office for Disaster Risk Reduction.

UNISDR (2013). *The Global Assessment Report on Disaster Risk Reduction (GAR2013)*. Geneva: UN Office for Disaster Risk Reduction.

Wu, Z., and Ma, T. (2014). The 2008 Wenchuan, China, earthquake. In: *Extreme Natural Hazards, Disaster Risks and Societal Implications*, eds. A. Ismail-Zadeh, J. U. Fucugauchi, A. Kijko, K. Takeuchi, and I. Zaliapin. Cambridge University Press, Cambridge, pp. 301–309.

Wyss, M., Nekraskova, A. and Kossobokov, V. (2012). Errors in expected human losses due to incorrect seismic hazard estimates. *Natural Hazards*, **62**, 927–935.

27

Geothermal Energy and a Future Earth

Ladislaus Rybach and Thomas Kohl

27.1 Introduction

Geothermal energy is available day and night, summer and winter; it is thermal energy generated and stored within the Earth in rock, trapped steam, or water. Ninety-nine per cent of the Earth's volume has temperatures greater than 1,000°C, with only 0.1 per cent at temperatures less than 100°C. The total heat content of the Earth is estimated to be about 10^{13} EJ and, thus, is immense. The main sources of geothermal energy are the residual energy from planet formation and the energy continuously generated by radionuclide decay. The Earth releases heat to the atmosphere, the thermal power being 40 million MW, without experiencing any noticeable cooling. This amount of heat is equivalent to the thermal power of about 13,000 nuclear power plants each of 1 GW_e capacity. Hence, the geothermal resource base is ubiquitous, sustainable, and sufficiently large to be a significant energy source.

Concerning the depth range of the energy source, two main categories are distinguished: shallow (<400 m; mainly utilized with ground-coupled heat pumps) and deep (>400 m). There are two main deep resource categories. In the first, the heat carrying geothermal fluid is locally present ("hydrothermal"). Such resources exist only in specific geological settings, such as volcanic regions or large, deep sediment basins, and are correspondingly rare. The second category of resources ("petrothermal") consists of the ubiquitous hot rocks with low permeability at depth. In these rocks, the heat carrier must be circulated artificially. In most cases (except for natural thermal springs), boreholes are needed to move the fluids with the heat to the surface.

Geothermal energy is utilized two ways: electricity generation and direct use. For electricity generation, the fluid from the ground must have a minimum temperature of 100°C –150°C. At lower temperatures, there is still a whole range of other possibilities: space heating, bathing and swimming, agricultural uses (greenhouses, fish farms).

In the following sections, the present status of geothermal utilization is summarized, first direct use and then power generation. Then, the development (growth) trends are presented, with emphasis on annual growth rates. More emphasis is placed on future, highly promising Enhanced Geothermal Systems (EGS). Further on, future prospects and implications of a Future Earth are highlighted.

27.2 Geothermal Direct Use

Reliable statistical data about global geothermal direct use are available from the World Geothermal Congresses, organized every five years. Data assembled for WGC 2015 are from 2014. World-wide direct use of geothermal heat in 2014 was 590 PJ in eighty-two countries, with a total installed capacity of 70 GW_{th} (Lund and Boyd, 2015). Table 27.1 shows the different use categories and their contributions; it clearly indicates that the main contribution was provided by geothermal heat pumps. This relatively new technology is one of the fastest-growing renewable energy branches (by about 20 per cent annually) and is by all means the real booming geothermal sector. Figure 27.1 shows the nearly exponential growth of geothermal heat pumps over the years 1995–2015. The second largest category, bathing and swimming in thermal waters, is certainly the oldest.

27.3 Geothermal Power Generation

Geothermal power plants convert heat from geothermal fluid into base-load electricity. The conversion efficiency depends mainly on the fluid's heat content as reflected by its temperature. Temperature commonly increases with depth and differs for young, geologically active regions in comparison to older, "cooled" regions.

Table 27.1 *Contributions of the Different Direct Use Categories*

Direct use category	Heat utilized (PJ/y)	% of total
Geothermal heat pumps	326.3	55.30
Space heating	88.5	15.01
Greenhouse heating	26.8	4.54
Aquaculture pond heating	12.0	2.03
Agricultural drying	2.0	0.35
Industrial uses	10.5	1.78
Bathing and swimming	119.8	20.31
Cooling / snow melting	2.6	0.44
Other	1.5	0.25

Compiled from data in Lund and Boyd (2015)

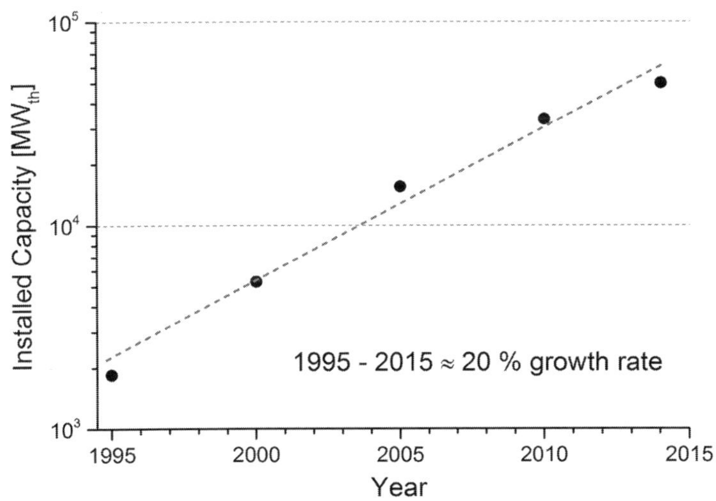

Figure 27.1 Rapid growth of geothermal heat pumps (installed capacity data of Lund and Boyd, 2015; plotted on logarithmic scale).

Consequently, the most attractive sites for this technology are the few geologically young areas with very high geothermal gradients. The most productive hydrothermal reservoirs with large volumes of stored geothermal fluids at abnormally high temperatures exist at a few hundred or even thousand meters' depth.

Geothermal power plants in twenty-two countries are spread around the globe; Figure 27.2 shows the situation in 2015. The globally installed gross capacity in early 2016 was 13.3 GW_e (GEA, 2016), which provided about 0.3 per cent of total electricity production.

So far, practically all geothermal power plants have been producing power from hydrothermal resources. Since these are highly unevenly distributed, many large countries such as Brazil cannot use hydrothermal

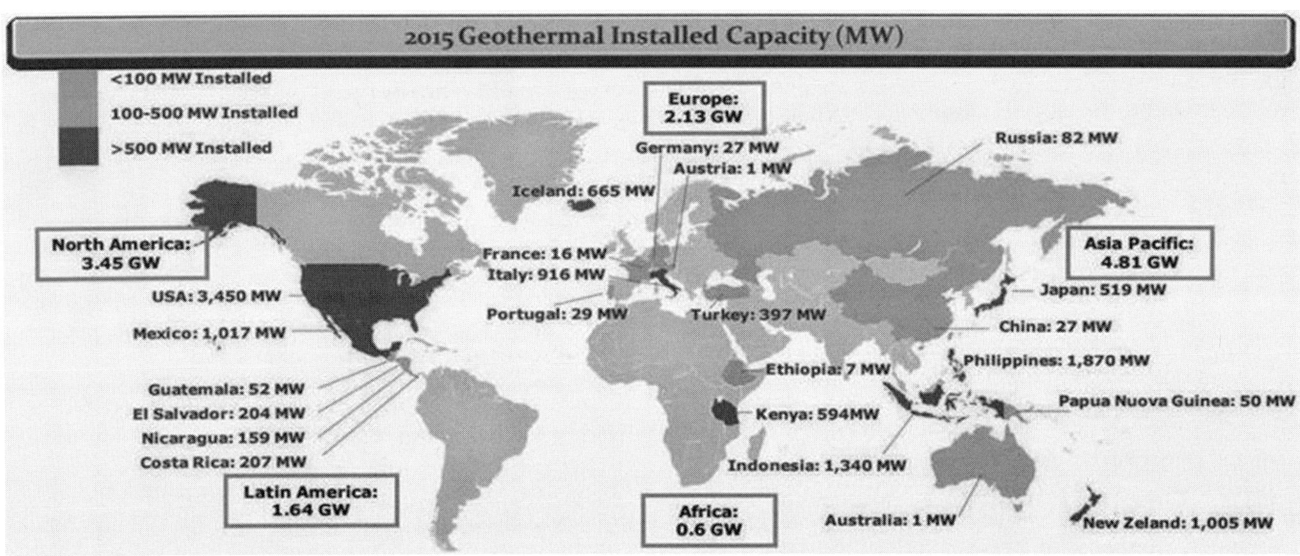

Figure 27.2 Global geothermal power in 2015: 22 countries with power plants in operation (from Bertani, 2015).

electricity. The development trends in power plant establishment will be presented in the following. Figure 27.3 shows the growth of globally installed geothermal capacity and electricity production since 1995 (Bertani, 2015).

The 5 percent growth seems to be rather constant, but how does it compare to other renewable technologies like solar photovoltaics (PV) or wind? Table 27.2 lists the potential estimates, according to the United Nations Development Program (UNDP).

Geothermal certainly has the lead in terms of potential, but how much has been realized? According to Bertani (2015) and Lund and Boyd (2015) the total geothermal production (electricity plus direct use) amounted to 0.850 EJ in 2014, which means that production is smaller by many orders of magnitude. And how does it relate to other renewables? For this, the annual growth numbers will be compared. Figure 27.3 shows that geothermal power growth in the years 1985–2015 was rather steady and amounted to 5 per cent per year.

REN21 (2016) reports the growth over time in the development of wind and solar PV power. It is nearly exponential with the annual growth rates being 35 per cent for solar and 25 per cent for wind. How quickly geothermal power growth falls behind is clearly evident from Figure 27.4: here the growth of solar PV and geothermal power is compared. It can be anticipated that both trends will continue in the next few years. Interestingly, statistical data show that geothermal power capacity was well ahead of solar PV until about 2007, but this is nearly invisible on the scale of Figure 27.4.

Of course, annual availability of geothermal power is much better than that of solar PV (and also of wind energy). This is why the electricity production numbers (GWh per year) should be compared as well. Table 27.3 shows this comparison. In terms of earlier REN21 data, solar PV production overtook geothermal power production in 2011.

The globally averaged availability of geothermal power plants is only 65 per cent (some plants are operating with greater than 90 per cent availability, i.e. 7784 hours a year). Among the reasons for the lower value are that geothermal power plant statistics report the nameplate (size) and not the operating capacity, there are production stops for maintenance, etc.

The slow geothermal power growth has its consequences on the market: investments are much smaller than for solar or wind energy. Figure 27.5 shows this very clearly – for this diagram it must be considered that the bars "Developed countries" and "Developing countries" should be added to illustrate the real relations.

Table 27.2 *Estimated Technical Potential of Renewable Energies*

Energy source	Capacity (EJ/yr)*
Geothermal	5000
Solar	1575
Wind	640
Biomass	276
Hydro	50
Total	7541

From Rogner (2000)
*1 EJ = 1018 J

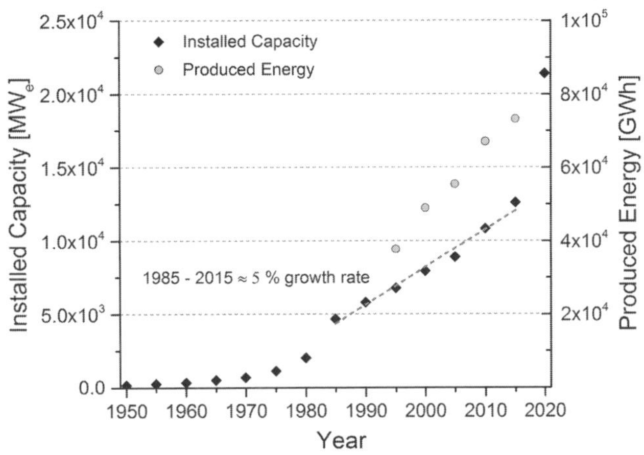

Figure 27.3 Global development of geothermal power: installed capacity (MW_e), produced electricity (GWh), and estimated capacity in 2020. The line through the data points 1985–2015 corresponds to 5 per cent growth per year (modified from Bertani, 2015).

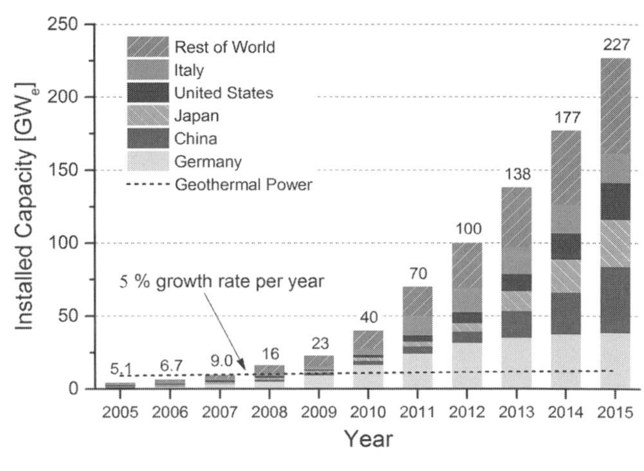

Figure 27.4 Global photovoltaic power development (installed capacity, GW_e), compared to geothermal power growth (figure from REN21 (2016) and geothermal addition from Figure 27.3, this volume).

Table 27.3 *Global Comparison of Installed Capacity and Annual Power Production of Renewables in 2015*

Technology	Installed capacity GW$_e$	Installed capacity %	Annual production GW$_e$	Annual production %	Availability %
Hydropower	1,064	57.7	3,940	70.9	42
Biomass	106	5.8	464	8.4	50
Wind	433	23.5	796	14.3	21
Geothermal	13.2	0.7	75	1.4	65
Solar PV	227	12.3	278	5.0	14
Total	**1,843.2**	**100**	**5,553**	**100**	

Compiled from data in REN21 (2016)

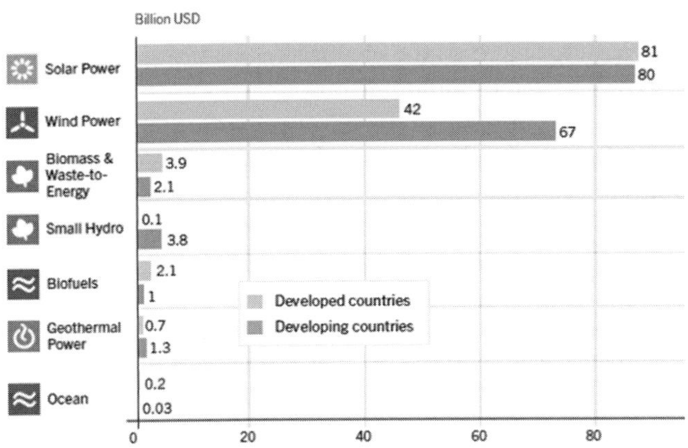

Figure 27.5 Global investments in renewable energies for power generation in 2015. (Compiled from data in REN21, 2016).

From all this it is more than evident: geothermal power growth is increasingly falling back and it should be accelerated soon and strongly. Otherwise geothermal power will become negligible! So far, practically all geothermal power plants have used hydrothermal resources. These are really rare, and it cannot be expected that their development rate can be increased significantly. So the only hope for the future is petrothermal resources, and this means EGS = Enhanced/Engineered Geothermal Systems. In other words, to accelerate the growth of geothermal power a principal shift from hydrothermal to petrothermal resources should be made.

27.4 EGS Status and Problems

The concept of Enhanced Geothermal Systems, EGS, was originally developed from the Hot Dry Rock (HDR) technology and some thirty years ago it was adapted to petrothermal resources (e.g. Armstead, 1983). Essentially, HDR attempts to create a new

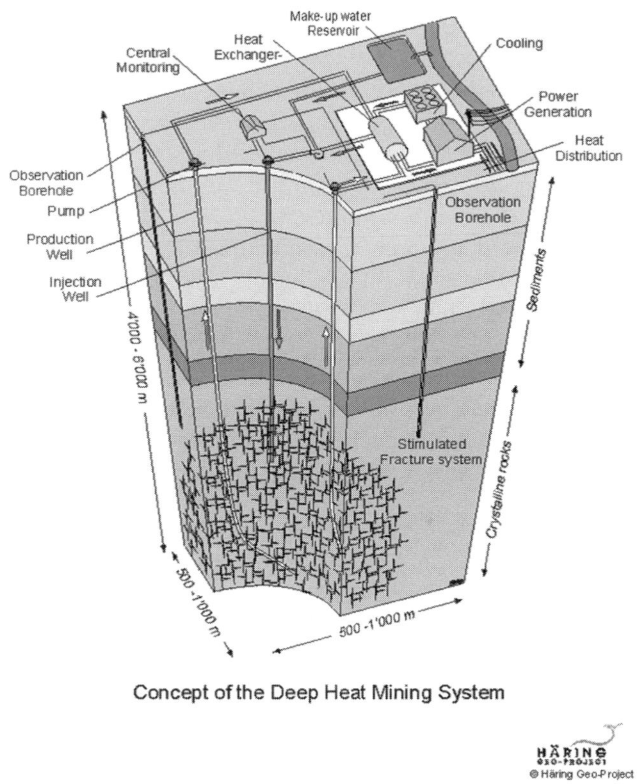

Figure 27.6 Schematics of an EGS installation for power generation and district heating.

artificial fracture network in host rock through massive fluid injection under high pressure to tap resources in the deep subsurface, where temperatures are high enough for power generation (150°C–200°C). Water from the surface is circulated through this several-kilometers-deep heat exchanger, using injection and production wells, and is recovered as steam/hot water. Further surface installations complete the circulation system. Figure 27.6 shows the schematics of an EGS installation. The extracted heat can be used for district heating and/or for power generation. International experience with this HDR concept often revealed unfavorable productivity and fluid losses, leading to a conceptual extension by the EGS principle. Whereas the HDR scheme is based on applying "fracking" technologies to create new flow pathways, EGS rather considers stimulation and extension of pre-existing fracture networks in order to enhance permeability.

EGS systems bridge the technology between conventional geothermal utilization in hydrothermal systems (typically developed in sedimentary rocks) and future utilization of petrothermal resources, which are based on stimulation technologies to enhance natural fluid circulations in crystalline rocks. Meanwhile, the aspect of stimulation that is often inevitable for most geothermal systems has been connected to the term "Engineered Geothermal Systems," EGS. Today, EGS technologies are typically applied to low-enthalpy areas. But operators of conventional high-enthalpy systems make efforts to apply EGS technology also at the low-temperature borders of their reservoirs. Numerous impressive EGS potentials have been reported for numerous countries, e.g. the United States (100,000 MW_e; Tester et al., 2006) or Germany (475,000 MW_e; Jain et al., 2015).

In terms of the EGS vision "Future of Geothermal Energy" (Tester et al., 2006), EGS power plants in principle could be built in large numbers all over the globe, since deep hot rocks are practically ubiquitous. There is general agreement about the required characteristics of deep EGS heat exchangers ("reservoirs"). Table 27.4 shows the numerical figures; the most important reservoir properties are volume, heat exchanger surface area, and flow impedance.

According to the basic idea of EGS, a standardized reservoir with the specified properties would be needed at any desired location. How such a structure could be established, irrespective of the local subsurface conditions, is not yet known: the necessary construction technology has not yet been developed. This is the key challenge for the future.

Table 27.4 *Requirements for a Technically Feasible and Economically Viable, Standard EGS Heat Exchanger, for a Target 5 Mw_e per Module*

Life time of the system	~20 years
Temperature of the wells	~200°C
Separation between the wells	~600 m
Production flow rate	~75 kg s^{-1}
Flow impedance	~0.1 MPa l^{-1} s^{-1}
Water loss	~10% max
Thermal drawdown	~10%
Contact surface area	~10 million m^2
Reservoir rock volume	~300 million m^3
Interest rate for the capital	~5%
Support	No CO_2 levy support

From EGEC (2012)

27.5 Future Research and Development (R&D) Needs Relating to EGS

As shown by the most detailed experimental monitoring of EGS at Soultz-sous-Forêts, current EGS development for controlled enhancement of the reservoir is hindered by lacking fundamental understanding of processes occurring at high volumes and rates of flow in fractured reservoir rocks (Schill et al., 2015). Additional complexity results from high differential stresses, the related mechanical behavior of the reservoir rock, and the chemical response of the fluid. Although experimental progress has been made, fundamental constitutive laws of flow and thermo-hydraulic-mechanical-chemical (THMC) coupling presently are not being considered adequately in reservoir engineering and plant operation in fractured environments. It is therefore most important to carry out large-scale in situ tests under hydrogeological conditions and comprehensively study these aspects (Freeze and Javandel, 2008). Underground research laboratories (URLs) are suited best to monitor and quantify the interrelated processes during large-volume injection into fractured rock with the help of large-scale in situ observational methods to complement present-day laboratory-scale experiments. Such experimental facilities for application and monitoring of experiments should address key aspects of a safe and economically efficient use of geothermal energy. Also, as was pointed out by OECD (2013) for the field of nuclear waste disposal research, URLs offer an excellent opportunity to integrate multiple disciplines (e.g. geology, hydrology, and engineering), build technical teams, and gain practical experience that will be invaluable for the next generation of researchers.

Contrary to hydrothermal resources, where prospective drilling sites must be located by geothermal exploration, the sites of future EGS plants will already be fixed, according to the demand. This relies on the fact that deep hot rocks are ubiquitous. Here the questions are how to extract heat locally, and how to get it to the surface. For this, a deep EGS heat exchanger should be established at the requested locations, with the properties mentioned. Then, the heat carrier has to be circulated from the surface through the deep heat exchanger and then back.

In this endeavour the local subsurface conditions must be known, especially the properties and conditions

Table 27.5 *The Most Important Performance Factors of EGS Heat Exchangers; For Comparison, Two Hydrothermal Fields, Dixie Valley and Beowawe, Are Added*

Project	AE[2] volume (km^3)	Tracer-swept volume (10^4 m^3)	Recovery factor (%)	Tracer-swept/ AE (10^{-5})	Impedance[4] (MPa s/kg)
Fenton Hill	0.12	0.1	2	1	25–30
Rosemanowes	0.6	0.12	0.7	0.2	0.1
Hijiori	0.25	0.02	0.2	0.1	0.2
Cooper Basin	0.4	2	1.6	5	0.2
Soultz-s.F.	2.4	1.6–3.0		0.7–1.3	0.1–0.25
Coso	0.15	4.1		27	
Dixie Valley	11[3]	400–1400	8–21	36–130	
Beowawe	5[3]	700–7000		140–1400	
Target[1]		30	20		0.1

From Grant (2016)
[1]Cf. Table 27.4
[2]AE: Volume with acoustic emissions (=microseismic signals); the tracer-swept volme is determind by tracer tests
[3]Reservoir volume
[4]Maximum feasible flow impedance

at depth. Site exploration for EGS development should therefore clarify the local temperature field (isotherm pattern), lithology (granite? gneiss? schist?), stress field (component magnitudes, orientation, anisotropy), kind and degree of the existing fracture network (orientation, density, extension, apertures, fillings, mechanical properties), and natural seismicity. And then means are needed to establish a proper EGS heat exchanger, with these local characteristics.

For heat extraction, the foreseen volume (=reservoir volume/tracer-swept volume) should host a network of permeable fractures. To create this network, hydraulic stimulation (possibly combined with chemical stimulation) will be applied in most cases. Experience shows that the stimulation can induce felt seismic events; the case of the Deep Heat Mining project in Basel, Switzerland, became widely known (Deichmann and Giardini, 2009). The risk of induced seismicity is decisive for EGS acceptance, which is why the event magnitudes need to be limited. In any case, large-magnitude events are not desired in reservoir development: high-magnitude events mean large fractures that can lead to hydraulic short-circuiting during heat extraction. The EGS heat exchangers should consist of many fine, well-connected heat exchange surfaces. This will enable the circulated heat carrier fluid to access a preferably large rock volume and to extract the heat efficiently.

Hence the main challenges in EGS development and the corresponding future R&D needs are:

- Establishing standard EGS heat exchangers, at any desired location, in suitable temperature/depth ranges, independent of local subsurface conditions;
- clarifying the long-term performance of EGS heat exchangers in terms of productivity and environmental effects;
- determining recovery factors (=extractable heat / heat in place).

Production sustainability is another open question. So far, no experience has been gained with respect to possible changes in the behaviour of an EGS heat exchanger with time. Permeability enhancement (e.g. by new fractures generated by cooling cracks, mineral dissolution) could increase the recovery factor, whereas permeability reduction (e.g. by mineral deposition, short-circuiting) could decrease recovery. Without having field-scale experience with long-term EGS production, the economic estimates of production and maintenance costs remain unsubstantiated.

Finally, up-scaling of the EGS power plant size to a few megawatts (MW_e) is needed. But in order to play a significant role in a future electricity supply a system capacity of at least several tens of megawatts would be essential. One of the main future R&D goals will be to see whether and how the EGS power plant size could be up-scaled, possibly by unconventional measures (still to be developed).

To become a significant player in future electricity supply systems, EGS still has a long way to go. Currently available data from field experiments (Grant, 2016) show that the target figures have not yet been reached; see Table 27.5.

One thing is certain: the heat is down there, in immense amounts – we just have to learn how to take it out! But:

- How? (sufficient deep heat exchanger realization ➔ proper, site-dependent stimulation, limiting magnitudes of induced seismicity)
- With what efficiency? (recovery factor ➔ enhancement, production sustainability)
- How economic? (investments needed ➔ private/municipality/private-public partnership financing)

These are the basic EGS questions and corresponding R&D tasks. Besides, up-scaling EGS power plant size (from a few megawatts to dozens of megawatts) will be decisive. And EGS pilot plants are badly needed, as well as long-term experience.

27.6 Geothermal in a Future Earth

What will the Earth look like in the future? Global emission of CO_2 and greenhouse gases will, in the short term, first increase and then ideally stabilize or even decrease. This, in turn, will influence global warming (surface temperature variations, changes in precipitation patterns, etc.). Rainfall changes may affect discharges of some hot springs; the availability of cooling water from surface water supplies for power plants could also be affected, which is why air-cooled condensers might be required (Goldstein et al., 2011). The latter could lose efficiency as a result of rising temperatures.

Unlike the composition and processes in the lower atmosphere and directly at the Earth's surface (mainly due to external or man-made influences), the situation in the Earth's deeper interior is not expected to change significantly in the future. The only exception: shallow geothermal resources (details in the following); deep geothermal resources remain practically unaffected by climate changes experienced at the Earth's surface.

For geothermal energy utilization in a Future Earth, long-term sustainability of heat production will be indispensable at all times. Contrary to mining operations, where excavated ore will never return, geothermal resources recover! They are completely self-sustainable: heat/fluid extraction creates heat sinks/hydraulic minima, which trigger strong temperature/pressure gradients and a corresponding heat/fluid inflow to refill the deficits. Depletion and recovery exhibit an asymptotic behaviour: strong at the beginning and slowing down subsequently (Rybach et al., 1999). Figure 27.7 shows examples of typical production and recovery curves in shallow and deep systems.

In earlier times, geothermal extraction systems and power plants were designed for thirty years of operation, and production was forced to achieve a quick return of investment. Production decline due to hydrothermal reservoir depletion was the result; see e.g. the case of the Geysers field, California (Bertani, 2015). There are two main ways to counteract and prevent decline: reinjecting the fluid produced after heat extraction and applying sustainable production rates. The latter must be set to comply with the natural heat/fluid recharge rate; see Rybach and Mongillo (2006). Another way is to guarantee production sustainability by applying lower, sustainable production rates. Some details are given in the following.

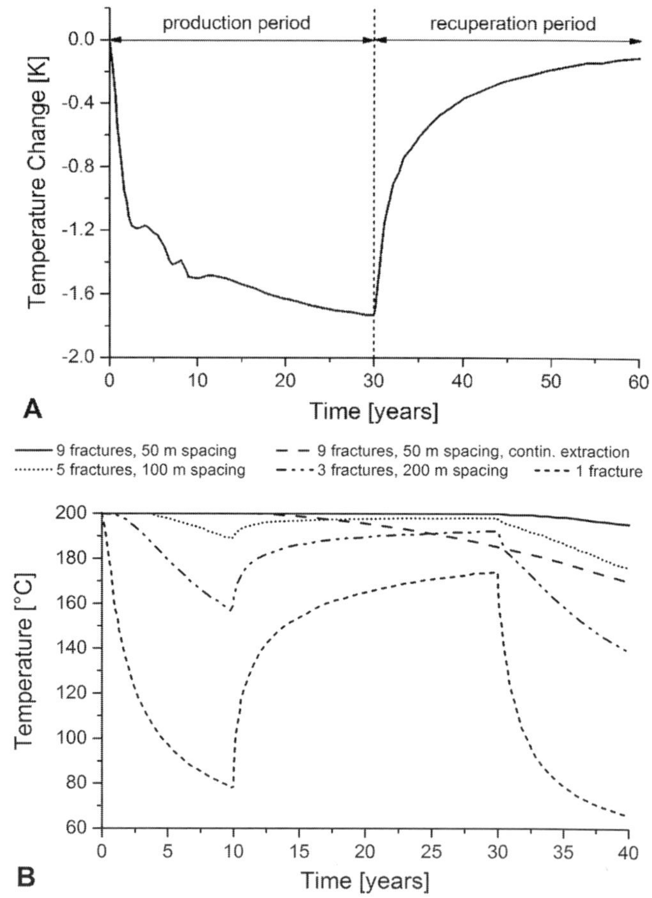

Figure 27.7 Typical depletion and recovery curves. A: ground temperature changes at 50-m depth and 1-m distance from an operating 100-m- long borehole heat exchanger – the first 10 years measured, the rest modelled (from Rybach et al., 2000). B: EGS fluid production temperatures with various heat exchanger fracture configurations (after Fox et al., 2013). Heat transfer by conduction in both cases. The similarity of the curve shapes is remarkable.

When producing from a geothermal source, sustainability will depend on the initial heat and fluid content and their regeneration rates (Wright, 1995). The reaction of the resource to production will largely depend on the rate of heat/fluid extraction. With high extraction rates the energy yield will be correspondingly high at the beginning (and with it the economic reward) but will decrease significantly with time, potentially terminating a commercially viable operation.

Lower production rates can ensure the longevity of production: i.e. relatively constant production rates can be sustained. In addition, sustainable production rates can provide total energy yields similar to those achieved with high extraction rates. To demonstrate this, the results of a study comparing high- and low-level production from an EGS model are summarized here (for details see Sanyal and Butler, 2005). The modeled reservoir had an area of 3.66 km x 3.66 km, with a vertical extension between 1.22 km and 2.74 km in depth. The average initial reservoir temperature was 210°C. A three-dimensional double-porosity, finite-difference numerical scheme was used to calculate power generation from this hypothetical EGS reservoir. A five-spot borehole array (injector at the model centre and production well at each corner of a square) with high (1800 t/hr) and low (475 t/hr) production rates was considered (injection flow rate = production flow rate). Production at the high rate yielded higher power generation capacity at the beginning (45 MW_e). A parasitic load of nearly 10 MW_e was needed to pump the high fluid circulation rate through the system. The fluid production temperature decreased with time, and reservoir depletion resulted in the stop of production after 20 years. The total energy produced amounted to 245 MW_e-year. At the lower circulation rate the starting capacity was only 12 MW_e, but the pumping load was nearly negligible. The temperature decline also was much smaller and the power generation capacity prevailed well beyond 30 years. The total energy produced over 30 years, 250 MW_e-year, was very similar to that of excessive production.

This example demonstrates that with lower extraction rates longevity of the resource and, thus, sustainable production can be achieved, while long-term energy generation is practically the same as in the case of excessive production. The level of sustainable production depends on the utilization technology as well as on the local geological conditions and resource characteristics. Its determination requires specific studies, especially model simulations of long-term production strategies.

27.7 Surface Conditions Influence Shallow Geothermal Resources

Surface temperature changes represent the principal boundary condition in thermal processes in the subsurface. The heat input at the surface will penetrate down into the ground; the duration of propagation increases with the square of the depth affected. The ground therefore is a prominent archive of past surface temperature changes (Bodri and Cermák, 2007). For example, cooling of the last ice age is reflected by decreased temperatures, mainly at the depth range of one to three kilometers at the German KTB site (Rybach, 1992).

Other, external sources can also influence the thermal conditions, especially in the shallow ground. Nowadays a significant temperature anomaly can be observed at many places in borehole temperature profiles in the top hundred meters of urban areas (Rybach et al., 2016). Temperature-depth profiles (= temperature logs) clearly and significantly differ from a linear temperature-depth trend: A systematic deviation starts at a depth of about one hundred meters with a tendency to increase gradually towards the surface. The so-called Urban Heat Island (UHI; man-made heat input due to urbanization) manifests itself in the subsurface of many cities and can also significantly increase groundwater temperatures (e.g. Menberg et al., 2013, 2014).

Figure 27.8 shows such an anomalous temperature log from the locality Meilen in the Zurich area in Switzerland (A). The interpretation of the anomalous form of the temperature log reveals that the surplus temperature has two components (B): input by global warming (dashed line) plus urban heating (plus signs). Urban heating is caused by sun-warmed pavements and heated buildings. At the Meilen site, at twenty-meter depth, total warming amounts to 2.2°C. Of this amount, the share of global warming is 0.9°C, while warming due to urbanization has a share of 1.3°C (59 per cent of total). For details, especially about the interpretation methodology, see Rybach et al. (2016).

It can be expected that the anthropogenic heat input into the ground, especially in megacities, will play an increasingly important role. Heat extraction from the ground by borehole heat exchanger–coupled geothermal heat pump systems will certainly profit from this additional heat source that may be sustainable for the time being.

27.8 Further Geothermal Possibilities in the Future

Although predictions of future developments are ambiguous, some considerations for the developments are given in the following, subdivided into short-, medium-, and long-term categories.

27.8.1 In the Short Term

The use of shallow (<400 m) geothermal resources will continue to grow at a respectable rate, possibly slowing down somewhat in the coming decades. The installation of geothermal heat pumps started with individual

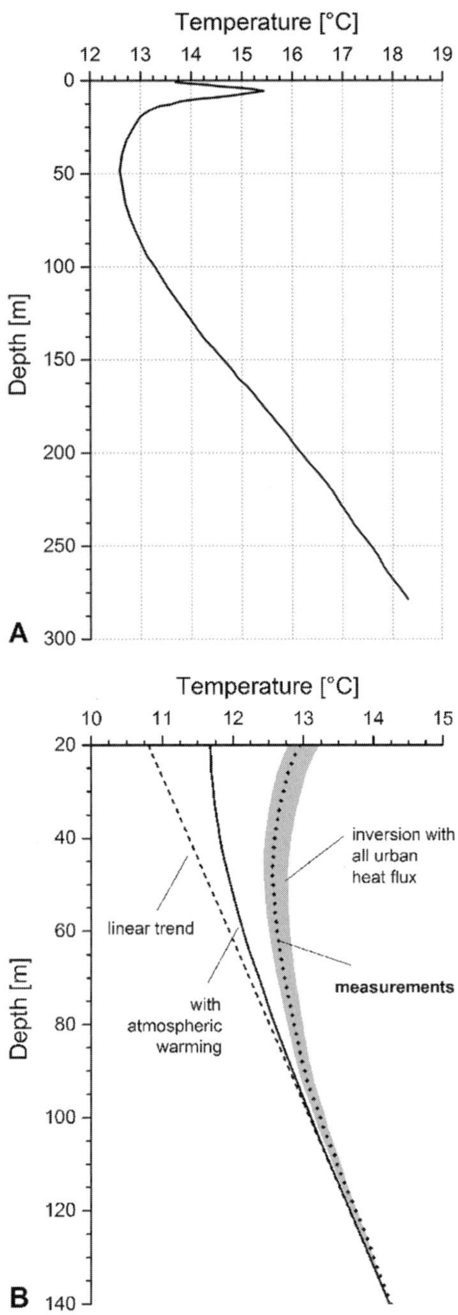

Figure 27.8 A: Temperature log at the locality of Meilen (measured on 27 October 2005, after thermal equilibration). B: Measured temperature profile at a depth of 20–140 m in Meilen (+ signs) with modelling results (grey error band for RMSE = ±0.1°C). Also shown are the undisturbed linear trend (dotted line, 29°C/km) and the profile due to climate forcing alone (dashed line).

heating systems in single-family houses. Nowadays, fields with hundreds of borehole heat exchangers are deployed for large-scale complexes with habitations, services, administration buildings, factories. The establishment of such set-ups shows a tendency to increase and this development will certainly continue. This is especially important in view of the generally increasing urbanization. Combined operation of the same system for heating and cooling is also increasing, even in zones of moderate climate. The near-surface surplus heat by global and urban warming will be beneficial for ground-coupled heat pump systems.

Direct use of deep, hydrothermal resources will also continue. In the future, a large fraction (up to 50 per cent) in the energy balance of many countries will be needed for heating. Large district heating systems could serve many users; the problem here is that – contrary to electricity, for which large utility companies supply whole regions – there are hardly any correspondingly extended heat supply networks and providers and the heat market is much too fragmented.

In addition, geothermal power development might fall behind the rapidly advancing wind and solar electricity generation, and geothermal progress will concentrate on direct heat applications. There are already tendencies of developing EGS installations to provide heat for industrial users rather than electricity; see the case of Rittershoffen in France (Baujard et al., 2016).

Geothermal fluids have a special advantage: they often have temperatures <150°C. This is warm enough for most industrial and agricultural applications. The much higher temperatures provided by the burning of fossil fuels result only in great losses.

27.8.2 In the Medium Term

Here, the deep resources will be the focus, in view of the immense amounts of heat stored at greater depths. A number of options are feasible:

Combined heat and power (CHP): The relatively low efficiency of converting heat to power at wellhead delivery temperatures below 150°C leaves geothermal power plants with lots of "waste fluid" that could be used further, e.g. for district heating. Provided that heat customers are available within reasonably short distances, the economic efficience of a CHP facility would be much better than that of a power generation station alone. A typical CHP installation is being operated successfully in the town of Unterhaching near Munich, Germany (Richter, 2010).

Cascading use: It is based on the fact that various direct heat applications operate at different temperature levels. In this arrangement, a chain of coupled applications is established in a succession of decreasing inlet temperatures. Starting with the fluid outlet of geothermal power plants, the chain can consist fully or partly of industrial, space heating, diverse balneological, and bathing and wellness uses, greenhouse and soil heating; de-icing of parking lots and pavements, production of domestic hot water, air-conditioning, heat pump applications, assorted low-temperature washing, low-temperature drying, and fish farming (Lund and Boyd, 1999; Rybach and Kohl, 2004).

Hybrid systems: in principle, biomass, wind, or solar photovoltaic systems can be combined with a geothermal source to produce electricity, by superheating the turbine inlet. Recently, Thain and DiPippo (2015) evaluated various design solutions for geothermal-biomass hybrid plants, and the MIT energy group (Manente et al., 2011) carried out several conceptual studies for solar-geothermal designs. Practical applications of this kind of solution are still rather rare, among them being the geothermal-biomass plant in Honey Lake, California, United States (GHL, 1988), and the geothermal-solar plant at Stillwater, Nevada, United States (DiMarzio et al., 2015). Since September 2015, a biomass-geothermal hybrid plant has been operated with eighteen megawatts at Cornia in the southern part of the Larderello geothermal field, Italy (Dal Porto et al., 2016). Kohl and Speck (2004) developed a hybridization concept for EGS power generation. On the basis of existing data from the Soultz-sous-Forêts (France) project, they demonstrated that rather high efficiency factors (~15 per cent) are achievable for geothermal systems of 150°C reservoir temperatures when superheating the already pure water-vapor cycle.

Use of abandoned oil/gas wells: there are several hundred thousand abandoned oil/gas wells all over the world. For hydrocarbon exploration and production the occurrence of hot formation waters has always been a nuisance; they have ever hardly been tested or utilized properly. For a number of years, there have been regular, annually repeated conferences at the Geothermal Lab, Southern Methodist University, in Dallas, Texas, United States, about geothermal energy in nonproductive oil and gas fields (www.smu.edu/Dedman/Academics/Programs/GeothermalLab/Conference). There the latest developments are presented and discussed. Currently, activities are evolving in China, Canada, and Nebraska in the United States. It is foreseeable that there will be increasing interest and activity in this promising field of geothermal development.

Break-through of EGS: finally, it is hoped that numerous successful pilot and demonstration facilities in various contrasting geological settings will prove the EGS concept in the foreseeable future. Until then, very significant efforts will be needed to secure very substantial funding for all the activities needed in EGS technology development (described in detail in Section 27.5, Future Research and Development (R&D) Needs Relating to EGS).

27.8.3 In the Long Term

It is imaginable that sometime in the future, a revolutionary new, cheap, and efficient deep drilling technology will be invented, which would greatly reduce the cost (often the majority of new geothermal development project expenditure) of drilling the necessary boreholes. Currently, there are various new drilling technologies in stages of experimental testing: laser drilling (www.foroenergy.com/technology), electro pulse boring (Schiegg et al., 2015), hydrothermal spallation drilling (Kant et al., 2015), and several others including hyperdrill, thermodrill, and radial jet drilling. So far, none of them has reached full field-size testing. It is hoped that one (or some) of these technologies will be operational in the future, and ready for the market later on.

Efficient and affordable desalination of seawater is one of the future challenges and hopes for agriculture in many dry-climate countries on seashores. Most technologies need heat to operate an evaporation/condensation loop desalination device. A multi-stage distillation system, supplied by geothermal heat, was tested and demonstrated in the Kimolos island (Greece) project (an Intelligent Energy Project of the EU; details at www.energia.gr/geofar/articlefiles/geothermalinnovative-desalination.pdf). Date et al. (2015) describe equipment to desalinate thermal spring water for drinking on the western coast of India. A desalination project for desert areas in Mexico is described in Gutierrez and Espindola (2010); another from Saudi Arabia in Chandrasekharam et al. (2016).

At present, steam or water/brine is the common heat carrying geothermal fluid. Various efforts are being undertaken to use CO_2 instead of water. When successful, this could lead to numerous applications with great potential, for example: EGS with CO_2 (enhanced geothermal systems (EGS) using CO_2 as working fluid (Pruess and Spycher, 2010) and geothermal energy production combined with geological CO_2 sequestration (Randolph and Saar, 2011).

There are offshore possibilities for geothermal power prodution, not only for wind energy: the black smokers in mid-oceanic ridges have great potential. These spectacular hydrothermal vents emit not only lots of heavy metals, but also extremely large amounts of heat: Hiriart et al. (2010) report 100 to 300 MW_{th} thermal power output per axial meter at places on the Juan de Fuca Ridge – and all this without any drilling! Of course, special submarine power plants (unmanned) would be needed (Shnell et al., 2015) as well as long seabed cables to carry the electric power to land. Nevertheless, this unique possibility should be further investigated in more detail.

27.9 Conclusions

At the World Geothermal Congress 2015, geothermal direct heat use was reported from eighty-two countries, with a total of 590 PJ for 2014, achieved with an installed capacity of 70 GW_{th}. The majority of production (55.3 per cent) is by geothermal heat pump systems, which is the only booming geothermal sector – its average annual growth rate amounted to 20 per cent over the period 1995–2014. This trend is expected to continue in the future, albeit at a somewhat smaller rate. The geothermal heat pump systems for space heating and cooling rely on shallow (<400 m depth) resources.

For power generation, deep hydrothermal resources are used. The total installed capacity in twenty-two countries was reported to be 13.3 GW_e in early 2016, which provided 75 TWh, or about 0.3 per cent of total electricity production. The average growth rate of geothermal electricity over the years 1985–2015 was rather constant and amounted to 4 per cent annually. This was much less than growth rates in solar PV or wind energy, although geothermal electricity production was higher than by solar PV until 2011. Investments in the other renewable sources show a corresponding pattern.

Acceleration of future geothermal electricity production can only be achieved by developing petrothermal resources, since these are ubiquitous in principle. High potentials are reported from various countries. So far, attempts have been made to use these resources by EGS systems. The general technical feasibility and economic viability of the EGS concept still need to be demonstrated, with successfully operating pilot and demonstration facilities, in different geologic settings. This will need very substantial pilot and demonstration funding.

Concerning other geothermal possibilities in the future, a number of options exist, with production sustainability being of paramount importance. In the short term, shallow systems will benefit from increasing urbanization, since this will lead to constant heat input into the ground. Effects of global warming will not reach deep hydrothermal resources for quite some time. Large geothermal district heating (and also cooling) systems may become more common, when financed by municipalities. In the medium term, combined heat and power, cascading use, hybrid systems, and use of abandoned oil/gas wells could be developed. A break-through of the EGS concept can also be envisaged. In the long term, a number of innovative developments could take place: revolutionary and much cheaper deep drilling methods, geothermal seawater desalination, combined CO_2 disposal and geothermal energy production, supercritical resources, and offshore geothermal power.

In any future energy mix, geothermal heat and power will play a role – ideally a significant one.

Acknowledgements

Jörg Meixner's (Institute of Applied Geosciences, Karlsruhe Institute of Technology) help in various editorial matters is gratefully acknowledged. Two anonymous reviewers provided helpful comments and suggestions, which have been duly considered.

References

Armstead, H. (1983). *Geothermal Energy: Its Past, Present and Future Contributions to the Energy Needs of Man*, 2nd ed.: E. & F.N. Spon, New York.

Baujard, C., Genter, A., Dalmais, E., Maurer, V., Hehn, R., Rosillette, R. (2016). Temperature and hydraulic properties of the Rittershoffen EGS reservoir, France. Proceedings, European Geothermal Congress 2016 Strasbourg, France, 19–24 September 2016 Available at: www.geothermal-energy.org/pdf/IGAstandard/EGC/2016/EGC2016-T-DU-112.pdf.

Bertani, R. (2015). Geothermal Power Generation in the World – 2010–2015 Update Report. Proceedings, World Geothermal Congress 2015 Melbourne, Australia, 19–25 April 2015. Available at: www.geothermal-energy.org/pdf/IGAstandard/WGC/2015/01001.pdf.

Bodri, L., Čermák, V. (2007). *Borehole Climatology – a New Method How to Reconstruct Climate*. Elsevier Science, Amsterdam.

Chandrasekharam, D., Lashin, A., Al Arifi, N., Bassam, A., Varun, C. (2016). Desalination of Seawater using Geothermal Energy to Meet Future Fresh Water Demand of Saudi Arabia Water Resources Management 30, 1–16. DOI: 10.1007/s11269-016-1419-2.

Dal Porto, F., Pasqui, G., Fedeli, M. (2016). Geothermal Power Plant Production Boosting by Biomass Combustion: Cornia 2 Case Study. Proceedings, European Geothermal Congress 2016 Strasbourg, France, 19–24 Sept 2016 Available at: www.geothermal-energy.org/pdf/IGAstandard/EGC/2016/EGC2016-T-PO-175.pdf.

Date, A., Ahmadi, M., Akbarzadeh, A., Gabguly, S., Mohan Kumar, M. (2015). Experimental Performance Investigation and Case Study of Combined Desalination and Power Generation. Proceedings, World Geothermal Congress 2015 Melbourne, Australia, 19–25 April 2015. Available at: www.geothermal-energy.org/pdf/IGAstandard/WGC/2015/26061.pdf.

Deichmann, N., and Giardini, D. (2009). 'Earthquakes Induced by the Stimulation of an Enhanced Geothermal System below Basel (Switzerland). *Seismological Research Letters*, 80(5), 784–798. DOI:10.1785/gssrl.80.5.784.

DiMarzio, G., Angelini, L., Price, W., Chun, C., Harries, S. (2015). The Stillwater Triple Hybrid Power Plant: Integrating Geothermal, Solar Photovoltaic and Solar Thermal Power Generation. Proceedings, World Geothermal Congress 2015 Melbourne, Australia, 19–25 April 2015. Available at: https://pangea.stanford.edu/ERE/db/WGC/papers/WGC/2015/38001.pdf.

EGEC (2012). *Strategic Research Priorities for Geothermal Technology - European Technology Platform on Renewable Heating and Cooling*. European Communities, Brussels. Available at: www.rhc-platform.org/fileadmin/Publications/Geothermal_SRA.pdf.

Fox, Don B., Sutters, D., Beckers, K., Maciej, Z., Koch, D., Anderson, B., Tester, J. (2013). Sustainable heat farming: Modeling extraction and recovery in discretely fractured geothermal reservoirs. *Geothermics* 46, 42–54. Available at: http://dx.doi.org/10.1016/j.geothermics.2012.09.001.

Freeze, R., Javandel, I. (2008). An interview with Paul Witherspoon, distinguished hydrogeologist from the USA. *Hydrogeol. J.* 16, 811–815. DOI: 10.1007/s10040–008–0308-z.

GEA (2016). Geothermal Energy Accociation 2016 Annual U.S. & Global Geothermal Power Production Report. Available at: http://geoenergy.org/reports/2016/2016%20Annual%20US%20Global%20Geothermal%20Power%20Production.pdf.

GHL (1988).' Honey Lake Power Facility under Construction', Geothermal Hot Line; (USA), 18, No. 2 pp. 61–62 Available at: http://pubs.geothermal-library.org/lib/journals/hotline_18_2_Dec_1988.pdf.

Goldstein, B., Hiriart, G., Bertani, R., Bromley, C., Gutierrez-Negrin, L., Huenges, E., Muraoka, H., Ragnarsson, A., Tester, J., Zui, V. (2011). Geothermal Energy. In IPCC Special Report on Renewable Energy Sources and Climate Change Mitigation [O. Edenhofer, R. Pichs-Madruga, Y. Sokona, K. Seyboth, P. Matschoss, S. Kadner, T. Zwickel, P. Eickemeier, G. Hansen, S. Schlomer, C. von Stechow (eds)], Cambridge University Press, Cambridge and New York. Available at: www.ipcc.ch/pdf/special-reports/srren/Chapter%204%20Geothermal%20Energy.pdf.

Grant, M.A. (2016). Physical performance indicators for HDR/EGS projects. *Geothermics* 63, 2–4. Available at: http://dx.doi.org/10.1016/j.geothermics.2015.01.004.

Gutierrez, H., Espindola, S. (2010). Using Low Enthalpy Geothermal Resources to Desalinate Sea Water and Electricity Production on Desert Areas in Mexico. Proceedings, World Geothermal Congress 2010 Bali, Indonesia, 25–29 April 2010. Available at: www.geothermal-energy.org/pdf/IGAstandard/WGC/2010/2808.pdf.

Hiriart, R., Prol-Ledesma, R.M., Alcocer, S., Espindola, S. (2010). Submarine Geothermics; Hydrothermal Vents and Electricity Generation Proceedings, World Geothermal Congress 2010 Bali, Indonesia, 25–29 April 2010. Available at: www.geothermal-energy.org/pdf/IGAstandard/WGC/2010/3704.pdf.

Jain, C., Vogt, C., Clauser, C. (2015). Maximum potential for geothermal power in Germany based on engineered geothermal systems. *Geothermal Energy* 3, 1–20. Available at: https://geothermal-energy-journal.springeropen.com/articles/10.1186/s40517-015-0033-5.

Kant, M., Becker, D., Brkic, D., Meier, T., Schuler, M., Philipp Rudolf von Rohr, P. (2015). Investigation of a Novel Drilling Technology – Influence of the Surface Temperature for Hydrothermal Spallation Drilling. Proceedings World Geothermal Congress 2015 Melbourne, Australia, 19–25 April 2015. Available at: www.geothermal-energy.org/pdf/IGAstandard/WGC/2015/21015.pdf.

Kohl, T., Speck, R. (2004). Electricity Production by Geothermal Hybrid Plants in Low-Enthalpy Areas. Proceedings, Twenty-Ninth Workshop on Geothermal Reservoir Engineering Stanford University, Stanford, California, January 26–28, 2004 Availabe at: www.geothermal-energy.org/pdf/IGAstandard/SGW/2004/Kohl.pdf.

Lund, J., Boyd, T. (1999). Small geothermal power project examples. *Geothermal Resources Council Bulletin, June* 1999, 9–26. Available at: https://geothermal.org/bulletin.html.

Lund, J., Boyd, T. (2015). Direct Utilization of Geothermal Energy 2015 Worldwide Review. Proceedings, World Geothermal Congress 2015 Melbourne, Australia, 19–25 April 2015. Available at: www.geothermal-energy.org/pdf/IGAstandard/WGC/2015/01000.pdf.

Manente G., Field R., DiPippo R., Tester J.W., Paci M., Rossi N. (2011). *Hybrid Solar-Geothermal Power Generation to Increase the Energy Production from a Binary Geothermal Plant, Proceedings 2011 ASME International Mechanic Engineering Congress and Exposition, Paper no.* IMECE 2011–63665. Denver, CO. Available at: https://inldigitallibrary.inl.gov/sti/6330916.pdf.

Menberg, K., Blum, P., Schaffitel, A., Bayer, P. (2013). Long-Term Evolution of Anthropogenic Heat Fluxes into a Subsurface Urban Heat Island. *Environmental Sci. Technol.* 47(17), 9747–9755. DOI: 10.1021/es401546u.

Menberg, K., Blum, P., Kurylyk, B., Bayer, P. (2014). Observed groundwater temperature response to recent climate change. *Hydrol. Earth Syst. Sci.* 18, 4453–4466. Available at: www.hydrol-earth-syst-sci.net/18/4453/2014/hess-18-4453-2014.pdf.

NEA (2013). Underground Research Laboratories (URL). Radioactive Waste Management. Nuclear Energy Agency OECD, Paris, 52 p. Available at: www.oecd-nea.org/rwm/reports/2013/78122-rwm-url-brochure.pdf.

OECD (2013) Radioactive Waste Management: Underground Research Laboratories (URL), NEA/RWM/R (2013)2. OECD, Nuclear Energy Agency.

Pruess, K., Spycher, N. (2010). Enhanced Geothermal Systems (EGS) with CO_2 as Heat Transmission Fluid – a Scheme for Combining Recovery of Renewable Energy with Geologic Storage of CO_2. Proceedings, World Geothermal Congress 2010 Bali, Indonesia, 25–29 April 2010. Available at: www.geothermal-energy.org/pdf/IGAstandard/WGC/2010/3107.pdf.

Randolph, J.B., Saar, M.O. (2011). Combining geothermal energy capture with geologic carbon dioxide sequestration, *Geophys. Res. Lett.* 38, L10401. DOI: 10.1029/2011GL047265.

REN21 (2016). Renewables 2016 Global Status Report 2016. REN21 Secretariat, Paris. Available at: www.ren21.net/wpcontent/uploads/2016/10/REN21_GSR2016_FullReport_en_11.pdf.

Richter, B. (2010). Geothermal Energy Plant Unterhaching, Germany. Proceedings World Geothermal Congress 2010 Bali, Indonesia, 25–29 April 2010. Available at: www.geothermal-energy.org/pdf/IGAstandard/WGC/2010/0408.pdf.

Rogner, H.H. (2000). Energy Resources. In: *World Energy Assessment: Energy and the Challenge of Sustainability*. United Nations Development Programme, New York, 135–172.

Rybach, L. (1992). An attempt to interpret the temperature profile of the KTB pilot drrillhole (Germany) by paleoclimatic considerations. *Paleogeogr. Paleoclimatol. Paleoecol. (Global Planet. Change Sect.)* 98, 193–197.

Rybach, L., Megel, T., Eugster, W. (1999). How renewable are geothermal resources? *Geothermal Resources Council Transactions* 23, 563–566. Available at: www.geothermal-library.org/index.php?mode=pubs&action=view&record=1016527.

Rybach, L., Megel, T., Eugster, W. (2000). At What Timescale Are Geothermal Resources Renewable? Proceedings World Geothermal Congress 2000, Kyushu, Tohoku, Japan, May 28 - June 10, 2000. Available at: www.geothermal-energy.org/pdf/IGAstandard/WGC/2000/R0900.PDF.

Rybach, L., Kohl, T. (2004). Waste heat problems and solutions in geothermal energy. In: Energy, Waste and the Environment, R. Gieré, P. Stille (eds.), *Geological Society, London, Special Publications* 236, p. 369–380. DOI: 10.1144/GSL.SP.2004.236.01.21.

Rybach, L., Mongillo, M. (2006). Geothermal sustainability – a review with identified research needs. *Geothermal Resources Council Transactions* 30, 1083–1090. Available at: www.geothermal-library.org/index.php?mode=pubs&action=view&record=1025179.

Rybach, L., Bayer, P., Rivera, J., Blum, P. (2016). Influence factors in the depth domain of borehole heat exchangers – global warming and urban heating. Proceedings, European Geothermal Congress 2016 Strasbourg, France, 19–24 September 2016. Available at: www.geothermal-energy.org/pdf/IGAstandard/EGC/2016/EGC2016-S-GP-267.pdf.

Sanyal, S., Butler, S. (2005): An Analysis of Power Generation Prospects from Enhanced Geothermal Systems. Proceedings World Geothermal Congress, Antalya, Turkey, 24–29 April 2005. Availabel at: www.geothermal-energy.org/pdf/IGAstandard/WGC/2005/1632.pdf.

Schiegg, H.O., Rødland, A., Zhu, G., Rødland, A., Yuen, D. (2015) Electro-pulse-boring (EPB): Novel super-deep drilling technology for low cost electricity. *J. Earth Sci.* 1, 37–46. DOI: 10.1007/s12583–015–0519-x.

Schill, E., Nicolas Cuenot, N., Genter, A., Kohl, T. (2015). Review of the Hydraulic Development in the Multi-Reservoir / Multi-Well EGS Project of Soultz-sous Forêts. Proceedings, World Geothermal Congress 2015 Melbourne, Australia, 19–25 April 2015. Available at: www.geothermal-energy.org/pdf/IGAstandard/WGC/2015/31064.pdf.

Shnell, J., Hiriart, G., Nichols, K., Orcutt, J. (2015). Energy from Ocean Floor Geothermal Resources. Proceedings World Geothermal Congress 2015 Melbourne, Australia, 19–25 April 2015. Available at: www.geothermal-energy.org/pdf/IGAstandard/WGC/2015/37011.pdf.

Tester, J.W., Anderson, B.J., Batchelor, A.S., Blackwell, D.D., DiPippo, R., Drake, E.M., Garnish, J., Livesay, B., Moore, M.C., Nichols, K., Petty, S. , Toksoz, M.N. and Veatch, Jr. R.W. (2006). The Future of Geothermal Energy: Impact of Enhanced Geothermal Systems (EGS) on the United States in the 21st Century, *Massachusetts Institute of Technology*, Cambridge, MA (November 1, 2006). Available at: https://energy.mit.edu/wp-content/uploads/2006/11/MITEI-The-Future-of-Geothermal-Energy.pdf.

Thain, I., DiPippo, R. (2015): Hybrid Geothermal-Biomass Power Plants: Applications, Designs and Performance Analysis. Proceedings World Geothermal Congress 2015 Melbourne, Australia, 19–25 April 2015. Available at: www.geothermal-energy.org/pdf/IGAstandard/WGC/2015/26020.pdf.

Wright, P.M. (1995). The Sustainability of Production from Geothermal Resources. Proceedings, World Geothermal Congress 1995, p. 2825–2836. Available at: www.geothermal-energy.org/pdf/IGAstandard/WGC/1995/4-Wright.pdf.

PART VIII

Climate Change and Global Change

28

Future Earth, Climate Change, and Global Change: Future Earth's Ocean

Martin Visbeck and Anke Schneider

28.1 Motivation

The ocean covers 71% of the Earth surface, provides more oxygen than all rainforests combined and holds vast and in some cases untouched resources. It has always played an important role for humanity. It provides us with food, materials, energy and transportation and plays a major role in stabilising our climate. In addition to those provisioning and regulating services the ocean offers cultural services – humans enjoy the seascape and use it for religious or recreational practices. At the same time storm surges, tsunamis, hurricanes and other natural disasters are intimately linked with the ocean. Over the last several decades an exponentially growing population and their development needs have put additional stresses on the ocean and coastal space. For most of human times more than half of the population lived within three hundred kilometres from the coast. Coastal build up and an increasing number of human settlements have put the coastal environment at potential risk. Today twelve out of fifteen mega cities are located near the ocean. In addition, growing levels of ocean use (fishing, mining, transport) and dramatic increases in ocean pollution from carbon dioxide (CO_2) and global warming to land-based nutrient runoff, oil and plastics are additional stressors on the ocean and its ecosystems. Despite the growing challenges and the opportunity to safeguard prosperity, currently the ocean space is weakly and not always consistently governed. Thus the sustainability of the ocean's services to humanity and its own ecosystem is at risk. The ocean space will benefit from international cooperation in discovering and observing the ocean, understanding its mechanisms and connections and improving ocean system models for assessment and predictions. Then the new and existing knowledge needs to be better shared and synthesised and optimal or reasonable pathways for sustainable ocean development need to be proposed and agreed on by all actors. More coherent, inclusive and effective ocean governance arrangements are needed as well as more international policies and cooperation at the regional to global scale.

28.2 Ocean Prosperity

Humans have always benefitted from the ocean and its diverse ecosystem services, which are of varying relevance for individuals, groups or stakeholders. Sometimes we speak of a "healthy and productive" ocean, referring to the desire to maintain the ocean in a prosperous state. More generically ecosystem services can be classified into four categories: provisioning services, supporting services, regulating services and cultural services.

Provisioning ecosystem services comprise products and goods for direct human sustenance but also spaces and areas that are needed to ensure sustainable productivity. Alongside transportation routes the most important provisioning service is food from the sea by fish. Globally fish and seafood are existentially important for the nutrition of many millions of people, and the fishery industry is of significant economic importance. Particularly the coastal populations in developing countries are often directly dependent on the fish catch, providing food and income for less developed coastal communities. Further, the ocean is a great source for freshwater – on the one hand, through evaporation and subsequent rainfall over land, which is important for agriculture and food security especially in tropical regions, and on the other hand, also increasingly through man-made desalinisation plants. Other provisioning services are non-renewable resources from the ocean. Sand, ores and fossil fuels such as oil and gas have been exploited for many years. Mining for coastal phosphate depositions or minerals on the sea floor could begin in the near future. A sustainable but so far rarely used resource from the ocean is the energy latent in tides, waves, vertical temperature gradients and strong offshore winds. There are increasing efforts to mobilise this renewable energy for electric power production. For the food and cosmetic industries and for the development of new medicines, the genetic information and biogeochemical substances from marine organisms seem to be of great potential.

Cultural ecosystem services encompass diverse functions, which serve the non-material well-being of humankind. Beside the cultural services linked to traditions, religion and social aspects, the recreational function and leisure value of the ocean and its landscapes are of major importance as they are closely linked to tourism. Every year a large number of tourists spend their holiday either directly at the beach or at the sea to pursue diving or other sporting activities.

Supporting ecosystem services are not of direct use for humans but rather can be regarded as the essential basis for the other categories of services. Among supporting services are biological, chemical and physical processes; the dynamics of the marine food web; the primary production of phytoplankton and the diversity of marine habitats itself. These services either form the basis of life on Earth or they balance and stabilise it. Thus, ultimately they are indispensable for humans.

Regulating ecosystem services comprise the many advantages and benefits gained from the regulating effect of the sea. Mangrove forests, sand dunes and coral reefs protect the coasts from erosion by storms and floods. Single-celled organisms and bacteria biodegrade excrements, other pollutants and even toxic substances we discharge into rivers and coastal waters and thereby maintain water purity. Furthermore the ocean has a controlling influence on the climate – regionally and globally. During summer the sea can store heat and then carry warmer air onto cold and dry land in coastal areas in winter. Because of evaporation over the sea, in many regions the ocean also is responsible for rainfall over the land. Water can store heat for a long period. Seawater in the tropics absorbs large quantities of solar radiation, which is then transported over many thousands of kilometres to the poles. The ocean also regulates gases. The surface ocean is permanently exchanging gases with the atmosphere. The uptake, for example, of large amounts of the greenhouse gas CO_2 since the Industrial Revolution prevented our atmosphere from far greater warming than has actually occurred.

These services collectively ensure a healthy and productive ocean and thus are fundamental for societal well-being and human survival. While many of them can be used sustainably, this is not always the case. A good example of overuse of a potentially sustainable service is ocean fishing.

Overfishing is probably one of the best known ocean issues. It can pose a serious threat to marine ecosystems and thus reduce the ocean's ability to provide food sustainably. Fish is the major source of animal protein in many poorly developed coastal regions. Economically fishing is a global business responding to the increasing demand for fish and seafood. Globally around 80 million tons of fish are caught per year; 25% of the stocks are overfished or close to overfishing and around half of the stocks are fished to maximum capacity with little safety margin (FAO 2016). Illegal fisheries but also poor management plus the absence of compliance with existing regulations pose a major threat. To allow fish stocks to recover, they have to be managed sustainably and the control and enforcement of regulations are needed. Another example of overuse is coastal build up due to transport or tourism needs.

28.3 Ocean Pollution

A rapidly growing, affluent and more technologically advanced humanity is using its surrounding environment heavily. In many cases the desirable outcome of more energy, food, material or water produces unwanted chemicals, substances or simply wasted material. The amount of pollution is especially high in densely populated coastal regions, settlements or mega cities. Coastal regions are among the mostly populated areas in the world and the dependence on coastal resources is even increasing (Visbeck, 2014).

CO_2 pollution: The continuous increase of atmospheric CO_2 by emissions from burning fossil fuels causes significant changes in the planet's radiation budget. Thus the planet's surface is warming. Ocean warming accounts for more than 90% of the extra energy that the Earth has been receiving (IPCC Report, 2013). The results are increased surface water temperatures, stronger ocean stratification, rising sea levels (due to thermal expansion and ice melting) and changes in ocean currents.

As the ocean is a major sink for the increased atmospheric carbon dioxide concentrations, it is partly buffering global warming. About 30% of anthropogenic CO_2 emissions have been taken up by the ocean (IPCC Report, 2013). However, the downside is a gradual acidification of the seawater (e.g. Doney, 2010) because when CO_2 dissolves in seawater it reacts with the water molecules to form carbonic acid (H_2CO_3), which quickly dissociates into HCO^{3-} and CO_3^{2-} leaving free H^+ ions and thereby decreasing the pH. Current research is exploring whether a reduction of the ocean pH may pose a threat to marine species, especially calcifying organisms such as corals, bivalves and some zoo- and phytoplankton species at the bottom of the food web.

Together global warming and ocean acidification have a large potential impact on marine ecosystems and coastal habitats. Another more indirect effect of increased temperatures and stratification is the loss of oxygen in the ocean. This yields, for example, an expansion of the subtropical oxygen-minimum zones (Stramma et al., 2008).

Eutrophication is an indirect effect of pollution and the result of heavy fertilisation and poor wastewater management. It is a serious problem for coastal waters, in particular big river deltas, where the nutrient-rich waters enter the sea. Phosphorus and nitrogen compounds lead to enhanced algal growth, resulting in algal blooms (WOA 2016, chapter 6). Firstly, algal blooms may change the turbidity of seawater and limit light penetration. Secondly, the degradation process of the algal biomass by bacteria consumes oxygen and in extreme cases the oxygen can be depleted. In these anoxic or dead zones, fish, crustaceans or bivalves can no longer survive. Furthermore, some algae have the potential for producing toxins with impacts on humans and animals (Glibert et al., 2005). This is a clear example of the terrestrial side to marine threats and necessary protection. The general need for policies and water pollution control in the fields of agriculture, urbanisation and sewerage services is obvious.

Chemical pollution of the environment caused by human activity is another major threat to marine ecosystems. A broad range of chemicals reaches the ocean. Pollutants are contained in untreated waste water, exhaust air from industry, oil from drilling platforms or tanker accidents and plastic litter. Transported by currents, pollutants can be found far away from their source in quite remote areas. In spite of two intergovernmental conventions for internationally mandatory marine protection (London Convention, 1972 and London Protocol, 1996), the situation continues to be very poor.

Plastic litter is mainly land-born, reaching the ocean via rivers or wind. Incredible amounts of plastic accumulate in the middle of the ocean basins, where currents converge and get caught in rotating swirls, such as the Great Pacific Garbage Patch. In addition the smallest plastic particles, called microplastics, and associated toxic substances such as plasticiser and solvents can accumulate in the food chain (WOA 2016, chapter 25).

Invasive species are another form of pollution that can have significant consequences for local ecosystems (WOA 2016, chapter 17). Global shipping but also changing environmental conditions such as warming or loss of oxygen have led to changes in species abundance. Ships use water ballast tanks to replace the loss of mass when the fuel is used. Thus they transport water from one ocean region to another. International rules have been established to require ships to kill organisms in the ballast water.

In many regions of the ocean the effects of the diverse factors of ocean pollution add up and cause enormous and sometimes dramatic changes. A visible example is the strong reduction of summer sea ice in the Arctic. Another prominent example is the decline of coral reefs. Coral reefs are one of the largest and most diverse ecosystems on Earth and they provide important ecosystem services for millions of people. Besides ocean acidification, other stress factors pose major threats to coral reefs. For instance, increasing ocean temperatures and extreme temperature events such as El Niño lead to bleaching and coral death if lasting too long. Further, viral attacks, overfishing, sedimentation, pollution and other local threats endanger the reefs. It has been estimated that about 20% of the global coral reefs already have been destroyed and that a further 35% have severely suffered (WOA 2016, chapter 43). When combining all threats, around 75% of reefs are endangered. But it is not only the coral reefs that are under threat in coastal regions. Coasts tend to be affected by a multitude of problems at the same time; the United Nations estimates that half of the world's population lives within sixty kilometres of the sea and are dependent on a healthy ocean and its coastal services. Interactions with negative impacts on the coastal zone are overfishing and destructive fishing methods, coastal construction projects, the discharge of pollutants and sediments from rivers or ships as well as tourism and recreational activities. The destruction of coral reefs, wetlands, mangrove forests and salt marshes by human activities has resulted and will continue to result in major alterations of the coasts, with consequences ranging from increased erosion to decrease in drinking water and loss of breeding grounds and restoration zones for biodiversity, and food supply to changes in coastal morphological dynamics and "natural beauty" of landscapes.

28.4 Ocean Resilience

Humans are benefiting from the ocean while ocean hazards can be devastating for coastal communities. Throughout human evolution civilisation has developed strategies to increase the resilience to threats from the ocean. Dikes have been built to protect whole coastlines from storm surges; critical infrastructure is built away from low-lying coastal areas. However, the rush for coastal recreation and access to the sea has led to newly built infrastructure that is very vulnerable to any types of ocean extreme events. Destruction of natural protection systems such as mangroves or healthy reef system and the decline of protecting coastal sea ice due to ocean warming have increased the vulnerability of coasts to strong winds and associated storm surges. Another dimension of resilience is marine biodiversity. A species-rich ecosystem is expected to be more resilient against pathogens, warming or temporary changes in ocean chemistry. Marine protected areas, natural coastal defences and regionally to globally agreed fishing quotas, mining codes or regulations about ocean pollution are all critical elements to safeguard ocean resilience. A specific challenge is the existing complex and often weak governance system of the ocean.

There have been several developments in the past to govern the marine environment regionally and internationally in a responsible and sustainable manner with some good examples. However, the existing system of ocean governance has failed to safeguard ocean resilience in several areas. Three factors have been identified: first, insufficient regulations; second, the lack of compliance and third, weak coherence at a global scale. The establishment of a consistent and comprehensive ocean governance regime is hampered through different types of ocean governance institutions, international treaties (conventions) and agreements involved. Most of the existing ocean governance regimes are sectorial and cover issues such as deep sea mining, fisheries in the open ocean, biodiversity, the conduct of science, international pollution, shipping and labour law. At the international level of the United Nations (UN) alone more than ten organizations are dealing with some aspects of ocean governance, and none is in charge of the whole architecture. For example, the International Maritime Organization (IMO) regulates shipping while the International Seabed Authority (ISA) is in charge of deep sea mining outside the economic zones of coastal states. UN organisations such as the United Nations Environmental Programme (UNEP) or the Food and Agricultural Organization (FAO) both have multinational activities related to the ocean. Possibly the most significant instrument is the United Nations Convention of the Law of the Sea (UNCLOS), which since 1994 sets out the ground rules for global marine policy. It is treated as an international constitution for the ocean and obliges all signatory states to protect and conserve the marine environment and to make use of marine resources in a sustainable way – nevertheless without being specific on the definition of sustainability or environmental protection standards. According to the UNCLOS the ocean is subdivided into various zones. The Territorial Sea is the 12-nautical-mile zone and is part of a state's sovereign territory – all activities are subject to the legislation of the specific states. Ships from all countries can pass though these waters. The Exclusive Economic Zone (EEZ) includes the sea floor and the water column and extends farther offshore to a distance of 200 nautical miles. The EEZ does not belong to the coastal state's sovereign territory, but only the coastal state is allowed to extract resources in its own EEZ. Other nations need to ask for permission and negotiate benefit sharing to do so. The high seas (or international waters), adjoining the 200-nautical-mile zone, are available for use by all countries. Their use is regulated, for instance, by Regional Fisheries Management Organizations (related to fisheries) or the International Seabed Authority (related to mining of sea bed resources). An exception to this zoning is the continental shelf, which is the sloping sea floor off the coast. In case this natural geological extension of the mainland stretches out over the 200-nautical-mile limit, a country can exploit the seafloor on the extended continental shelf up to a maximum of 350 nautical miles or even more. Especially in areas with a high potential for possible exploitation of mineral resources and with more than one country claiming an extended continental shelf, this evokes dispute. One example is the North Pole, where Canada, Norway, Russia, Denmark (via Greenland), and the United States claim areas.

However, as fish stocks or toxic substances as well as the threats of climate change do not follow any borders, the zoning of UNCLOS is not always encouraging a comprehensive ocean governance regime. Further, the UNCLOS has no provision for the implementation of marine protection areas, an oversight that has been recognised and is currently being addressed. In addition to the UN authorities and global agreements, there are a multitude of regional ocean organisations and non-governmental institutions regulating different aspects such as fishing, pollution and cooperation at the regional level.

Marine Protected Areas (MPAs) can be established in international and in territorial waters and generally have individual protection objectives and only a few enforce comprehensive protection. Thus, for instance, banning sea floor trawling whereas fishing in the water column is still allowed can conserve sensitive habitats on the sea floor but does not reduce overfishing. The legal designation of an MPA is straight forward for national since the coastal state has juridiction. For the establishment of MPAs in international waters, an agreement on the protection objectives between countries that make use of the sea area must be reached. However, there is a process ongoing to derive a legal frame for that. Consensus cannot always be achieved and other states cannot be obliged to respect the MPA. At the biodiversity conference in Nagoya in 2010 it was agreed that 10% of the global ocean surface should be placed under protection by the year 2020. Up to now, there are too few protected areas, and often they are very isolated from each other. To link protected areas in trans-regional networks would help to keep up with habitat conservation and protection of species with wide-ranging distribution.

Another mechanism for multinational ocean governance is the Regional Seas Programme (RSP) initiated in the 1970s by the UNEP. This global initiative concentrates on the protection of particular sea regions based on agreements between the respective neighbouring states. Today the following eighteen regions are covered within the RSP: Antarctic, Arctic, Baltic, Black Sea, Caspian, Eastern Africa, East Asian Seas, Mediterranean, North-East Atlantic, North-East Pacific, Northwest Pacific, Pacific, Red Sea and Gulf of Aden, Regional Organisation for Protection of the Marine Environment (ROPME) Sea Area, South Asian Seas, South-East Pacific, Western Africa and Wider Caribbean. Most of the RSPs function through Action Plans and the majority of countries are participating. However, because of financial and institutional challenges and the lack of political will, the implementation and the outcome of the RSPs vary considerably among states and programmes. The Helsinki Commission (HELCOM), responsible for the protection of the Baltic Sea, and the Oslo-Paris Convention (OSPAR), covering the North Sea and the Northeast Atlantic, are examples for good ocean governance within the framework of the Regional Seas Programme. The Charlie Gibbs Marine Protection Area in the Atlantic is one of the few protected areas in international waters and has been designated and agreed on by the OSPAR contracting parties even though it is located outside their exclusive economic zones. In contrast, the Coordinating Body on the Seas of East Asia (COBSEA), in authority for the East Asian Seas, has been confronted with major problems in the implementation of the Action Plan. The reasons can be found in the variety of political, economic and social systems of the states involved and the many environmental pressures and threats in the region.

28.5 Ocean Sustainable Development

The original concept of sustainability arose from the need for a sustainable supply of wood for mining operations, and the ensuing concept in a nutshell refers to using natural resources intelligently so that the supply never runs out. This long-term dimension of sustainable development thus ties in very closely with intergenerational justice, i.e. that the needs of current and future generations are both equally considered. In the classic model of sustainability, the environment, the economy and the society are represented as three pillars equally supporting sustainability (Bruntland, 1987). However, in reality this is seldom the case. Numerous modifications of this model have been developed, but even now, the critics point out in many cases economic concerns still take precedence over environmental and social aspects. Today an important distinction is made between strong and weak sustainability. According to the model of weak sustainability, natural capital that has been consumed can in principle be replaced without limit by real capital and human capital, i.e. full substitution of benefits. According to the idea of strong sustainability, in turn, natural capital can only be consumed if it can be replaced by equivalent natural capital: i.e. no substitutions are allowed.

Another way to consider sustainable development stems from the concept of planetary boundaries, which defines nine essential (and exemplary) environmental dimensions such as biodiversity loss or ocean acidification with corresponding planetary thresholds or guardrails (Steffen, 2009). Exceeding a critical limit could result in sometimes-irreversible environmental changes and unforeseen consequences for human life on Earth. The main challenges are to quantify meaningful (global) boundaries and to formulate comprehensive solutions. Furthermore, humankind is facing several societal problems such as hunger, poverty, lack of health and education systems, and a desire for social justice exists. Thus, the revised version of the planetary boundaries concept has been supplemented with these social aspects. The linkage between the environmental and the societal dimensions creates a safe operating space bordered by the essential human needs on the one side and the planetary boundaries on the other side. A critical issue in this combined concept is the consideration of local, regional and global scales. Only where the social dimensions at all scales have been fulfilled but the critical environmental limits at all scales have not been crossed does a safe and just space for humankind emerge.

However, the reported concepts all lack a clear and agreed development path that ensures sustainability. Furthermore, damage to the marine environment is often not directly visible or measurable and continues far beyond national borders and into international areas. Compared to critical issues on land, this makes it more challenging to apply and implement the concept of sustainability to the ocean and its governance arrangements.

In the year 2000 eight Millennium Development Goals (MDGs) were established by the United Nations in order to achieve improvements in social conditions in the developing countries by 2015. Several goals have been reached. The criticism, however, was that neither marine resources nor governance aspects were mentioned in the MDGs and that they only applied to developing countries. In 2010 it was agreed that governance aspects for the period from 2015 to 2030 should be addressed and that the goals should be universal and apply to all countries. Further it was recognised that living conditions could not be improved if neglecting the environment. In order to define new development goals that are taking these considerations into account for the time after 2015, an Open Working Group (OWG) was established in 2012 under the auspices of the UN. The purpose of the Open Working Group was to involve a large range of stakeholders. Interest groups, businesses and individuals could feed into the process via an Internet portal and submit proposals on the new goals. Experts from the scientific community and civil society groups were invited to share ideas and experiences on several aspects of sustainability. The consultation process lasted eight months, after which the OWG with representatives for all member countries developed the new goals. The report was published in spring 2014 and consisted of 17 Sustainable Development Goals (SDGs) and 169 targets to be achieved until 2030. Projects to combat poverty or hunger were no longer seen as isolated from climate actions and one Sustainable Development Goal (SDG 14) was explicitly dedicated to the ocean. In September 2015 the Agenda 2030 for sustainable development and its 17 SDGs were approved at the United Nations General Assembly and thus for the first time a framework for action towards comprehensive sustainable global development existed, although financing of the implementation of 2030 Agenda has up to now not been clarified.

Critique: Already in spring 2015 the International Council for Science (ICSU) reviewed the new SDGs and especially criticised that the SMART criteria were not always met. The SMART criteria is a concept from business and project management stating that goals can only be achieved when they are **s**pecific, **m**easurable, **a**ttainable (and ambitious), **r**elevant, and **t**ime-bound. The reason for this might be the fact that the UN negotiations are a political process and consensus has to be found among all countries. However, even if not all of the goals will be reached by 2030, the purpose of overcoming the limitations of the MDGs and developing a sustainable agenda that covers environmental, social and economic aspects as comprehensively as possible was reached. The first set of indicators has been released but scientists are not sure that the current indicator framework meets the ambition for sustainable ocean development.

In practice several coastal states have realised the challenge of sustainable ocean development and decided to embark on a space-based planning and regulatory framework. The idea for national implementation is to realise an ecosystem approach to ocean management and balance the various human demands on the ocean space, such as shipping, fisheries, extraction, energy production, tourism and marine protected areas. In other words, Marine Spatial Planning (MSP), as a means of coordinating the various coastal and marine interests, should always be realised in a way that prevents negative impacts on marine habitats and the provision of ecosystem services.

28.6 The Need for More Integrated and International Ocean Research

Comprehensive and sustainable use and governance of the ocean are a major challenge for the future. During the negotiations of the Sustainable Development Goals, the implementing agreement to establish marine protected areas in the open ocean or for the Paris climate agreement, it became clear that there are many remaining scientific questions about how the ocean system works, how it will evolve in the future under different development and protection scenarios and what any development will mean for sustained ocean prosperity.

Fundamentally it is not clear whether the world will or even can achieve the Sustainable Development Goals in general and the ocean goal (SDG 14) in particular. Thus there is a need to explore a range of regional specific development pathways and their appropriate governance strategies. Such a process will provide new insight but also call out areas where lack of scientific understanding or ocean information hinders the selection of a preferred or most beneficial development pathway. While there are a number of integrated research priorities and inherent uncertainties, there are some promising targets to be tackled to facilitate ocean sustainability:

1. The establishment of sufficient MPAs within the economic zones of coastal states and in the open ocean: MPAs are an instrument for placing particular areas under protection and can be established both in international and in territorial

waters. Generally they are used to pursue individual protection objectives. Thus the establishment of an MPA does not mean that the sea area may no longer be used at all and is protected in every respect.

2. The reduction of marine pollution: Marine pollution embraces all areas of discharge (toxic substances, nutrients and litter) as well as noise pollution and the uptake of anthropogenic CO_2. A reduction is essential to counteract the many negative implications for the marine environment.
3. Encouragement of national and regional consistent MSP: To protect the marine environment effectively, there must be better coordination between its conservation and use, and MSP is an important tool in achieving this goal by meeting social, economic, and ecological objectives.
4. The development of a more coherent and adequate ocean governance: In Europe, for example, the Joint Communication (an international ocean governance agenda for the future of our ocean) is an integral part of the EU's response to the United Nations' 2030 Agenda for Sustainable Development, in particular SDG 14. It sets out detailed actions to shape international governance.
5. Improvement of the capacity to observe, understand, predict and manage the ocean at the local, regional and global levels by education and training.

However, in order to accomplish SDG 14 and its interaction with the other SDGs fully, more research is needed to identify possible and promising ocean development scenarios on the global, regional and local levels. There is an increasing need for more global cooperation in trans-disciplinary ocean research. There are several internationally coordinated research activities such as WCRP, ICSU-SCOR and IOC-GOOS and in addition several core projects, formerly IGPB, IHDP and Diversitas, now merged into Future Earth. Future Earth is an initiative to advance knowledge and build capacity in the rapidly expanding area of global sustainability science and to provide an international research agenda to guide natural and social scientists working around the world. It is also a platform for international engagement to foster integrated global environmental change research, designed and conducted in partnership with society to support transformation towards global sustainability in the coming decades.

In order to facilitate rapid and effective ocean knowledge generation a balanced approach at the global, regional and local levels is needed. One way to think about the challenge is to conceptualise ocean information gathering by an ocean value chain (Figure 28.1) that has five critical elements: ocean discovery and observing, ocean understanding, ocean modelling and

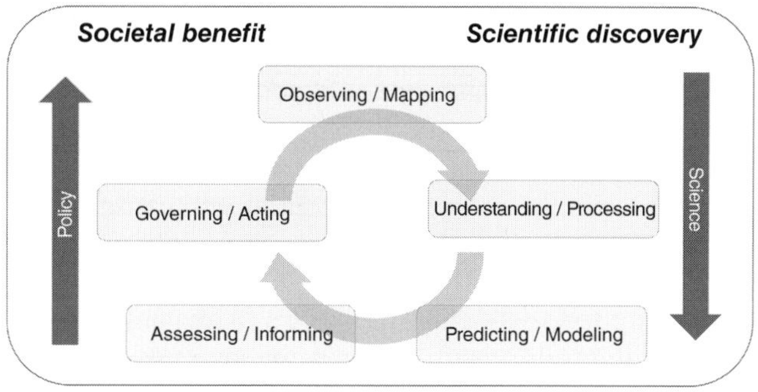

Figure 28.1 Ocean value chain: Increased need for ocean and climate information to meet a growing range of societal needs can only be fully realised if all elements of the value chain are resourced adequately.

projection, ocean assessment and finally ocean policy and governance.

1. Ocean discovery and observing: The vast volume of the ocean and its complex coastlines are neither fully observed nor adequately understood. In particular the deep sea is a frontier of ocean sciences and many discoveries are still to be made. Sometimes internationally coordinated teams can discover new areas of our planet, new phenomena of profound importance or new organisms and substances. In addition sustained and systematic ocean observing is needed to document ocean change, initialise ocean system models and provide critical information for improved ocean understanding.
2. Ocean understanding: The ocean is a very complex and connected global system. For centuries ocean scientists have been trying to understand the ocean dynamics, chemistry, biology, ecosystems and the geology of the sea floor and the meteorology above the ocean. Internationally several disciplinary core projects are conducting innovative and coordinated research to improve understanding of the ocean and the interactions among the various systems. Two areas canbe highlighted as in need of more coordinated research:
   a) Ocean extremes: Extreme events are challenging in the context of risk management. Human interactions with the ocean are growing. However, the full spectrum of marine risks has yet to be discovered, understood and quantified. Ocean risk covers topics beyond the hydro-meteorological domain and includes geohazards, ecosystem changes and tipping points.
   b) Ocean pollution: Increased human activities on the ocean but also in coastal settings more often than not yield increased input of materials, chemicals and other substances. Many of them have the potential to alter the ocean ecosystems.

3. Ocean modelling and projections: Ocean system models tend to focus on specific challenges such as the physical climate, ocean biogeochemistry, the ocean sediment interface, coastal regions or fisheries. Population increase, economic wealth and an increasing environmental footprint will cause more profound changes in the future, raising the question, How will the ocean change in the Anthropocene? What effects will climate change, increasing fishing, coastal development and growing levels of environmental pollution have on the ocean in the next 20, 50, 100 and 200 years?
4. Ocean assessment: Ocean observation, ocean system modelling and improved ocean understanding lead to a wealth of new scientific information. However, for the political system to react there is a need for improved ocean assessment. The ocean's role in the climate system is part of the climate assessment (IPCC – Intergovernmental Panel on Climate Change) while the changes in the marine biodiversity are somewhat covered by the IPBES (Intergovernmental Science-Policy Platform on Biodiversity and Ecosystem Services) process. At the UN level the First World Ocean Assessment was released in 2016. However, no assessment covers all aspects of the ocean regime.
5. Ocean policy and governance: The legal regime of the ocean and coastal zone is complex. And the connectivity of the global ocean by its circulation means that ocean governance will yield the best outcome when it is regionally or globally coherent. Authority, transparency and the ability to encourage good behaviour are particular challenging in the ocean domain. Equitable access, burden and benefit sharing as well as transnational cooperation warrant innovative approaches to ocean governance.

Once the value chain is implemented and the information flow across the interfaces is well established, one would expect that a whole range of innovative and new ocean solutions would emerge in the context of sustainable development of the marine sphere. Areas such as sustainable fisheries, least destructive exploration, renewable energy, effective and equitable marine spatial planning, marine carbon management, transparent governance and sustainable tourism are just some examples of ocean solutions for a more sustainable world.

28.7 Conclusion

In order to promote more sustainable interactions between humans and our ocean there is an urgent need to integrate existing ocean knowledge and share that with all stakeholders in a solutions-oriented manner. In many areas knowledge exists already but is not easily available; in other areas more in-depth insights are needed. Overall it is important to recognise the full ocean information value chain, which includes ocean discovery and observation, ocean understanding and where possible simulation, ocean predictions and future scenarios. All this information needs to be assessed in order to inform specific ocean policy and governance challenges.

A modern approach to obtain this ocean information in an international, more efficient, more integrated, and fit-for-purpose way is the concept of co-design, co-production and co-dissemination of knowledge. It is based on the idea that research towards global sustainability will be both designed and conducted in partnership between science and society and that integration is an iterative process that involves reflection among all stakeholders. A similar approach is needed for the underlying ocean observations that should be made in a more integrated (cross-issue), more international (sharing) and fit-for-purpose (systems approach) way as articulated in the framework for ocean observing (A Framework for Ocean Observing, 2012). Moreover, an effective ocean information assessment framework is needed, building on the IPCC and IPBES and a significantly improved World Ocean Assessment.

The UN's 2030 Agenda for sustainable development has for the first time explicitly articulated a focussed ocean goal. However, from a scientific perspective this can only be a start and a hook into the global development policy arena. Science needs to define smart targets for what one might call "ocean health" as a goal for ocean sustainability. It needs to establish the knowledge base to develop policies and actions to maintain and improve the productivity and "health" of ocean systems. Science is challenged to offer more integrated information that allows predicting and responding to shocks or disasters to and from ocean systems. Our capacity to chart a course from knowledge of ocean systems to changes in policies, practices, governance and behaviours that will assist in sustaining those systems will depend on our ability to develop scalable and integrated systems approaches to these issues. Future Earth together with its science partner organisations (WCRP, IOC-UNESCO, ISCU-SCOR, SDSN and UNEP) is well positioned to facilitate the formation of trans-disciplinary teams that can address these challenges through an integrative Knowledge-Action Network (KAN). Future Earth Knowledge–Action Networks are collaborative frameworks stimulating highly integrative sustainability research. Their aim is to generate the multifaceted knowledge needed to inform solutions for complex societal issues.

Future Earth and its ocean partners are in the process of establishing an Ocean Knowledge–Action Network (Ocean KAN) responding to demands that have been elaborated and articulated by a range of meetings and conferences. The initial Ocean KAN partners are Future Earth and its marine core projects, WCRP-CLIVAR, IOC-UNESCO and ICSU-SCOR. Following the first international scoping workshop of the Ocean KAN in late 2016, the process of articulating a Research and Engagement Perspective as well as an Implementation and Resource Strategy was set in place. Starting in 2017 over an eighteen-month period a Development Team composed of the research and the stakeholder communities with balanced representation of disciplines, gender, age and regional diversities developed the ambition for the Ocean KAN.

The Ocean KAN is expected to focus on solutions-oriented trans-disciplinary research and the engagement of a broad range of researchers, societal partners and stakeholders from diverse sectors and regions to provide new knowledge, innovation and transformation pathways for societies to develop towards more sustainable interactions with the ocean. Firstly, research priorities in support of achieving SDG 14 will be defined in a co-design process in which the Ocean KAN functions as facilitator for the creation of new integrative initiatives and collaboration of the scientific community, users of knowledge and decision makers around the world. Secondly, the Ocean KAN aims at building scientific capacity worldwide to assess the state of the oceanic and coastal domains, provide new knowledge and expert advice in order to improve the management and governance of the ocean.

References

Bruntland et al. (1987). *Our Common Future World, Commission on Environment and Development*. Oxford: Oxford University Press, p. 27.

Doney S.C. (2010). The Growing Human Footprint on Coastal and Open-Ocean Biogeochemistry. *Science*, **328**, 1512–1516.

FAO (2016). *The State of World Fisheries and Aquaculture 2016: Contributing to Food Security and Nutrition for All*. Rome: FAO.

A Framework for Ocean Observing (2012). By the Task Team for an Integrated Framework for Sustained Ocean Observing, UNESCO, IOC/INF-1284, doi: 10.5270/OceanObs09-FOO.

Glibert P.M., Anderson D.M., Gentien P., Graneli, E., Sellner K.G. (2005). The global complex phenomena of harmful algae blooms. *Oceanography*, **18**, 136–147.

IPCC (2013). Climate Change 2013: The Physical Science Basis. Contribution of Working Group I to the Fifth Assessment Report of the Intergovernmental Panel on Climate Change [Stocker, T.F., D. Qin, G.-K. Plattner, M. Tignor, S.K. Allen, J. Boschung, A. Nauels, Y. Xia, V. Bex and P.M. Midgley (eds.)]. Cambridge University Press, Cambridge and New York, doi:10.1017/CBO9781107415324.

Steffen, W. et al. (2009). Planetary boundaries: Guiding human development on a changing planet. *Science*, **347**.

Stramma, L., Johnson, G.C., Sprintall, J., Mohrholz, V. (2008). Expanding oxygen-minimum zones in the tropical oceans. *Science*, **320**, 655–658.

Visbeck, M., Kronfeld-Goharani, U., Neumann, B. et al. (2014). Securing blue wealth: The need for a special sustainable development goal for the ocean and coasts. *Marine Policy*, **48**, 184–191.

World Ocean Assessment I: The First Global Integrated Marine Assessment (2016). Online: www.un.org/depts/los/global_reporting/WOA_RegProcess.htm (16.07.2016).

29

Asia's Sustainability Challenges and Future Earth

Tetsuzo Yasunari, Hein Mallee, and Reiichiro Ishii

29.1 Introduction

The notion that we have entered a new geological era, the Anthropocene, in which humanity has become the dominant driver of Earth system change (Crutzen, 2002; Steffen et al., 2007) is rapidly gaining currency. Asia's sheer size and dramatic development are increasingly making it a major contributor to this trend. About 60% of the global population lives in Asia, and its total GDP has exceeded one-third of the global GDP. The region is experiencing tremendous demographic change, rapid economic growth and urbanization. Associated with these trends of human activity, Asia has become a hotspot for greenhouse gas emissions and air and water pollution, which have begun to affect regional and even global hydro-climate change and air quality.

These changes are accelerating the decline of ecosystem services, with biodiversity loss in the terrestrial and marine ecosystems of Asia increasing most rapidly in the world. At the same time, economic development trends give rise to great disparities of wealth both within and between countries. Under these social and environmental conditions, the vulnerability of societies and ecosystems to the potential impacts of climate change is being aggravated.

On the other hand, the so-called Asian Green Belt, with its humid monsoon climate and rich biodiversity, has historically offered many examples of long-term social and ecological sustainability. Here, traditional systems of agriculture and livelihoods, with their own unique cultures, have supported large numbers of people through time. In meeting its formidable contemporary sustainability challenges, Asia will need to develop novel approaches in science, technology and governance, building on the wisdom embodied in traditional livelihoods and cultures. Designing sustainable interactions between humanity and nature in Asia is a global challenge, and the ultimate goal of Future Earth, global sustainability, cannot be achieved without sustainability in Asia (Yasunari et al., 2013; Manton et al., 2015).

In this chapter, we first provide an overview of the natural system of Asia, which underpins its burgeoning human society. We then document the current environmental issues, and some of the impacts on human health, caused by active and rapid industrial and urban development in recent decades, as indicated by selected literature. At the end of the chapter, we emphasize the importance and urgent necessity of interdisciplinary and transdisciplinary research and engagement in Asia as part of Future Earth in Asia (or "Future Asia"), to cope with the environmental issues and to achieve long-term sustainability.

29.2 Changes in Climate and Water

29.2.1 Geophysical and Climatic Conditions

The topography of Asia is characterized by an active tectonic zone with its center the Himalaya-Tibetan Plateau (HTP), surrounded by a subduction zone along the periphery of the continent. The HTP and all the mountain ranges that surround it form the highest (average elevation more than 4000 m above sea level) and most extensive highland in the world. The Asian monsoon is a large-scale atmospheric circulation system induced by seasonal differential heating between the Eurasian continent and the Indian and Pacific Oceans. Many previous climate modeling studies have shown the critical role of the HTP (e.g., Hahn and Manabe, 1975; Kutzbach et al., 1993) in the formation of the Asian monsoon. Recent studies using coupled ocean–atmosphere general circulation models have also suggested that the uplift of the HTP over the geological time of the Tertiary to the Quaternary era has greatly impacted the formation of the Asian monsoon system and the atmosphere–ocean coupled system (e.g., Abe et al., 2003; 2004). Because of the far-ranging impact of the monsoon on climate, agriculture and ultimately human society, the affected region is often referred to as "Monsoon Asia".

The HTP contains the largest amount of glacially stored fresh water (snow and ice) outside the Arctic and the Antarctic (Yao et al., 2012), and is sometimes called the "Asian water tower" (Immerzeel et al., 2010). Asia's major river systems are formed from this water tower, with many alluvial plains and valleys formed under these tectonic and geomorphological conditions, combined with the monsoon climate. These alluvial valleys, plains and delta systems (e.g., of the Ganges, the Mekong, the Yangtze and the Pearl River) have provided ideal conditions for paddy rice cultivation and have given rise to some of the most densely populated areas in the world. These geophysical and climatological conditions, on the other hand, cause frequent and serious natural disasters (e.g., earthquakes, landslides, tsunamis and floods) in Monsoon Asia. Particularly, the major river delta areas are highly vulnerable to shifts or shocks originating upstream or in neighboring seas and oceans (Syvitski et al., 2009).

29.2.2 Changes in Hydro-Climate and Floods in Asia

Typhoons, heavy rains, floods and droughts, landslides, earthquakes and tsunamis are the cause of regular natural disasters in Asia and the region needs to improve its capacity for risk management related to such extreme hydro-climatological and tectonic events (ADRC, 2011; Krechowicz and Venugopal, 2010). In particular, an increase in frequency of extremely heavy rainfall associated with the monsoon and typhoons has been reported in Asia in the past few decades (Fujibe et al., 2005; Goswami et al., 2006; Endo et al., 2005). The IPCC concluded from state-of-the-art climate models that this tendency could be further enhanced by global warming, particularly in this region (Field et al., 2012). In contrast to extreme rainfall events, the frequency of moderate rainfall events over central India during the monsoon season has decreased over the period 1950–2000 (Goswami et al., 2006). For the East Asia summer monsoon, studies have found a clear weakening of the monsoon over the last three decades, exhibiting a drying trend (decrease of rainfall amount) in northern China and a wetting trend (increase of rainfall amount) in the Yangtze River basin; such changes may give rise to simultaneous floods and droughts over eastern Asia (Endo et al., 2005). Strong increasing trends of the very wet day index are observed in both Southeast Asia and Korea, where heavy rainfall events occur during the monsoon period. This contrasts with decreasing trends in northern China and south eastern Australia. It has been shown that the increase in extreme events is linked to global warming and that this development is larger and faster than previously expected (Salinger et al., 2013). These extreme heavy rainfall events frequently cause severe floods in the region. For example, one study of the observed and projected trends in human exposure to tropical cyclones (typhoons in Asia) and floods showed that more than 90% of the global population exposed to such events are found in Asia, both currently and in projections for 2030 (Handmer et al., 2012). Disasters expose different dimensions of social-ecological vulnerability and therefore disaster risk reduction is an important field of sustainability research and action (UNISDR, 2011). In the event of disaster, it is crucial to address immediate recovery needs while also developing learning systems that will improve social-ecological resilience to potential future disasters and build sustainability in the long term (Takeuchi, 2002).

29.2.3 Water Cycles, Water Security and Management

Water is at the nexus of the coupled nature and human systems, connecting human consumption, agricultural irrigation and hydropower generation through important geophysical and bio-geochemical processes in the atmosphere, oceans and land. Water is one of the most important elements for human development and therefore human sustainability. Under changing climate and rapid human development, water security has been threatened in numerous ways that require immediate attention in order to ensure a sustainable Future Earth (Manton et al., 2015). The grand challenge of sustainable water resources management is to secure the access to high-quality water necessary to meet basic human needs, such as health and sanitation, food production and renewable energy. To do this, the linkages among ecological change, land use, urbanization, industrial activities and water systems need to be taken into account (Pahl-Wostl et al., 2008).

Local and regional-scale water cycles in Asia are changing, especially through changes of precipitation associated with changes in the monsoon climate and evapotranspiration processes associated with land use/cover changes. Continued glacier retreat in the Himalaya has significant implications for the water supply to Asia's major rivers and will affect the dynamics of water resources and water availability for human use. The importance of groundwater processes in the water cycle and water resources management is discussed separately in Chapter 13 of this book (Taniguchi, 2018).

Aiming at flood control, water conservation, power generation and crop irrigation, hydrological processes are altered in various ways, including diversion of water to other watersheds and construction of hydropower dams. Measures of agricultural intensification such as irrigation and control of water flows through dams as well as the pollution and contamination of water associated with intensification make the already scarce water less accessible and useable. Urbanization also puts pressure on water resources. Moreover, these human alterations also alter the hydrological processes, which in turn affect water distribution and ultimately the availability to different communities and nations.

A major global study of human water security and river biodiversity (Vörösmarty et al., 2010) found that nearly 80% of the world's population lives in areas of high threat. In Asia, this includes zones of intensive agriculture and dense population settlement, such as large parts of Central Asia, India, and eastern China, Korea and Japan, as well as drylands such as those of Central Asia, where water scarcity is the main factor.

29.3 Changes in Biodiversity and Ecosystems (Services)

29.3.1 Biodiversity and Ecosystems in Asia

Monsoon Asia has a unique and rich biosphere with great diversity, running contiguously from the tropics to the boreal region, forming the "Asian Green Belt" along the east coast of the continent. Recent climate model studies have also suggested that the Asian Green Belt plays an active role in the monsoon climate, by controlling the energy, water and material cycles through vegetated land surface cover (Yasunari et al., 2006; Takata et al., 2009). The Asian Green Belt is rich in biodiversity: it contains seven out of seventeen megabiodiverse countries (Australia, China, India, Indonesia, Malaysia, Papua New Guinea and the Philippines), and fourteen of the world's thirty-five biodiversity hotspots.

The ecosystems with the richest biodiversity per area can be observed in Southeast Asia. The land is naturally covered by dense and tall tropical rain forests dominated by Dipterocarp trees, while the mangrove forests and seagrass beds covering the coastal zone exhibit high productivity and species richness. Vascular plants exhibit higher species richness in Asia than in other regions (Barthlott et al., 2005) and the Borneo lowlands contain more vascular plant species than any other ecoregion on Earth, with ten thousand species (Kier et al., 2005). In Lambir Hills National Park, Sarawak, 1,175 tree species were recorded in a 52-ha plot, the highest documented tree diversity in Borneo (Ashton, 2005). In the sea, the Coral Triangle is known to host the highest species richness, where 76% (605) of the world's coral species (798) are found (Hoegh-Guldberg et al., 2009). These complex structures of autotrophs in turn foster the diversity of inhabiting plants and animals. Besides the tropic regions, rich biodiversity is observed in temperate and boreal regions. The warm ocean current and high precipitation enable the ecosystems to have relatively high productivity and large biomass.

The richness of biodiversity in Asia cannot be attributed only to the current broad spectrum of biomes, but also the geographical properties and history of this area are considered to be essential for the development of this rich biodiversity. Being located in the vast Eurasian continent enabled the floral and faunal elements from adjacent regions to be moved in and mixed, while the mostly continuous terrestrial extent along the latitudinal gradient helped terrestrial species to avoid extinction and enhanced speciation under the dual impact of geological and global climatic cycles (Lohman et al., 2011). For example, consider the period during the Quaternary glacial episodes to see the effect of glacial cycles. The terrestrial continuity prevented extinction of some temperate species in northern Asia by allowing them to migrate southwards to reach refuges and they retained their presence thereafter (Petit et al., 2008). Meanwhile in the Southeast Asian region, the present-day islands of Borneo, Sumatra, Java and Bali and the Malay Peninsula and Sunda shelf were connected to form a large peninsula called Sundaland when sea levels dropped by 120 m at the time of the last glacial maximum (LGM; 21 k.y.a.). The extent of conducive climatic conditions at the LGM is considered to have been large enough for Dipterocarp tree species to sustain the forest biota with their high species diversity (Raes et al., 2014). In contrast, during the interglacial period, the sea level rose and it converted mountains into geographically isolated islands, creating conditions to facilitate speciation. The episodic sea-level changes repeatedly connected the islands of Sundaland to the Asian mainland, enabling biotic migrations from the mainland to the archipelago (Sodhi et al., 2004; Meijaard, 2004). Thus, the natural biodiversity in Southeast Asia took a very long time to become as rich as we are currently observing and can be regarded as the gift of the dynamics and history of the geology and climate of this region. This is the case for the biodiversity richness of Asia as a whole.

The Asian monsoon climate system thus contributes to the rich biodiversity and high ecosystem productivity of the Asian Green Belt, which also means it is eminently suitable for human habitation. The ecosystems in

Monsoon Asia, including the agro-ecosystem created mainly in lowland and deltas, have provided people with rich food and other substantial biological resources allowing the population size to exceed 4.1 billion within this region, almost 60% of the world population (Population Reference Bureau, 2016). In the course of population growth, people dispersed over the region and formed diverse traditional systems of agriculture and livelihoods adapted to local biodiversity. The complex mosaic of cultures exhibited by Asia today is the result of this process.

29.3.2 Changing Interactions between Humans and Ecosystems

Recently, however, ecosystems in Asia have been undergoing drastic and sometimes irreversible changes induced by increasing anthropogenic pressures. This is basically due to rapid growth of human population, which increased threefold in 50 years; the increased outbound trade of ecosystem products has also strongly boosted this trend, with its underlying driver being the growth of the global economy. The growing demand for Asian ecosystem products induced the vast land use change through logging, plantation agriculture, aquaculture and overexploitation, especially in Southeast Asia, and these are regarded as the primary threats to its rich biodiversity (UNEP, 2011a). How the biodiversity in Asia is under threat can be recognized, for example, on the maps of globally threatened mammal and bird species (Baillie et al., 2004).

Significant transformations of forest ecosystems in the past two decades can be recognized from the forest cover change from 1990 to 2010. During the first decade, there was a slight decline in forest cover at the rate of some 600 000 ha per year in Asia as a whole. However, in the latter decade, 2000–2010, there was an increase in forest cover with a net gain of more than 2.2 million hectares per year in the period (FAO, 2011). Hidden in this crude trend, however, there exist considerable differences in the forest states and their trends among countries. In the decade of 2000–2010, for example, the bulk of forest cover increase was observed in China because of large-scale afforestation activities, while severe forest loss took place in Indonesia, mainly due to clearing of natural tropical rain forests with the corresponding loss of their rich biodiversity. In Southeast Asia, agriculture is the largest industry (in terms of GDP) for most countries and agricultural production has increased significantly since the 1960s, with most countries more than doubling their cultivated area (Gasparatos et al., 2010).

Threats to coral reefs in the Coral Triangle are also becoming more and more severe. More than 85% of reefs within the Coral Triangle Region are currently threatened by local stressors (such as overfishing, pollution and coastal development), which is substantially higher than the global average of 60%. Nearly 45% are at high or very high threat levels. When the influence of recent thermal stress and coral bleaching is combined with these local threats, the proportion of reefs rated as threatened increases to more than 90%, which is substantially greater than the global average of 75% (Burke et al., 2012).

What do these trends in biodiversity degradation mean in terms of ecosystem services for Asian people? The ecosystem services humans depend on are not only the provision of ecological resources used for food and other raw materials. There are other classes of fundamental ecosystem services, often categorized into regulating, cultural and supporting services (Millennium Ecosystem Assessment, 2005). Rich forest stands, for example, play substantial roles in stabilizing soil layers and in regulating the regional water cycle, which in turn contributes to supply stable water resource and avoidance of disasters. Rich cultural diversity intertwined with local biodiversity of course is essential for the wellbeing and identities of Asian people and their societies. The extensive forests in the Asian Green Belt, expanding from tropics to boreal regions, together with other vegetation and coral reefs, also have the crucially important role of carbon sequestration to regulate the global carbon cycle. These ecosystem services all derive from biodiversity and are intrinsically linked with the environmental issues occurring in the Asian region. Unsustainable tree removal practices, such as clear-felling, prompt erosion and soil salinity, as well as disturbance of the groundwater table. In regions with less precipitation, deforestation facilitates the transformation of fertile areas into barren land, a process known as desertification. Rapid expansion of cultivated area by clearing natural forest is the easiest way to bring about more provisioning services at least in the short term, but will negatively affect services of the other classes that people can enjoy in the longer term. In other words, the Asian countries are, by overly relying on the increase of provisioning services, more and more disturbing the balance of the entirety of ecosystem services for regional people. Further biodiversity decline will threaten hundreds of millions of Asian people in the future.

While the provisioning services are easy to recognize or evaluate economically, the other classes of ecosystem services are generally less visible and therefore their importance for the sustainability of human society is not highly appreciated in comparison with that of the provisioning services; nor is the biodiversity behind them. The lack of balanced evaluation of ecosystem services has resulted in poor regulation of ecosystem management and, in some cases, unsustainable practices could not be stopped. It has become increasingly apparent throughout the region that the enduring costs from unsustainable land use ultimately overwhelm the more immediate gains even in economic evaluation.[1] However, the values of ecosystem services vary and cannot always be evaluated economically. Therefore, there is an urgent need for developing methodologies to understand what an ecosystem means to the people with different stakes at different scales from local to global, as well as effective governance mechanisms for conservation and sustainable use of biodiversity and ecosystem services.

We do not fully understand the properties of biodiversity and ecosystem services. At least, however, it is certain that they are not sufficiently resilient against or self-sustainable under current anthropogenic pressures *over long periods.* Furthermore, once severely degraded, ecosystems containing rich forests or coral reefs will never recover as they were. Even if partially possible, it requires a long time and large expense. Their loss means the elimination of potential ecosystem resources and this leads to other environmental problems induced by adverse effects of changes in water, soil and climate regimes.

29.4 Development, Environmental Change and Human Health

29.4.1 Urbanization

Asia is home to almost 60% of the global population and exerts a major global impact, both economically and ecologically: it accounts for one-third of global GDP, consumes more than 60% of global materials, is a growing emitter of greenhouse gases and is undergoing rapid demographic change and economic growth (UNEP, 2011). Urbanization and urban growth are an inseparable element of this development process. Historically, the monsoon climate enabled agriculture to thrive in Asia's deltas and plains and this, in turn, supported large populations and thriving cities – sixty-four of the one hundred largest cities were in Asia in 1800 (Satterthwaite 2007). However, from a worldwide perspective, the region has been and still is relatively less urbanized. The urbanization level was 43.1% in 2011, compared to 50.9% for the entire world. Asia is engaged in an unprecedented catch-up race, however, having added a billion urban dwellers since 1980 with an urbanization increase rate that was above the global average. Asia is also home to twelve of the world's twenty-three megacities and this is expected to grow to twenty-one out of thirty-seven by 2050 – thus becoming more in line with Asia's global population share. It should also be noted that, in spite of these powerful urbanizing trends, Asia still contains 69% of all rural people in the world (Manton et al., 2015).

This transformation from a predominantly rural to an urban-dominated society in Asia has therefore been highly compressed in time in comparison with the experiences in the West, and this high-speed change has been accompanied by rapid and profound changes in most aspects of life, as well as the environment. Between 1980 and 2005, 80% of the growth in world materials use happened in Asia and the Pacific – this was not only due to population growth but also to higher per capita use (UNEP, 2011), as urbanization proceeded. A related outcome has been the growth of air pollution problems (discussed in Section 29.4.2), and urban heat island effects are becoming more frequent. Some of these adverse environmental outcomes are likely to be aggravated by climate change. Many urban centers are found in coastal areas and will be vulnerable to sea level rise (World Bank 2010).

29.4.2 Air Pollution

Outdoor air pollution in Asia has grown to such proportions that international media reports about Beijing or New Delhi schools having to close to protect children from the harmful effects of smog have become common.[2] Such media reports are increasingly confirmed by scientific studies such as Chapter 17, this volume. Brauer et al. (2016) report that the highest concentrations of fine particles with a diameter of less than 2.5 micrometers ($PM_{2.5}$) worldwide are found in South and East Asia, especially in northern India and eastern China,[3] resulting from combustion emissions related to such sources as household solid fuel use, coal-fired power plants, landscape fires, industry and transportation. A separate study of $PM_{2.5}$ pollution in forty-five megacities worldwide in 2013 reports that the most polluted cities were found in east-central China and the Indo-Gangetic Plain (Cheng et al., 2016). Likewise, ambient nitrogen dioxide (NO_2), mostly related to the burning of fuel, is a major pollutant in Asia. A recent satellite-based study found that while the population-weighted annual mean NO_2 in Asia Pacific (mainly Japan and South Korea) had peaked in the mid-1990s and then declined, the concentrations in East Asia (China, North Korea and Taiwan) had almost tripled between 1996 and 2012 (Geddes et al., 2016).

The drivers behind this dramatic worsening of the quality of Asia's skies are complex, but they are broadly connected to the fast pace of economic development. This development is driven by and at the same time drives the escalation of fuel use, including the burning of more polluting coal. The increase in income and urbanization has given rise to lifestyles that depend more on fossil fuels, with the use of private cars becoming much more common, while the increase in urban sprawl has given rise to longer commutes.

In parts of Asia, especially Southeast Asia, an important form of air pollution, the haze from forest and peat land fires, relates to development in a different way. It is part of a larger pattern of tropical deforestation due to logging and conversion to palm oil plantations and to some extent the expansion of subsistence farming. In Sumatra and Kalimantan in Indonesia, fire is commonly used in land clearing and may get out of control in the dry season, especially in drained peat lands, which are highly inflammable (Howes and Wyrwoll, 2012). In years of serious drought due to a combination of El Niño and positive Indian Ocean Dipole conditions, such as 1997, 2006 and 2015, severe haze developed and was carried by the prevailing winds across Indonesia and to Singapore and Malaysia (Koplitz et al. 2016). As discussed later, such haze episodes present considerable health threats to the population in the areas affected.

Finally, while air pollution has serious local and regional consequences, developments in Asia increasingly leave their mark on a global scale. Between 1990 and 2013, global $PM_{2.5}$ concentrations increased by 20.4% and this change was mostly due to trends in South Asia, Southeast Asia and China (Brauer et al. 2016).

29.4.3 Population, Health and Environmental Change

Asia is also characterized by great diversity in terms of demography and health (Manton et al., 2015). The population of Asia and the Pacific was almost 4.5 billion people in 2016 (2016 ESCAP Population Data Sheet),[4] 60% of the world's total. Overall, human health as reflected in average life expectancy has increased (72.2 years on average in twenty-two Asian countries in 2010) and fertility levels have gone down (from an average of four children per woman in 1990 to three in 2010) (OECD, 2012). The resulting population structure is young, but with great variation – ranging from very "young" countries such as Afghanistan and Pakistan to the rapidly aging societies of East Asia. In view of this population composition and fertility and mortality levels, it is expected that Asia's population will rise to 5 billion by 2050 and eventually decline to about 4 billion at the end of the century (UNDESA, 2012). Japan is at the head of the Asian (and the world's) demographic curve, with its population now declining, but other countries such as Korea, China and Singapore are also rapidly ageing. It is expected that the dependency ratio (ratio of the population older than 65 to those aged 15 to 64) in Japan, Korea and Singapore will range from 58% to 70% by 2050 (Chomik and Piggott, 2015). The trends in both total population and its composition will have profound implications for sustainability in Asia and beyond.

The region has seen very rapid economic development (on average GDP growth in developing Asia between 1990 and 2010 was 7.0%;Manton et al., 2015) and urbanization has taken place at a very high rate, leading to an urbanization level of 43% in 2011, as described previously. Broadly speaking, as development progresses, disease patterns are shifting from infectious to chronic diseases, but in developing countries a "double burden" is manifested where lifestyle-related chronic diseases are becoming more prominent while infectious diseases still are taking an important toll. Many aspects of health and disease in Asia are directly or indirectly related to human interference with ecosystems. The following sections briefly discuss the linkages of health to air pollution and the threat of emerging infectious diseases in Asia.

29.4.4 Air Pollution and Health

The state of ambient air pollution in Asia and its impact beyond the region have been discussed in Section 29.4.2. The World Health Organization (WHO) has recently sounded the alarm over the threat of air pollution to human health, linking some 3 million deaths a year globally to exposure to outdoor air pollution and a similar number to indoor air pollution. Two in three such deaths occur in Asia and the Pacific, mostly because of non-communicable diseases that result from long-term exposure to pollutants, such as cardiovascular diseases, stroke, chronic obstructive pulmonary disease and lung cancer.[5] $PM_{2.5}$ particulate matter has in recent years received much attention as being particularly dangerous to human health.

It is estimated that 70% of premature deaths due to $PM_{2.5}$ (and ozone) occur in the Asia and the Pacific region (2.3 million deaths) with pollution hotspots being particularly prominent in eastern China and the north of India (Lelieveld et al., 2015). $PM_{2.5}$ pollution in Asian megacities shows clear seasonality with populations being exposed to very heavy pollution episodes during winters (Cheng et al., 2016). The health impacts of air pollution increasingly have also become a cross-boundary issue in Asia, and researchers, for example, estimate that haze from Indonesian forest fires in 2015 resulted in 100,300 excess deaths across Indonesia, Malaysia and Singapore (Koplitz et al., 2016).

29.4.5 Emerging Infectious Diseases

Asia is a key area in the emergence of new infectious diseases (Jones et al., 2008; Coker et al., 2011). The emerging infectious diseases (EIDs) that have surfaced in Asia in the past two decades have caused much disruption and human and economic costs, but followed diverse trajectories. The Nipah virus that emerged in Malaysia in 1998 was effectively brought under control by policies that separated pig raising activities from the fruit bats that carried the virus. SARS, which rapidly spread from southern China to the corners of the world in 2002–2003, infecting more than eight thousand people and killing almost eight hundred, mysteriously disappeared. Avian influenza H5N1, which from 2003 rapidly spread out of southern China into most of Asia, Europe and Africa, has settled into an endemic existence in poultry in a number of countries with high human and livestock densities, occasionally claiming human victims.

A major study of the emergence of 335 EIDs worldwide between 1940 and 2004 found that 60.3% were caused by zoonotic pathogens and of these 71.8% had a wildlife origin (Jones et al., 2008). It also identified EID "hotspots," among which South and Southeast Asia and China stand out prominently.

While for wildlife-derived diseases the species diversity of wild hosts is an important contributing factor, the emergence and spread of these diseases ultimately are an outcome of economic development and resulting societal and environmental change. Land use change has reduced the natural habitats of wildlife, increasing the likelihood, for example, of migratory wild birds' having contact with domestic poultry. Rising incomes have led to a precipitous increase in the consumption of meat, provided by a burgeoning livestock sector, including the proliferation of largely unregulated intermediate-size commercial farms, connected to live markets by informal trading networks (Scoones, 2010). High densities of rural human and livestock populations continuously place people in close contact with animals. Such factors facilitate pathogens' exploitation of new ecological niches, and once new diseases of humans have emerged, increased mobility and globalization foster their rapid spread (Wu et al., 2017).

Outbreaks of diseases in domestic poultry and related counter measures and market responses have major economic consequences also for small and poor farm families. It is, however, the human pandemic potential of the EIDs that has been the main source of concern and policy attention. The struggle with avian influenza, for example, is a never-ending battle, as the virus rapidly mutates, continuously giving rise to new varieties, some able to infect and kill humans. The spread in 2003 and subsequent continued simmering of avian influenza H5N1 have been referred to earlier. In 2013 a new variety, H7N9, began to give rise to human infections in China, with more than thirteen hundred human cases reported, about one-quarter of which were lethal. Infection largely was due to contact with live poultry. Even more recently, H5N8 has spread from Korea in wild and domestic birds across Asia, Europe and Africa (with even some cases in North America) without infecting people, while the H5N6 strain from southern China is currently widespread in Asia and has had some human victims. It is therefore likely that Asia's countryside will continue to give rise to new infectious diseases with pandemic potential for the foreseeable future.

29.5 Conclusion: A Role for Future Earth in Asia

Asia's sustainability challenges as recounted in this chapter range from greenhouse gas emissions to emerging infectious diseases and from biodiversity loss to societal ageing. It has recently become common to refer to such challenges as "wicked problems," problems that defy close definition, have multiple causes and unforeseeable consequences, affect many stakeholders and lack suitable governance structures (Brown et al. 2012). They are problems that emerge from the solutions to earlier problems and need to be understood from multiple valid, but sometimes contradictory, perspectives. Research that addresses such problems therefore cannot remain confined to specific disciplines, but needs to adopt an interdisciplinary perspective that integrates diverse research fields (across the natural and social sciences, humanities, health and engineering) around shared and synthetic research questions. If research is not only to address wicked problems, but also to contribute to exploring solutions to them, a transdisciplinary approach is called for (Funtowicz and Ravetz, 1993; Pohl and Hirsch Hadorn, 2007; Brandt et al., 2013); in other words, research needs to be co-designed and co-produced in close collaboration with societal stakeholders. It was to support and facilitate such research that Future Earth was established (Future Earth, 2013).

The problems discussed here contribute to and, at the same time, are manifestations of global environmental change. They are, however, also the product of specifically Asian natural conditions and historical trends. We have described how the topography of the region, dominated by the Himalaya-Tibetan Plateau, in combination with seasonal patterns of differential heating of the continent and the surrounding oceans, has given rise to a humid monsoon climate. This benign climate fostered rich biodiversity in the uplands and enabled highly productive rice-based agriculture in the coastal deltas and plains. Such agriculture, in turn, supported large populations including those in thriving towns and cities. Traditional agriculture-based livelihood systems evolved as part of societies with distinct cultures and in some cases were sustained for several thousand years. Therefore, in grappling with the wicked problems of sustainability, not only does research need to involve and engage societal stakeholders, but it should also recognize their regional embeddedness and historic dynamics.

We argue that there is a need to develop an Asian community of research, practice and policy that contributes to solving problems through co-designed and co-produced research. Such research would be embedded in contemporary international understandings of global environmental change, while adopting a regional perspective. Such a community would also take responsibility for fostering a new generation of researchers and practitioners. In collaborating across national borders, it might eventually catalyze the creation of an "Asian Environmental Community" for the sustainable future of Asia.

Internet Resources and Notes

[1] See, for example, the Economics of Ecosystems and Biodiversity initiative, which aims at "making nature's values visible" (www.teebweb.org).

[2] For example: www.theguardian.com/world/2016/dec/17/beijing-smog-pollution-red-alert-declared-in-china-capital-and-21-other-cities and www.bbc.com/news/world-asia-india-37887937.

[3] $PM_{2.5}$ concentrations in northern Africa and the Middle East were also very high, but this is due to mineral dust blown by the wind.

[4] www.unescap.org/sites/default/files/SPPS%20PS%20data%20sheet%202016%20v15-2.pdf.

[5] www.who.int/mediacentre/news/releases/2016/air-pollution-estimates/en/.

References

Abe M., Kitoh, A., and Yasunari, T. (2003). An evolution of the Asian summer monsoon associated with mountain uplift -simulation with the MRI atmosphere-ocean coupled GCM. *Journal of the Meteorological Society of Japan*, 81(5), 909–933.

Abe M., Yasunari, T., and Kitoh, A. (2004). Effects of large-scale orography on the coupled atmosphere-ocean system in the tropical Indian and Pacific oceans in boreal summer. *Journal of the Meteorological Society of Japan*, 82(2), 745–759.

ADRC (Asian Disaster Reduction Center) (2011). Natural Disaster Data Book 2009. http://174.129.193.82/mes_files/pdf/

Ashton, P. S. (2005). Lambir's forest: the world's most diverse known tree assemblage? In *Pollination Ecology and the Rain Forest: Sarawak Studies*, eds. D. W. Roubik, S. Sakai and A. A. Hamid Karim, New York: Springer, pp. 191–216.

Baillie, J., Hilton-Taylor, C., & Stuart, S. (eds.) (2004) *IUCN Red List of Threatened Species. A Global Species Assessment*. Gland, Switzerland: IUCN.

Barthlott, W. et al. (2005). Global centers of vascular plant diversity. *Nova Acta Leopoldina*, 92(342), 61–83.

Brandt, P., Ernst, A., Gralla, F., et al. (2013). A review of transdisciplinary research in sustainability science. *Ecological Economics*, 92, 1–15

Brauer, M., Freedman, G., Frostad, J., et al. (2016). Ambient air pollution exposure estimation for the global burden of disease 2013. *Environmental Science and Technology*, 50(1), 79–88.

Brown, V. A., Harris, J. A., and Russell, J. W. (eds.) (2010). *Tackling Wicked Problems, through the Transdisciplinary Imagination*. London: Earthscan.

Burke, L., et al. (2012) *Reefs at Risk Revisited in the Coral Triangle*. Washington, DC: World Resource Institute.

Cheng, Z., Luo, L., Wang, S. et al. (2016). Status and characteristics of ambient PM2.5 pollution in global megacities. *Environment International*, 89, 212–221.

Chomik, R., and Piggott, J. (2015). Population ageing and social security in Asia. *Asian Economic Policy Review*, 10, 199–222.

Coker, R. J., Hunter, B. M., Rudge, J. W., Liverani, M., and Hanvoravongchai, P. (2011). Emerging infectious diseases in Southeast Asia: regional challenges to control. *The Lancet*, January 25, DOI:10.1016/S0140–6736(10) 62004–1.

Crutzen, P. J. (2002). Geology of mankind: the anthropocene. *Nature*, 415, 23.

Endo, N., Ailikun, B., and Yasunari, T. (2005). Trends in precipitation amounts and the number of rainy days and heavy rainfall events during summer in China from 1961 to 2000. *Journal of the Meteorological Society of Japan*, 83, 621–631.

FAO (Food and Agriculture Organization of the United Nations) (2011). *State of the World's Forests 2011*,

Rome: Food and Agriculture Organization of the United Nations.

Field, C. B., Barros, V., Stocker, T. F., et al. (2012). *Managing the Risks of Extreme Events and Disasters to Advance Climate Change Adaptation*. Cambridge and New York: Cambridge University Press, 582.

Fujibe, F., Yamazaki, N., Katsuyama, M., and Kobayashi, K. (2005). The increasing trend of intense precipitation in Japan based on four-hourly data for a hundred years. *Scientific Online Letters on the Atmosphere*, 1, 41–44.

Funtowicz, S, O., and Ravetz, J. R. (1993). *Science for the Post-Normal Age*. *Futures*, Sept, 739–755.

Future Earth (2013). Future Earth Initial Design: Report of the Transition Team. Paris: International Council for Science (ICSU).

Gasparatos, A. Subramanian, S. M., Elliott, W., and Braimoh, A. (2010). Unraveling the Drivers of Southeast Asia's Biodiversity Loss. In *UNU-IAS Policy Report Climate and Human-Related Drivers of Biodiversity Decline in Southeast Asia*. United Nations University.

Geddes, J. A., Martin, R. V., Boys, B. L., and van Donkelaar, A. (2016). Long-term trends worldwide in ambient NO_2 concentrations inferred from satellite observations. *Environmental Health Perspectives*, 124(3), 281–289.

Goswami, N., Venugopal, V., Sengupta, D., Madhusoodanan, M. S., and Xavier, P. K. (2006). Increasing trend of extreme rain events over India in a warming environment. *Science*, 314, 1442–1445.

Hahn, D. G., and Manabe, S. (1975). The role of mountains in the south Asian monsoon circulation. *Journal of the Atmospheric Sciences*, 32, 1515–1541.

Handmer, J., Honda, Y., Kundzewicz, Z. W. et al. (2012). Changes in impacts of climate extremes: human systems and ecosystems. In Field, C. B., Barros, V., Stocker, T. F., et al. *Managing the Risks of Extreme Events and Disasters to Advance Climate Change Adaptation*. Cambridge and New York: Cambridge University Press.

Hoegh-Guldberg, O., Hoegh-Guldberg, H., Veron, J. E. N. et al. (2009). *The Coral Triangle and Climate Change: Ecosystems, People and Societies at Risk*. Brisbane: WWF Australia.

Howes, S., and Wyrwoll, P. (2012). Asia's Wicked Environmental Problems. ADBI Working Paper 348. Tokyo: Asian Development Bank Institute www.adbi.org/working-paper/2012/02/28/5009.asia.wicked.environmental.problems/.

Immerzeel, W. W., van Beek, L. P. H., and Bierkens, M. F. P., (2010). Climate change will affect the Asian water towers. *Science*, 328, 1382–1385.

Jones, K. E., Patel, N. G., Levy, M. A. et al. (2008). Global trends in emerging infectious diseases. *Nature*, 451, 990–994.

Kier, G., Mutke, J., Dinerstein, E., et al. (2005). Global patterns of plant diversity and floristic knowledge. *Journal of Biogeography*, 32, 1–10.

Koplitz, S. N., Mickley, L. J., Marlier, M. E. et al. (2016). Public health impacts of the severe haze in Equatorial Asia in September–October 2015: demonstration of a new framework for informing fire management strategies to reduce downwind smoke exposure. *Environmental Research Letters*, 11, 094023, DOI:10.1088/1748-9326/11/9/094023

Krechowicz, D., and Venugopal, S. (2010). *Analyzing Environmental Trends: Taking the Pulse of Asia's Financial Community*. *World Resources Institute Working Paper*. Washington, DC: World Resources Institute.

Kutzbach, J. E., Prell, W. L., and Ruddiman, W. F. (1993). Sensitivity of Eurasian climate to surface uplift of the Tibetan Plateau. *Journal of Geology*, 101, 177–190.

Lelieveld, J., Evans, J. S., Fnais, M., Giannadaki, D., and Pozzer, A. (2015). The contribution of outdoor air pollution sources to premature mortality on a global scale. *Nature*, 525, 367–371.

Lohman, D. J. de Bruyn, M., Page, T., et al. (2011). Biogeography of the Indo-Australian Archipelago. *Annual Review of Ecology, Evolution, and Systematics*, 42, 205–226.

Manton, M., Yasunari, T., Ailikun, Mallee, H., Lasco, R., and Ramesh, R. (2015). *Initial Strategic Research Plan for Future Earth in Asia*. Beijing, China: Meteorological Press.

Meijaard, E. (2004). Biogeographic history of the Javan leopard Panthera pardus based on a craniometric analysis. *Journal of Mammalogy*, 85, 302–310.

Millennium Ecosystem Assessment (2005). *Millennium Ecosystem Assessment*. Washington, DC: Island Press.

OECD Health at a Glance: Asia Pacific 2012. Life expectancy at birth. www.oecd-ilibrary.org/sites/9789264183902-en/01/01/index.html;jsessionid=hqnpt1peah2.x-oecd-live-01?contentType=&itemId=/content/chapter/9789264183902-4-en&containerItemId=/content/serial/23054964&accessItemIds=/content/book/9789264183902-en&mimeType=text/html. Retrieved 18 Aug 2013

Pahl-Wostl, C., Kabat, P., and Möltgen, J. (eds.)(2008). *Adaptive and Integrated Water Management: Coping with Complexity and Uncertainty*. Heidelberg, Springer.

Petit, R. J., Hu, F. S., and Dick, C. W. (2008). Forests of the past: A window to future changes. *Science*, 320, 1450–1452.

Pohl, C., and Hirsch Hadorn, G. (2007). *Principles for Designing Transdisciplinary Research*. Munich, Oekom Verlag.

Raes, N., Cannon, C. H., Hijmans, R. J., et al. (2014). Historical distribution of Sundaland's Dipterocarp rainforests at Quaternary glacial maxima. *Proceedings of the National Academy of Sciences of the United States of America*, 111, 16790–16795.

Salinger, M. J., Bell, J. D., Evans, K., et al. (2013). Climate and oceanic fisheries: recent observations and projections and future needs. *Climatic Change*, 119, 213–221.

Satterthwaite, D. (2007). *The Transition to a Predominantly Urban World and Its Underpinnings*. London: International Institute for Environment and Development.

Scoones, I. (ed.) (2010). *Avian Influenza—Science, Policy and Politics*. London: Earthscan.

Sodhi, N. S., Koh, L. P., Brook, B. W., and Ng, P. K. (2004). Southeast Asian biodiversity: an impending disaster. *Trends in Ecology & Evolution*, 19, 654 – 660.

Steffen, W., Crutzen, P. J., and McNeill, J. R. (2007). The Anthropocene: are humans now overwhelming the great forces of nature? *Ambio*, 36, 614–621.

Syvitski, J. P. M., Kettner, A. J., Overeem, I. et al. (2009). Sinking deltas due to human activities. *Nature Geoscience*, 2, 681–686.

Takata, K., Saito, K., and Yasunari, T. (2009). Changes in the Asian monsoon climate during 1700–1850 induced by preindustrial cultivation. *Proceedings of the National Academy of Sciences of the United States of America*, 106, 9586–9589.

Takeuchi, K. (2002). Flood management in Japan – from rivers to basins. *Water International*, 27, 20–26.

Taniguchi, M. (2018). Asian Groundwater Perspectives on Global Change and Future Earth - *Chapter 13 of this book*.

UNEP (United Nations Environment Programme) (2011a). *Global Biodiversity Outlook 3 Regional Summaries: State of Biodiversity in Asia and the Pacific*. Bangkok, United Nations Environment Programme Regional Office for Asia and the Pacific.

UNEP (2011b). *Resource Efficiency: Economics and Outlook for Asia and the Pacific*. Bangkok: United Nations Environment Programme.

UNISDR (2011). *Global Assessment Report on Disaster Risk Reduction*. Geneva: United Nations International Strategy for Disaster Reduction.

UNDESA (United Nations Department of Economic and Social Affairs/Population Division) (2012). World Urbanization Prospects 2011: The 2011 Revision.

Vörösmarty, C. J., McIntyre, P. B., Gessner, M. O., et al. (2010). Global threats to human water security and river biodiversity. *Nature*, 467, 555–561.

World Bank (2010). *Climate Risks and Adaptation in Asian Coastal Megacities: A Synthesis Report*. Washington, DC: World Bank.

Population Reference Bureau (2016). *2016 World Population Data Sheet*. Washington, DC: Population Reference Bureau.

Wu, T., Perrings, C., Kinzig, A., et al. (2017). Economic growth, urbanization, globalization, and the risks of emerging infectious diseases in China: a review. *Ambio*, 46(10), 18–29.

Yasunari, T., Niles D. N., Taniguchi, M., and Chen, D. (2013). Asia: proving ground for global sustainability. *Current Opinion in Environmental Sustainability*, 5(3–4), 288–292.

Yasunari T., Saito, K., and Takata, K. (2006). Relative roles of large-scale orograghy and land surface processes on global hydroclimate. Part I: Impacts on monsoon systems and the tropics. *Journal of Hydrometeorology*, 7, 626–641.

Yao, T., Thompson, L., Yang, W., et al. (2012). Different glacier status with atmospheric circulations in Tibetan Plateau and surroundings. *Nature Climate Change*, 2, 663–667.

30

Looking Back to Move Forward: Institutional Capacity Required by Global Governance Changes

*Jane E. Rovins** *and Sarah Beaven*

30.1 Introduction

In order to understand current global governance arrangements for disasters and development, it's important to start with a look back to see how they have evolved. International recognition of the need to manage global disaster risk and global poverty began to gather momentum in the late eighties. In 1987, the United Nations General Assembly declared 1990–1999 to be the International Decade for National Disaster Reduction (United Nations General Assembly 1987, 1989). In the same year, this body also accepted the Brundtland report definition of sustainable development as "development which meets the needs of the present without compromising the ability of future generations to meet their own needs" (Drexhage and Murphy 2010, p. 2). In 1988, the Intergovernmental Panel on Climate Change (IPCC) was set up by the United Nations Environment Programme (UNEP) and World Meteorological Organization (WMO) to provide policymakers with the scientific basis of climate change, its impacts and future risks and ways to mitigate and adapt (Intergovernmental Panel on Climate Change [IPCC] 2013). Since then, a series of global initiatives focused on reducing the risk of rising disaster losses, climate change and global poverty have continued to evolve largely in parallel.

Increased global interconnectivity and interdependence have been widely acknowledged to be contributing factors in the growing scale and complexity of transnational threats such as climate change, disaster risk, economic crisis, biodiversity loss and global poverty (Duke 2006; Jasanoff 2011; Patrick 2014). These developments reflect and are part of increased globalisation, as rapid technological change, economic liberalisation and international regulation shift decision making increasingly into the international sphere (Skogstad 2003; Castells 2008; Jasanoff 2011; Jensen et al. 2015). It has proved difficult to achieve multilateral action to address these complex global problems through traditional international forums, however, since the international political arena is made up of 'independent sovereign units that recognise no higher authority' (Patrick 2014, p. 59). Growing need for collective action has driven the emergence of new international collaborative mechanisms alongside traditional institutions, as those involved turn to more convenient, ad hoc venues, and a wide range of tailored, informal cooperative networks (Fordham 2007; Patrick 2014; Jensen et al. 2015). Collective action of this kind 'to address challenges and seize opportunities that transcend national frontiers' has been defined by Patrick (2014) as global governance (Patrick 2014, p. 62).

These alterations in international governance practice have given rise to a series of multi-layered globalised social systems, each able 'to work as a unit on a planetary scale in real or chosen time' (Castells 2008, p. 81; Jensen et al. 2015). Although interlinked, global governance systems in disaster risk management, sustainable development and climate change have evolved largely in parallel. Since these systems are decentralised, they rely on distributed institutional capacity to link and connect the many organisations and networks involved (Patrick 2014; Jensen et al. 2015). At the global level, this has driven a move away from attempts to drive and administer comprehensive solutions towards arrangements that provide institutional monitoring, coordinating, guiding and integrating capacity (Castells 2008; Patrick 2014; Jensen et al. 2015). Required to link diverse networks within each larger system, institutional capacity is also needed to create greater alignment and cohesion between science and practice, and across sectors and stakeholders

* Jane E. Rovins was the inaugural Executive Director of the Integrated Research on Disaster Risk (IRDR) program.

(McNie 2007; Jensen et al. 2015). As changes in global governance practice gather pace, and global governance systems continue to diverge, demand continues to grow for integration between increasingly distinct global governance systems in sustainable development, climate change and disaster risk reduction (Fordham 2007).

The following discussion of integration efforts over the last twenty years aims to increase understanding of global governance changes in the sustainable development and disaster risk reduction areas. The focus is on the associated need for increased institutional capacity, and the difficulty of providing it. Recently established science bodies, Integrated Research on Disaster Risk (IRDR) and Future Earth, are used as case studies, to illustrate two different approaches to integrating science and practice in disaster risk reduction and sustainable development areas, respectively.

30.2 Global Integration Efforts

Five years after the United Nations accepted the Brundlandt definition of sustainable development, world leaders agreed on a series of broad sustainable development principles at the 1992 United Nations Conference on Environment and Development in Rio de Janeiro, Brazil (Drexhage and Murphy 2010). In 2000, the United Nations declared eight broad Millennium Development Goals; two of these ('Protecting the Environment' and 'Protecting the Vulnerable') included brief mentions of disaster risk. In the same year, the United Nations General Assembly established a secretariat (United Nations International Strategy for Disaster Reduction;UNISDR)[1] to address disaster risk reduction, and to acknowledge the impact it has on sustainable development (United Nations General Assembly 2000).

Then in 2002 the World Summit on Sustainable Development explicitly identified natural disasters as a significant threat to sustainable development and world safety (United Nations 2002a, 2002b; United Nations Conference on Sustainable Development Secretariat 2012). The associated Implementation Plan proposed that disaster risk reduction, mitigation and preparedness as well as response and recovery were key to addressing impacts on environment, development and climate (United Nations 2002b). Dealing 'effectively with natural disasters and conflicts, including their humanitarian and environmental impacts' was identified as essential for sustainable development, on the basis of many cases in which 'both the gains and efforts aimed at sustainable development' had been 'obliterated' by natural disasters and conflicts (United Nations, 2002b, paragraph 65).

In 2004, further progress on the integration front was indicated by complementary global reports produced by the UNISDR and the United Nations Development Program (UNDP) (United Nations 2004; United Nations Development Program [UNDP] 2004). Written in collaboration, both reports are concerned with disaster risk reduction and call for greater alignment between disaster risk reduction and sustainable development efforts. These reports were shared with the sustainable development community at the 2002 Johannesburg Sustainable Development Summit.

A year later, the gains made through this collaborative activity were evident in the *Hyogo Framework for Action 2005–2015: Building the Resilience of Nations and Communities to Disasters*. The opening paragraphs draw from a similar paragraph in the UNISDR *Living with Risk* (2004) report to align the Hyogo Framework explicitly with the Millennium Development Goals:

> There is now international acknowledgement that efforts to reduce disaster risks must be systematically integrated into policies, plans and programmes for sustainable development and poverty reduction, and supported through bilateral, regional and international cooperation, including partnerships. Sustainable development, poverty reduction, good governance and disaster risk reduction are mutually supportive objectives, and in order to meet the challenges ahead, accelerated efforts must be made to build the necessary capacities at the community and national levels to manage and reduce risk. Such an approach is to be recognized as an important element for the achievement of internationally agreed development goals, including those contained in the Millennium Declaration.
>
> (United Nations 2005, paragraph 4).

The Hyogo Framework includes several references to the integration of disaster risk reduction into sustainable development planning, programmes and reporting. There are also references to building resilience, which is an essential component of sustainable development.

While this marked something of a high point in efforts to integrate these two fields, in retrospect the focus appears to be somewhat uni-directional. Integration is a two-way street, as Fordham (2007) suggests, requiring that both fields expand the focus of their policy and practice in the direction of the other. At the 2002 World Summit and in the UNDP *Reducing Disaster Risk* report, sustainable development leaders explicitly acknowledged the need to expand their field to incorporate disaster risk reduction into their own policy and practice (UNDP 2004). The Hyogo Framework agreed that expansion of sustainable development is necessary. But it does not explicitly acknowledge the need to incorporate sustainable development into every aspect of disaster risk reduction policy and practice and to this extent fails to make the reciprocal integration gesture.

30.3 Global Summits

The years 2015–2016 were watershed years for global issue–based agendas and frameworks, as disaster risk reduction, sustainable development, climate change and humanitarian affairs communities joined within their respective communities to develop new and updated global agendas. These summits were all convened to address the same broader issue, the impacts of gross environmental degradation due to human development. The results included the goals provided in the *Sendai Framework for Disaster Risk Reduction 2015–2030* (Sendai Framework), the *Transforming Our World: The 2030 agenda for Sustainable Development* (17 seventeen Sustainable Development Goals), the Paris Agreement (under the United Nations Framework Convention on Climate Change) and the World Humanitarian Summit Commitments to Action, respectively. As responses to the impacts of human development, these frameworks underline the extent to which global poverty, development, disaster risk and impacts and climate change issues are profoundly interlinked. However, little progress appears to have been made on integrating the global systems focused on addressing them. It may even be that these efforts are now less integrated than they were under the Hyogo Framework and the Millennium Development Goals.

The first Sendai Framework goal does state that disaster risk reduction can only be achieved

> through the implementation of integrated and inclusive economic, structural, legal, social, health, cultural, educational, environmental, technological, political and institutional measures that prevent and reduce hazard exposure and vulnerability to disaster, increase preparedness for response and recovery, and thus strengthen resilience.
> (UNISDR 2015, paragraph 17)

This goal is indicative of a "new perspective" beyond disaster-specific activities, and points to a more explicit emphasis on the importance of reconstruction, rehabilitation, resilience and recovery. The emphasis on integrated disaster risk reduction measures reflects a growing awareness, among disaster risk management communities, of the effects of disasters on development and sustainability. The Sendai Framework has repeated references to building the resilience of infrastructure, and health and education provisions implicitly acknowledge the role of sustainable development in disaster risk reduction. Like the Hyogo Framework, however, the Sendai Framework stops short of considering sustainable development as a key component of disaster risk management (UNISDR 2015).

The United Nations General Assembly Resolution 70/1 (2015), *Transforming Our World: The 2030 Agenda for Sustainable Development* (SDG) replaces the Millennium Development Goals. Like the 2002 United Nations Johannesburg Declaration on Sustainable Development, this resolution acknowledges that disasters and natural hazards are a significant threat to sustainable development (United Nations General Assembly 2015). The SDGs do acknowledge the role disasters play in instability and development. But disaster reduction is not one of the seventeen goals, and the SDGs do not include strategies and initiatives designed to integrate disaster risk reduction into sustainable development approaches and strategies.

These summits continued to state the need to better align sustainable development and disaster risk reduction but failed actually to do so. This reiteration of earlier statements to the same effect underlines both the lack of progress on this front and the continuing difficulty of bringing this about. Yet a comparison of the 17 Sustainable Development Goals and the Sendai Framework disaster risk management targets point to an increasing overlap between these fields. This overlap indicates the potential for a much more integrated approach to progressing disaster risk and sustainable development. When the Sendai priorities are also considered, the SDGs and Sendai Framework targets are clearly working towards the same ends (Table 30.1).

30.4 Integrating Science, Policy and Practice

The International Council for Science (ICSU) and the International Social Science Council (ISSC) share a longstanding responsibility to advance and strengthen international science.[2] Both have been centrally involved in the provision of science support for the United Nations–led disaster risk management and sustainable development efforts. Leading up to the 2002 World Summit, for example, ICSU and ISSC members were involved in researching and producing

Table 30.1 *Sustainable Development Goals Compared to Sendai Targets*

Sustainable Development Goals (SDGs)	**Sendai Framework Targets**
Goal 1 - End poverty in all its forms everywhere 1.5 By 2030, build the resilience of the poor and those in vulnerable situations and reduce their exposure and vulnerability to climate-related extreme events and other economic, social and environmental shocks and disasters	(a) Substantially reduce global disaster mortality by 2030, aiming to lower average per 100,000 global mortality between 2020–2030 compared to 2005–2015. (b) Substantially reduce the number of affected people globally by 2030, aiming to lower the average global figure per 100,000 between 2020–2030 compared to 2005–2015. (c) Reduce direct disaster economic loss in relation to global gross domestic product (GDP) by 2030.
Goal 2 - End hunger, achieve food security, and improved nutrition and promote sustainable agriculture 2.4 By 2030, ensure sustainable food production systems and implement resilient agricultural practices that increase productivity and production, that help maintain ecosystems, that strengthen capacity for adaptation to climate change, extreme weather, drought, flooding and other disasters and that progressively improve land and soil quality	(b) Substantially reduce the number of affected people globally by 2030, aiming to lower the average global figure per 100,000 between 2020–2030 compared to 2005–2015.
Goal 3 - Ensure healthy lives and promote well-being for all at all ages 3.9 By 2030, substantially reduce the number of deaths and illnesses from hazardous chemicals and air, water and soil pollution and contamination	(d) Substantially reduce disaster damage to critical infrastructure and disruption of basic services, among them health and educational facilities, including through developing their resilience by 2030.
Goal 4 - Ensure inclusive and equitable education and promote lifelong learning opportunities for all	(d) Substantially reduce disaster damage to critical infrastructure and disruption of basic services, among them health and educational facilities, including through developing their resilience by 2030. (g) Substantially increase the availability of and access to multi-hazard early warning systems and disaster risk information and assessments to the people by 2030.
Goal 5 - Achieve gender equality and empower all women and girls	
Goal 6 - Ensure availability and sustainable management of water and sanitation for all 6.3 By 2030, improve water quality by reducing pollution, eliminating dumping and minimizing release of hazardous chemicals and materials, halving the proportion of untreated wastewater and substantially increasing recycling and safe reuse globally	(d) Substantially reduce disaster damage to critical infrastructure and disruption of basic services, among them health and educational facilities, including through developing their resilience by 2030.
Goal 7 - Ensure access to affordable, reliable, sustainable and modern energy for all	
Goal 8 - Promote sustained, inclusive and sustainable economic growth, full and productive employment and decent work for all	
Goal 9 - Build resilient infrastructure, promote inclusive and sustainable industrialization and foster innovation	(d) Substantially reduce disaster damage to critical infrastructure and disruption of basic services, among them health and educational facilities, including through developing their resilience by 2030.
Goal 10 - Reduce inequality within and among countries	

Table 30.1 (*cont.*)

Sustainable Development Goals (SDGs)	**Sendai Framework Targets**
Goal 11 - Make cities and human settlements inclusive, safe, resilient and sustainable 11.5 By 2030, significantly reduce the number of deaths and the number of people affected and substantially decrease the direct economic losses relative to global gross domestic product caused by disasters, including water-related disasters, with a focus on protecting the poor and people in vulnerable situations 11.b By 2020, substantially increase the number of cities and human settlements adopting and implementing integrated policies and plans towards inclusion, resource efficiency, mitigation and adaptation to climate change, resilience to disasters, and develop and implement, in line with the Sendai Framework for Disaster Risk Reduction 2015–2030, holistic disaster risk management at all levels	(a) Substantially reduce global disaster mortality by 2030, aiming to lower average per 100,000 global mortality between 2020–2030 compared to 2005–2015. (b) Substantially reduce the number of affected people globally by 2030, aiming to lower the average global figure per 100,000 between 2020–2030 compared to 2005–2015. (c) Reduce direct disaster economic loss in relation to global gross domestic product (GDP) by 2030. (e) Substantially increase the number of countries with national and local disaster risk reduction strategies by 2020.
Goal 12 - Ensure sustainable consumption and production patterns	
Goal 13 - Take urgent action to combat climate and its impacts 13.1 Strengthen resilience and adaptive capacity to climate-related hazards and natural disasters in all countries 13.2 Integrate climate change measures into national policies, strategies and planning 13.3 Improve education, awareness-raising and human and institutional capacity on climate change mitigation, adaptation, impact reduction and early warning 13.a Implement the commitment undertaken by developed-country parties to the United Nations Framework Convention on Climate Change to a goal of mobilizing jointly $100 billion annually by 2020 from all sources to address the needs of developing countries in the context of meaningful mitigation actions and transparency on implementation and fully operationalize the Green Climate Fund through its capitalization as soon as possible 13.b Promote mechanisms for raising capacity for effective climate change-related planning and management in least developed countries and small island developing States, including focusing on women, youth and local and marginalized communities	(a) Substantially reduce global disaster mortality by 2030, aiming to lower average per 100,000 global mortality between 2020–2030 compared to 2005–2015. (b) Substantially reduce the number of affected people globally by 2030, aiming to lower the average global figure per 100,000 between 2020–2030 compared to 2005–2015. (c) Reduce direct disaster economic loss in relation to global gross domestic product (GDP) by 2030. (e) Substantially increase the number of countries with national and local disaster risk reduction strategies by 2020.
Goal 14 - Conserve and sustainably use the oceans, seas and marine resources for sustainable development 14.2 By 2020, sustainably manage and protect marine and coastal ecosystems to avoid significant adverse impacts, including by strengthening their resilience, and take action for their restoration in order to achieve healthy and productive oceans	
Goal 15 - Protect, restore and promote sustainable use of terrestrial ecosystems, sustainably manage forests, combat desertification, and halt and reverse land degradation and halt biodiversity loss 15.3 By 2030, combat desertification, restore degraded land and soil, including land affected by desertification, drought and floods, and strive to achieve a land degradation-neutral world	(g) Substantially increase the availability of and access to multi-hazard early warning systems and disaster risk information and assessments to the people by 2030.

Table 30.1 (*cont.*)

Sustainable Development Goals (SDGs)	Sendai Framework Targets
Goal 16 - Promote peaceful and inclusive societies for sustainable development, provide access to justice for all and build effective, accountable and inclusive institutions at all levels	
Goal 17- Strengthen the means of implementation and revitalize the Global Partnership for Sustainable Development	(f) Substantially enhance international cooperation to developing countries through adequate and sustainable support to complement their national actions for implementation of this framework by 2030.

both the UNDP *Living with Risk* and UNISDR *Reducing Disaster Risk* 2004 reports. ICSU and ISSC were also active in the production of Intergovernmental Panel on Climate Change reports.

These collaborative activities built relationships between ICSU and ISSC as well as with various United Nations entities involved in coordinating disaster risk management and sustainable development efforts. Both activities and relationships increased shared awareness of the need for much greater integration of the different scientific disciplines and traditions contributing to the disaster risk management and sustainable development efforts.

30.5 Case Study: Integrated Research on Disaster Risk and Future Earth

In 2010, the IRDR program was launched by co-sponsors ICSU, ISSC and UNISDR. Five years later, ICSU and ISSC collaborated with United Nations and other bodies focused on sustainable development to establish Future Earth. Both IRDR and Future Earth programmes are explicitly tasked with growing the global institutional capacity required to increase connections and consistency across emerging multilateral issue–based networks. Together they exemplify distinct contemporary approaches to engaging with emerging global governance arrangements, in order to integrate, monitor and coordinate the way science is brought to bear on complex global issues. Considering IRDR and Future Earth in this context provides the opportunity to explore a few broad questions around the integrative role of global science bodies in an increasingly networked and distributed global governance environment. How are these organisations increasing integration of researchers and practitioners? What aspects of this role are exemplified by Future Earth and IRDR? What institutional capacity might they contribute in future? How might these two global bodies work more closely together, and so contribute to greater integration of global disaster risk management and sustainable development efforts?

As a global initiative, IRDR was tasked with "seeking to address the challenges brought by natural hazard events, mitigate their impacts, and improve related policy-making mechanisms" by increasing integration within the science domain, and across sectors and policy and practice (International Council for Science (ICSU) n.d.b). IRDR was set up to develop international disaster risk management networks by weaving a network of links across existing disaster risk management science programs and projects, as well as policy and practice communities at all levels.

The programme is small. A complex governance arrangement involves multiple reporting lines, one to ICSU, one to the IRDR science committee and others to those funding the program, the Chinese Academy of Sciences/Institute for Remote Sensing and Digital Earth (host organization) and the China Association of Science and Technology (CAST). The initial focus was on face-to-face networking to identify and join together relevant research programs, national and regional organisations, United Nations and other global bodies, international policy communities and funding agencies, with the aim of coordinating the implementation of multidisciplinary disaster risk research that answered the needs of decision makers.

By 2015, robust outreach activities had built an international network of two thousand-plus stakeholder organizations, and resulted in the establishment of eight IRDR National Committees, and five International Centres of Excellence (ICoE).[3] Two large international conferences (in 2011 and 2014) drew a wide range of disciplines together with global leaders and a range of practitioners, stimulating international cross-disciplinary and cross-sector debate and relationship building. Alongside these networking activities, a series of projects are led by science committee members. Two aim to improve understanding, of the Risk Interpretation and Action (RIA project) under conditions of uncertainty, and of the global disaster risk reduction research effort (AIRDR project). Another two set out to provide guidelines and standardised frameworks, for forensic post-disaster assessments (FORIN project), and for gathering and managing human and economic loss data (DATA project).

The connections and reputation developed through these activities contributed to IRDR's capacity to influence the declaration from the Asian Ministerial Conferences for Disaster Risk Reduction in 2012, and again in 2014, leading the thematic group on science, technology and academia. Subsequently, significant input from IRDR is reflected in the increased emphasis on scientific integration in the 2015 Sendai Framework (Aitsi-Selmi et al. 2016), and in a lead role in formalising the UNISDR Asia Science, Technology and Academia Advisory Group (ASTAAG) in May 2015.

As a small programme with limited resources, IRDR has been able to draw on support from global science and United Nations bodies to operate with agility. Size, comparative autonomy and UNISDR co-sponsorship facilitate the ongoing international face-to-face networking at global, regional and national levels that has helped draw other science programmes and policy and practitioner communities closer together. On the other hand, the same attributes have also limited IRDR potential to challenge the status quo more actively. Size and resourcing constraints largely restrict the programme to monitoring and contributing to other programmes rather than driving unique research projects of its own. Regular unplanned turnover in leadership on both the IRDR science committee in years 1–3 and in executive leadership in years 4–7 were also a limiting factor. These constraints have affected the organisation's capacity to grow its international profile, and effectively promote and disseminate international guidelines and standardised approaches.

Launched five years after IRDR, 'Future Earth is a global platform for international scientific collaboration, providing the knowledge required for societies in the world to face risks posed by global environmental change and to seize opportunities in a transition to global sustainability' (Future Earth n.d.b). It is co-sponsored by ICSU, ISSC and United Nations development–related bodies, and like IRDR offers standardising guidelines (for research engagement, and for the development of integrated Knowledge-Action Networks). Three distinct work streams divide the organisation according to focus. Knowledge-Action Networks are collaborative frameworks that facilitate highly integrative sustainability research, aimed at generating 'the multifaceted knowledge needed to inform solutions for complex societal issues' (Future Earth n.d. c). Medialab is focused on science communication initiatives (Future Earth n.d.a.), while the large federation of projects and other initiatives at 'the core' of Future Earth makes up Global Research Projects (Future Earth n.d.b). Scarcely a year old at the time this chapter is being written, Future Earth is still at such an early stage it is too soon to consider its performance. Drawing from the IRDR experience, however, it is possible to make some useful observations about the very different approach to integration taken in Future Earth.

Whereas IRDR is a small, lean organisation in comparison with other global science bodies, Future Earth is at the other end of the size spectrum, a large body by any standards. This is a consequence of what appears to be a more direct and incorporative approach to providing the institutional capacity required to increase integration at the global level. Where IRDR was set up to build networks among existing programmes and across sectors, Future Earth joins a range of (twenty-eight) existing initiatives together within a single broad platform. Larger global change research programmes absorbed during the development of Future Earth include the International Geosphere Biosphere Programme (IGBP), the International Human Dimensions Programme on Global Environmental Change (IHDP) and DIVERSITAS as well as twenty other projects (Future Earth n.d.d; ICSU 2013). These three well-established, long-running programs (founded 1987, 1990 and 1991, respectively) had proven track records (DIVERSITAS 2011; IHDP 2014; World Climate Research Project 2016; IGBP n.d). Future Earth also builds on the World Climate Research Project, one of the global change programmes. The World Climate Research Project was not, however, absorbed into Future Earth, remaining within the World Meteorological Organization. Each of these pre-existing programmes was co-sponsored by ICSU and ISSC, with partners from the World Meteorological Organization and various United Nations organizations. Since these are many of the same sponsor organizations that created Future Earth, and many of the funding agencies are the same as well, it is clear that Future Earth marks an apparent shift in approach to providing the institutional capacity required to integrate global change research efforts.

There are several obvious advantages to this incorporative approach. Future Earth joins United Nations players and existing research programmes from both sustainable development and climate change areas. This approach to science integration also offers to give the new body the cumulative authority and resourcing previously dispersed across its component programmes. Fulfilling this potential would allow Future Earth to generate the momentum, profile and influence required to contribute to global-level change. Using a single platform has associated potential when it comes to disseminating robust standardised research approaches across global sustainable development and climate change efforts. Future Earth provides guidelines for engagement and for the development of knowledge-action networks to generate 'high quality actionable scientific knowledge through the integration of research and the involvement of societal partners' (Future Earth 2016). Both these guidelines are informed by current research understandings of best practice.

Although Future Earth has yet to be fully under way, it is clear that in order to fulfil its potential this new organisation will need to mitigate the risks associated with its size and complexity actively. Future Earth is the largest and most potentially powerful issue-based global science body to date. As such, it risks being perceived to have achieved research dominance at the expense of its own constituent programmes, smaller issue-based organisations and those of opposing or alternative views, and to be actively inhibiting the diversity of research and integrated networks currently active across the globe. Perceptions of this kind have been found to erode the legitimacy of integrating arrangements at all scales rapidly, irrespective of their validity (Cash and Moser 2000; Cash et al. 2003).

High levels of transparency around funding and research relationships, decision-making structures and criteria will be required to minimise the risk of generating such perceptions (Cash et al. 2003). Since this risk increases at the same rate as the authority and size of the initiative, the levels of transparency required of Future Earth are high. The Future Earth website and associated materials attempt to ensure that this organisation's structure, governance and research approaches are as transparent as possible. However, the enormous complexities of this very large amalgam of organisations and projects works against these efforts at transparency.Collecting a wide range of funders, United Nations bodies and global programmes has resulted in complicated, nested structural and governance arrangements that remain difficult to understand, even when clicking through each nested layer on the website to trace the constituent elements of these complex networks of affiliation. IRDR is a smaller organisation, a characteristic that should make it easier to grasp the programme's basic structure, projects and overall 'shape.' Yet the complexity of the IRDR governance structure and multiple reporting lines means that they, too, are hard to grasp from information provided on the IRDR website.

Future Earth and IRDR both struggle to achieve transparency when it comes to decision making. Both websites fail to make the criteria used to guide funding, organisational and research decisions explicit. Neither organisation provides detailed information on its website concerning decision-making processes or the outcomes of individual decisions. Consequently collaborative decision making occurs to a large extent 'behind the scenes' in both organisations, making it difficult to find out how decisions have been made, why they were made, who was responsible for making them and how to be involved in future decision making.

Another area of risk for Future Earth arises out of the integration between climate change and sustainable development science initiatives. This integration reflects common disciplinary and disciplinary political ground. As Fordham (2007) noted, the coalescence of networks from across disciplines and global efforts around common disciplinary approaches and ideas offers promise for increased integration both within the science domain and between issue-based global efforts going forward. There is no doubt that extending Future Earth to include global governance networks focused on both climate change and sustainable development represents a significant step forward with respect to integration across issue-based governance systems.

Yet joining sustainable development and climate change scientific initiatives together in this way appears, at this early stage at least, to have involved a shift in integrating focus away from disaster risk management. Rising disaster risk does not feature in the eight major societal challenges to sustainability identified in the *Future Earth 2025 Vision* (Future Earth 2014). Future Earth's turn away from disaster risk management to date appears to result in a research focus that includes only those hazards associated with climate change and does not include hazard-related risk. This is surprising, in view of the science that has established that climate change is linked to the increased severity and frequency of disasters (Centre for Research on the Epidemiology of Disasters (CRED) and UNISDR 2015). The exclusion of hazard-related risk is also inconsistent with the SDGs and the Sendai Framework, which both emphasise a multi-hazard approach linked to risk.[4] Gains in connecting climate change and sustainable development research made by Future Earth may have required a step backward in terms of integration with both the larger disaster risk management effort and the sustainable development practice and policy fields that produced the SDGs and the Sendai Framework. It is also in contrast with the IRDR emphasis on integrating natural hazards research traditions with those focused on risk, which reflects the current prevalence of this approach in scientific contributions to global disaster reduction efforts. The IRDR emphasis is closer to that of both sustainable development and disaster risk management practice fields, as evidenced in the Sendai Framework.

But perhaps the greatest challenge faced by Future Earth arises out of its position within the science domain. Like IRDR, Future Earth is required to foster integration between scientific and practice domains, from within the science domain. The practice domain comprises those who are involved in disaster risk management implementation and decision making in government, business or industry. The Knowledge-Action Networks work stream is focused on developing 'greater interactions among disciplines and between research and societal partners, in both public and private sectors and civil society,' in order to 'improve the role of research to identify solutions for societal challenges' (Future Earth n.d.c). This emphasis on cross-sector collaboration is not apparent, however, in the governance structure, or at the design and implementation stage. The decision to give responsibility for designing and setting up Future Earth to a transition team and ex officio members recruited from climate change, sustainable development and biodiversity research fields represented a missed opportunity to ensure input from practice communities and organisations (ICSU n.d.c). The Governing Council retains responsibility for the strategic direction of Future Earth, and does not include representatives from practice organisations, instead giving decision-making authority at this level to global and United Nations science bodies and science funders. This is a top-down approach. Below this, parallel science and engagement advisory committees provide advice and support to the governance committee. Information and materials currently available on the website imply a traditional higher-level emphasis on supply-driven research, generated by science as a solution to end-user problems, rather than the co-creation of research, policy and practice through collaborations involving stakeholders/end-users in all research processes and stages to resolve end-user issues.

It has been well established that the integration of science into other spheres of activity requires a fundamental trade-off concerning time (Parker and Crona 2012; Sarkki et al. 2014). Research processes and the consensus building required for peer assessment are time-consuming, meaning that time frames in this domain tend to be medium- to long-term. By contrast, most business, practice and policy processes aim to respond to developments and achieve outcomes as close to real time as possible. The resulting tension has been found to be fundamental, representing a major challenge for initiatives requiring collaborations between science and policy or practice domains; high degrees of collaboration and compromise from both sides are required to make such cross-sector initiatives effective (Sarkki et al. 2014; Rovins, McAdoo and Weiss 2016).

Achieving the necessary levels of compromise from inside a single domain, whether science or practice, is difficult. Initiatives based in the practice domain risk making decisions that are not based on current evidence, since it has not yet had time to become established. Conversely, initiatives largely based in the science domain often lag behind policy and practice developments, in part because of the 'closed-loop' yet byzantine flow of information from field to lab, lab to publications and publications back to another lab, and investors that lack the facility to discuss concerns that are emerging from disaster researchers' (Rovins, McAdoo and Weiss 2016). This process lag can mean that by the time science and research reach a consensus, business and practice are likely to have moved on, without the benefits provided by a robust evidence base (Rovins, McAdoo and Weiss 2016). In the disaster risk reduction field, for example, this effect can be seen in ICSU publications concerning the impact of hazards and disasters on development including the *Priority Area Assessment on Environment and Its Relation to Sustainable Development* (ICSU 2003) as well as the *ICSU Foresight Analysis* (ICSU 2004), where the role of hazard and disaster risk in sustainable development is described as an "emerging" issue (ICSU 2008, p. 7). In the scientific sphere, ten years is not long ago. For practice communities, ten years can be a lifetime, and this issue was no longer new by 2004; the International Decade for National Disaster Reduction was over, the UNISDR had been launched and the Hyogo Framework was only a year away. Practice had moved on, while science governance was just getting organised.

Like those of Future Earth, IRDR governance arrangements are science-heavy. But the small size of the earlier organisation, the strong emphasis on face-to-face networking with practice organisations, initial staff with both practice and research qualifications and support from UNISDR gave IRDR some room to manoeuvre when attempting to integrate science and practice from within the science domain. The comparatively large size of Future Earth, the strong emphasis on facilitating research solutions and lack of sponsorship and involvement from practice organisations and practitioners are likely to make it more difficult for this new organisation to negotiate this tension.

30.6 Conclusion

IRDR and Future Earth exemplify different contemporary approaches to engaging with emerging global governance arrangements. Such arrangements are needed in order to integrate, monitor and coordinate the way science is brought to bear on complex global issues. Small by comparison with most global standards, and established five years earlier, IRDR has used face-to face networking and conferences to link and connect the diverse range of research programs and stakeholder organisations across an increasingly networked and distributed global governance environment. The recently established Future Earth, by contrast, is large by any standards, joining a range of related but diverse science programs in a single broad integrating platform. Each strategy carries both promise and risk. Limited size and resourcing have made it difficult for IRDR to build a profile and sustain momentum. Future Earth is at the other end of the scale, with much greater authority and scope. Although it is too early to tell whether this organisation will be able to wield these to fulfil its considerable potential, it is likely that this outcome will depend on the way this global body manages the risk associated with its size and incorporative approach. Currently both bodies remain science-heavy, requiring more integration at both governance and research decision-making levels. Neither appears to be moving towards more integration with disaster risk management and sustainable development practice.

How might these two global bodies work more closely together, and so contribute to greater integration of global disaster risk management and sustainable development efforts? This goal seems unlikely to be achieved by a further merger. Where sustainable development and climate change science communities share common ground, it would be much more difficult to integrate IRDR effectively into Future Earth, particularly in light of the current lack of Future Earth focus on disaster risk management. Such integration, moreover, would be likely to increase the complexity and size of Future Earth even further, and heighten the associated risks.

It is, of course, too early to judge the performance of Future Earth, and IRDR has recently undergone a review.[5] It is not yet clear whether Future Earth's approach marks a step change forward in efforts to increase integration between larger global governance systems or a step back towards earlier more monolithic global institutions traditionally tasked with providing comprehensive solutions to transnational issues. Only time will tell.

However, it is clear that their success will depend on ability to balance science, policy and practice influence at all levels of organisation, and in all decision making relating to research, practice and policy. More than a decade of research focused on integrating initiatives has found that involvement from all those impacted by the initiative is required to ensure that decisions and outcomes strike the appropriate balance between diverse and sometimes opposing interests (Cash and Moser 2000; Cash et al. 2003; Parker and Crona 2012; Sarkki et al. 2014; Bryson, Crosby and Stone 2015; Beaven et al. 2017). Achieving – and remaining focused on – this on-going balancing act at all levels is likely to be the most useful first step on the path to moving disaster risk management and sustainable development, and the larger global change efforts that have been established closer together (Rovins in press). The bottom line is that there is a need for much more explicit and detailed mechanisms for networking processes at the higher level between organizations. But these must include both science and practice.

Internet Resources

[1] UNISDR has since become the United Nations Office for Disaster Risk Reduction.

[2] ICSU was founded in 1931, with antecedents going back to 1899; ISSC was founded in 1950 (ICSU, n.d.a; ISSC, n.d.). ISSC and ICSU merged in 2018 to form the International Science Council (ISC).

[3] Since this watershed year – 2015 – two additional National Committees and seven ICOEs have been added, bringing the total to ten and twelve, respectively.

[4] The SDGs and the Sendai Framework are concerned with different hazards and risks, but both are concerned with multiple linked hazards and risks.

[5] The IRDR Midterm review is available on the ICSU website – www.icsu.org/publications/review-of-the-integrated-research-on-disaster-risk-irdr-programme-2016

References

Aitsi-Selmi, A., Murray, V., Wannous, C. et al. (2016). Reflections on a Science and Technology Agenda for 21st Century Disaster Risk Reduction. *Int. J Disaster Risk Science*, 7(1), pp. 1–29. DOI 10.1007/s13753–016–0081-x.

Beaven, S., Wilson, T., Johnston, L., Johnston, D., and Smith, R. (2017). Role of Boundary Organization after a Disaster: New Zealand's Natural Hazards Research Platform and the 2010 – 2011 Canterbury Earthquake Sequence. *Natural Hazards Review*, **18**(2), pp. 1–13. https://doi.org/10.1061/(ASCE)NH.1527–6996.0000202.

Bryson, J. M., Crosby, B. C., and Stone, M. (2015). Designing and Implementing Cross-Sector Collaborations: Needed and Challenging. *Public Administration Review*, **75**(5), pp. 647–663. http://doi.org/10.1111/puar.12432.

Cash, D. W., and Moser, S. (2000). Linking Global and Local Scales: Designing Dynamic Assessment and Management Processes. *Global Environmental Change*, **10**(2), pp. 109–120.

Cash, D. W., Clark, W. C., Alcock, F. et al. (2003). Knowledge Systems for Sustainable Development. *PNAS - Proceedings of the National Academy of Sciences*, **100** (14), pp. 8086–8091.

Castells, M. (2008). The New Public Sphere : Global Civil Society, Communication Networks, and Global Governance. *American Academy of Political and Social Science*, **616**(1), pp. 78–93.

Centre for Research on the Epidemiology of Disasters (CRED) and UNISDR. (2015). *The Human Cost of Weather Related Disasters 1995–2015*. Geneva, UNISDR and CRED.

DIVERSITAS. (2011). *Mission and History*. Available online: www.diversitas-international.org/about/mission-and-history [Accessed 15/09/2016].

Drexhage, J., and Murphy, D. (2010). *Sustainable Development : From Brundtland to Rio 2012*. New York: United Nations Headquarters.

Duke, C. (2006). Prosperity, complexity and science. *Nature Physics*, **2**, 426–428.

Fordham, M. (2007) Disaster and Development: A Necessary Eclecticism. In *Handbook of Disaster Research*, ed. H. Rodriguez, E. L. Quarantelli, and R. R. Dynes, New York: Springer, pp. 335–346.

Future Earth. (2014). *Future Earth 2025 Vision*. Paris: ICSU.

Future Earth. (2016). *Knowledge-Action Networks Framing and Operationalisation*. Available online: www.futureearth.org/media/knowledge-action-networks-framing-and-operationalisation [Accessed 1/10/2016].

Future Earth. (n.d.a). Future Earth Media Lab Science, Design, Storytelling. Available online: http://medialab.futureearth.org/ [Accessed 15/10/2016].

Future Earth. (n.d.b). Global Research Projects. Available online: www.futureearth.org/projects [Accessed 15/10/2016].

Future Earth. (n.d.c). Knowledge-Action Networks. Available online: http://futureearth.org/knowledge-action-networks [Accessed 15/10/2016].

Future Earth. (n.d.d). Who We Are. Available online: www.futureearth.org/who-we-are [Accessed 30/09/2015].

International Council for Science (ICSU). (2004). *Foresight Analysis: Report of the CSPR*. Paris: ICSU.

International Council for Science (ICSU). (2008). *A Science Plan for Integrated Research on Disaster Risk: Addressing the Challenge of Natural and Human-Induced Environmental Hazards*. Paris: ICSU.

International Council for Science (ICSU). (2013). *Future Earth Initial Design: Report of the Transition Team Paris: International Council for Science (ICSU)*. Paris: ICSU.

International Council for Science (ICSU). (n.d.a). *A Brief History of ICSU*. Available online: www.icsu.org/about-icsu/about-us/a-brief-history [Accessed 15/03/2017].

International Council for Science (ICSU). (n.d.b). *Disaster Risk (IRDR)*. Available online: www.icsu.org/what-we-do/interdisciplinary-bodies/irdr [Accessed 15/09/2016).

International Council for Science (ICSU). (n.d.c). *Transition Team*. Available online: www.icsu.org/future-earth/who/transition-team [Accessed 15/09/2016].

International Council for Science (ICSU). (n.d.d). *World's top bodies representing the social and natural sciences vote to pursue a merger, forming a single organization representing all social and natural sciences by 2018*. Available online: www.icsu.org/news-centre/news/top-news/world2019s-top-bodies-representing-the-social-and-natural-sciences-vote-to-pursue-a-merger-forming-a-single-organization-representing-all-social-and-natural-sciences-by-2018 [Accessed 14/03/2017].

Intergovernmental Panel on Climate Change (IPCC). (2013). *IPCC Factsheet: What Is the IPCC?* Geneva: IPCC Secretariat.

International Council for Science (ICSU). (2003). *Priority Area Assessment on Environment and its Relation to Sustainable Development*. Paris: ICSU.

International Geospere-Biospere Programme. (n.d.). History of global-change research. Available online: www.igbp.net/about/history.4.1b8ae20512db692f2a680001291.html [Accessed 15/09/2016].

International Human Dimension Programme on Global Environmental Change. (2014). *About IHDP*. Available online: www.ihdp.unu.edu/pages/?p=about [Accessed 15/09/2016].

International Social Science Council (ISSC). (n.d.). *History and Mission*. www.worldsocialscience.org/about/history-mission/ [Accessed 15/03/2017].

Jasanoff, S. (2011). Constitutional Moments in Governing Science and Technology. *Science and Engineering Ethics*, 17(4), pp. 621–638. https://doi.org/10.1007/s11948–011–9302–2.

Jensen, S., Feldmann-Jensen, S. J., Johnston, D. M., and Brown, N. A. (2015). The Emergence of a Globalized System for Disaster Risk Management and Challenges for Appropriate Governance. *International Journal of Disaster Risk Science*, **6**(1), pp. 87–93.

McNie, E. C. (2007). Reconciling the supply of scientific information with user demands: an analysis of the problem and review of the literature. *Environmental Science and Policy*, 10(1), pp. 17–38. https://doi.org/10.1016/j.envsci.2006.10.004.

Parker, J., and Crona, B. (2012). On Being All Things to All People: Boundary Organizations and the Contemporary Research University. *Social Studies of Science*, **42**(2), pp. 262–289.

Patrick, S. (2014). The Unruled World: The Case for Good Enough Global Governance. *Foreign Affairs*, **93**(1), pp. 58–73.

Rovins, J. (in press). A Wicked Problem: Making DRR and CCA work for resilience. In *Disaster Resilience: An Integrated Approach*, ed. D. Paton and D.M. Johnston, 2nd edn. Springfield, IL: Charles C. Thomas.

Rovins, J.E., McAdoo, B., and Weiss, R. (2016). *South East Asia Risk Collaboration Hub (SEARCH)*, Middle East Institute, Available online: www.mei.edu/content/map/south-east-asia-risk-collaboration-hub-search (Accessed 8/8/2016).

Sarkki, S., Niemelä, J., Tinch, R., van den Hove, S., Watt, A., and Young, J. (2014). Balancing Credibility, Relevance and Legitimacy: A Critical Assessment of Trade-Offs in Science-Policy Interfaces. *Science and Public Policy*, **41**(2), pp. 194–206.

Skogstad, G. (2003). Who Governs? Who Should Govern?: Political Authority and Legitimacy in Canada in the Twenty-First Century. *Canadian Journal of Political Science/Revue Canadienne de Science Politique*, 36, pp. 955–973. https://doi.org/10.1017/S0008423903778925

United Nations Conference on Sustainable Development Secretariat. (2012). Disaster Risk Reduction and Resilience Building, RIO 2012 Issues Briefs No. 8.

United Nations Development Programme (UNDP). (2004.) *Reducing Disaster Risk a Challenge for Development*. New York: UNDP.

United Nations General Assembly. (1987). International Decade for Natural Disaster Reduction (11 December 1987, A/RES/42/169).

United Nations General Assembly. (1989). The Rule of Law at the National and International Levels (22 December 1989, A/RES/44/236).

United Nations General Assembly. (2000). International Decade for Natural Disaster Reduction: Successor Arrangements (3 February 2000, A/RES/54/219).

United Nations General Assembly. (2015). Transforming out World: the 2030 Agenda for Sustainable Development (25 September 2015, A/RES/70/1).

United Nations International Strategy for Disaster Reduction. (2015). Sendai Framework for Disaster Risk Reduction 2015–2030. Available online: www.wcdrr.org/uploads/Sendai_Framework_for_Disaster_Risk_Reduction_2015-2030.pdf [Accessed 15/8/2016].

United Nations. (2002a). Johannesburg Declaration on Sustainable Development (2 September 2002, A/Conf. 199/20).

United Nations. (2002b). Plan of Implementation of the World Summit on Sustainable Development (4 September 2002, A/Conf. 199/20).

United Nations. (2004). *Living with Risk: A Global Review of Disaster Reduction Initiatives*. Geneva: United Nations.

United Nations. (2005). Hyogo Framework for Action 2005–2015: Building the Resilience of Nations and Communities to Disasters (22 January 2005,A/CONF.206/6)

World Climate Research Programme (WCRP). (2016). WCRP History. Available online: www.wcrp-climate.org/about-wcrp/about-history [Accessed 14/9/ 2016].

Name Index

Subject Index